T0338245

SEMICONDUCTOR TERAHERTZ TECHNOLOGY

SEMICONDUCTOR TERAHERTZ TECHNOLOGY

DEVICES AND SYSTEMS AT ROOM TEMPERATURE OPERATION

Editors

Guillermo Carpintero

Universidad Carlos III de Madrid, Spain

Luis Enrique García Muñoz

Universidad Carlos III de Madrid, Spain

Hans L. Hartnagel

Technische Universität Darmstadt, Germany

Sascha Preu

Technische Universität Darmstadt, Germany

Antti V. Räisänen

Aalto University, Finland

This edition first published 2015

Registered office
John Wiley & Sons Ltd, The Atrium, Southern Gate, Chichester, West Sussex, PO19 8SQ, United Kingdom

For details of our global editorial offices, for customer services and for information about how to apply for permission to reuse the copyright material in this book please see our website at www.wiley.com.

Library of Congress Cataloging-in-Publication Data applied for.

A catalogue record for this book is available from the British Library.

ISBN: 9781118920428

Set in 9/11pt, TimesLTStd by SPi Global, Chennai, India
Printed and bound in Singapore by Markono Print Media Pte Ltd

1 2015

Contents

Acknowledgments

Prof. Enrique García Muñoz, Prof. Daniel Segovia-Vargas, and Prof. Guillermo Carpintero acknowledge the support received from the European Science Foundation (ESF) in the framework of the ESF activity entitled "New Frontiers in Millimetre/Sub-Millimetre Waves Integrated Dielectric Focusing Systems" and from Proyecto de investigación "DiDaCTIC: Desarrollo de un sistema de comunicaciones inalambrico en rango THz integrado de alta tasa de datos," TEC2013-47753-C3 and CAM S2013/ICE-3004 "DIFRAGEOS" projects.

Prof. Guillermo Carpintero also acknowledges support from the European Commission through the iPHOS FP7 STREP project (Integrated photonic transceivers at sub-terahertz wave range for ultra-wideband wireless communications), which he coordiated. Special thanks go to the Project Officer, Henri Rajbenbach, and the project reviewers (Beatrice Cabon and Adolfo Cartaixo) for their encouragement to fully achieve the challenging project objectives.

Prof. Ramón Gonzalo would like to thanks all coworkers and collaborators involved in the research on THz technology; in particular, Iñigo Ederra, Jorge Teniente, and Juan Carlos Iriarte at the Antenna Group, UPNa. Furthermore, he specially acknowledges the support received by the Spanish Project funded by the Spanish Minister of Economics and Competitiveness "DiDaCTIC: Desarrollo de un sistema de comunicaciones inalambrico en rango THz integrado de alta tasa de datos," TEC2013-47753-C3-R1.

S.Preu and C. Damm acknowledge the Hessian priority programme "LOEWE-Sensors Towards Terahertz" for funding.

The authors of Chapter 8, would like to thank all coworkers and collaborators involved in the research on liquid crystals for millimeter-wave and THz application; in particular Atsutaka Manabe, Dr. Michael Wittek, and Dagmar Klass at Merck KGaA, Darmstadt; Carsten Fritzsch, Alexander Gäbler, Matthias Höfle, Matthias Jost, Prof. Dr.-Ing. Rolf Jakoby, Dr.-Ing. Onur Karabey, Matthias Maasch, Dr.-Ing. Andreas Penirschke, and Sebastian Strunck at the Institute for Microwave Engineering and Photonics, TU Darmstadt.

Frédéric van Dijk would like to thank colleagues from the III–V Lab and non-permanent staff who contributed to the work on photonic integrated circuits on InP; in particular Marco Lamponi, Mourad Chtioui, Gaël Kervella, François Lelarge, Yohann Moustapha-Rabault, Eric Vinet, Dallila Make, Bernadette Duval, Yannick Robert, and Alvaro Jimenez.

ACST GmbH acknowledges all the engineers and technical staff of the Technical University of Darmstadt (Germany) and Radiometer Physics GmbH (Germany), who contributed to the development work addressed in Sections 6.4 and 6.5. This activity was partially supported by the European Space Agency under several ESA/ESTEC contracts.

Prof. A.V. Räisänen's work during the writing of Chapter 4 was supported by "Cátedras de Excelencia," based on an agreement between Banco Santander and Universidad Carlos III de Madrid. He would like to thank Diby Jeo Raol Amara, Mazidul Islam, Subash Khanal, Mohammad Suzan Miah, Mohammad Sajjad Mirmoosa, Jinsong Song, and Dimitrios Tzarouchis, who all attended the Aalto University doctoral course on THz techniques in the fall of 2014, for helping in proof reading the full manuscript of this book.

E.R. Brown acknowledges support by, or in part by, the U.S. Army Research Office under contract number W911NF-11-10024, and an Endowed Chair at Wright State University under the Ohio Research Scholars Program.

Finally, we would like to thank the patient and exceptional support that we have received from Wiley, especially from Ella Mitchell and Prachi Sinha Sahay.

The authors, would especially like to acknowledge Helen Hartnagel for the precise corrections throughout the document.

Preface

"Midway upon the journey of our life, I found myself within a forest dark, for the straightforward pathway had been lost." These words from Dante Aliguieri can summarize the main motivation that we had for writing this book: trying to cover and explain recent advances in THz frequencies from a semiconductor technology perspective. Our motivation started with a meeting in Nürnberg (Germany) in October 2013, where we (all authors of this book) had agreed to meet to discuss the objectives and structure of the book in its present form.

THz frequencies have been investigated for a long time, since the middle of the twentieth century, having coined popular terms such as the "THz gap" to indicate the underdevelopment of this part of the electromagnetic spectrum. However, the technological challenge to develop efficient devices for both generation and detection of THz waves has only recently started to be addressed. The mechanisms that govern the propagation, emission, and characterization at THz frequencies have been studied and developed by both physicists and engineers. We have realized that, very frequently, there is a huge lack of knowledge about the different perspectives that are involved. Let us illustrate this point with an example: the photomixing technique for the generation of THz waves. Photomixing uses two lasers emitting at slightly different wavelengths that beat in a semiconductor device, obtaining the difference of the two lasers as the THz emitted frequency. It must thus be studied, taking semiconductor physics into account. Conversely, the study of the antennas and/or lenses attached for extracting the THz wave generated (or for the propagation and guiding of THz signals) is typically done by the classical approach of macroscopic electrodynamics. Engineers typically treat the photomixing process as a "black box" where they have to insert their antennas to get the best matching. On the other hand, the physicists who developed the device sometimes forget about the technology limits or have modest knowledge about antenna radiation mechanisms, and use classical topologies on the basis that they worked previously, not being aware that more efficient approaches are available. The result is that both research approaches, although totally valid and rigorous, have a lack of knowledge in the other area. In order to optimize the devices, the researchers need to address design issues that lead them into areas of expertise that they do not master. That is what we call entering into "the forest dark" of Dante.

We believe that partnership among researchers is the best approach to explore the new regions. With this spirit in mind, we organized the meeting at Nürnberg, contacting specialists from all the regions of the THz forest. The main objective for this book in the beginning was to shed some light into the different parts of the THz forest. We can affirm that the book is a self-contained manual for both physicists and engineers who are working or starting their research in semiconductor THz technology. The international team of authors, which comes from both areas of knowledge, that is, Physics and Engineering, wrote all the contributions with extreme care, explaining the basic concepts of their areas up to the current state-of-the-art. For this reason, an expert in an area could find a few pages of the book "rather elementary". These are, however, strictly necessary to provide total consistency and the self-contained aspect of the book, and to cover fully the current state-of-the-art.

In Chapter 2, the theoretical background of terahertz generation by photomixing is discussed in detail. Basic design rules are specified for obtaining highly efficient optical to THz power conversion for both photoconductive and high frequency p-i-n diodes, considering pulsed as well as continuous-wave operation. State-of-the-art realizations of photomixers at 800 and 1550 nm laser wavelengths are shown. Limiting electrical and thermal constraints to the achievable THz power are also addressed. Finally, this chapter gives an overview of electronic means for THz generation, such as Schottky diodes, negative differential resistor oscillators, and plasmonic effects that are used in THz generation. The chapter starts with a quick overview of the most relevant THz generation schemes based on nonlinear media, accelerating electrons, and actual THz lasers. This serves to place in context the two schemes discussed in detail thereafter: photomixing and electronic generation. The chapter goes through the theoretical frameworks, principles of operation, limitations, and reported implementations of both schemes for pulsed and continuous-wave operation when applicable. The chapter also covers to a lesser extent the recently explored use of plasmonics to improve the efficiency of THz generation in photomixing, nonlinear media, and laser schemes.

Chapter 3 presents the theoretical background of antenna theory, tailored to terahertz applications. A general discussion is provided on the issues of THz antennas, especially for matching to the photomixer. Array theory is presented, together with an exhaustive and precise analysis of one of the most promising and new solutions for generating THz emission with high power levels, that is, the large area emitter concept.

In Chapter 4 we first briefly introduce Maxwell's equations and derive the Helmholtz equation, that is, a special case of the wave equation, and introduce its different solutions for fields that may propagate in THz waveguides. The second section describes different waveguides operating at THz frequencies, and the third section is devoted to the beam waveguide and quasi-optics. Material issues related to waveguides and quasi-optical components are also discussed. The chapter concludes with THz wave propagation in free space.

Chapter 5 is a comprehensive review of the physical principles and engineering techniques associated with contemporary room-temperature THz direct detectors. It starts with the basic detection mechanisms: rectification, bolometric, pyroelectric, and plasma waves. Then it addresses the noise mechanisms using both classical and quantum principles, and the THz coupling using impedance-matching and antenna-feed considerations. Fundamental analyses of the noise mechanisms are provided because of insufficient coverage in the popular literature. All THz detectors can then be described with a common performance formalism based on two metrics: noise-equivalent power (NEP) and noise-equivalent temperature difference (NETD). The chapter concludes with a comparison of the best room-temperature THz detector experimental results to date above ~300 GHz, and suggests that as none of these detector types are operating very close to fundamental theoretical limits there is room for significant performance advances.

Several key topics in THz electronics are discussed in Chapter 6. We describe operating principles, limitations, and state-of-the-art of resonant-tunneling diodes (RTDs) and THz RTD oscillators. Furthermore, THz fundamental or sub-harmonic flip-chip Schottky diode mixer configurations are described. Different measurement techniques are commented and their properties outlined. The chapter also describes the use of advanced mixer configurations. Fabrication technologies for Schottky-diode based structures for THz wave applications are included, together with the low-barrier Schottky diode characterization for millimeter-wave detector design. Finally, low noise amplifiers (LNAs) for sub-millimeter waves are discussed, including up-to-date design approaches and resulting performance, with emphasis on the necessary technological modification to extend monolithic microwave integrated circuit (MMIC) approaches toward the THz region.

The THz spectral range has not yet been fully exploited to its full potential due to the current limitations in sources and detectors. To open the THz frequency range for applications, photonic solutions have been at the technological forefront. For instance, the advances of time domain spectroscopy techniques using short pulse lasers have enabled the provision offull spectroscopy data across the range. There are different types of systems and their development should be governed by the requirements of the potential application. Photonic techniques are desirable solutions for millimeter wave and THz generation in terms of their energy efficiency and, above all, their tuning range. Recent developments in this area target the improvement of optical-to-THz converters as well as enhancing the level of integration of semiconductor laser sources in order to address their main drawbacks, cost, and spectral purity. The purpose of Chapter 7 is to describe the main types of photonically enabled THz systems and the expected performances from their components. A description is provided of the key elements in designing each of the components and their limitations. The final part of the chapter is a discussion on potential future development and the importance of integration.

Finally, Chapter 8 summarizes and explains some novel approaches and applications of THz, such as liquid crystals, graphene technology, or resonator theory based on a nonlinear up-conversion process. This makes the approach very appealing for its use as highly-sensitive receivers.

We expect that the reader will find in the book not only answers but also at least some hints for continuing the advances in THz technology. Of course, we hope that the reader shares the same feeling of satisfaction experienced by the authors when writing and discussing the present book.

Prof. Dr. Guillermo Carpintero
Prof. Dr. Luis Enrique García Muñoz
Madrid, Spain, November 2014

Prof. Dr. Hans Hartnagel
Prof. Dr. Sascha Preu
Darmstadt, Germany, November 2014

Prof. Dr. Antti V. Räisänen
Espoo, Finland, November 2014

Foreword

Terahertz technology is one of the most exciting areas being explored today by researchers interested in high frequency electromagnetic systems and associated solid-state devices, antennas, quasi-optical components, and related techniques. Although tantalizing in its potential, terahertz technology has also been frustratingly slow to develop to maturity. This book represents the most detailed and up-to-date exposition of the current state of the art for terahertz technology, and an exposition of the reasons for its slow progress. It contains discussions of terahertz sources, including the scaling up of microwave and millimeter wave oscillators and frequency multipliers, as well as the mixing of infrared or optical lasers to produce difference frequencies in the terahertz band. Radiation of terahertz signals is described, including the use of antennas as well as the direct radiation from emitting sources. Terahertz propagation in various types of waveguides is discussed, in addition to propagation characteristics of the atmosphere. Detection of terahertz signals by direct detection methods using diodes, bolometers, and pyroelectric devices is presented, along with noise characteristics of these techniques. Electronic devices useful for detection, mixing, generation, amplification, and control of terahertz signals are thoroughly described.

This book includes seven chapters written by the leading researchers in terahertz technology research today. Chapter authors represent research groups from around the world, including workers from Spain, Germany, Italy, the United Kingdom, Japan, Finland, Ireland, and the United States. Again reflecting the uniqueness of terahertz technology as an interdisciplinary effort, contributors come from the fields of both engineering and physics, as do the subjects reported here. For these reasons, this work should prove to be of interest and value to researchers, teachers, and students.

The terahertz frequency band (100 GHz to 10 THz) is located between microwave frequencies at the low end, and far-infrared frequencies at the high end. Although initial work at terahertz frequencies was done as early as the 1970s, major progress on systems using such frequencies has been slow due to the difficulty in fabricating sources, detectors, amplifiers, antennas, and other components at such high frequencies. On the other hand, terahertz technology, being positioned between microwave technology and optical technology, can benefit from the unique methods and techniques of each of these domains. A good example of this is presented in Chapter 2, where the generation of terahertz power is described using solid state oscillators and frequency multipliers (scaling up of techniques and devices commonly used at microwave and millimeter wave frequencies), as well as generation using photomixing of two lasers (a common technique in the optical domain). Several other examples of how terahertz technology capitalizes on both electronics technology and optical technology are described in other chapters. As most new technology is built “on the shoulders of giants”, terahertz technology has two sets of “giants” to build upon.

Applications of terahertz technology include sensors, imaging for medical and security purposes, and short-range communications. One of the earliest applications of terahertz frequencies was for astronomical spectroscopy, as many of the more complex molecules being searched for in space have resonances at these frequencies. Terahertz frequencies also provide unique features for communications systems, providing extremely high bandwidth, and built-in resistance to eavesdropping due to extremely high rates of atmospheric attenuation. As more progress is made with solid state components for terahertz frequencies, it is certain that more applications will be found for this unique technology. Such progress seems to be accelerating: it was just recently announced that the first MMIC amplifier operating above 1 THz has been developed in the US. Developments like this mean that we can look forward to seeing terahertz technology move from the research lab into practical and affordable commercial and science applications in the near future. This book will provide the background that workers will need to be productive in this exciting field.

David Pozar
Professor Emeritus
Electrical and Computer Engineering
University of Massachusetts Amherst
Amherst, MA USA
November 2014

List of Contributors

Miguel Beruete, Departamento de Ingeniería Eléctrica y Electrónica, Universidad Pública de Navarra, Campus Arrosadia s/n, 31006 Pamplona, Navarra, Spain

Elliott R. Brown, Wright State University, 3640 Colonel Glenn Hwy, Dayton, OH 45435, USA

Guillermo Carpintero, Departamento de Tecnología Electronica, Universidad Carlos III de Madrid, Av. de la Universidad, 30, Leganés 28911, Madrid, Spain

Oleg Cojocari, ACST GmbH, Josef-Bautz-Straße 15, D-63457 Hanau, Germany

Christian Damm, Terahertz Sensors Group, Dept. of Electrical Engineering and Information Technology, Technische Universität Darmstadt, Merckstraße 25, 64283 Darmstadt, Germany

Gottfried H. Döhler, Max Planck Institute for the Science of Light, Guenther-Scharowsky-Str. 1, Bldg. 24, 91058 Erlangen, Germany

Michael Feiginov, Department of Electronics and Applied Physics, Tokyo Institute of Technology, Ookayama, Meguro-ku, 2-12-1-S9-3, Tokyo 152-8552, Japan (Presently with Canon Inc., Frontier Research Centre, THz Imaging Division, Tokyo, Japan)

Alejandro García-Lampérez, Departamento de Teoría de la Señal y Comunicaciones, Universidad Carlos III de Madrid, Avda. de la Universidad 30, 28911 Leganés, Madrid, Spain

Luis Enrique García Muñoz, Departamento Teoría de la Señal y Comunicaciones, Universidad Carlos III de Madrid, Av. de la Universidad, 30, 28911 Leganes, Madrid, Spain

Andrey Generalov, Aalto University, School of Electrical Engineering, Department of Radio Science and Engineering, P.O. Box 13000, FI-00076 AALTO, Espoo, Finland

Thorsten Göbel, Terahertz Group/Photonic Components, Fraunhofer Heinrich Hertz Institute, Einsteinufer 37, 10587 Berlin, Germany

David González Ovejero, University of Siena, Department of Information Engineering and Mathematics, Palazzo San Niccolo, Via Roma 56, 53100 Siena, Italy

Ramón Gonzalo, Departamento de Ingeniería Eléctrica y Electrónica, Universidad Pública de Navarra, Campus Arrosadia, 31006 Pamplona, Navarra, Spain

Marcin L. Gradziel, Maynooth University Department of Experimental Physics, National University of Ireland, Maynooth, Co. Kildare, Ireland

Hans Hartnagel, Technical University Darmstadt, Institut für Hochfrequenztechnik, Merckstr. 23, 64283 Darmstadt, Germany

Matthias Hoefle, ACST GmbH, Josef-Bautz-Str. 15, 63457 Hanau, Germany, and Institute for Microwave Engineering and Technologies, Technische Universität Darmstadt, Merckstr. 25, 64283 Darmstadt, Germany

Ernesto Limiti, Dipartimento di Ingegneria Elettronica, Università degli Studi di Roma Tor Vergata, Via del Politecnico 1, 00133 Roma, Italy

Dmitri Lioubtchenko, Department of Radio Science and Engineering, School of Electrical Engineering, Aalto University, P.O. Box 13000, FI-00076 AALTO, Finland

Itziar Maestrojuán, Anteral S.L., edificio Jerónimo de Ayanz, Campus Arrosadía s/n, 31006 Pamplona, Navarra, Spain

Stefan Malzer, Friedrich-Alexander Erlangen-Nürnberg Universität, Applied Physics, Staudtstr. 7, Bldg. A3, 91058 Erlangen, Germany

Javier Montero-de-Paz, Departamento Teoría de la Señal y Comunicaciones, Universidad Carlos III de Madrid, Av. de la Universidad, 30, 28911 Leganes, Madrid, Spain

J. Anthony Murphy, Maynooth University Department of Experimental Physics, National University of Ireland, Maynooth, Co. Kildare, Ireland

Miguel Navarro-Cía, Optical and Semiconductor Devices Group, Dept. Electrical and Electronic Engineering, Imperial College London, South Kensington Campus, London SW7 2AZ, UK

Créidhe O'Sullivan, National University of Ireland, Maynooth, Co. Kildare, Ireland

Sascha Preu, Terahertz Systems Technology Group, Dept. of Electrical Engineering and Information Technology, Technische Universität Darmstadt, Merckstr. 25, 64283 Darmstadt, Germany

Antti V. Räisänen, Aalto University, School of Electrical Engineering, Department of Radio Science and Engineering, P.O. Box 13000, FI-00076 AALTO, Espoo, Finland

Cyril C. Renaud, Department of Electronic & Electrical Engineering, University College London, Torrington Place London WC1E 7JE, UK

Vitaly Rymanov, Universität Duisburg-Essen, ZHO/Optoelektronik, Lotharstr. 55, 47057 Duisburg, Germany

Magdalena Salazar-Palma, Universidad Carlos III de Madrid, Departamento de Teoría de la Señal y Comunicaciones, Avenida Universidad 30, 28911 Leganés, Madrid, Spain

Harald G. L. Schwefel, Max Planck Institute for the Science of Light, Günther-Scharowsky-Str. 1, Bldg 24, 91058 Erlangen, Germany and Friedrich-Alexander University Erlangen-Nürnberg (FAU), Institute for Optics, Information and Photonics, Günther-Scharowsky-Str. 1, Bldg 24, 91058 Erlangen, Germany; presently at University of Otago, Department of Physics, Dunedin, New Zealand

Florian Sedlmeir, Max Planck Institute for the Science of Light, Günther-Scharowsky-Str. 1, Bldg 24, 91058 Erlangen, Germany, and Friedrich-Alexander University Erlangen-Nürnberg (FAU), Institute for Optics, Information and Photonics, Günther-Scharowsky-Str. 1, Bldg 24, 91058 Erlangen, Germany

Daniel Segovia-Vargas, Departamento Teoría de la Señal y Comunicaciones, Universidad Carlos III de Madrid, Avda Universidad 30, 28911 Leganés, Spain

Andreas Stöhr, Universität Duisburg-Essen, ZHO/Optoelektronik, Lotharstr. 55, 47057 Duisburg, Germany

Neil Trappe, Maynooth University Department of Experimental Physics, National University of Ireland, Maynooth, Co. Kildare, Ireland

Frédéric van Dijk, Alcatel-Thales III-V Lab, Campus de Polytechnique, 1, Avenue Augustin Fresnel, 91767 Palaiseau, Cedex, France

Christian Weickhmann, Technische Universität Darmstadt, Institut für Mikrowellentechnik und Photonik, Merckstraße 25, 64283 Darmstadt, Germany

1

General Introduction

Hans Hartnagel[1], Antti V. Räisänen[2], and Magdalena Salazar-Palma[3]
[1]*Technical University Darmstadt, Institut für Hochfrequenztechnik, Darmstadt, Germany*
[2]*Aalto University, School of Electrical Engineering, Department of Radio Science and Engineering, Espoo, Finland*
[3]*Universidad Carlos III de Madrid, Departamento de Teoría de la Señal y Comunicaciones, Avenida Universidad Madrid, Spain*

In this book, we define TeraHertz (THz) waves as the part of the electromagnetic spectrum with wavelengths ranging from 3 mm down to 30 μm, that is, from 100 GHz to 10 THz covering the upper part of millimeter waves (30–300 GHz), the whole range of submillimeter waves (300 GHz to 3 THz), and the lower end of infrared waves (from 3 THz to visible light).

THz science and technology is a relatively young area both in research and applications. THz applications started from radio astronomy in the 1970s. This was based on the property that molecules and atoms can be identified by their radiation spectrum caused by their rotational and vibrational resonances. Since then the THz band has found many other potential applications because it provides unprecedented bandwidth and opportunities for completely new sensor applications. It is feasible and potential for many ground-based commercial applications as well as for Earth science applications: remote sensing of the Earth's surface and atmosphere, broadband high data-rate indoor (e.g., smart home) and short-range outdoor wireless communications, short-range, long-range, and multi-function automotive radars and ultra wide band (UWB) high-resolution radars, telematics for road traffic and transport, both between vehicles and between vehicles and infrastructures, imaging for security, medical, and other purposes.

In free space, transmission of THz waves typically requires line-of-sight between the transmitter and the receiver. The length of a terrestrial communication hop cannot be very long since the water vapor of the atmosphere is highly absorbent, ranging from not more than a few meters to hundreds of meters. On the other hand, this high attenuation enables one to limit THz-communication distances to secure distances. Depending on the relative humidity (between 25 and 100%), the atmospheric attenuation at sea level and at a temperature of 25 °C ranges from 0.3 to 1 dB km^{-1} at 100 GHz and from 50 to 250 dB km^{-1} at 700 GHz. At water vapor absorption peaks, for example, at 557 GHz, the attenuation may reach values over 10^4 dB km^{-1}.

So far the employment of THz systems for applications has been slow because of the immature technology. The frequency range from 0.1 to 10 THz is often called the terahertz gap, because technologies for generating and detecting this radiation are much less mature than those at microwave or infrared

Semiconductor Terahertz Technology: Devices and Systems at Room Temperature Operation, First Edition.
Edited by Guillermo Carpintero, Luis Enrique García Muñoz, Hans L. Hartnagel, Sascha Preu and Antti V. Räisänen.

frequencies. Mass production of devices at the terahertz gap frequencies and operation at room temperature have mostly not yet been realized. High-power THz waves (beyond the kilowatt levels) with microwave concepts can be generated by using vacuum tube generators, such as backward wave oscillators (BWOs) and free electron lasers (FELs), and lower THz power waves by multiplying semiconductor oscillator frequencies from the microwave or low millimeter wave bands by Schottky diode multipliers to the THz waves. On the other hand, THz radiation with photonics means may be generated by using optically pumped gas lasers or photomixers pumped with two infrared semiconductor lasers. In the latter case, the power of the THz signal is typically only microwatts. Additionally it is possible to generate THz signals of very low power values (nanowatt levels) by ultrashort bias pulsing, and this technique has been very successful in THz characterization of chemical elements.

However, recent strong advances in the development of semiconductor components and their manufacturing technology are making THz systems and applications more feasible and affordable. The terahertz gap is shrinking slowly but definitely due to strong developments from both directions, from microwaves and from photonics.

New approaches to the design and manufacturing of THz antennas are another indispensable basis for developing future applications. THz frequencies require integration of antennas with the active electronics. In most applications, electronic focusing and beam steering is needed or is at least a very valuable asset. One of the main components of a radar sensor is a beam-steering device that scans surroundings for hidden objects. Another application that requires implementation of the same technology is point-to-point wireless communication systems. If beam steering is implemented in receiving and transmitting antennas, a self-adapting mechanism can be elaborated for innovative performance of the future secure high-capacity communication links. Because the RF (radiofrequency) spectrum in the microwave region is highly populated with different communication standards, a possible frequency range for high-capacity communication systems, which require a wide bandwidth, can be found at millimeter wavelengths, for example, at 86, 150, and 250 GHz.

This book is intended to be valuable for researchers in this field, for industrialists looking to open new markets, and particularly also for teaching at university level. Therefore it concentrates on those parts of the subject which are important for these aspects. Other important scientific areas such as THz spectroscopy are not treated in detail, except for the possibility of its use in security by identification of explosive materials.

2

Principles of THz Generation

Sascha Preu[1], Gottfried H. Döhler[2], Stefan Malzer[3], Andreas Stöhr[4], Vitaly Rymanov[4], Thorsten Göbel[5], Elliott R. Brown[6], Michael Feiginov[7,*], Ramón Gonzalo[8], Miguel Beruete[8], and Miguel Navarro-Cía[9]

[1]*Technische Universität Darmstadt, Dept. of Electrical Engineering and Information Technology, Darmstadt, Germany*
[2]*Max Planck Institute for the Science of Light, Erlangen, Germany*
[3]*Friedrich-Alexander Universität Erlangen-Nürnberg, Applied Physics, Erlangen, Germany*
[4]*Universität Duisburg-Essen, ZHO/Optoelektronik, Duisburg, Germany*
[5] *Fraunhofer Heinrich Hertz Institute, Terahertz Group/Photonic Components, Berlin, Germany*
[6]*Wright State University, Dayton, OH, USA*
[7] *Tokyo Institute of Technology, Department of Electronics and Applied Physics, Japan*
[8]*Universidad Pública de Navarra, Departamento de Ingeniería Eléctrica y Electrónica, Pamplona, Navarra, Spain*
[9] *Imperial College London, Dept. Electrical and Electronic Engineering, London, UK*

2.1 Overview

The THz frequency range (100 GHz to 10 THz) is situated between microwaves and infrared optics. For a long time, it was called the "THz gap", since there were no efficient sources and detectors available, in contrast to the neighboring microwave and optical domains. In the meantime, a multitude of means to generate THz radiation has been developed in order to close this gap. For the highest THz power levels, that is, tens of watts average power and tens of microjoules pulse energies [1–3], factory hall sized free electron lasers (FELs) and synchrotrons have been constructed. A heavily accelerated, relativistic electron beam is guided into an undulator, a structure where the electron is deflected back and forth by alternating magnetic fields. The acceleration of the relativistic electrons perpendicular to their main direction of propagation results in dipole radiation along the propagation axis. In FELs, the undulator is

* Presently with Canon Inc., Frontier Research Centre, THz Imaging Division, Tokyo, Japan

Semiconductor Terahertz Technology: Devices and Systems at Room Temperature Operation, First Edition.
Edited by Guillermo Carpintero, Luis Enrique García Muñoz, Hans L. Hartnagel, Sascha Preu and Antti V. Räisänen.

usually situated inside a THz cavity [4, 5]. The cavity field acts back on the electron beam, resulting in microbunching, that is, dicing the electron beam into packets. Constructive interference of all electron packets occurs, resulting in a laser-like behavior.

Another source is the backward wave oscillator (BWO). This could be considered as the small brother of a synchrotron or FEL. It consists of a vacuum tube that houses an electron source and a slow wave circuit (instead of an undulator). At resonance, a wave is built up in the slow wave circuit that propagates in the opposite direction to the electron beam, hence the name "backward wave" oscillator. In contrast to FELs, BWOs are table-top sources. They provide power levels in the range of milliwatts (at 1 THz) to 100 mW (at ~100 GHz) [6].

True THz lasers are gas lasers, p-germanium (p-Ge) lasers and quantum cascade lasers (QCLs). Gas lasers use polar molecules as the laser medium. Many spectral resonances of molecule rotations of basically all polar molecules (e.g., methanol) are situated in the THz range and therefore suited as THz laser levels. The gas cell, set up within a THz cavity, is usually pumped with an infrared laser, in most cases a CO_2 laser [7]. Only the THz wavelength that is in resonance with both the cavity and a rotational resonance of the molecule is amplified. Average power levels in the milliwatt range can easily be achieved with record continuous-wave (CW) levels above 1 W [8]. However, the frequency is limited to the molecule-specific resonance frequencies. The p-Ge laser consists of p-type germanium mounted in a crossed E (electric) and B (magnetic) field. The laser transitions are either transitions from light to heavy hole valence subbands or transitions between cyclotron resonances [9, 10]. The laser cavity is the crystal itself or an external cavity [11]. So far, p-Ge lasers have to be operated in a cryogenic environment. QCLs consist of a superlattice of quantum wells which is electrically biased [12]. Each superlattice period consists of several quantum wells of different width and barrier thickness. The upper and lower laser levels are extended over several suitably designed wells. Strong coupling through a narrow barrier allows filling of the upper laser level. In order to achieve population inversion, fast depletion of the lower level is ensured by resonant longitudinal optical (LO) phonon-assisted transitions to lower states and strong coupling to the neighboring "ground state" through a narrow barrier. This ground state represents the injector into the upper laser level of the next period of the "quantum cascade" structure, which is comprised of the order of $N \approx 100$ periods. Ideally, a single electron produces 1 THz photon for each sequence in the cascade. As the *spontaneous* radiative transitions between the upper and lower laser levels are competing with by orders of magnitude more efficient non-radiating phonon-assisted transitions, the dark current density in QCLs is high. With sophisticated design, it has become possible to achieve very high THz photon densities in high-Q laser cavities, such that ultimately the stimulated laser transitions largely outnumber the non-radiative processes. Although the threshold currents are high, the quantum efficiency above threshold becomes high and CW power levels in the tens of milliwatts are available in the upper part of the THz range [13]. However, the tunability is limited by the linewidth of the THz resonator (usually the QCL chip itself), and it is very challenging to operate a QCL below 1 THz [14]. Both p-Ge lasers and THz QCLs require cryogenic operation. This often hinders commercial applications.

Another very important optical method of THz generation is nonlinear frequency conversion [15]. Many materials show nonlinear components of the electric susceptibility, χ. The polarization, $P(t)$, of a crystal caused by incident light with field strength $E(t)$ is given by

$$P(t) = \varepsilon_0 \left(\chi^{(1)} E(t) + \chi^{(2)} (E(t))^2 + \chi^{(3)} (E(t))^3 + \ldots \right), \tag{2.1}$$

where $E(t)$ is the sum of all incident electric fields. The strength of the *nonlinear* components (i.e., $\chi^{(n)}$ with $n > 1$) depends strongly on the crystallographic structure of the material. For THz generation, the second-order nonlinearity, $\chi^{(2)}$, is most commonly used. For two incident lasers that differ slightly in frequency, this term contains a component of the difference frequency. $P(t)$ then oscillates and emits photons at the difference frequency that can easily be chosen to be in the THz range. Since $P^{(2)}(t) \sim (E(t))^2$, high fields are required for efficient generation. Therefore, mostly pulsed laser sources are used, with a few CW examples also demonstrated. It also works for single to few THz cycle pulses, since the pulses contain a broad frequency spectrum. The optical laser signal and the THz signal have to co-propagate through some length, $l \sim 0.5$ mm, of the nonlinear crystal. The THz (phase) refractive index and the optical (group) refractive index have to be matched in order to assure propagation at the same speed of light ("phase-matching"). Due to the Manley–Rowe criterion (simply speaking, one pair of optical photons can only generate 1 THz photon), $\chi^{(2)}$ THz generation is very inefficient the further the THz frequency and the optical frequencies are apart (if secondary photons are recycled, however, efficiencies above the Manley–Rowe limit are possible). Typical power efficiency values for THz generation with near-infrared (NIR) lasers are in the range of 10^{-5} [15]. However, the high available power of state of the art pulsed laser systems, the high THz peak power (50 kW in Ref. [16]), the large frequency coverage, and room-temperature operation make nonlinear THz generation very attractive for many applications.

A detailed description of all these approaches is beyond the scope of this book. This chapter will therefore focus only on THz generation by photomixing and by electronic means.

Much like nonlinear THz generation, photomixers down-convert optical photons to THz photons. In contrast to polarization of a crystal, where *one pair* of optical photons generates *one THz photon*, a pair of optical photons generates *one electron–hole pair* in a semiconductor. Each electron–hole pair can emit *many THz photons*. Therefore, THz generation via photogenerated electron hole pairs is typically much more efficient than nonlinear generation, particularly at the lower end of the THz spectrum. This approach works well at room temperature and under ambient conditions, and offers an extremely large tuning range. These sources can be operated with a single, few cycle pulsed laser (broadband operation), two pulsed lasers with pulse widths much longer than the inverse THz frequency (quasi-continuous-wave operation), and in the CW mode (requiring two CW lasers). They are very versatile and are used in many applications. In this chapter, we will develop the theoretical framework, the limitations, and provide various realizations of photomixers.

In Section 2.3 we will discuss the fundamentals of electronic generation of THz radiation. This includes electronic high frequency oscillators such as negative resistance devices (resonant tunneling diodes, and Esaki diodes) and electronic up-conversion of microwave radiation to the THz domain using Schottky diodes and hetero-barrier varactors. Since microwave circuits offer high power levels and high efficiencies, electronic THz generation is very successful and offers many applications. Such applications range from astronomic applications on satellites to table-top applications in the laboratory.

2.2 THz Generation by Photomixers and Photoconductors

2.2.1 Principle of Operation

For simplicity, we start with CW or quasi-continuous-wave photomixing. The formalisms are simpler to understand since all equations have to be derived only for a single THz frequency. However, most results will also be valid and applicable for pulsed operation. A photomixer consists of a semiconductor that is excited with a pair of lasers with powers P_1 and P_2, and optical frequencies $\upsilon_{1,2} = \overline{\upsilon} \pm f_{\mathrm{THz}}/2$, that is, they differ in frequency by the THz frequency. The frequency of the lasers must be sufficiently high in order to generate electron–hole pairs in the semiconductor by absorption, that is $\upsilon_{1,2} > E_{\mathrm{G}}/h$, where E_{G} is the band gap energy of the semiconductor (e.g., $E_{\mathrm{G}} = 1.42$ eV for GaAs) and h is the Planck constant. The lasers with electric field strengths $E_{1,0} \sim \sqrt{P_1}$ and $E_{2,0} \sim \sqrt{P_2}$, are heterodyned, resulting in a total optical field strength of

$$\vec{E}(t) = \vec{E}_1(t) + \vec{E}_2(t) = \vec{E}_{1,0}\mathrm{e}^{\mathrm{i}(\overline{\omega}+\omega_{\mathrm{THz}}/2)t} + \vec{E}_{2,0}\mathrm{e}^{\mathrm{i}(\overline{\omega}-\omega_{\mathrm{THz}}/2)t-\mathrm{i}\varphi}, \tag{2.2}$$

where $\omega_i = 2\pi\upsilon_i$ are angular frequencies and φ is the relative phase. The optical intensity is

$$I_{\mathrm{L}}(t) \sim |\vec{E}(t)|^2 = E_{1,0}^2 + E_{2,0}^2 + 2|\vec{E}_{1,0} \circ \vec{E}_{2,0}| \cos(\omega_{\mathrm{THz}}t + \varphi), \tag{2.3}$$

as illustrated in Figure 2.1a–d.

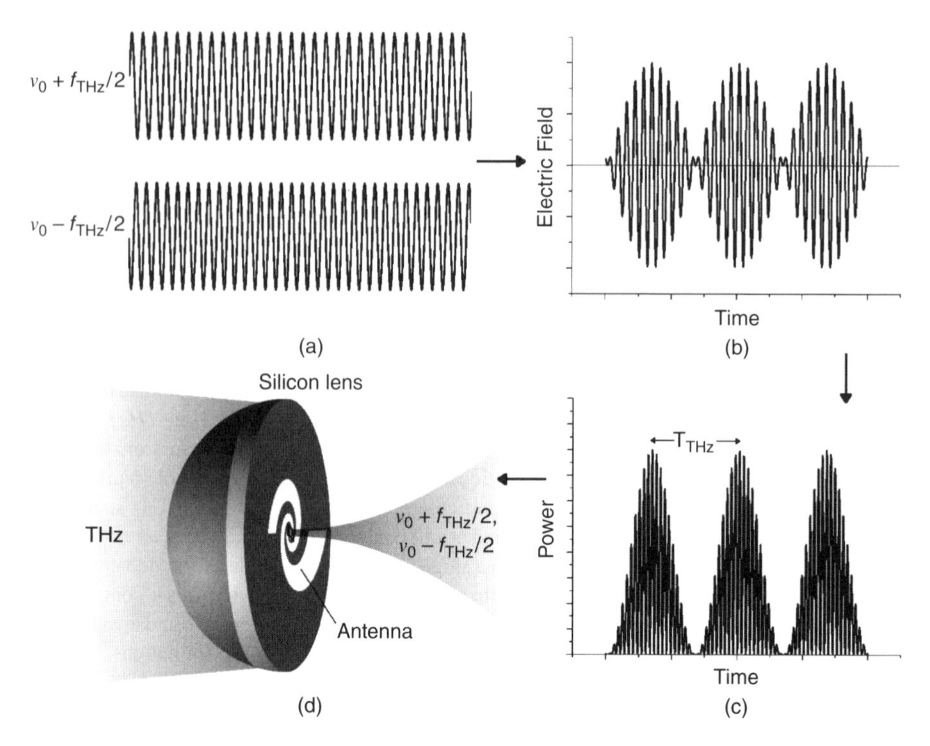

Figure 2.1 Schematic of the photomixing process. (a) Two lasers are heterodyned, (b) resulting electric field, (c) resulting power modulation with a beat node at the difference frequency and (d) the heterodyned lasers are focused on a photomixer. It is connected to an antenna to radiate the THz signal. A silicon lens assists to couple out the THz beam to free space.

Expressing Eq. (2.3) in terms of power yields

$$P_L(t) = P_1 + P_2 + 2\sqrt{P_1 P_2} \cos\beta \cdot \cos(\omega_{THz} t + \varphi), \tag{2.4}$$

where β is the angle between the electric fields (polarizations) of the lasers. An *ideal* semiconductor device (i.e., all light is absorbed, no losses) will generate a photocurrent

$$I_{Ph}^{id}(t) = \frac{eP_L(t)}{h\overline{\upsilon}} = \frac{e(P_1 + P_2)}{h\overline{\upsilon}} + 2\frac{e\sqrt{P_1 P_2}}{h\overline{\upsilon}} \cos\beta \cdot \cos(\omega_{THz} t + \varphi), \tag{2.5}$$

with a DC component of $I_{DC}^{id} = e(P_1 + P_2)/h\overline{\upsilon}$ and an AC amplitude of $I_{THz}^{id} = 2e\sqrt{P_1 P_2}\cos\beta/h\overline{\upsilon}$. The AC current is maximized for $P_1 = P_2 = P_L = ½\, P_{tot}$ and $\beta = 0$, that is, the two lasers have identical power and polarization, yielding $I_{THz}^{id} = eP_{tot}/h\overline{\upsilon} = I_{DC}^{id} = I^{id}$, where $P_{tot} = 2P_L$ is the total laser power. The total current reads

$$I_{Ph}^{id}(t) = I^{id}[1 + \cos(\omega_{THz} t + \varphi)]. \tag{2.6}$$

The AC current is usually fed into some kind of antenna with radiation resistance R_A and an (ideal) THz power is radiated,

$$P_{THz}^{id} = \frac{1}{2} R_A (I_{THz}^{id})^2 = \frac{1}{2} R_A \left(\frac{e}{h\overline{\upsilon}}\right)^2 P_{tot}^2 \tag{2.7}$$

To summarize, two laser beams that differ in frequency by the THz frequency are absorbed by a semiconductor device. The device produces an AC current at the difference frequency of the lasers, namely the THz frequency. This current is fed into an antenna for THz emission. At first glance, this seems to be a complicated scheme. However, the photomixing concept has several outstanding advantages:

- It is very simple to tune a laser by 1 THz (see Chapter 7 for details). For 1550 nm (800 nm) lasers, this corresponds to a wavelength offset of 8 nm (2.2 nm). Most lasers are tunable by several nanometers to tens of nanometers, so photomixers are inherently tunable over an extremely wide range. This is particularly important for spectroscopic applications.
- The linewidth of the THz radiation from a CW photomixer is determined by the linewidth of the lasers. Typical values are a few MHz to a few tens of MHz. This is sufficient for most applications. Long cavity lasers can offer linewidths in the 100 kHz range. To obtain even narrower linewidths, it is also possible to stabilize the lasers [17].
- The interaction length of the laser beam with the semiconductor is short compared to the THz wavelength. In contrast to nonlinear mixing there are no phase matching problems.
- Most importantly, photomixers can be operated at room temperature, but are not limited to that.

For pulsed operation, the device absorbs a single, short optical pulse. By using some kind of radiating structure, such as an antenna, the photocurrent yields THz radiation. The emitted THz field is proportional to the *first time derivative* of the photocurrent,

$$E_{\mathrm{THz}}(t) \sim \frac{\partial I_{\mathrm{Ph}}(t)}{\partial t} \tag{2.8}$$

In order to obtain the frequency spectrum of the emitted signal, the THz field has to be Fourier transformed, yielding

$$E_{\mathrm{THz}}(\omega) \sim \mathrm{FT}\left[\frac{\partial I_{\mathrm{Ph}}(t)}{\partial t}\right] = \mathrm{i}\omega I_{\mathrm{Ph}}(\omega), \tag{2.9}$$

where $I_{\mathrm{Ph}}(\omega)$ is the spectrum of the photocurrent. In the *ideal* case, the photocurrent spectrum is proportional to the optical pulse spectrum. According to the Fourier theorem, the spectral width, $\Delta\omega$, is inversely proportional to the temporal width (current pulse duration), $\Delta\tau$, that is, $\Delta\omega\Delta\tau = 0.5$. Therefore, the optical pulse duration must be (much) shorter than the period of the maximum THz frequency to be obtained. In *real* devices and in contrast to the CW case, the relation between photocurrent and optical power is more complicated: The photocurrent density at time t of a carrier density generated at time t' is given by $j_{\mathrm{Ph}}(t,t') = en(t')v(t-t')$, with the carrier velocity $v(t-t')$. When an optical photon is absorbed close to the band edge, it generates an electron–hole pair that is at rest. The carriers are subsequently accelerated, for instance by an applied DC bias. Thus, not only the carrier concentration (generation rate $\partial n(t)/\partial t \sim P_{\mathrm{L}}(t)$) but also the carrier velocity is time dependent. The THz pulse is thus temporally broader than the optical pulse and depends on the details of the carrier transport in the semiconductor. These broadening mechanisms will be discussed in the next section.

2.2.2 Basic Concepts and Design Rules

There exist two kinds of photomixers: *p-i-n* diode-based mixers and *photoconductive* mixers. In order to develop the limiting factors of photomixers at high THz frequencies, we will start with p-i-n diodes and discuss photoconductive mixers subsequently.

2.2.2.1 p-i-n Diode-Based Photomixers

At DC and at low and intermediate RF frequencies, p-i-n diodes are implemented to generate photocurrent from incident light in many applications. For instance, p-i-n diodes are used in solar cells and also as receivers in communication electronics, where they act as optical to RF converters. A p-i-n diode

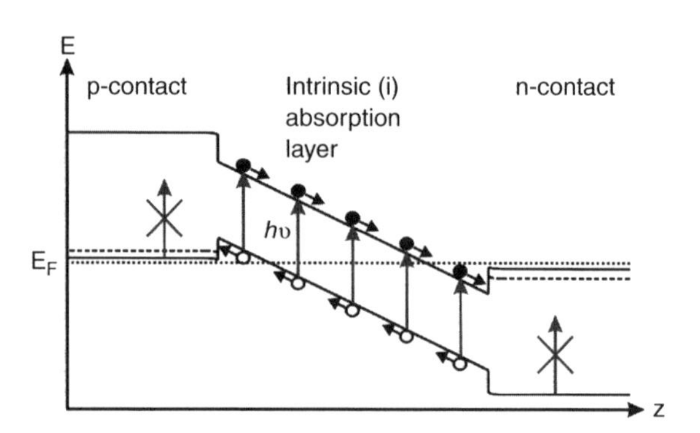

Figure 2.2 Double hetero p-i-n diode structure as used in RF electronics or solar cells. The vertical arrows indicate the photon energy, $h\upsilon$. Optically generated holes move toward the p-contact, electrons to the n-contact. Due to the larger band gap, absorption in the field-free contact layers is not possible.

consists of a p-doped semiconductor, followed by an intrinsic (undoped) part, i, and an n-doped semiconductor. For moderately n/p-doped samples, an electric field of strength $E_i = E_G/d_i$ builds up in the intrinsic layer of length d_i, where E_G is the band gap of the semiconductor. Electron–hole pairs generated in the i-region are efficiently separated by the built-in field. In order to suppress slow contributions due to diffusion of photogenerated electrons from the field-free p-layer into the i-layer and the corresponding contributions due to holes generated in the field-free n-layer, the band gap of the doping layers is increased to energies exceeding the photon energy, for example, by adding aluminum to GaAs or to $In_{0.53}Ga_{0.47}As$. These "double-hetero" (DH) p-i-n diodes with carrier generation restricted to the i-region (see Figure 2.2) can – in principle – also be used at THz frequencies.

Although each absorbed photon fully contributes an elementary charge, e, to the photocurrent, there are several mechanisms that limit the efficiency of the device. First, the semiconductor surface will reflect some of the laser power. This reflection (R) can be minimized by an anti-reflection coating (ARC). Second, not all optical power is absorbed within the i-layer. A real device has a finite absorption length (p-i-n diodes in communication applications typically ~1 μm) for an absorption coefficient α in the range of $10^4\ \mathrm{cm}^{-1}$. Typical materials used for 800 nm (GaAs) and 1550 nm ($In_{0.53}Ga_{0.47}As$), exhibit an absorbance $A = 1 - \exp(-\alpha d_i) = 67\ \%\ (55\ \%)$ for a 1 μm absorption length and an absorption coefficient (close to the band edge) of $1.1 \times 10^4 \mathrm{cm}^{-1}$ for GaAs ($0.8 \times 10^4 \mathrm{cm}^{-1}$ for $In_{0.53}Ga_{0.47}As$) . The reduction of both the AC and DC photocurrent with respect to the ideal photocurrent is summarized in the external quantum efficiency,

$$\eta_{\mathrm{ext}}^{\mathrm{I}} = (1 - R) \cdot [1 - \exp(-\alpha d_i)]. \tag{2.10}$$

Since the THz power is proportional to the square of the photocurrent, the external THz quantum efficiency is

$$\eta_{\mathrm{ext}} = (1 - R)^2 \cdot [1 - \exp(-\alpha d_i)]^2. \tag{2.11}$$

At high frequencies, two further effects reduce the AC current amplitude: Any electronic device has a certain capacitance. In the case of a p-i-n diode, the capacitance is simply that of a plate capacitor with a plate spacing of d_i, the intrinsic layer thickness:

$$C_{\mathrm{pin}} = \varepsilon_0 \varepsilon_r \frac{A}{d_i}, \tag{2.12}$$

where A is the cross-section of the diode. This capacitance is parallel to the radiation resistance, R_A, of the antenna, as illustrated in Figure 2.3. At high frequencies, the antenna is shorted by the capacitance,

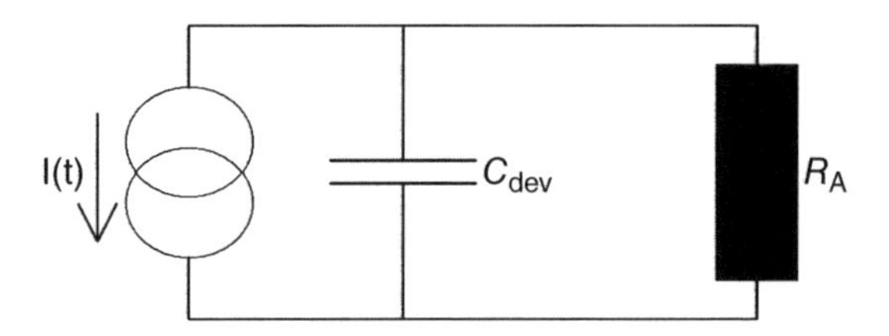

Figure 2.3 Simplified equivalent circuit of a THz emitter. The device capacitance, $C_{dev} = C_{pin}$ is parallel to the radiation resistance of the antenna, R_A. Real devices may further feature a finite conductance parallel to the current source.

reducing the power delivered to the antenna according to

$$\eta_{RC} = \frac{1}{1 + (2\pi R_A C_{pin} f_{THz})^2}. \tag{2.13}$$

This roll-off is called the ***RC roll-off***. At high frequencies, the THz power decreases as f_{THz}^{-2}, the 3 dB frequency (i.e., the frequency where the THz power is reduced by a factor of 2) is $f_{3dB}^{RC} = (2\pi R_A C_{pin})^{-1}$.

Another roll-off is attributed to the carrier transport inside the diode. Carriers that are generated at different times cause currents while being transported to the respective n- or p-contact. These currents interfere. We will discuss two extreme cases of diodes, namely a diode with a strongly confined absorption region and a diode with homogeneous absorption along the intrinsic layer. For a diode where a narrow absorption region is confined close to the p-contact, hole contributions are negligible since they remain stationary and only electrons are transported. For simplicity, we assume for now that electrons are instantaneously accelerated to the saturation velocity, v_{sat}, and then transported at that constant velocity. This is a reasonable approximation for diodes operated at high transport fields. More realistic cases will be treated later. The total current is

$$I(t) = \frac{1}{\tau_{tr}} \int_0^{\tau_{tr}} I(t, \tau')\mathrm{d}\tau', \tag{2.14}$$

where $\tau_{tr} = d_i/v_{sat}$ is the transport time of the carriers and $I(t, \tau') = I^{id}\{\cos[\omega(t-\tau')] + 1\}$ is according to Eq. (2.6) the current at time t that has been generated at time τ'. Integration yields

$$I(t) = I^{id}\left[1 + \cos\left(\omega t - \frac{\omega\tau_{tr}}{2}\right) \cdot \mathrm{sinc}\frac{\omega\tau_{tr}}{2}\right], \tag{2.15}$$

where $\mathrm{sinc}\, x = (\sin x)/x$. The AC *current* amplitude is thus reduced by sinc$(\omega\tau_{tr}/2)$, whereas the DC amplitude remains the same. The 3 dB frequency of the THz *power*, $P_{THz} \sim I^2$, is given by $\mathrm{sinc}^2(\omega\tau_{tr}/2) = 0.5$. This yields a 3 dB frequency of $f_{3dB}^{tr} = 0.44/\tau_{tr} \approx 1/(2\tau_{tr})$. The reduction of the THz power with increasing frequency due to the transport of the carriers is called ***transit-time roll-off***. At higher frequencies, the THz power rolls off quickly, as illustrated in Figure 2.4. The nodes are attributed to our simplifying assumptions of identical carrier velocity and the same transit-time for all carriers. This is, of course, an unrealistic case for real diodes. The envelope function of the roll-off can be estimated by

$$\eta_{tr} \approx \frac{1}{1 + (2\tau_{tr} f_{THz})^2}. \tag{2.16}$$

This envelope is illustrated in Figure 2.4. A saturation velocity of $\sim 10^7$ cm/s (GaAs) and a transport length of 300 nm yield a 3 dB frequency of only 147 GHz. The THz power at higher frequencies will be strongly reduced. The envelope function decreases as f_{THz}^{-2}, similar to the RC roll-off. Obviously, the 3 dB frequency must be increased for efficient devices. One way would be to reduce the transport length.

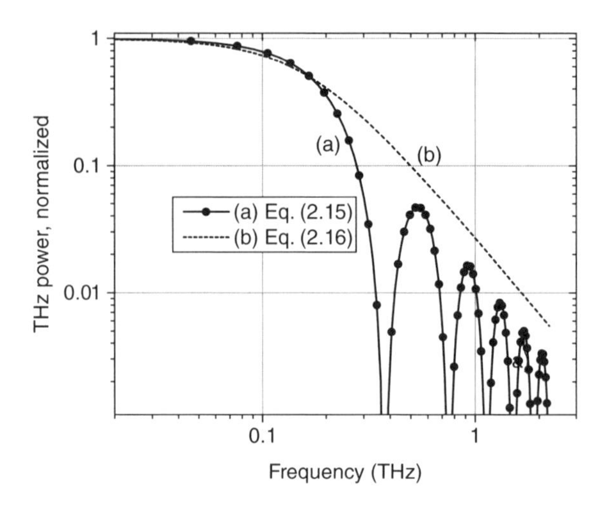

Figure 2.4 (a) THz power versus frequency for a diode with a narrow absorption region close to the p-contact where carriers are quickly accelerated to the saturation velocity, v_{sat}, and then transported across the intrinsic layer with v_{sat} as calculated in Ref. [18]. The nodes are due to neglecting the required acceleration period of the carriers and to the infinitely narrow absorption region. (b) Approximate roll-off according to Eq. (2.16). Below the 3 dB frequency, Eq. (2.16) reproduces the roll-off well; above, the envelope shows the same power law dependence. The exact frequency dependence, however, depends on the details of the carrier transport [18].

According to Eq. (2.12), however, this increases the capacitance and therefore increases the effect of the RC roll-off. Our example of $d_i = 300$ nm yields for a fairly small device of 8 μm × 8 μm, an antenna with a radiation resistance of 70 Ω, and $\varepsilon_r = 13$ an RC 3 dB frequency of only 93 GHz. Such a device is already strongly RC-limited at THz frequencies and thicker intrinsic layers are required, opposing the transit-time limitation. Reducing the device cross-section may be an alternative. In this case, the power density within the device will increase for a given optical power. This limits the maximum photocurrent due to thermal and electrical issues occurring at high power densities. Since the THz power is proportional to the square of the photocurrent, smaller cross-sections strongly reduce the maximum THz power. Thermal limitation will be discussed in detail in Section 2.2.3. The transit-time performance can be improved without affecting the RC-time by optimizing the charge carrier transport within the diode. If the (average) velocity of the carriers can be increased, the same intrinsic layer length results in shorter transit-times and, hence, higher transit-time 3 dB frequencies. This requires a closer look on high-field transport in semiconductors on a short-time scale. In Figure 2.5a, the band structure of GaAs, a typical semiconductor with a direct band gap at the Γ-point, E_Γ, is shown. Apart from the Γ-valley there are additional minima in the conduction band at the L- and the X-points of the Brillouin zone. As they are higher in energy by several 100 meV, the L- and X-valleys are normally not occupied by electrons and, hence, do not contribute to transport. The absorption of photons with energy $h\upsilon$ larger than the band gap, E_G, results in the generation of electrons in the Γ-valley, (and of holes at the top of the corresponding valence bands). Due to the very small effective mass (e.g., $m_{eff} = 0.067m_0$ for GaAs) the acceleration, a, of electrons by electric fields E, $a = eE/m_\Gamma$, is very efficient. For electrons at the bottom of the Γ-valley, there exist no efficient mechanisms for scattering. Therefore, they are "quasi-ballistically" accelerated, that is, their velocity increases with time according to $v = at = \frac{eE}{m_\Gamma}t$. This quasi-ballistic motion persists until the electrons reach the energy of the first side valley, $E_{\Gamma L}$. Here, extremely efficient inter-valley phonon scattering from the Γ- into the L-side valley sets in. The effective masses in these valleys (reflecting the much

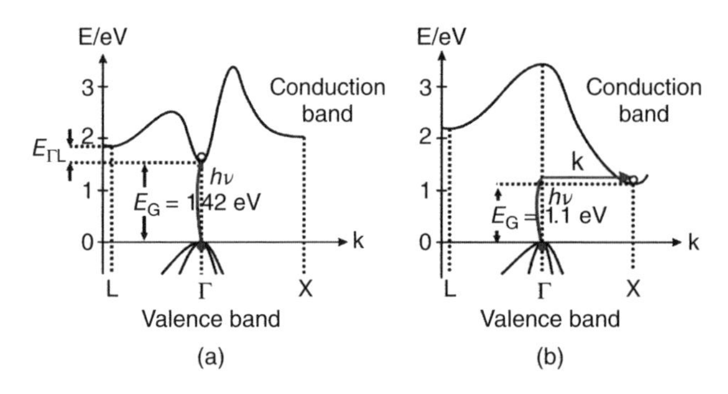

Figure 2.5 Bandstructure of GaAs (a) versus silicon (b).

smaller curvature of the band near the respective conduction band minima) are much larger. Further, very efficient phonon scattering between various side valleys randomizes their direction of motion, limiting the (average) drift velocity to the nearly field-independent saturation velocity $v_{sat} \approx 10^7$cm/s.

Particularly, InGaAs-based devices benefit from ballistic effects since the inter-valley energy of InGaAs of $E_{\Gamma L} = 0.46$ eV [19, 20] is more than 50% higher than that of GaAs[1] ($E_{\Gamma L} = 0.29$ eV [20, 21]) and the effective mass in the Γ-valley is smaller ($m_\Gamma = 0.04m_0$ vs. $m_\Gamma = 0.067m_0$ for GaAs). For InGaAs, the maximum ballistic velocity, $v_{bal} \approx 2 \times 10^8$ cm/s ($\approx 10^8$ cm/s in GaAs) is about ten times higher than the saturation velocity, v_{sat}. Results of Monte-Carlo simulations of the transient velocity $v(t)$ of an ensemble of electrons generated at $t = 0$ for InGaAs are depicted in Figure 2.6 for different electric fields.

With increasing field, the "velocity overshoot" approaches the maximum ballistic velocity. The left panel of Figure 2.6b, depicting the distance covered by an electron within the time, t, is particularly instructive. At a field of 20 kV/cm the electron travels over a distance of about 250 nm within less than 300 fs. At a field of 40 kV/cm the ballistic transport turns into propagation with saturation velocity already at about 150 fs and it takes several picoseconds (!) before the electron will have covered the same distance of 250 nm. With increasing field the situation becomes even worse.

In semiconductors with an indirect band gap, like silicon (Si), the bandstructure in momentum space is illustrated in Figure 2.5b. X-valleys are situated at the bandgap energy above the top of the valence bands, whereas the Γ-valley is found at higher energies. Therefore, the electrons always occupy these valleys, exhibiting a rather large effective mass, high scattering rates (particularly between the X-valleys), and, hence, much lower mobilities ($\mu < 1400$ cm^2/Vs for intrinsic Si) compared with direct gap semiconductors. The short-time high-field transport exhibits no velocity overshoot due to scattering between all X-valleys. The high-field drift velocities do not exceed the value $v = \mu E$ at any time. For Si, for example, the stationary velocity is $<10^7$cm/s, even at 100 kV/cm. Apart from the unfavorable transport properties, indirect bandgap semiconductors exhibit very low absorption coefficients near the bandgap as absorption takes place only via second order phonon-assisted processes. For these reasons, the "standard semiconductor" Si is practically not used for THz photomixing; all work focuses on (direct band gap) III/V semiconductors.

Unfortunately, holes do not show any (remarkable) velocity overshoot. If electrons are quasi-ballistically transported, holes are much slower. There is a great benefit to generate holes close to or in the p-contact such that their transport length remains short. This is achieved by implementing a semiconductor with a band gap $E_G < h\bar{\upsilon}$ at the p-contact where absorption and carrier generation

[1] Materials parameters for various semiconductors taken from www.ioffe.ru\SVA\NSM\Semicond\index.html and references thereof.

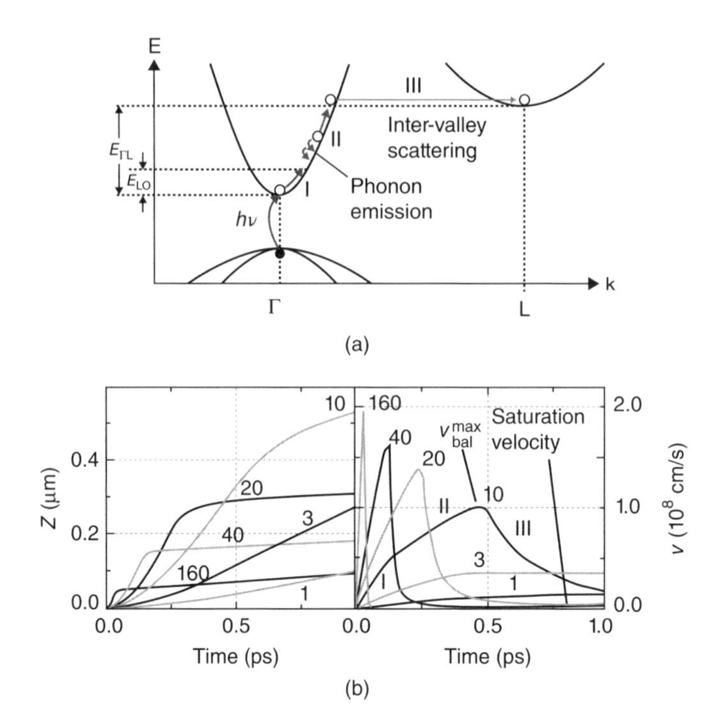

Figure 2.6 (a) Illustration of the electron transport in InGaAs (energy axis not to scale); in region I the electrons are accelerated by an applied field, E, with $a = eE/m_\Gamma$. In region II, the electrons have surpassed a kinetic energy of E_{LO}. Longitudinal optical (LO) phonon emission results in a somewhat reduced acceleration of the electrons. III: The electrons reached the inter-valley energy $E_{\Gamma L}$. Scattering in one of the 6 equivalent L-valleys strongly reduces the average velocity of the electron ensemble. (b) Monte-Carlo simulation of the covered distance and the electron velocity for InGaAs for electric field strengths from 1 to 160 kV/cm. The regions I–III are indicated for the accelerating field of 10 kV/cm. Reproduced and adapted with permission from Ref. [18] S. Preu *et al.*, J. Appl. Phys. 109, 061301 (2011) © 2011, AIP Publishing LLC.

can take place, followed with another, intrinsic semiconductor with $E_G > h\bar{\upsilon}$ such that no carriers are generated there. The latter is called the *transport region* or *transport layer*. Since only electrons are transported, such devices are called *Uni-Traveling-Carrier (UTC)* photodiodes. A typical band structure is illustrated in Figure 2.7. Another way to restrict the hole transport to a small fraction of the intrinsic layer is just modifying the intrinsic layer of a DH – p-i-n diode (see Figure 2.2) by a gradually increasing Al-content toward the n-layer.

In order to get the optimum transport field, that is, the field in the intrinsic layer where the electrons are transported ballistically as far as possible (see Figure 2.6), a small external DC bias can be applied to the diode. Since quasi-ballistic electrons can reach peak velocities that are about a factor of 10 higher than the saturation velocity, their average velocity is up to a factor of 3–5 higher. For our example with an intrinsic layer length of 300 nm, this yields a transit-time 3 dB frequency of about 0.75 THz and increases the THz power at high frequencies by a factor of 25 according to Eq. (2.16). The RC roll-off can be improved by using smaller devices (which will ultimately limit the maximum photocurrent) or by using a series connection of p-i-n diodes, as realized by the n-i-pn-i-p superlattice photomixer concept [22]. Various layouts of p-i-n diode-based photomixers will be discussed in Section 2.2.6.

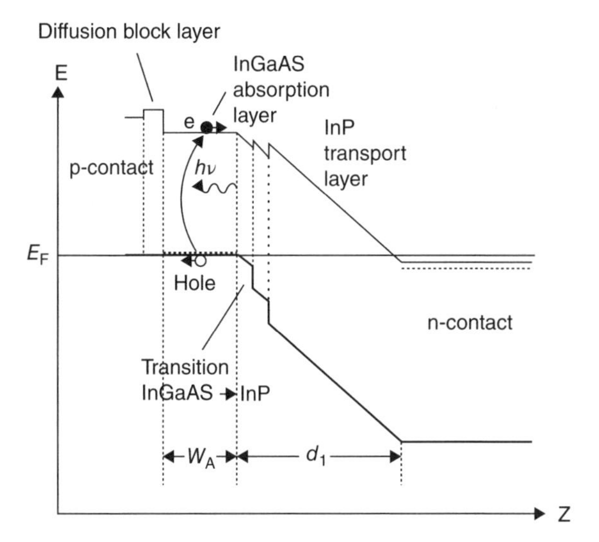

Figure 2.7 Typical band structure of a uni-traveling-carrier (UTC) photomixer. Photons are absorbed in the (thin) InGaAs absorption region and generate electron–hole pairs. The electrons are transported across the InP transport layer to the n-contact, Holes drift to the p-contact. Reproduced and adapted with permission from Ref. [18] S. Preu *et al.*, J. Appl. Phys. 109, 061301 (2011) © 2011, AIP Publishing LLC.

2.2.2.2 Photoconductive Mixers

The second type of photomixer is the photoconductive mixer. It consists of a highly resistive semiconductor that is covered with metal contact electrodes, as illustrated in Figure 2.8. The heterodyned laser beams are absorbed by the semiconductor in the electrode gap, leading to the generation of electron–hole pairs. Since there is no built-in field as in p-i-n diodes, an external DC bias is required in order to separate the carriers and generate a photocurrent. Typical bias levels range from a few volts to ~100 V, depending on the electrode gap, w_G, and the semiconductor break-down field strength.

Figure 2.8a shows the side view of the electrode structure. The electric field in the gap is inhomogeneous. The field lines become longer the deeper we look in the photoconductive material. The highest field strengths are at the surface in the vicinity of the electrodes. In addition, the electron–hole generation rate is highest at the surface. The absorption follows Lambert–Beer's law according to Eq. (2.10). Typical values of the absorption coefficient of 11 000 cm^{-1} (GaAs at 840 nm wavelength) yield a $1/e$ penetration depth of 0.9 μm. This value is roughly the required thickness of the photoconductive material for high absorption. The thickness of the active photoconductive layer and the inhomogeneity of the absorption may be reduced by growing a Bragg reflector below the photoconductor. Often, an ARC (a transparent dielectric layer of width $d = \lambda_0/(4n)$, where $n \approx \sqrt{n_{sc}}$ is the refractive index of the dielectric) is applied on top of the electrode structure in order to reduce reflection and to passivate the structure. The latter is very important since both carrier generation and electric fields are highest at the surface. The generated heat can lead to oxidation and finally to destruction of the device.

In most cases, an antenna is connected to the electrodes in order to emit THz radiation. Similar to p-i-n diode-based photomixers, photoconductors also show an RC roll-off toward high frequencies. The capacitance of a finger-like electrode structure as illustrated in Figure 2.8b is [19]

$$C = \frac{K(k)}{K(k')} \cdot \frac{\varepsilon_0(1+\varepsilon_r)}{w_e + w_g} A, \tag{2.17}$$

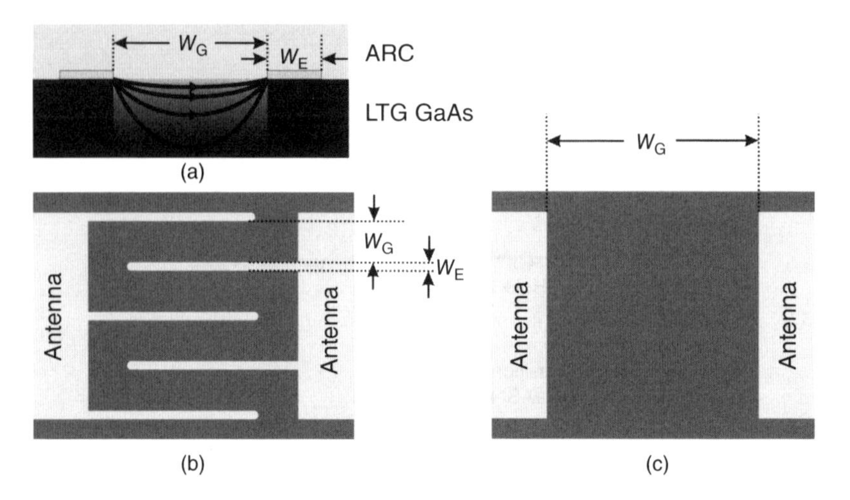

Figure 2.8 (a) Cross-section of a photoconductor. The electric DC fields, due to biasing the electrodes, are indicated by the bowed lines. The gradient illustrates the amount of photo-generated carriers in the gap (logarithmic scale; bright: many carriers, dark: few carriers). (b) Top view of a photoconductive mixer with fingers for CW operation. (c) Fingerless gap for pulsed operation. See plate section for color representation of this figure.

with $k = \tan^2(\pi w_E/[4(w_E + w_G])$, $k' = \sqrt{1 - k^2}$, and $K(k) = \int_0^{\pi/2} \frac{d\theta}{\sqrt{1-k^2 \sin^2\theta}}$ is the complete elliptic function of the first kind, w_G is the gap width, w_E is the electrode width, and A is the total (square) active area. In order to assure efficient coupling of the lasers, the diameter of the electrode structure must be of the same order as the laser beam diameter. Typical values are $10 \times 10\ \mu m^2$. An electrode width of $w_E = 0.2$ μm and an electrode gap of 1.8 μm, requires five electrodes in order to cover 10 μm. This yields a capacitance of 1.7 fF. For an antenna with 72 Ω, the RC 3 dB frequency is 1.3 THz, already leading to an RC roll-off-free performance at lower frequencies. We see that much smaller electrode gaps are not beneficial for RC-roll-off-free operation below 1 THz. Further, the area covered by the metal electrodes cannot be optically excited. Smaller gaps at a fixed, finite electrode width therefore reduce the optically active area.

CW Operation

Similar to p-i-n diodes, the transport of carriers also results in a roll-off. For CW operation, the semiconductor is always populated with optically generated carriers. In p-i-n diodes, transport lengths of a few 100 nm already result in a transit-time 3 dB frequency below 1 THz. Typical gaps of (CW operated) photoconductors, however, are in the range of 2 μm. In order to speed up the carrier transport, high biases have to be applied to the electrodes. For a close-to breakdown field of ~300 kV/cm for GaAs at the surface, a maximum bias of 60 V can be applied for a 2 μm electrode gap. For high quality GaAs with a mobility of $\mu = 8000\ \text{cm}^2/\text{Vs}$, carriers can easily reach the saturation velocity of $\sim 10^7$ cm/s at such accelerating fields. This results in a transport time of $\tau_{tr} \sim 16$ ps where the time for accelerating the carrier was even neglected. Velocity overshooting effects can be neglected at these time scales since they only play a role in the first few 100 nm of the transport and require lower fields. For intrinsic GaAs, the transit-time 3 dB frequency is only $f_{3dB}^{tr} = 1/(2\tau_{tr}) = 31$ GHz. Such a low value would result in a severe roll-off at THz frequencies since most carriers would need too much time to cross the electrode gap. Therefore, (CW) photomixers use materials with a short carrier lifetime, where carriers are trapped (and subsequently recombine with their counterparts) on their way to the electrode and no longer contribute to the displacement current. In order to calculate the displacement current, we assume transport at the (constant) saturation velocity without any velocity overshoot (which does not play any role in short-lifetime material). We further simplify the transport by assuming that the transport length equals the electrode gap since most electrons are generated close to the surface. The number of charges generated at time t' within a time interval $\Delta t'$ by the beating of the lasers is $\Delta N(t') = \eta^{\mathrm{I}}_{\mathrm{ext}} I^{\mathrm{id}}[1 + \cos(\omega_{\mathrm{THz}} t')]\Delta t'$, according to Eq. (2.6). They undergo exponential damping due to fast trapping and recombination within τ_{rec}. At a time $t > t'$, only $\Delta N(t') \exp(-[t - t']/\tau_{\mathrm{rec}})$ carriers have survived. In order to calculate the current, all remaining carriers (i.e., non-trapped) have to be considered to contribute to the current. The current *density* is given by $j = env_{\mathrm{dr}}$, where v_{dr} is the drift velocity, and $en = N/V$ is the charge density per volume. The volume is spanned by the gap width, w_{G}, along the x direction, the $1/e$ penetration depth of the laser, z_0, and the length of the electrode, l_y, yielding $V = z_0\, l_y\, w_{\mathrm{G}}$. The area $A = z_0\, l_y$ is the cross-section of the current, $I = jA$. At time t, the generated current is $\Delta I(t, t') = \Delta N(t')\, v_{\mathrm{dr}}/w_{\mathrm{g}} \exp(-[t - t']/\tau_{\mathrm{rec}})$ for carriers generated at time t'. The term $w_{\mathrm{g}}/v_{\mathrm{dr}} = \tau_{\mathrm{tr}}$ is the transport-time of the carriers. The current at time t is the integration over all generation times, t',

$$I(t) = \frac{1}{\tau_{\mathrm{tr}}} \int_{t-t'=0}^{t-t'=\tau_{\mathrm{tr}}} \eta^{\mathrm{I}}_{\mathrm{ext}} I^{\mathrm{Id}}[1 + \cos(\omega_{\mathrm{THz}} t')] \exp(-[t - t']/\tau_{\mathrm{rec}}) \mathrm{d}t', \tag{2.18}$$

where $t - t'$ is the transport time of a carrier after its generation. Since most of the carriers should be trapped on their way to the contact, we can assume that the transport time is much larger than the trapping/recombination time, $\tau_{\mathrm{tr}} \gg \tau_{\mathrm{rec}}$, and therefore we can shift the upper integration boundary toward infinity with marginal error. For the same reason, we can also ignore the dependence of the transit-time with respect to the place of generation of the carriers relative to the electrodes. Substituting $t - t' = \tau$ and integrating Eq. (2.18) yields

$$I(t) = \eta^{\mathrm{I}}_{\mathrm{ext}} I^{\mathrm{id}} \frac{\tau_{\mathrm{rec}}}{\tau_{\mathrm{tr}}} \left[\frac{\sin\left(\omega_{\mathrm{THz}} t - \varphi\right)}{\sqrt{1 + (\omega_{\mathrm{THz}} \tau_{\mathrm{rec}})^2}} + 1 \right], \tag{2.19}$$

with $\varphi = \arctan(\omega_{\mathrm{THz}} \tau_{\mathrm{rec}})$. We see that both the time-dependent (AC) and the DC part of the current are damped due to trapping by a factor of $g = \tau_{\mathrm{rec}}/\tau_{\mathrm{tr}}$. Although this quantity is usually (much!) smaller than 1, it is called the ***photoconductive gain***. The time-dependent (AC) current is further reduced by a factor of $\left(\sqrt{1 + \left(\omega_{\mathrm{THz}} \tau_{\mathrm{rec}}\right)^2}\right)^{-1}$. Since the THz power is $P \sim I^2$, the power is reduced at high frequencies by

$$\eta_{\mathrm{LT}} = \frac{1}{1 + (2\pi f_{\mathrm{THz}} \tau_{\mathrm{rec}})^2}, \tag{2.20}$$

where we replaced $\omega_{\mathrm{THz}} = 2\pi f_{\mathrm{THz}}$. Mathematically, this ***lifetime roll-off*** has the same form as the transit time roll-off for p-i-n diodes in Eq. (2.16), however, with a 3 dB frequency of $f^{\mathrm{LT}}_{3\mathrm{dB}} = 1/(2\pi\tau_{\mathrm{rec}})$ that is formally a factor of π smaller than the transit-time 3 dB frequency. Compared to an ideal device, the emitted THz power of a photoconductor is reduced by

$$\eta_{\mathrm{PC}} = \frac{g^2}{1 + (2\pi f_{\mathrm{THz}} \tau_{\mathrm{rec}})^2}. \tag{2.21}$$

In order to obtain a flat frequency response of the device, short lifetimes, τ_{rec}, are required. However, they should not be too small since the photoconductive gain is also proportional to the lifetime (the latter becomes important when we discuss thermal limitation of a photoconductor in Section 2.2.3). A low lifetime reduces the mobility of the carriers, requiring higher fields in order to accelerate them. Particularly for CW operation, engineering of the carrier lifetime of the photoconductive material is of key importance for efficient operation. For a flat frequency response up to 1 THz, a lifetime of 160 fs is required. Examples of materials with such a short carrier lifetime will be given in Section 2.2.5. Optimum values for both lifetime and electrode gap for photoconductors have been calculated in Ref. [23] for both a typical resonant antenna and a typical broadband antenna.

Pulsed Operation

For pulsed operation, a short lifetime is not so crucial for the photoconductor (or Auston switch, named after David Auston who first used photoconductors for generation of THz pulses). According to Eq. (2.9), the (ideal) THz spectrum is proportional to the spectral width of the photocurrent. Even if the decay of the current is slow (i.e., on the time scale of 20 ps as for the case of intrinsic GaAs), the turn-on can be fast, providing high frequency components in the Fourier spectrum. Further, the material can relax on long time scales between two subsequent pulses – in contrast to the CW case where carriers are continuously generated.

In order to derive the emitted THz spectrum under pulsed operation, the details of the photocurrent generation have to be investigated. The generated field by a radiating structure (such as an antenna or by any current) is proportional to the time-derivative of the current [24–26]. The photocurrent density at time t resulting from a carrier density, $n(\vec{r}, t, t')$ at position $\vec{r}$, generated at time t' at depth z with respect to the surface is given by

$$\partial \vec{j}_{\mathrm{Ph}}(\vec{r}, t, t') = e\vec{v}(\vec{r}, t - t')\, \partial n(\vec{r}, t, t'), \tag{2.22}$$

where $\vec{v}(\vec{r}, t - t')$ is the velocity of the carriers at time t. The photoconductor may feature a short carrier lifetime, yielding $n(\vec{r}, t, t') = n_0(\vec{r}, t') \exp(-[t - t']/\tau_{\mathrm{rec}})$ as for the CW case. For the total current, holes and electrons have to be considered. Since the formulas are formally the same for both carrier types, we will restrict our treatment to electrons only. However, holes feature a larger effective mass and do not show any velocity overshoot, resulting in smaller values for $\vec{v}(\vec{r}, t - t')$. Consequently, the contribution of hole currents is usually smaller. Further, we will for now only treat the small-signal limit, neglecting any saturation effects which will be treated later in Section 2.2.4. The electron–hole generation *rate* is proportional to the optical power,

$$\frac{\partial n_0(\vec{r}, t')}{\partial t'} = \alpha \frac{P_{\mathrm{L}}(\vec{r}, t')}{h\bar{\upsilon}A} = \alpha \frac{\Phi_{\mathrm{L}}(\vec{r}, t')}{h\bar{\upsilon}}, \tag{2.23}$$

where A is the device cross-section where the optical power, $P_{\mathrm{L}}(\vec{r}, t - \tau) = \exp(-\alpha z) P_{\mathrm{L},0}(x, y, t - \tau)$, is distributed, and α is the absorption coefficient of the photoconductor. $\Phi_{\mathrm{L}}(\vec{r}, t') = P_{\mathrm{L}}(\vec{r}, t')/A$ is the optical intensity. Using the same formalism for Eq. (2.23) for the current as in Eq. (2.18) and integrating over the photoconductor cross-section yields for the local current

$$I_{\mathrm{Ph}}(x, t) = \int \int_{t-t'=0}^{t-t'=\tau_{\mathrm{tr}}(x)} e\alpha \frac{\Phi_{\mathrm{L}}(\vec{r}, t')}{h\bar{\upsilon}} \exp(-[t - t']/\tau_{\mathrm{rec}}) v_{\mathrm{T}}(\vec{r}, t - t') \mathrm{d}t'\, \mathrm{d}z\, \mathrm{d}y, \tag{2.24}$$

where $v_{\mathrm{T}}(\vec{r}, t - t')$ is the surface-tangential component of the velocity. In order to get the total current, we have to sum over all components generated at all coordinates x,

$$I_{\mathrm{Ph}}(t) = \frac{1}{w_{\mathrm{G}}} \int_0^{w_{\mathrm{G}}} I_{\mathrm{Ph}}(x, t) \mathrm{d}x \tag{2.25}$$

The same formula applies to the current generated by holes. The two contributions have to be summed in order to provide the total current, and hence, the emitted THz field which is proportional to the time derivative of the current.

Since the carrier velocity depends on both the field strength of the accelerating DC field (see Figure 2.6b) and the carrier dynamics in the semiconductor (ballistic acceleration vs transport at the saturation velocity), Eq. (2.25) can only be solved analytically by making a few simplifying assumptions. We are only interested in the temporal evolution. The majority of the carriers will be generated close to the surface. The fields at small penetration depths are very similar, resulting in little spatial variation of the carrier velocities. Since the electric field is proportional to the time derivative of the current, we are only interested in time-varying components. This means that either the pulse duration or the carrier lifetime (or both) should be much shorter than the transit-time. We further assume a homogeneous illumination. Gaussian beams could also be considered, however, they make the derivations more complicated. Under these simplifying assumptions, the integrand no longer possesses any spatial dependence. The spatial integrals result in the total absorbed power. Dropping all other constants and substituting $t-t'=\tau$ yields

$$E_{\mathrm{THz}}(t) \sim \frac{\partial}{\partial t}\int_{0}^{\tau_{\mathrm{tr}}} P_{\mathrm{L}}(t-\tau)\mathrm{e}^{-\tau/\tau_{\mathrm{rec}}}v(\tau)\mathrm{d}\tau. \tag{2.26}$$

In order to investigate the influence of the optical pulse width and the lifetime of the carriers, we will now focus on a few important special cases. The first case is that of a photoconductor with a long lifetime. The exponential damping due to the lifetime can be neglected. The transit-time should be long compared to the pulse duration. The transport shall be non-ballistic providing $v(\tau)=v_{\mathrm{sat}}=\mathrm{const}$. The device is excited with a very short Gaussian pulse, $P_{\mathrm{L}}(t)=P_0\exp(-t^2/\tau_{\mathrm{pls}}^2)$, where τ_{pls} is the pulse duration. The THz-field becomes

$$E_{\mathrm{THz}}(t) \sim \frac{\partial}{\partial t}\int_{0}^{\infty} P_{\mathrm{L}}(t-\tau)v_{\mathrm{sat}}\mathrm{d}\tau \sim v_{\mathrm{sat}}P_{\mathrm{L}}(t). \tag{2.27}$$

The generated electric field is then simply proportional to the optical pulse, $E_{\mathrm{THz}}(t) \sim P_0\exp(-t^2/\tau_{\mathrm{pls}}^2)$ as illustrated in Figure 2.9a with a power spectral density of

$$P_{\mathrm{THz}}(\omega) \sim |E_{\mathrm{THz}}(\omega)|^2 \sim v_{\mathrm{sat}}^2P_{\mathrm{L}}^2(\omega), \tag{2.28}$$

with $P_{\mathrm{L}}(\omega)=P_0\exp(-\omega^2\tau_{\mathrm{pls}}^2/4)$. This calculated field, however, is not the emitted THz field. It is strictly positive and therefore contains also a DC component which cannot be radiated. In addition, any kind of THz set-up has a cut-off frequency: At lower frequencies, larger implemented antennas and apertures must be used (size of the order of the wavelength at least) which ultimately will hit experimental limits. Often, low frequency components may be suppressed by the detection technique due to a detector cut-off or similar. Therefore, low frequency components are always filtered. Removing the low frequency components results in a single cycle THz pulse, as obtained in experiments and shown in Figure 2.9a by the black line. Its frequency spectrum is shown in Figure 2.9b. It is of Gaussian shape, dropping fast at frequencies above $f_{3\mathrm{dB}}=\sqrt{\ln 2}/(\sqrt{2}\pi\tau_{\mathrm{pls}})$. Here, no RC roll-off was taken into account. Similar to CW operation, the RC roll-off will reduce the power at higher THz frequencies.

For most practical cases, the transit-time, τ_{tr}, is orders of magnitude longer than all other involved time scales. For the following cases, the upper integration boundary of Eq. (2.25) will therefore be shifted to infinity and explicit dependences on the transit time of the carriers will be neglected. By continuing the velocity, $v(\tau)$ toward negative values of τ by adding zeroes, that is, setting $v^*(\tau)=v(\tau)\theta(\tau)$, where $\theta(\tau)$ represents the Heaviside step function, the lower integral boundary of Eq. (2.26) can be set to $-\infty$. With

$$h(\tau)=\exp(-\tau/\tau_{\mathrm{rec}})v(\tau)\theta(\tau) \tag{2.29}$$

Equation (2.26) becomes a convolution of the pulse shape and the function $h(\tau)$,

$$E_{\mathrm{THz}}(t) \sim \frac{\partial}{\partial t}[P_{\mathrm{L}}(t) * h(t)], \tag{2.30}$$

where "*" represents the convolution operator. If both functions are integrable, the Fourier transformation of Eq. (2.26) yields the simple result [27]

$$E_{\mathrm{THz}}(\omega) \sim \mathrm{i}\omega P_{\mathrm{L}}(\omega)H(\omega), \tag{2.31}$$

with $H(\omega)$ being the Fourier transform of $h(\tau)$. The THz power spectrum is [27]

$$P_{\mathrm{THz}}(\omega) \sim \omega^2|P_{\mathrm{L}}(\omega)|^2 \cdot |H(\omega)|^2, \tag{2.32}$$

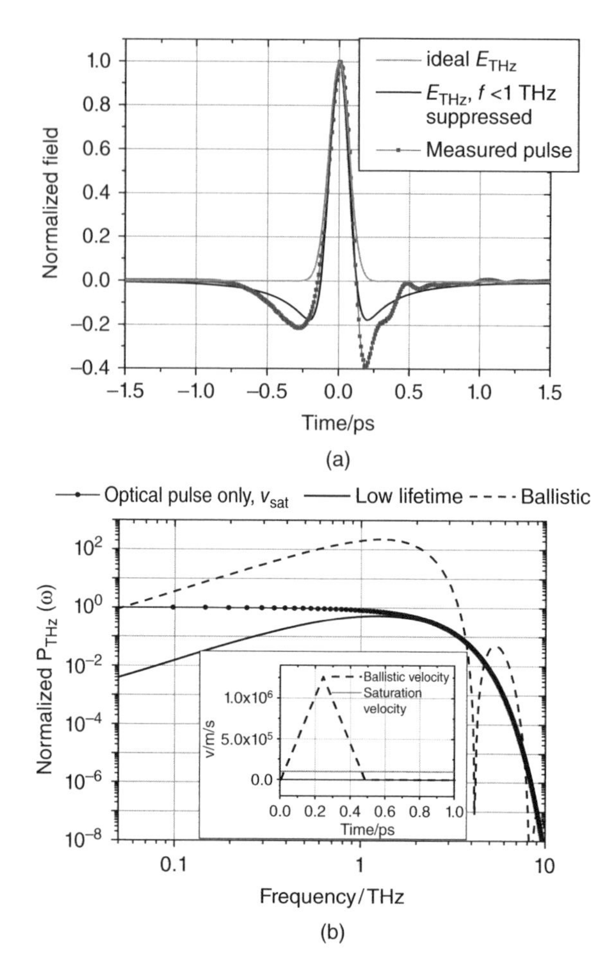

Figure 2.9 (a) Bright line: Calculated field according to Eq. (2.28) for $\tau_{pls} = 100$ fs. Solid black line: pulse resulting from the bright line by suppressing frequency components below 1 THz as $\sim\omega^2$. Line with symbols: An experimentally obtained THz pulse with a comparable pulse duration. (b) Fourier spectra of the fields: solid line with points according to Eq. (2.28) for semi-insulating (SI) GaAs. Gray solid line: Response of a material with short lifetime according to Eq. (2.34). Dashed line: Spectrum of a material with ballistic transport according to Eq. (2.36). Inset: Assumed velocity trace for ballistic and non-ballistic transport. Reproduced and adapted from Ref. [27] S. Preu, J. Infrared Milli Terahz Waves 35, 998 (2014) with permission from Springer, © 2014 Springer.

that is, the product of the power spectrum of the laser, the power spectrum of the carrier dynamics (represented by the function $H(\omega)$) and the term ω^2. The term $\sim\omega^2$ can be identified as the Hertzian dipole radiation resistance $R_A^H \sim \omega^2$ [28]. It needs to be replaced by the frequency dependence of the antenna radiation resistance if any is used.

Equation (2.32) is a powerful tool in order to calculate the emitted spectra since it separates the influence of the pulse shape and the carrier dynamics. If the device shows an RC roll-off, it can simply be multiplied to Eq. (2.32).

We now focus on two further important cases by using Eq. (2.32): first, a photoconductive material with a short lifetime, $\tau_{rec} < 1$ ps. Since scattering dominates the transport in a low lifetime material, no velocity overshoot has yet been reported. Therefore, the carrier velocity can be assumed to be the saturation velocity, $v(\tau) = v_{\mathrm{sat}} = \mathrm{const}$. For typical accelerating fields, the time required to accelerate the carriers is very short. Therefore, the carriers are assumed to be moving at the saturation velocity for the whole transport, neglecting slower velocities right after carrier generation. Furthermore, the short lifetime assures that the dependence of the point of generation does not play any dominant role and the transit-time can be set to infinity. This results in the integrable function $h(\tau) = \exp(-\tau/\tau_{\mathrm{rec}})v_{\mathrm{sat}}\theta(\tau)$. The power spectrum is

$$|H(\omega)|^2 = \frac{v_{\mathrm{sat}}^2 \tau_{\mathrm{rec}}{}^2}{1 + (\omega\tau_{\mathrm{rec}})^2}. \tag{2.33}$$

The function $|H(\omega)|$ can be identified as the effective dipole responsible for THz generation, namely the covered distance of separated electrons and holes, $l_{\mathrm{D}} = v_{\mathrm{sat}}\tau_{\mathrm{rec}}$, and a transit-time roll-off.

The power spectrum according to Eq. (2.26) is

$$P_{\mathrm{THz}}(\omega) \sim \frac{(\omega v_{\mathrm{sat}}\tau_{\mathrm{rec}})^2}{1 + (\omega\tau_{\mathrm{rec}})^2} P_{\mathrm{L}}^2(\omega). \tag{2.34}$$

It is also illustrated in Figure 2.9. Note that the transit-time roll-off term in $|H(\omega)|$ does not prevent generation of high frequency components since $P_{\mathrm{L}}(\omega)$ contains these frequencies. Above $f > (2\pi\tau_{\mathrm{rec}})^{-1}$, the frequency dependences of the prefactor in Eq. (2.34) cancel, resulting in $P_{\mathrm{THz}}(\omega) \sim v_{\mathrm{sat}}^2 P_{\mathrm{L}}^2(\omega)$ as in the case of semi-insulating (SI) GaAs with transport at the saturation velocity.

The second case is that of a material with ballistic electron transport. Ballistic transport is not possible in a short lifetime material because of excessive scattering at trap states. Therefore, the lifetime must be long, the exponential term in Eq. (2.26) has to be dropped. The electron velocity distribution shown in Figure 2.6b is quite complex. We approximate the onset of the transport by a ballistic acceleration of $a = e/m_\Gamma E_{\mathrm{DC}}$, where E_{DC} is the accelerating field resulting from the DC bias to the structure. This acceleration prevails until the electron reaches enough energy to scatter into the side valley. This happens after a ballistic flight time of

$$\tau_{\mathrm{bal}} = \sqrt{\frac{2m_\Gamma E_{\Gamma\mathrm{L}}}{e^2 E_{\mathrm{DC}}^2}} \tag{2.35}$$

Equation (2.35) was obtained by equating the inter-valley energy $E_{\Gamma\mathrm{L}}$ to the kinetic energy of the electron. For simplicity, we assume that the deceleration due to scattering happens with an acceleration of $a' = -a$ until the electron is at rest. According to Figure 2.6b, right panel, this is a rather realistic assumption. Of course, in the experiment the electron density will not be at rest but rather travel with the saturation velocity. Due to the time derivative and the Fourier transformation, transport at a constant, comparatively small saturation velocity does not play any role. For an optimum accelerating field of 20 kV/cm and an effective electron mass of $m_\Gamma = 0.067 m_0$, the acceleration is $a = 5.25 \times 10^{18}$ m/s^2 and the ballistic flight time is $\tau_{\mathrm{bal}} = 0.24$ ps. The power spectral density of the carrier dynamics, $|H(\omega)|^2$ is then

$$|H(\omega)|^2 = (v_{\mathrm{bal}}^{\max}\tau_{\mathrm{bal}})^2 \left(\mathrm{sinc}\frac{\omega\tau_{\mathrm{bal}}}{2}\right)^4, \tag{2.36}$$

where $v_{\mathrm{bal}}^{\max} = a\tau_{\mathrm{bal}}$ is the maximum ballistic velocity of the carriers. The power spectrum according to Eqs. (2.32) and (2.36) generated by the electrons is shown in Figure 2.9b. Since holes do not show any remarkable velocity overshoot, their contribution will follow Eq. (2.28) and has to be summed with the electron contribution.

The derived spectra are the idealized cases only. We neglected contributions of carriers that are further away from the surface. The accelerating fields are lower there, such that lower frequency components will be generated. Further, any device specific roll-offs and the frequency dependence of the antennas will alter the generated spectra. The latter, however, can simply be multiplied to the respective spectra (Eqs. (2.28), (2.32), (2.34), and (2.36)).

Comparing ballistic SI GaAs (Eq. (2.36)) with non-ballistic low-temperature grown (LTG) GaAs (Eq. (2.28)) shows that the emitted power is proportional to the square of the product of the relevant transport time and the transport velocity, $d_{\mathrm{bal}} = v_{\mathrm{bal}}^{\max}\tau_{\mathrm{bal}}$ for the case of ballistic transport and $d_{\mathrm{LT}} = v_{\mathrm{sat}}\tau_{\mathrm{rec}}$ for a short lifetime material. This is the length of the respective dipole that is responsible for THz generation. For the ballistic case, the dipole length can be as large as $d_{\mathrm{bal}} = 300$ nm, whereas the dipole length for the short lifetime material is only of the order of 20 nm (for a lifetime of $\tau_{\mathrm{rec}} = 200$ fs). For the specific example, SI GaAs produces more than 200 times higher output power for the same biasing conditions and same electrode layout. In the next section, we will see that SI material has some drawbacks in terms of damage threshold and biasing which levels the performance of both materials.

We note that Eq. (2.32) is so general that it can also be used to derive the frequency response of the material under CW operation for a Hertzian antenna and for p-i-n diodes as long as the respective transport kinetics are taken into account. The spectral answer of the CW photomixer becomes [27]

$$P_{\mathrm{THz}}(\omega) \sim r_{\mathrm{A}}^{\mathrm{H}}(\omega)|P_{\mathrm{L}}(\omega)|^2 \cdot |H(\omega)|^2, \tag{2.37}$$

where we used $r_{\mathrm{A}}^{\mathrm{H}}(\omega) \sim R_{\mathrm{A}}^{\mathrm{H}}(\omega)/l_{\mathrm{D}}^2 = \omega^2/(6\pi\varepsilon_0\varepsilon_{\mathrm{r}}c^3)$ for the radiation resistance of a Hertzian dipole normalized to its length. For CW lasers with a (narrow) Gaussian linewidth, $\Delta\omega$, the optical power is $P_{\mathrm{L}}(\omega) \approx P_0 \exp(-\omega^2/\Delta\omega^2)$. Typical values for the linewidth are in the few MHz range. The term $|H(\omega)|^2$ for all discussed transport mechanisms is spectrally much broader, that is, in the range of a few THz. The generated THz power spectrum is therefore dominated by the spectral characteristics of the lasers, the actual carrier transport only plays a minor role but decreases the amplitude at high frequencies. From a physics picture, the non-harmonic components of the current under CW operation interfere away, only the components at the difference frequency of the lasers survive.

2.2.2.3 Photoconductors as THz Detectors

Photoconductors can also be used as THz detectors. When no bias is applied, optically generated electron–hole pairs just recombine without generation of a current. If, however, an antenna is coupled to the electrodes in order to receive a THz signal, a bias at the THz frequency will be generated. This bias is proportional to the THz field strength. For a CW signal, the bias is $U_{\mathrm{THz}}(t) \sim E_{\mathrm{THz}} \cos(\omega_{\mathrm{THz}}t + \varphi)$. If the electrode gap is excited with the same laser signal that was used to generate the THz signal, the carrier generation rate is modulated at the same THz frequency, $n(t) \sim P_{\mathrm{L}}(t) = P_{\mathrm{L}}(1 + \cos\omega_{\mathrm{THz}}t)$. The carrier velocity is modulated by the received THz field as well, since $v(t) \sim \mu U_{\mathrm{THz}}(t) \sim E_{\mathrm{THz}} \cos(\omega_{\mathrm{THz}}t + \varphi)$. The resulting current is proportional to the product of carrier velocity and carrier concentration, $I(t) \sim n(t)v(t) \sim \cos(\omega_{\mathrm{THz}}t)\cos(\omega_{\mathrm{THz}}t + \varphi)$ and the device behaves like a homodyne mixer generating a DC component of

$$< I > \sim < U_{\mathrm{THz}}(t)P_{\mathrm{L}}(t) > \sim E_{\mathrm{THz}}P_{\mathrm{L}} \cos\varphi, \tag{2.38}$$

where the brackets stand for the time average of the quantities. The phase, φ, is due to a finite delay, for instance by different optical path lengths, d, between the optical and the THz wave, that is, $\varphi = k_{\mathrm{THz}}d$. The photoconductor is therefore a homodyne mixer where both field amplitude and phase are detectable.

In a similar fashion, photoconductors can be used as detectors in pulsed systems. The optical pulse is used to scan the THz pulse by including a delay time, τ. The resulting DC signal is the convolution of the optical signal and the THz field,

$$I(\tau) \sim \int E_{\mathrm{THz}}(t)P_{\mathrm{L}}(\tau - t)\mathrm{d}t. \tag{2.39}$$

The optical pulse therefore "samples" the electric field THz strength. Its Fourier transform yields the THz power spectrum. This technique is frequently applied in terahertz spectroscopy and is called "time domain spectroscopy" (TDS). Often, the understanding of the detection mechanism is that the THz pulse is sampled with a much narrower optical pulse, similar to electronic sampling. Eq. (2.28), however, shows that the THz field can have similar spectral shape and similar temporal width to the optical beam. It is easy to show that the same Gaussian pulse shape of optical and THz beam, Eq. (2.39), results in a broadening of the detected signal $I(t)$ by a factor of $\sqrt{2}$. Other typical pulse shapes like $\mathrm{sech}^2(t/\tau_{\mathrm{pls}})$ result in similar factors. The measured signal appears temporally wider and spectrally narrower than the actual THz signal. At a first glance, this result may render spectroscopic data of TDS systems useless, since the measured spectrum differs from the spectrum of the actual THz signal. In spectroscopy, however, it is not the individual spectrum that matters but rather the ratio of the spectrum of a transmitted or reflected signal by a device under test, and the reference signal. The Fourier transform of Eq. (2.39) is

$$I(\omega) \sim E_{\mathrm{THz}}(\omega)P_{\mathrm{L}}(\omega) \tag{2.40}$$

The transmission (or alternatively, the reflection) coefficient for a transmitted (reflected) signal $I_{\mathrm{M}}(\omega)$ and a reference signal $I_{\mathrm{R}}(\omega)$ is

$$t(\omega) = \frac{I_{\mathrm{M}}(\omega)}{I_{\mathrm{R}}(\omega)} = \frac{E^{\mathrm{M}}_{\mathrm{THz}}(\omega)}{E^{\mathrm{R}}_{\mathrm{THz}}(\omega)}. \tag{2.41}$$

The dependence of the spectrum on the optical pulse (and other experimental parameters) simply cancels out. The time domain data result in the correct THz spectra, independent of the optical pulse spectrum.

2.2.3 *Thermal Constraints*

The THz power of photomixers increases with the square of the laser power, $P_{\mathrm{THz}} \sim P_{\mathrm{L}}^2$. Therefore, the highest possible laser power should be used, if reasonably large THz power is desired. This, however, leads to excessive heating with several unfavorable effects on the device performance. First, the increased phonon scattering rate at higher temperature reduces carrier mobility as well as ballistic benefits. The transit time increases, reducing the THz power at high frequencies, ($f_{\mathrm{THz}} > f_{\mathrm{tr}}^{3\mathrm{dB}}$). Second, the device ages faster, since high temperatures allow (exponentially) increased dopant and trap migration, finally leading to breakdown of the device. To estimate the effects due to thermal dissipation, we look at the thermal conductivity of a semiconductor, which can be approximated by [29]

$$\lambda_{\mathrm{th}}(T) \approx \lambda_{\mathrm{th}}(T_0) \cdot (T_0/T)^c. \tag{2.42}$$

For most semiconductors, the exponent c is larger than 1, resulting in a reduced thermal conduction at higher temperatures. As an example, $c = 1.375$ for InGaAs lattice matched to InP or $c = 1.55$ for InP [29]. Further, the band gap shrinks with increased temperature, leading to a higher absorption coefficient and increased optical power dissipation. Both effects may lead to thermal runaway of the device and, finally, its destruction. In this section, the maximum optical power and electric fields for safe device operation will be estimated. Furthermore, we will discuss device design issues in order to reduce thermal effects.

In the previous section, the lifetime roll-off for photoconductors was introduced in Eqs. (2.20) and (2.21). An expansion of η_{PC} with respect to τ_{rec} yields $\eta_{\mathrm{PC}} \sim (\tau_{\mathrm{rec}})^2$ for $f < f_{3\mathrm{dB}}^{\mathrm{LT}}$ and $\eta_{\mathrm{PC}} = f_{\mathrm{THz}}^{-2}$ for $f \gg f_{3\mathrm{dB}}^{\mathrm{LT}}$, independent of τ_{rec}. So it seems that short lifetimes reduce the output power at low frequencies and are of no use at high frequencies. The picture changes when we take thermal limitation into account. Any photomixer experiences two heat sources: (i) current under an external bias results in Joule heating and (ii) absorbed laser power. Photoconductors with small gain, $g \ll 1$, generate very little external current, $I_{\mathrm{ext}} = gI_{\mathrm{id}}$, where $I_{\mathrm{id}} = eP_{\mathrm{L}}/h\upsilon$ is the ideal photocurrent due to *absorbed* optical power. The Joule heat is $P_{\mathrm{H}}^{\mathrm{J}} \approx I_{\mathrm{ext}}U_{\mathrm{DC}} = gI_{id}U_{\mathrm{DC}}$, with the external bias U_{DC}. Joule heating can be reduced by using small gain values in order to reduce the external current. This fact explains why materials with a short lifetime are so successful in the THz range as compared to materials with a long lifetime: they can tolerate much higher optical power levels due to reduced Joule heating. For p-i-n diodes, Joule heating is less dramatic than for photoconductors, since little external bias, U_{DC}, is required during operation. Heating by absorbed laser power is more important. P-i-n diode-based photomixers are typically operated under reverse bias. Therefore, most of the optical power is transformed into heat, since carriers emit phonons while relaxing to the band edge during transport to the respective contact. For photoconductors, just a small fraction of the laser power is extracted as current due to the low gain. In both cases, the heat produced by the laser is approximately given by the absorbed optical power, $P_{H}^{L} \approx P_{L}$. The total heat dissipated by the photomixer is given by [23]

$$P_{H} = P_{L}\left(1 + g\frac{e}{h\upsilon_{L}}U_{DC}\right), \tag{2.43}$$

where $g = 1$ for p-i-n diodes. A maximum temperature change of $\Delta T_{max} = 120$ K has been determined empirically for GaAs photoconductors before damage occurs [30]. Given that there is sufficient lateral heat spreading (i.e., a constant temperature in the cross-section), heat is transported across a mesa of height, d, as

$$P_{H} = \Delta T_{max}\lambda A/d, \tag{2.44}$$

with the thermal conductance, λ. In most cases, however, the heat source due to absorption of the lasers can be identified as a thin disk (i.e., the absorbing semiconductor) on a thermal conductor (i.e., the substrate) [19]. Taking the spreading resistance into account yields

$$P_{H} \approx \Delta T_{max}\lambda\sqrt{\pi A}. \tag{2.45}$$

For GaAs with a thermal conductance of $\lambda = 0.43$ W/cmK, a device cross-section of $A = 50\ \mu m^2$ results in a maximum thermal power of $P_{H} = 65$ mW according to Eq. (2.45) for $\Delta T_{max} = 120$ K.

Since $P_{L} < P_{H}$ this is also the maximum limit for the optical power. However, devices are illuminated with a laser that often features approximately a Gaussian shape. The non-homogeneous optical power distribution may lead to a failure at (much) lower optical power levels. Typically, an absorbed optical power of the order of 50 mW for a photoconductor ($A \sim 50\ \mu m^2$) is considered to be safe for long term use in photoconductors. For UTC diodes, fabricated of mainly InP with a higher thermal conductance of $\lambda = 0.68$ W/cmK, maximum current levels in the range of 100–150 kA/cm^2 have been reported [31, 32]. For the area of $A = 50\ \mu m^2$, as used in the above calculation, these current densities correspond to an absorbed optical power in the range of 70 mW at 1550 nm (note that usually only a fraction of the optical power is absorbed in p-i-n diodes; the used laser powers are typically much higher than the presented value). Equation (2.45) predicts 100 mW for InP, very close to the experimental value. For long term use, the optical power must be smaller in order to prevent rapid degradation of the device. This implies very similar values for the maximum optical power for photoconductors and p-i-n diodes. In order to obtain the maximum THz power from the given limit of optical power, three equations have to be optimized simultaneously: RC-roll off, Eq. (2.13), transit-time (CW p-i-n diodes only, Eq. (2.16)) or lifetime roll-off (CW photoconductors only, Eq. (2.21)), and thermal dissipation (summarized by Eq.(2.43)–(2.45)). These equations are not independent. The thermal transport depends either linearly on the device cross-section, A ((Eq. 2.44)), or on $\sqrt{A}$ (Eq. (2.45)). The RC-roll-off depends on the device cross-section and the transport distance (intrinsic layer length, d_i, for p-i-n diodes and electrode gap, w_G, and number of electrodes for photoconductors). The transit-time or lifetime depends on the transport distance and the recombination time of the semiconductor. For photoconductors, the thermal dissipation and the lifetime roll-off further depend on the gain. The optimization has to be performed for a specific THz frequency, f_{max}, since transport- and RC roll-off are frequency-dependent. Brown calculated the ideal parameters for photoconductors for both a resonant antenna with $R_A = 215\ \Omega$ and a broadband antenna with $R_A = 72\ \Omega$ in Ref. [23]. For operation at 1 THz with a broad band photomixer, the optimum parameters have been determined to be $w_G = 0.99\ \mu m$, $\tau_{rec} = 0.3$ ps, $A = 73\ \mu m^2$, eight electrodes (in total). A minimum finger width due to processing limitations of 0.2 μm was assumed.

Strategies to improve heat transport from the active area are of key importance for improving the maximum power. This is particularly difficult for devices operated at 1550 nm. The absorption layer is typically made from $In_{0.53}Ga_{0.47}As$, a ternary compound with very low thermal conductance of 0.05 W/cmK [21]. InP, a standard substrate for telecom devices, features a 13 times higher thermal conductance of 0.68 W/cmK [21]. In UTC diodes, the layers with low thermal conductance are kept as small as possible. The n-contact, p-contact, and the transport layer are InP layers. Only the absorption layer with a length in the range of 50–100 nm is InGaAs. GaAs-based devices benefit from a relatively high thermal conductance of 0.43 W/cmK [20] as compared to $In_{0.53}Ga_{0.47}As$. Incorporation of AlAs heat spreader layers [33] further improves the performance. Transfer onto a substrate with higher thermal conductance has also been demonstrated [34].

Thermal effects are one of the main constraints of CW systems. For pulsed systems, the thermal power is mainly generated during the pulse duration (typically much below $\tau_{\mathrm{pls}} < 1$ ps). It scales with the average thermal power, P_{H}, not the peak power, $P_{\mathrm{H,pk}}$. Thermal limitations can therefore be mitigated by using lasers with low repetition rates, f_{rep}, since $P_{\mathrm{H}} = P_{\mathrm{H,pk}}\tau_{\mathrm{pls}}f_{\mathrm{rep}}$. Electrical limitations, as being discussed next, are more detrimental and challenging for pulsed systems.

2.2.4 *Electrical Constraints*

Besides thermal limitations, the carrier transport and other electrical effects also show a power dependence. These effects play a particular role in pulsed operation where a high charge density is generated within the semiconductor but only a moderate average thermal power is dissipated. In order to get insight into the physical limits, some simplifying approximations are necessary in order to get analytical expressions. For exact values, simulations are required, taking the time evolution of the carrier transport into account. Since p-i-n diodes and photoconductors perform very differently, the two cases will be treated separately.

2.2.4.1 p-i-n Diodes

We will see soon that p-i-n diodes are typically used in CW systems only. Therefore, this section specifically focuses on CW operation. The main results, however, are also valid for pulsed excitation. In order to minimize the transit-time roll-off for CW generation, and to increase the effective dipole length for pulsed operation, carrier transport has to be optimized for operation in the THz range. This is especially the case for THz frequencies where transport at the saturation velocity already causes a roll-off, that is, $f_{\mathrm{THz}} > (2\tau_{\mathrm{tr}}^{\mathrm{sat}})^{-1} = v_{\mathrm{sat}}/(2d_{\mathrm{i}})$. For all p-i-n diodes, there exists an optimum transport field, E_{T}, since they benefit from ballistic transport: on the one hand, too low fields result in small carrier acceleration, $a = eE_{\mathrm{T}}/m_{\Gamma}$. It simply takes too long to accelerate the carriers. On the other hand, too high transport fields (>40 kV/cm, see Figure 2.6) are detrimental for ballistic transport. For efficient operation, it is of key importance to maintain the ideal transport field throughout the whole structure. The ideal field can either be the built-in field of the diode, E_{bi}, or being fine-tuned with an additional external field, E_{ext}, via a DC bias. However, photo-carriers will alter the local fields within the diode. When optically generated carriers, ($n_{\mathrm{ph}}(z)$ electrons and the same amount of holes, $h_{\mathrm{ph}}(z)$) are separated in an electric field, $E_{\mathrm{DC}} = \mathrm{E}_{\mathrm{bi}} + \mathrm{E}_{\mathrm{ext}}$, they generate a field, E_{ph}, that opposes E_{DC}. According to the first Maxwell equation,

$$\frac{\partial}{\partial x}E_{\mathrm{ph}}(z) = -\frac{e[n_{\mathrm{ph}}(z) - h_{\mathrm{ph}}(z)]}{\varepsilon_0\varepsilon_{\mathrm{r}}}. \tag{2.46}$$

Subsequently generated carriers experience a smaller acceleration of

$$a(z) = \pm\frac{e}{m_{\Gamma}}[E_{\mathrm{DC}} - E_{\mathrm{ph}}(z)], \tag{2.47}$$

Where the "+" sign is for electrons and the "−" sign for holes. In the small signal limit of low optical powers (i.e., small $n_{\mathrm{ph}}(\mathrm{z})$, $h_{\mathrm{ph}}(z)$), the screening field is much smaller than the original accelerating field, $|E_{\mathrm{ph}}(z)| \ll |E_{\mathrm{DC}}|$ and the small perturbation of the carriers to the constant field E_{DC} can

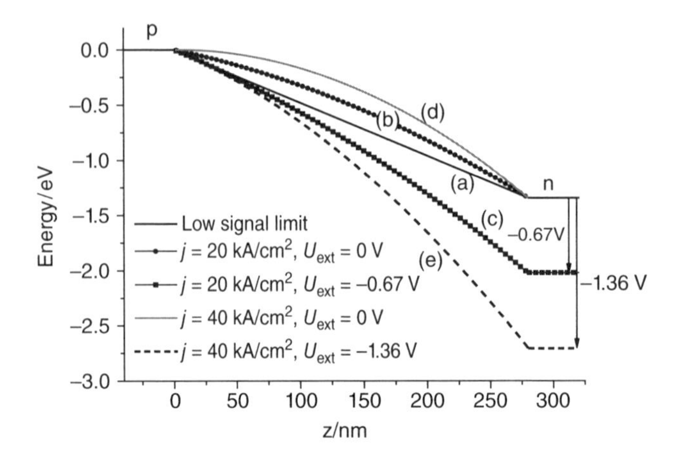

Figure 2.10 (a) Band profile of an ideal UTC p-i-n diode with $d_i = 280$ nm in the small signal limit (black solid line). The band profile is unperturbed and linear. (b) Intermediate optical power level: The carriers are mildly screening the built-in field (shown here for 20 kA/cm^2), particularly at the p-side of the transport layer. A small external bias can restore the original field strength (dotted, (c)) at the beginning of the i-layer, however, leading to higher fields close to the n-contact. (d) High power limit (approximate solution according to Eqs. (2.51) and (2.52), shown for $j = 40$ kA/cm^2). Without external bias, the carriers screen the built-in field so strongly that a region with very low accelerating fields is formed at the p-side of the intrinsic layer. In this limit, a self-consistent calculation would be required to calculate the exact band profile. A strong reverse bias has to be applied (e) to restore the transport field at the p-side of the intrinsic layer to its original value but the transport field is no longer homogeneous and shows strong warping.

be neglected. For intermediate power levels, the perturbation by the photogenerated carriers can, at least partially, be compensated with an external DC bias, as shown in Figure 2.10 for a photocurrent density of $j = env = 20$ kA/cm^2. A small deviation from the linear band profile is visible but the accelerating field strength is still sufficient throughout the structure. At high power levels, the optically generated carriers can create field kinks in the range of the built-in field, $E_{bi} = E_G/d_i$, or even larger. Therefore, the band profile gets strongly distorted, even regions of a flat band profile between the space charge accumulation can occur. In such situations, transport will be heavily restrained, leading to saturation or even a decline of the THz power with increasing optical power.

We will estimate the critical current densities and the corresponding optical power where the strong screening sets in with a few simplifying assumptions. For exact solutions, self-consistent calculations are required. This goes beyond the scope of this book.

From Figure 2.6 we can extract that efficient ballistic transport can be maintained for fields within $E_{pin} = E_{bal}^{opt} \pm \Delta E_{bal} = E_{bal}^{opt} \pm 10$ kV/cm, where E_{bal}^{opt} is the optimum ballistic field. E_{bal}^{opt} is slightly dependent on the intrinsic layer length and on the material of the transport layer, with typical values around $E_{bal}^{opt} = 20$ kV/cm. A variation of the electric field *within* the intrinsic layer cannot be compensated with an external bias since the external field just adds to the internal field. In order to allow for analytical solutions, the p-i-n diode is subject to the following assumptions:

- Absorption takes place in a very narrow area close to the p-contact (ideal UTC diode). Holes remain stationary and only electrons move.
- For CW operation, the carriers within one THz wave are assumed to be generated instantaneously, that is, the CW signal is approximated by a pulse train instead of a cosine shape. For pulsed operation, all carriers shall be generated at once (i.e., the pulse duration is much shorter than the transit-time).

For a device operated at the transit-time 3 dB frequency where only a single electron bunch propagates, the electron density that has propagated to a distance $s(t)$ after generation at time t is

$$n(z) = \frac{\bar{j}_{\mathrm{bal}}^{\max} \tau_{\mathrm{tr}}}{e} \delta(z - s(t)), \tag{2.48}$$

where $\bar{j}_{\mathrm{bal}}^{\max}$ is the maximum current density for ballistic transport that will be derived in the following. Since holes are stationary, they do not produce any current. The induced field kink according to Eq. (2.46) is

$$\Delta E_{\mathrm{Ph}}(z) = \frac{\bar{j}_{\mathrm{bal}}^{\max} \tau_{\mathrm{tr}}}{\varepsilon_0 \varepsilon_{\mathrm{r}}}. \tag{2.49}$$

For a device with a transit-time 3 dB frequency of $f_{\mathrm{tr}}^{3\mathrm{dB}} = 1/(2\tau_{\mathrm{tr}}) = 1$ THz, and a maximum kink of $\Delta E_{\mathrm{Ph}}(z) = 10$ kV/cm in order to achieve ballistic transport before and after the kink, the maximum current density for $\varepsilon_{\mathrm{r}} = 13$ is $\bar{j}_{\mathrm{bal}}^{\max} = 23\ \mathrm{kA/cm^2}$. For the 50 μm^2 device that was used for the estimates regarding the thermal limitations, the maximum current is 11.5 mA, and the *absorbed* optical power for operation at 1550 nm is 9.2 mW. For current densities above $\bar{j}_{\mathrm{bal}}^{\max}$, only part of the transport region can be covered ballistically. This increases the transit-time of the carriers and reduces the transit-time 3 dB frequency. Since we assumed that all carriers are generated at once, this value represents only a lower limit. Continuous carrier generation will lead to slightly higher values.

For low frequencies, where transport at the saturation velocity is sufficient in order to achieve roll-off free operation, the THz power will still increase quadratically with the optical power. The screening can be compensated with an external bias as long as this bias does not exceed the breakdown voltage of the semiconductor. The latter case is the ultimate power limit for the photomixer. The exact photo-induced screening of the field in the intrinsic layer depends strongly on the transport properties. The maximum current density will, however, certainly be higher than $\bar{j}_{\mathrm{bal}}^{\max}$. In order to provide an analytical estimate for the maximum current density (optical power), we assume that the majority of the transport takes place with the saturation velocity, v_{sat}, and the device is operated above its transit-time 3 dB frequency under CW operation in the quasi-steady state, such that the time-dependent quantities can be replaced with their time-averages (indicated by bars). The following calculation is not valid for pulsed operation since there is no steady state. The space charge that builds up during carrier propagation can relax slowly between the pulses, recovering the original internal fields as without photogenerated carriers. For the case of pulsed excitation, time-dependent quantities would have to be used. Since p-i-n-diodes are typically used in CW experiments, we will restrict the discussion on the CW case and use again the case of an ideal UTC diode with a very narrow absorption region close to the p-contact. The average carrier density in the device is given by the continuity equation,

$$e\bar{n}(z) = \bar{j}/v_{\mathrm{sat}} = \text{const.} \tag{2.50}$$

The average photo-generated bias at position z according to the first Maxwell equation is

$$\overline{U}_{\mathrm{Ph}}(z) = -\iint \frac{en(z'')}{\varepsilon_0 \varepsilon_{\mathrm{r}}} \mathrm{d}z'' \mathrm{d}z' = -\frac{nez^2}{2v_{\mathrm{sat}} \varepsilon_0 \varepsilon_{\mathrm{r}}} = -\frac{\bar{j}z^2}{2v_{\mathrm{sat}} \varepsilon_0 \varepsilon_{\mathrm{r}}}, \tag{2.51}$$

as illustrated in Figure 2.10. For the boundary condition $j = \text{const}$ (Eq. (2.50)) an external bias has to be applied to compensate for the photo-voltage. We allow a small additional external bias, U_{ext} for fine-tuning the field in the intrinsic layer in order to compensate the photo bias and to allow a minimum transport field, $E_{\min}$. The external DC bias becomes

$$U_{\mathrm{ext}} = U_0 + \overline{U}_{\mathrm{Ph}}(d_{\mathrm{i}}) = U_0 - \frac{\bar{j}d_{\mathrm{i}}^2}{2v_{\mathrm{sat}} \varepsilon_0 \varepsilon_{\mathrm{r}}}, \tag{2.52}$$

where U_0 is the bias at low optical power. The total (average) field in the transport layer is

$$E_{\mathrm{tot}}(z) = E_{\mathrm{bi}} + E_0 - \frac{\partial}{\partial z} \overline{U}_{\mathrm{Ph}}(z) = E_{\mathrm{bi}} + E_0 + \frac{\bar{j}z}{v_{\mathrm{sat}} \varepsilon_0 \varepsilon_{\mathrm{r}}}, \tag{2.53}$$

with E_{bi} the built-in field and $E_0 = U_0/d_i$. The field is lowest at the p-contact, where the carriers are generated, and highest at the n-contact, $z = d_i$. In order to sustain efficient carrier transport, the field inside the diode should not fall below a minimum value of $E_{min} = 5$ kV/cm. Below this value, carriers are accelerated too slowly and may not even reach the saturation velocity. By setting $E_{tot}(0) = E_{min}$, Eq. (2.53) yields $E_0 = E_{min} - E_{bi}$ and

$$U_0 = d_i(E_{bi} - E_{min}). \tag{2.54}$$

Note that typically $U_0 > 0$, that is, in the small signal limit, even a small forward bias can be tolerated. On the other hand, the field must not exceed the breakdown field of the semiconductor. For InP (InGaAs), the breakdown field is $E_B = 500$ kV/cm (200 kV/cm) [19]. By setting $E_{tot}(d_i) = E_B$ in Eq. (2.53) the maximum current is

$$\bar{j}_{Ph}^{max}(E_{ext}) = \frac{v_{sat}\varepsilon_0\varepsilon_r(E_B - E_{bi} - E_0)}{d_i} = \frac{v_{sat}\varepsilon_0\varepsilon_r(E_B - E_{min})}{d_i}. \tag{2.55}$$

For a device with an InP transport layer, an InGaAs absorption region, (safe) operation at $E_B = 200$ kV/cm, a band gap energy of the transport layer of $E_G = 1.34$ eV, $\varepsilon_r = 12.5$, $v_{sat} = 1.5 \cdot 10^7$ cm/s and an intrinsic layer length of $d_i = 280$ nm the maximum current is $\bar{j}_{max} = 115\ \text{kA/cm}^2$. This current density is five times higher than the maximum current density for ballistic transport. The assumption that the carriers mainly move with the saturation velocity is satisfied. If there is still a ballistic contribution, the maximum current can be somewhat higher since $\bar{j}_{max} \sim v_e$. Substituting U_0 from Eq. (2.54) and the maximum current for Eq. (2.55) into Eq. (2.52), the required external bias is $\bar{U}_{ext} = -1.6$ V. These values are in excellent agreement with data found in the literature [31] where $\bar{j}_{max} = 60 - 150\ \text{kA/cm}^2$ (area 13 μm^2, maximum current 8–20 mA) and a DC bias between −1.5 and −2 V are reported for diodes with $d_i \sim 280$ nm. Equation (2.55) shows that the maximum current scales with the inverse intrinsic layer length. Higher current densities can be achieved with shorter intrinsic layers without electrically destroying the device. This is also beneficial for the transit-time roll-off but deteriorates the RC optimization. However, the above approximations of using the average values are only valid if the device is operated far above the transit-time 3 dB frequency, requiring rather thick intrinsic layers. For operation below the transit-time limit, a rigorous simulation using the time-dependent quantities must be performed. Further cases of other types of p-i-n diodes have been calculated in Ref. [18].

In order to reduce the photo-induced field according to Eq. (2.53), the transport layer can be slightly n-doped. The positive space charge of the ionized donors results in a band bending opposite to that of the free electron space charge and less external bias is required to maintain the transport field in the intrinsic layer. An example is illustrated in Figure 2.11.

For comparison of electrical and thermal limitations, we use the same geometry as in the last section. The maximum optical power for a 50 μm^2 device with an electrically limited current density of 115 kA/cm^2 is $P_{Abs} = 75$ mW. This power level is similar to the maximum thermal power (70–100 mW had been estimated with Eqs. (2.43)–(2.45)). CW operated p-i-n diodes with intrinsic layers in the range of 200–300 nm are both electrically and thermally limited at the same time with little room for improvement. For pulsed operation, however, electrical limitations dominate the performance since the carrier concentration during the pulse is much larger. Much shorter intrinsic layers would be required to achieve both electrically and thermally limited performance. However, this would degrade the RC roll-off drastically. Therefore, p-i-n diodes are typically used in CW applications only.

So far, we have treated an idealized p-i-n diode with a very narrow absorption region (of width w_A), much narrower than the transport layer. In real devices, this is not practicable since the absorbed optical power (and, consequently, the photocurrent) would be very small according to Lambert–Beer's law (see Eq. (2.10)). As an example, even a 200 nm absorber layer with an absorption coefficient of $\alpha = 8000\ \text{cm}^{-1}$ (a typical value for InGaAs) absorbs a laser power of only $P_{abs}/P_L = 1 - \exp(-\alpha w_a) = 14.8\ \%$, if no power was reflected. However, a 200 nm thick absorber layer cannot be considered as very narrow compared to a 280 nm transport layer. The thickness of the absorber must be thinner in order to prevent large hole transport lengths, which would deteriorate the transport performance of the device. A further problem

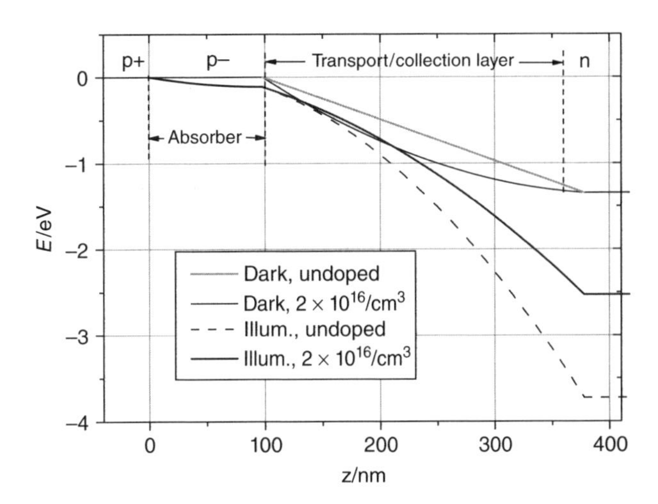

Figure 2.11 Approximate conduction band diagram of a UTC diode with an absorber doping of $p^- = 10^{17}/cm^3$, and a transport layer doping of $n = 2 \times 10^{16}/cm^3$. The band diagram is depicted for an optically generated current density of $80\,kA/cm^2$. The curvature of the transport layer due to photo-generated carriers under strong illumination is much smaller than for the undoped case. The doping also substantially reduces the required DC bias for maintaining the transport field at the beginning of the transport layer (at the p^--layer). The self-biasing effect in the p^- region is also shown.

is the structure of the absorber layer. As discussed in Section 2.3.1, in terms of thermal conduction, InP is more preferable than InGaAs. However, InP does not absorb 1550 nm light. InGaAs layers have to be used for the absorbing layer. Therefore, many THz p-i-n diodes consist of an InGaAs absorber layer with a length in the range of 50–120 nm, followed by an InP transport layer in the range of 200–300 nm [31, 35, 36]. A key challenge in sample growth is the transition from InGaAs to InP, particularly concerning the exchange of the Group V element (As, P). Smooth transitions (such as graded layers) are difficult to grow, therefore most diodes with InP transport layers consist of a sequence of two to three $InGaAs_yP_{1-y}$ layers. Each transition results in a step in the band edge that gives rise to carrier accumulation and scattering. In order to reduce the step in the conduction band, the transition layers are slightly n-doped [36] in most cases. Since the InGaAs absorber is at the p-side of the junction, the n-doping could result in formation of a junction with a large field drop right in the absorber layer instead of the transport layer. Therefore, the InGaAs absorber layer must be p^--doped and the transition length should be kept short. The p^--doping results in very low accelerating fields in the absorber layer (see Figure 2.7). This deteriorates the transit-time performance since the electron transit-time from their point of generation to the transport layer through the absorber layer can be very long if it is diffusion-dominated. For pure diffusion, the transit-time through the absorber layer is [37]

$$\tau_A^{diff} = \frac{w_A^2}{2D} + \frac{w_A}{v_{th}}, \qquad (2.56)$$

where $D = k_B T \mu_e / e$ is the diffusion constant and v_{th} is the electron thermal velocity. For an *average* drift length of $w_A = 50$ nm (total absorber length 100 nm), the diffusion transit-time through the absorber is in the range of $\tau_A^{diff} \approx 0.5$ ps, depending on the doping level which alters the carrier mobility. The same accounts for holes. Fortunately, holes are much slower than electrons and generate a space charge field in the absorber layer similar to Eqs. (2.51) and (2.53), but with opposite sign. The amplitude of the space charge field generated by stationary holes, however, is much smaller due to the p^- doping of the absorber that is usually higher than the photogenerated carrier density. It is illustrated in Figure 2.11. The small space charge field due to holes supports field-driven acceleration of the electrons in the p^- -layer toward the transport layer. For high optical powers, the acceleration can be high enough to allow drift-limited transport [37]. Still, the average transit-time through the InGaAs absorber layer, τ_A, increases the total transit-time, $\tau_{tr} = \tau_A + \tau_T$, where τ_T is the transit-time through the InP transport layer.

In order to mitigate the dilemma between increased absorption (requiring a long absorber) and small transit-time (requiring a short absorber), the propagation direction of the laser light can be adapted. So far, we have considered direct illumination from the top. Several groups employ a waveguide design, where the absorption takes place along the p-i-n diode's lateral dimension [36, 38]. Narrower absorption regions can be used by elongated diodes, still achieving large optical responsivities in the range of 0.36 A/W [36] (the maximum responsivity at 1550 nm is $e/h\upsilon = 1.25$ A/W).

2.2.4.2 Photoconductors

Space Charge Screening

For a material with short lifetime, such as LTG GaAs, carriers are trapped soon after their generation. The average distance an electron or hole can move with a drift velocity v_{dr} before it is trapped is

$$l_D = v_{dr}\tau_{rec}, \tag{2.57}$$

as illustrated in Figure 2.12a,b. The maximum drift velocity is the saturation velocity, v_{sat}. However, the mobilities for electrons in LTG GaAs are much smaller than in undoped GaAs, with values around 400 cm^2/Vs. At high optical fluences, the large amount of optically generated carriers can additionally result in increased carrier–carrier scattering, further reducing the mobility [39]. To accelerate electrons to the saturation velocity, fields in the range of 25 kV/cm are necessary in LTG GaAs and it takes already tens of femtoseconds. In the following calculation, we will use $v_{dr} = v_{sat}$. The results will only be upper limits for space charge effects. We will first treat the CW case.

CW Operation: A recombination time of $\tau_{rec} \sim 200$ fs and transport with v_{sat} results in a dipole length of $l_D \sim 20$ nm. Electron and hole clouds are displaced, as illustrated in Figure 2.12b. The electric field

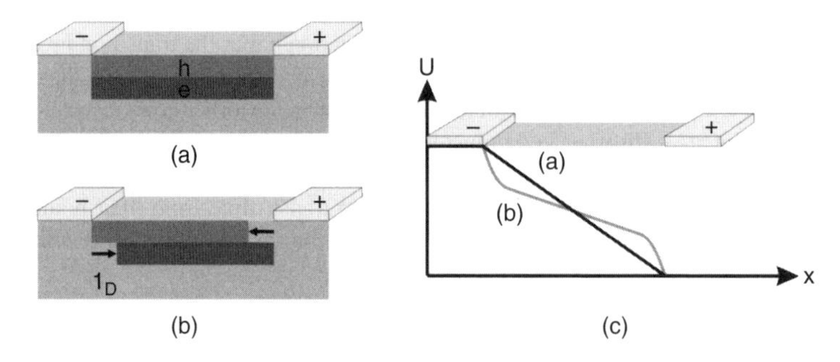

Figure 2.12 Illustration of the screening effect in photoconductors for a short pulse. (a) Electron and hole distribution right after their generation. For better visibility, electron and hole clouds are vertically offset. (b) The charge clouds can only move a length l_D before they get trapped. Carriers at the contact are withdrawn. In the central part of the structure, there are both electrons and holes available. The carriers can quickly recombine without leaving any net charge. (c) The external bias is mainly screened close to the contacts, lowering the effective transport field throughout most of the structure. The electric field (slope in (c)) right at the contact is higher than the average transport field.

in the gap is

$$E_{\mathrm{tot}}(x) = \begin{cases} E_0 - \dfrac{en_0}{\varepsilon_0\varepsilon_r}x & for \quad 0 \le x < l_D \\ E_0 - \dfrac{en_0}{\varepsilon_0\varepsilon_r}l_D & for \quad l_D \le x < w_G - l_D \\ E_0 - \dfrac{en_0}{\varepsilon_0\varepsilon_r}\left(w_G - x\right) & for \quad w_G - l_D \le x < w_G \end{cases} \tag{2.58}$$

with E_0 the field right at the electrode, as illustrated in Figure 2.12c. For CW, we assume that the carrier density, n_0, is the average space charge carrier density at the electrodes and is time-independent. The external bias source can deliver a constant average bias of U_{DC} to the electrodes, maintaining the average potential. The boundary conditions are therefore chosen such that $U_{\mathrm{tot}} = \int_0^{w_G} E_{\mathrm{tot}}(x)\mathrm{d}x = U_{DC} = E_{DC}/w_G$, with E_{DC} the unperturbed field in the gap. This yields for the field at the electrodes $E_0 = E_{DC} + \frac{en_0}{\varepsilon_0\varepsilon_r}\frac{l_D}{w_G}(w_G - l_D) \approx E_{DC} + \frac{en_0}{\varepsilon_0\varepsilon_r}l_D$. The field in the gap center (Eq. (2.58), center) is reduced by screening due to photo-generated carriers to

$$E_{\mathrm{scr}} = E_{DC} - \Delta E_{Ph} = E_{DC} - \frac{n_0 e}{\varepsilon_0\varepsilon_r}\frac{l_D^2}{w_G}, \tag{2.59}$$

The total screened bias at the electrodes is

$$\Delta U_{ph} = \frac{n_0 e l_D^2}{\varepsilon_0\varepsilon_r}. \tag{2.60}$$

For CW operation, it is necessary to clearly distinguish between trapping and recombination. Historically, the *trapping time* of carriers in photoconductors is termed the *recombination time*, τ_{rec}. Since the carrier is captured, it no longer contributes to the current and the trapping time enters in the lifetime roll-off Eqs. (2.20) and (2.21). However, this does not necessarily mean that the carrier has already *recombined* with another carrier of opposite sign. Sufficiently far away from the electrodes, the photo-generated electron and hole densities are equal. Therefore, fast recombination due to alternate electron and hole capture by a sufficiently large number of deep traps is a very fast process. However, close to the electrodes within a drift length l_D the carriers of opposite sign are missing and the deep defects can no longer act as recombination centers but only as traps. Once a trap is occupied, it cannot trap more carriers of the same charge. But these carriers will remain trapped for very long times compared to the recombination time: thermal "detrapping" requires a thermal activation energy of about $\frac{1}{2}E_G$ to get back into the respective band. Overall relaxation or detrapping times, τ_{DT}, between 100 ps [25] and several ns have been reported [40]. Therefore, under CW conditions, nearly all the recombination centers within the drift length l_D from the electrodes will become charged traps with space charge density $-en_0$ and $+en_0$, respectively. The number of occupied deep traps, n_0, can largely exceed the photogenerated electron/hole density created on the recombination time scale, that is, $n_0 \gg \alpha P_{abs}\tau_{rec}/(h\upsilon A)$. This justifies the initial assumption of a time-independent, averaged space charge carrier density, n_0. But the number of trapped charges, n_0, cannot exceed the number of trap states, which is estimated to be in the range of $2 \times 10^{18}/\mathrm{cm}^3$ [41], roughly representing an upper limit to the space charge. This value, however, depends strongly on the growth and annealing conditions of the photoconductor and can vary within a wide range. A trap density of $2 \times 10^{18}/\mathrm{cm}^3$ leads to a reduction of the field strength of $\Delta E_{Ph} = 5.5$ kV/cm for a gap of 2 μm and $l_D = 20$ nm, a screened bias of $\Delta U_{ph} = 1.1$ V, and an increase of the field at the electrode of $\Delta E_0 = E_0 - E_{DC} = 550$ kV/cm which is in the range of the breakdown field strength. In summary, bias field screening for CW excitation is small, but space charge effects can cause early breakdown close to the electrodes due to drastic increase in the local field to values clearly above those derived from gap spacing and DC bias.

A calculation of the maximum number of optically generated carriers (and, therefore, the maximum optical power for CW generation) requires precise knowledge of the de-trapping time and may require a dynamic calculation. This calculation goes beyond the scope of this book.

Pulsed Operation: Under pulsed operation, the material can relax between pulses such that only the carriers generated in a single pulse contribute to the space charge. There is no accumulation due to long de-trapping times, as long as the repetition rate of the laser is slower than the inverse detrapping time. The carrier generation and separation happens on the (sub-) ps time scale without formation of a constant, average space charge carrier density, n_0 at the electrodes. The external source that provides the bias, U_{DC}, will be too slow to react on the THz time scale. Various capacitances, inductances and resistances (including the device resistance) hinder supplying charges from the bias source that compensate photo-generated carriers on such a short time scale – in contrast to the CW case where it needed to supply an average bias only. It is therefore inappropriate to assume a constant potential of U_{DC} at the electrodes. It is more realistic to assume that the constant DC bias is screened by optically generated carriers, leading to a reduction in the electrode potential during the THz pulse. Therefore, Eqs. (2.58)–(2.60) need modification and have to be calculated time-dependent. In order to get an estimate of the maximum screening effects, we assume an extremely short optical pulse and quick charge trapping in order to calculate the fields in the junction right after generation and trapping where screening effects are maximal. The electric field in the gap becomes

$$E_{tot}(x) = \begin{cases} E_{DC} - \dfrac{en_0}{\varepsilon_0\varepsilon_r}x & for \quad 0 \le x < l_D \\ E_{DC} - \dfrac{en_0}{\varepsilon_0\varepsilon_r}l_D & for \quad l_D \le x < w_G - l_D \\ E_{DC} - \dfrac{en_0}{\varepsilon_0\varepsilon_r}\left(w_G - x\right) & for \quad w_G - l_D \le x < w_G \end{cases} \tag{2.61}$$

with $E_{DC} = U_{DC}/w_G$. Note that Eq. (2.61) is almost identical to Eq. (2.58). Only the (CW steady state) field at the electrode, E_0, is replaced by the external DC field, E_{DC}, since we assumed that the structure can relax between the pulses. Further, n_0 is the optically generated carrier density per pulse. The field in the gap center is reduced by screening due to photo-generated carriers by

$$\Delta E_{Ph} = \frac{n_0 e}{\varepsilon_0\varepsilon_r}l_D, \tag{2.62}$$

The electrode potential is screened by

$$\Delta U_{ph} = \frac{n_0 e l_D}{\varepsilon_0\varepsilon_r}(w_G - l_D). \tag{2.63}$$

Comparing the screened field of Eq. (2.62) with that of p-i-n diodes (with $z = d_i$ and $j = en_0 v_{sat}$) in Eq. (2.53) of $\Delta E_{ph} = \frac{n_0 e d_i}{\varepsilon_0\varepsilon_r}$, the maximum screened field is a factor of $d_i/l_D \sim 10$ smaller for photoconductors, where d_i is the transport length for p-i-n diodes. Most pulsed systems therefore use photoconductors instead of p-i-n diodes. Much higher carrier densities, n_0, are possible in pulsed operation before reaching any space charge limit. For the same values as above ($l_D = 20$ nm, $w_G = 2$ μm, $n_0 = 2 \times 10^{18}/\text{cm}^3$), however, the screened bias is $\Delta U_{ph} = 110$ V, that is, a factor of $l_D/(w_G - l_D) = 100$ larger than for the CW case. This value is larger than typical biases of 30–60 V that can be used for a gap width of 2 μm before breakdown occurs. Therefore, saturation certainly plays a role at such high carrier concentrations. But bias field screening saturation is a dynamic process: the space charge builds up on a time scale of τ_{rec}. This is the same time scale where THz radiation is efficiently emitted. In low field regions, carriers can no longer reach the saturation velocity, resulting in a slower response and a reduced spectral width. Therefore, not only the THz amplitude will be altered by bias field screening but also the spectral shape. The change in the THz spectra requires simulations [42]. The maximum bias and breakdown voltage, however, can be estimated by a quasi-static approach since de-trapping takes much longer than trapping. The bias due to space charge and the space charge field are given by Eqs. (2.61) and (2.62). In contrast to CW operation, $n_0 = \alpha E_{pls}/(h\upsilon A)$ is the carrier concentration generated by *one* pulse. The external applied bias must compensate the space charge field in Eq. (2.62). To sustain the transport in the gap, a minimum field of about 25 kV/cm is necessary. Nowhere in the gap, particularly close to the electrodes, the breakdown voltage ($E_B = 500$–1000 kV/cm for LTG GaAs [23, 43]) may be exceeded. This leads to a maximum carrier density of

$$n_0^{max} = \frac{\varepsilon_0 \varepsilon_r}{e l_D}(E_B - E_{min}) \tag{2.64}$$

For $\tau_{rec} = 200$ fs, $v_{sat} = 10^7$ cm/s, and $E_B = 1000$ kV/cm the maximum optically generated carrier density is $n_0 = 3.6 \times 10^{18}$ cm^{-3} (pulse energy for GaAs of 40 pJ for $A = 50$ μm^2, or pulse fluences of 80 mJ/cm^2). On the one hand, this calculation is fairly optimistic: carriers are generated within a thickness of $1/\alpha$ in the semiconductor. But the electrodes are planar. Carriers from the depth of the photoconductor have to drift toward the electrodes, increasing the local space charge density and reducing the maximum pulse energy. On the other hand, space charge screening will reduce the local field inside the gap, therefore reducing the carrier velocity and the dipole length, l_D. In turn, higher optical densities are allowed according to Eq. (2.64). The device saturates and the THz pulses become wider. Estimates on the trap state density are of the same order of magnitude (2×10^{18}/cm^3 [41]). The photoconducting material will not be able to trap more charges, resulting in longer effective trap times and temporal broadening of the pulse. The calculated order of magnitude for saturation is in excellent agreement with experimental data [39, 41, 44].

For a photoconductor without traps, such as SI GaAs, the carriers drift toward the contacts due to the applied DC bias. This is similar to a p-i-n diode with absorption throughout an intrinsic layer of length w_G. SI GaAs is very inefficient as a CW source due to the extremely long transit-time, resulting in a severe roll-off, and the high electrical power consumption, resulting in early thermal breakdown. In pulsed systems, however, SI GaAs has proven to be competitive with LTG GaAs. Thermal limitation is alleviated since practically no heat is generated between pulses where the device is in its high resistive dark state. Due to a sharp onset of the current when the pulse illuminates the sample, a broadband spectrum is obtained, as illustrated in Figure 2.9. However, due to a lower breakdown voltage of the order of 300 kV/cm as compared to 500–1000 kV/cm for LTG GaAs, the maximum bias voltage is lower. In addition, the carriers drift longer since they are not trapped, resulting in larger space charge effects. The devices saturate earlier than LTG GaAs devices. An early experimental comparison of SI GaAs with LTG GaAs from Ref. [44] is shown in Figure 2.13.

Impedance Matching and Radiation Field Screening

If an antenna is used to emit the radiation resulting from the changes in the photocurrent generated by the device, the radiation impedance, $Z_A = R_A + iX_A$, and the impedance matching to the photoconductor determine the radiated power. The radiated THz power for CW operation is given by [23, 45]

$$P_{THz} = \frac{1}{2} I^2 \frac{G_A(\omega)}{(G_A(\omega) + G_P)^2 + (B_A(\omega) + \omega C)^2} \tag{2.65}$$

with $G_A(\omega) = \Re[Z_A^{-1}(\omega)]$ and $B_A(\omega) = \Im[Z_A^{-1}(\omega)]$, C is the capacitance and $G_P = I/U_{DC}$ is the conductance of the device. Note that the emitted THz power in Eq. (2.65) becomes maximal for a fixed radiation resistance for $G_P \to 0$. For CW operation, the source resistance of p-i-n diodes can be in the range of the radiation resistance (typical values are between 27 and 100 Ω for an antenna on a GaAs/air or InP/air interface) and impedance matching is possible [46]. The resistance of photoconductors is much higher than the radiation resistance of any practical antenna for CW operation. Typical dark resistances for GaAs-based devices are in the MΩ to GΩ range. Even with strong pumping, it is difficult to get the device resistance below ~1 kΩ in CW operation, with very few exceptions (such as in Ref. [47]). The conductance of the device, $G_P \ll G_A(\omega)$, can be omitted in Eq. (2.65). For low and intermediate power levels this remains the same for pulsed operation. For the highest optical power levels, however, the

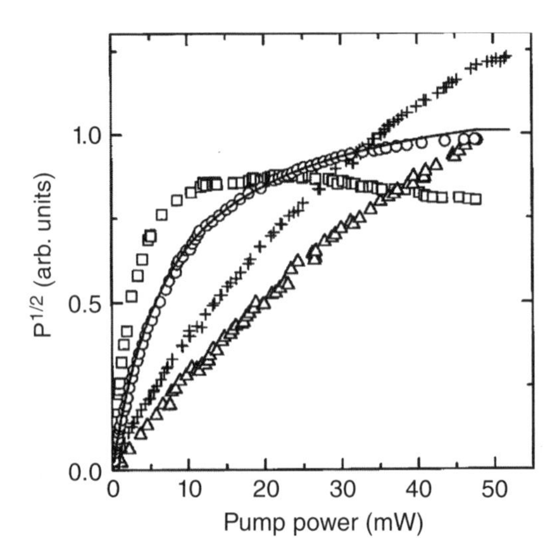

Figure 2.13 Integrated powers for pulsed operation ($\upsilon_{\text{rep}} = 82$ MHz, $\tau_{\text{pls}} = 80$ fs) of an LTG GaAs photomixer ($A \sim 100\,\mu\text{m}^2$) equipped with a dipole antenna (open circles) at $E_{\text{DC}} = 20$ kV/cm, and a SI GaAs photomixer with the same antenna (open squares) at $E_{\text{DC}} = 36$ kV/cm. The other graphs represent data with LTG GaAs using other types of antennas. A linear increase with the square root of the pump power represents the non-saturated regime. Despite the 80% higher bias, the SI GaAs saturates earlier (pulse fluence ~25 mJ/cm^2 compared to ~100 mJ/cm^2 for LTG GaAs). Saturation occurs probably due to both bias screening and radiation field screening. Under similar biases and excitation conditions the order of magnitude of the emitted power was the same. Figure reprinted from Ref. [44] with permission from the Optical Society of America, © 1997 Optical Society of America.

conductance may become higher than the antenna admittance $G_{\text{A}}(\omega)$. This decreases the THz power according to Eq. (2.65). The antenna is short circuited by the device. However, Eq. (2.65) is difficult to use in pulsed operation in the limit of low device conductance: the frequency dependence of the excitation, $I(\omega) \sim P_{\text{L}}(\omega)$ is contained in the current. The device conductance $G_{\text{P}}(t)$ is time varying, for example, by trapping of the carriers. Its Fourier transform also shows a frequency dependence. A similar approach as in Eq. (2.32) could be used. In order to estimate the power level where G_{P} becomes comparable to the antenna admittance we will use a simplified approach. For an exciting optical pulse that is much shorter than τ_{rec}, the instantaneous conductivity of a photoconductor right after absorption of the optical pulse is

$$G_{\text{P}} = en_0\mu \cdot \frac{Nl_{\text{y}}}{\alpha w_{\text{G}}}, \tag{2.66}$$

for a finger structure with fingers of length l_{y} and N gaps of width w_{G}. α is the absorption coefficient, and

$$n_0 = \alpha \frac{E_{\text{pls}}/N}{h\upsilon w_{\text{G}} l_{\text{y}}} \tag{2.67}$$

with an absorbed pulse energy of E_{pls}. Substituting Eq. (2.67) into Eq. (2.66) yields

$$G_{\text{P}} = \mu \frac{eE_{\text{pls}}}{h\upsilon} \cdot \frac{1}{w_{\text{G}}^2}. \tag{2.68}$$

For an antenna with a frequency-independent radiation resistance of $R_{\text{A}} = 72\,\Omega$ (e.g., a self-complementary broadband antenna), negligible reactance, X_{A}, a mobility of 400 cm^2/Vs, $h\upsilon = 1.5$ eV

(830 nm, LTG GaAs) and a gap width of 2 μm, the instantaneous device resistance becomes equal to the radiation resistance of the antenna at an optical pulse energy of 2.1 pJ. This is more than an order of magnitude smaller than the laser power where bias field screening becomes apparent, as derived in the previous section for a 50 μm^2 device. For SI GaAs, the mobility can be up to a factor of 20 higher than for LTG GaAs, resulting in impedance matching at a pulse energy of only 0.1 pJ. Equation (2.68) scales inversely quadratic with the gap width. It is therefore reasonable to use larger gaps for pulsed operation, in order to allow for higher optical powers and consequently higher THz powers. On the one hand, larger gaps reduce the photoconductive gain, $g = \tau_{rec}/\tau_{tr}$, which reduces the current fed to the antenna (Eq. (2.21)). On the other hand, larger gaps show less capacitance, improving the RC performance particularly above 1 THz, compensating the smaller current. For a single gap of $w_G = l_y = 7\,\mu m$ (area $\sim 50\,\mu m^2$), impedance matching is achieved at 50 pJ absorbed energy for LTG GaAs (optical pulse fluence of 100 $\mu J\,cm^2$), corresponding to an optical carrier density of $n \approx 4.2 \cdot 10^{18}/cm^3$. In this range, trap saturation and field screening may also influence the device performance in LTG GaAs, since the trap density is estimated to be in the range of $2 \times 10^{18}/cm^3$ [41]. The carrier density (pulse energy) for impedance matching does not represent a limit. At higher pulse energies, the device impedance will be lower than the antenna impedance right after absorption of the optical pulse. It will cross the point of impedance matching within the time scale of the recombination time, and become highly resistive again afterwards. Broadening of the pulse was reported [41] and a sub-linear increase of the emitted THz field with increasing fluence [39]. In many cases, τ_{rec} is in the same range as the pulse duration, τ_{pls}, such that many carriers are trapped before the optical pulse is over. The presented values are therefore underestimates.

At very high optical power levels where $G_P \gg G_A$, it will not make sense to further increase the optical power, since the device will saturate. Equation (2.65) becomes in this limit $P = \frac{1}{2} I^2 \frac{G_A(\omega)}{G_P^2} = \frac{1}{2} G_A(\omega) U_{DC}^2 \rightarrow 0$ since $G_A(0\,Hz) = 0$ and U_{DC} possesses only the DC component. The frequency dependences of current and conductance cancel, there is no longer noticeable AC modulation of the current fed into the antenna, and no power will be generated during the time where $G_P(t) \gg G_A$. Note, however, that for $G_P(t) \sim G_A$ increasing the bias, U_{DC}, still quadratically increases the THz power (particularly during pulse times where $G_P(t) \leq G_A$). This leads to the name of the device: "photoconductive switch". The laser pulse modulates the resistance of the photoconductor from highly resistive to low-ohmic and therefore switches the current due to the external DC bias on and off.

A term that is more frequently used to describe the saturation with respect to increasing optical power is *radiation field screening*. At high fluences, the emitted THz field can reach the same order of magnitude as the accelerating DC field, $E_{DC} = U_{DC}/w_G$, supplied to the electrodes. In the following, we will show that radiation field screening is indeed described by Eq. (2.65). For simplicity, we neglect the reactance of both the antenna and the device and discuss only the 1D problem of carrier transport along the x-direction between the electrodes where all z-dependences have been integrated. We further assume long electrodes without any y-dependence on fields and currents. Due to the induction law, the THz field reduces the accelerating field, [48, 49]

$$E_{loc}(x, t) = E_{DC} - \Delta E_{Ph}(x, t) - E_{THz}(x, t), \tag{2.69}$$

where $\Delta E_{Ph}(x, t)$ is the (time-dependent) screened bias by charges in the photoconductor according to Eq. (2.62), and $E_{THz}(x,t)$ is the radiated THz field that back-acts on the carriers. The bias field screening by $\Delta E_{Ph}(x, t)$ will be neglected in the following since it was discussed in the last section. The radiated field is proportional to the time derivative of the 2D photocurrent density, $E_{THz}(x, t) \sim \partial/\partial t\, j^{(2D)}(x, t)$ (with $j^{(2D)}$ in units A/m due to z-integration). The current density is x-dependent since we have to calculate the total displacement current which is the sum of all currents generated within the device. The 2D current density can be understood as an effective surface current. In the frequency domain, this reads

$$E_{THz}(x, \omega) = \beta \cdot i\omega j^{(2D)}(x, \omega) \tag{2.70}$$

where β is a proportionality constant defined by the radiation resistance of the antenna, $Z_A(\omega) = U_{THz}(\omega)/I_{THz}(\omega) = \left|\int E_{THz}(x, \omega) dx\right| / \left|\int j(x, \omega) dA\right| = i\beta\omega$, where $j^{(2D)}(x, \omega)$ is the surface current density generated by the device. The 2D current density is related to the local accelerating field by $j^{(2D)}(x,\omega)=G_P(x,\omega)E_{loc}(x,\omega)$. The radiated field at the antenna in Eq. (2.70) can then be expressed in terms of the local field as

$$E_{THz}(x,\omega) = Z_A(\omega)G_P(x,\omega)E_{loc}(x,\omega) \tag{2.71}$$

Substituting the local field from Eqs. (2.69) into Eq. (2.71) and solving for $E_{THz}(x,t)$ yields

$$E_{THz}(x,\omega) = \frac{Z_A(\omega)G_P(x,\omega)E_{DC}}{1+G_P(x,\omega)Z_A(\omega)} = \frac{Z_A(\omega)j_{id}^{(2D)}(x,\omega)}{1+G_P(x,\omega)Z_A(\omega)}, \tag{2.72}$$

with $G_P(x,\omega)E_{DC}=j_{id}^{(2D)}(x,\omega)$, the ideal current density generated by the device with $\frac{1}{w_G}\int j_{id}^{(2D)}(x,\omega)\,dxdy = I(\omega)$. For a device conductance comparable to the inverse radiation resistance, the current fed into the antenna in order to generate THz radiation is reduced by the denominator in Eq. (2.72). Replacing the radiation resistance by the admittance, the THz field becomes

$$E_{THz}(x,\omega) = \frac{j_{id}^{(2D)}(x,\omega)}{G_A(\omega)+G_P(x,\omega)} \tag{2.73}$$

Using $P_{THz}(\omega)=\frac{1}{2}G_A(\omega)\left|\int E_{THz}(x,\omega)\,dx\right|^2$ and replacing the local conductance, $G_P(x,\omega)$, by the externally accessible value of the lumped element, $G_P(\omega)$, yields the impedance matching formula in Eq. (2.62), however, without imaginary parts since they were neglected in this calculation. These can easily be accommodated in Eq. (2.73) by replacing $G_P(x,\omega)$ and $G_A(\omega)$ by their respective complex quantities.

Equation (2.72) is frequently used to describe radiation field screening for large area or large aperture emitters [24, 50] where the spatial dependence can be omitted for homogeneous illumination. In large aperture emitters, the carriers situated at a semiconductor/air interface directly emit the THz power into free space. The radiation resistance therefore has to be replaced by the wave impedance, $Z_0/(1+n)$, where n is the refractive index of the substrate. Further, the lumped element conductance must be replaced by the surface conductivity (i.e., the conductivity integrated over depth, z), $\sigma^{(2D)}$, since the device cannot be considered as a lumped element but rather as an array of distributed emitters (see also Section 3.6.2). The current responsible for radiation reads [24]

$$j^{(2D)}(t) = \frac{\sigma^{(2D)}(t)E_{DC}}{1+\frac{\sigma^{(2D)}(t)Z_0}{1+n}} \tag{2.74}$$

The numerator is the ideal current density. Since the THz power is proportional to $P_{THz}\sim[j^{(2D)}(t)]^2$, radiation field screening reduces the THz power as $\eta_{RFS}=[1+\sigma^{(2D)}Z_0/(1+n)]^{-2}$.

From large area emitters based on SI GaAs, pulse fluences in the range of ~1 mJ/cm^2 for accelerating fields of 2–2.5 kV/cm [50, 51] to about 10 mJ/cm^2 for 20 kV/cm [52] have been used at the onset of radiation field screening (i.e., sub-linear increase of the emitted THz field with increasing fluence). Higher DC accelerating fields allow (linearly) higher saturation fluences since a higher radiated THz field is required for screening. From the presented values, each mJ/cm^2 requires at least 2 kV/cm of DC field in order to prevent radiation saturation of the field screening for GaAs. However, early data [39] suggest later saturation at fluences in the range of 80 mJ/cm^2 for a bias field of 4 kV/cm (extrapolated from Figure 8 of Ref. [39]).

For LTG GaAs, Loata *et al.* [49] found radiation field screening saturation of the device for optical pulse fluences in the range of 50–100 mJ/cm^2 (n_0~few 10^{18} cm^{-3}) for an accelerating field of 20 kV/cm. It saturates at higher optical fluences than SI GaAs.

For pulsed operation, trap saturation, bias field screening, and impedance matching/radiation field screening are becoming important at similar optical power levels for both low lifetime material and SI GaAs. It is often difficult to distinguish the different contributions.

2.2.5 *Device Layouts of Photoconductive Devices*

The development of THz photoconductive devices is mainly electrical engineering and materials science. For CW operation, the material must feature (i) a low carrier lifetime in order to improve the transit-time performance, (ii) a high dark resistance for reducing the DC electrical load, (iii) a high breakdown field in order to allow for high DC biases, and, last but not least, (iv) an acceptable carrier mobility. For pulsed operation, a low carrier lifetime (i) is not desperately necessary, however, it helps to increase the breakdown field and the dark resistance. It is fairly simple to accomplish (i)–(iii) for instance by destroying or damaging the semiconductor lattice, however, at the cost of mobility (iv). Materials with low mobility, μ, do not allow transport at the saturation velocity, if $v = \mu E_{DC} < v_{sat}$ for the applied maximum DC field. This results in low photocurrents and little emitted THz power. Materials need to be engineered in order to optimize items (i)–(iv) at the same time. Several examples for both 800 and 1550 nm operation are discussed below.

2.2.5.1 Photoconductors at 800 nm

Photoconductors based on GaAs are the work horses in many THz laboratories. GaAs can be grown with a very small amount of unintentional doping. Due to the sufficiently large band gap of 1.42 eV, there are practically no thermally generated carriers and the dark resistance is high (~GΩ). These photoconductors require laser wavelengths shorter than 870 nm. They are often excited with Ti:sapphire lasers, particularly for pulsed operation. CW systems often consist of distributed feedback (DFB) diodes that are amplified with (tapered) semiconductor optical amplifiers (SOAs). Typical optical power levels for CW operation are 50 mW. A difference frequency of 1 THz requires 2.4 nm wavelength offset at 800 nm center frequency. Typical DFB diodes can be tuned by more than 2.5 nm, and a tuning range of ~5 nm (~2.1 THz) can easily be achieved. In the GHz range, SI GaAs can be used for fast optical receivers [53]. In the THz range, however, intrinsic or SI GaAs features a very pronounced transport-time roll-off for CW operation due to the lack of trapping centers. Space charge saturation occurs at fairly low optical power levels (see Eqs. (2.59) and (2.60) for $l_D = w_G$) and the electrical power dissipation is high. The lower breakdown voltages and the lack of traps lead to nonlinearities in the *IV*-characteristics at substantially lower biases than in materials with low carrier lifetime [44]. In pulsed operation, however, SI GaAs devices perform very well: The space charge limitations occur much earlier than for LTG GaAs, but the THz power emitted by an electron–hole pair is much higher, as described in Section 2.2.2.2, Figure 2.9 and discussion thereof. Since thermal dissipation scales with the average power but THz power scales with the peak power, thermal limits are suppressed by a factor of $\upsilon_{rep}\tau_{pls}$. Therefore, pulsed SI GaAs devices perform as well as devices based on materials with a short lifetime [44]. In the following, however, we focus on low lifetime material since such materials can be used in both CW and pulsed applications.

Low Temperature Grown GaAs (LTG GaAs)

LTG GaAs is typically fabricated by molecular beam epitaxy (MBE) at low growth temperatures (~200 °C) under As overpressure on SI GaAs substrates [23]. At this low temperature, up to 1.5% excess arsenic is incorporated. By annealing at temperatures up to 500–580 °C the excess As forms antisite defects, which act as deep (double) donors, and quasi-metallic clusters. A very low lifetime in the range of 100 fs [54] and a reasonable mobility around 400 cm^2/Vs can be achieved. Carrier lifetimes and resistance can be widely engineered by the growth temperature and the subsequent annealing temperature (Figure 2.14). Carrier trapping and scattering by the As clusters prevents any velocity overshoot. The *IV* characteristics remain linear. The breakdown field is about two to three times higher than for SI GaAs. For operation at 800 nm, the 1/*e* absorption length is about 0.8 μm. A typical sample structure for CW operation is 1–2 μm thick and consists of two electrodes with three to four fingers each that are deposited on top of an LTG GaAs layer (see Figure 2.8). The finger width is in the range of 0.2 μm, the gap width, w_G~2 μm, depending on the design frequency. The area covered by the

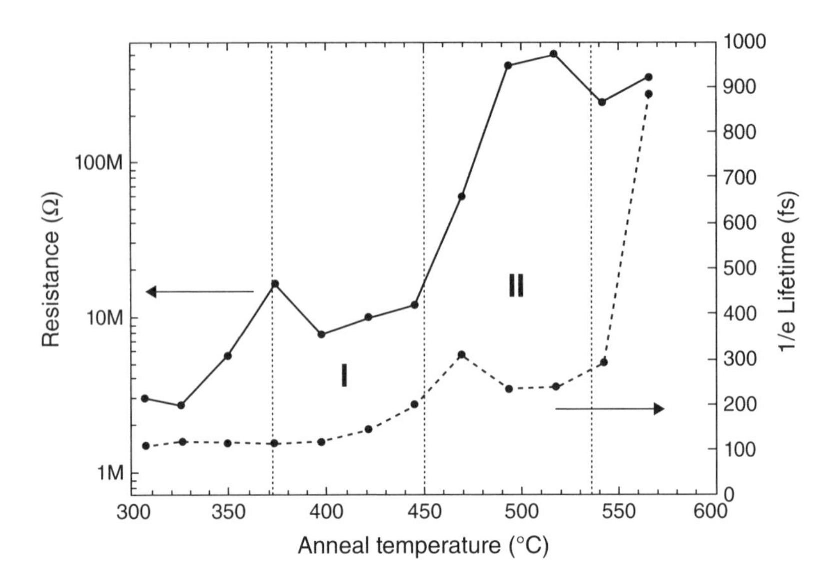

Figure 2.14 Carrier life time of LTG GaAs versus annealing temperature. The growth was performed at 230 ± 10 °C. At low annealing temperatures (region I and below), the material is dominated by defects, showing a comparatively low resistance and also a short lifetime. In region II, the lattice begins to relax; doping defects are reduced, increasing the resistance but the lifetime remains on a similar level. Region II is ideal for producing photoconductive switches. Above region II, point defects are cured, the lattice constant relaxes toward that of the substrate. The annealing also reduces the trap state concentration, resulting in an increase in the lifetime to ~1 ps or even more. Reproduced from Ref. [54], © 2003, AIP Publishing LLC.

fingers is in the range of 100 μm² [23], resulting in an RC 3 dB frequency around 1.3 THz for a 72 Ω load. No mesa etching is required due to the high dark resistance in the GΩ/square range [23], since a noteworthy conductance only occurs in optically excited areas. Usually, an ARC (thickness ~140 nm for an SiO_2-film at 800 nm) is deposited on top of the active structure; this also prevents oxidation of the semiconductor at high optical powers. In order to improve thermal conduction, often an AlAs heat spreader layer is deposited below the active material. A Bragg mirror, consisting of AlGaAs-AlAs $\lambda/2$ layers, can be grown underneath the active material to reflect the incoming laser and enhance the absorption [55]. This allows thinner LTG GaAs absorber layers. It also improves the electrical performance since the electric field applied by the finger electrodes is more homogeneous close to the surface. Several groups also reported improvement in the thermal management by sample transfer onto a highly thermally conductive material, such as sapphire or silicon [43, 47]. Industrial diamond and SiC have also been proposed as thermal substrates.

To the knowledge of the authors, the highest reported CW THz power in the low THz-frequency range is around 1.8 mW at 252 GHz [47]. A tunability range from <100 GHz to 3.8 THz with a single device has been reported [55] with still microwatt-level powers at 1 THz. LTG GaAs-based photoconductors have not only been used as antenna coupled devices but also as large area emitters (see Section 3.6.2) [43].

LTG GaAs photoconductors have to be operated with laser wavelengths shorter than 870 nm due to their band gap. Although absorption can also take place via the mid-gap trap state [56] or by two photon processes (only in pulsed operation) [57], the absorption cross section is so low that very little has been reported with 1550 nm excitation.

ErAs:GaAs at 800 nm

ErAs:GaAs is a promising alternative to LTG GaAs. It consists of continuous growth of a superlattice sequence of GaAs-ErAs-GaAs with up to two monolayers (ML) of ErAs [58] and a thickness for the GaAs layers in the range of 10–30 nm. Erbium tends to precipitate and form ErAs nanoparticles with a size in the range of ~2 nm [58] and a density in the range of 7×10^{12} cm^{-2} [59]. Bulk ErAs is a semi-metal with a Fermi energy slightly above the band gap center of GaAs. Due to quantization effects, small particles have been reported to show a transition from semi-metallic to semi-conducting [60]. In contrast to LTG GaAs, ErAs:GaAs is grown at only slightly lower temperatures than standard GaAs, with typical values around 535–580 °C [59, 61]. Due to the position of the Fermi energy in the band gap of GaAs, the ErAs particles are very efficient in capturing carriers. The trapping times mainly depend on the drift/diffusion time from the point of generation (by optical absorption) to the next ErAs particle. Higher particle density results in shorter trapping times, however, also in lower mobilities due to increased scattering. Trapping times in the range of $\tau_{rec} \sim$ 120–250 fs [33, 62, 63] and a significantly higher mobility than LTG GaAs [23] have been reported. The incomplete surface coverage of ErAs allows continuous overgrowth with defect-free GaAs, because of the available GaAs surface where growth can continue. The electrical performance is comparable to that of LTG GaAs. A breakdown voltage of 200–500 kV/cm has been reported [33, 59]. The dark resistance is about 100 times lower than LTG GaAs [23], however, still high enough in order to produce useful photoconductors with a dark resistance in the several MΩ range [23]. The comparable electrical properties and lifetimes result in comparable performance as LTG GaAs photoconductors. Interestingly, a special growth mode of ErAs:GaAs allows noticeable absorption at 1550 nm, allowing implementation of ErAs:GaAs photomixers at telecom wavelengths. This will be discussed in the next section.

2.2.5.2 Photoconductor Concepts at 1550 nm

1550 nm compatible photomixers benefit from the huge variety and low cost of telecom components. A wavelength difference of 8 nm is required for a difference frequency of 1 THz. CW systems often consist of two DFB diodes that are commercially available with a mode hop free tuning range of about 4.5 nm. Two diodes can therefore cover a frequency range of about 1.2 THz. For a larger tuning range, one of the DFB diodes can be replaced with a grating tuned source. In both cases, the laser linewidth is in the low MHz range. DFB diodes are also very stable (on the same scale as the linewidth, see Chapter 7). If more power is required, erbium-doped fiber amplifiers (EDFAs) are used. The systems are often completely fiber-based (with polarization maintaining fibers), allowing simple handling and very little alignment effort. Only the free space THz path needs alignment. In 1550 nm CW systems, photoconductors are mainly used as coherent detectors. Although they can also generate THz radiation, p-i-n diode-based mixers are typically more efficient.

Pulsed systems frequently use 1550 nm fiber lasers. Due to dispersion, the fiber laser length, l, is usually kept fairly short, resulting in high repetition rates in the range of $\upsilon_{rep} = c/2l \sim 70$ MHz. Photoconductors are used both as sources and coherent detectors. If coherent detection by electro-optic sampling (EOS) is desired, the 1550 nm signal can be frequency-doubled such that EOS crystals with phase-matching at ~800 nm (e.g., ZnTe) can be used.

In contrast to well-established 800 nm photomixers, devices for 1550 nm operation suffer intrinsically by the ~2 times lower band gap (~0.74 eV). It is very challenging to obtain deep traps and high dark resistances while maintaining a sufficiently high carrier mobility and absorption coefficient. LTG $In_{0.53}Ga_{0.47}As$, for instance, is strongly n-type [64], in contrast to LTG GaAs. Further, the breakdown voltage, E_B, is lower. From semiconductor physics, the breakdown field tends to vary with band-gap energy superlinearly and studies have yielded universal empirical relationships such as $E_B[\mathrm{V/cm}] = 1.73 \times 10^5 (U_G[eV])^{2.5}$ for low-doped direct-band-gap materials [65]. Hence, the difference between the GaAs E_B ($U_G = 1.42$ eV) and the E_B of InGaAs is predicted to be a factor of 4.9, which is rather close to the observed difference in the maximum bias voltage between homogeneous GaAs and InGaAs ultrafast PC devices. Several strategies have been demonstrated to fulfill the needs and to overcome the discussed problems of telecom-compatible photomixers.

Extrinsic Photoconductivity at 1550 and 1030 nm of ErAs:GaAs

A promising alternative to InGaAs is to use GaAs with 1030 or 1550 nm drive lasers and utilize sub-band-gap photon absorption mechanisms via the high concentration of defect- or impurity-levels that ultrafast materials generally have. In pulsed-mode, for example, attempts have been made to utilize two-photon absorption and sub-picosecond recombination via the mid-gap states associated with As precipitates in LTG GaAs [66, 67]. This was then used to demonstrate a PC switch, but the resulting photoconductivity was found to be impractically weak compared to the intrinsic cross-gap effect. In CW mode, attempts have been made to overcome the weak 1550 nm absorption by embedding the LTG GaAs in a dielectric-waveguide, distributed p-i-n photodiode [68]. But the waveguide length required for strong absorption reduces the electrical bandwidth because of difficulties in velocity matching the photonic and RF waves.

To date, the most attractive extrinsic photoconductive material has been co-doped (homogeneous) Er:GaAs. Test structures such as the PC switch in Figure 2.15 have yielded valuable data pertaining to DC photoconductive and THz performance [69]. First, the DC photoconductivity was characterized at 1550 and 1030 nm with up to $P_0 = 140$ mW of average laser power at each wavelength. This limitation is necessary on the one hand to guard against damaging the PC switch, and on the other hand for consistency between the 1550 and 1030 nm measurements, since this was the maximum power available from the 1550 nm laser but not the 1030 nm laser. The 1550 nm laser was an EDFA mode-locked laser with a repetition frequency of $f_{rep} = 36.7$ MHz and pulse width of $\tau_{pls} \sim 300$ fs. The 1030 nm laser was an YDFA (ytterbium-doped fiber amplifier) mode-locked laser with a repetition frequency of $f_{rep} = 31.1$ MHz and pulse width of $\tau_{pls} \sim 190$ fs. Laser power measurements were taken with a commercial thermopile-type detector. The laser beam at both wavelengths was focused into the active (central) gap of the PC switch in Figure 2.15 with a standard 10× microscope objective. The bias voltage was fixed at 77 V – a value found low enough to provide reliable operation at the maximum 140 mW laser power level. The DC photoconductivity results are plotted in Figure 2.16a,b for 1550 and 1030 nm light, respectively. The two curves are substantially different at low P_0, but similar at high P_0 where they both approach a linear asymptote. This is in contrast to the continuous quadratic-up behavior displayed by LTG GaAs switches with 1550 nm drive [67]. At low P_0, Figure 2.16a,b also show nonlinear behavior, but concave-down in Figure 2.16a and concave-up in Figure 2.16b, which is not yet understood. At the low end of the linear asymptotic region of Figure 2.16a, the current responsivity is $\mathfrak{R} \approx 5\ \mu$A/mW, and at the high end it is

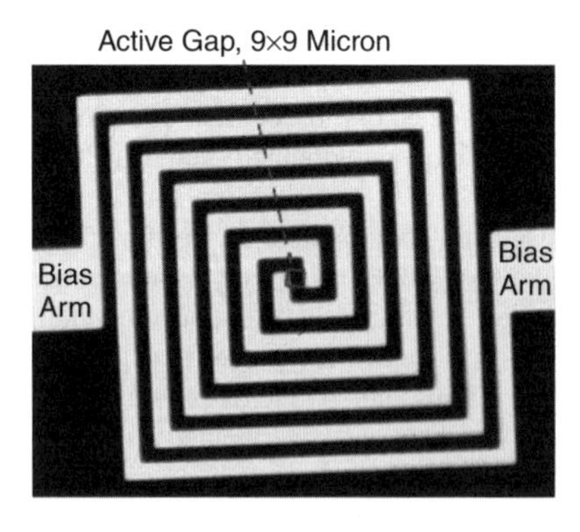

Figure 2.15 ErAs:GaAs self-complementary square-spiral photo-conductive switch.

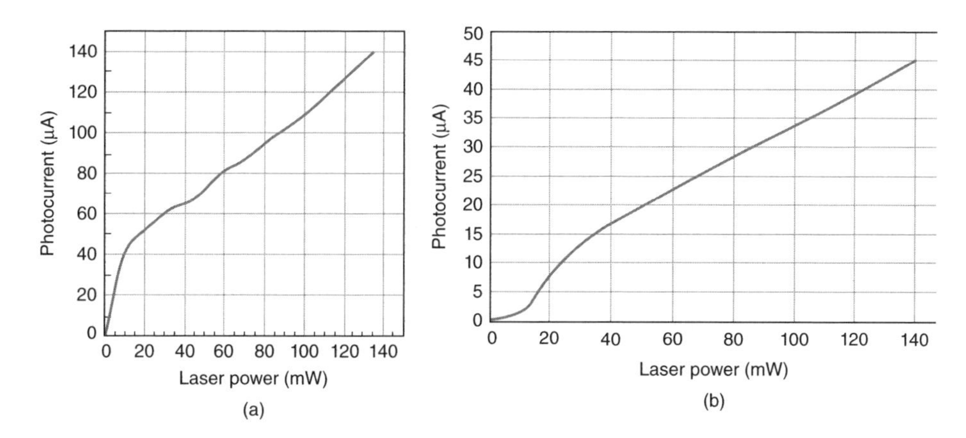

Figure 2.16 (a) DC photocurrent versus average 1550 nm laser power at 77 V bias. (b) Same as (a) except with 1030 nm laser.

$\Re \approx 1.0\,\mu$A/mW. The former is about four times less than the $\Re$ from an identical type of PC switch (same ErAs:GaAs material) measured at the same U_{DC} with a 780 nm sub-picosecond pulsed laser.

In contrast, the photocurrent for the 1030 nm laser in Figure 2.16b reaches a maximum value of 44.9 μA, corresponding to $\Re \approx 0.32\,\mu$A/mW, which is about three times less than the 1550 nm result.

However, it still supports our interpretation of the new effect as extrinsic photoconductivity rather than two-photon absorption. This is because 1030 nm photons would not match the energy-momentum conservation criteria for two-photon transitions across the GaAs band gap, so would experience a far weaker absorption coefficient than 1550 nm photons. The promising 1550 nm photocurrent suggests that this PC switch should produce measurable THz power, assuming of course that the photocarrier lifetime associated with the new photoconductive mechanism is comparable to that of the traditional intrinsic, cross-gap effect (typically $\ll$ 1 ps). The THz output power was first measured with a broadband, calibrated pyroelectric detector having a crude low-pass filter (0.010-inch-thick black polyethylene) to block the mid-IR-to-visible radiation. Its resulting THz responsivity is roughly 5000 V/W. The pyroelectric signal versus bias, U_{DC} , (Figure 2.17a) was measured at a fixed 1550 nm laser power of 140 mW, and the signal versus P_0 (Figure 2.17b) was measured at a fixed bias voltage of 77 V. The vertical scale

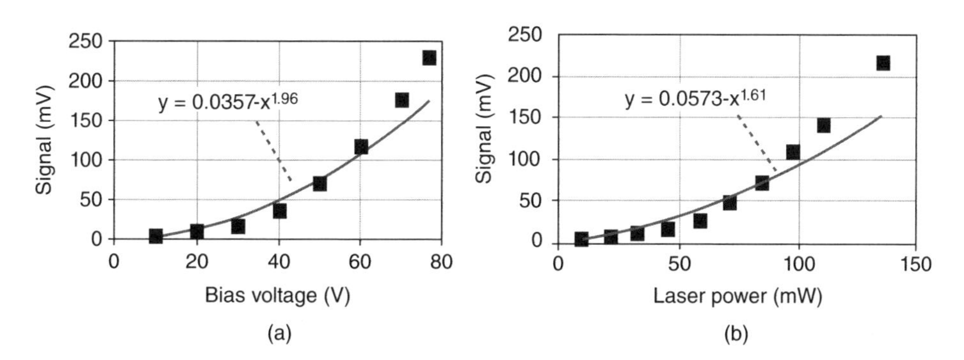

Figure 2.17 (a) RMS output from THz pyroelectric detector versus bias voltage with a constant 1550-nm laser power of 140 mW. (b) RMS output from THz pyroelectric detector versus 1550-nm average laser power at a constant bias of 77 V.

in both plots is RMS lock-in amplifier reading; when adjusted for this, the maximum reading in both Figure 2.17a,b corresponds to a peak-to-peak reading of approximately 520 mV (confirmed on an oscilloscope). Hence the maximum power measured from the switch is ≈105 μW [69]. In stark contrast, no THz power was measured from the same PC switch driven by the 1030 pulsed laser under the same operating conditions. This suggests that the photocarrier lifetime associated with the extrinsic photoconductivity is quite sensitive to laser drive wavelength, but for reasons that are not yet understood.

Inspection of Figure 2.17a,b provides an interesting contrast to the operating characteristics usually displayed by intrinsic photoconductive switches driven at 780 nm [70]. The nearly quadratic dependence on bias voltage in Figure 2.17a is stronger than at 780 nm, whereas the sub-quadratic $[P_{THz} \approx (P_0)^{1.6}]$ dependence on drive power in Figure 2.17b is weaker than at 780 nm [71]. The maximum 1550 nm-driven THz power is comparable to the broadband THz power measured from an identical type of switch (same ErAs:GaAs material) at the same U_{DC}, but driven with 25 mW of average 780-nm power. Hence, the new 1550-nm-driven photoconductive mechanism is about five-times less efficient in terms of THz-to-laser power conversion. An obvious way to improve this efficiency would be to increase the thickness of the active epitaxial layer, which is approximately 1.0 μm in all the devices tested to date. To estimate the power spectrum and bandwidth of the 1550-nm-driven PC switch, we carried out spot-frequency power measurements using a set of Schottky-diode zero-bias rectifiers mounted in a rectangular waveguide and operating in three distinct bands centered around 100 (W band), 300 and 500 GHz, as shown in Figure 2.18a [69]. These rectifiers act as band-limited filters with very sharp low-frequency turn-on (waveguide cut-off) and more gradual high-frequency roll-off. This enables a discrete estimate of the THz switch power spectrum knowing the external responsivity of the rectifiers and their noise equivalent bandwidth. The resulting data are plotted in Figure 2.18b, normalized to the signal from the 100 GHz rectifier. The bandwidth is obtained by fitting the discrete spectrum to a single-pole Lorentzian function, $S(f) = A/[1 + (2\pi f\tau)^2]^{-1}$, where A is a constant and τ is the photocarrier lifetime. This has been found to be a good fit to the THz power spectrum of PC switches whose photocarrier lifetime is significantly longer than the RC electrical time constant – a likely condition in the present case since the gap capacitance of the switch is ≪1 fF. For the experimental data in Figure 2.18b, the best fit to the data occurs when $A = 1.08$ and $\tau = 0.45$ ps. This corresponds to a 3 dB frequency-domain bandwidth of $B = (2\pi\tau)^{-1} = 354$ GHz, which is comparable to the bandwidth deduced from 780-nm time-domain measurements for the identical type of switch (same ErAs:GaAs material and antenna) with a 780-nm femtosecond laser [70]. However, the laser pulse in the present experiments (300 fs) is considerably

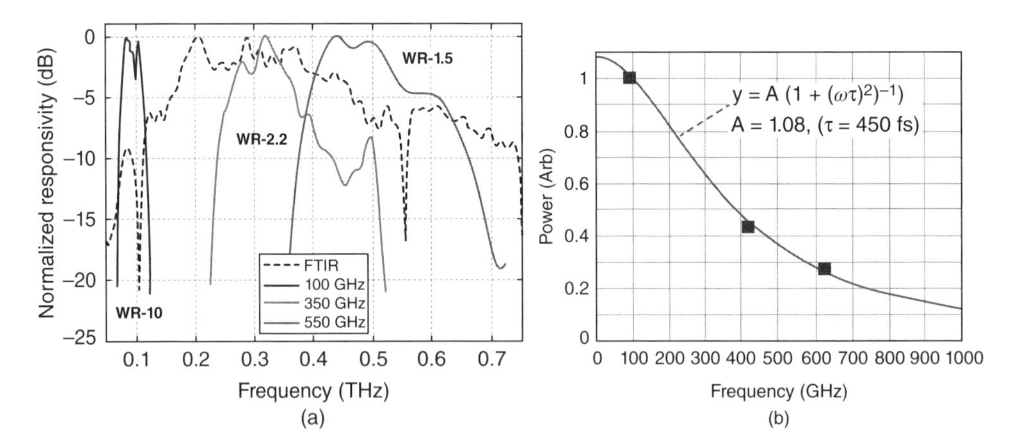

Figure 2.18 (a) Broadband spectrum from 780-nm-driven PC switch, and responsivity spectra of three waveguide Schottky rectifiers used for spot-frequency measurements. (b) Spot-frequency spectrum from 1550-nm-driven PC switch and curve fit.

longer than that used at 780 nm, so that the lifetime-limited bandwidth of the extrinsic PC switch could be even higher than 354 GHz.

ErAs:InAlAs/InGaAs Devices

The fairly low absorption of ErAs:GaAs at 1550 nm can only be overcome if the inter-band transition is used for absorption: for $In_{0.53}Ga_{0.47}As$, the absorption coefficients are in the range of 8000/cm. However, the typical problems for InGaAs-based photoconductors have to be faced: the Fermi energy of semi-metallic ErAs in InGaAs is close to (or even above) the conduction band edge [72]. Although carriers are still trapped by the semi-metal, the high density of states of ErAs result in a strong n-type background that has to be compensated with a p-dopant. Typically, either beryllium or carbon is used. The large p-doping (of the order of $10^{20}/cm^3$) and scattering by the ErAs particles results in rather low mobilities in the range of 900 cm^2/Vs [73] for material with sufficiently low lifetime, getting lower with higher Er concentration and stronger doping. In most cases, the n-background cannot be completely compensated, leading to fairly low resistances in the range of 10–100 Ω cm (100 kΩ/square for a 1 μm thick sample) [74]. In Ref. [75] a resistivity of 343 Ω cm was specified, however, without specifying the mobility. Furthermore, exact compensation is difficult: a small error in the doping concentration around the point of perfect compensation can push the Fermi energy from the center of the band gap right to the conduction- or valence band edge if no deep traps are present that pin the Fermi energy. In contrast to ErAs:GaAs, a compromise between short carrier lifetime (requiring high ErAs concentration), low dark conductivity (low ErAs concentration and high p compensation doping beneficial) and mobility (low ErAs concentration and low p-doping) has to be found. The problem of low breakdown voltage cannot be overcome. However, functional photomixers have been fabricated from ErAs:InGaAs with lifetimes in the range of 200–300 fs [75, 76]. Low-temperature operation can help to freeze out carriers for overcoming thermal heating limitations [77] by excess dark conductance.

In order to reduce the large n-background, a superlattice of InGaAs-ErAs:InAlAs has been implemented, as illustrated in Figure 2.19a,b [78]. ErAs is a deep trap in InAlAs. If the InAlAs layer is chosen only a few nm thick, carriers can tunnel into the ErAs states in InAlAs and are efficiently trapped. Some p-doping is still necessary in order to compensate for the n-background by carriers tunneling out of the ErAs states. Since carriers have to tunnel into the ErAs states (with limited tunnel probability), the carrier-lifetime can be quite long. However, it is already sufficient to grow the ErAs between an InAlAs and an InGaAs layer (Figure 2.19b), as a compromise between lifetime and resistance. Generally, the resistance of such a structure can be much higher than that of the ErAs:InGaAs material. The absorption takes place in the adjacent InGaAs layers between the InAlAs layers. The band structures shown in Figure 2.19a,b resemble quantum wells. Shifts of the absorption edge toward longer wavelengths due to quantization effects remain small if the InGaAs layer is chosen fairly wide, with thicknesses in the range of 15 nm. The absorption coefficient is just slightly lowered at 1550 nm as compared to bulk InGaAs. Besides the higher resistance, another advantage is the much higher carrier mobility due to the absence of ErAs in the absorber material and less p-background doping. Since carriers first have to drift toward the InAlAs barriers and then have to tunnel into the ErAs trap states, the carrier lifetime is longer than that of ErAs:InGaAs. The layer thicknesses of both the InAlAs tunnel barriers and the InGaAs absorber layers are a compromise between high resistance, absorption, and low lifetime. Typical values are 2.5 nm for the p-doped InAlAs layers, followed by a 15 nm InGaAs layer. In order to achieve an absorber thickness in the range of 1 μm, a sequence of $N = 70$ superlattice periods has been used, resulting in a total thickness of 1400 nm. Such InGaAs-ErAs:InAlAs photomixers have been used for pulsed measurements [78]. A representative measurement with a large area emitter is shown in Figure 2.19c. The material showed a resistivity of 670 Ω cm, a mobility of 1100 cm^2/Vs and a carrier lifetime of 1.2 ps.

LTG InAlAs:InGaAs Devices

A further concept for the realization of suitable photoconductors for the operation at 1.5 μm optical wavelength is a LTG InGaAs/InAlAs multilayer structure, which is additionally doped with beryllium.

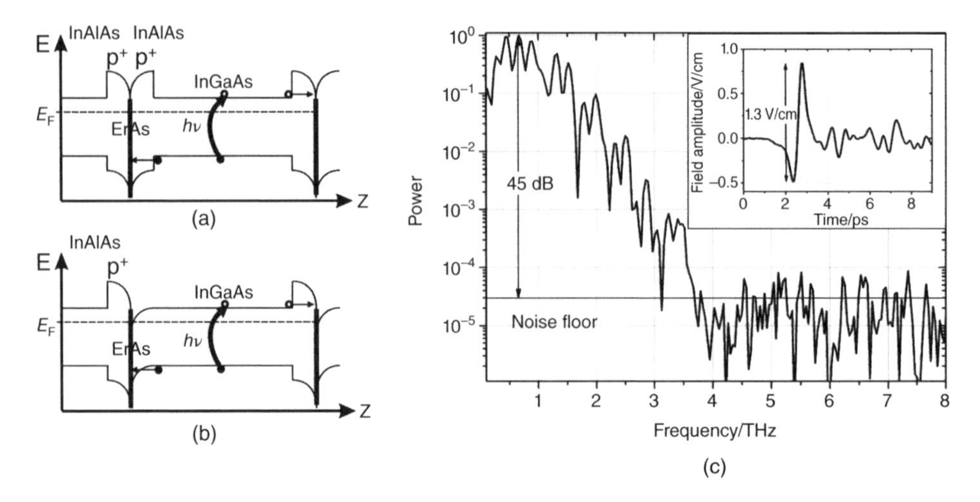

Figure 2.19 (a) InGaAs-ErAs:InAlAs photomixer ideal band diagram. The ErAs-layer is sandwiched between InAlAs tunnel barriers in order to increase the dark resistance of the material. (b) Device with only one tunnel barrier. The sheet resistance is still strongly increased. (c) Typical power spectrum obtained with an InGaAs-ErAs:InAlAs large area emitter (design as in (b)) under pulsed operation (laser wavelength 1550 nm, 100 mW average power, pulse duration 100 fs, repetition rate 78 MHz). Electro-optic sampling with ZnTe was used for detection. This results in the stronger roll-off above 3 THz where the detection crystal absorbs. The inset shows the time domain pulse. The ringing is due to reflections by an attached heat spreader.

Grown at temperatures below 200 °C, excess arsenic is incorporated as point defects on Ga lattice sites (antisite defect, As_{Ga}), which form deep donor levels in the material. In LTG GaAs these defects are mid-bandgap, whereas As_{Ga} defects in LTG InGaAs lie energetically close to the conduction band (CB) edge with activation energies around 30–40 meV. The reduction in the electron lifetime in LTG material compared to standard temperature grown (STG) semiconductors is attributed to electron capture by ionized arsenic antisites (As_{Ga}^{+}). In LTG GaAs the arsenic antisites are ionized by Ga vacancies, whereas ionization in LTG InGaAs is mainly thermal due to the low activation energy of the defect. This shift of the Fermi level toward the conduction band leads to highly n-conductive material at room temperature. To increase the resistivity of the material, the InGaAs-layers are sandwiched between InAlAs-layers barriers which trap the residual carriers. These layers are high bandgap and therefore transparent for the incident optical signal. By using the p-dopant beryllium, the resistivity is further increased.

The structure illustrated in Figure 2.20 (i.e., 12 nm InGaAs, 8 nm InAlAs), typically features a Hall mobility of 600 cm^2/Vs, a resistivity of 300 Ω cm and a carrier lifetime of 500 fs for a Be-doping concentration of 2×10^{18} cm^{-3}. This multilayer photoconductor successfully combined the required material properties for THz generation at 1.5 μm optical wavelength for the first time and enabled the first fully fiber-coupled pulsed THz-system in 2008, featuring a bandwidth of roughly 3 THz and a dynamic range of 30 dB [79].

A general disadvantage of photoconductive antennas with on-top metal contacts results from the decreasing in-plane electrical field component in the depth of the photoconductor. In the depicted multilayer structure, this problem is enhanced due to the many interfaces at the intermediate InAlAs layers with higher bandgap. The interaction of photocarriers with the electric field is reduced, which results in a low mobility limiting the obtainable THz power. This problem was solved by mesa-type structures with electrical side contacts. Here, the electrical field is directly applied, even to deeper layers, and the current in the receiver does not need to cross heterostructure barriers. As compared to their planar counterpart,

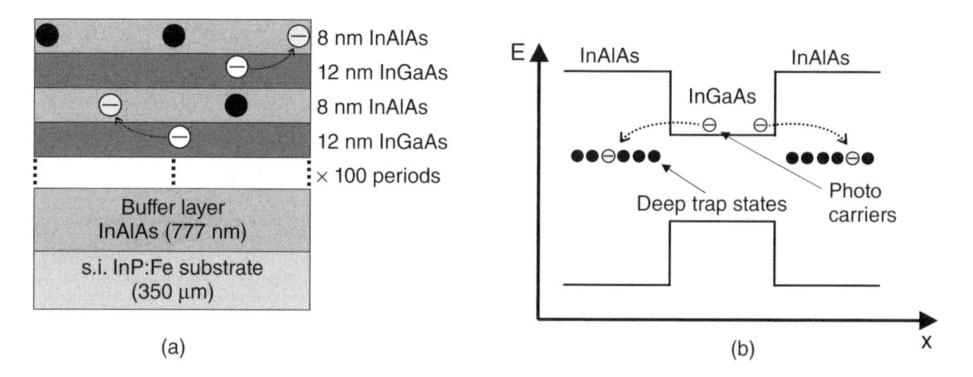

Figure 2.20 (a) Schematic of InGaAs/InAlAs heterostructure, with 100 periods of a 12 nm InGaAs layer followed by an 8 nm InAlAs layer with cluster-induced defects acting as electron traps. (b) Schematic of the respective band-diagram in real space with deep cluster-induced defect states. Reproduced and adapted with permission from Ref. [83], © 2013 Optical Society of America.

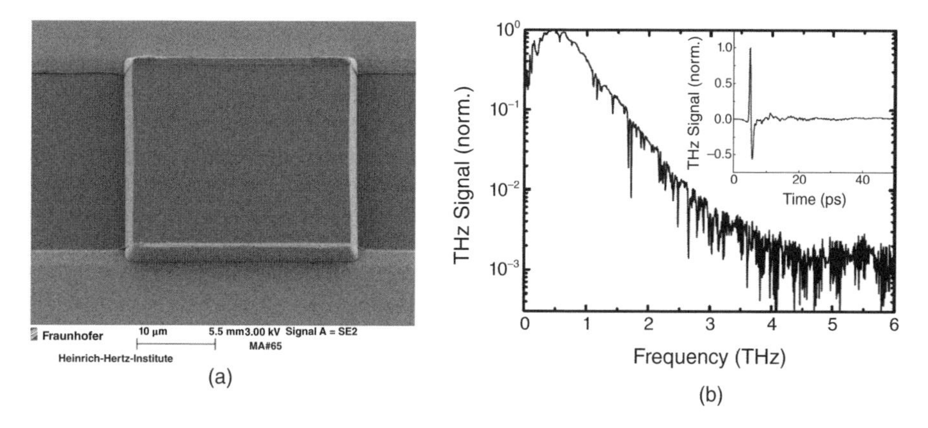

Figure 2.21 Mesa-structured InGaAs/InAlAs photoconductor (a) and corresponding THz-spectrum (b). The photoconductor features a stripline geometry with 25 μm gap, which was biased with 5 V and illuminated with 10 mW optical power. Reproduced and adapted with permission from Ref. [83], © 2013 Optical Society of America.

the mesa-type photoconductor increased the peak–peak amplitude of the THz-pulse by a factor of 28 and the bandwidth of the system exceeded 4 THz [80] (Figure 2.21).

A quick comparison of the mobility of the multilayer structure (600 cm^2/Vs) and the mobility of bulk InGaAs (10 000 cm^2/Vs) reveals that the photoconductor does not profit from the high InGaAs mobility, which limits the obtainable THz power. This is due to the high defect concentration of the material causing a strongly reduced carrier mobility due to elastic and inelastic (i.e., trapping) scattering of carriers at defect sites. A device which better matches the mentioned requirements can be realized by MBE growth of InGaAs/InAlAs multilayer heterostructures by utilizing a special characteristic of MBE growth of InAlAs. Within a substrate temperature range $T_S = 300$–$500\,°C$ the growth of InAlAs shows strong alloy clustering effects with InAs-like and AlAs-like regions featuring clusters sizes of several nanometers, with a maximum cluster density for $T_S \approx 400\,°C$.

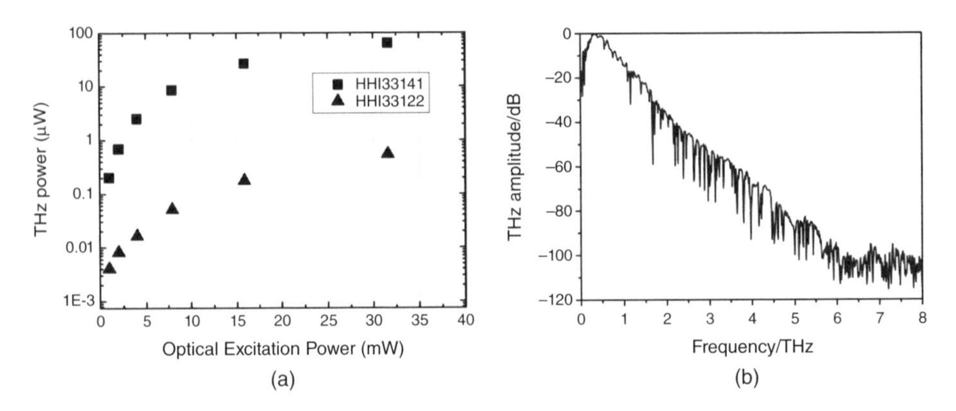

Figure 2.22 Emitted THz power of a 100 μm stripline antenna from a sample grown at 400 °C (HHI33141) and 200 °C (HHI33122) in dependence on the optical excitation power at a bias field of 15 kV/cm. In combination with an optimized receiver, a dynamic range of 100 dB was obtained, with a bandwidth of 6 THz. Reproduced and adapted with permission from Ref. [84], © 2013 AIP Publishing LLC.

The activation energies of these cluster defects have been measured to be in the region of $E_A = 0.6–0.7$ eV, which leads to semi-isolating InAlAs. By exploiting the above characteristics it is possible to obtain InAlAs/InGaAs photoconductors with low defect density and high mobility InGaAs layers adjacent to InAlAs layers with high defect densities at the same growth temperature for efficient carrier trapping. In such a way, the mobility of the material can be increased up to 5000 cm^2/Vs with a resistivity of 300 Ω cm. With this highly mobile material, THz powers up to 64 μW have been obtained for an optical illumination power of 32 mW [81] (cf. Figure 2.22).

In addition, photoconductors are also excellent detectors for CW THz systems operating with 1550 nm drive lasers. These detector applications will be discussed in detail in Chapter 7.

Ion-Implanted Photoconductors

Another option to generate deep traps in InGaAs is by means of ion damage. Ions with energies in the high kiloelectronvolt to megaelectronvolt range irradiated onto a semiconductor create many defect states by penetrating through the lattice, for example, by kicking out atoms or pushing atoms into interstitials. This results in a variety of trap states of different flavors, reducing both the carrier mobility and lifetime. A variety of ions has been used such as Br [82], Fe [83, 84], H, Au [85], and others. The optimum dose depends strongly on the implanted ions (weight) and implantation energy. For Be at 11 MeV, a dose around $10^{11}–10^{12}/cm^2$ was used [86]. The type of ion is secondary: In most cases, ion energies are chosen high enough to ensure that most of the ions come to rest far below the THz-active zone (~1–2 μm below the surface, defined by the absorption depth of the laser light). As-irradiated material is usually so strongly distorted that it cannot be used for THz applications; the carrier mobility is extremely low. An annealing step at temperatures between 500 and 800 °C [83, 84] is required to partially restore the lattice and allow accumulation of trap centers. Mobilities up to 3600 cm^2/Vs with lifetimes of 0.3 ps have been reported [87]. An average THz power of 0.8 μW was obtained under pulsed operation [87]. A pulse is shown in Figure 2.23a and the respective spectrum is shown in Figure 2.23b. The breakdown field increased to $E_B \approx 7.5$ kV/cm [86] by about a factor of 4 compared to non-implanted samples. The resistivity of that specific device was not reported. In Ref. [88], the same group reported on a value of ~5 Ω cm only. The authors pointed out that the residual carrier density is fairly invariant with respect to the ion dose. Some compensation doping would be required to reduce the dark conductivity. For further reading on ion-implanted materials, we refer to Ref. [86].

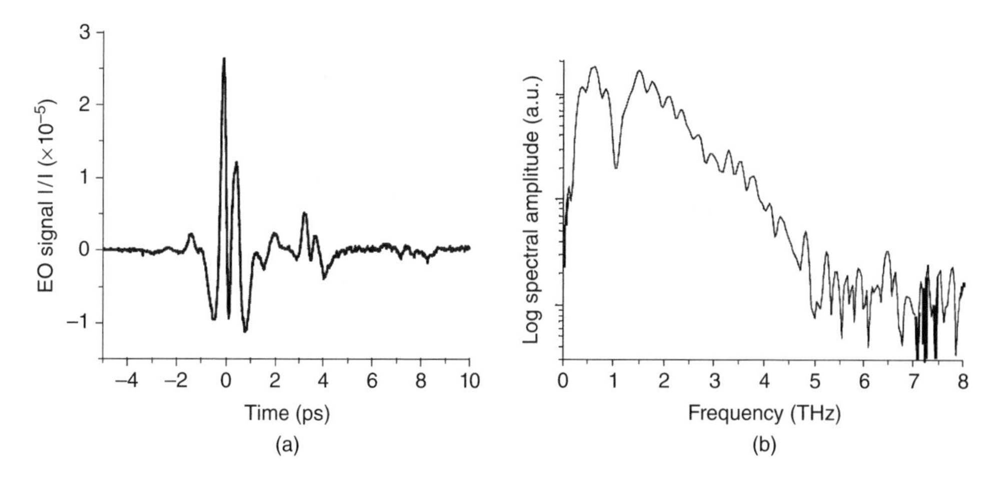

Figure 2.23 (a) THz pulse generated by Be^+ ion-damaged InGaAs excited with a $\tau_{pls} = 80$ fs 1550 nm laser and recorded with a DAST crystal. (b) Fourier spectrum. The dip at 1 THz is due to absorption of DAST. Figure reprinted from Ref. [86] with permission from Springer, © 2012 Springer.

2.2.5.3 Plasmonic Enhancement for Photoconductors

Plasmonic effects have recently been explored for improving the performance of photoconductors [89–91]. There exist two types of plasmonic enhancements that have been implemented: (i) Optical absorption enhanced structures: Well designed, nanometer-sized, metallic objects can feature plasmonic resonances that are excited by the incident laser power. The plasmons strongly increase the absorption close to the metallic obstacle. This allows much shorter penetration depths of the laser light, where the DC field by the electrodes is fairly strong. Further, the reduced absorption depth allows thinner InGaAs layers with low thermal conductance. Often, the nanometer-sized plasmonic structures are implemented in the electrodes. One carrier type (typically the electron) has very short transit lengths, resulting in high quantum efficiency. (ii) THz field enhanced structures: Due to sharp tips or nanometer-sized gaps and structures, the THz field is locally strongly enhanced. This improves, for instance, the performance of a photoconductive detector where the enhancement factor due to plasmonics is directly proportional to the detected photocurrent. In the following, we will briefly discuss some examples for plasmonic enhancement of photoconductors from the literature. Plasmonic enhancement for other THz generation concepts will be discussed in detail in Section 2.3.3.

Park *et al.* [91] reported on a photoconductive gap that was modified with silver nanoparticles with a diameter of ~170 nm. The lasers induced a plasmonic response, resulting in enhanced absorption close to the nanoparticles. This increased the THz power by a factor of 2 for the same excitation conditions. However, the authors pointed out that the breakdown DC voltage may be reduced by the particles, leading to lower maximum THz power.

Tanoto *et al.* [90] have designed a nano-gap electrode structure, as illustrated in Figure 2.24, which benefits from both aforementioned effects. The strong field enhancement at each electrode tip (particularly in the gap) results in efficient carrier extraction. Further, the capacitance of the structure is fairly small since it obtains interdigitated electrodes. Although the active area (i.e., the nano-gap area) is small, a strong improvement of the emitted THz power of up to a factor of ~100 as compared to a standard design, was reported.

Another approach by Berry *et al.* [89, 92] that applies both types of plasmonic enhancements uses plasmonic grating contact electrodes, as illustrated in Figure 2.25. The lasers illuminate the gratings that are attached to the antenna arms. A fairly large gap is used that is not illuminated. Typical parameters

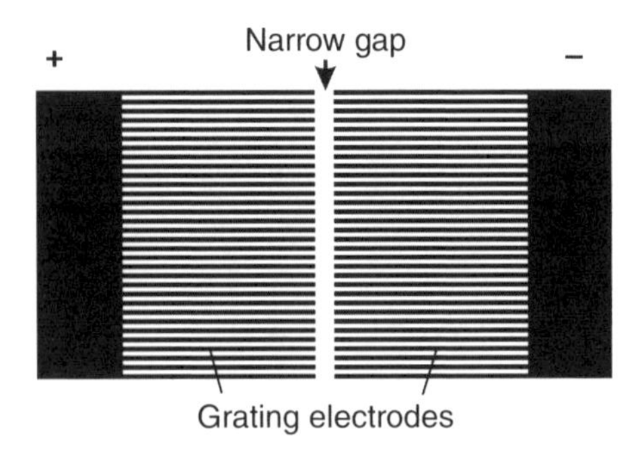

Figure 2.24 Nano-gap electrode structure as reported by Tanoto *et al.* [90]. A THz antenna is connected to the electrodes.

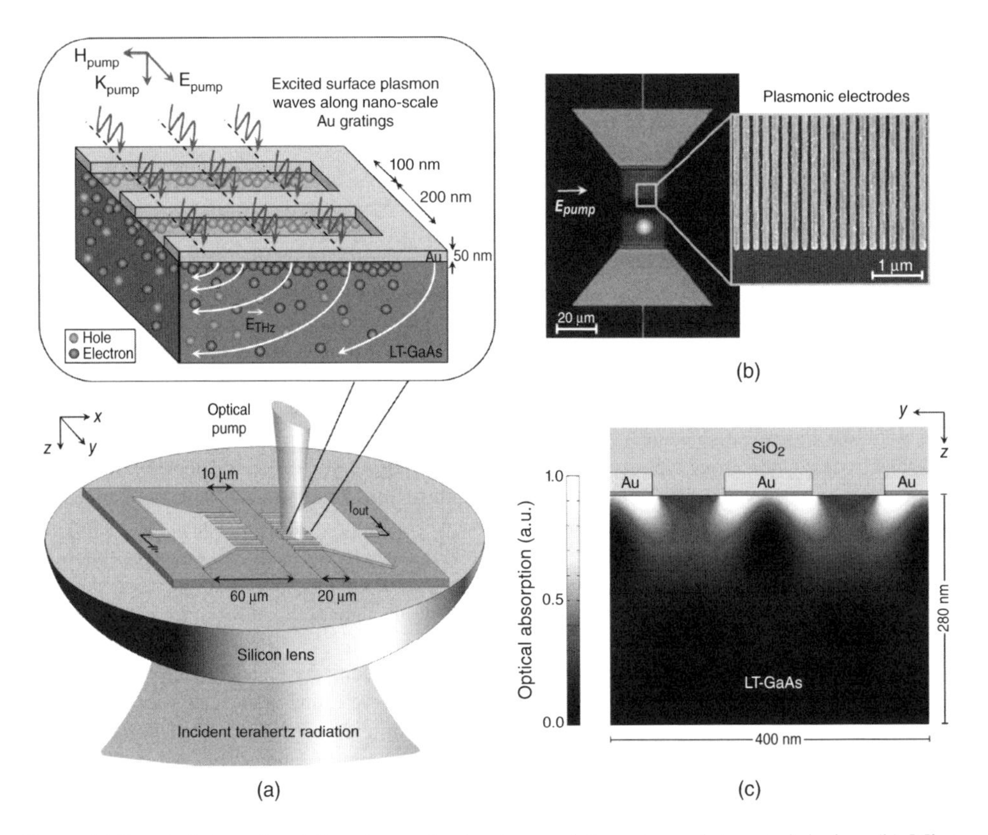

Figure 2.25 (a) Schematic of the plasmonic structure and the antenna-integrated device. (b) Micrograph of the structure. (c) Simulated optical absorption. Figures reproduced from Ref. [92] with permission from the Optical Society of America, © 2013 Optical Society of America. See plate section for color representation of this figure.

for the grating are gold electrodes with 200 nm pitch, 100 nm spacing, and 50 nm height. In order to improve coupling to the plasmons in the electrodes, an ARC with 150 nm SiO_2 is deposited on top of the grating. The thickness is optimized for maximizing the plasmonic resonance for 800 nm. Due to the small grating spacing of only 100 nm and to plasmonically enhanced absorption close to the electrodes, electrons feature a very short transit-time resulting in a large gain for electrons. Holes, however, are either trapped or have to move several microns to the opposite electrode. They produce a space charge background and do not contribute to the THz response. The large distance of the electrodes results in a low capacitance and a high DC resistance. The small electron transit-time results in a fast response. A 30 times improvement of the THz detection sensitivity has been reported [92].

2.2.6 Device Layouts of p-i-n Diode-Based Emitters

There are very few examples for p-i-n diodes at 850 nm in the literature. Most results are for 1550 nm-based operation using InGaAs/InP-based devices.

2.2.6.1 UTC Diodes

UTC diodes are, up to now, the most successful p-i-n diode-based photomixers for CW operation. They were invented by researchers at Nippon Telegraph and Telephone (NTT, Japan). A typical band structure is depicted in Figure 2.7. In UTC diodes, the p^--doped absorber layer is attached to the p-contact. The transport and current is mainly due to electrons with at least partial ballistic transport. This is achieved by using a transport or collection layer with a larger band gap as compared to the absorption layer. A few groups have reported on ~830 nm UTC diodes based on (Al)GaAs [93]. The majority of UTC diodes, however, are designed for 1550 nm operation with InGaAs absorbers, followed by two to three transition steps with InGaAsP layers to the InP transport (or collection) layer. Typical lengths for the absorber and transport layers are 100 and 300 nm, respectively [31, 36]. Besides telecom compatibility, InGaAs/InP UTC diodes also benefit from the larger inter-valley energy of InGaAs ($E_{\Gamma L} = 0.46$ eV) and InP ($E_{\Gamma L} = 0.59$ eV) as compared to GaAs ($E_{\Gamma L} = 0.29$ eV). The electrons can gain much higher energy before being scattered which allows larger ballistic transport lengths. An increase in the average transport velocity by a factor of 3 as compared to the saturation velocity, v_{sat}, has been reported in Ref. [94], indicating partial ballistic transport.

The original design by NTT, Japan [31] features an output power of 2.6 μW at 1.04 THz with a broadband logarithmic-periodic antenna and even 10.9 μW with a resonant antenna. The device cross-section was 13 μm^2, with photocurrent densities in the range of 100 kA/cm^2 (i.e., photocurrents of 13 mA). The photocurrent responsivity was 0.02–0.03 A/W, the transit-time 3 dB frequency was 170 GHz, the RC 3 dB frequency of the device with broadband antenna 210 GHz. With a design optimized for 100 GHz, 20 mW of output power has been reported [46].

Other designs, including backside illumination and side illumination by a waveguide design have been realized [36, 38]. In Ref. [36], Renaud *et al.* describe a traveling wave UTC diode with a resonant antenna, delivering 24 μW at 914 GHz (at 100 mW optical power). The optical responsivity was in the range 0.14–0.36 A/W. The absorber layer was 70 nm and the transport layer was 330 nm long.

Beck *et al.* [32] have designed a TEM-horn-coupled UTC diode with a diameter of only 2 μm allowing a high RC 3 dB frequency. The transport layer was also chosen short (only 137 μm) in order to increase the transit-time 3 dB frequency. With a photocurrent of only 2.75 mA (optical power 50 mW) they achieved 1.1 μW at 940 GHz.

Waveguide coupled 1550 nm UTC diodes by the Heinrich-Hertz Institute, Germany, so-called waveguide integrated photodiodes with a THz antenna (WIN-PDA), are commercially available. An output power of 5 μW at 500 GHz has been achieved at an optical power of only 25 mW [95].

2.2.6.2 Triple Transit Region Diodes

The UTC diode has been demonstrated to have the highest performance within the THz frequency range, as reported in the sections above. In contrast to the conventional p-i-n diodes, where both types of the photogenerated carriers contribute to the overall photocurrent, there is only electron drift in the UTC diode, which is a key feature for high-frequency operation and for overcoming saturation effects such as the space charge effect. In a conventional UTC diode, this is achieved by using an undepleted p-type absorber instead of a depleted non-intentionally-doped (n.i.d.) absorber as is the case in the p-i-n diode. Thus, one can assume hole relaxation and neglect hole diffusion in the absorption region of the UTC diode. However, the transit-time-limited bandwidth of a UTC diode still suffers from the slow electron diffusion in the highly p-doped (InGaAs) absorber layer. This drawback has been tackled by further inventions that have led, for example, to the development of the so-called modified uni-traveling-carrier (MUTC) diodes which also greatly overcome the saturation effects [96, 97]. In contrast to the original UTC diode, the absorption region of a MUTC not only consists of an undepleted p-doped absorption region but also of an additional depleted InGaAs absorption layer. This, however, implies that there is a substantial hole drift in the depleted absorber of MUTC PDs, which is somewhat in contradiction to the expression "uni-traveling-carrier."

To overcome the limiting low electron diffusion occurring in UTC and also in MUTC diodes, a new structure called a triple transit region (TTR) photodiode was invented and reported in Ref. [98] by

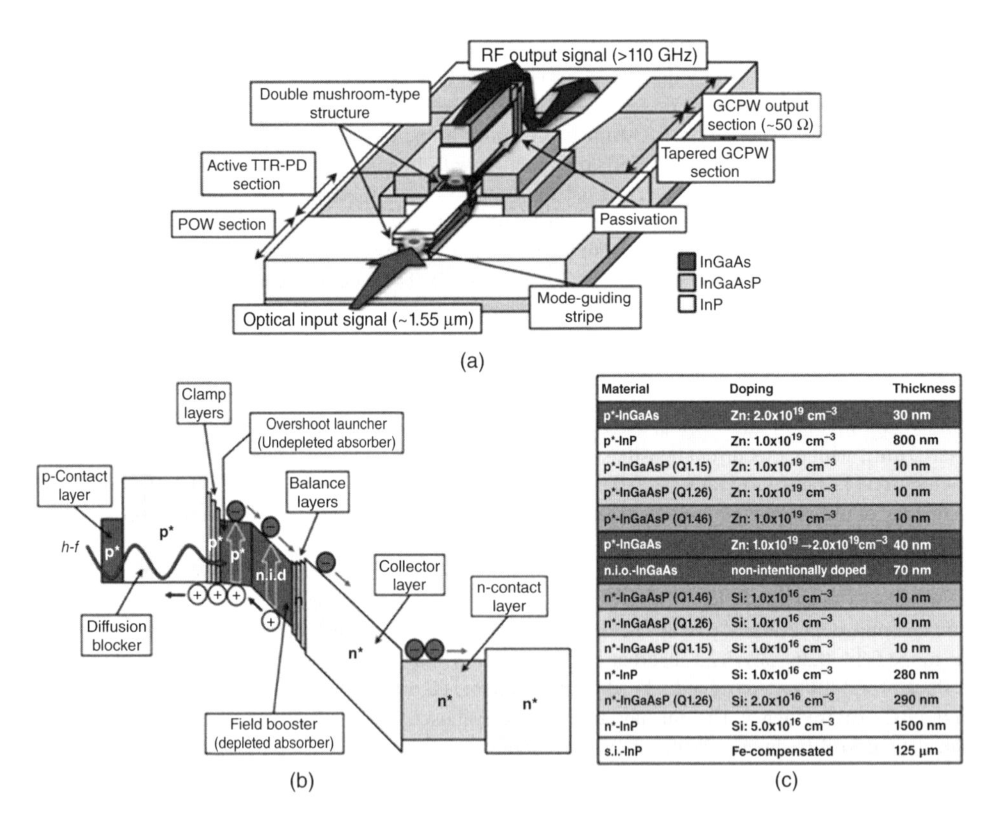

Material	Doping	Thickness
p*-InGaAs	Zn: $2.0x10^{19}$ cm^{-3}	30 nm
p*-InP	Zn: $1.0x10^{19}$ cm^{-3}	800 nm
p*-InGaAsP (Q1.15)	Zn: $1.0x10^{19}$ cm^{-3}	10 nm
p*-InGaAsP (Q1.26)	Zn: $1.0x10^{19}$ cm^{-3}	10 nm
p*-InGaAsP (Q1.46)	Zn: $1.0x10^{19}$ cm^{-3}	10 nm
p*-InGaAs	Zn: $1.0x10^{19} \rightarrow 2.0x10^{19} cm^{-3}$	40 nm
n.i.o.-InGaAs	non-intentionally doped	70 nm
n*-InGaAsP (Q1.46)	Si: $1.0x10^{16}$ cm^{-3}	10 nm
n*-InGaAsP (Q1.26)	Si: $1.0x10^{16}$ cm^{-3}	10 nm
n*-InGaAsP (Q1.15)	Si: $1.0x10^{16}$ cm^{-3}	10 nm
n*-InP	Si: $1.0x10^{16}$ cm^{-3}	280 nm
n*-InGaAsP (Q1.26)	Si: $2.0x10^{16}$ cm^{-3}	290 nm
n*-InP	Si: $5.0x10^{16}$ cm^{-3}	1500 nm
s.i.-InP	Fe-compensated	125 μm

Figure 2.26 (a) Schematic view of the developed TTR-PD (b) with band diagram and (c) layer structure. See plate section for color representation of this figure.

Rymanov *et al.* A schematic view of developed TTR diodes, the band diagram and the entire TTR diode layer structure, including a monolithically integrated lower InGaAsP/InP passive optical waveguide (POW) grown on a semi-insulating (SI) InP substrate (thinned to 125 μm), is shown in Figure 2.26. The key innovation of the TTR diode manifests in the active waveguide section of the diode. Top-down, it consists of a highly p-doped InP upper cladding layer (800 nm) which also serves as a diffusion blocker

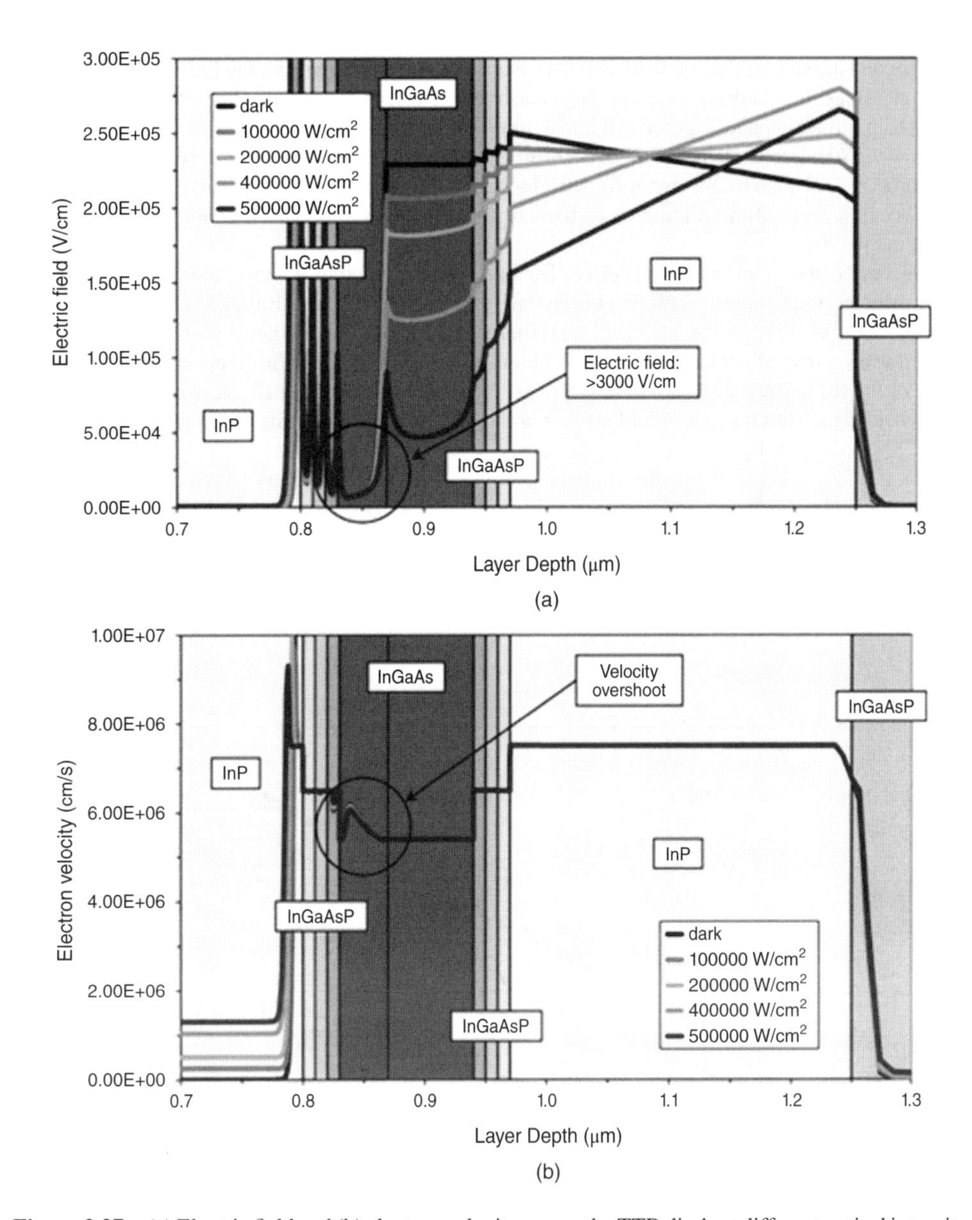

Figure 2.27 (a) Electric field and (b) electron velocity across the TTR diode at different optical intensity levels for a reverse bias of 8 V. Figure reproduced from Ref. [98] with permission from the Optical Society of America, © 2014 Optical Society of America. See plate section for color representation of this figure.

for electrons. This is followed by three 10 nm thick highly p-doped electric field clamp layers, which were introduced between the upper InP cladding layer and an undepleted InGaAs absorber (40 nm). Furthermore, there is a 70 nm thick depleted InGaAs absorber that acts as an electric field booster for the graded p-doped absorber. This is to ensure that the electric field in the graded p-doped absorption layer is always higher than the critical electric field. Thus, electron drift in the p-doped absorber determines the transit time rather than slow electron diffusion as is the case in UTC and MUTC diodes. Thanks to the three clamp layers between the upper InP cladding and the undepleted InGaAs absorber, strong electric fields beyond the critical electric field, and thus overshoot velocity, are maintained, even for high optical input intensities. In addition, the clamp layers also reduce carrier trapping at the InP/InGaAs interface [99], which further enhances the overall transit time.

Below the 70 nm thick depleted InGaAs absorber, three 10 nm thick slightly n-doped InGaAsP layers function as electric field balancing layers between the InGaAs field booster layer and the slightly n-doped InP collector layer (280 nm). Furthermore, these bridging InGaAsP layers are used for handling the band discontinuity [99].

As discussed above, and as can be derived from the name "triple transit region", there are three transit sections contributing to the drift motion of electrons for enhancing the transit-time-limited bandwidth. In contrast to MUTC PDs [96] and UTC-PDs [100], there are strong electric fields even in the undepleted absorption layer (overshoot launcher) of the TTR diode, which exceed the critical electric field in InGaAs of 3 kV/cm [101]. Thus, drift motion at overshoot velocity of about 6.1×10^6 cm/s instead of electron diffusion distinguishes the overshoot launcher of the TTR structure from the doped absorber in MUTC or UTC diodes [97, 100].

For a theoretical study of the electric fields and velocities in the different layers of the TTR diode, numerical simulations based upon the drift-diffusion-model (DDM) were performed [98]. Figure 2.27 shows the electric field and the resulting carrier velocities across the TTR diode at a reverse bias of 8 V.

As can be seen in Figure 2.27b, the three InGaAsP clamp layers ensure that the electric field strength in the absorber always exceeds the critical electric field, even at high optical excitation. Thus, even for very high optical intensities up to 500 kW/cm^2, the electron velocity remains in the regime of ballistic

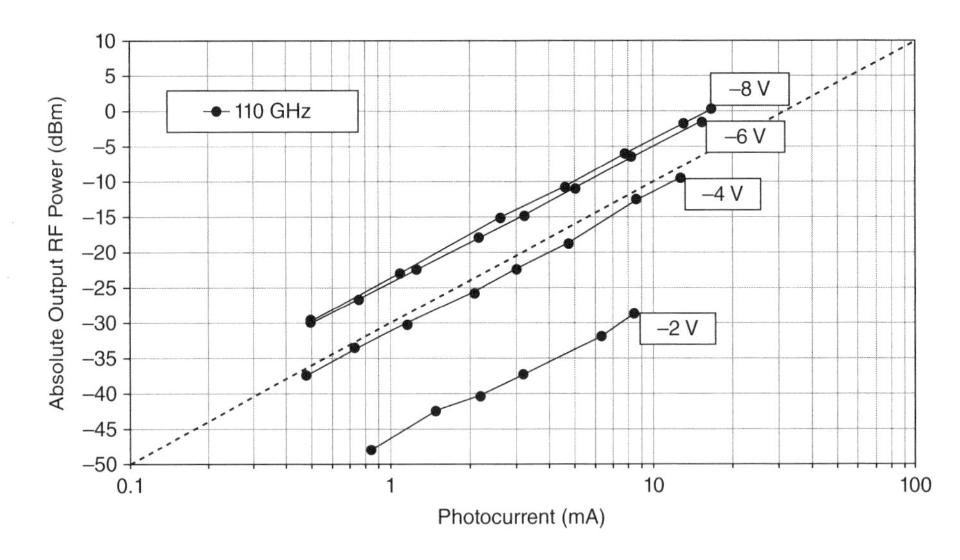

Figure 2.28 Output RF power at 110 GHz versus the photocurrent measured for different drive voltages. Figure reproduced from Ref. [98] with permission from the Optical Society of America, © 2014 Optical Society of America.

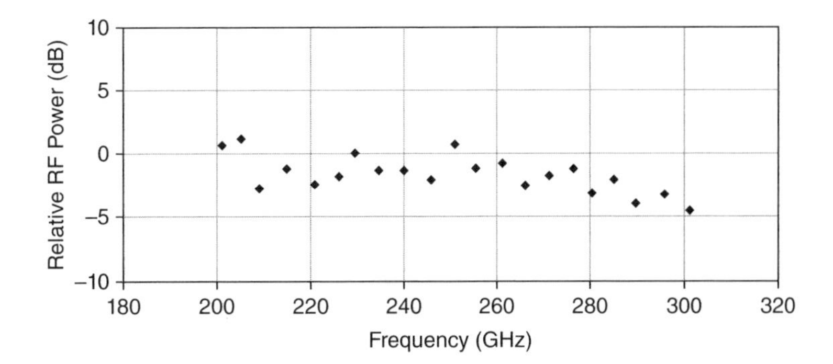

Figure 2.29 200–300 GHz operation of TTR diodes with integrated planar semi-circular bow-tie antennas. Figure reproduced from Ref. [103] with permission from VDE VERLAG, Berlin, © 2014 VDE VERLAG.

transport (>6.1×10^6 cm/s). Also in the intrinsic and quasi-intrinsic InGaAs and InP regions, electrons drift at the saturation velocity of 5.4 and 7.5×10^6 cm/s, respectively [102]. For the holes, relaxation can be assumed for the undepleted absorption layer, whereas in the depleted absorber, holes will drift at a somewhat slower velocity of ~4.8×10^6 cm/s. Only at optical intensity levels exceeding 500 kW/cm^2 and due to the resulting accumulation of holes and electrons in the depleted absorber, will the electric field fall below the critical electric field level, leading to saturation effects.

Experimentally, TTR diodes achieved broadband operation with a 3 dB bandwidth beyond 110 GHz as well as output power levels exceeding 0 dBm. Related output RF power measurements at 110 GHz are plotted in Figure 2.28 and were reported in Ref. [98].

Furthermore, TTR diodes were integrated on-chip with planar semi-circular bow-tie antennas (SCBTAs) as reported in Ref. [103]. Here, a wireless operation within 200–300 GHz, along with a 6 dB bandwidth, was demonstrated for the SCBTA-integrated TTR diodes, as shown in Figure 2.29.

2.2.6.3 n-i-pn-i-p Superlattice Diodes

The previous designs of p-i-n diode-based photomixers are optimized for high THz powers at lower THz frequencies. The n-i-pn-i-p superlattice diodes, in contrast, are explicitly optimized for transit-time-free operation up to 1 THz. For operation above a few 100 GHz, all previous concepts have a trade-off between transit-time and RC limitation: The transit-time is given by the transport layer length divided by the average transport velocity$\tau_{tr} = d_i/\overline{v}$. The transit-time can be reduced by making use of ballistic transport, allowing peak velocities of the order of 10^8 cm/s and average velocities in the region of 5×10^7 cm/s for InGaAs. Therefore, a maximum intrinsic layer length of 250 nm can be used for $f_{tr}^{3dB} = 1/(2\tau_{tr}) =$ 1 THz. At the same time, the RC roll-off 3 dB frequency, scales with $(d_i)^{-1}$. For $f_{RC}^{3dB} = 1/(2\pi R_A C_{pin}) =$ 1 THz, and $C_{pin} = \varepsilon_0 \varepsilon_r A/d_i$, a device with a cross-section of 50 μm^2 and an antenna resistance of only 27 Ω (resonant half wave dipole on InP/air interface), an intrinsic layer length of $d_i = 975$ nm would be required. This is in conflict with the previous value of 250 nm for the transit-time 3 dB frequency. Smaller device cross-sections, A, would allow reduction of the capacitance and, hence, increase the RC 3 dB frequency accordingly, however, at the cost of maximum current, $I_{max} = j_{max}A \sim A$, and therefore THz power. A way out of this dilemma is the n-i-pn-i-p superlattice photomixer concept: The capacitance can be reduced without affecting the intrinsic layer length and the device cross-section by connecting a number N of p-i-n diodes in series. The capacitance decreases as $C_{SL} = C_{pin}/N$. Only $N = 4$ periods are required in order to shift the RC 3 dB frequency above 1 THz for the same device parameters as above. The serial connection is obtained by stacking N p-i-n diodes during growth. A band diagram is depicted in Figure 2.30.

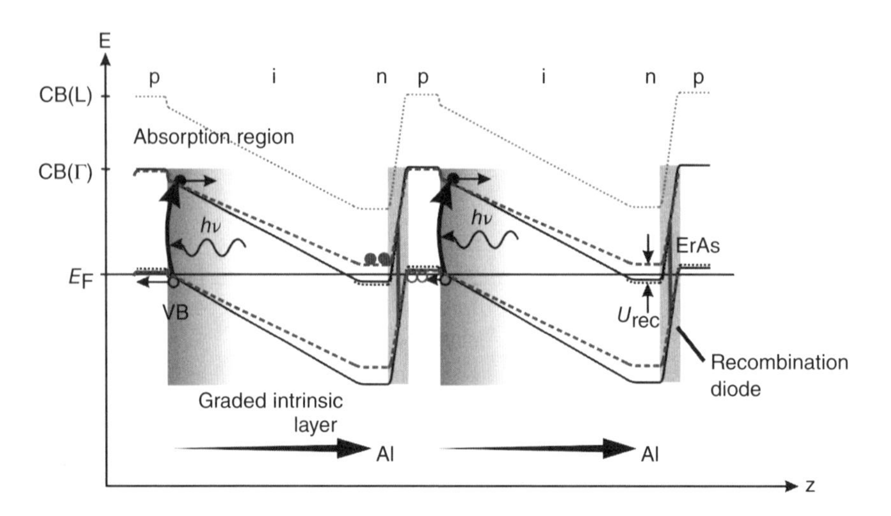

Figure 2.30 Band diagram of a n-i-pn-i-p superlattice photomixer ($N = 2$). Conduction band, valence band and L-sidevalley (dotted) are shown. The dashed line depicts the band diagram under illumination. A small forward bias, U_{rec}, evolves at the recombination diodes due to charge accumulation. The bias is only a fraction of the bandgap voltage. Reproduced and adapted with permission from Ref. [18] S. Preu *et al.*, J. Appl. Phys. 109, 061301 (2011) © 2011, AIP Publishing LLC. See plate section for color representation of this figure.

Between the p-i-n diodes, an np junction is formed. Both optically generated electrons and holes accumulate left and right of the resulting barrier. In order to prevent a flat band situation by charge accumulation, the np junction is designed as a highly efficient recombination diode that allows recombination of electrons and holes. High recombination current densities at low forward bias (~tens of kA/cm^2 at a fraction of the band gap voltage for InAlGaAs-based devices [104]) are achieved by implementing one to two monolayers of semi-metallic ErAs between the highly doped n and p layers. In order to supply the same optical power to each p-i-n diode, N times higher optical power is required for the same photocurrent as in a single p-i-n diode. However, optical power at 1550 nm is usually not a problem due to the availability of high power laser diodes and EDFAs.

The n-i-pn-i-p superlattice photomixing concept has been demonstrated both for 850 nm operation (AlGaAs-based devices [22]) and 1550 nm operation (InAlGaAs based devices [35]). The approximately twice larger band gap in AlGaAs-based devices, however, reduces the current density of the recombination diodes exponentially. Sufficiently high recombination currents in the range of 6 kA/cm^2 at 1 V bias could be still demonstrated [105]. Further, the n-i-pn-i-p superlattice makes full use of ballistic enhancement. Due to the larger inter-valley energy and higher recombination diode currents, InGaAs-based devices are more successful than GaAs devices. In order to prevent scattering, no transitions from InGaAs to InP are implemented, the whole structure consists of In(Al)GaAs. The absorption region is confined to the p-side by smoothly increasing the aluminum content of the InAlGaAs intrinsic (transport) layer. The maximum aluminum concentration is kept below ~10% since the inter-valley energy (and therefore the maximum ballistic velocity) decreases with increasing Al-content. For a transport length of 200 nm, a transit-time 3 dB frequency of 0.85 ± 0.15 THz has been determined.

The quaternary compound, however, features a fairly small thermal conductance ($\lambda_{InAlGaAs} < \lambda_{InGaAs} = 0.05$ W/K cm), limiting the current density to about 15 kA/cm^2. The n-i-pn-i-p superlattice photomixer therefore produces less output power at frequencies below the transit-time 3 dB frequency of UTC diodes (typically of the order of $f^{3dB}_{tr,UTC} \sim 200$ GHz) but is an excellent emitter at frequencies a few times above where the smaller current density is compensated by the excellent transit-time performance. With a

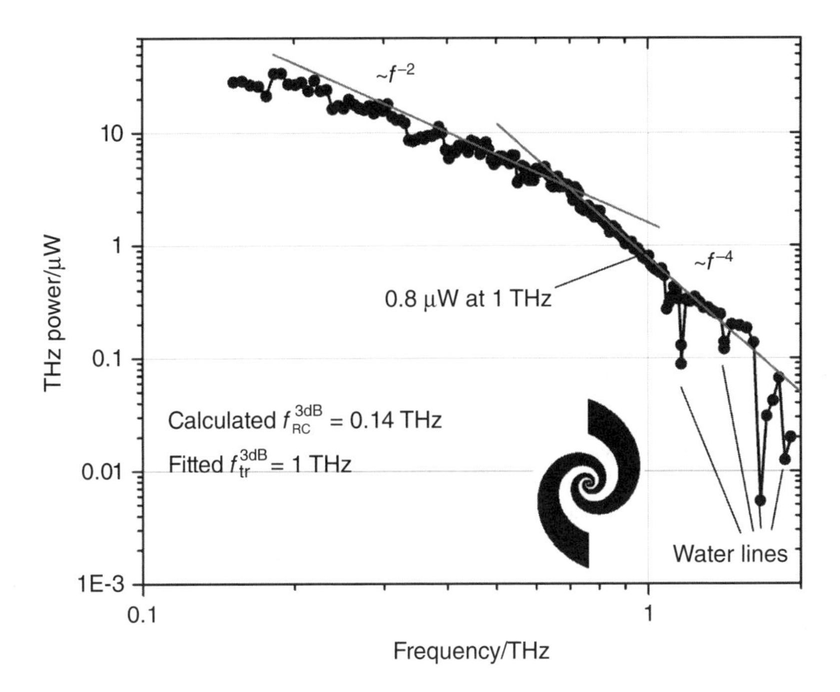

Figure 2.31 Power spectrum emitted by a logarithmic spiral by a three period n-i-pn-i-p superlattice photomixer. A fairly large device with a cross section of 82 μm^2 was used, resulting in an RC 3 dB frequency of 140 GHz. The represented values are corrected for 30% reflection of the silicon lens. The inset shows a schematic of the spiral antenna.

self-complementary broadband logarithmic spiral, a THz power of 0.65 μW at a photocurrent of 9.5 mA has been obtained with a $N = 3$ period device. Under extreme operation conditions close to thermal saturation, the power could be increased to 0.8 μW (17 mA photocurrent), as illustrated in Figure 2.31. With a resonant antenna, more than 3 μW can be estimated. At low frequencies (~75 GHz), a larger device with a broadband logarithmic-periodic antenna produced about 0.2 mW without any specific optimization.

2.3 Principles of Electronic THz Generation

Generation of radiation by electronic means is a fairly advanced and well-studied topic, however, with still a lot of active research, particularly in engineering. Electric and electronic generation and detection of electromagnetic radiation dates back to the early experiments by Heinrich Hertz who succeeded in transmitting electromagnetic signals for the first time. In the meantime, a rich spectrum of electronic sources and detectors has been developed, covering frequencies up to the THz range. In the microwave range, a large variety of powerful oscillators and amplifiers has been developed, with W-level output powers [106, 107]. Oscillators include Gunn diodes, impact ionization avalanche transit time (IMPATT) diodes, or tunnel-injection transit-time (TUNNETT) devices still with milliwatt power levels at 300 GHz [108]. Monolithic millimeter wave integrated circuits (MMICs) can be realized with transistor-based oscillators and amplifiers; the active elements are, for instance, high electron mobility transistors (HEMTs; based on III–V semiconductors), heterojunction-bipolar transistors (HBTs; SiGe or III-V devices), and field effect transistors (FETs). There are efforts to shift the operation frequency of these devices toward Terahertz frequencies. In the European DOTFIVE project [109], the development of SiGe HBTs with a maximum frequency of 0.5 THz was successfully demonstrated [110]. Its follow up, DOTSEVEN [111] aims for

SiGeC HBT development with an $f_{max} = 0.7$ THz. III–V-based transistors are also able to reach similar frequencies. Maximum frequencies $f_{max} = 0.8$ THz and 1.1 THz have been reported in Refs. [112, 113]. The improvement of these devices is mainly incremental by trying to overcome technological limits such as minimum structure size, reduction of access resistance and unwanted capacitances and inductances, and so on. The main design issues and concepts are already covered by other books [114]. Scaling laws are presented in Ref. [115]. In the following, we will give a brief overview of other room-temperature operating approaches that are used for Terahertz systems. Details on these systems follow in Chapter 6.

2.3.1 Oscillators with Negative Differential Conductance

Let us assume that we have a resonant circuit consisting of an inductor (L) and capacitor (C) connected in parallel, see Figure 2.32a. In the ideal case, such a resonant circuit is lossless and, if we excite some oscillations in the circuit, the circuit will oscillate forever, see Figure 2.32b. In reality of course, all resonant circuits have losses. We can represent the losses by a resistor R connected in parallel to the circuit, as shown in Figure 2.32c. Due to the resistor, the oscillations in the resonant circuit will attenuate with time, see Figure 2.32d. However, if we have a magic device, a negative resistor (R_D), we can use it to compensate for the losses in the resonant circuit, as illustrated in Figure 2.32e. If the negative resistor, R_D, has sufficiently low value (sufficiently high negative conductance), then the total combined value (R') of the resistors R and R_D will be negative, see conditions in Figure 2.32e. The amplitude of the oscillations excited in the resonant circuit will grow with time in this case. Small-amplitude initial seed oscillations are always present in the electronic circuits, particularly noise could be the source

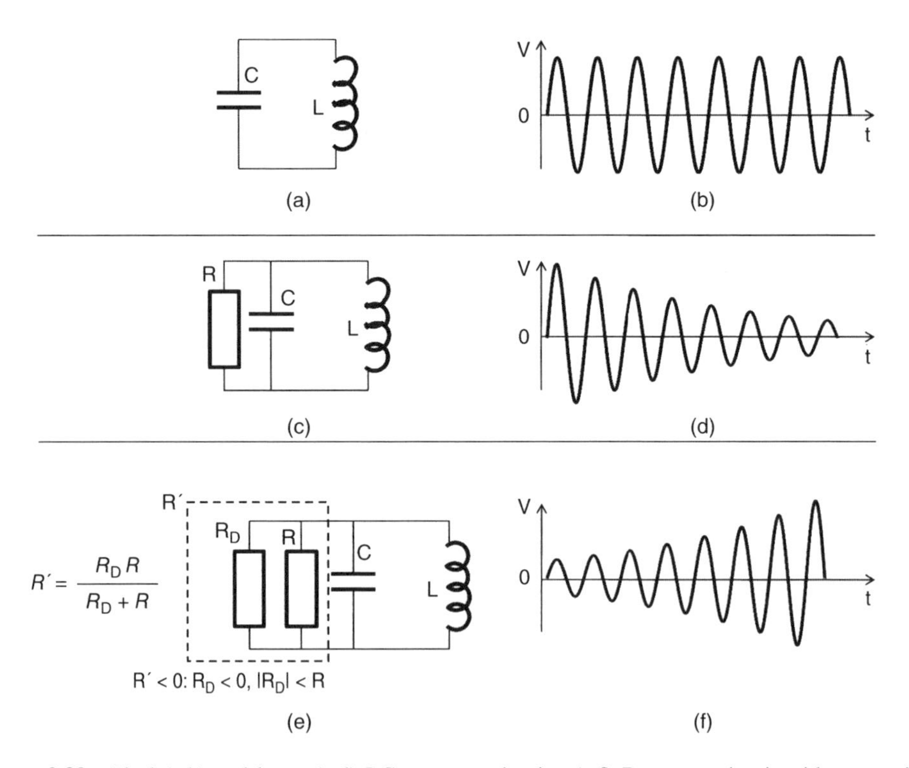

Figure 2.32 Ideal (a,b) and lossy (c,d) LC resonant circuits. (e,f) Resonant circuit with a negative resistor.

of such oscillations. The seed oscillations are then amplified by the negative resistor and, following transient oscillations, the circuit comes to a regime with stable large-amplitude oscillations. In such a way, a negative resistor can turn a simple lossy resonant circuit into an oscillator. Such resonant circuits with negative resistors are, probably, the simplest existing oscillators. The simplicity of the oscillators allows one to reduce their dimensions and to use the concept at very high and even THz frequencies. Instead of an LC resonant circuit with lumped-element inductor and capacitor, one usually uses resonators based on hollow waveguides or planar resonant antennas, like slot-, patch-, and other types of antennas. At high and particularly THz frequencies, the resonators have to be miniaturized, their characteristic length scale is roughly determined by the wavelength at the given oscillation frequency. The free-space half wavelength at 1 THz is 150 μm. If the resonator is immersed in a dielectric medium or the device with the negative resistor has a high inherent capacitance, then the characteristic dimensions of the THz oscillators should be reduced and they are usually on the scale of several tens of micrometers. Since such oscillators include all the necessary elements within their resonators, they are one of the most compact, if not the most compact, type of THz sources.

To make such oscillators, we need to be able to realize the magic devices with negative resistance in practice. Luckily, this is possible. There are few tunnel semiconductor devices with very particular *I–V* curves, sketched in Figure 2.33. The *I–V* curves have regions with negative differential conductance (NDC). If we apply a DC bias to such devices, so that they are biased in the NDC region, then their small-signal AC resistance will be negative and they could be used to make oscillators.

Resonant-tunneling diodes (RTDs) [116, 117] are double-barrier tunnel semiconductor structures. Their two barriers are very close one to another so that the barriers form a quantum well (QW) between them, see Figure 2.33a. The mechanism of the electron transport through the barriers is resonant tunneling

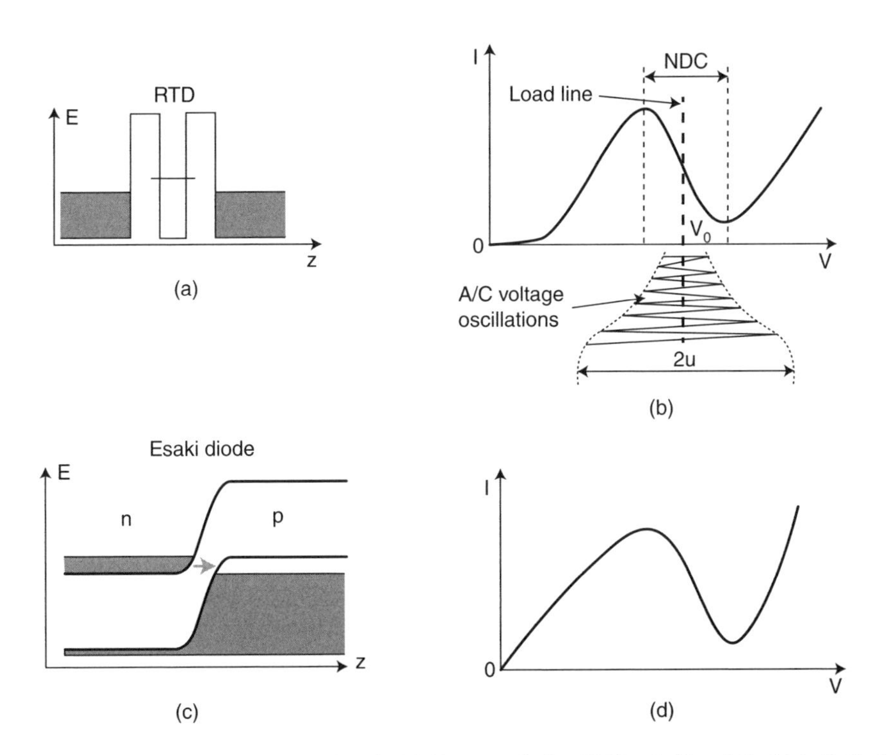

Figure 2.33 Band diagrams (a,c) and typical I-V curves (b,d) of RTDs (a,b) and Esaki (c,d) diodes.

through the quantized subbands in the QW. As a consequence, their *I–V* curve has an N-type shape with NDC region, see Figure 2.33b. RTDs will be discussed in detail in Chapter 6.

Another type of device with NDC is the Esaki diode [118]. The diode includes a pn junction with very high doping level in the adjacent p and n regions, see Figure 2.33c. The electrons can tunnel between p and n regions due to inter-band tunneling, since the transition layer between the p and n regions is very thin in such diodes. At low biases, the electrons can tunnel from the n region into the empty states in the p region above the quasi-Fermi level there. The tunnel current grows with increase in bias in this regime, as shown in Figure 2.33d. At higher biases, the bottom of the conduction band in the n region gets higher than the top of the valence band in the p region. The elastic inter-band tunneling becomes impossible in this regime and the current drops. This leads to appearance of the NDC region in the *I–V* curve. With further increase in bias, the electrons can flow from the n region to the p-region due to thermal excitation and the diode current starts to grow again. That explains the particular form of the diode's *I–V* curve sketched in Figure 2.33d.

The above devices rely on the tunnel mechanism of electron transport. Tunneling can be, in principle, an extremely fast process. Therefore the devices should be able to operate at very high frequencies. Indeed, the oscillators with RTDs work at THz frequencies nowadays and RTDs are the highest-frequency active electronic devices which exist presently [119, 120, 121].

When the amplitude of the initial oscillations in an RTD (or Esaki diode) is growing with time, the AC voltage oscillation amplitude becomes larger than the NDC region, indicated in Figure 2.33b by the dashed lines. The voltage swing extends into positive-differential-conductance regions of the RTD *I–V* curve. The resulting averaged AC differential conductance of an RTD becomes less negative and the amplitude of the oscillations grows slower and eventually stabilizes at a certain level. The behavior of RTDs is essentially nonlinear in the regime of stable oscillations.

There are not many known tunnel devices with NDC, apart from RTDs and Esaki diodes. Superlattices [122] are the best known example of other structures, although superlattice oscillators can operate only at sub-THz frequencies so far. There are also a few other novel concepts, like tunneling between two graphene layers [123], tunnel Schottky structures with a two-dimensional channel [124] and a few other, although applicability of the structures for sub-THz or THz generation still remains to be proven.

Other examples for NDC devices are Gunn diodes and IMPATT diodes. These devices perform excellently in the microwave range where they are frequently used as power sources. However, they are typically not used as fundamental oscillators above 100–150 GHz.

2.3.2 Multipliers (Schottky Diodes, Hetero-Barrier Varactors)

In recent years and due to the increasing number of applications in the THz band, the need for power sources operating at high frequencies is becoming crucial for the success of such applications. Up to ~150 GHz this need can be satisfied with fundamental all-solid-state local sources, such as Gunn or IMPATT oscillators. Output powers of the order of several hundreds of milliwatts to a few watts can be achieved with these devices. Nevertheless, when the frequency increases up to several THz, these solid-state fundamental oscillators are not able to provide power and therefore other alternatives have to be implemented.

In this case, frequency multipliers are one of the best options. Frequency multipliers are nonlinear devices that generate a specific harmonic of an input signal. In principle, any nonlinear impedance can be used to generate such frequency harmonics; however, variable capacitance or varactor diodes present better efficiency and power handling capabilities than variable resistance or varistor diodes. An input and output matching network is required at the desired harmonic to maximize the transferred power from the input signal. This requirement becomes difficult to comply with when the order of the harmonic is high; in fact, rarely over four or five times the source frequency [125].

The nonlinear behavior is achieved through the use of non-symmetrical devices like Schottky diodes [126], or symmetrical devices like heterostructure barrier varactors (HBVs) [127]. Note that Schottky structures are usually not used for multiplication factors higher than $N = 4$. In both cases, a nonlinear capacitance or varactor is used to generate the higher harmonics. These multipliers generally take the form of doublers (×2) or triplers (×3). Nevertheless when higher frequencies are needed, the solution is a combination of several of them (cascades of multipliers) to achieve higher multiplication factors.

Schottky diodes are the preferred device for building frequency multipliers, mainly due to the simplicity in the fabrication process. In fact, when the operating frequency increases, it is possible to reduce the size of the active area of the diode during the fabrication process in order to improve its main parameters, such as the junction capacitance, parasitic capacitances, and series resistance. The value of these parameters needs to be reduced to operate at THz frequencies.

On the other hand, HBV structures present current–voltage symmetry which simplifies the circuit topology since only odd harmonics ($(2n+1)f_c$) are produced and no bias is needed [128]. Nevertheless, HBV diodes require a complex material engineering process to reduce the active area and thus achieve good performance at THz frequencies.

In general, there is an important drawback in the THz multiplier systems; with each multiplication step, the output power is much less than the input power due to the very low conversion efficiency (typically 0.5–30% [129]). This efficiency is reduced as the operational frequency is increased. Furthermore, the matching requirements between different multiplier stages limit the operational bandwidth of the multiplier chain. A possible solution to short out the low efficiency problem is the use of balanced doublers that share the input power between several diodes. This configuration is able to provide efficiencies on the order of 50% [130]. On the other hand, HBV diodes, due to the cancelling of even harmonics, allow one to work with higher harmonics, such as the 5th and 7th. Besides, HBV devices provide the handling of higher input powers when compared with Schottky diode technology, which is an important advantage in order to achieve higher output powers.

In general, two types of multipliers have to be distinguished; broadband and narrow band. The first are able to cover a complete waveguide band, as WR10 (70–110 GHz), WR8 (90–140 GHz), and so on, while the second operate in a narrow frequency band. Narrow band multipliers always offer a higher efficiency and output power levels than broadband configurations.

The current state of the art of multipliers using planar GaAs Schottky diode technology (the most typical used in current applications) for output power levels and efficiency values is described in Table 2.1

Table 2.1 Status of multipliers: narrow band operation.

Frequency	Output power	Efficiency	Typical 3 dB bandwidth
250 GHz	80–100 mW	20–25%	≅8%
400 GHz	≅10 mW	<10%	≅4%
500 GHz	<70 mW	<7%	<4%

Table 2.2 Status of multipliers: broadband operation.

Frequency	Output power	Efficiency
260–400 GHz	<5 mW	<4%
400–600 GHz	<3 mW	<4%
600–900 GHz	<300 μW	<5%
750–1100 GHz	<200 μW	<2%
1100–1700 GHz	<80 μW	<1%
1400–2200 GHz	<50 μW	<0.6%

for the case of narrow band multipliers and in Table 2.2 for the case of broadband multipliers. Note that Table 2.2 values are for multipliers based on triplers. Fabrication techniques for Schottky diodes will be discussed in Section 6.5.

HBVs technology is not yet able to operate at frequencies around 1 THz. Some reported results in the literature are 180 mW output power at the W-band (90–110 GHz) with efficiencies of 23% [131] or 1 mW at 450 GHz with efficiencies below 2% [132].

RF multipliers are known and have been developed for a long time, with a manifold of excellent books on this topic. We refer to Ref. [133] for further reading.

2.3.3 Plasmonic Sources

In Section 2.2.5.3 we have already shown examples where plasmons are used to enhance the efficiency of photoconductors. In the following, we will discuss more general applications of plasmons for THz generation.

2.3.3.1 Plasmons

Plasmonics refers to a variety of techniques to generate, control, and detect electromagnetic radiation using metals [134–136]. The term plasmonics derives from plasmon waves, more commonly called surface plasmons (SPs), which, from a microscopic viewpoint, can be understood as collective electron oscillations that appear at a metal/dielectric interface under external p-polarized electromagnetic illumination.

Alternatively, plasmons can be identified as transversal magnetic (TM) surface waves (bounded modes) that propagate along the interface of two media with opposite signs of the real part of the permittivity. Imposing this boundary condition and forcing bounded surface waves (i.e., evanescent decay toward both media) plasmons can be directly derived from Maxwell's equations, obtaining their dispersion relation as well as complex propagation constants (the reader is referred to [137] for a simple and elegant explanation). With this macroscopic perspective, plasmons are classical waves, members of the more general family of complex waves, which also encompasses leaky waves, surface waves, and, closely related to plasmons, Zenneck waves.

Although plasmonics is a time-honored topic (the first studies can be found in the works of Zenneck – more than 100 years ago – about electromagnetic propagation on imperfectly conducting planes [138]) it has been revisited in the last decades due to the exciting properties and promising applications envisioned with them, leading to the so-called SP resurrection [139]. A key property of SPs is that they are confined to the metal/dielectric interface and thus wave–matter interaction is enhanced. Moreover, the wavelength of plasmons is smaller than the free-space wavelength beating the diffraction limit, a property that makes plasmons valuable for light manipulation at the nanoscale. In fact, plasmons have been proposed as prominent candidates for merging photonics and electronics, which could enable the long-desired all optics processing in plasmonic chips. All these features make them ideally suited for a variety of applications in fields like biology, optics, material science, nanoelectronics and nanophotonics and, recently, terahertz (THz) waves [136]. In fact, placed between microwaves and infrared, THz is the natural playground where ideas from both regimes can be successfully applied.

One of the main actors in the resurgence of plasmonics research was the discovery of extraordinary optical transmission (EOT) through subwavelength hole arrays by Ebbesen *et al.* in 1998 [140] and other periodic structures on metals [141], where the explanation for the enhanced transmission was put in terms of light–plasmon coupling. Soon after, similar phenomena were found in other regions of the electromagnetic spectrum such as microwaves/millimeter waves [142, 143] where metals are classically modeled as highly conductive, with a non-dispersive conductivity, rather than plasmonic, which implies a dispersive permittivity modeled with a Drude equation, demonstrating the generality of the phenomenon.

Under the high-frequency plasmonic perspective, this was interpreted as a generalization and extension of plasmons to lower frequencies, called "spoof plasmons", in diluted (i.e., perforated) metals, where periodicity of the array served to tune the plasma frequency [144]. Alternatively, evolving from low frequency concepts, EOT was explained in terms of leaky waves excited by the periodic structure [145]. Both explanations are complementary and describe the same phenomenon. The Drude model at low frequency is practically equivalent to a finite conductivity model [146]. On the other hand, a leaky wave excited by a periodic structure becomes a leaky plasmon at sufficiently high frequencies [147]. No matter which explanation is preferred, the important conclusion is that, aside from subtle differences mainly related to the exact metallic model, plasmonics research can benefit from mature techniques borrowed from optics or microwaves to obtain interesting technological realizations [146].

Another important actor in the resurgence of plasmonics is the advance in nanofabrication, both bottom-up and top-down processes [148]. Current nanofabrication techniques allow the realization not only of texturized surfaces supporting propagating SPs, but also particles of almost any shape with sub-nanometer precision. The latter support localized surface plasmons (LSPs), that is, collective oscillation of conduction electrons experiencing a restoring force due to the induced surface charges at the boundaries of the particle. LSPs hold promise for similar applications as SPs but, specifically, they are becoming key for single molecule spectroscopy [149], local heat generation [150], and nonlinearities [151] as a result of their high field enhancement in sub-wavelength volumes.

A variety of techniques have been proposed for the generation of THz waves where plasmons play a prominent role: nonlinear THz generation based on optical techniques where the use of metals allows relaxation of the constraint of phase matching; THz generation based on photoconductive materials benefits from plasmonic resonances and their strong field concentration to enhance the absorption and conversion of photons into electrons, as already discussed in Section 2.2.5.3; in modern quantum cascade lasers (QCLs) plasmons play a key role for the performance at THz, both for waveguiding and also beaming capabilities. In the following sections, we will review the main contributions of plasmons to the topic of THz generation.

2.3.3.2 Plasmonic THz Generation Based on Optical Nonlinear Effects

One of the most popular methods for THz radiation generation is the excitation of nonlinearities employing a pulsed laser of higher frequency, usually in the NIR. Two main mechanisms have been identified in the nonlinear process: optical rectification (OR) and multiphoton photoelectron emission (MPE), which have many common aspects but also subtle differences. OR is a second-order nonlinear effect in which broadband THz radiation is generated in a material with second-order susceptibility owing to the difference-frequency mechanism among the frequencies components contained in the femtosecond pulse. Consequently, the THz fluence generated depends on the square of the incident laser pump intensity. If the material with second-order susceptibility is illuminated with two frequency-offset continuous wave lasers, the difference-frequency mechanism leads to a continuous wave THz radiation similar to the case of photomixers discussed in Section 2.2.1. On the other hand, within the MPE framework, the dependence is of higher order (third-, fourth-, or even higher). In this case, THz radiation is mainly linked with multiphoton ionization and ponderomotive[2] acceleration of electrons. Although the debate is not yet closed, it seems that both mechanisms are usually present in the nonlinear-pulsed THz generation, with the main difference being the incident power and/or the repetition rate of the source which can mask either of the mechanisms [152, 153], as will be elucidated at the end of this section.

Historically, the first evidence of THz emission based on OR was reported in the seminal paper of Yang *et al.* in 1971 [154] where a picosecond pulsed laser was used to illuminate a $LiNbO_3$ slab used as the nonlinear material to produce THz radiation. Afterwards, the accuracy of the method was enhanced

[2] A ponderomotive force is a nonlinear force exerted by an inhomogeneous oscillating electromagnetic field on a charged particle.

by Auston *et al.* [155] by using a femtosecond pulsed laser and lithium tantalate as the electro-optic medium and was extended later to THz generation using the depletion layer of semiconductors [156]. More efficient techniques, such as coherent detection (i.e., amplitude and phase) of the THz radiation [157] using a variety of electro-optic crystals [157, 158] were proposed later, launching definitely the THz generation based on OR. The efficiency of OR and electro-optic detection depends decisively on the so-called phase matching, which currently is achieved by coincidental, birefringent, and non-collinear schemes. Recently OR in organic salt crystals has been explored for extremely intense THz pulses (on the order of MV/cm) covering the full THz gap (0.1–10 THz) [159].

More recently, with the explosion of plasmonics, metals have also been proposed as candidates for nonlinear THz generation, in which the constraint of phase matching is relaxed. The first realization of OR employing metals was reported by Kadlec *et al.* [160] with flat silver and gold slabs illuminated by a NIR pulsed femtosecond laser, obtaining a peak electric field of 200 V/cm and paving the way for the investigation of nonlinearities in metals [161] and the development of novel THz emitters, potentially more compact than previous solutions. In their experiments a second-order dependence between the emitted THz fluence and the incident laser pump intensity was obtained and thus OR was identified as the underlying mechanism in THz generation.

In Ref. [162] a gold grating instead of a flat metallic surface was investigated. The gold grating and a reference ZnTe slab were illuminated with a high power pulsed laser source to excite nonlinearities. Experiments at different angles were done and it was observed that maximum THz-pulse fluence appeared near the angle where SPs were excited by the periodic structure under p-polarized illumination, demonstrating the importance of SPs in the efficiency of the nonlinear process. Interestingly, a third-, fourth-, or even higher order dependence of the emitted THz fluence on the incident light was noticed, in sharp contrast to the results of [161]. Therefore, MPE was invoked as the underlying mechanism. It was argued that nanoscale structures are able to interact strongly with light, strengthening the field concentration in very small volumes and enhancing nonlinearities. The study was extended in Ref. [163] where the involvement of SPs was again explicitly mentioned as well as evanescent-wave acceleration of photoelectrons in the THz emission. The coupling of incident photons to SPs causes a high confinement of electromagnetic energy in a very small (subwavelength) area, accompanied by a strong field confinement (i.e., high gradient) at the metal/dielectric interface favoring the MPE process [164].

The strategy of structuring metals was extended in the analytical and numerical study of Ref. [165]. There, the performances of several nanostructured metallic surfaces were compared: flat gold films, metal gratings, single nanoparticle and nanoparticle arrays with circular (disks), coaxial and pyramidal shapes. The study demonstrated that THz emission can be optimized by judiciously choosing the nanoparticle parameters, that is, by tuning the SP resonance of the array. An experimental proof was presented in Ref. [166] by comparing semicontinuous (percolated) silver films and ordered arrangements of silver nanoparticles, arrayed as a honeycomb of triangular particles. The SP role in the THz emission was investigated here by modifying the thickness of the silver deposited, which shifts the plasmon resonance, demonstrating that the excitation of the plasmon resonance is directly related to an enhancement of the THz emission process. The dependence of THz emission on pump laser intensity was found to be between second and sixth order. Thus MPE was proposed as the underlying mechanism.

However, in Ref. [167] percolated silver films were analyzed and a clear second-order dependence was found, giving evidence of OR. Even more intriguing, for a flat metal surface, no THz emission was observed, seemingly contradicting the results of [161]. These apparent inconsistencies were finally resolved in Refs. [152, 153]: when low peak-power laser pulses are used as sources, high-order nonlinearities are faintly excited. However, with a high repetition rate a sensitive lock-in detection of relatively weak signals is possible, and thus second-order nonlinear optical processes, like OR, can be detected. On the other hand, high peak power amplified lasers can excite high-order nonlinearities to generate THz pulses. However, these high-power sources usually have a low repetition rate which makes it difficult to detect weak THz pulses from a second-order nonlinear process. In general, both OR and MPE will be present but depending on the characteristics of the sources one of them will dominate in the THz generation process.

Very recently, metamaterials have been proposed for the generation of broadband THz radiation from 0.1 to 4 THz [168]. Screens of split-ring resonators (SRRs) [169] are illuminated with infrared radiation and THz emission emerges through the excitation of the magnetic resonance.

2.3.3.3 Applications of Plasmonics in Quantum Cascade Lasers

Since their first experimental demonstration at 70 THz [170] by Faist *et al.*, QCLs have become by their own merits one of the most promising sources in THz technology. At difference with the sources discussed in the previous section, QCLs provide direct and powerful CW THz radiation.

The fundamentals of QCLs can be found in the theoretical works of Kazarinov and Suris in 1971 [171]. The operation principle relies on the emission of THz waves by relaxation of electrons between subbands of quantum wells. The first experimental realization had to wait for more than 20 years [170] and, after formidable efforts, the operation frequency was finally reduced to 4.4 THz [172] in 2002 by Köhler *et al.*, more than 40 years after the first theoretical study. Since then, research on QCL has evolved spectacularly [173].

A severe hurdle in THz-QCLs is the guidance and confinement of the THz signal. Although metallic waveguides operate with relatively low loss in microwaves and millimeter-waves, it is a fact that metallic losses increase with frequency, deterring their use at THz frequencies [174]. Similarly, dielectric waveguides usually present high loss, since many polymers have absorption bands at these frequencies [174]. An alternative to circumvent these issues is to take advantage of plasmonic concepts developing plasmonic waveguides. The key idea is that by high doping of semiconductors a negative permittivity can be synthesized. In this way, doped semiconductors behave like metals at infrared but with a plasma frequency in the mid- or far-IR instead of ultraviolet. In fact, in the seminal paper of Köhler *et al.* [172], a plasmonic waveguide with superior characteristics than previous designs was synthesized by doping a GaAs semiconductor layer surrounded by an undoped GaAs substrate. Soon after, a DFB QCL design was proposed [175] to improve the single mode operation of the source. This was accomplished by introducing a grating into the doped GaAs layer, imitating infrared lasers, where corrugations are etched on metals. In Ref. [176] a different topology for DFB QCL is analyzed, with a metal–metal ridge waveguide. Moreover, in Ref. [177] a periodic array of thin slits etched on a metallic slab is used as a grating for SPs. This is further extended in Ref. [178] where the semiconductor is also removed from the slit to avoid coupling of the SP inside the slit. Further developments of THz-QCL relying on SPs guiding can be found in Ref. [179].

Plasmons are also employed to enhance the directivity (beaming) of QCL sources. In Ref. [180] a corrugated structure was placed at the output of a mid-IR QCL, imitating antennas initially developed in microwaves [181, 182] and later evolved at THz frequencies [183]. A similar design was implemented in Ref. [184] where a double periodicity structure was proposed: a surface waveguide was designed using a small period corrugated structure; to enhance radiation through leaky waves [147], a larger periodicity was superimposed.

References

[1] DIFFER. FELIX2, Nijmegen, http://www.differ.nl/en/felix/specifications (accessed 25 November 2014).

[2] HZDR. FELBE, Helmholtz-Zentrum Dresden-Rossendorf, Germany, https://www.hzdr.de/db/Cms?pNid=205 (accessed 25 November 2014).

[3] UCSB-FEL. University of California, Santa Barbara, CA, http://sbfel3.ucsb.edu/ (accessed 25 November 2014).

[4] Madey, J.M.J. (1971) Stimulated emission of Bremsstrahlung in a periodic magnetic field. *Journal of Applied Physics*, **42**, 1906.

[5] Ramian, G. (1992) The new UCSB free-electron laser. *Nuclear Instruments and Methods in Physics Research, Section A*, **318**, 225.

[6] RPC Istok, Fryazino, Russia, http://www.istokmw.ru/vakuumnie-generatori-maloy-moshnosti/ (accessed 26 November 2014).

[7] Jacobsson, S. (1989) Optically pumped far infrared lasers. *Infrared Physics*, **29**, 853.

[8] Dodel, G. (1999) On the history of far-infrared (FIR) gas lasers: thirty-five years of research and application. *Infrared Physics and Technology*, **40**, 127.

[9] Bründermann, E., Linhart, A.M., Reichertz, L. *et al.* (1996) Double acceptor doped Ge: a new medium for inter-valence-band lasers. *Applied Physics Letters*, **68**, 3075.

[10] Pfeiffer, P., Zawadzki, W., Unterrainer, K. *et al.* (1993) p-type Ge cyclotron-resonance laser: theory and experiment. *Physical Review B*, **47**, 4522.

[11] Muravjov, A.V., Nelson, E.W., Peale, R.E. *et al.* (2003) Far-infrared p-Ge laser with variable length cavity. *Infrared Physics and Technology*, **44**, 75.

[12] Williams, B.S., Kumar, S., and Hu, Q. (2005) Operation of terahertz quantum-cascade lasers at 164 K in pulsed mode and at 117 K in continuous-wave mode. *Optics Express*, **13**, 3331.

[13] Ajili, L., Scalari, G., Faist, J. *et al.* (2004) High power quantum cascade lasers operating at λ=87 and 130 μm. *Applied Physics Letters*, **85**, 3986.

[14] Scalari, G., Walther, C., Fischer, M. *et al.* (2009) THz and sub-THz quantum cascade lasers. *Laser & Photonics Reviews*, **3**, 45.

[15] Vodopyanov, K.L. (2006) Optical generation of narrow-band terahertz packets in periodically-inverted electro-optic crystals: conversion efficiency and optimal laser pulse format. *Optics Express*, **14**, 2263.

[16] Minamide, H., Hayashi, S., Nawata, K. *et al.* (2014) Kilowatt-peak Terahertz-wave generation and sub-femtojoule Terahertz-wave pulse detection based on nonlinear optical wavelength-conversion at room temperature. *Journal of Infrared, Millimeter, and Terahertz Waves*, **35**, 25.

[17] Pine, S., Suenram, R.D., Brown, E.R., and McIntosh, K.A. (1996) A terahertz photomixing spectrometer: application to SO2 self broadening. *Journal of Molecular Spectroscopy*, **175**, 37.

[18] Preu, S., Döhler, G.H., Malzer, S. *et al.* (2011) Tunable, continuous-wave Terahertz photomixer sources and applications. *Journal of Applied Physics*, **109**, 061301.

[19] Goldberg, Y.A. and Schmidt, N.M. (1999) in *Handbook Series on Semiconductor Parameters*, 2nd edn (eds M. Levinshtein, S. Rumyantsev, and M. Shur), World Scientific, London, p. 62.

[20] Adachi, S. (1994) *GaAs and Related Materials*, World Scientific, Singapore.

[21] Adachi, S. (1992) *Physical Properties of III-V Semiconductor Compounds*, John Wiley & Sons, Inc., New York.

[22] Döhler, G.H., Renner, F., Klar, O. *et al.* (2005) THz-photomixer based on quasi-ballistic transport. *Semiconductor Science and Technology*, **20**, 178.

[23] Brown, E.R. (2003) THz generation by photomixing in ultrafast photoconductors. *International Journal of High Speed Electronics and Systems*, **13**, 497.

[24] Benicewicz, P., Roberts, J.P., and Taylor, A.J. (1994) Scaling of terahertz radiation from large-aperture biased photoconductors. *Journal of the Optical Society of America B*, **11**, 2533.

[25] Jepsen, P.U., Cooke, D.G., and Koch, M. (2011) Terahertz spectroscopy and imaging-modern techniques and applications. *Laser & Photonics Reviews*, **5**, 124.

[26] Ashida, M. (2008) Ultra-broadband Terahertz wave detection using photoconductive antenna. *Japanese Journal of Applied Physics*, **47**, 8221.

[27] Preu, S. (2014) A unified derivation of the Terahertz spectra generated by photoconductors and diodes. *Journal of Infrared, Millimeter, and Terahertz Waves*, **35**, 998.

[28] Balanis, C.A. (2005) *Antenna Theory: Analysis and Design*, 3rd edn, John Wiley & Sons, Inc., Hoboken, NJ.

[29] Bhattacharya, P. (ed) (1993) *Properties of Lattice-Matched and Strained Indium Gallium Arsenide*, Inspec, New York.

[30] Verghese, S., McIntosh, K.A., and Brown, E.R. (1997) Optical and terahertz power limits in the low-temperature-grown GaAs photomixers. *Applied Physics Letters*, **71**, 2743.

[31] Ito, H., Nakajima, F., Futura, T., and Ishibashi, T. (2005) Continuous THz-wave generation using antenna-integrated uni-travelling-carrier photodiodes. *Semiconductor Science and Technology*, **20**, 191.

[32] Beck, A., Ducournau, G., Zaknoune, M. *et al.* (2008) High-efficiency uni-travelling-carrier photomixer at 1.55 μm and spectroscopy application up to 1.4 THz. *Electronics Letters*, **44**, 1320.

[33] Bjarnason, J.E., Chan, T.L.J., Lee, A.W.M. *et al.* (2004) ErAs:GaAs photomixer with two-decade tunability and 12 W peak output power. *Applied Physics Letters*, **85**, 3983.

[34] Latzel, P., Peytavit, E., Dogheche, E., and Lampin, J.-F. (2011) Improving properties of THz photoconductors by bonding to a high thermal conductivity substrate. 36th International Conference on Infrared, Millimeter and Terahertz Waves (IRMMW-THz), p. 1

[35] Preu, S., Renner, F.H., Malzer, S. *et al.* (2007) Efficient Terahertz emission from ballistic transport enhanced n-i-p-n-i-p superlattice photomixers. *Applied Physics Letters*, **90**, 212115.
[36] Rouvalis, E., Renaud, C.C., Moodie, D.G. *et al.* (2010) Traveling-wave uni-traveling carrier photodiodes for continuous wave THz generation. *Optics Express*, **18**, 11105.
[37] Ishibashi, T., Furuta, T., Fushimi, H. *et al.* (2000) InP/InGaAs uni-traveling-carrier photodiodes. *IEICE Transactions on Electronics*, **E83-C**, 938.
[38] Stanze, D., Bach, H.-G., Kunkel, R. *et al.* (2009) Coherent CW terahertz systems employing photodiode emitters. 34th International Conference on Infrared, Millimeter, and Terahertz Waves, IRMMW-THz 2009, p. 1.
[39] Darrow, J.T., Zhang, X.-C., Auston, D.H., and Morse, J.D. (1992) Saturation Properties of large-aperture photoconducting antennas. *IEEE Journal of Quantum Electronics*, **28**, 1607.
[40] Loata, G.C., Löffler, T., and Roskos, H.G. (2007) Evidence for long-living charge carriers in electrically biased low-temperature-grown GaAs photoconductive switches. *Applied Physics Letters*, **90**, 052101.
[41] Segschneider, G., Jacob, F., Löffler, T. *et al.* (2002) Free-carrier dynamics in low-temperature-grown GaAs at high excitation densities investigated by time-domain terahertz spectroscopy. *Physical Review B*, **65**, 125205.
[42] Kim, D.S. and Citrin, D.S. (2006) Coulomb and radiation screening in photoconductive terahertz sources. *Applied Physics Letters*, **88**, 161117.
[43] Awad, M., Nagel, M., Kurz, H. *et al.* (2007) Characterization of low temperature GaAs antenna array terahertz emitters. *Applied Physics Letters*, **91**, 181124.
[44] Tani, M., Matsuura, S., Sakai, K., and Nakashima, S.-I. (1997) Emission characteristics of photoconductive antennas based on low-temperature-grown GaAs and semi-insulating GaAs. *Applied Optics*, **36**, 7853.
[45] Berry, C.W. and Jarrahi, M. (2012) Principles of impedance matching in photoconductive antennas. *Journal of Infrared, Millimeter, and Terahertz Waves*, **33**, 1182.
[46] Ito, H., Nagatsuma, T., Hirata, A. *et al.* (2003) High-power photonic millimetre wave generation at 100 GHz using matching-circuit integrated uni-travelling-carrier photodiodes. *IEE Proceedings: Optolectronics*, **150**, 138.
[47] Peytavit, E., Latzel, P., Pavanello, F. *et al.* (2013) CW source based on photomixing with output power reaching 1.8 mW at 250 GHz. *IEEE Electron Device Letters*, **34**, 1277.
[48] Liu, J., Zou, S., Yang, Z. *et al.* (2013) Wave shape recovery for terahertz pulse field detection via photoconductive antenna. *Optics Letters*, **38**, 2268.
[49] Loata, G.C., Thomson, M.D., Löffler, T., and Roskos, H.G. (2007) Radiation field screening in photoconductive antennae studied via pulsed terahertz emission spectroscopy. *Applied Physics Letters*, **91**, 232506.
[50] Hattori, T., Tukuamoto, K., and Nakatsuka, H. (2001) Time-resolved study of intense Terahertz pulses generated by a large-aperture photoconductive antenna. *Japanese Journal of Applied Physics*, **40**, 4907.
[51] Rodriguez, G. and Taylor, A.J. (1996) Screening of the bias field in terahertz generation from photoconductors. *Optics Letters*, **21**, 1046.
[52] Beck, M., Schäfer, H., Klatt, G. *et al.* (2010) Impulsive terahertz radiation with high electric fields from an amplifier-driven large-area photoconductive antenna. *Optics Express*, **18**, 9252.
[53] Rogers, D.L. (1991) Integrated optical receivers using MSM detectors. *Journal of Lightwave Technology*, **9**, 1635.
[54] Gregory, I.S., Baker, C., Tribe, W.R. *et al.* (2003) High resistivity annealed low-temperature GaAs with 100 fs lifetimes. *Applied Physics Letters*, **83**, 4199.
[55] Brown, E.R., McIntosh, K.A., Nichols, K.B., and Dennis, C.L. (1995) Photomixing up to 3.8 THz in low-temperature-grown GaAs. *Applied Physics Letters*, **66**, 285.
[56] Hisatake, S., Kitahara, G., Ajito, K. *et al.* (2013) Phase-sensitive terahertz self-heterodyne system based on photodiode and low temperature-grown GaAs photoconductor at 1.55 μm. *IEEE Sensors Journal*, **13**, 31.
[57] Tani, M., Lee, K.-S., and Zhang, X.-C. (2000) Detection of terahertz radiation with low-temperature-grown GaAs-based photoconductive antenna using 1.55 μm probe. *Applied Physics Letters*, **77**, 1396.
[58] Kadow, C., Johnson, J.A., Kolstad, K. *et al.* (2000) Growth and microstructure of self-assembled ErAs islands in GaAs. *Journal of Vacuum Science & Technology B*, **18**, 2197.
[59] Kadow, C., Jackson, A.W., Gossard, A.C. *et al.* (2000) Self-assembled ErAs islands in GaAs for optical-heterodyne THz generation. *Applied Physics Letters*, **76**, 3510.
[60] Scarpulla, M.A., Zide, J.M.O., LeBeau, J.M. *et al.* (2008) Near-infrared absorption and semimetal-semiconductor transition in 2 nm ErAs nanoparticles embedded in GaAs and AlAs. *Applied Physics Letters*, **92**, 173116.
[61] Kadow, C., Johnson, J.A., Kolstad, K., and Gossard, A.C. (2003) Growth-temperature dependence of the microstructure of ErAs islands in GaAs. *Journal of Vacuum Science & Technology B*, **21**, 29.

[62] Kadow, C., Fleischer, S.B., Ibbetson, J.P. *et al.* (1999) Self-assembled ErAs islands in GaAs: Growth and subpicosecond carrier dynamics. *Applied Physics Letters*, **75**, 3548.
[63] Griebel, M., Smet, J.H., Driscoll, D.C. *et al.* (2003) Tunable subpicosecond optoelectronic transduction in superlattices of selfassembled ErAs nanoislands. *Nature Materials*, **2**, 122.
[64] Metzger, R.A., Brown, A.S., McCray, L.G., and Henige, J.A. (1993) Structural and electrical properties of low temperature GaInAs. *Journal of Vacuum Science & Technology B*, **11**, 798.
[65] Hudgins, J.L., Simin, G.S., Santi, E., and Khan, M.S. (2003) An assessment of wide bandgap semiconductors for power devices. *IEEE Transactions on Power Electronics*, **18**, 907.
[66] Grenier, P. and Whitaker, J.F. (1997) Subband gap carrier dynamics in low-temperature-grown GaAs. *Applied Physics Letters*, **70**, 1998.
[67] Erlig, H., Wang, S., Azfar, T. *et al.* (1999) LT-GaAs detector with 451 fs response at 1.55-micron via two-photon absorption. *Electronics Letters*, **35**, 173.
[68] Chiu, Y.-J., Zhang, S.Z., Fleischer, S.B. *et al.* (1998) GaAs-based 1.55- μm high speed, high saturation power, low-temperature grown GaAs pin photodetector. *Electronics Letters*, **34**, 1253.
[69] Middendorf, J.R. and Brown, E.R. (2012) THz pulse generation using extrinsic photoconductivity at 1550 nm. *Optics Express*, **20**, 16504.
[70] Taylor, Z.D., Brown, E.R., Bjarnason, J.E. *et al.* (2006) Resonant-optical-cavity photoconductive switch with 0.5% conversion efficiency and 1.0W peak power. *Optics Letters*, **31**, 1729.
[71] Suen, J.Y., Li, W., Taylor, Z.D., and Brown, E.R. (2010) Characterization and modeling of a THz photoconductive switch. *Applied Physics Letters*, **96**, 141103.
[72] Driscoll, D.C., Hanson, M., Kadow, C., and Gossard, A.C. (2001) Electronic structure and conduction in a metal–semiconductor digital composite: ErAs:InGaAs. *Applied Physics Letters*, **78**, 1703.
[73] Driscoll, D.C., Hanson, M.P., Gossard, A.C., and Brown, E.R. (2005) Ultrafast photoresponse at 1.55 μm in InGaAs with embedded semimetallic ErAs nanoparticles. *Applied Physics Letters*, **86**, 051908.
[74] Sukhotin, M., Brown, E.R., Gossard, A.C. *et al.* (2003) Photomixing and photoconductor measurements on ErAs/InGaAs at 1.55 μm. *Applied Physics Letters*, **82**, 3116.
[75] Ospald, F., Maryenko, D., von Klitzing, K. *et al.* (2008) 1.55 μm ultrafast photoconductive switches based on ErAs:InGaAs. *Applied Physics Letters*, **92**, 131117.
[76] Brown, E.R., Driscoll, D.C., and Gossard, A.C. (2005) State-of-the-art in 1.55 μm ultrafast InGaAs photoconductors, and the use of signal-processing techniques to extract the photocarrier lifetime. *Semiconductor Science and Technology*, **20**, 199.
[77] Williams, K.K., Taylor, Z.D., Suen, J.Y. *et al.* (2009) Toward a 1550 nm InGaAs photoconductive switch for terahertz generation. *Optics Letters*, **34**, 3068.
[78] Preu, S., Mittendorff, M., Lu, H. *et al.* (2012) 1550 nm ErAs:In(Al)GaAs large area photoconductive emitters. *Applied Physics Letters*, **101**, 101105.
[79] Sartorius, B., Roehle, H., Künzel, H. *et al.* (2008) All-fiber terahertz time-domain spectrometer operating at 1.5 μm telecom wavelengths. *Optics Express*, **16**, 9565.
[80] Roehle, H., Dietz, R.J.B., Hensel, H.J. *et al.* (2010) Next generation 1.5 μm terahertz antennas: mesa-structuring of InGaAs/InAlAs photoconductive layers. *Optics Express*, **18**, 2296.
[81] Dietz, R.J.B., Globisch, B., Gerhard, M. *et al.* (2013) 64 μW pulsed terahertz emission from growth optimized InGaAs/InAlAs heterostructures with separated photoconductive and trapping regions. *Applied Physics Letters*, **103**, 061103.
[82] Delagnes, J.C., Mounaix, P., Němec, H. *et al.* (2009) High photocarrier mobility in ultrafast ion-irradiated $In_{0.53}Ga_{0.47}As$ for terahertz applications. *Journal of Physics D: Applied Physics*, **42**, 195103.
[83] Carmody, C., Tan, H.H., Jagadish, C. *et al.* (2003) Ion-implanted $In_{0.53}Ga_{0.47}As$ for ultrafast optoelectronic applications. *Applied Physics Letters*, **82**, 3913.
[84] Suzuki, M. and Tonouchi, M. (2005) Fe-implanted InGaAs terahertz emitters for 1.56 m wavelength excitation. *Applied Physics Letters*, **86**, 051104.
[85] Tousley, B.C., Mehta, S.M., Lobad, A.I. *et al.* (2003) Femtosecond optical response of low temperature grown $In_{0.53}Ga_{0.47}As$. *Journal of Electronic Materials*, **22**, 1477.
[86] Mangeney, J. (2012) THz photoconductive antennas made from ion-bombarded semiconductors. *Journal of Infrared, Millimeter, and Terahertz Waves*, **33**, 455.
[87] Mangeney, J., Chimot, N., Meignien, L. *et al.* (2007) Emission characteristics of ion-irradiated $In_{0.53}Ga_{0.47}As$ based photoconductive antennas excited at 1.55 μm. *Optics Express*, **15**, 8943.
[88] Mangeney, J. and Crozat, P. (2008) Ion-irradiated $In_{0.53}Ga_{0.47}As$ photoconductive antennas for THz generation and detection at 1.55 μm wavelength. *Comptes Rendus Physique*, **9**, 142.

[89] Berry, C.W., Wang, N., Hashemi, M.R. *et al.* (2013) Significant performance enhancement in photoconductive terahertz optoelectronics by incorporating plasmonic contact electrodes. *Nature Communications*, **4**, 1622.
[90] Tanoto, H., Teng, J.H., Wu, Q.Y. *et al.* (2013) Nano-antenna in a photoconductive photomixer for highly efficient continuous wave terahertz emission. *Science Reports*, **3**, 2824.
[91] Park, S.-G., Choi, Y., Oh, Y.-J., and Jeong, K.-H. (2012) Terahertz photoconductive antenna with metal nanoislands. *Optics Express*, **20**, 25530.
[92] Wang, N., Hashemi, M.R., and Jarrahi, M. (2013) Plasmonic photoconductive detectors for enhanced terahertz detection sensitivity. *Optics Express*, **21**, 17221.
[93] Shi, J.-W., Li, Y.-T., Pan, C.-L. *et al.* (2006) Bandwidth enhancement phenomenon of a high-speed GaAs–AlGaAs based unitraveling carrier photodiode with an optimally designed absorption layer at an 830nm wavelength. *Applied Physics Letters*, **89**, 053512.
[94] Ito, H., Kodama, S., Muramoto, Y. *et al.* (2004) High-speed and high-output InP–InGaAs uni-traveling-carrier photodiodes. *IEEE Journal of Selected Topics in Quantum Electronics*, **10**, 709.
[95] Stanze, D., Roehle, H., Dietz, R.J.B., *et al.* (2010) Improving photoconductive receivers for 1.5 μm CW THz systems. 35th International Conference on Infrared Millimeter and Terahertz Waves (IRMMW-THz) 2010, p. 1.
[96] Chtioui, M., Lelarge, F., Enard, A. *et al.* (2012) High responsivity and high power UTC and MUTC GaInAs-InP photodiodes. *IEEE Photonics Technology Letters*, **24**, 318.
[97] Pan, H., Beling, A., Chen, H., and Campbell, J.C. (2008) Characterization and optimization of high-power InGaAs/InP photodiodes. *Optical and Quantum Electronics*, **40**, 41.
[98] Rymanov, V., Stöhr, A., Dülme, S., and Tekin, T. (2014) Triple transit region photodiodes (TTR-PDs) providing high millimeter wave output power. *Optics Express*, **22**, 7550.
[99] Wey, Y.-G., Giboney, K., Bowers, J. *et al.* (1995) 110-GHz GaInAs/InP double heterostructure p-i-n photodetectors. *Journal of Lightwave Technology*, **13**, 1490.
[100] Pasalic, D. and Vahldieck, R. (2008) Hybrid drift-diffusion-TLM analysis of high-speed and high-output UTC traveling-wave photodetectors. *International Journal of Numerical Modelling: Electronic Networks, Devices and Fields*, **21**, 61.
[101] Menon, P.S., Kandiah, K., Ehsan, A.A., and Shaari, S. (2011) Concentration-dependent minority carrier lifetime in an $In_{0.53}Ga_{0.47}As$ interdigitated lateral PIN photodiode model based on spin-on chemical fabrication methodology. *International Journal of Numerical Modelling: Electronic Networks, Devices and Fields*, **24**, 465.
[102] Williams, K.J. (2002) Comparisons between dual-depletion-region and uni-travelling-carrier p-i-n photodetectors. *IEE Proceedings: Optolectronics*, **149**, 131.
[103] Rymanov, V., Dülme, S., Babiel, S., and Stöhr, A. (2014) Compact photonic millimeter wave (200–300 GHz) transmitters based on semicircular bow-tie antenna-integrated 1.55 μm triple transit region photodiodes. German Microwave Conference (GeMiC 2014), Aachen, Germany, p. 1.
[104] Preu, S., Malzer, S., Döhler, G.H. *et al.* (2010) Efficient III–V tunneling diodes with ErAs recombination centers. *Semiconductor Science and Technology*, **25**, 115004.
[105] Pohl, P., Renner, F.H., Eckardt, M. *et al.* (2003) Enhanced recombination tunneling in GaAs pn junctions containing low-temperature-grown-GaAs and ErAs layers. *Applied Physics Letters*, **83**, 4035.
[106] Mader, T.B., Bryerton, E.W., Markovic, M. *et al.* (1998) Switched-mode high-efficiency microwave power amplifiers in a free-space power-combiner array. *IEEE Transactions on Microwave Theory and Techniques*, **46**, 1391.
[107] Micovic, M., Kurdoghlian, A., Margomenos, A., *et al.* (2012) 92–96 GHz GaN power amplifiers. IEEE MTT-S International Microwave Symposium Digest (MTT) 2012, p. 1.
[108] Eisele, H. (2010) State of the art and future of electronic sources at terahertz frequencies. *Electronics Letters*, **46**, 8.
[109] 5^{THz} dotfive. Towards 0.5 TeraHertz Silicon/Germanium Heterojunction Bipolar Technology. http://www.dotfive.eu/ (accessed 26 November 2014).
[110] Heinemann, B., Barth, R., Bolze, D. *et al.* (2010) SiGe HBT technology with fT/fmax of 300GHz/500GHz and 2.0 ps CML gate delay. IEEE International Electron Devices Meeting (IEDM) 2010, p. 30.5.1.
[111] 7_{THz} dotseven DOTSEVEN: Towards 0.7 Terahertz Silicon Germanium Heterojunction Bipolar Technology. http://www.dotseven.eu/ (accessed 26 November 2014).
[112] Jain, V., Lobisser, E., Baraskar, A. *et al.* (2011) InGaAs/InP DHBTs in a dry-etched refractory metal emitter process demonstrating simultaneous f_τ/f_{max} ~430/800 GHz. *IEEE Electron Device Letters*, **32**, 24.
[113] Urteaga, M., Pierson, R., Rowell, P. *et al.* (2011) 130 nm InP DHBTs with f_τ >0.52 THz and f_{max} >1.1 THz. Proceedings of the 69th IEEE Device Research Conference, p. 281.

[114] Rodwell, M.J.W. (2001) *High-Speed Integrated Circuit Technology: Towards 100 GHz Logic*, World Scientific Publishing, Singapore.
[115] Mark Rodwell, E., Lind, Z., Griffith, A.M. *et al.* (2007) On the feasibility of few-THz bipolar transistors. IEEE Bipolar/BiCMOS Circuits and Technology Meeting 2007, p. 17.
[116] Tsu, R. and Esaki, L. (1973) Tunneling in a finite superlattice. *Applied Physics Letters*, **22**, 562.
[117] Chang, L.L., Esaki, L., and Tsu, R. (1974) Resonant tunneling in semiconductor double barriers. *Applied Physics Letters*, **24**, 593.
[118] Esaki, L. (1958) New phenomenon in narrow germanium p-n junctions. *Physical Review*, **109**, 603.
[119] Kanaya, H., Sogabe, R., Maekawa, T. *et al.* (2014) Fundamental oscillation up to 1.42 THz in resonant tunneling diodes by optimized collector spacer thickness. *Journal of Infrared, Millimeter, and Terahertz Waves*, **35**, 425.
[120] Feiginov, M., Sydlo, C., Cojocari, O., and Meissner, P. (2011) Resonant-tunnelling-diode oscillators operating at frequencies above 1.1 THz. *Applied Physics Letters*, **99**, 233506.
[121] Feiginov, M., Kanaya, H., Suzuki, S., and Asada, M. (2014) Operation of resonant-tunneling diodes with strong back injection from the collector at frequencies up to 1.46 THz. *Applied Physics Letters*, **104**, 243509.
[122] Schomburg, E., Henini, M., Chamberlain, J.M. *et al.* (1999) Self-sustained current oscillation above 100 GHz in a GaAs/AlAs superlattice. *Applied Physics Letters*, **74**, 2179.
[123] Britnell, L., Gorbachev, R.V., Geim, A.K. *et al.* (2013) Resonant tunnelling and negative differential conductance in graphene transistors. *Nature Communications*, **4**, 1794.
[124] Feiginov, M. and Kotelnikov, I.N. (2007) Evidence for attainability of negative differential conductance in tunnel Schottky structures with two-dimensional channels. *Applied Physics Letters*, **91**, 083510.
[125] Maestrini, A., Thomas, B., Wang, H. *et al.* (2010) Schottky diode based terahertz frequency multipliers and mixers. *Comptes Rendus Physique*, **11**, 480.
[126] Crowe, T.W. (1989) GaAs Schottky barrier diodes for the frequency range 1-10 THz. *International Journal of Infrared and Millimeter Waves*, **10**, 765.
[127] Kollberg, L. and Rydberg, A. (1989) Quantum-barrier-varactor diode for high efficiency millimeter-wave multipliers. *Electronics Letters*, **25**, 1696.
[128] Dillner, L., Ingvarson, M., Kollberg, E., and Stake, J. (2001) High efficiency HBV multipliers for millimeter wave generation, in *Terahertz Sources and Systems*, NATO Science Series, vol. **27**, Klumer Academic Press, p. 27.
[129] Räisänen, A.V. (1992) Frequency multipliers for millimeter and sub millimeter wavelengths. *Proceedings of the IEEE*, **80**, 1842.
[130] Porterfield, D., Hesler, J., Crowe, T. *et al.* (2003) Integrated terahertz transmit/receive modules. European Microwave Conference, Munich, Germany 2003, p. 1319.
[131] Malko, A., Bryllert, T., and Vukusic, J. (2013) Silicon integrated InGaAs/InAlAs/AlAs HBV frequency tripler. *IEEE Electron Device Letters*, **34**, 843.
[132] Schumann, M., Duwe, K., Domoto, C. *et al.* (2003) High-performance 450-GHz GaAs-based heterostructure barrier varactor tripler. *IEEE Electron Device Letters*, **24**, 138.
[133] Faber, M.T., Chramiec, J., and Adamski, M.E. (1995) *Microwave and Millimeter-Wave Diode Frequency Multipliers*, Artech House, Norwood, MA.
[134] Maier, S.A. (2007) *Plasmonics: Fundamentals and Applications*, Springer, New York.
[135] Ozbay, E. (2006) Plasmonics: merging photonics and electronics at nanoscale dimensions. *Science*, **311**, 189.
[136] Dragoman, M. and Dragoman, D. (2008) Plasmonics: applications to nanoscale terahertz and optical devices. *Progress in Quantum Electronics*, **32**, 1.
[137] Ishimaru, A. (1991) *Electromagnetic Wave Propagation, Radiation, and Scattering*, Prentice Hall, Inc., Englewood Cliffs, NJ.
[138] Zenneck, J. (1907) Über die Fortpflanzung ebener elektromagnetischer Wellen längs einer ebenen Leiterfläche und ihre Beziehung zur drahtlosen Telegraphie. *Annalen der Physik*, **328**, 846.
[139] Editorial Surface plasmon resurrection, Nature Photonics **6**, 707 (2012).
[140] Ebbesen, T.W., Lezec, H.J., Ghaemi, H.F. *et al.* (1998) Extraordinary optical transmission through sub-wavelength hole arrays. *Nature*, **391**, 667.
[141] Lezec, H.J., Degiron, A., Devaux, E. *et al.* (2002) Beaming light from a subwavelength aperture. *Science*, **297**, 820.
[142] Beruete, M., Sorolla, M., Campillo, I. *et al.* (2004) Enhanced millimeter-wave transmission through subwavelength hole arrays. *Optics Letters*, **29**, 2500.
[143] Beruete, M., Campillo, I., Dolado, J.S. *et al.* (2004) Enhanced microwave transmission and beaming using a subwavelength slot in corrugated plate. *IEEE Antennas and Wireless Propagation Letters*, **3**, 328.

[144] Pendry, J.B., Martín-Moreno, L., and Garcia-Vidal, F.J. (2004) Mimicking surface plasmons with structured surfaces. *Science*, **305**, 847.
[145] Lomakin, V. and Michielssen, E. (2005) Enhanced transmission through metallic plates perforated by arrays of subwavelength holes and sandwiched between dielectric slabs. *Physical Review B*, **71**, 235117.
[146] Torres, V., Ortuño, R., Rodríguez-Ulibarri, P. *et al.* (2014) Mid-infrared plasmonic inductors: enhancing inductance with meandering lines. *Science Reports*, **4**, 1.
[147] Jackson, D.R., Oliner, A.A., Zhao, T., and Williams, J.T. (2005) Beaming of light at broadside through a subwavelength hole: leaky wave model and open stopband effect. *Radio Science*, **40**, 1.
[148] Zheng, W., Chiamori, H.C., Liu, G.L. *et al.* (2012) Nanofabricated plasmonic nano-bio hybrid structures in biomedical detection. *Nanotechnology Reviews*, **1**, 213.
[149] Lakowicz, J.R. and Fu, Y. (2009) Modification of single molecule fluorescence near metallic nanostructures. *Laser & Photonics Reviews*, **3**, 221.
[150] Baffou, G. and Quidant, R. (2013) Thermo-plasmonics: using metallic nanostructures as nano-sources of heat. *Laser & Photonics Reviews*, **7**, 171.
[151] Aouani, H., Rahmani, M., Navarro-Cía, M., and Maier, S.A. (2014) Third-harmonic-upconversion enhancement from a single semiconductor nanoparticle coupled to a plasmonic antenna. *Nature Nanotechnology*, **9**, 290.
[152] Ramakrishnan, G., Kumar, N., Planken, P.C.M. *et al.* (2012) Surface plasmon-enhanced terahertz emission from a hemicyanine self-assembled monolayer. *Optics Express*, **20**, 4067.
[153] Polyushkin, D.K., Márton, I., Rácz, P. *et al.* (2014) Mechanisms of THz generation from silver nanoparticle and nanohole arrays illuminated by 100 fs pulses of infrared light. *Physical Review B*, **89**, 125426.
[154] Yang, K.H., Richards, P.L., and Shen, Y.R. (1971) Generation of far-infrared radiation by picosecond light pulses in LiNbO3. *Applied Physics Letters*, **19**, 320.
[155] Auston, D., Cheung, K., Valdmanis, J., and Kleinman, D. (1984) Cherenkov radiation from femtosecond optical pulses in electro-optic media. *Physical Review Letters*, **53**, 1555.
[156] Zhang, X.-C., Hu, B.B., Darrow, J.T., and Auston, D.H. (1990) Generation of femtosecond electromagnetic pulses from semiconductor surfaces. *Applied Physics Letters*, **56**, 1011.
[157] Zhang, X.-C., Jin, Y., and Ma, X.F. (1992) Coherent measurement of THz optical rectification from electro-optic crystals. *Applied Physics Letters*, **61**, 2764.
[158] Rice, A., Jin, Y., Ma, X.F. *et al.* (1994) Terahertz optical rectification from $\langle 110 \rangle$ zinc-blende crystals. *Applied Physics Letters*, **64**, 1324.
[159] Vicario, C., Ruchert, C., and Hauri, C.P.P. (2013) High field broadband THz generation in organic materials. *Journal of Modern Optics*, **39**, 2660.
[160] Kadlec, F., Kuzel, P., and Coutaz, J.-L. (2004) Optical rectification at metal surfaces. *Optics Letters*, **29**, 2674.
[161] Kadlec, F., Kuzel, P., and Coutaz, J.-L. (2005) Study of terahertz radiation generated by optical rectification on thin gold films. *Optics Letters*, **30**, 1402.
[162] Welsh, G., Hunt, N., and Wynne, K. (2007) Terahertz-pulse emission through laser excitation of surface plasmons in a metal grating. *Physical Review Letters*, **98**, 026803.
[163] Welsh, G.H. and Wynne, K. (2009) Generation of ultrafast terahertz radiation pulses on metallic nanostructured surfaces. *Optics Express*, **17**, 2470.
[164] Aeschlimann, M., Schmuttenmaer, C.A., Elsayed-Ali, H.E. *et al.* (1995) Observation of surface enhanced multiphoton photoemission from metal surfaces in the short pulse limit. *Journal of Chemical Physics*, **102**, 8606.
[165] Gao, Y., Chen, M.-K., Yang, C.-E. *et al.* (2009) Analysis of terahertz generation via nanostructure enhanced plasmonic excitations. *Journal of Applied Physics*, **106**, 074302.
[166] Polyushkin, D.K., Hendry, E., Stone, E.K., and Barnes, W.L. (2011) THz generation from plasmonic nanoparticle arrays. *Nano Letters*, **11**, 4718.
[167] Ramakrishnan, G. and Planken, P.C.M. (2011) Percolation-enhanced generation of terahertz pulses by optical rectification on ultrathin gold films. *Optics Letters*, **36**, 2572.
[168] Luo, L., Chatzakis, I., Wang, J. *et al.* (2014) Broadband terahertz generation from metamaterials. *Nature Communications*, **5**, 3055.
[169] Marqués, R., Baena, J.D., Beruete, M. *et al.* (2005) Ab initio analysis of frequency selective surfaces based on conventional and complementary split ring resonators. *Journal of Optics A: Pure and Applied Optics*, **7**, 38.
[170] Faist, J., Capasso, F., Sivco, D.L. *et al.* (1994) Quantum cascade laser. *Science*, **264**, 553.
[171] Kazarinov, F. and Suris, R.A. (1971) Possibility of the amplification of electromagnetic waves in a semiconductor with a superlattice. *Soviet Physics – Semiconductors*, **5**, 707.
[172] Köhler, R., Tredicucci, A., Beltram, F. *et al.* (2002) Terahertz semiconductor-heterostructure laser. *Nature*, **417**, 156.

[173] Kumar, S. (2011) Recent progress in terahertz quantum cascade lasers. *IEEE Journal on Selected Topics in Quantum Electronics*, **17**, 38.

[174] Mitrofanov, O., James, R., Fernandez, F.A. *et al.* (2011) Reducing transmission losses in hollow THz waveguides. *IEEE Transactions on Terahertz Science and Technology*, **1**, 124.

[175] Mahler, L., Köhler, R., Tredicucci, A. *et al.* (2004) Single-mode operation of terahertz quantum cascade lasers with distributed feedback resonators. *Applied Physics Letters*, **84**, 5446.

[176] Williams, B., Kumar, S., Hu, Q., and Reno, J. (2005) Distributed-feedback terahertz quantum-cascade lasers with laterally corrugated metal waveguides. *Optics Letters*, **30**, 2909.

[177] Mahler, L., Tredicucci, A., Köhler, R. *et al.* (2005) High-performance operation of single-mode terahertz quantum cascade lasers with metallic gratings. *Applied Physics Letters*, **87**, 181101.

[178] Demichel, O., Mahler, L., and Losco, T. (2006) Surface plasmon photonic structures in terahertz quantum cascade lasers. *Optics Express*, **14**, 5335.

[179] Tanvir, H., Rahman, B.M.A., Kejalakshmy, N. *et al.* (2011) Evolution of highly confined surface plasmon modes in terahertz quantum cascade laser waveguides. *Journal of Lightwave Technology*, **29**, 2116.

[180] Yu, N., Fan, J., Wang, Q.J. *et al.* (2008) Small-divergence semiconductor lasers by plasmonic collimation. *Nature Photonics*, **2**, 564.

[181] Beruete, M., Campillo, I., Dolado, J.S. *et al.* (2006) Dual-band low-profile corrugated feeder antenna. *IEEE Transactions on Antennas and Propagation*, **54**, 340.

[182] Beruete, M., Campillo, I., Dolado, J.S. *et al.* (2005) Very low-profile 'Bull's Eye' feeder antenna. *IEEE Antennas and Wireless Propagation Letters*, **4**, 365.

[183] Beruete, M., Beaskoetxea, U., Zehar, M. *et al.* (2013) Terahertz corrugated and bull's-eye antennas. *IEEE Transactions on Terahertz Science and Technology*, **3**, 740.

[184] Yu, N., Wang, Q.J., Kats, M.A. *et al.* (2010) Designer spoof surface plasmon structures collimate terahertz laser beams. *Nature Materials*, **9**, 730.

3

Principles of Emission of THz Waves

Luis Enrique García Muñoz[1], Sascha Preu[2], Stefan Malzer[3], Gottfried H. Döhler[4], Javier Montero-de-Paz[5], Ramón Gonzalo[5], David González-Ovejero[6], Daniel Segovia-Vargas[1], Dmitri Lioubtchenko[7], and Antti V. Räisänen[7]

[1]*Departamento Teoría de la Señal y Comunicaciones, Universidad Carlos III de Madrid, Av. de la Universidad, Leganes-Madrid, Spain*
[2]*Dept. of Electrical Engineering and Information Technology, Technische Universität Darmstadt, Darmstadt, Germany*
[3]*Friedrich-Alexander University Erlangen-Nürnberg, Applied Physics, Erlangen, Germany*
[4]*Max Planck Institute for the Science of Light, Erlangen, Germany*
[5]*Departamento de Ingeniería Eléctrica y Electrónica, Universidad Pública de Navarra, Navarra, Spain*
[6]*University of Siena, Department of Information Engineering and Mathematics, Palazzo San Niccolo, Siena, Italy*
[7]*Aalto University, School of Electrical Engineering, Department of Radio Science and Engineering, Finland Espoo, Finland*

3.1 Fundamental Parameters of Antennas

In this section, the main parameters of antennas are briefly reviewed. These parameters are explained in a simple way, but there exists a vast literature where the reader can go into further detail [1–6].

3.1.1 Radiation Pattern

The radiation pattern is a representation of the radiation properties of an antenna as a function of the different directions of space at a fixed distance. Normally, spherical coordinates (r, θ, φ) are used and the

Semiconductor Terahertz Technology: Devices and Systems at Room Temperature Operation, First Edition.
Edited by Guillermo Carpintero, Luis Enrique García Muñoz, Hans L. Hartnagel, Sascha Preu and Antti V. Räisänen.

representation is referred to the electric field. As the electric field is a vector magnitude, two orthogonal components, normally θ and φ, must be calculated at each point of the constant radius sphere defined by the spherical coordinates. Field radiation pattern or power radiation pattern have the same information, since the power density is proportional to the square of the electric field module.

It is common to represent the radiation pattern in a three-dimensional plot or in a two-dimensional one showing different cuts of the radiation pattern. In linearly polarized antennas (see Section 3.1.6) the E-plane and the H-plane can be defined. The E-plane is the plane containing the direction of propagation and the E-field, while the H-plane is the plane containing the direction of propagation and the H-field. Both planes are perpendicular and their intersection represents a line which defines the direction of propagation.

Bi-dimensional cuts of the radiation pattern can be represented either in polar or Cartesian coordinates. An example of both types of representation can be seen in Figure 3.1. In the first case, the radius represents the intensity of the radiated electric field or the power density, while the angle represents the space direction. In the second case, the angle is represented on the abscissa axis while electric field or power density is represented on the ordinate axis. Field or power density can be represented in an absolute or a relative way (normalized to the maximum), with natural or decibel scales. From a radiation pattern representation, several parameters can be extracted. The direction of space where the radiation is maximum is called the *main lobe,* whereas the *secondary lobe* is the lobe with more amplitude than the rest of the lobes.

3 dB *beamwidth* ($\Delta\theta_{3dB}$) or *half power beamwidth* (HPBW) is the angular distance in the directions where the density power radiation pattern is equal to half the maximum.

Beam-width between zeros ($\Delta\theta_c$) or *first null beamwidth* (FNBW) is the angular distance in the directions where the main lobe has a minimum.

Side lobe level (SLL) is the relation between the value of the radiation pattern in the direction of maximum radiation and the value of the radiation pattern in the direction of the secondary lobe. It is normally expressed in decibel.

Front to back radiation is the relation between the value of the radiation pattern in the direction of maximum radiation and the value of the radiation pattern in the opposite direction.

Depending on the shape of the radiation pattern three types of radiation patterns can be identified: isotropic, omnidirectional, and directive (Figure 3.2).

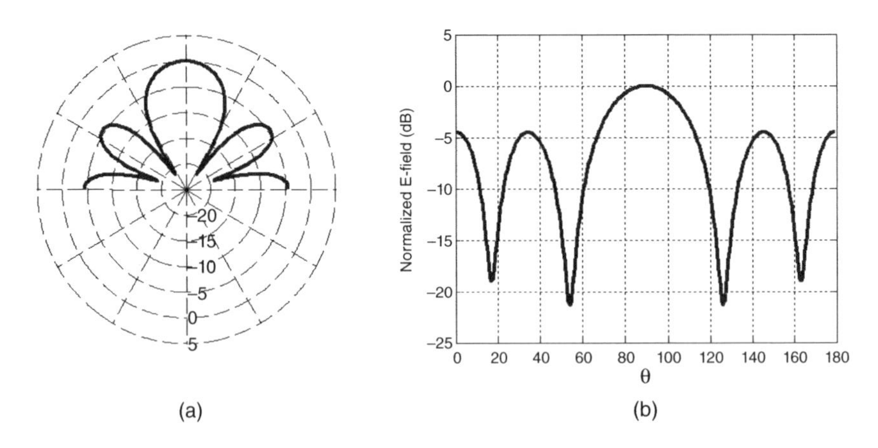

Figure 3.1 Radiation patterns. (a) Polar and (b) Cartesian representation.

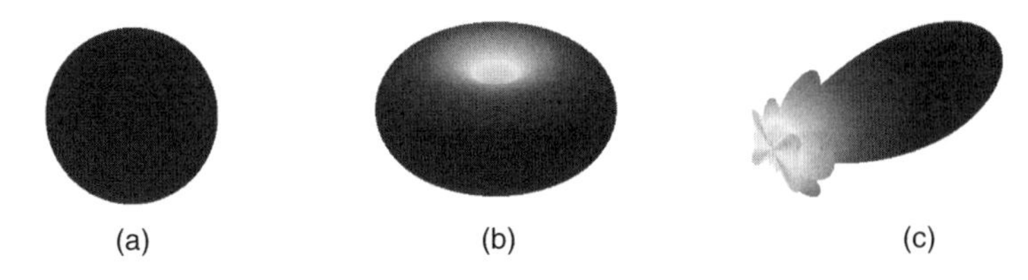

Figure 3.2 Radiation patterns. (a) Isotropic, (b) omnidirectional, and (c) directive.

3.1.2 Directivity

Directivity of an antenna is defined as the relationship between power density radiated in a direction at a fixed distance and the power density that will be radiated at that distance by an isotropic antenna that radiates the same power as the antenna:

$$D(\theta,\varphi) = \frac{P(\theta,\varphi)}{P_r/(4\pi r^2)} \tag{3.1}$$

If no angular distribution is specified, it is understood that directivity refers to the maximum radiation direction:

$$D = \frac{P_{max}}{P_r/(4\pi r^2)} \tag{3.2}$$

Directivity can be obtained from the expression:

$$D = \frac{4\pi}{\iint_{4\pi} t(\theta,\varphi)\mathrm{d}\Omega} \tag{3.3}$$

where

$$t(\theta,\varphi) = \frac{P(\theta,\varphi)}{P_{max}} \tag{3.4}$$

is the normalized radiation pattern.

In directive antennas, a good approximation to calculate the directivity is given by

$$D = \frac{4\pi}{\Delta\theta_1 \cdot \Delta\theta_2} \tag{3.5}$$

where $\Delta\theta_1$ and $\Delta\theta_2$ are the 3 dB beamwidth in the two main planes of the radiation pattern.

Once the maximum directivity D and the normalized radiation pattern $t(\theta,\varphi)$ are known, it is easy to obtain the directivity at any direction:

$$D(\theta,\varphi) = D \cdot t(\theta,\varphi) \tag{3.6}$$

3.1.3 Gain and Radiation Efficiency

3.1.3.1 Gain

Gain (G) is directly related to directivity. Its definition is the same as that of directivity, but instead of comparing with the radiated power, delivered power is used. Then, it is possible to take into account the losses of the antenna. Directivity and gain are related by *radiation efficiency* (ε_{rad}):

$$G(\theta,\varphi) = \frac{P(\theta,\varphi)}{P_{delivered}/(4\pi r^2)} = \frac{P_r}{P_{delivered}}\frac{P(\theta,\varphi)}{P_r/(4\pi r^2)} = \varepsilon_{rad} \cdot D(\theta,\varphi) \tag{3.7}$$

3.1.3.2 Radiation Efficiency

The existence of losses in the antenna means that not all the power delivered from the transmitter to the antenna is radiated to the free space. The amount of power radiated to the free-space is defined by the *radiation efficiency* (ε_{rad}), which takes values between 0 (no power is radiated to the free-space) and 1 (all power delivered to the antenna is radiated to the free-space).

3.1.4 Effective Aperture Area and Aperture Efficiency

Effective aperture or *effective area* is defined as the relation between the power that the antenna delivers to its load (suppose antenna working in reception) and the power density of the incident wave. It is related to the gain by the following formula:

$$\frac{A_{\text{eff}}}{G} = \frac{\lambda^2}{4\pi} \tag{3.8}$$

The effective aperture cannot be larger than the physical dimension of the antenna, so an *aperture efficiency* ($\varepsilon_{aperture}$) that relates the effective area and the physical area of the antenna can be defined:

$$\varepsilon_{\text{aperture}} = \frac{A_{\text{eff}}}{A_{\text{phy}}} \tag{3.9}$$

3.1.5 Phase Pattern and Phase Center

Under some circumstances it is desirable to plot not the amplitude of the electric field (as in the radiation pattern) but the phase of the electric field. Such a representation is called the *phase pattern*.

When observing an antenna at a great distance, its radiation can be seen as coming from a single point. In other words, its wave front is spherical. This point, the center of curvature of the surfaces with constant phase, is called the *phase center* of the antenna.

3.1.6 Polarization

Polarization represents the field vector orientation in a fixed point as a function of time. It can be identified by the geometric figure described, as time goes, by the end of the electric field vector in a fixed point of the space in the plane perpendicular to the direction of propagation. Three figures can be generated: an ellipse (which is the most common one), a segment (linear polarization), and a circle (circular polarization).

The direction of rotation of the electric field, either in circularly or elliptically polarized waves, is called right-hand polarization if it is clockwise and left-hand if it is counterclockwise.

The axial ratio of an elliptically polarized wave is defined as the ratio between the major and minor axes of the ellipse. It takes values between 1 and infinity. For a circularly polarized wave the axial ratio is 1, whereas for a linearly polarized wave it is infinity.

3.1.7 Input Impedance and Radiation Resistance

3.1.7.1 Input Impedance

The input impedance of an antenna is the relation between the voltage and the current at the input of the antenna. It is normally an imaginary number that depends on the frequency:

$$Z_{\text{in}}(\omega) = R(\omega) + \mathrm{j}X(\omega) \tag{3.10}$$

If $X(\omega) = 0$ at some specific frequency, it is said that the antenna is resonant at that frequency. Knowing the input impedance of an antenna is a key factor because normally the antenna is connected to a transmission line or to an active device (a field-effect transistor (FET), diode, etc.). If a mismatch occurs between the antenna and the device, then not all the power transmitted through the device will be delivered to the antenna (transmitter) or not all the power received by the antenna will be delivered to the device (receiver).

3.1.7.2 Radiation Resistance

When delivering power to an antenna, a part of it is radiated through the free-space. This quantity can be defined by a *radiation resistance*, R_r, which is defined as the resistance value that will dissipate the same power as that radiated by the antenna.

$$P_{\text{radiated}} = I^2 R_r \tag{3.11}$$

Not all the power delivered to an antenna is radiated into free-space. Associated to this we can define the *loss resistance* R_Ω. This resistance refers to the losses that appear in the antenna and is defined as the resistance value that will dissipate the same power as that not radiated by the antenna.

$$P_{\text{delivered}} = P_{\text{radiated}} + P_{\text{losses}} = I^2 R_r + I^2 R_\Omega \tag{3.12}$$

This is related to the radiation efficiency (ε_{rad}) defined previously:

$$\varepsilon_{\text{rad}} = \frac{P_{\text{radiated}}}{P_{\text{delivered}}} = \frac{P_{\text{radiated}}}{P_{\text{radiated}} + P_{\text{losses}}} = \frac{R_r}{R_r + R_\Omega} \tag{3.13}$$

3.1.8 Bandwidth

The bandwith represents the frequencies at which the antenna works. It is the frequency interval at which the antenna does not overcome some prefixed limits. It can be represented as the absolute value ($f_{\max} - f_{\min}$) or relative value (fractional bandwidth, FBW):

$$\text{FBW} = \frac{f_{\max} - f_{\min}}{f_0} \tag{3.14}$$

In broadband antennas, it is common to represent the bandwidth in the form:

$$\text{BW} = \frac{f_{\max}}{f_{\min}} : 1 \tag{3.15}$$

The criteria used to determine the bandwidth of an antenna are related to the radiation pattern (directivity, polarization purity, beamwidth, SLL) or to the impedance (input impedance, reflection coefficient, or standing wave ratio).

3.2 Outcoupling Issues of THz Waves

THz antennas are typically placed lying on the interface of two media, one of them being a semiconductor, which can be sometimes considered as a semi-infinite substrate. The use of typical commercial software for the analysis of this electromagnetic problem may provide results far away from the actual solution. This is the reason why a solution that exploits the Green's function of the structure will be presented.

The analysis of antennas embedded or just lying on the interface between two electrically different media has been compiled from a broad spectrum of research [7–13]. The interest in this kind of antenna and topology has increased in the last years due to, for example, the new devices for THz generation based on the use of the substrate's semiconductor properties, which avoids the use of antennas. The equivalent electromagnetic model of these new devices called large area emitters (LAEs) is as simple as an array of Hertzian dipoles lying on the interface of the semiconductor and air [14]. The lattice and geometry of this array will be parameterized as a function of the excitation of the device and the geometry of the semiconductor device. Knowledge of the radiation pattern of such arrays is required by the scientific community.

Antennas lying on a dielectric medium have some special features that distinguish them from antennas in free-space. In free-space, the power radiated by a planar antenna divides equally above and below the plane by symmetry, whereas an antenna on a dielectric radiates mainly into the dielectric. This results both from the way elementary sources radiate and from the way waves propagate along metals at a dielectric interface [8]. The phenomenon can be explained by a reasoning based on the different impedances of both media. Let us consider a dipole placed at the air/dielectric interface. If the dipole is used as a receiving antenna, it measures the electric field parallel to it. This electric field is the field transmitted through the interface. In the transmission-line model of wave propagation [15], a wave from a high impedance material (air) is incident on a low-impedance material (dielectric). The transmitted electric field is small, so the dipole response is small. The effect is the same as that of the voltage across a coaxial cable terminated with a low resistance. When the wave is incident from the dielectric side, the transmitted electric field is large and the dipole response is large, since the material has a lower impedance than air. The analogous situation in a coaxial cable corresponds to the voltage seen at a high-impedance load. The result is that the response is larger when the wave is incident from the dielectric. By reciprocity, the power transmitted from the dipole is larger in the dielectric. Since any antenna can be seen as a collection of elementary Hertzian dipoles, one can state that the antenna will ordinarily radiate more strongly into the dielectric [8]. Another factor that favors the radiation into the dielectric is the way waves propagate along metals at an air/dielectric interface. The waves tend to propagate at a velocity that is between the velocity of waves in the air and the velocity of waves in the dielectric [9, 16]. Indeed, the wave propagates with a velocity close to the characteristic velocity of a material with a dielectric constant equal to the mean of the two dielectric constants. The effect is that the wave is slow as far as the air is concerned and mainly evanescent waves are excited. On the other hand, for the dielectric the wave is fast, and the radiation is strong there. The effect is that of a leaky-wave antenna, with radiation primarily in the dielectric [8]. So it seems clear that the energy should be focused in from the dielectric side.

Another issue that has to be taken into account is the thickness of the substrates, which is of the order of some wavelengths at THz and sub-THz frequencies. Under these conditions, the substrate is very thick and its effect has to be properly taken into account. From a ray tracing point of view (Figure 3.3), the rays incident at an angle larger than the critical angle are completely reflected and trapped as surface waves. The issue of power coupled into surface waves will be discussed in Section 3.2.2.2; this is a key issue

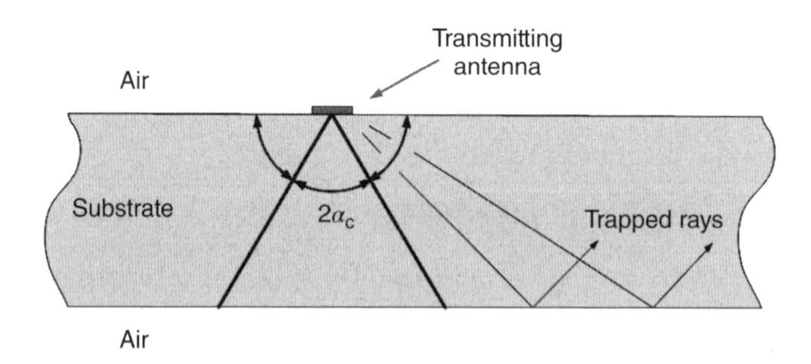

Figure 3.3 Transmitting antenna on a dielectric substrate showing the rays trapped as surface waves. The critical angle is represented by α_c.

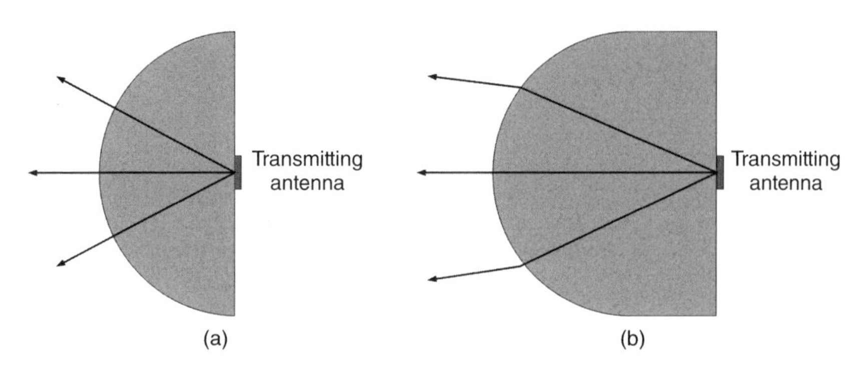

Figure 3.4 Substrate lenses. (a) Hemisphere and (b) extended hemisphere.

since it can dominate the radiated power under some circumstances. Nevertheless, the presence of surface waves can be controlled by the inclusion of a lens on the back side of the substrate with the same dielectric constant. This eliminates the problem, since, as shown in Figure 3.4, the rays' angle of incidence is now nearly normal to the surface and, hence, rays do not suffer total internal reflection. The substrate lens takes advantage of the sensitivity of an antenna to radiation from the substrate side and eliminates the surface waves. In addition, directivity is improved. The disadvantages of the substrate lens are those of any system using refractive optics: absorption and reflection loss [8]. Among the available dielectric lenses, two are particularly attractive because they are aplanatic, adding no spherical aberration or coma [17]. These are the hemispherical lens (Figure 3.4a) and the extended hemispherical lens (Figure 3.4b). A thorough analysis of the latter is carried out in Section 3.4.

3.2.1 Radiation Pattern of a Dipole over a Semi-Infinite Substrate

In some cases, a good approximation to obtain the radiation pattern of antennas lying over or on a dielectric substrate is to consider the thickness of such a substrate as infinite. At THz frequencies, available substrates have a thickness of the order of several λ (typical 350 μm thickness silicon wafers correspond to $\sim 4\lambda$). Thus, commercial full-wave electromagnetic simulators would need a lot of CPU time and memory resources to solve the problem due to its large electrical size. For the reasons above, it seems reasonable to consider the substrate's thickness as infinite. In some applications, such as the design of hyperhemispherical dielectric lenses (Section 3.4) this radiation pattern over a semi-infinite substrate is needed. In this section, we provide the radiation pattern of the most elementary antenna, a Hertzian dipole, over a semi-infinite substrate. To do so, analytical expressions for the radiation pattern of an electric dipole located in a dielectric medium 1 at distance z_0 from the interface to a different dielectric medium 2 have been derived, based on Lukosz's work [10–12], and compared to those obtained using the structure's Green's function [18]. In particular, the radiation patterns of dipoles located on the interface ($z_0 = 0$) and oriented perpendicular or parallel to it are examined. The results are similar to those presented in Refs. [7–12] and the rigorous mathematical methodology can be followed in Refs. [10–12]. Both media, 1 and 2, are assumed to be isotropic, homogeneous, and non-absorbing dielectrics with refractive indices n_1 and n_2, respectively. The methodology presented in Refs. [10–12] can be summarized as follows. First, the unperturbed dipole field is represented as a superposition of s- and p-polarized plane and evanescent waves. Let us consider a plane of incidence $z = \text{constant}$, the waves are transverse electric (TE) or s-polarized waves when the electric field is perpendicular to the plane of incidence. On the other hand, a transverse magnetic (TM) wave or p-polarized wave has the magnetic field perpendicular to the plane of incidence. This representation is rigorously valid both in the dipole's far-field and near-field. Then, the reflection and refraction of each of these plane and evanescent waves incident onto the interface are

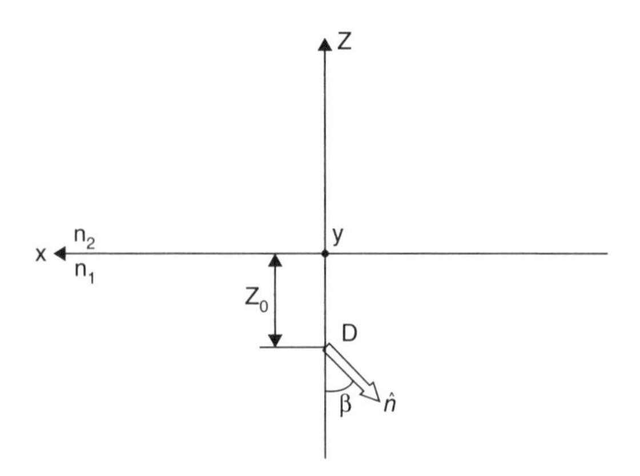

Figure 3.5 Dipole located in medium 1 at distance z_0 from the interface to medium 2. The unit vector $\vec{n}$ in the direction of the dipole moment lies in the x-z-plane.

considered separately. In Figure 3.5 a schematic of the problem is depicted. In the following subsections the horizontal dipole and vertical dipole cases are treated separately.

3.2.1.1 Vertical Dipole

In Figure 3.6, the radiation pattern of a vertical dipole lying on the interface between two semi-infinite media is depicted. The upper medium is vacuum and the lower medium is GaAs ($\varepsilon_r = 12.9$). One can observe that the radiation pattern does not depend on the angular coordinate φ (as for a vertical dipole on an unbounded medium) and that the maximum in the radiation pattern appears at the critical angle ($\alpha_c = \sin^{-1}(1/12.9) = 16.16° \rightarrow \theta = 180° - \alpha_c = 163.84°$). These results are equivalent to those presented in the literature [7–12].

Figures 3.7 and 3.8 depict the angular distribution of the power emitted by a vertical electric dipole, normalized to the unbounded medium ($n_1 = 1$ and $n_2 = 1$) case. The radiation patterns only depend on the distance z_0 and the refractive indices n_1 and n_2. The maximum of the radiation pattern occurs

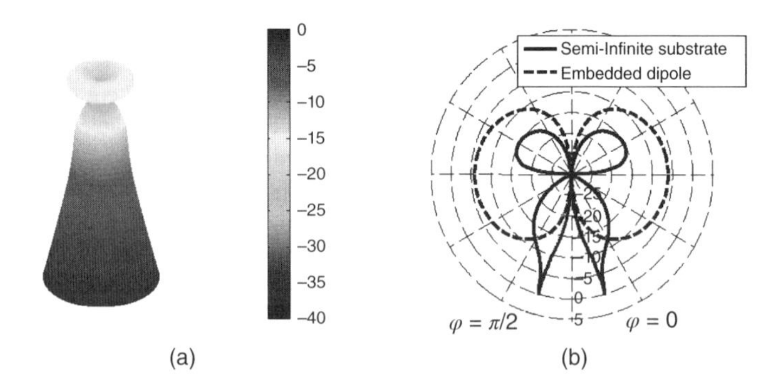

Figure 3.6 Radiation pattern of a dipole lying in the interface between two media with $n_1 = \sqrt{12.9}$ (GaAs), $n_2 = 1$, and $\beta = 0$. (a) 3D plot and (b) polar plot comparing with the case of a dipole embedded in a dielectric medium ($n_1 = n_2 = \sqrt{12.9}$).

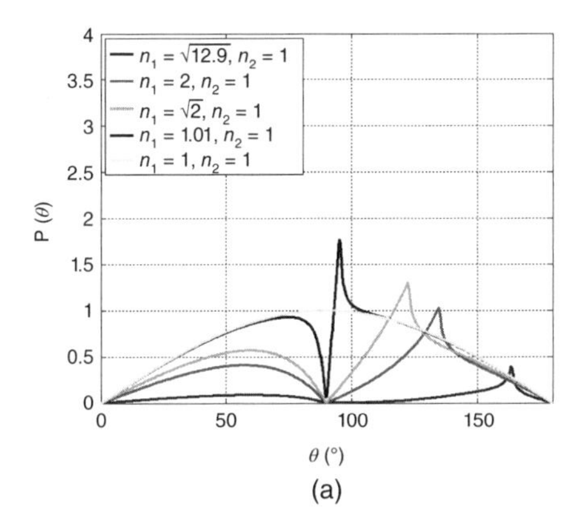

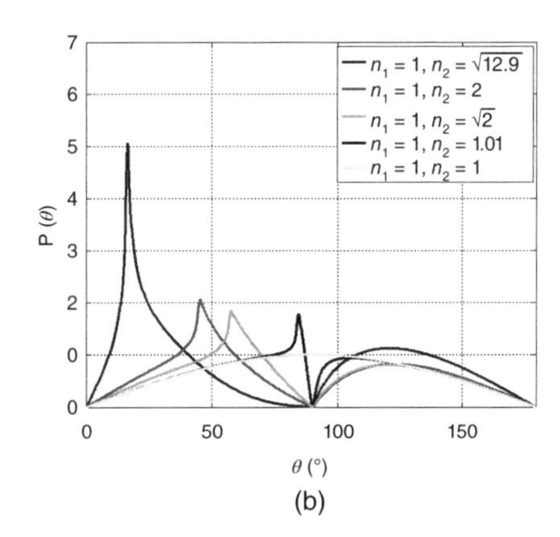

Figure 3.7 Angular distribution $P(\theta)$ of emitted power normalized to $n_1 = 1$ and $n_2 = 1$ case versus angle θ for vertical electric dipoles located on the interface ($z_0 = 0$). (a) $n_1 > n_2$ and (b) $n_1 < n_2$. See plate section for color representation of this figure.

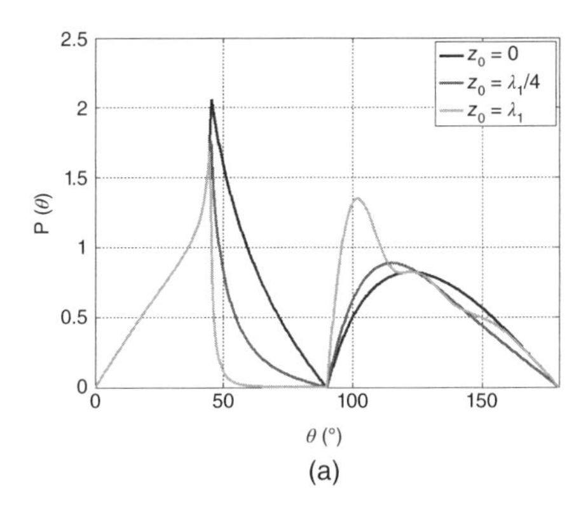

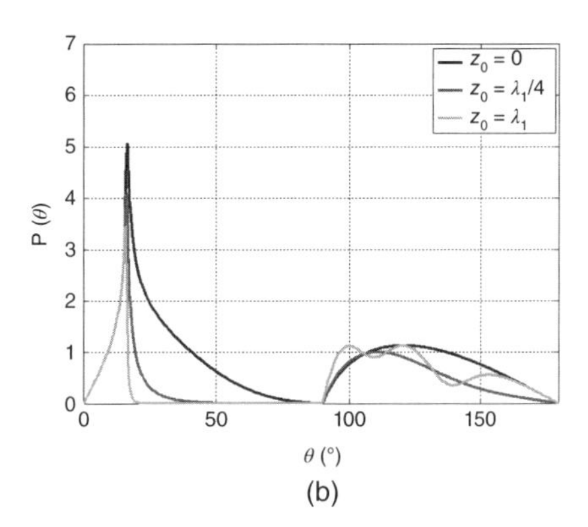

Figure 3.8 Angular distribution $P(\theta)$ of emitted power normalized to $n_1 = 1$ and $n_2 = 1$ case versus angle θ for vertical electric dipoles located at different distances z_0. (a) $n_1 = 1$ and $n_2 = \sqrt{2}$ and (b) $n_1 = 1$ and $n_2 = \sqrt{12.9}$. See plate section for color representation of this figure.

at the critical angle ($\alpha_c = \sin^{-1}(n_2/n_1)$), where the radiation pattern has an uncertainty and it remains at the same angular point regardless of the distance z_0. The power radiated in a direction beyond the critical angle, α_c, of the medium decreases exponentially with the dipole's distance z_0 from the interface. Here, $\Delta z(\alpha_2)$ is the penetration depth of that evanescent wave in the dipole field whose refraction at the interface gives rise to the transmitted plane wave with angle of refraction α_2. For distances $z_0 \gg \Delta z(\alpha_2)$ the evanescent wave does not reach the interface, and no transmitted light appears in medium 2 at $\alpha_2 > \alpha_c$.

3.2.1.2 Horizontal Dipole

In this section the case of the radiation pattern of a horizontal dipole ($\beta = 90°$) lying on the interface of two media (Figure 3.9) is analyzed. These results are also equivalent to those presented in the literature [7–12].

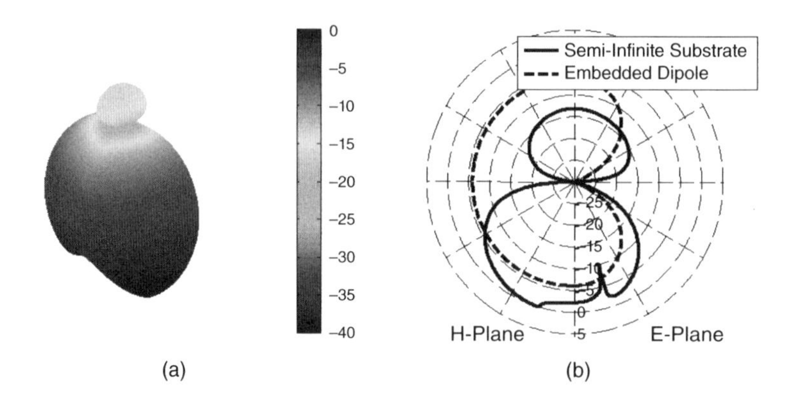

Figure 3.9 Radiation pattern of a dipole lying in the interface between two media $n_1 = \sqrt{12.9}$ (GaAs), $n_2 = 1$, with $\beta = 90°$. (a) 3D plot and (b) polar plot comparing with embedded dipole ($n_1 = n_2 = \sqrt{12.9}$).

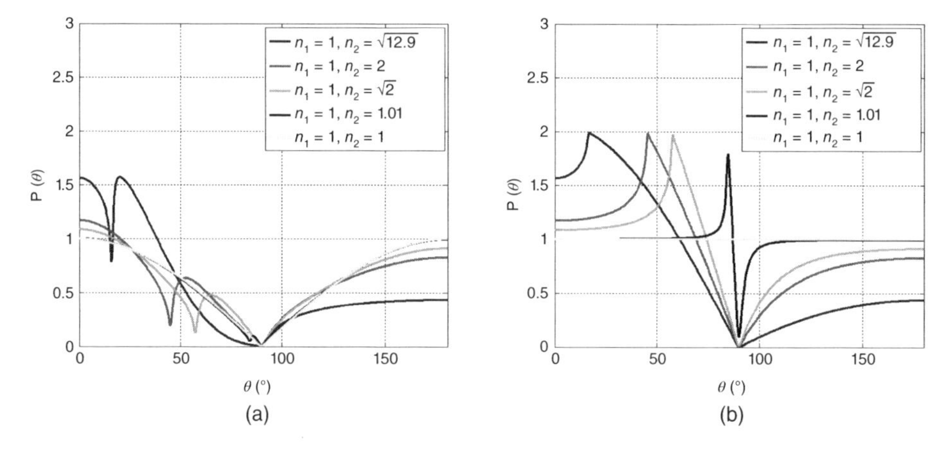

Figure 3.10 Angular distribution $P(\theta)$ of emitted power normalized to the free-space ($n_1 = 1$ and $n_2 = 1$) case versus angle θ for horizontal electric dipoles located on the interface ($z_0 = 0$). (a) E-plane and (b) H-plane. See plate section for color representation of this figure.

Figures 3.10 and 3.11 depict the angular distribution of the power emitted by a horizontal electric dipole normalized to the unbounded medium ($n_1 = n_2 = 1$) case. The radiation patterns only depend on the distance z_0 and refractive indices n_1 and n_2. The maximum of the radiation pattern occurs at the critical angle ($\alpha_c = \sin^{-1}(n_2/n_1)$) for the H-plane and a null appears at the same angle in the E-plane, where the radiation pattern has an uncertainty, and it remains at the same angular point regardless of the distance z_0. The power radiated in a direction beyond the critical angle, α_c, of the medium decreases exponentially with the dipole's distance z_0 from the interface. Here $\Delta z(\alpha_2)$ is the penetration depth of that evanescent wave in the dipole field whose refraction at the interface gives rise to the transmitted plane wave with angle of refraction α_2. For distances $z_0 \gg \Delta z(\alpha_2)$ the evanescent wave does not reach the interface, and no transmitted light appears in medium 2 at $\alpha_2 > \alpha_c$.

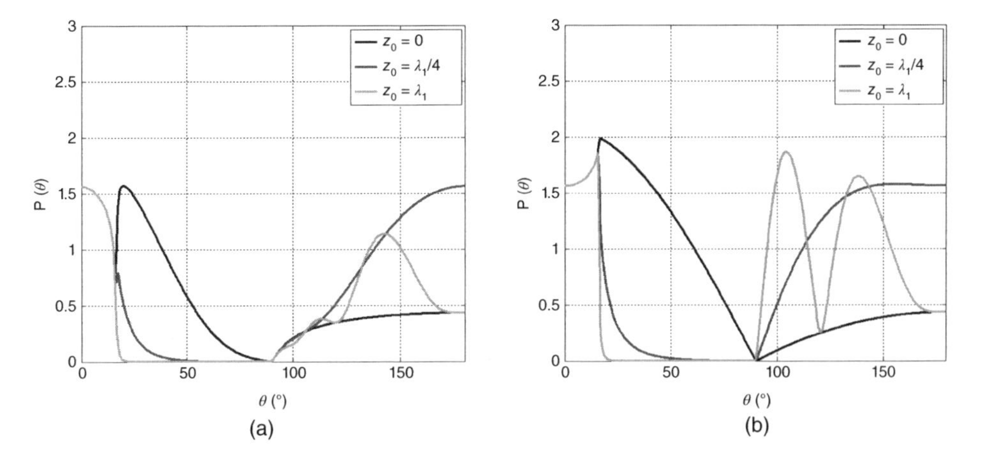

Figure 3.11 Angular distribution $P(\theta)$ of emitted power normalized to the free-space ($n_1 = 1$ and $n_2 = 1$) case versus angle θ for horizontal electric dipoles located at different distances z_0 with $n_1 = 1$ and $n_2 = \sqrt{12.9}$. (a) E-plane and (b) H-plane. See plate section for color representation of this figure.

3.2.2 *Radiation Pattern of a Dipole in a Multilayered Medium*

This section describes the radiation properties of a dipole (either vertical or horizontal) lying above or in a planar multilayered medium. In the previous section the substrate beneath the antenna was considered to have semi-infinite thickness, this case does not correspond to a realistic situation and is useful only under certain circumstances. Next, it will be shown that this finite thickness substrate changes the radiation pattern of antennas placed above or inside the substrate.

First, the case of both horizontal and vertical Hertzian dipoles placed on a typical 350 μm GaAs wafer is analyzed at different frequencies, and compared with the radiation patterns obtained in the previous section. Then, the Hertzian dipole is placed inside the InP wafer and the radiation pattern as well as the radiated power is calculated. It will be shown that when a dipole is in the dielectric slab, a Fabry-Perot behavior appears and part of the power is kept through the dielectric and is not radiated to the air.

3.2.2.1 Hertzian Dipole over a Finite Thickness Substrate

A 350 μm GaAs ($n = \sqrt{12.9}$) substrate and a Hertzian dipole lying on it is considered in this section. For $z > 350$ μm and $z < 0$ air ($\varepsilon_r = 1$) is set as the background material while for $0 < z < 350$ μm GaAs is considered. Two positions of the Hertzian dipole are considered: placed along the z-direction (vertical dipole) or along the x-direction (horizontal dipole), in both cases the dipole lies in the $z = 350$ μm plane.

Vertical Dipole

Figure 3.12 shows the radiation patterns of a vertical Hertzian dipole lying on a 350 μm GaAs substrate and the results are compared with the semi-infinite case calculated previously. One can see that, while in the semi-infinite case the radiation pattern does not depend on the frequency, in this finite case frequency plays an important role. Although the physical dimensions of the GaAs wafer are the same, the electrical thickness changes with frequency, so an effect on the radiation pattern is expected. A maximum at the critical angle no longer exists and it is translated to angles close to $\theta = 90°$ (interface between air and GaAs). So it appears to be that considering the substrate thickness as infinite is not a good option in most cases, no matter how thick the substrate is.

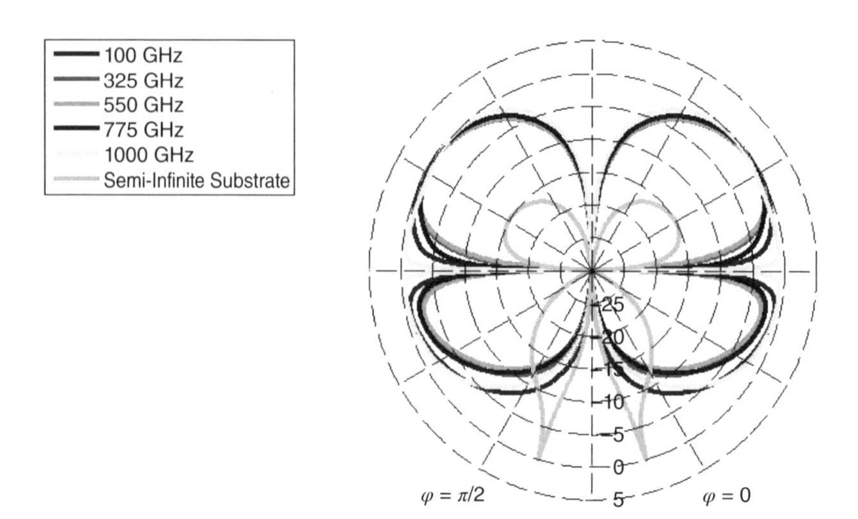

Figure 3.12 Radiation patterns of a vertical Hertzian dipole lying on a 350 μm GaAs substrate obtained at different frequencies. The case of a dipole on a semi-infinite GaAs substrate is shown as a reference. See plate section for color representation of this figure.

Horizontal Dipole

Figure 3.13 shows the radiation patterns for a horizontal (x-direction, $\varphi = 0$) Hertzian dipole lying on 350 μm GaAs substrate and the obtained results are compared with the patterns on a semi-infinite substrate. One can observe that, while in the semi-infinite case the radiation pattern does not depend on the frequency, in the finite thickness case frequency plays an important role. Although the physical dimensions of the GaAs wafer are the same, the wafer's electrical thickness changes with frequency, so an effect in the radiation pattern is expected. Neither maximum nor null exist at the critical angle and the maximum is displaced to angles close to $\theta = 90°$ (interface between air and GaAs). Thus, it seems that considering the substrate thickness as infinite is not a good option in most cases, no matter how thick the substrate is.

3.2.2.2 Hertzian Dipole Inside Finite Thickness Substrate

Recently, owing to the improvements in the CMOS (complementary metal oxide semiconductor) manufacturing process, antennas integrated with other circuits are receiving great attention at THz frequencies [19]. In this technology, the antennas are included between different dielectric media so it is a key point to correctly calculate their radiation pattern. Another important application where antennas are placed inside a finite thickness substrate are LAEs. This antenna-free scheme is explained in detail in Section 3.6.2, where an electromagnetic radiation pattern equivalent circuit is obtained by discretizing the continuous currents into an array of Hertzian dipoles. As an example of the effect of the finite thickness of a dielectric on the radiation pattern, both vertical and horizontal Hertzian dipoles are analyzed.

When a Hertzian dipole oriented across the z-direction (a vertical dipole) is placed in the GaAs wafer 25 μm away from the top air/dielectric interface, the radiation pattern displayed in Figure 3.14 is obtained. In this configuration, the radiation pattern is no longer symmetric and, depending on the frequency, more power is radiated toward the upper ($z > 0$) or lower ($z < 0$) semispace.

When a Hertzian dipole oriented across the y-direction (a horizontal dipole) is placed in the GaAs wafer 25 μm away from the top air/dielectric interface, the radiation pattern displayed in Figure 3.15 is obtained. Also in this configuration, where the dipole is horizontally oriented, the radiation pattern is no longer symmetric and, depending on the frequency, more power is radiated toward the upper ($z > 0$) or lower ($z < 0$) semispace.

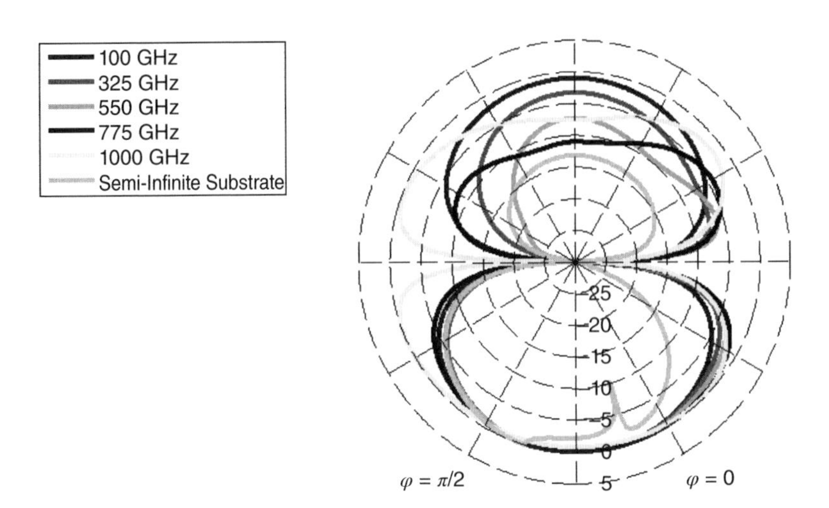

Figure 3.13 Radiation patterns of a horizontal Hertzian dipole lying on a 350 μm GaAs substrate obtained at different frequencies. The case of a dipole on a semi-infinite GaAs substrate is shown as a reference. See plate section for color representation of this figure.

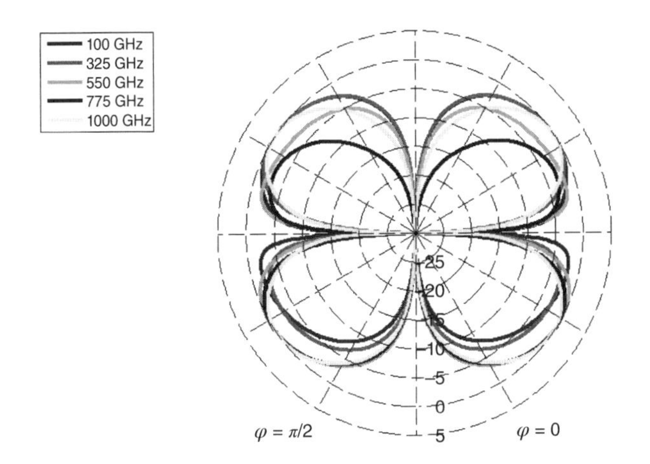

Figure 3.14 Radiation patterns of a vertical Hertzian dipole inside a 350 μm GaAs substrate obtained at different frequencies. The dipole is placed 25 μm away from the upper part of the wafer. See plate section for color representation of this figure.

Another key issue is to estimate the amount of power that is radiated outside the wafer and how much power is retained in it, thus reducing the radiation efficiency. Figure 3.16 depicts the power radiated by the dipole in the wafer, normalized to the power radiated by a dipole embedded in GaAs, as a function of frequency. One can see that the power radiated by the horizontal dipole is much higher than the power radiated by the vertical dipole (in the plot the radiated power for the vertical dipole is multiplied by a factor of 20 for clarity). This is because the vertical position of the dipole makes it easier to excite a guided dielectric mode, thus reducing the radiated amount of power. In addition, there are certain frequencies at which the radiated power is much higher than others. At these frequencies the Fabry-Perot behavior can be appreciated. Those radiated power peaks are separated by 240 GHz approximately ($\lambda_{eff} \approx 350$ μm when considering GaAs) and alternate maxima with minima every 120 GHz approximately ($\lambda_{eff}/2 \approx 175$ μm

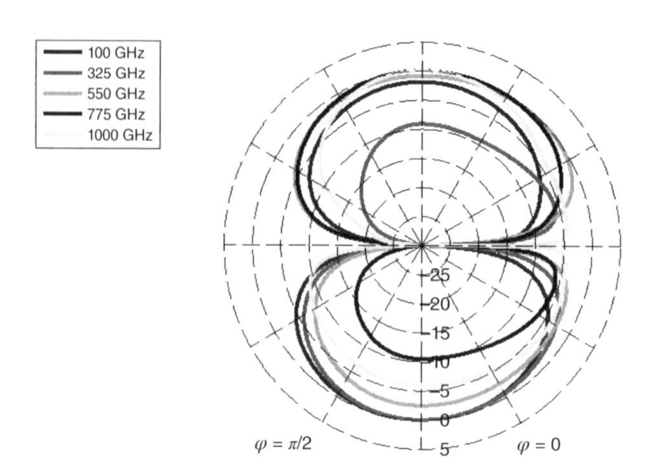

Figure 3.15 Radiation patterns of a horizontal Hertzian dipole inside a 350 μm GaAs substrate obtained at different frequencies. The dipole is placed 25 μm away from the upper part of the wafer. See plate section for color representation of this figure.

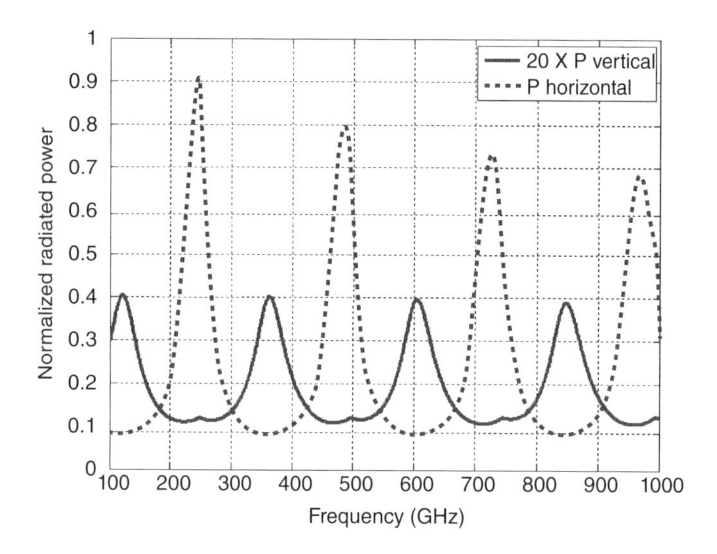

Figure 3.16 Fabry-Perot behavior on the radiation pattern. Radiated power versus frequency normalized to the case of a dipole embedded in a homogeneous medium with $\varepsilon_r = 12.9$ for: horizontal (dotted) and vertical (full line) Hertzian dipoles lying in the middle of a 350 μm GaAs wafer.

when considering GaAs). The obtained results are exactly the same as those predicted by Brueck in his paper related to the radiation pattern of dipoles embedded in a dielectric slab [13].

3.2.3 *Anomalies in the Radiation Pattern*

Several approaches have been applied for solving the problem of calculating the emitted power and the radiation pattern of dipoles and slots lying at the interface of two media, as mentioned in the above sections [7–12]. However, even though all of them present pretty similar results based on the use of the method of moments (MoMs) [7] or a rigorous mathematical methodology [10–12], there exists an issue in the radiation pattern: The radiation pattern from a dipole lying on the interface of two media presents some potential singularities that are discussed in this section and, to the author's best knowledge, there is no explanation in the literature for them.

If one tries to give an explanation for the radiation patterns shown before it is possible to extract some interesting conclusions. As explained by Pozar, Brewitt-Taylor, and Rutledge in their seminal papers [7–9], for the dipole, the H-plane pattern in the dielectric has a maximum at the critical angle $\theta_c = \pi - \sin^{-1}(1/\sqrt{\varepsilon_r})$ and the E-plane pattern has a minimum there. Both patterns have a null at the interface except the H-plane pattern for $\varepsilon_r = 1$, as discussed by Rutledge *et al.* [8]. A maximum and/or minimum at the critical angle does not occur for slot antennas, since the conducting plane effectively isolates the two media, as is well explained by Pozar in Ref. [7].

However, it is interesting to have a look at the maximum of the radiation patterns. Coming back to the results obtained from the semi-infinite case, one can plot the two cases, vertical and horizontal dipoles, on the interface of a medium with relative refractive index $n = \sqrt{12.9}$ in rectangular coordinates (Figure 3.17).

Let us consider now Bernstein's theorem, well-known in Signal Theory and with direct application to Antenna Theory, as discussed in detail in Ref. [20]. To apply this theorem to our radiation patterns, one defines F_1 as an upper limit of the radiation function $F(\tau)$, where $\tau = \sin\theta$, and D is the dimension of the aperture. Supposing that the radiation pattern $F(\tau)$ possesses a bounded spectrum $(-\upsilon_0, +\upsilon_0)$, it follows from Bernstein's theorem that the relative slope of the pattern is limited by the spatial frequency bandwidth of the antenna:

$$\frac{1}{F_1}\frac{\mathrm{d}F}{\mathrm{d}\tau} \leq 2\pi v_o = \pi\frac{D}{\lambda} \tag{3.16}$$

The slope limitation plays an important role, such as with the angular sensitivity of tracking antennas. This is also the case with radar antennas where it is desired that the gain falls rapidly to zero below the horizon to avoid ground clutter. More generally, Bernstein's theorem implies a limit to all derivatives of bounded signals with bounded spectra:

$$\frac{1}{F_1}F^{(n)}(\tau) \leq (2\pi v_o)^n \tag{3.17}$$

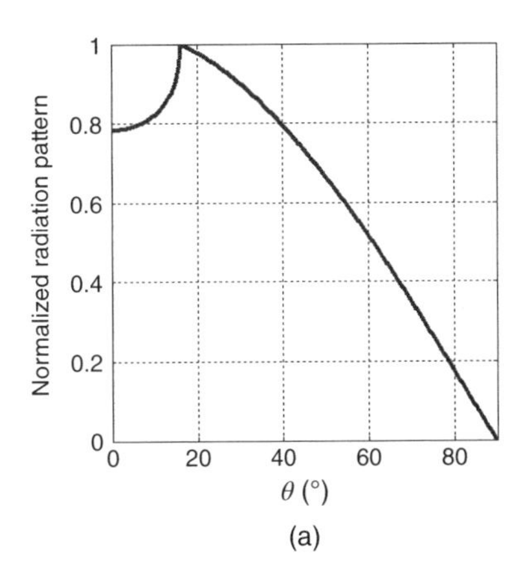

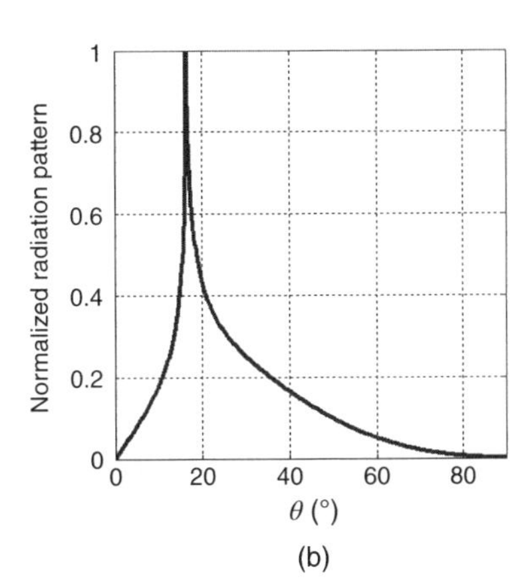

Figure 3.17 Radiation pattern of a dipole from the interface. Relative refractive index $n = \sqrt{12.9}$. (a) Horizontal dipole H-plane and (b) vertical dipole E-plane.

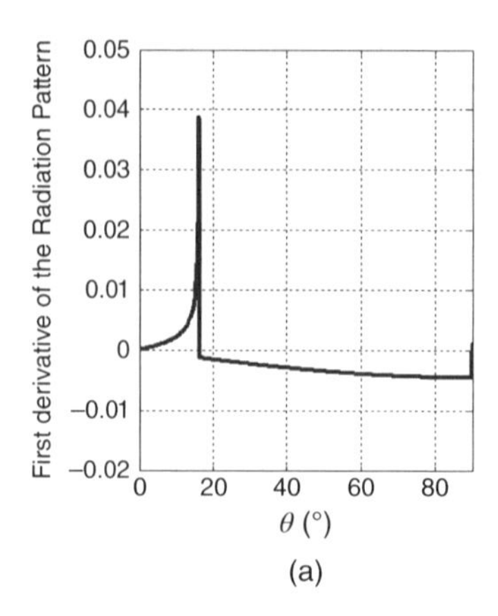

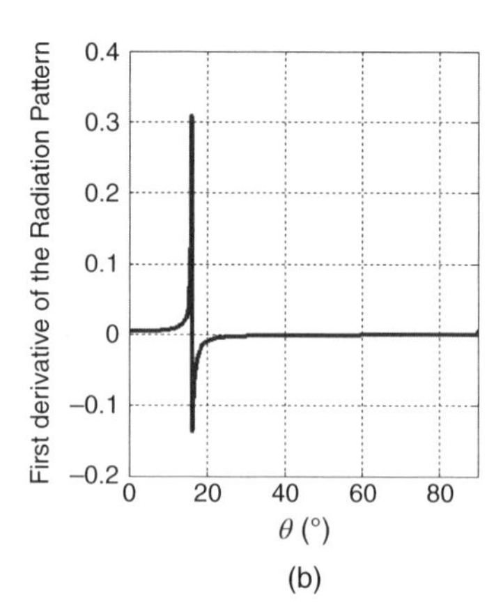

Figure 3.18 First derivative from the radiation pattern of a dipole from the interface. Relative refractive index $n = \sqrt{12.9}$. (a) Horizontal dipole H-plane and (b) vertical dipole E-plane.

So, it is possible to deduce a relationship between the dimension D of the aperture and the half-power beamwidth (the so-called θ_{3dB}) of the main lobe of the radiation pattern. This, easily manipulated, implies the Rayleigh limit for the accuracy resolution in far-field.

Calculating the derivative of the radiation pattern of a dipole (horizontal and vertical) from Figure 3.17 results in the curve in Figure 3.18. In any event, the result of Bernstein's theorem says that the first derivative is bounded, which could be right for our pattern of interest since the pattern has a slope discontinuity, but the slopes are finite. But then the second derivative of the pattern versus angle will be infinite at the point of the slope discontinuity, and so Bernstein's theorem will not be satisfied. So if these results apply to any size antenna, the patterns we are looking at seem to violate this theorem.

An alternative argument is that the far-field cannot have a slope discontinuity because the field must satisfy Maxwell's equations (in the far-field), and taking the curl of an E field with a slope discontinuity will produce a step function in H, and doing a curl on H to find E would lead to a delta function, and so on, which is clearly incorrect.

This discontinuity in the H-plane pattern occurs because the stationary phase (SP) approximation (which has been used to derive the far-field expressions in most of the references) breaks down at the critical angle. The SP method, in simple form, has an integrand of the form $F(x)e^{-jkf(x)}dx$ where $F(x)$ is assumed to have a well-behaved phase near the stationary point. Examination of the integrand for the half-space problem, however, shows that $F(x)$ goes from real to complex at the critical angle, thus violating this assumption, and producing the non-physical change in slope. The problem is apparently with the application of the SP method.

This anomaly in the radiation pattern studied in this section is also presented in many other antenna structures, as can be read in Ref. [21], so, one should be careful while applying the SP method.

3.3 THz Antenna Topologies

In this section, the most common antenna topologies used in the THz range are presented. Among these topologies one finds resonant antennas such as dipoles or slot antennas, and self-complementary antennas like spirals, log-spirals, log-periodic, and bow-tie antennas.

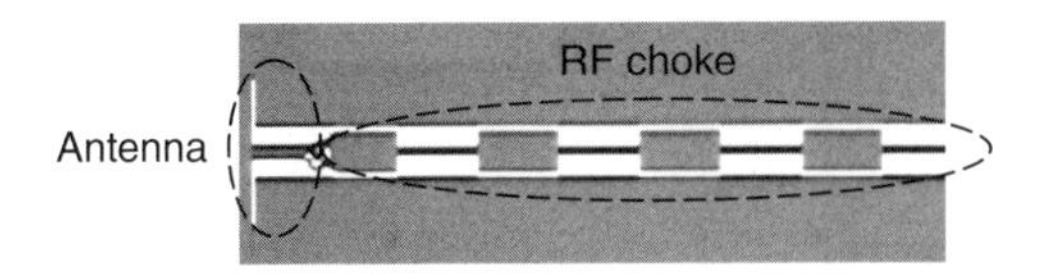

Figure 3.19 SEM picture of a dipole antenna with a stepped impedance low pass filter acting as RF choke.

3.3.1 Resonant Antennas

Regarding resonant antennas in the THz band, the most widely used are dipoles or related topologies. Either detector or emitter devices possess a capacitive part in their input impedance that has a strong influence at higher frequencies. Traditionally, this capacitive part is compensated by including a RF (radio frequency) choke (normally a low pass filter) [22]. In Figure 3.19 an example of a dipole antenna with a RF choke is plotted. In this case the RF choke is a stepped impedance low-pass filter.

Once this capacitive part is compensated, an antenna with the real part of its input impedance equal to the real part of the device's impedance is desired to maximize the power delivered to/from the antenna. Depending on the class of device one topology or another will be more suitable. For instance, photoconductors [14] possess a very high input impedance, so an antenna with a very high input impedance will be the best option. On the other hand, devices such as Schottky diodes or FET have a lower real part of the input impedance, so different antennas may be used. In Ref. [23] a detailed numerical study on some of these resonant antennas at THz frequencies is presented.

3.3.1.1 Dipole Antennas

A dipole antenna is one of the most commonly used and simple antennas [1]. It consists of two conducting arms that are fed at their junction. Depending on the shape of these two arms different topologies can be identified.

λ/2 and λ Dipoles

The geometry and design parameters ($L_{D,}$ w_d) of a single dipole are shown in Figure 3.20. Depending on the total length (L_D), two kinds of dipoles can be identified: $\lambda/2$ and λ dipoles. The main difference between them is the input impedance they provide at the feeding point. Indeed, in $\lambda/2$ dipoles the voltage is a minimum and the current a maximum at the feeding point (unfilled arrow), thus, a minimum is obtained for the input impedance. Such impedance can be modified changing the gap and width of the dipole. For a silicon substrate an impedance around 50 Ω is typically obtained at resonance. On the other hand, λ dipoles have a maximum of voltage and a minimum of current at the feeding point (unfilled arrow), thus a maximum is obtained for the input impedance. Again, this behavior can be modified by changing the gap and width of the dipole. For a silicon substrate around 250 Ω are usually obtained at the resonance frequency. A typical $\lambda/2$ dipole antenna operating at 1 THz and lying on a silicon substrate ($\varepsilon_r = 11.7$) has $L_d \approx 45$ μm and $w_d \approx 3$ μm, whereas a typical λ dipole antenna would have $L_d \approx 90$ μm and $w_d \approx 3$ μm.

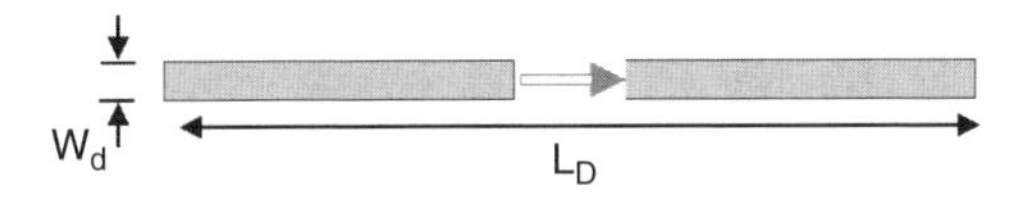

Figure 3.20 Geometry and design parameters of a dipole antenna.

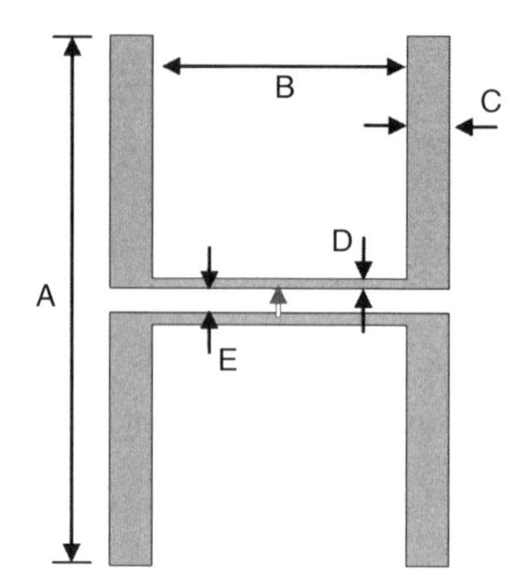

Figure 3.21 Geometry and design parameters of a dual-dipole antenna.

Dual-Dipoles

Dual antenna elements present several advantages over single antenna element designs. These advantages include more symmetric beam patterns, which leads to higher Gaussian beam efficiency, and a higher radiation resistance [24]. The geometry and design parameters of a dual-dipole are shown in Figure 3.21. The feeding point (unfilled arrow) is placed in the middle of the antenna and two symmetric arms form the dipole antenna. The input impedance achieved with this configuration is similar to that of a λ dipole, but the radiation pattern is improved. The design parameters of a typical dual-dipole antenna on a silicon substrate ($\varepsilon_r = 11.7$) working at 1 THz are $A \approx 70\ \mu m$, $B \approx 50\ \mu m$, $C \approx 3\ \mu m$, $D \approx 1\ \mu m$, and $E \approx 5\ \mu m$. This topology has been widely used in THz and sub-THz resonant systems design [22, 25].

Meander Dipoles

A meander dipole antenna is similar to a dipole in which the radiating arms have been bent to a meander shape in order to reduce the antenna size [26]. The geometry and main design parameters of a meander dipole are shown in Figure 3.22. Traditionally, this antenna is used to reduce the antenna size (antenna miniaturization) and in systems that need a very high value of the input impedance, since it provides higher values of input impedance at its resonance than λ-dipoles or dual-dipoles [26]. With this antenna topology, the input impedance achieved at its resonance can be higher than $1000\ \Omega$. This antenna is also very convenient for biasing the active device (photomixer, diode, etc.) placed in the middle of it. This is because almost no current is excited at higher frequencies in the borders of the arms, so biasing circuitry does not affect the high frequency behavior of the antenna. On the other hand, lower values of radiation efficiency are achieved compared to conventional dipoles or dual-dipoles. The design parameters

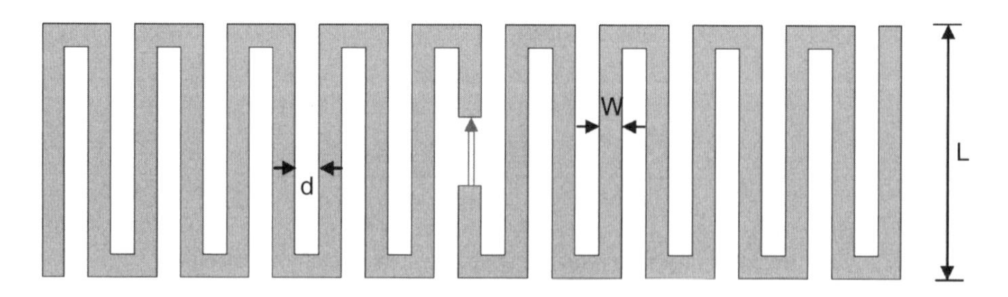

Figure 3.22 Geometry and design parameters of a meander dipole antenna.

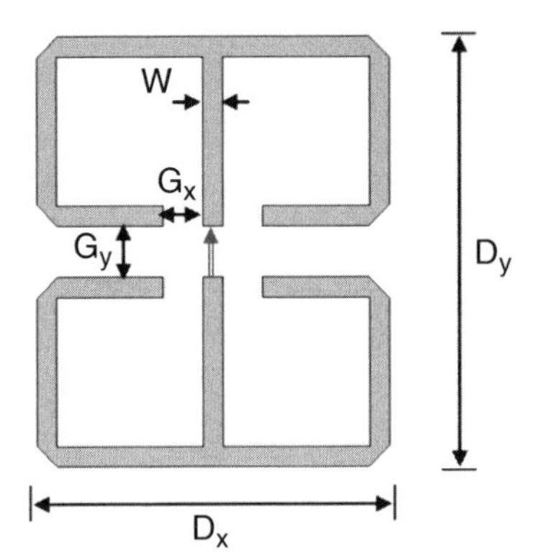

Figure 3.23 Geometry and design parameters of a full-wavelength four-leaf-clover antenna.

of a typical meander dipole antenna on a silicon substrate ($\varepsilon_r = 11.7$) working at 1 THz are $L \approx 45$ μm, $w \approx 3$ μm, and $d \approx 3$ μm.

Full-Wavelength Four-Leaf-Clover

The full-wavelength four-leaf-clover antenna was proposed in Ref. [27] and its geometry is shown in Figure 3.23 along with its design parameters. This antenna provides an input impedance similar to that of the meander dipole (more than 1 kΩ). Moreover, since this design is a dual antenna, it presents the advantage of improving the radiation pattern with respect to meander dipoles. A typical full-wavelength four-leaf-clover antenna over silicon substrate ($\varepsilon_r = 11.7$) working at 1 THz has $D_x = D_y \approx 37$ μm, $G_x = G_y \approx 2$ μm, and $w \approx 3$ μm.

3.3.1.2 Slot Antennas

The geometry and design parameters of a rectangular slot antenna are shown in Figure 3.24, where the gray area represents the metallic part and the white area the aperture. Slot antennas are complementary to dipole antennas, that is, while dipoles have an electric moment, slots present a magnetic one (Babinet's principle [1]). Accordingly, a single slot antenna has a maximum on the input impedance at $L_s = \lambda/2$. All the dipoles mentioned above can also be manufactured in a slot topology. A similar behavior regarding the input impedance is expected. A typical single slot antenna operating at 1 THz and lying on a silicon substrate ($\varepsilon_r = 11.7$) has $L_s \approx 45$ μm and $w_s \approx 3$ μm dimensions.

3.3.2 Self-Complementary Antennas

Self-complementary antennas are ones where the metallic and non-metallic areas have the same shape and can be superimposed on each other by rotation. An antenna with a self-complementary structure has a constant input impedance, independently of the source frequency and shape of the structure. Self-complementary antennas are broadband antennas and can achieve a bandwidth of more than one octave [28].

According to [28], the input impedance of self-complementary antennas is equal to:

$$Z_{in} = \frac{Z_0}{2} \tag{3.18}$$

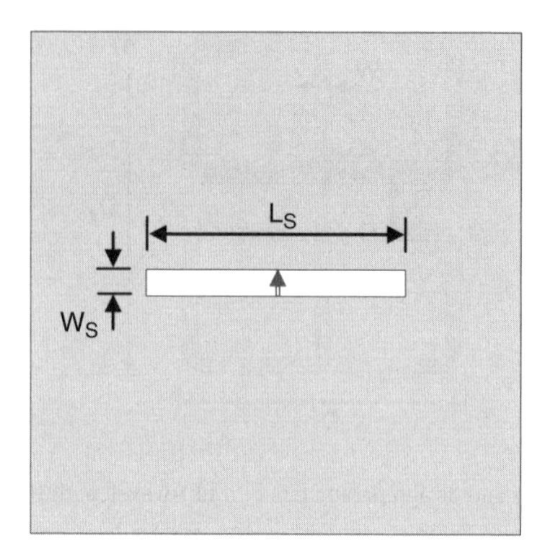

Figure 3.24 Geometry and design parameters of a single slot antenna.

where Z_0 is the intrinsic impedance of the medium. In the case of free-space, $Z_0 = \eta = 120\pi\ \Omega$. It can be seen that this impedance is always real and constant with frequency.

Self-complementary antennas are very attractive because they have excellent wideband radiation characteristics and because predicting resonant behavior of a source is extremely difficult in the high-frequency region. Therefore, the use of antennas with a frequency-independent response allows one to conduct tests over a wide spectral range with a single device. In applications such as spectroscopy, self-complementary antennas are the preferred option due to their broadband nature.

Among the available self-complementary topologies [28], the most widely used are: log-spiral, log-periodic, and bow-tie antennas. In the following sections each will be briefly described. In Ref. [29] a detailed numerical study of these three antennas at THz frequencies is presented.

3.3.2.1 Log-Spiral Antennas

Log-spiral antennas are self-complementary antennas when designed with specific parameters such as the proper arm length and width. The most important characteristic of these antennas is the constant input impedance (theoretical) not depending on frequency [28], when both the width of the arms and spacing between them are equal. Obviously, a log-spiral should be of infinite extent to be rigorously self-complementary. Nevertheless, for practical reasons they are always truncated and the input impedance cannot be constant over an infinite bandwidth, both in the high and the low band. The truncation has effects on the low frequency limit, where the size of the element is comparable to the wavelength and the energy spreads throughout all the structure. At high frequencies, the limitation is due to the source feed itself because of the roll-off and also to the truncation of the spiral on the inner side where the feed is placed.

The log-spiral design parameters are:

$$r_1 = ke^{a\varphi}, r_2 = ke^{a(\varphi-\delta)} \tag{3.19}$$

where r_1 and r_2 represent the inner and outer radius of the spiral, respectively. The angle δ determines the arm width, and a and k are constants, which control the growth rate of the spiral and the size of the terminal region, respectively. In Figure 3.25 the log-spiral geometry is depicted. Its lower frequency

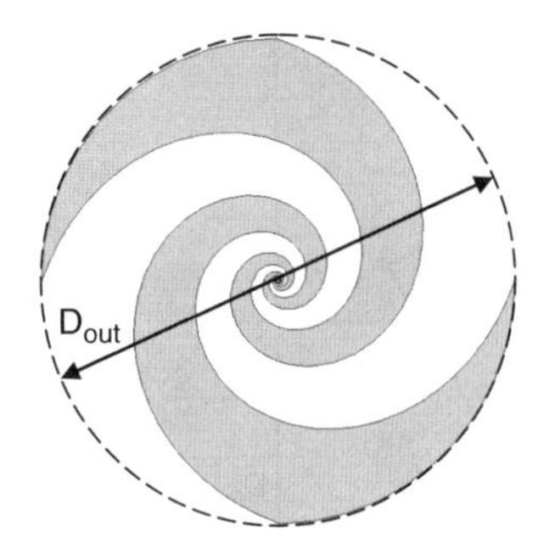

Figure 3.25 Geometry of a log-spiral antenna.

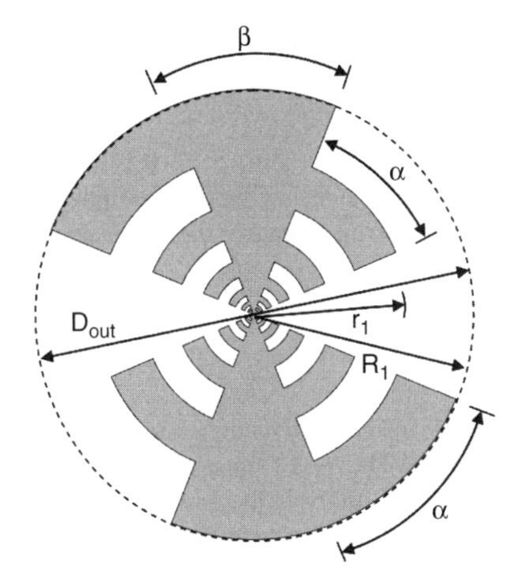

Figure 3.26 Geometry and design parameters of a log-periodic antenna.

is determined approximately by $\lambda = D_{out}$ while its higher frequency is determined by the gap in the center.

Log-spiral antennas provide a relatively constant input impedance and circular polarization. Their radiation efficiency is high and they provide an almost constant radiation pattern.

3.3.2.2 Log-Periodic Antennas

The log-periodic antenna has a log-periodic circular-toothed structure with tooth and bow angles of $\alpha = \beta = 45°$ (for the self-complementary case). The ratio of the successive teeth $(R_{n+1}/R_n) = 0.5$ while the size ratio of the tooth and anti-tooth (r_n/R_n) is equal to $\sqrt{0.5}$. Figure 3.26 depicts the geometry of this antenna. Its lower frequency of operation is determined approximately by D_{out}, whereas its higher frequency is determined by the gap in the center.

Log-periodic antennas possess a non-constant impedance and their polarization is not constant with frequency. In addition, their radiation efficiency decreases with frequency. On the other hand, they have a good radiation pattern and can achieve a high directivity over a wide bandwidth. This topology is one of the most used at THz frequencies.

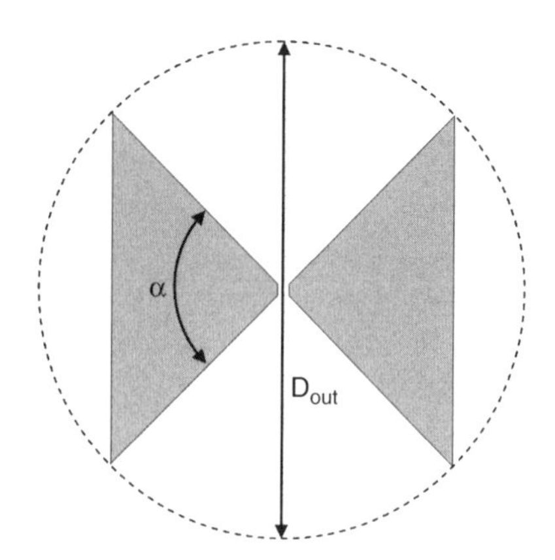

Figure 3.27 Geometry of a bow-tie antenna.

3.3.2.3 Bow-Tie Antennas

The bow-tie antenna has the shape depicted in Figure 3.27. It is similar to a dipole where its arms are wider at their extremes. A self-complementary bow-tie has an angle $\alpha = 90°$. This self-complementary topology provides the most constant input impedance of the three analyzed cases but its radiation pattern deteriorates with frequency.

3.4 Lenses

Dielectric lenses have been one of the most used elements as substrates for planar antennas in the millimeter and sub-millimeter frequency range. The use of substrate lenses allows one to improve the radiation pattern of planar antennas, since the directivity is increased as well as the gaussicity, while the back radiation is reduced [8, 25]. As mentioned before, at these frequencies available substrates are too thick in terms of the wavelength. This results in an increase in the losses due to unwanted modes generated inside the substrate [30], as has been shown in the above sections. Using a dielectric lens as substrate, these modes are not generated because of the lens geometry.

The ideal shape of a dielectric lens is ellipsoidal [3, 5]. However, it is very difficult to manufacture and using it is not practical. An elliptical lens can be approximated by an extended hemispherical one [25] (Figure 3.28) or a hyper-hemispherical one (basically a truncated sphere), both of which are much easier to fabricate. Hence they are the most commonly used lenses. Both extended hemispherical and hyper-hemispherical lenses behave in a similar manner and the methodology explained here is valid for both.

In order to design these lenses, the computational strength of current full-wave electromagnetic simulators (CST [31], HFSS [32], FEKO [33], etc.) is combined with ray tracing techniques and conventional physical optics. The problem in designing planar antennas with dielectric lenses is that the dimensions of the dielectric lenses are on the order of several wavelengths (7–15λ), while typically planar antennas are smaller than λ. For that reason, simulating the whole structure (planar antenna + dielectric lens) will require a large consumption of memory and CPU time and, in some cases, the amount of computational resources needed will exceed those available.

3.4.1 Lens Design

A planar antenna placed on a high permittivity substrate radiates most of the power into it (as shown in Section 3.2), so a unidirectional radiation pattern is obtained [25]. An extended hemispherical geometry

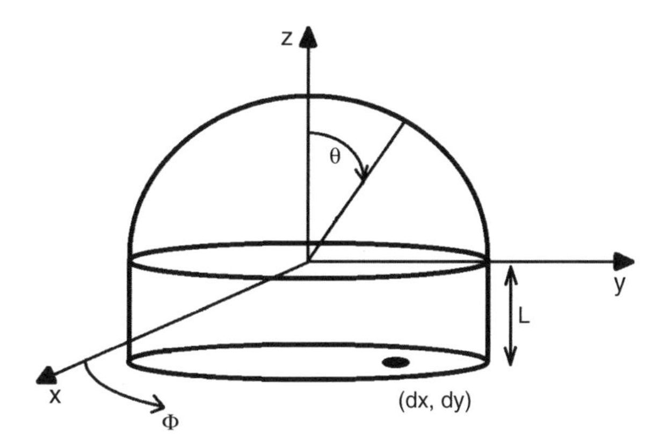

Figure 3.28 Extended hemispherical dielectric lens geometry.

is used for the dielectric lens since it can approximate the elliptical lens behavior and is easier to manufacture.

The dielectric lens consists of a hemisphere with radius R and a cylindrical block with length L and the same radius. The planar antenna is placed on the bottom of the lens with a displacement dx, dy from the axis origin.

Dielectric extended hemispherical lenses satisfy the sine condition, which guarantees circular coma absence, and they do not have spherical aberrations. This implies that if an optical system is designed in such a way that all rays are focused to a point, an extended hemispherical lens can be included in the system and the rays will be focused to the same point [25].

With this lens, the directivity of the planar antenna is increased and the radiation pattern fits well to a Gaussian beam system. The parameter that most influences the system directivity is the length of the cylindrical block (L), as will be shown in the next subsection. This is due to the fact that the antenna must be placed at the focus of the lens, but in this case the lens is hemispherical so a cylindrical block must be included to place the antenna in the focus of the equivalent elliptical lens.

The methodology followed for the design of planar antennas on dielectric extended hemispherical lenses can be divided into two parts. The first consists in designing the planar antenna with a commercial full-wave electromagnetic simulator (CST [31], HFSS [32], FEKO [33], etc.) or with an in-house software (as the one used in previous sections). To do this, the antenna is placed over a semi-infinite dielectric substrate with the same permittivity as the lens. With this simulation, the radiated fields are obtained as a function of the spherical coordinates θ and φ.

Once the radiated fields are obtained, a ray tracing technique is carried out from a point (antenna position) in the same direction as the propagation. The lens geometry can be divided into four parts (Figure 3.29, [34]). The first (Figure 3.29a) is limited by the total reflection angle. In this region both transmitted and reflected waves appear and it is the main contributor to the radiation pattern of the lens. In the developed software, this region is the only one considered to obtain the radiation pattern. The second region (Figure 3.29b) is the one between the critical angle and the end of the hemisphere. Here no transmitted waves appear since the incident angle is higher than the critical one and all incident waves are reflected. Evanescent waves appear outside the lens but do not contribute to the radiated field. The third region (Figure 3.29c) is limited by the end of the hemisphere and the critical angle in the slab. Here the incident angle is higher than the critical angle, so only reflected waves appear. Even lower evanescent waves than in the second region appear outside the lens because directive antennas are used. The contribution to the far-field calculation can be neglected. Finally, the fourth region (Figure 3.29d) is limited by the critical angle in the slab and the end of it (both transmitted and reflected waves).

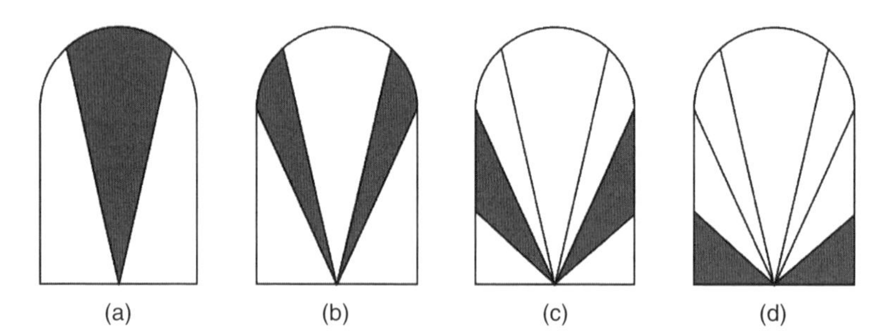

Figure 3.29 Regions into which the extended hemispherical lens can be divided. (a) Spherical region where transmitted and reflected rays appear, (b) spherical region where total reflection occurs, (c) planar region where total reflection occurs, and (d) planar region where transmitted and reflected rays appear.

Here the lowest radiation intensity of the antenna is obtained and the contribution to the far-field can also be neglected. Reflected rays among the different regions can be reflected again and impinge on the lens/air interface, thus contributing to the generated final field, but their contribution will be of low significance [35].

After that, physical optics is applied to generate the currents in the surface of the first region of the lens (Figure 3.29a). Finally, the radiation pattern generated by those currents is obtained. The steps of this second part are summarized below.

The first consists in decomposing the electric and magnetic fields into TE and TM components at the lens/air interface. Then, we can use the formulas that provide the reflection on a dielectric with permittivity ε_r [36]:

$$\Gamma_{TE} = \frac{\sqrt{\varepsilon_r}\cos\theta_i - \sqrt{1-\varepsilon_r\sin^2\theta_i}}{\sqrt{\varepsilon_r}\cos\theta_i + \sqrt{1-\varepsilon_r\sin^2\theta_i}} \tag{3.20}$$

$$\tau_{TE} = 1 + \Gamma_{TE} \tag{3.21}$$

$$\Gamma_{TM} = \frac{\sqrt{\varepsilon_r}\sqrt{1-\varepsilon_r\sin^2\theta_i} - \cos\theta_i}{\sqrt{\varepsilon_r}\sqrt{1-\varepsilon_r\sin^2\theta_i} + \cos\theta_i} \tag{3.22}$$

$$\tau_{TM} = (1 + \Gamma_{TM})\frac{\cos\theta_i}{\sqrt{1-\varepsilon_r\sin^2\theta_i}} \tag{3.23}$$

where ε_r is the dielectric constant of the lens material, θ_i is the incidence angle with respect to the normal lens surface, and Γ and τ are TE and TM reflection and transmission coefficients.

Once the electric and magnetic fields have been defined (primary fields), electric ($\vec{J}_s$) and magnetic ($\vec{M}_s$) the current densities are obtained outside the lens surface [36]:

$$\vec{J}_s = \hat{n} \times \vec{H} \tag{3.24}$$

$$\vec{M}_s = \hat{n} \times \vec{E} \tag{3.25}$$

where $\hat{n}$ is the vector normal to the interface.

From this magnetic and electric density currents, θ and φ components of the electric field in far-field are calculated [36]:

$$E_\theta \cong -\frac{jke^{-jkr}}{4\pi r}(L_\varphi + \eta N_\theta) \tag{3.26}$$

$$E_\varphi \cong +\frac{jke^{-jkr}}{4\pi r}(L_\theta + \eta N_\varphi) \tag{3.27}$$

where $\vec{N}$ and $\vec{L}$ are the radiation vectors defined by [36]:

$$\vec{N} = \iint_{S'} \vec{J}_s e^{jkr' \cdot \cos\psi} ds' \tag{3.28}$$

$$\vec{L} = \iint_{S'} \vec{M}_s e^{jkr' \cdot \cos\psi} ds' \tag{3.29}$$

where S' is the external surface of the lens, r' is the distance from origin to where the equivalent magnetic and electric density currents are calculated, r is the distance from origin to the observation point, and ψ is the angle between r and r'.

3.5 Techniques for Improving the Performance of THz Antennas

Some novel and promising advances are presented in this section, which deals with the improvement of the radiation efficiency presenting solutions from the antenna and RF point of view.

3.5.1 Conjugate Matching Technique

From the antenna point of view only two parameters can be optimized in order to improve the performance of both an emitter and a receiver: the radiation efficiency (ε_{rad}) and the mismatching factor (M). The mismatching factor is defined with the M-factor given by [37]:

$$M = \frac{4R_a R_p}{(R_a + R_p)^2 + (X_a + X_p)^2} \tag{3.30}$$

where $Z_a = R_a + jX_a$ is the input impedance of the antenna and $Z_p = R_p + jX_p$ is the input impedance of the active device (either acting as an emitter or a receiver).

The antenna efficiency can then be defined as the product of these two efficiencies and the polarization efficiency (ε_{pol}) [1]:

$$\varepsilon_{ant} = \varepsilon_{rad} \cdot M \cdot \varepsilon_{pol} \tag{3.31}$$

This efficiency expresses how far the system is from the ideal behavior of the active device. Maximizing this efficiency will enhance the performance of the device and it is the key factor that one should use to optimize the design of the antenna. This design is based on the Active Integrated Antenna concept [38, 39] and tries to maximize the antenna efficiency as defined in Eq. (3.31). In the design phase, both the mismatching factor and the radiation efficiency must be considered.

From now on, we will focus on the design of an antenna emitter (AE) based on a photomixer, but the same methodology can be followed no matter what photomixer or active device is used. CW THz generation by photomixing consists of generating a THz current in a semiconductor device using two heterodyned laser beams of photon energies $\hbar(\omega_0 \pm \omega_{THz}/2)$ (with the same power, $P_L/2$, and polarization), differing in photon frequency by the THz frequency ω_{THz} (where $\hbar$ is the reduced Planck constant and ω_0 the central frequency). In a first step, the heterodyned laser signal is absorbed on typical length

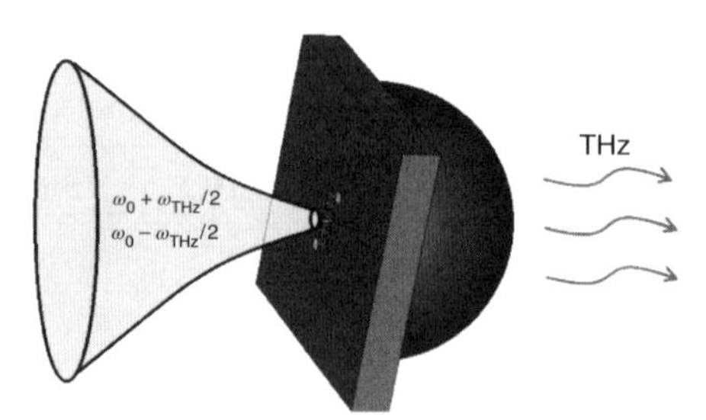

Figure 3.30 Schematic of the CW THz photomixing antenna emitter. Two optical lasers impinge in the photomixer device which is integrated with an antenna over a GaAs substrate. A silicon lens is included to avoid substrate reflections and increase directivity.

scales shorter than 1 μm, that is, much shorter than the THz wavelength. In the second step the resulting photocurrent is fed into an antenna, which then emits THz radiation [14]. A typical schematic of the CW THz generation by photomixing device, can be seen in Figure 3.30.

An interdigitated low temperature (LT)-GaAs-based photomixer is used as the nonlinear device to generate the THz signal. It consists of eight interdigitated fingers, with a finger gap equal to 1 μm, and finger width of 200 nm. Figure 3.31a shows a SEM (scanning electron microscope) photograph of the device. The small-signal equivalent circuit of the interdigitated photomixer is shown in Figure 3.31b [22], where $C = 2$ fF is the parasitic capacitance from the fringing fields between the 200 nm wide fingers and $G = (10\,\text{k}\Omega)^{-1}$ is the conductance given by I_{DC}/V_{DC}. The antenna admittance ($Y_a(\omega)$) is also included in the equivalent circuit of Figure 3.31b.

With the photomixer parameters mentioned before, the input impedance ($Z_p(\omega) = 1/Y_p(\omega) = 1/(G + j\omega C)$, Figure 3.31b) is found to be $0.57 - j75.8\,\Omega$ at 1.05 THz. Thus, an antenna with an input impedance of $0.57 + j75.8\,\Omega$ at that frequency will maximize Eq. (3.30). This low value of the real part and relatively high value of the imaginary part is very difficult to obtain simultaneously with a simple resonant antenna.

A meander dipole antenna is similar to a dipole in which the radiating arms have been bent to a meander shape in order to reduce the antenna size [26]. The typical geometry of a meander dipole is shown in Figure 3.32a along with its main design parameters. Traditionally, this antenna has been used to reduce the antenna size (antenna miniaturization) and in systems that need a very high value of the input impedance, since this antenna provides higher values of input impedance at its resonance than λ-dipoles or dual-dipoles [26], as shown in (Figure 3.32b).The novelty here stems from using the meander dipole out of its main resonance region [39], since at frequencies lower than the resonant frequency its input

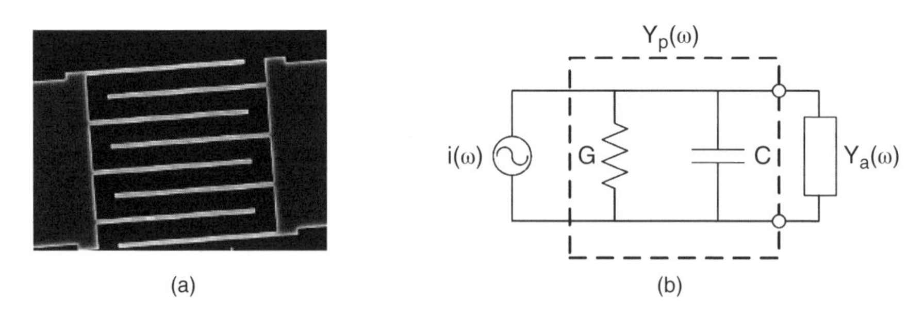

Figure 3.31 (a) SEM photograph of the interdigitated LT-GaAs photomixer and (b) its equivalent circuit.

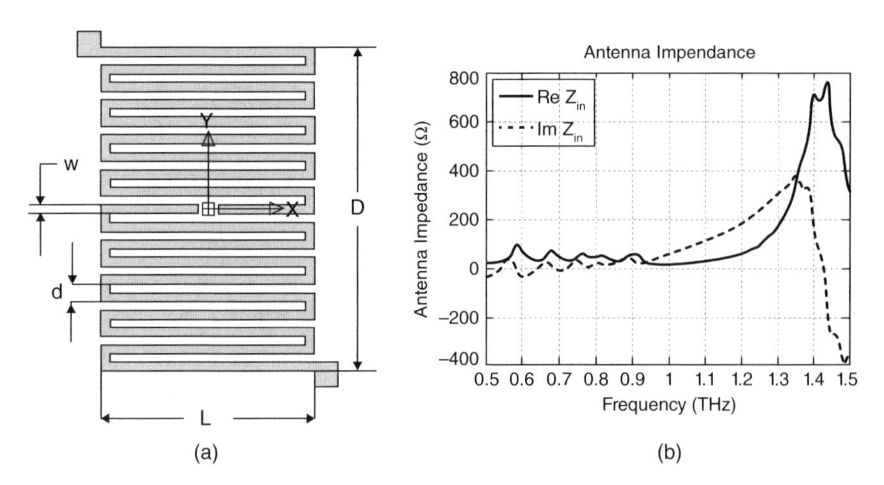

Figure 3.32 Meander dipole antenna. (a) Geometry and main parameters. (b) Simulated input impedance of a meandered dipole antenna over a semi-infinite silicon substrate with parameters: $L = 37\ \mu m$, $w = 3.5\ \mu m$, $d = 2\ \mu m$, $D = 105.5\ \mu m$, and nine bends per arm.

impedance exhibits a low real part and an inductive behavior (as shown in Figure 3.32b in the frequency range from 0.9 to 1.3 THz) suitable for matching it with the photomixer capacitive behavior. With such a design, we intend to achieve a conjugate matching between the antenna and the photomixer without any additional element, thus increasing the mismatching factor. The antenna will show the optimal input impedance to the photomixer, so an integrated active antenna is obtained [38]. According to Refs [38, 39], this design philosophy will improve the system parameters and the figure of merit of the overall transmitter.

3.5.1.1 Parametric Study

In order to analyze the effect of the different parameters on the antenna efficiency, several simulations have been carried out and the results are shown in Figure 3.33. The full-wave electromagnetic simulator CST Microwave Studio was used to carry out the simulations. A silicon semi-infinite substrate was set as the background material in order to reduce computational time [23]. Both the mismatching factor and the radiation efficiency are shown, as well as the antenna efficiency (Eq. (3.31), assuming $\varepsilon_{pol} = 1$).

The design parameters of the meander dipole antenna are: the length (L) and the width (w) of the dipole arms, the separation between arms (d) and the number of bends (see Figure 3.32a). Some conclusions can be extracted from these simulations. The parameter L changes the resonant frequency of the meander antenna: the higher the value of L, the lower the resonant frequency (larger size of the antenna). In addition, the closer the lines of the meander dipole antenna (lower value of d), the lower the radiation efficiency. This is because the currents have opposite directions so they cancel each other [26]. It is important to keep enough space between lines in order to have a relatively high value of the radiation efficiency. Regarding the number of bends, it has been noticed that for a number of bends higher than 3 (total length of the dipole $>2\lambda$), neither the input impedance nor the radiation efficiency change significantly. Finally, w does not have an important role in either of the efficiencies and can be kept constant during the optimization process.

It is important to highlight that the radiation efficiency keeps a relatively high value up to frequencies lower than 2/3 of its main resonance frequency. In all the above simulations, the main resonance is located somewhere close to 1.5 THz (see Figure 3.32b) and our design frequency is 1.05 THz.

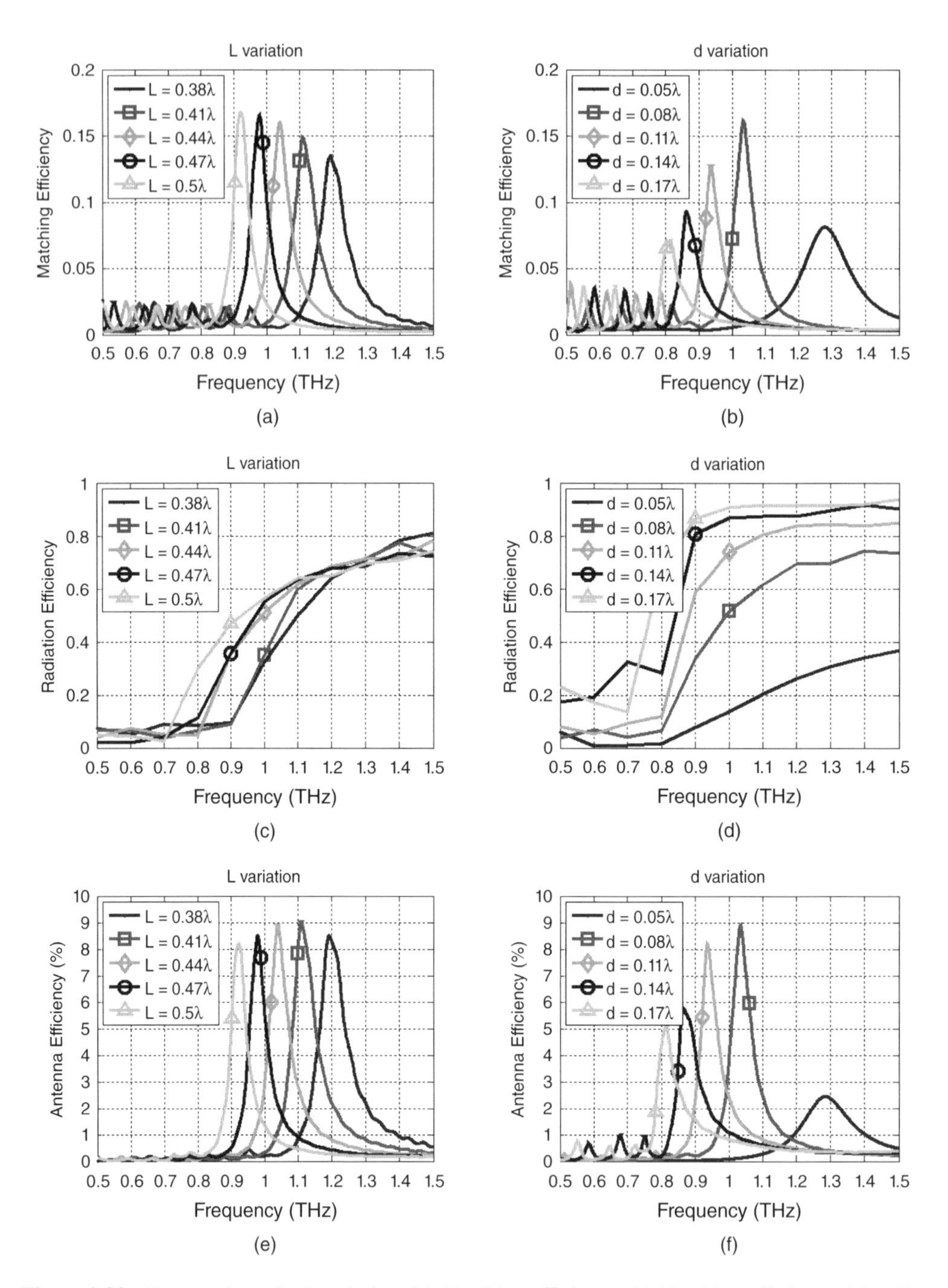

Figure 3.33 Parametric study. L variation. (a) Matching efficiency, (b) Matching efficiency, (c) radiation efficiency, and (d) radiation efficiency, (e) antenna efficiency. $w = 0.044\lambda$, $d = 0.08\lambda$. d variation and (f) antenna efficiency. $w = 0.044\lambda$, $L = 0.44\lambda$.

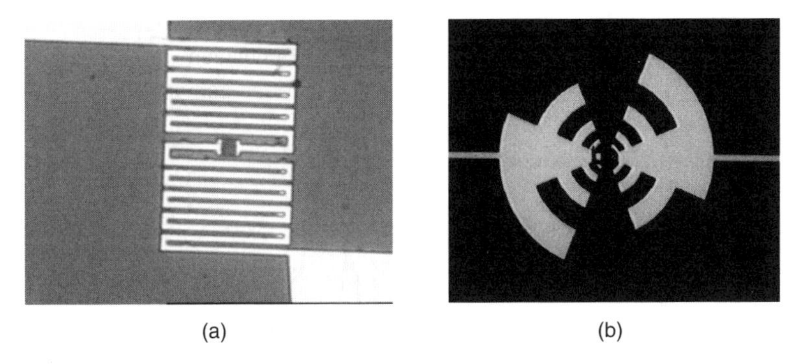

Figure 3.34 Photograph of the manufactured antennas integrated with the LT-GaAs photomixer device. (a) Meander dipole antenna and (b) log-periodic antenna.

3.5.1.2 Prototype at 1.05 THz

A prototype working at 1.05 THz as a local oscillator (LO) for a radioastronomy heterodyne detector [40] was designed and manufactured. The final design parameters are $L = 37\,\mu m$, $w = 3.5\,\mu m$, $d = 2\,\mu m$, $D = 105.5\,\mu m$, and nine bends per arm (Figure 3.34). The separation between the middle arm and the following ones (up or down) was fixed to 3.5 μm to keep enough space for the photomixer device. The simulated input impedance is shown in Figure 3.32b. With such a structure an antenna efficiency of 7.05% at 1.05 THz is estimated, with matching and radiation efficiencies equal to 14.73 and 47.83%, respectively. For the sake of comparison, a self-complementary log-periodic antenna [1] was designed and its antenna efficiency calculated. Its maximum and minimum dimensions were 200 and 10 μm, respectively. In contrast with the meander dipole design, broadband antennas exhibit an almost constant impedance with frequency (self-complementary antennas: $60\pi/\sqrt{\varepsilon_r}$ for the real part and 0 for the imaginary part [1]), so little improvement can be achieved in the matching efficiency. In this second AE, an antenna efficiency of 1.73% is estimated with matching efficiency equal to 2.01% and radiation efficiency equal to 86.03%. Figure 3.35 shows the simulated power improvement obtained with the meander dipole AE given by $\varepsilon_{ant}^{meander}/\varepsilon_{ant}^{log-per}$. A 6 dB improvement at 1.05 THz is expected with the meander dipole antenna.

A typical CW THz photomixing measurements set-up was utilized to characterize the emitted power. It consists of two 850 nm lasers that illuminate the photomixer device. The THz frequency is tuned and the THz emitted power is measured from 80 GHz up to 2 THz. The photomixer output THz frequency depends on the frequency of the two lasers impinging over it. In the measurements shown in Figure 3.36, the frequency of one of the lasers remains constant while the other is changed to obtain the desired output THz frequency. This frequency change can be done either by modifying the current or the temperature of the laser. The THz radiation emitted by the antenna with the lens is first chopped with an optical chopper and then beam focused with two parabolic mirrors. The THz signal impinges a diamond window Golay cell connected to a lock-in amplifier which uses the chopper signal as the reference. Measured output power with both meander and log-periodic AEs are shown in Figure 3.36.

It can be observed that at 1.1 THz a maximum in the emitted power is obtained with the meander AE (0.26 μW). If we compare both power measurements (meander and log-per AE) the plot in Figure 3.35 is obtained. A small shift from 1.05 to 1.1 THz can be appreciated but the power improvement is more or less the same (6 dB). In addition, the simulated quality factor (Q-factor) is 18.33, while the measured one is 10.11. This Q-factor decrease is due to the fact that additional losses not contemplated in the simulations, such as losses in the silicon dielectric, are occurring.

The meander dipole antenna has been presented as a solution to match both the capacitive susceptance and the conductance of photomixers, while maintaining a relatively high value of radiation efficiency. To achieve such a goal, the antenna is forced to work out of its main resonance, at lower frequencies, where it exhibits an inductive behavior.

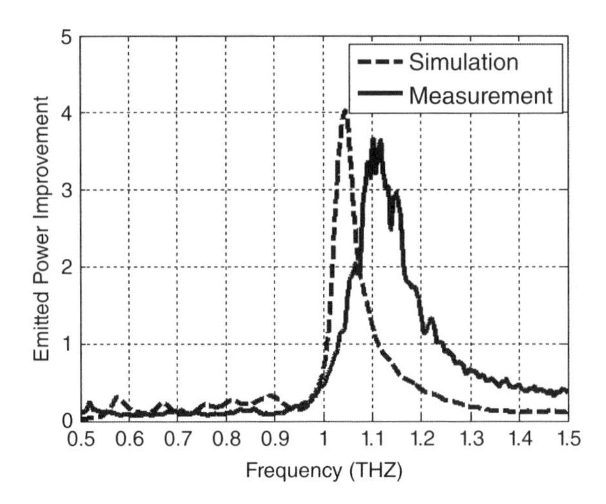

Figure 3.35 Emitted power improvement when comparing meander dipole antenna with log-periodic antenna.

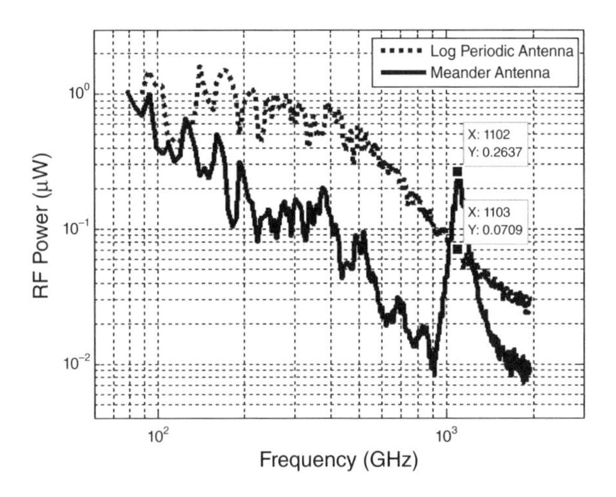

Figure 3.36 Measured THz emitted power of both antenna emitters: meander dipole antenna and log-periodic antenna.

With this approach, there is no need to include a RF choke, so the design of an antenna for CW THz based on photomixing is simplified to just the design of the meander dipole antenna. A joint design process of the antenna is proposed where both matching and radiation efficiencies (AIA concept) are taken into account. The proposed antenna design can be followed independently of the photomixer used and it consists in trying to maximize the radiated power obtained from it.

A prototype working at 1.05 THz was designed, manufactured, and measured, and the obtained results show a 6 dB improvement in THz output power when compared with the log-periodic prototype. This result is comparable with previously reported ones [22] at 1 THz but with the novelty of not using the RF choke.

3.5.2 Tapered Slot Antenna on Electromagnetic Band Gap Structures

The goal of this section is to develop a class of antennas that allows us to get rid of the Si lens, while achieving a high gain and symmetric radiation patterns. Moreover, these antennas are printed on a planar substrate in order to reduce the cost and to facilitate their integration with devices such as photomixers or diodes. One of the best candidates to meet all of these requirements is a tapered slot antenna [41–43]. Tapered slot antennas (TSAs) have been widely used as ultra-wide band radiating elements since they were first introduced in Ref. [41]. Their use as the individual elements in ultra-wide band arrays has been underpinned by their ease of fabrication in either microstrip or coplanar strip technology. The square kilometer array (SKA) project is one of the most recent examples where TSAs have been employed in a dense array for radio astronomy in low frequency bands (200 MHz to 1 GHz [44, 45]). TSAs are also versatile candidates for application as antennas in the low THz band (0.3 to 3 THz). Their use can reduce the effect of the increasing transmission losses and low efficiency in planar antenna systems due to the use of dielectric lenses. Studies of TSAs printed on a substrate have been reported in Refs. [46–48], in which the MoM yielded a comprehensive prediction of the TSAs properties. Other numerical methods, such as FDTD (finite-difference time-domain method) and FEM (finite element method), have also been developed to study the TSAs in a dense array configuration [49–51].

One of the main limitations in the manufacture of these antennas for use in the THz regime is the thin substrate. The goal of this section is to study the possibility of new topologies that allow the construction of TSAs in the THz range. In Refs. [42, 43] guidelines based on experimental results are provided for the design of linear tapered slot antennas (LTSAs), that is, TSA antennas with a linear shape aperture, by employing coplanar-strip technology with a substrate of permittivity ε_r. Moreover, the so-called Yngvesson limit is introduced, which provides an experimental maximum thickness of the dielectric for which the LTSA operates efficiently. If the effective thickness of the substrate is given as $t_{eff} = t(\sqrt{\varepsilon_r} - 1)$, where t is the actual substrate height, the range for acceptable operation in terms of beam symmetry of a TSA has been found experimentally to be $0.005 \leq t_{eff}/\lambda_0 \leq 0.03$. For a substrate thickness above or below these bounds the performance of the LTSA in terms of efficient endfire radiation deteriorates [42, 43]. This limit is especially critical at very high frequencies, especially in the THz regime, where the antenna dimensions are sometimes electrically larger than the Yngvesson limit because of manufacturing considerations [42]. In Refs. [52] and [53] the authors discuss LTSAs suspended on membranes, printed over EBGs (electromagnetic band gaps), as well as novel configurations allowing thicker substrates.

In this section, an application of the previous theory is developed by means of cutting off modes that propagate on a thick substrate too, but with an EBG approximation. The theory for this application can be understood from two perspectives: the reduction of the effective ε_r of the substrate, or the behavior as an EBG of the structure, cutting off the vertical component of the modes on the substrate. It is important to note that these kinds of structures should have their cut-off frequency within the operating frequency of the antenna, so the structures are tuned. The drawback of this solution is the extremely narrow band operation of these kinds of structures, that only operate in a very restrictive range of frequencies when tuned.

The proposed structure consists of several dielectric layers with different permittivities and thicknesses stacked on top of each other. As shown in Figure 3.37, two types of layers with different effective thickness are alternated in a periodic structure. In this figure, there are considered combinations of ε_r and thickness, having therefore a different Z_0 on each layer, that repeat periodically [54, 55]. Several structure types are studied with the aim of characterizing the behavior of the structure. To this end, the reflection coefficient is calculated according to Eq. (3.32).

$$\Gamma = \frac{E^-}{E^+} \tag{3.32}$$

The simulation was performed with CST Microwave Studio and the electric field results in the illumination of the structure with a plane wave, as illustrated in Figure 3.37. The electric field probes (one on each axis) are located in the Z^+ direction, in order to compute the reflection coefficient of the structure [54, 55].

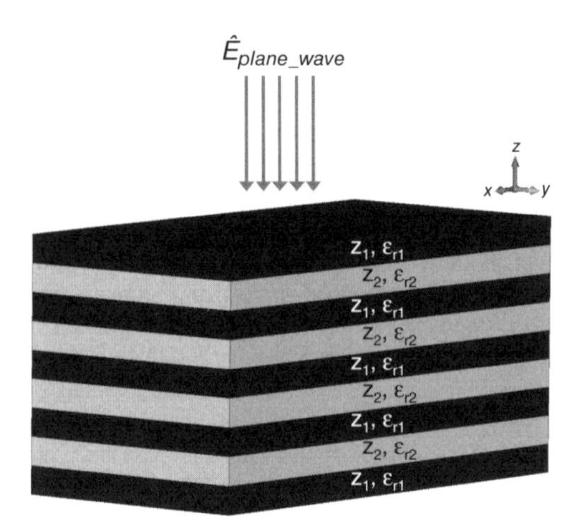

Figure 3.37 Example of an EBG structure.

The higher order modes with a strong transversal component are those responsible for the non-endfire radiation of the TSA antenna. With this EBG, these components are eliminated, having no modes with this field configuration. The transversal direction agrees with the direction where the EBG is composed ($\hat{z}$). This component is suppressed, so all the power is concentrated again on the fundamental mode, allowing end-fire radiation.

Different parametric studies have been carried out to predict the behavior of the structure as an EBG, especially to determine accurately its working frequency band. The design parameters are Z_0, which can be tuned by changing the thickness, and ε_r of each of the layers, and the number of periods. It is expected that the narrowband properties of the structure will be strengthened by the increasing periodicity [56], so the selectiveness of the resulting filter is higher and also with more resonances within the band, while the main operating frequencies will be set by a trade-off between the Z_0 of each layer and the number of periods.

Figure 3.38 shows an example with a period of six sequences, that is, seven layers with $\varepsilon_r = 12.9$ and six layers with $\varepsilon_r = 2.2$. For these conditions, an additional resonance appears compared to a three period structure (if we focus on the same band as for the three period case) and the operation bands are reduced in bandwidth: the first is between 65 and 80 GHz and the second between 95 and 105 GHz.

These simulations are an example of how the reflection coefficient behaves with frequency and with the variation of the layer thickness. The selection of these two substrates is due to the ease in finding a feed for the TSA manufactured in a very high permittivity substrate and to the several Teflon substrates available for microwave applications. These materials can be replaced by similar ones with analogous properties.

Another tuning parameter is the variation of the thickness ratio between the high and low permittivity substrate. In the previous example, this ratio was 1 : 1, but it can be modified to achieve the desired impedance on each layer and the reflection coefficient on the bands. All these results are obtained for frequencies centered at 80 GHz, when the thickness of each of the layers is comparable to the wavelength, so the results can be applied to practical cases, as described in the next section.

Since the objective of this section is to design a TSA on a substrate with a higher overall dielectric thickness, the EBG structure from previous studies will be employed as the substrate for several TSAs, including array configurations and different shapes of the antenna to compare performances.

For manufacturing purposes, the chosen working frequencies are in the (60–100) GHz band. Therefore, simulations are also carried out within this frequency band and the antenna is fed with a 50 Ω coplanar stripline (CPS).

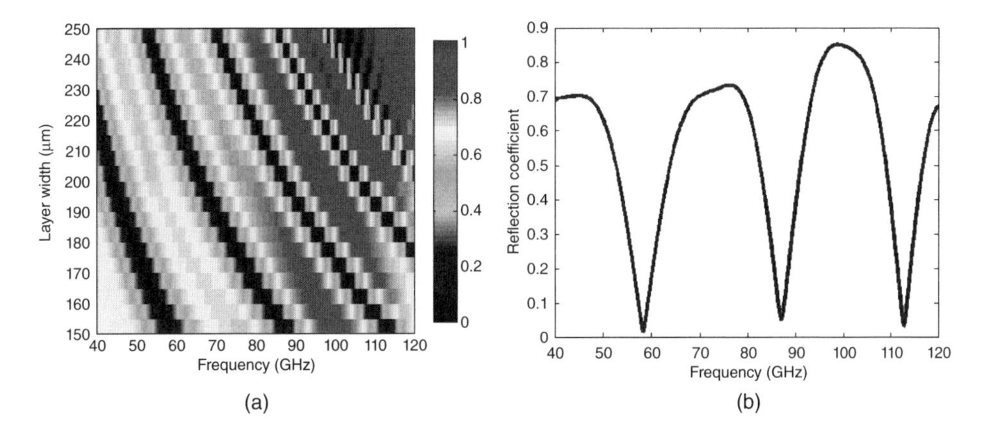

Figure 3.38 Reflection coefficients for six periods of stacking layers with $\varepsilon_{r1} = 12.9$ and $\varepsilon_{r2} = 2.2$ in terms of frequency and substrate width. See plate section for color representation of this figure.

Two types of TSA are used: Vivaldi (with an exponential profile of the slot) and LTSA. The substrates chosen to accommodate the selected frequencies are Rogers 3010 and Rogers 4450F. These substrates are 0.127 and 0.202 mm thick and have $\varepsilon_{r1} = 10.2$ and $\varepsilon_{r2} = 3.52$, respectively. These laminates can be found on the market and are easy to process. Besides, the characteristics of Rogers 4450F (Prepeg) allows the stacking and gluing of all the layers. The antennas were printed on copper on the outer layers of the dielectric.

The reflection coefficient for this structure between 60 and 100 GHz is shown in Figure 3.39. The operating frequencies for this particular stacking are between 65 and 85 GHz where the reflection coefficient is 1 or close to this value (at least more than 0.5). If the frequencies are such that $\Gamma > 0.6$, the obtained bands are 62–86 and 108–120 GHz. Beyond 120 GHz the antenna is not simulated due to the upper limit bandwidth of the Schottky diode.

According to previous work [42], the aperture of a TSA should not be larger than λ_{eff}(λ_{eff} being the effective wavelength on the substrate) to obtain the best performances in terms of directivity and symmetrization of the beam. Moreover, the length of the TSA should not be larger than $12\lambda_0$, where λ_0 is the free-space wavelength. Several designs have been simulated to obtain the best performance in the (62–86) GHz band. To calculate λ_{eff} on the stacked layer substrate, ε_{eff} and t_{eff} are also calculated, taking into account both substrates as in Eqs. (3.33) and (3.34) [42], respectively. For the stacking used in this work, the resulting relative permittivity is $\varepsilon_{\text{eff}} = 6.58$. If one simulates the same antenna on a substrate with the same width and the equivalent ε_r, it does not radiate in any of the bands under study. For conciseness the simulations are not shown here.

$$\varepsilon_{\text{eff}} = \varepsilon_{r1}\frac{4 \times \text{thickness}_1}{\text{total thickness}} + \varepsilon_{r2}\frac{3 \times \text{thickness}_2}{\text{total thickness}} \tag{3.33}$$

$$t_{\text{eff}} = \left(\sqrt{\varepsilon_r} - 1\right) t/\lambda_0 \tag{3.34}$$

As the final implementation of the prototypes is based on a homodyne scheme with a Schottky detector, a discrete port is used to feed the CPS in the simulations. The input impedance of the line is designed to obtain 50 Ω on the high permittivity substrate. The manufacturing method employed is micromachining.

Going further with the design and taking advantage of the isolation and component suppression of the EBG, two TSA antennas can be printed on the substrate, one on each side. In this way, an array structure can be implemented in the same device and the size of the element would be lower. This configuration is implemented in a linear array, having therefore a $2 \times N$ array device. The behavior of this structure is studied and compared with the single element in the following.

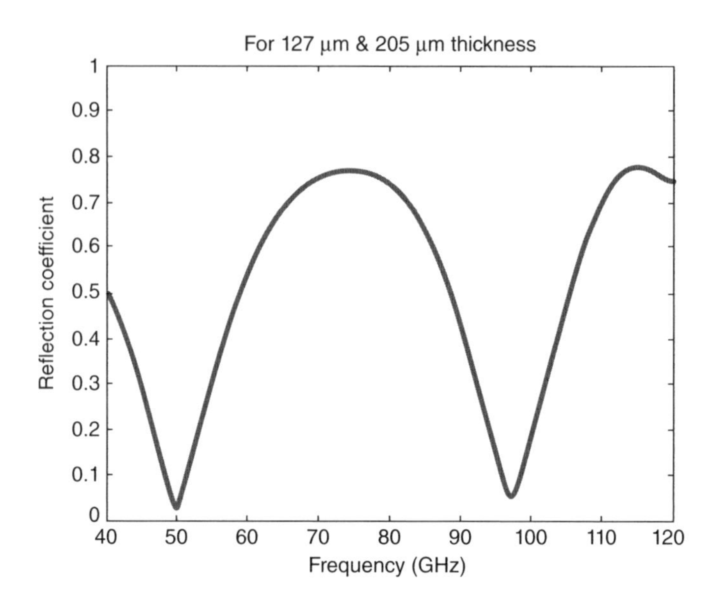

Figure 3.39 Reflection coefficient for the EBG of seven layers (three periods).

3.5.2.1 Single Element One-Sided

In order to illustrate the performance of the structure, a LTSA has been designed and printed on a stack made of several layers of the above-mentioned high frequency commercial dielectrics, from 50 to 100 GHz. The aperture of the LTSA is $0.8\lambda_0$ at 90 GHz and the antenna is $3.38\lambda_0$ long. The simulations have been carried out with CST Microwave studio and the fields obtained in the aperture are shown in Figure 3.40, where the EBG is suppressing the vertical component or not, respectively.

The radiation patterns of the structure in Figure 3.41a at 70 and 90 GHz are shown in Figure 3.41b,c, respectively. The radiation pattern at 70 GHz is end-fire with a single main lobe, whereas at 90 GHz the radiation is no longer end-fire. The dimensions of the EBG are within the limits of the typical design parameters, but not the overall effective thickness of the substrate which is well beyond the limits predicted in Ref. [42] for the resulting ε_{eff}.

The first attempt to implement the described structure consists in a single antenna over the EBG studied above. Several prototypes have been manufactured with the aim of validating the simulations made for both linear and Vivaldi TSAs. This element will be the baseline for all other devices composed of more than one antenna, in an array configuration.

The same Schottky diode as mentioned before is employed. The feeding region of the antenna must be designed in order to accommodate the diode and to obtain the conjugate input impedance, reducing the mismatch as much as possible. The most optimal feeding of the TSA antenna is the CPS when functioning with a diode of around 50 Ω impedance. The minimum spacing between the lines is fixed by the diode dimensions of 265 μm to be able to bond the terminals to both of the lines. As studied by Simons [57], the lower the gap between the lines, the higher the impedance. There is also a manufacturing limitation of 100 μm for the thickness of the copper lines. With these constraints and bearing in mind the diode's impedance, the resulting CPS line is depicted in Figure 3.42a.

As a transition between the TSA antenna and the CPS line, a curve based on a tangent function is used, due to the need for a smooth transition to both conductors of the antenna. The curve equation is defined in Eq. (3.35), where C and the smooth factor are constants based on the transition smoothness and offset

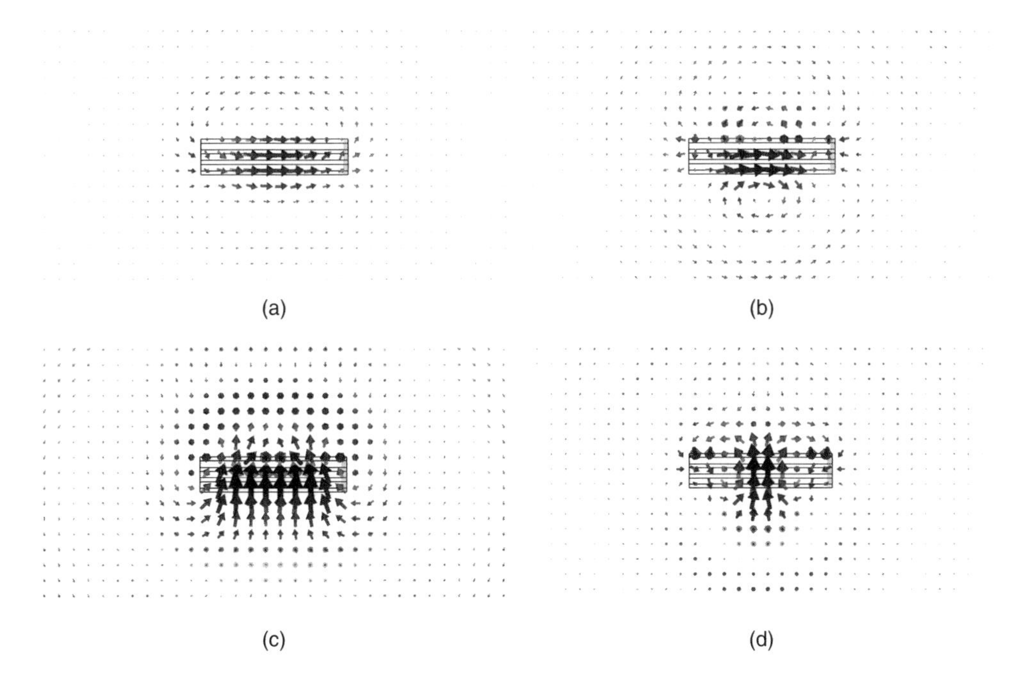

Figure 3.40 Fields on the aperture of a LTSA printed over an EBG. 70 GHz. (a) E-field and (b) E-field. (c) H-field. 90 GHz and (d) H-field.

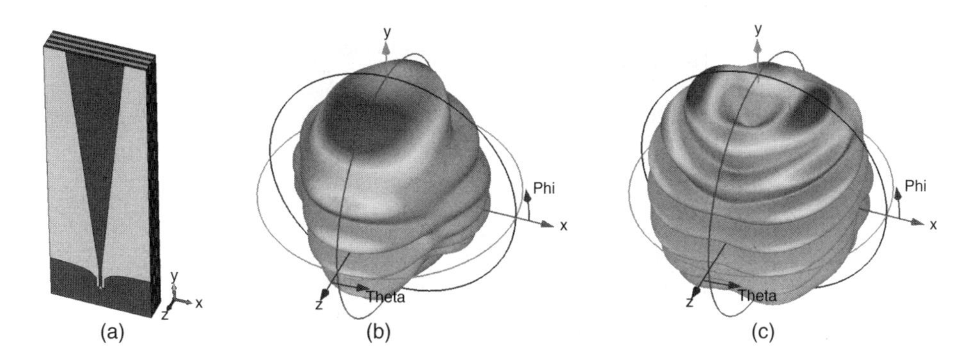

Figure 3.41 (a) Simulated structure and far-field at (b) 70 GHz and (c) 90 GHz.

is related to the antenna total width.

$$Y = C \cdot \tan^{-1}\left(\frac{X - \text{offset}}{\text{smooth factor}}\right) - \text{offset} \tag{3.35}$$

For the single element, some measurements have been performed to validate the simulations. First, the change in the shape of the antenna strongly influences the performance of the EBG. The end-fire radiation is only achieved at some of the working frequencies of the EBG. The range employed for the TSA length is between $2\lambda_c$ and $4\lambda_c$, while the aperture is between $0.7\lambda_c$ and $1.2\lambda_c$, where $\lambda_c = c/90$ GHz.

The simulated input impedance of the antenna in Figure 3.41a is shown in Figure 3.42b with a discrete port simulating the Schottky diode employed as excitation at the beginning of the CPS line. This antenna

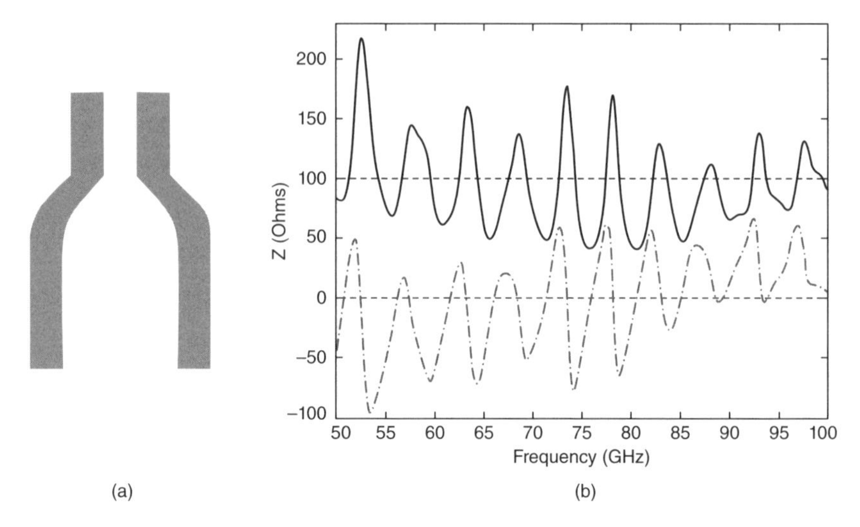

Figure 3.42 (a) Feeding CPS line transition for TSA antennas and (b) real (solid) and imaginary (dashed) part of input impedance.

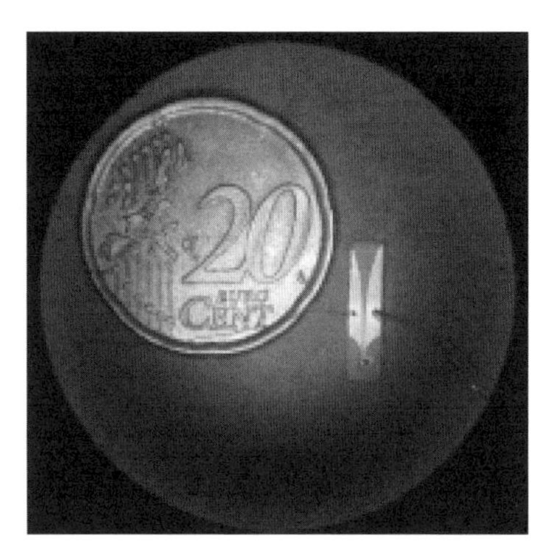

Figure 3.43 Manufactured Vivaldi over an EBG at 80 GHz.

is the same as for the fields in Figure 3.40. The real part of the impedance is oscillating around 100 Ω while the imaginary part is centered at 0. This is in accordance with the characteristic impedance of traveling wave and ultra-wide band antennas [58].

Some examples of the simulated radiation patterns are shown in Figure 3.41 for the antenna with λ_0 aperture and $3.5\lambda_0$ length, with λ_0 being the free-space wavelength at 100 GHz. Figure 3.43 shows a picture of one of the manufactured antennas, in this case a Vivaldi TSA with an exponential profile, and a reference to compare the size. The two wires connected to each of the arms of the antenna are the contacts for the DC bias of the Schottky diode.

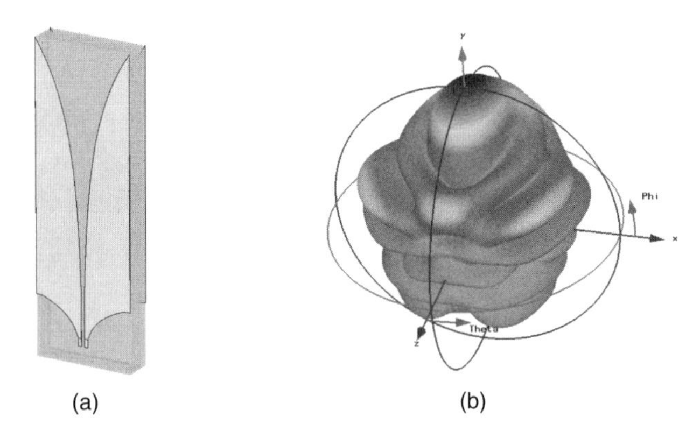

Figure 3.44 (a) Example of a double Vivaldi antenna over an EBG substrate and (b) simulated far-field 70 GHz.

3.5.2.2 Single Element Double-Sided

The next design consists of two antennas printed on each side of a dielectric EBG stacked structure. The antennas are both excited with the Schottky diode employed in the previous section. In the measurement set-up, only one of the antennas is in receiving configuration, while the other is passive and not biased, connected to the equivalent resistance of the diode at the terminals of the antenna. Thus, it behaves as a resistance that terminates the CPS in the TSA antenna. The same EBG composite structure used for the single antenna has been employed in this simulation. Figure 3.44a shows the geometry of a Vivaldi element, whereas Figure 3.44b presents the radiation pattern obtained at 80 GHz. The pattern obtained biasing the diode of the other face is identical, since the structure is symmetrical. There is a small asymmetry on the pattern regarding the z-axis, the EBG is acting as a high order mode suppressor and the properties for wave propagation in the EBG are not the same as in free-space.

3.5.2.3 Linear Array

In order to increase the radiated power in the THz range, the structure in Section 3.5.2.1 is arranged in an array configuration. In this experiment, the antennas are working with a homodyne detection scheme. Nevertheless, the antenna concept of a TSA array with an EBG can be checked due to the reciprocity theorem of antennas.

The first arrayed configuration is the single-sided system in Section 3.5.2.1. For this test Vivaldi and linear TSAs have also been considered for simulation, measurements, and manufactured prototypes. The structure, shown in Figure 3.45a, consists of three elements arranged in a linear array and printed on the EBG substrate. Figure 3.45b,c shows the radiation patterns for one of the corner elements and the central element in the 1×3 array. The elements of the array are connected (side by side). The beam of the corner element is tilted toward the direction of the array center, which also happens in a classical array configuration with conventional substrates. This effect is due to the strong mutual coupling [49]. The elements are excited one by one, therefore only one of the antennas is receiving, while the other two are loaded with the lumped port impedance (50 Ω). The simulations are shown at 80 GHz, since there is a shift in frequency with respect to the single element, as usually observed in this type of composition [49].

Two pictures of arrays are shown in Figure 3.46 with a reference for size comparison, for five and nine elements systems.

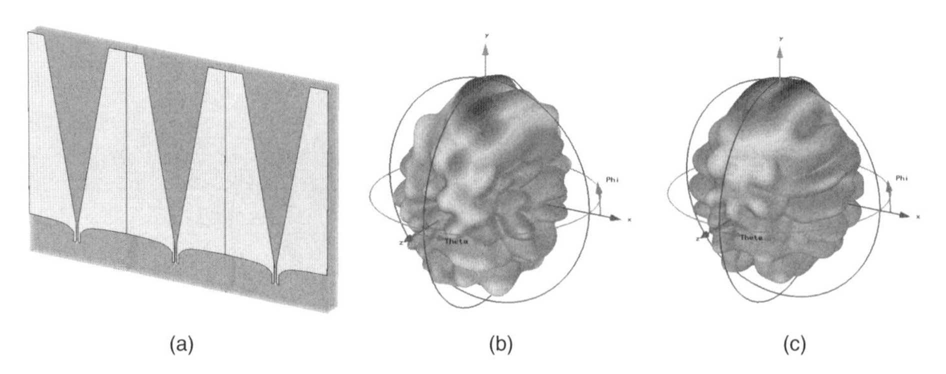

Figure 3.45 1 × 3 array of LTSAs and radiation patterns at 80 GHz: (a) structure simulated, (b) corner element, and (c) central element.

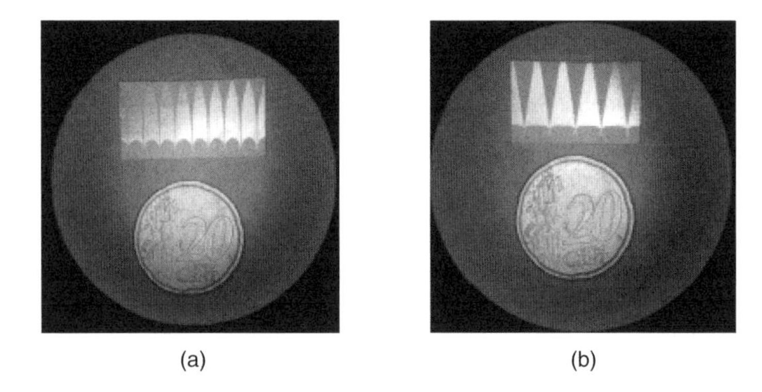

Figure 3.46 Manufactured arrays at 80 GHz (a) Vivaldi and (b) LTSA.

3.5.2.4 Measurements

The measurements in this section are intended to validate the EBG substrate designed before. Two antennas have been measured, one single- and one double-sided. In such a way, the performance of the EBG substrate can be checked as well as the isolation between the two antennas in the double-sided configuration. The feed for the elements is a GaAs Beam Schottky diode MA4E2037 in a homodyne receiving configuration. With this device, it is possible to measure the radiation pattern of the structure, checking therefore the suitability of the EBG for these types of antennas. The diode is biased with 0.6 V, and connected to a lock-in amplifier. The measurement set-up Figure 3.47b and polarizing circuit scheme for this experiment is illustrated in Figure 3.47a, where the lock-in amplifier is connected in parallel to the diode.

Two different antennas have been measured. The first is an isolated, single sided antenna with a width of 3.15 mm and length of 15.10 mm, that corresponds to Figure 3.48a. The two curves represent the E-plane for two different frequencies 70 and 75 GHz. The difference in the level as well as the noise in the measurement is due to the measurement set-up in an optical table more suitable for higher frequencies. A broadside radiation, without deformation of the beam is obtained, as was the intention to demonstrate according to the previous theory.

The second antenna measured is the double-sided Vivaldi, with a width of 3.92 mm and length of 13.07 mm, corresponding to Figure 3.48b. For this case only one measurement on the E-plane at 80 GHz was taken due to the time limitation of the anechoic chamber. There is a maximum in the broadside direction, which demonstrates once more the fundamental mode operation of the TSA.

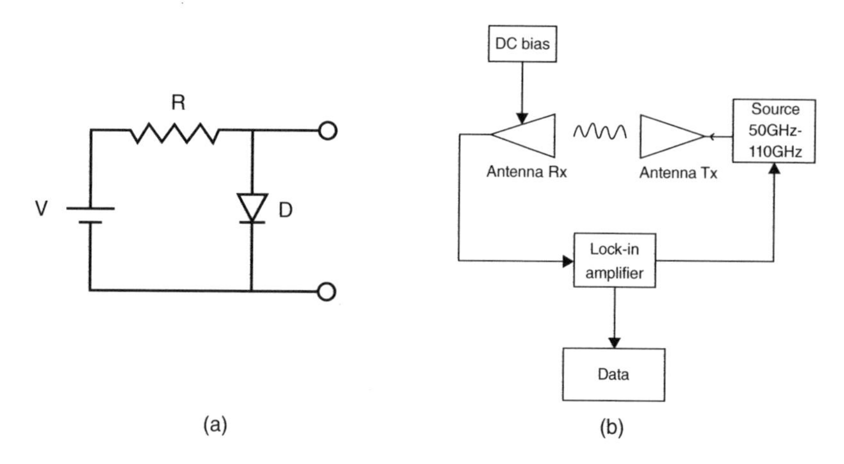

Figure 3.47 (a) Polarizing circuit and (b) set-up for pattern measurement.

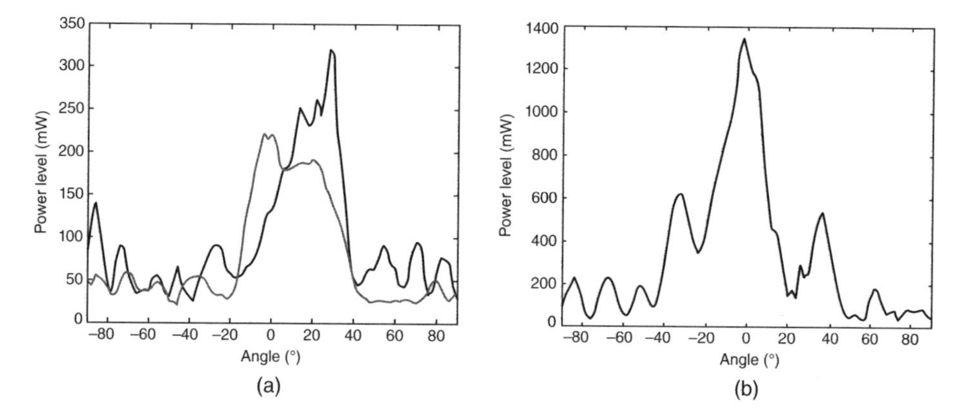

Figure 3.48 Measurements in E-plane for EBG antennas: (a) single sided at 70 GHz (Maximum level around 300 mW) and 75 GHz (Maximum level around 200 mW). (b) Double sided at 80 GHz.

3.6 Arrays

Some novel and promising advances are presented in this section. First, a short introduction to array theory is presented, then some state-of-the-art experiments which exploit an array configuration for increasing the THz emitted power and/or the sensibility and the beam-steering properties. Finally, as a direct application, the LAE concept is presented as a promising photomixing technique.

3.6.1 *General Overview and Spectral Features of Arrays*

In order to overcome the power limitations in THz AEs, the use of (planar) arrays of AEs has been proposed as a promising approach. In addition, beam-steering capabilities can be achieved by means of antenna arrays. In this section, basic array theory will be presented and an example of an array of AEs working in the THz range is shown.

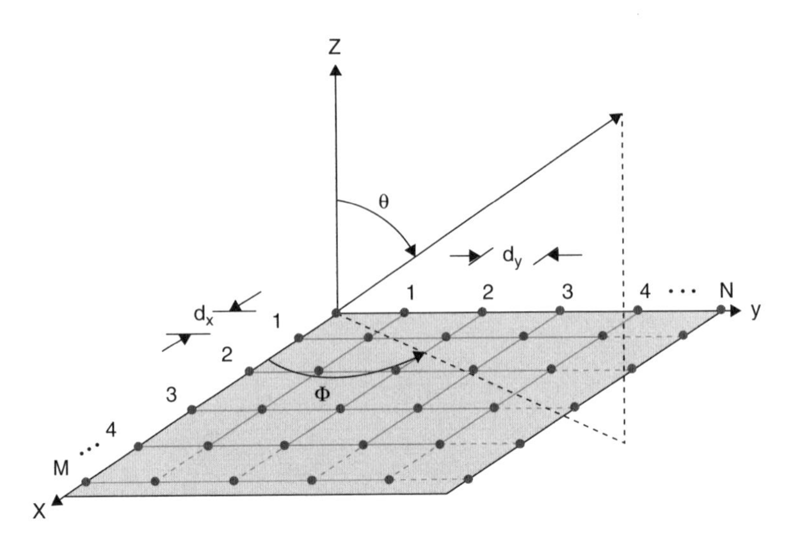

Figure 3.49 Geometry of a planar array.

The radiation pattern of a planar rectangular array of antennas (Figure 3.49) can be obtained with the radiation pattern of an isolated element and the so-called array factor (AF):

$$\mathrm{AF}(\theta, \varphi) = \sum_{m=1}^{M} \sum_{n=1}^{N} I_{mn} \cdot \mathrm{e}^{\mathrm{j}(m-1)(k \cdot d_x \cdot \sin\theta \cdot \cos\varphi + \beta_x)} \cdot \mathrm{e}^{\mathrm{j}(n-1)(k \cdot d_y \cdot \sin\theta \cdot \cos\varphi + \beta_y)} \tag{3.36}$$

$$\boldsymbol{E}_{\mathrm{array}}(\theta, \varphi) = \boldsymbol{E}_{\mathrm{isolated}}(\theta, \varphi) \cdot \mathrm{AF}(\theta, \varphi) \tag{3.37}$$

where θ and φ are the spherical coordinates where the radiation pattern is calculated, k is the wavenumber, I_{mn} is the amplitude of the current of element at the position $n \times m$, d_x is the separation between elements in the x-direction, d_y is the separation between elements in the y-direction, and β_x and β_y are the phase difference in the x- and y-directions between feeding currents in adjacent elements. With the amplitude distribution (uniform, triangular, binomial, etc.) of the feeding currents one can control the SLL, whereas with the phase of these currents chosen as in Eq. (3.38) the direction of the main lobe is fixed at (θ_0, φ_0):

$$\begin{aligned} \beta_x &= -kd_x \sin\theta_0 \cos\varphi_0 \\ \beta_y &= -kd_y \sin\theta_0 \sin\varphi_0 \end{aligned} \tag{3.38}$$

Changing the phase of the array elements, the main lobe direction can be changed and beam-steering capabilities can be obtained.

Photomixers are ideally suited for coherently radiating arrays. The power of the two lasers used for the generation of the mixing input signal can be coherently amplified and distributed to an (in principle) unlimited number of array elements by using fiber amplifiers and beam splitters (including the possibility of beam steering, which was discussed elsewhere [59, 60]).

First, we consider a planar array of AEs positioned at the points $\vec{l} = \vec{n}a = (n_x\hat{x}, n_y\hat{y})a$, with n_x, $n_y = 0, \pm 1, \ldots \pm n_{\max}$ (or n_x, $n_y = \pm 1/2, \ldots \pm (1/2 + n_{\max})$) and a pitch, a, in the lattice.

The radiation intensity pattern $U^{\mathrm{AEA}}(\theta, \varphi)$ of the antenna emitter array (AEA) is simply obtained by multiplying the radiation intensity of a single AE, $U_0^{\mathrm{AE}}(\theta, \varphi)$, by the square of the array factor:

$$U^{\mathrm{AEA}}(\theta, \varphi) = U_0^{\mathrm{AE}}(\theta, \varphi) |\mathrm{AF}(\theta, \varphi)|^2 \tag{3.39}$$

The array factor in Eq. (3.39) is defined, for elements radiating with the same phase and intensity, as

$$\mathrm{AF}(\theta,\varphi) = \sum_{n} \exp(i\vec{k}\circ\vec{l}) \quad (3.40)$$

where $\vec{k} = (2\pi n_{sc}/\lambda_0)(\cos\varphi\sin\theta, \sin\varphi\sin\theta, \cos\theta)$ is the THz wave vector in the semiconductor, with n_{sc} being the refractive index of the semiconductor. The intensity normal to the surface ($\theta = 0$) is then increased by a factor N^2, where $N = n^2$ for an $n \times n$ array. The pitch a of the array has to be of the order of or larger than $\lambda_0/2n_{sc}$, since the size of the individual antennas is, at least, in this range. Interference effects are, therefore, strongly affecting the intensity at angles θ for which $\vec{k} \circ \vec{l}_{max} = (2\pi a n_{max} n_{sc}/\lambda_0)\sin\theta$ becomes comparable with $\theta \sim \pi/2$. As a consequence, the radiation pattern becomes more directional with increasing n_{max} and increasing a/λ. It has to be noticed that antenna crosstalk (mutual effects) is also taken into account. At these frequencies and with these distances between antennas, the mutual coupling is extremely low. However, the devices also have to be electrically connected with bias lines resulting in electrical crosstalk through the bias lines. This can affect the radiation pattern of the antennas but this effect is not critical for the general results shown in this chapter.

Figure 3.50a shows the radiation intensity patterns $U^{AEA}(\theta,\varphi)$ in the E-plane and H-plane for 3×3, 5×5, and 7×7 AEAs along with the pattern of an isolated dipole. The patterns have been normalized with respect to the directivity of $U_0^{AE}(\theta = 0,\varphi)$. In this case, both the radiation into the substrate and the radiation into the air are plotted. The pitch of the arrays is $a = \lambda_0/(2n_{sc}) = 42\ \mu m$. The authors are well aware that this is extremely difficult to realize experimentally. The only way would require the use of microlenses. Even doing so, there would be too many problems for the alignment of the set-up [61]. However, the purpose of this section is to show the potential improvements or drawbacks for different topologies. The maximum radiation intensity $U_0^{AE}(\theta = 0,\varphi)$ is increased by a factor 9^2, 25^2, and 49^2, corresponding to 19, 28, and 34 dB, respectively, as expected. At the same time, the width of the lobes decreases. Nulls of the intensity pattern in an $n \times n$ array occur for those angles for which the path difference between the waves originating from neighboring elements is $(\lambda_0/n_{sc})/n$ or a multiple of it. For a pitch a this implies that the zeros for an $n \times n$ array are found at:

$$\theta^{m}_{nxn,0} = \arcsin\left[\left(\frac{m}{n}\right)\left(\frac{\lambda_0}{a n_{sc}}\right)\right],$$
$$(m = 1, 2, \ldots, n-1, n+1, \ldots, 2n-1, 2n+1, \ldots), \quad (3.41)$$

whereas main maxima occur at

$$\theta^{m}_{n\times n,\max} = \arcsin\left[m\left(\frac{\lambda_0}{a n_{sc}}\right)\right],$$
$$\left(m = 1, 2, \ldots, \frac{\mathrm{int}\left(a n_{sc}\right)}{\lambda_0}\right) \quad (3.42)$$

Specifically, for $a = \lambda_0/2n_{sc}$ this yields for the angles $\theta_{3\times 3,0} = \arcsin(2/3) = 41.8°$, $\theta_{5\times 5,0} = \arcsin(2/5) = 23.6°$, and $\arcsin(4/5) = 53.1°$, and $\theta_{7\times 7,0} = \arcsin(2/7) = 16.6°$, $\arcsin(4/7) = 34.8°$, and $\arcsin(6/7) = 59.0°$. These results confirm the expected increase of directivity with the size of the array. Figure 3.50b shows the radiation intensity patterns in the planes $U_{2mm}{}^{AEA}(\theta,\varphi = 0)$, $U_{2mm}{}^{AEA}(\theta,\varphi = \pi/2)$ for 3×3, 5×5, and 7×7 AEAs, composed of the same AEs but with one lens of radius $R_l = 2$ mm. Again, the patterns have been normalized to the directivity of $U_0^{AE}(\theta = 0,\varphi)$ and the radiation intensity of the isolated dipole ($U_0^{AE}(\theta,\varphi = 0)$, $U_0^{AE}(\theta,\varphi = \pi/2)$) is also presented. Figure 3.50c depicts the same results for a lens with radius $R_l = 5$ mm.

Comparing the results with a lens to those without we notice a drastic additional increase in radiation intensity at $\theta = 0$. The increase is more pronounced for the smallest (3×3) array and the larger lens. For the larger arrays it is significant only for the larger lens. Nevertheless, in all the cases the increase in

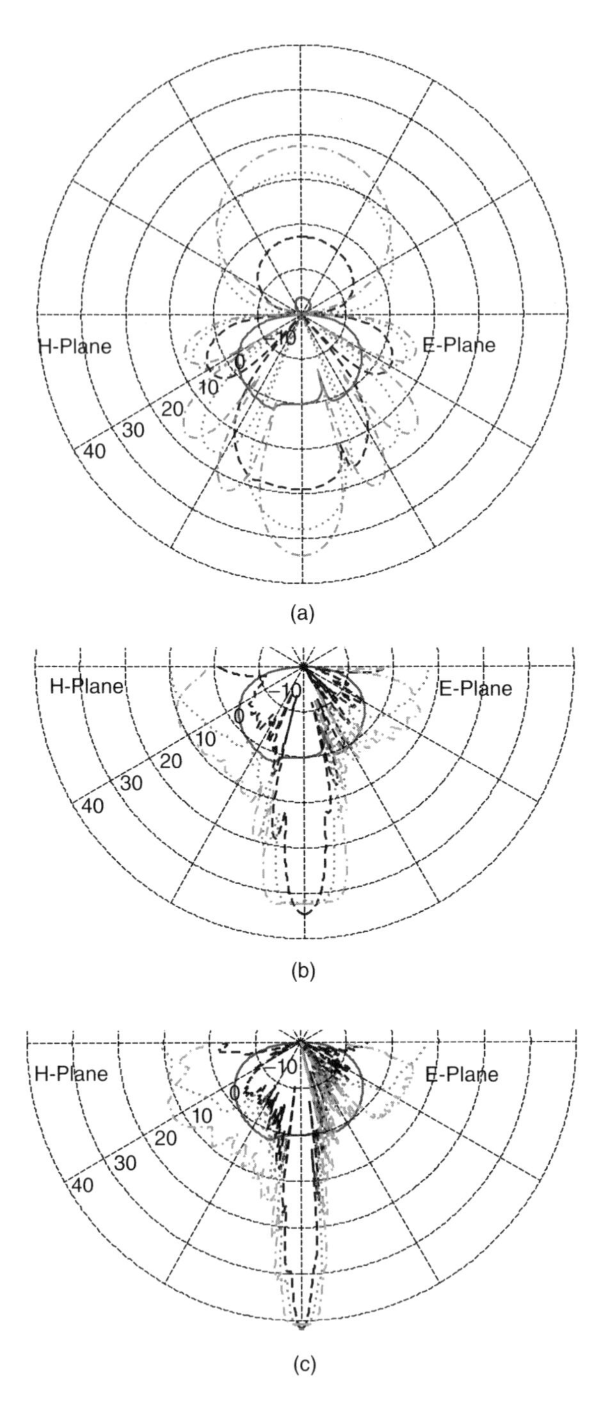

Figure 3.50 Radiation intensity patterns of a 3 × 3 array of isolated dipoles (--), a 5 × 5 array of isolated dipoles (··), and a 7 × 7 array of isolated dipoles (-··-) with pitch $a = \lambda_0/(2n_{sc})$. (a) Arrays without lens, (b) with a 2 mm radius and 0.69 mm slab hyperhemispherical lens, and (c) with a 5 mm radius and 1.725 mm slab hyperhemispherical lens. For the sake of comparison also the radiation intensity pattern for the single AE is depicted (-). See plate section for color representation of this figure.

radiation intensity is less than the naive sum of the decibel increases due to the lens and the array. On the other hand, we also observe a narrowing of the central lobe, which, however, is much less pronounced than the narrowing achieved by the lens for the single AE. This will be explained below. In summary, the results obtained for arrays without a lens appear very appealing, as an increase in radiation intensity of 34 dB can be achieved for a 7×7 AEA due to the lens focusing capability. The additional advantage of a lens becomes marginal for large arrays, even if the lens is large. Only 10 dB or less can be gained with a 7×7 array with a lens of 5 mm radius compared to either a 7×7 array without lens or a single AE with 5 mm lens. A major advantage of the lens, however, is the possibility of extracting radiation emitted under an angle exceeding the critical angle for total reflection $\theta_c \approx 16°$. Although the radiation pattern of the arrays is already rather narrow, it helps collecting a larger fraction of the emitted THz radiation.

There is, however, a problem regarding the implementation of arrays with small pitch $a = \lambda_0/(2n_{sc}) \approx 40\,\mu m$, as assumed so far. Although there are problems with designing and fabricating arrays of resonant $\lambda/2$-dipole antennas in combination with relatively large devices (of dimensions $> 10 \times 10\,\mu m^2$!) and with connecting them to a DC bias, the main problem is the coherent illumination of such arrays. Illumination by free space optics requires very sophisticated refractive or diffractive components and, in addition, extremely demanding alignment efforts [61]. A more promising approach is based on a fiber array with the same pitch as the AE array as they can be relatively easily aligned. The pitch of such an array, however, is limited by the diameter of the core of the fiber to about $a > 150\,\mu m$. Therefore, we present in Figure 3.51a the results corresponding to Figure 3.50a but with the pitch increased from $a = \lambda/2n_{sc} \approx 42\,\mu m$ to $a = 4\lambda/2n_{sc} \approx 168\,\mu m$. For the arrays without a lens (Figure 3.51a) the intensity pattern now exhibits $2n$ lobes with zeros at the angles $\theta^{(m)}_{n\times n,0} = \arcsin(m/2n)$ $(m = 1, 2, \ldots, n-1, n+1, n+2, \ldots, 2n-1)$ and main-maxima at $\theta^{n=m}_{n\times n,max} = \arcsin(1/2) = 30°$. Therefore, zeros for the 3×3 array occur at the angles $\theta^{(1)}_{3\times3,0} = \arcsin(1/6) = 9.6°$, $\theta^{(2)}_{3\times3,0} = \arcsin(2/6) = 19.5°$, $\theta^{(4)}_{3\times3,0} = \arcsin(4/6) = 41.8°$, and $\theta^{(5)}_{3\times3,0} = \arcsin(5/6) = 56.4°$, and a main maximum appears at $\theta^{(3)}_{3\times3,max} = \arcsin(3/6) = 30°$. Accordingly, the 5×5 array exhibits 8 zeros and the 7×7 array 12 zeros. The main maximum at 30° occurs for all the arrays.

The radiation intensity patterns for the low-density arrays with lenses (Figure 3.51b,c) differ strongly from the corresponding pattern found for the dense arrays. In particular, the radiation intensities at $\theta = 0$ are even lowered for the 7×7 array (by about 8 dB for $R_l = 2$ mm and about 3 dB for $R_l = 5$ mm). In order to explain this behavior let us consider Figure 3.52. In a good approximation, an optimized hyperhemispherical lens of radius R has a focal length $f = R_l/n_{sc}$. Waves originating from a pixel at one pitch length a away from the center position are directed to an angle:

$$\theta_a = \arctan(a/f) = \arctan(an_{sc}/R_l) \tag{3.43}$$

away from the optical axis (Figure 3.52). If this angle becomes comparable with the angle of the central diffraction-limited radiation lobe of the lens:

$$\theta_{Ai,l} = \arcsin(1.22\lambda_0/2R_l) \tag{3.44}$$

or even significantly larger, the field patterns originating from different pixels have a reduced or even negligible overlap. Hence, even for a coherent phase, the constructive interference is strongly reduced. As a result, the radiation intensity no longer reaches the naively expected maximum values, and, at the same time, the radiation pattern becomes broadened. For the dense arrays with $a = \lambda_0/n_{sc}$ the angles θ_a and $\theta_{Ai,l}$ differ only by the factor 1.22 in Eq. (3.44).

Therefore, there is still constructive interference but it is already significantly reduced. The values for the deviation from the optical axis are $\theta_a = \arctan(an_{sc}/R_l) = \arctan(\lambda_0/2R_l) = 0.3/4 = \arcsin 0.075 = 4.3°$ (for $R_l = 2$ mm) and $\theta_a = \arctan 0.030 = 1.72°$ (for $R_l = 5$ mm). This explains why the maximum intensity at $\theta = 0$ is smaller and why the radiation pattern is broader than expected.

For the low-density array, however, $\theta_a = \arctan(2\lambda_0/R_l) = 16.7°$ (for $R_l = 2$ mm) and $\theta_a = 6.8°$ (for $R_l = 5$ mm) turns out to be much larger than $\theta_{Ai,1}$. Therefore, constructive interference hardly exists anymore. As θ_a and $\theta_{Ai,1}$ exhibit the same dependence on the lens radius, this problem cannot be overcome

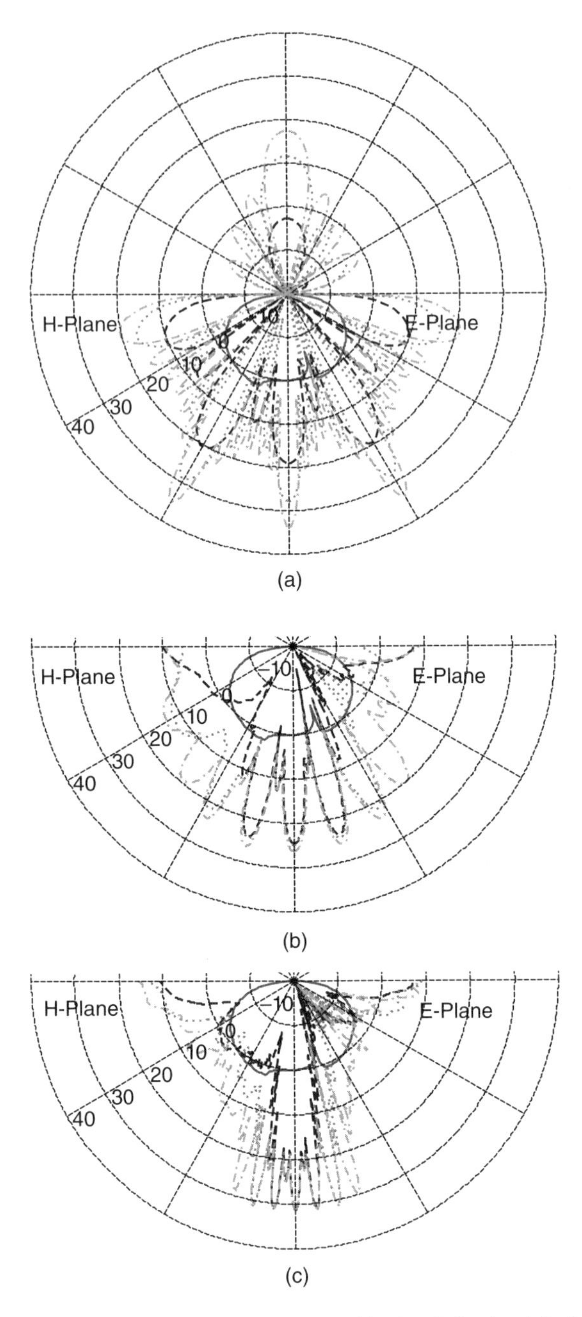

Figure 3.51 Radiation intensity patterns of 3 × 3 array of isolated dipoles (--), 5 × 5 array of isolated dipoles (··), and 7 × 7 array of isolated dipoles (-··) with pitch $a = 4\lambda_0/(2n_{sc})$. (a) Arrays without lens, (b) with 2 mm radius and 0.69 mm slab hyperhemispherical lens, and (c) with 5 mm radius and 1.725 mm slab lens. For comparison also the radiation intensity pattern for the single AE is depicted (-). The lobes in (b) and (c) are directly related to the off-center positions of the pixels (three lobes for the 3 × 3, five lobes for the 5 × 5, and seven lobes for the 7 × 7 array). The angle between neighboring lobes closely corresponds to the values obtained from Eq. (3.43). See plate section for color representation of this figure.

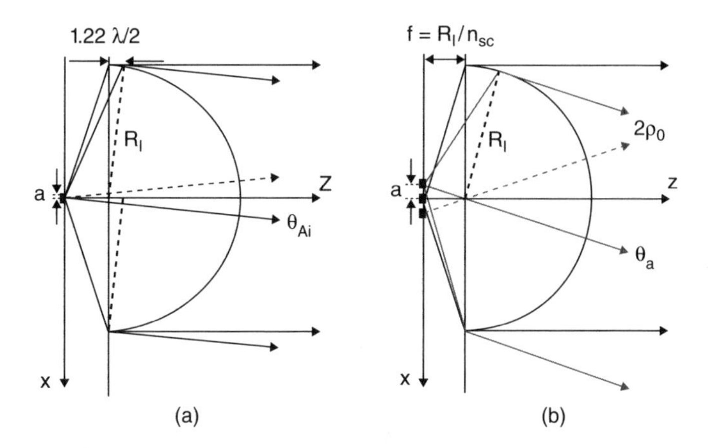

Figure 3.52 Illustration of the geometrical imaging and the diffraction aspects of an array with lens. (a) Angle θ_a between the collimated radiation originating from neighboring pixels and (b) angle $\theta_{Ai,1}$ corresponding to the width of the first Airy disk.

by increasing the lens radius. Thus it appears that low-density arrays do not appear to be very promising. This potentially would change if a very large lens without hyperhemisphericity (non-focusing) and a parabolic mirror for collimation were used. However, apart from the alignment issues, this would not be a compact and integrated system.

An alternative option is to use an array of AEs, each one having its own lens. This scheme implies that the minimum pitch becomes $a = 2R_l$. Due to the large period the angular spacing between neighboring zeros of the intensity becomes narrow and even the number of main maxima increases drastically, according to Eqs. (3.41) and (3.42), respectively.

In Figure 3.53 the radiation intensity pattern of a 3×3, 5×5, and 7×7 arrays, obtained with the analytical solution of a dipole lying on semi-infinite substrate and the PO analysis for the lenses, are presented. As expected, the radiation intensity $U_l^{AEA}(\theta = 0,\varphi)$ for the array is increased by a factor $N^2 = (n \times n)^2$, that is, by 19, 28, and 34 dB, for the 3×3, 5×5, and the 7×7 arrays, respectively, compared with the intensity from a single AE with lens, $U_l^{AE}(\theta = 0,\varphi)$. Apart from the central lobe the main interference maxima dominate the intensity pattern. They occur at the angles $\theta^{(m)}_{n\times n,max} = \arcsin[m(\lambda_0/a)] = \arcsin[m(0.3\text{ mm}/4\text{ mm})] = 4.3°, 8.6°, 13.0°, 17.5°, 22.0°, \ldots$ for the array of $R_l = 2$ mm lenses, that is, between the main lobes there are 1, 3, or 5 side lobes (with much weaker intensity maxima). Consequently, the width of the main lobes decreases with the size of the array. As the width of the radiation pattern of the single AE with a lens, $U_l^{AE}(\theta, \varphi)$, is narrow (determined by the Airy disk), the intensity of the side lobes is found to be strongly reduced, although the maxima of the array factor $|AF(\theta, \varphi)|^2$ in Eq. (3.45) are not decreasing with the order m of the main maxima.

For the calculation of the angular radiation intensity of the array the angular radiation intensity pattern of an AE without a lens, $U_0^{AE}(\theta, \varphi)$, has to be replaced by the angular radiation intensity pattern of an AE with a lens, $U_l^{AE}(\theta, \varphi)$, in Eq. (3.39), that is,

$$U_l^{AEA}(\theta, \varphi) = U_l^{AE}(\theta, \varphi)|AF(\theta, \varphi)|^2 \tag{3.45}$$

3.6.2 *Large Area Emitters*

In the previous section, the concept of "conventional" arrays was discussed as one of the possibilities to increase the intensity and the directivity of emitted THz beams. In addition, arrays provide the possibility

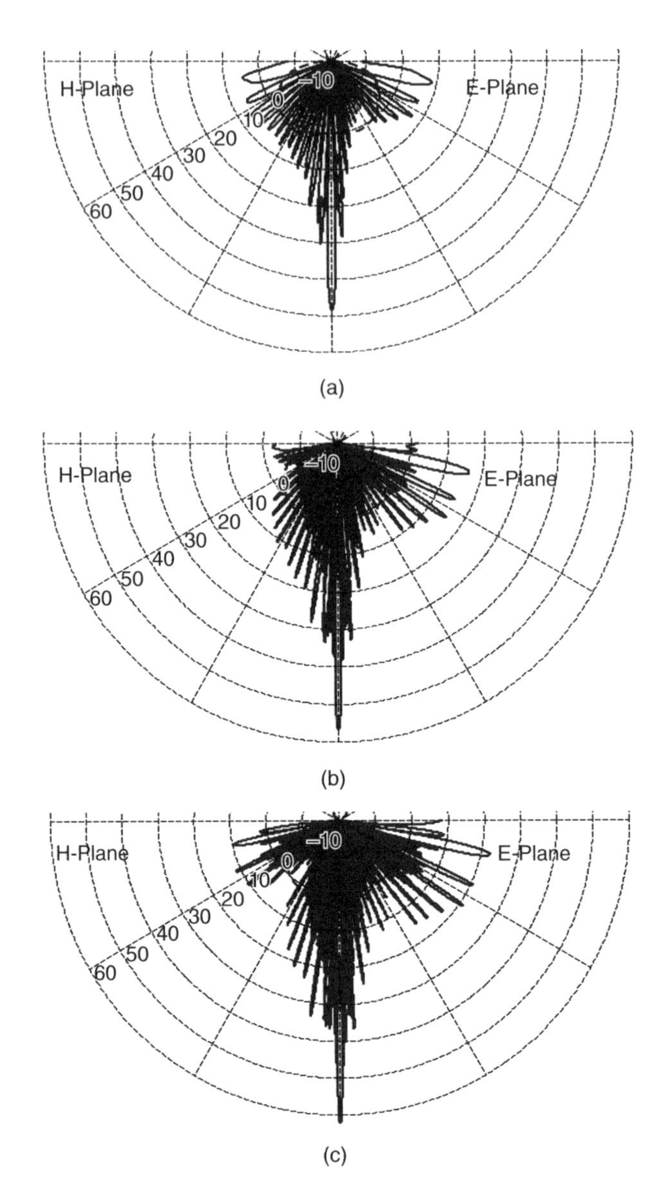

Figure 3.53 Radiation intensity pattern of (a) 3×3-, (b) 5×5-, and (c) 7×7-array consisting of $n \times n$ AEs with individual lens.

of beam steering by controlling the phase of the laser beams. Although very appealing results can be expected from two-dimensional arrays of lumped element photoconductive or p-i-n-diode THz emitters, this concept still suffers from several drawbacks:

1. The maximum tolerable optical power and, hence, the THz power emitted by the individual elements of the array is limited due to the small active device area required to keep the RC roll-off at a reasonable level.

2. (Hyperhemispherical) lenses are required for efficient collection of the radiation emitted into the substrate under different angles.
3. As the minimum pitch of the arrays is determined by the diameter of the individual antennas only one central element of the array can be positioned on the optical axis of the lens. Therefore, the off-axis positions of the other array elements result in the emission of the focused beams under quite different angles, unless the diameter of the lens is chosen unreasonably large. Even in this case, additional, high quality THz focusing elements are required to obtain a reasonably collimated beam exhibiting coherent superposition of the THz fields originating from the individual elements.
4. An alternate approach, avoiding these problems, is based on an array of emitters with individual lenses. The realization of this approach, however, is technologically extremely demanding.
5. Last, but not least, the illumination of the individual elements of the array by laser beams focused onto their small active areas requires a well-adjusted microlens array, a suitably designed diffractive element, or a fiber bundle. An even distribution of the optical power and the alignment are challenging.

To overcome such problems, large area emitters (LAEs) will be presented and discussed in the following sections. LAEs can be considered as high-density arrays, or, more correctly, as continuous arrays. It will be shown that they represent an elegant approach which avoids the previously mentioned problems.

3.6.2.1 The Basic Concept of LAEs

Up to here, it has always been assumed that the THz radiation emitted by photoconductive or p-i-n photodiode THz emitters is due to the temporal changes of a photocurrent, generated in the semiconductor device and fed into an antenna. However, antennas are not a sine qua non, since, according to basic electrodynamics, electromagnetic fields are generated by each photogenerated charge carrier, if accelerated by the DC electric field present in a photoconductor or a p-i-n photodiode. As typical charge carrier acceleration and deceleration times at intermediate and high fields in semiconductors are in the 100 fs range (see Section 2.2), these carriers will emit THz pulses. If many charge carries are generated simultaneously by femtosecond laser pulses within a large illuminated area, the superposition of the coherently emitted electromagnetic fields of these carriers will result in the emission of strong THz pulses. Limitations by local heating and saturation, as occurring in lumped element devices, can be circumvented by using larger areas. In fact, there are numerous earlier reports in the literature about such LAEs in which this approach was used. Early demonstrations date back to the 1990s where "large aperture photomixers" used a large photoconductive gap (in the range of $\sim$1 mm), biased with hundreds of volts in order to achieve accelerating fields in the low kV/cm range [62]. Higher fields were not possible due to breakdown at the electrodes. In the meantime, LAEs were used within the framework of pump and probe experiments as a tool for the investigation of high-field transport phenomena in (quasi-) bulk semiconductors [63], or surface depletion fields [64], or for the study of novel transport phenomena such as Bloch oscillations in semiconductor superlattices [65], or plasmon excitation [66]. In the mid-2000s [67, 68] a new layout of LAEs that allowed emission into the far-field with much shorter gap widths ($\sim$10 μm) were experimentally demonstrated. During the past few years the generation of THz current pulses by laser pulses in LAE photoconductors has therefore found increasing interest as an alternate method for the generation of THz pulses [69–74]. In fact, record THz fields generated by photomixers have been reported. Recently, this LAE approach has been used successfully also for the generation of CW-THz radiation by photomixing [75].

A detailed description of layouts of LAEs will be given in Section 3.6.2.6. The goal of the following sections is an evaluation of the microscopic mechanism responsible for THz emission and of the characteristic features of LAEs under pulse excitation and under CW photomixing conditions, with emphasis on the radiation pattern and the achievable THz power. Both photoconductive and p-i-n photodiode-based LAEs are investigated. We find that the fundamental as well as the technological problems associated with discrete arrays can be avoided in LAEs and that higher THz power in combination with a nearly ideal Gaussian beam profile can be obtained.

3.6.2.2 THz Generation by Accelerated Carrier in LAEs

In Section 2.2 the THz field emitted by an emitter with antenna ("AE") was introduced in terms of the time derivative of the photo current $I_{\mathrm{ph}}(t)$ in Eq. (2.8). The contribution of the electrons to the photocurrent density at the time t, $\vec{j}_{\mathrm{ph}}(t)$ can be expressed exactly as the sum of the contributions $e\vec{v}_i(t)$ of all electrons contained in the volume, V. In order to take the potential time dependence of the number of carriers moving at velocity $\vec{v}_i(t)$ into account, it can also be expressed in terms of a time-dependent, averaged electron density $n(t)$, and velocity $\vec{v}(t)$.

$$\vec{j}_{\mathrm{ph}}(t) = (e/V)\sum_i \vec{v}_i(t) = en(t)\vec{v}(t) = en\vec{v}f(t) \tag{3.46}$$

In Eq. (3.46) the function $f(t)$ concatenates all time dependences, v and n are the respective maximum values. Similar expressions apply for the photogenerated holes. The time derivative of the current density becomes

$$\frac{\mathrm{d}\vec{j}_{\mathrm{ph}}(t)}{\mathrm{d}t} = en\vec{v}\frac{\mathrm{d}f(t)}{\mathrm{d}t} = en\vec{a}(t). \tag{3.47}$$

In order to derive the characteristic features of the THz radiation emitted by the large number of charge carriers in a LAE we will first use the last expression in Eq. (3.47), which expresses the derivative of the photocurrent in terms of a constant carrier density n and the "quasi-acceleration" $\vec{a}(t) = \vec{v}\mathrm{d}f(t)/\mathrm{d}t$. The evaluation of the current changes in terms of the detailed generation, recombination, and transport processes will be postponed to Section 3.6.2.5.

According to fundamental electrodynamics, the electric field at the time t at the point $\vec{r}$, emitted by an elementary charge e at $\vec{\rho} = 0$ in a semiconductor and subjected to a time-dependent (quasi-) acceleration $\vec{a}(t)$, is [76]

$$\vec{E}_e(t, \vec{r}, \vec{\rho} = 0) = \frac{e}{4\pi\varepsilon_0 c_0^2}\frac{\vec{r}\times[\vec{r}\times\vec{a}(t_r)]}{r^3} \tag{3.48}$$

with a field amplitude of

$$E_e(t, \vec{r}, \vec{\rho} = 0) = \frac{e}{4\pi\varepsilon_0 c_0^2}\frac{a(t_r)\sin\theta_a}{r^3} \tag{3.49}$$

Here, θ_a is the angle between $\vec{a}(t_r)$ and $\vec{r}$ and

$$t_r = t - r/c_{\mathrm{sc}} = t - r\,n_{\mathrm{sc}}/c_0 \tag{3.50}$$

takes into account the retardation of the field at $\vec{r}$. The radiation intensity $U_e(t,\theta)$, emitted under the angle θ_a is defined as the power emitted per solid angle $\mathrm{d}\Omega = \sin\theta\mathrm{d}\theta\mathrm{d}\varphi$, that is,

$$U_e(t, \theta_a) = r^2|S_e(t, \vec{r}, \theta_a)| \tag{3.51}$$

where $\vec{S}_e(t, \vec{r}, \theta_a)$ is the Poynting vector

$$\vec{S}_e(t, \vec{r}, \theta_a) = c_{\mathrm{sc}}\varepsilon_0\varepsilon_{\mathrm{sc}}[E_e(t, \vec{r}, \theta_a)]^2\vec{r}/r. \tag{3.52}$$

This yields, with $c_{\mathrm{sc}} = c_0/n_{\mathrm{sc}} = c_0/\varepsilon_{\mathrm{sc}}^{1/2}$ for the radiation intensity

$$U_e(t, \theta_a) = \frac{\sqrt{\varepsilon_{\mathrm{sc}}}}{16\pi^2\varepsilon_0 c_0^3}[ea(t_r)]^2\sin^2\theta_a, \tag{3.53}$$

and for the total emitted power,

$$P_e(t) = \int U_e(t, \theta_a) \mathrm{d}\Omega = \frac{\sqrt{\varepsilon_{\mathrm{sc}}}}{6\pi^2 \varepsilon_0 c_0^3} [ea(t_r)]^2 = \frac{8\pi}{3} U_e(t, \theta_a = \pi/2) \tag{3.54}$$

If, instead of one, N photogenerated carriers are being accelerated, the emitted field increases by a factor N, whereas the radiation intensity and the total power increase by N^2, provided that the carriers are generated within an area or volume of dimensions sufficiently small compared with the THz wavelength to justify neglecting phase differences, that is,

$$U_{\mathrm{LAE}}(t, \theta_a) = \frac{\sqrt{\varepsilon_{\mathrm{sc}}}}{16\pi^2 \varepsilon_0 c_0^3} [Nea(t_r)]^2 \sin^2\theta_a = N^2 U_e(t, \theta_a), \tag{3.55}$$

and

$$P_{LAE}(t) = \frac{\sqrt{\varepsilon_{\mathrm{sc}}}}{6\pi^2 \varepsilon_0 c_0^3} [Nea(t_r)]^2 = N^2 P_e(t). \tag{3.56}$$

For calculation of the THz radiation intensity and power generated by real LAEs, three important factors, which modify the radiation intensity and power of LAEs with respect to Eqs. (3.55) and (3.56), have to be taken into account:

1. *Spatial interference yielding a "continuous array factor".* As the dimensions of typical LAEs are comparable or even larger than the THz wavelength, normally, the interference between the fields emitted by charge carriers generated at different points $\vec{\rho} \neq 0$ has to be taken into account. If all the carriers are coherently generated and accelerated, constructive interference is expected for electromagnetic fields propagating in the direction of the exciting laser beam (or, the corresponding directions of refraction or reflection). In Figure 3.54a,b the situation is illustrated for two important special cases. Case (a) depicts the typical case of a photoconductive LAE with a DC electric field $\vec{E} = (E_x, 0, 0)$ parallel to the surface, illuminated by a Gaussian laser beam of radius ρ_0 at normal incidence. In case (b) a DC electric field normal to the surface is assumed, $\vec{E} = (0, 0, E_z)$, corresponding to the typical scenario in LAEs based on a p-i-n photodiode scheme. According to Eq. (3.48), the THz field emitted in the direction of the accelerating DC-field is zero if a finite angle of incidence $\theta_{\mathrm{vac},0}$ for the laser beam is assumed. In case (a) the direction of fully constructive interference is normal to the surface and coincides with the incoming beam. In case (b) fully constructive interference happens for the radiation into the substrate under the refraction angle $\theta_{\mathrm{sc},0}$ and into the vacuum under the reflection angle $\theta_{\mathrm{vac,r}} = -\theta_{\mathrm{vac},0}$. It should be noted that the far-field THz *intensities* obtained in all these cases increase with the square of the absorbed laser power and, according to Eq. (3.55) with N^2, independent of the dimensions of the LAE and independent of the THz wavelength. This implies that the thermal limitations discussed in Section 2.2.3, which apply to any emitter with antenna (and arrays of those), are drastically relaxed for such large LAEs, as the incoming laser power can be distributed over a large area using the maximum thermally tolerable laser intensity. Destructive interference, however, will become increasingly significant at small angular deviations from the intensity maxima, if the dimensions of the LAE increase. Therefore, the THz *power* will no longer increase quadratically with the area (or laser power, respectively) and N^2. Hence, the increase in the THz *power* will be less than expected from Eq. (3.56), for LAEs with large dimensions. We will take the interference effects into account quantitatively by introducing a (continuous) array factor for LAEs in Section 3.6.2.3. The cases of typical photoconducting LAEs with in-plane DC-fields and of p-i-n-type LAEs with vertical DC-fields will be discussed in separate sections.
2. *Modifications of the angular radiation intensity $U_e(t,\theta_a)$ due to the proximity of the accelerated carriers to the semiconductor/air interface:* The THz fields are generated within the semiconductor and will be modified at the interface to the air. Only an angle-dependent fraction will cross the interface, whereas another part will be reflected, with an angle-dependent phase shift. In particular, total reflection affects all field components emitted under an angle exceeding the critical angle for total

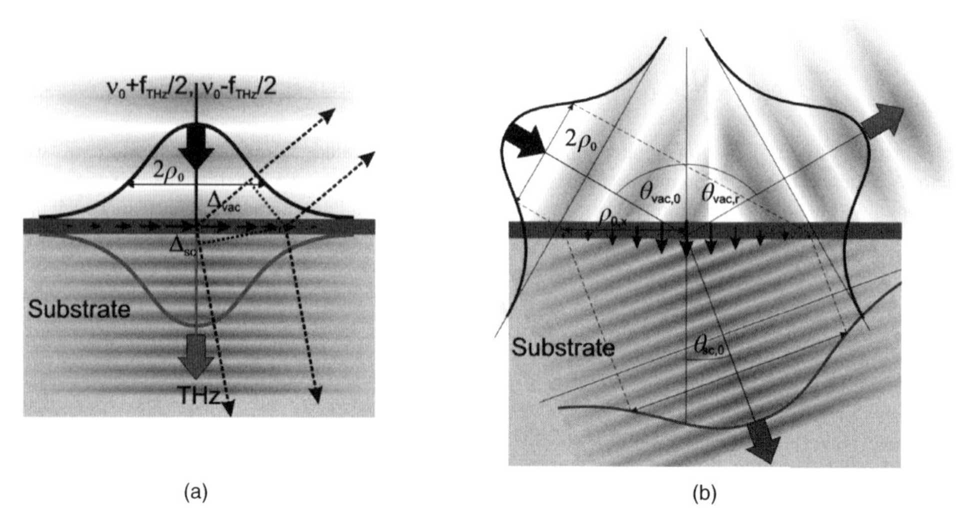

Figure 3.54 (a) Excitation conditions for a LAE with horizontally oriented dipoles. An excitation with a perpendicularly incident Gaussian beam results in maximum emission of a Gaussian THz wave in the same direction. Off-axis components emitted by different dipole elements experience a phase shift, Δ_{vac}, and Δ_{sc}, respectively, suppressing these components at large angles. (b) Excitation scenario for an LAE with vertical dipoles. The photoconductor has to be excited under an angle since the emission of the vertical dipole along its axis is zero.

reflection θ'_c. This implies that the intensity and power emitted into air correspond only to that within the small cone at angles smaller than θ'_c. Therefore, intensity and power emitted into air are significantly smaller than expected according to Eq. (3.56). These aspects are taken into account in Section 3.6.2.4.

3. *The dependence of the acceleration term on the transport properties and on CW or pulse excitation:* Similar to the case of the AEs discussed in Section 2.2.2, the THz currents at time t depend on the generation rates given by the spectrum of the laser excitation. We therefore introduce the macroscopic (quasi-)acceleration $A(t)$. It reflects the contributions of all the carriers generated at previous times t' having experienced a transient acceleration at the time t before reaching a constant velocity or disappearing by trapping, recombination with its counterpart, or at a contact electrode. This macroscopic (quasi-) acceleration term $A(t)$ also includes temporal changes of the carrier density and follows from integration of the generation rates, the acceleration, and the recombination rates over all previous times, as discussed in Section 2.2.2.1 for p-i-n diodes and Section 2.2.2.2 for photoconductors. Although $A(t)$ expresses temporal changes in the LAE current, this term differs from the time derivative of the displacement current relevant for AEs. In order to obtain quantitative information about the radiation intensity and THz power of the LAEs, the acceleration term will be evaluated in detail in Section 3.6.2.5. In the case of CW LAEs that are excited with a carrier density of $N(\omega_{\text{THz}}) \sim P_{\text{L}}(\omega_{\text{THz}})$ at the angular THz frequency $\omega_{\text{THz}} = 2\pi\upsilon_{\text{THz}}$, only the Fourier components at ω_{THz} will be relevant for the power and the radiation pattern. For pulsed excitation, the THz emission can also be expressed in terms of $N(t)$ and $a(t)$. However, all the Fourier components of the broadband THz spectrum are relevant. For these reasons, the discussion in Section 3.6.2.5 will be split into sections dealing with CW and with pulsed excitation.

Because of the size-, wavelength- and, hence, frequency-dependence of the interference, we will consider in the following a harmonic time dependence of the quasi-acceleration $\vec{a}(t)$

$$\vec{a}(t) = \vec{a}_0 \cos \omega t = \vec{v}_0 \omega \sin \omega t = -\vec{d}_0 \omega^2 \cos \omega t \tag{3.57}$$

With this periodic acceleration, each electron can be interpreted as an elementary radiating electronic dipole of length $d_0 = a_0/\omega^2$ and velocity amplitude $v_0 = a_0/\omega$. Here, $\omega = 2\pi f$ stands for a THz frequency corresponding to a fixed THz vacuum wavelength $\lambda_0 = c/f = 2\pi c/\omega$ and $\lambda_{sc} = \lambda_0/n_{sc}$ in the semiconductor. As $d_0 \ll \lambda_0$ for realistic values, we are dealing with Hertzian dipoles. For the non-periodic acceleration $a(t)$ and carrier generation, the Fourier components can be evaluated correspondingly.

3.6.2.3 Spatial Interference and "Continuous Array Factor"

a) Photoconductive Scenario

In this section we investigate the typical photoconductor scenario depicted in Figure 3.54a. This implies normal incidence of the laser beam, $\theta_{vac} = 0$, and an in-plane electric field assumed to be parallel to the x-direction, $\vec{E} = (E_x, 0, 0)$, which yields $\vec{a}_0 = (a_0, 0, 0)$ for the periodic acceleration in Eq. (3.57). The values of the amplitude a_0 at the (angular) frequency ω for realistic transport processes (non-ballistic or ballistic transport, including finite recombination lifetime) under photomixing or pulse excitation conditions will be evaluated in detail in Sections 3.6.2.3a and b respectively.

We further assume that the photogeneration of electrons and holes takes place closely below the surface of a semi-infinite semiconductor, within a depth small compared to the THz wavelength λ_{sc}, that is, approximately at $\vec{\rho} = (\rho_x, \rho_y, 0)$. The expression for the electric field from Eq. (3.48) becomes, for a periodically accelerated carrier at $\vec{\rho}$

$$\vec{E}_e(t, \vec{r}, \vec{\rho}) = \frac{e}{4\pi\varepsilon_0 c_0^2} \frac{(\vec{r} - \vec{\rho}) \times [(\vec{r} - \vec{\rho}) \times \vec{a}_0]}{|\vec{r} - \vec{\rho}|^3} \exp(\mathrm{i}[\vec{k} \circ (\vec{r} - \vec{\rho}) - \omega t]) \tag{3.58}$$

In Eq. (3.58) we have introduced the THz wave vector

$$\vec{k} = \vec{k}_0 = \frac{2\pi}{\lambda_0} \cdot \frac{(\vec{r} - \vec{\rho})}{|\vec{r} - \vec{\rho}|}, \tag{3.59}$$

for propagation in vacuum and

$$\vec{k} = \vec{k}_{sc} = \frac{2\pi}{\lambda_{sc}} \cdot \frac{(\vec{r} - \vec{\rho})}{|\vec{r} - \vec{\rho}|} = \frac{2\pi n_{sc}}{\lambda_0} \cdot \frac{(\vec{r} - \vec{\rho})}{|\vec{r} - \vec{\rho}|} = n_{sc}\vec{k}_0 \tag{3.60}$$

in semiconductors.

For the far-field ($r \gg \rho$) Eq. (3.58) becomes

$$\vec{E}_e(t, \vec{r}, \vec{\rho}) = \vec{E}_e(t, \vec{r}, 0) \exp(-\mathrm{i}\vec{k} \cdot \vec{\rho}) \tag{3.61}$$

with

$$\vec{E}_e(t, \vec{r}, 0) = \frac{e}{4\pi\varepsilon_0 c_0^2} \frac{\vec{r} \times (\vec{r} \times \vec{a}_0)}{r^3} \exp[\mathrm{i}(\vec{k} \cdot \vec{r} - \omega t)] \tag{3.62}$$

According to Eq. (3.61) the fields originating from different points $\vec{\rho}$ differ only by the phase factor $\exp(-\mathrm{i}\vec{k} \cdot \vec{\rho})$.

If the electrons are generated by a Gaussian laser beam of total power P_L with radius ρ_0 and photon energy $h\upsilon_0$, the photon flux distribution $\phi_{ph}(\vec{\rho})$ at the semiconductor surface is

$$\phi_{ph}(\vec{\rho}) = \frac{P_L}{\pi\rho_0^2} \exp\left(-\frac{(\rho_x^2 + \rho_y^2)}{\rho_0^2}\right) = \phi_{ph,0} \exp\left(-\frac{(\rho_x^2 + \rho_y^2)}{\rho_0^2}\right) \tag{3.63}$$

The (areal) density distribution of accelerated photo-excited electrons near the surface, $n(\vec{\rho})$, in general, will have the same Gaussian shape, that is,

$$n(\vec{\rho}) = n_0 \exp\left[-\left(\rho_x^2 + \rho_y^2\right)/\rho_0^2\right] \tag{3.64}$$

with $n_0 \propto \phi_{\mathrm{ph},0} = P_\mathrm{L}/(\pi\rho_0^{\,2})$. The total generated electric field at $\vec{r}$ is the sum of the fields coherently emitted by all the accelerated electrons. This yields with Eq. (3.61)

$$\vec{E}_{\mathrm{LAE}}(t,\vec{r}) = \int n(\vec{\rho})\vec{E}_\mathrm{e}(t,\vec{r},\vec{\rho})\mathrm{d}\rho_x\mathrm{d}\rho_y = \vec{E}_\mathrm{e}(t,\vec{r},0)\int n(\vec{\rho})\exp(-\mathrm{i}\vec{k}\circ\vec{\rho})\mathrm{d}\rho_x\mathrm{d}\rho_y, \tag{3.65}$$

with $n(\vec{\rho})$ from Eq. (3.64).

With Eqs. (3.51)–(3.53) adapted to harmonic acceleration according to Eq. (3.57) and $\vec{a}_0 = (a_0, 0, 0)$ the radiation intensity of the LAE becomes

$$U^{\mathrm{h}}_{\mathrm{LAE}}(\theta,\varphi) = U^{\mathrm{h}}_{\mathrm{e}}(\theta,\varphi)\left|\int n\left(\vec{\rho}\right)\exp(-\mathrm{i}\vec{k}\circ\vec{\rho})\mathrm{d}\rho_x\mathrm{d}\rho_y\right|^2 \tag{3.66}$$

with

$$U^{\mathrm{h}}_{\mathrm{e}}(\theta,\varphi) = \frac{1}{2}\cdot\frac{\sqrt{\varepsilon_{\mathrm{sc}}}}{16\pi^2\varepsilon_0 c_0^3}(ea_0)^2(1-\sin^2\theta\cos^2\varphi) \tag{3.67}$$

In Eq. (3.67), the $\sin\theta$-term in the expression for $U_\mathrm{e}(\theta_a)$ from Eq. (3.53) has been expressed in terms of conventional polar coordinates, with $\vec{e}_r = (\sin\theta\cos\varphi,\ \sin\theta\sin\varphi,\ \cos\theta)$ and $\vec{e}_a = (1,0,0)$. Moreover, the superscript "h" has been added to indicate the horizontal orientation of the DC electric field and the acceleration. The factor ½ in Eq. (3.67) is due to the time average of $\langle\vec{a}(t)^2\rangle = \langle a_0^2\cos^2\omega t\rangle = {}^1\!/_2\ a_0^2$.

Equation (3.66) expresses a crucial result for LAEs: The radiation intensity emitted into a given solid angle $\mathrm{d}\Omega = \sin\theta\ \mathrm{d}\theta\ \mathrm{d}\varphi$ can be split into a product of the angular dependent radiation intensity $U_e(\theta,\varphi)$ emitted under the polar angles (θ,φ) by a single emitter (i.e., the periodically accelerated electron, in our case) and the corresponding value of a "continuous array factor," AF(θ,φ) for the same angle, that is,

$$U^{\mathrm{h}}_{\mathrm{LAE}}(\theta,\varphi) = U^{\mathrm{h}}_{e}(\theta,\varphi)\mathrm{AF}(\theta,\varphi) \tag{3.68}$$

with

$$\mathrm{AF}(\theta,\varphi) = \left|\int n\left(\vec{\rho}\right)\exp(-\mathrm{i}\vec{k}(\theta,\varphi)\cdot\vec{\rho})\mathrm{d}\rho_x\mathrm{d}\rho_y\right|^2 \tag{3.69}$$

For an isotropic electron density distribution $n(\vec{\rho}) = n(\rho)$ the array factor obviously depends only on θ, but not on φ.

At this point a comparison between these continuous arrays and the familiar arrays, consisting of discrete elements, seems appropriate. The discrete counterpart of our continuous array is a coherently emitting $(m\times m)$ – array, consisting of discrete elements ij with the same angular-dependent radiation intensity, positioned at $\rho_{ij} = (x_i, y_j, 0)$, (with $(x_i,y_j) = (0,\pm1,\pm2\ldots\pm m/2)b$ and similar dimensions $mb \approx 2\rho_0$),

$$\mathrm{AF}_{mm}(\theta,\varphi) = \left|\sum_{ij} N_{ij}\exp\left(-\mathrm{i}\vec{k}\cdot\vec{\rho}_{ij}\right)\right|^2 \tag{3.70}$$

The quantity N_{ij} in Eq. (3.70) corresponds to the number of accelerated electrons in the ij-element of the array, given by

$$N_{ij} = n(\vec{\rho}_{ij})b^2. \tag{3.71}$$

With sufficiently small values of b ($\ll\lambda_{\mathrm{sc}}/2$), the differences between the discrete array factor, AF$_{mm}(\theta,\varphi)$, and its continuous counterpart, AF(θ,φ), become negligible. It should be noted, however, that the fabrication of discrete arrays, even satisfying the relaxed condition $b < \lambda_{\mathrm{sc}}/2$ represents a tremendous task, as discussed in Section 3.6.1, whereas the design and fabrication of LAEs will turn out to be even less demanding than in the case of photo-conducting or p-i-n-diode-based THz emitters with antenna (AEs).

Specifically, for the Gaussian carrier density form Eq. (3.63), the expression for the array factor, according to Eq. (3.69), becomes very simple. As the Gaussian $n(\vec{\rho})$ is isotropic in the (ρ_x, ρ_y)-plane, the only angular dependence is due to the factor $\vec{k} \cdot \vec{\rho}$. Without loss of generality we can choose the coordinate system such that $\vec{k}$ lies in the (xz) – plane and we obtain $\exp(i\vec{k} \cdot \vec{\rho}) = \exp(ik\rho_x \sin\theta)$. This yields for the array factor

$$AF(\theta, \varphi) = n_0^2 \left| \int \exp\left[-\left(\rho_x/\rho_0\right)^2\right] \exp(-ik\rho_x \sin\theta) \mathrm{d}\rho_x \cdot \int \exp[-(\rho_y/\rho_0)^2] \mathrm{d}\rho_y \right|^2. \quad (3.72)$$

The integrals can be solved analytically, with the result

$$\mathrm{AF}(\theta) = (n_0 \pi \rho_0^2)^2 \exp\left(-\frac{1}{2}\left[k\rho_0 \sin\theta\right]^2\right). \quad (3.73)$$

The θ-dependence varies strongly with the parameter $k\rho_0/2 = \pi\rho_0/\lambda$. For a small LAE with $k\rho_0 = 2\pi\rho_0/\lambda < \sqrt{2}$ the array factor is nearly isotropic. It drops by less than a factor $1/e = 0.37$ between $\theta = 0$ and $\pi/2$. Therefore, the array factor becomes approximately independent of θ for $\rho_0 \ll 1/k = \lambda/2\pi$, that is,

$$\mathrm{AF}(\theta) = (n_0 \pi \rho_0^2)^2 \text{ for } \rho_0 \ll 1/k = \lambda/2\pi \quad (3.74)$$

and, according to Eq. (3.66), the radiation intensity of the LAE becomes

$$U_{\mathrm{LAE}}^{\mathrm{h}}(\theta, \varphi) = U_{\mathrm{e}}^{\mathrm{h}}(\theta, \varphi) \cdot (n_0 \pi \rho_0^2)^2 \quad (3.75)$$

This implies that $U_{\mathrm{LAE}}^{\mathrm{h}}(\theta, \varphi)$ exhibits the same angular dependence as $U_e(\theta, \varphi)$ and for any angle it increases quadratically with the number of photogenerated carriers and, hence, with the laser power, that is,

$$U_{\mathrm{LAE}}^{\mathrm{h}}(\theta, \varphi)/U_{\mathrm{e}}^{\mathrm{h}}(\theta, \varphi) = (n_0 \pi \rho_0^2)^2 \sim P_{\mathrm{L}}^2 \text{ for all angles } (\theta, \varphi). \quad (3.76)$$

For a large LAE with $k\rho_0 = 2\pi\rho_0/\lambda > \sqrt{2}$ the array factor becomes increasingly directional. For large LAEs with $\rho_0 \gg 1/k = \lambda/2\pi$, the quadratic exponential decrease of the array factor from Eq. (3.73) sets in already at small angles θ. Therefore, $\sin\theta$ can be replaced by θ in the relevant range of angles and $\mathrm{AF}(\theta)$ becomes, in very good approximation,

$$\mathrm{AF}(\theta) \approx (n_0 \pi \rho_0^2)^2 \exp\left(-\frac{1}{2}\left[k\rho_0 \theta\right]^2\right) \quad (3.77)$$

The 3 dB angle, defined as $\mathrm{AF}(\theta_{3\mathrm{dB}}) = ½\mathrm{AF}(\theta = 0)$ is reached at

$$\exp\left(-\frac{1}{2}\left[k\rho_0 \theta_{3\mathrm{dB}}\right]^2 = \frac{1}{2}\right) \quad (3.78)$$

which yields

$$\theta_{3\mathrm{dB}} = \sqrt{2\ln 2}/(k\rho_0) = \sqrt{2\ln 2}\ [\lambda/(2\pi\rho_0)] = 0.1874(\lambda/\rho_0) \quad (3.79)$$

Note that for emission into the semiconductor substrate λ stands for $\lambda_0/n_{\mathrm{sc}}$, whereas for emission into the air $\lambda = \lambda_0$ applies. For $\rho_0 = \lambda_0$, for example, Eq. (3.79) yields

$$\theta_{3\mathrm{dB}} = 10.8°, \quad \text{for emission into the air,} \quad (3.80)$$

and for emission into a semiconductor with $n_{\mathrm{sc}} = 3.6$

$$\theta_{3\mathrm{dB}} = 3.0°, \quad \text{for emission into the substrate.} \quad (3.81)$$

Thus, the array factor and, hence, the THz emission becomes strongly directional already for relatively small diameters of the LAE. This is an appealing aspect of LAEs, since no high-index hyper-hemispheric

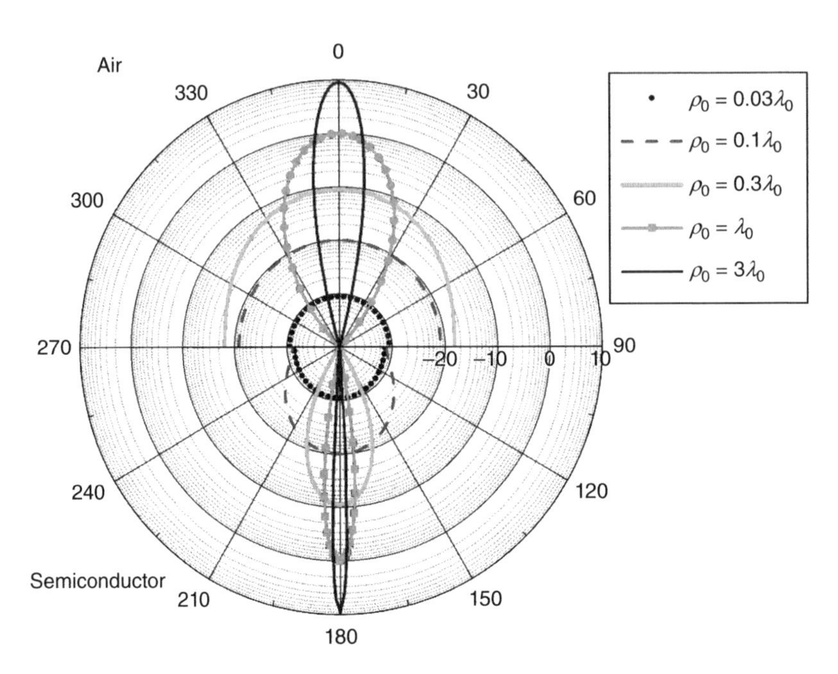

Figure 3.55 Array factor for horizontal LAEs for emission into air (top) and semiconductor (bottom) for various excitation spot radii, ρ_0, according to Eq. (3.73). For small excitation spot radii, $\rho_0 < \rho_c$ from Eq. (3.88), the LAE emits uniformly. At large radii $\rho_0 > \rho_c$, the emission becomes strongly directional. Due to the large refractive index of the semiconductor, n, the emission toward the substrate is more directional since $\rho_{c,sc} = \rho_{c,vac}/n$.

lens, like the familiar Si lens used for AEs, is required for efficient extraction of the THz radiation from the semiconductor substrate and a clean, small-aperture Gaussian beam shape is easily achieved.

In Figure 3.55 the array factor from Eq. (3.73) is shown for emission into air and into the semiconductor in a log-polar diagram for $\rho_0 = (0.03, 0.1, 0.3, 1.0, 3.0)\ \lambda_0$, normalized to its value at $\rho_0 = \lambda_0$ at 0°. It is evident that the directivity becomes pronounced already at lower values of ρ_0 for the emission into the semiconductor. With Eq. (3.73) inserted into Eq. (3.68), the radiation intensity $U^{h}_{LAE}(\theta, \varphi)$ of a LAE, excited by a Gaussian laser beam of radius ρ_0 is given by

$$U^{h}_{LAE}(\theta, \varphi) = U^{h}_{e}(\theta, \varphi) \cdot (n_0 \pi \rho_0^2)^2 \exp\left(-\frac{1}{2}\left(k\rho_0 \sin\theta\right)^2\right) \tag{3.82}$$

We notice that this quadratic increase applies only within the range of angles $\theta < \theta_{3dB}$, for large LAEs with $\rho_0 \gg 1/k = \lambda/2\pi$. For the case $\rho_0 \ll 1/k = \lambda/2\pi$, the quadratic increase with the square of the number of photogenerated carriers, according to Eq. (3.76), is valid for all angles (θ, φ),

$$U^{h}_{LAE}(\theta, \varphi)/U^{h}_{e}(\theta, \varphi) = (n_0 \pi \rho_0^2)^2 \sim P_L^2, \quad \text{for}\ \theta < \theta_{3dB} = 0.1874(\lambda/\rho_0). \tag{3.83}$$

Therefore, the size dependence of the array factor also strongly affects the size dependence of the total THz *power* emitted by an LAE. With Eq. (3.73) the expression for the total power emitted by an LAE becomes

$$P^{h}_{LAE} = \int U^{h}_{e}(\theta, \varphi)(n_0 \pi \rho_0^2)^2 \exp\left(-\frac{1}{2}\left(k\rho_0 \sin\theta\right)^2\right) d\Omega \tag{3.84}$$

This integral can be evaluated easily for the limiting cases $\rho_0 \ll \text{or} \gg 1/k = \lambda/2\pi$. As the angle dependence of the array factor is negligible for a small beam radius ρ_0, we obtain

$$P^{\mathrm{h}}_{\mathrm{LAE}} = (n_0\pi\rho_0^2)^2 \int U^{\mathrm{h}}_{\mathrm{e}}(\theta,\varphi)\mathrm{d}\Omega = (n_0\pi\rho_0^2)^2 \frac{8\pi}{3} U^{\mathrm{h}}_{\mathrm{e}}(\theta=0) = (n_0\pi\rho_0^2)^2 P_{\mathrm{e}} \sim P_{\mathrm{L}}^2,\ \text{for}\ \rho_0 \ll 1/k = \lambda/2\pi \tag{3.85}$$

with P_{e} adapted from Eq. (3.54).

For $\rho_0 \gg 1/k = \lambda/2\pi$ we obtain significant contributions to the integral only for small values of θ. Therefore, we can approximate $\sin^2\theta \approx \theta$ and $U_e(\theta,\varphi) \approx U_e(\theta=0)$ in Eq. (3.84). With $\mathrm{d}\Omega = 2\pi\theta\mathrm{d}\theta$ we obtain

$$P^{\mathrm{h}}_{\mathrm{LAE}} = (n_0\pi\rho_0^2)^2 U^{\mathrm{h}}_{e}(\theta=0) \int \exp\left(-\frac{1}{2}\left(k\rho_0 \sin\theta\right)^2\right) 2\pi\theta\mathrm{d}\theta, \tag{3.86}$$

and, finally,

$$P^{\mathrm{h}}_{\mathrm{LAE}} = (n_0\pi\rho_0^2)^2 U^{\mathrm{h}}_{e}(\theta=0) \cdot \frac{\lambda^2}{\pi\rho_0^2} \sim P_{\mathrm{L}},\ \text{for}\ \rho_0 \gg 1/k = \lambda/2\pi \tag{3.87}$$

From a comparison of results for small LAEs (Eq. (3.85)) with those of large ones (Eq. (3.87)), we notice a transition from a regime where the THz power increases with the square of the cross section of the laser beam, $(\pi\rho_0{}^2)$, and where it does not explicitly depend on the THz wavelength, λ, to a regime with linear increase in the THz power and a strong wavelength dependence, favoring longer wavelengths. The critical beam diameter ρ_c, characterizing the crossover from "small LAEs" to the "large LAEs" is obtained from the value which would (formally) yield the same value for the total THz power P_{LAE} in Eqs. (3.85) and (3.87), giving

$$\rho_{\mathrm{c}} = \sqrt{3\lambda^2/(8\pi^2)} = \sqrt{6}(\lambda/4\pi) = 0.195\lambda \tag{3.88}$$

The results of Eqs. (3.85) and (3.87) can be combined into

$$P^{\mathrm{h}}_{\mathrm{LAE}}(\rho_0) = P_e \frac{(n_0\pi\rho_0^2)^2}{1+(\rho_0/\rho_{\mathrm{c}})^2} \tag{3.89}$$

The conversion efficiency, defined as the ratio of emitted THz power to the laser power, increases linearly in the "small LAE" range for $\rho_0 < \rho_{\mathrm{c}}$ and saturates in the "large LAE" range.

$$\eta^{\mathrm{h}}_{\mathrm{LAE}}(\rho_0) = \eta_{\mathrm{sat}} \frac{(\rho_0/\rho_{\mathrm{c}})^2}{1+(\rho_0/\rho_{\mathrm{c}})^2}, \tag{3.90}$$

with $\eta_{\mathrm{sat}} \sim P_e\pi\rho_{\mathrm{c}}^2$. It should be noted that in Eq. (3.82)–Eq. (3.90) the values of the wavelength in vacuum or in the semiconductor have to be used for emission into air ($\lambda = \lambda_0$) or into the substrate ($\lambda = \lambda_0/n_{\mathrm{sc}}$), respectively. This applies, in particular also to the crossover radius ρ_{c}. Thus, the crossover from quadratic to linear increase in THz power for emission into air occurs at

$$\rho_{\mathrm{c,vac}} = \sqrt{3\lambda^2/(8\pi^2)} = \sqrt{6}(\lambda_0/4\pi) = 0.195\lambda_0, \tag{3.91}$$

whereas it happens already at

$$\rho_{\mathrm{c,sc}} = \sqrt{6}\,(\lambda_{\mathrm{sc}}/4\pi) = 0.195\lambda_0/n_{\mathrm{sc}} \approx 0.056\lambda_0 \tag{3.92}$$

for emission into the substrate with a refractive index of $n_{\mathrm{sc}} = 3.6$.

For 1 THz, for example, $\rho_{\mathrm{c,vac}} = 58.5\ \mu\mathrm{m}$ and $\rho_{\mathrm{c,sc}} = 16.25\ \mu\mathrm{m}$. Hence, the illuminated area at the crossover point is $\pi\rho_{\mathrm{c,vac}}^2 = 10\ 751\ \mu\mathrm{m}^2$ for emission into the air or $\pi\rho_{\mathrm{c,sc}}^2 = 887\ \mu\mathrm{m}^2$ into the semiconductor. These values indicate that the maximum tolerable *laser power* $P_{\mathrm{L,max}}^{\mathrm{LAE}} = \varphi_0^{\max}\pi\rho_0^2$ for LAEs can be increased by factors of about 20–200 compared to the maximum tolerable power $P_{\mathrm{L,max}}^{\mathrm{AE}} = \varphi_0^{\max}A_{\mathrm{AE}}$

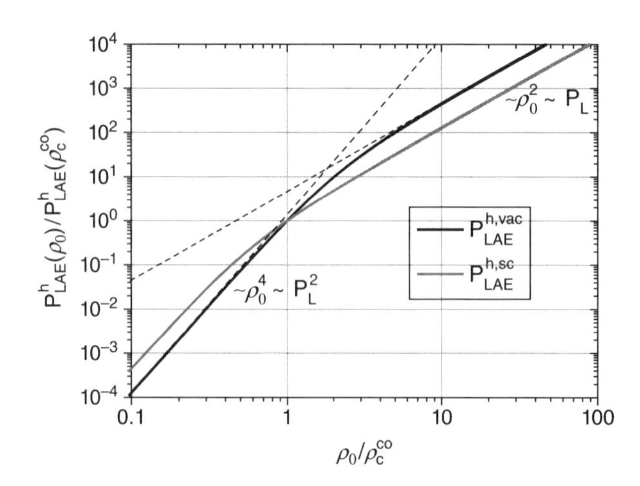

Figure 3.56 Emitted power of the horizontal LAE, normalized to the power emitted at the cross-over, $\rho_0 = \rho_{c,co}$. The power emitted into air shows a transition from quadratic laser power dependence to linear laser power dependence at higher spot radii. Therefore, the power emitted into air is higher at a larger radius. Adapted with permission from Ref. [14] S. Preu *et al.*, J. Appl. Phys. 109, 061301 (2011) © 2011, AIP Publishing LLC.

of AEs (requiring small areas $A_{AE} < 50\,\mu m^2$ for efficient operation at 1 THz). We assumed here that the maximum tolerable *laser intensity,* φ_0^{max}, is the same for LAEs and AEs. Due to the quadratic increase of P_{LAE} with P_L, the THz power could be increased by factors of the order of 400–40 000, provided the conversion efficiency of LAEs and AEs was comparable (this is generally not the case and will be discussed in Section 3.6.2.5).

Rewriting Eq. (3.89) separately for emission into air and into the semiconductor yields

$$P^{h}_{LAE,sc}(\rho_0) \approx P^{h}_{e,sc}\frac{(n_0\pi\rho_0^2)^2}{1+(\rho_0/\rho_c^{sc})^2} \tag{3.93}$$

$$P^{h}_{LAE,vac}(\rho_0) \approx P^{h}_{e,vac}\frac{(n_0\pi\rho_0^2)^2}{1+(\rho_0/\rho_c^{vac})^2}. \tag{3.94}$$

According to our present simplified considerations, we note that $P^{h}_{e,vac} = P^{h}_{e,sc}/\sqrt{\varepsilon_{sc}} = P^{h}_{e,sc}/n_{sc}$, whereas $\rho_{c,vac} = n_{sc}\rho_{c,sc}$. Therefore, we expect a crossover from $P^{h}_{e,sc}(\rho_0) > P^{h}_{e,sc}(\rho_0)$ to $P^{h}_{e,vac}(\rho_0) > P^{h}_{e,sc}(\rho_0)$, at $\rho_{c,co} = \sqrt{\rho_{c,vac}\rho_{c,sc}}$ as illustrated by the graphs of Eqs. (3.93) and (3.94) shown in Figure 3.56.

So far, we have neglected that the radiated intensity and power of an elementary emitter in proximity to the interface semiconductor/air are strongly affected by the boundary conditions for the THz fields. This implies it is necessary to determine the modified radiation intensities $U^{if,h}_{e,sc}(\theta,\varphi)$ and $U^{if,h}_{e,vac}(\theta,\varphi)$ from Eq. (3.68), in addition to the different array factors $AF_{sc}(\theta,\varphi)$ and $AF_{vac}(\theta,\varphi)$. These modifications will be discussed in Section 3.6.2.4.

b) p-i-n Photodiode Scenario

If acceleration of the photogenerated carrier takes place in p-i-n-type structures, the DC electric field is normal to the semiconductor surface, $\vec{E} = (0, 0, E_z)$ and, consequently the acceleration, $\vec{a}_0 = (0, 0, a_0)$ and the corresponding Hertzian dipole, $\vec{d}_0 = (0, 0, d_0) = (0, 0a_0/\omega^2)$, will be oriented normal to the semiconductor surface. As already discussed in Section 3.6.2.2, the radiation intensity is zero for the direction parallel to the acceleration. Therefore, for normal incidence of the laser beam, the maximum of the array factor would coincide with the zero of radiation intensity in Eq. (3.53), which obviously represents a very unfavorable scenario. As Eq. (3.51) also applies for $\vec{a}_0 = (0, 0, a_0)$, an angle θ close to $\pi/2$ would be desirable to obtain large THz field amplitudes and radiation intensities (see Figure 3.54b). The maximum (critical) angle θ_{sc} for $\theta_{vac} = \pi/2$, however, is $\theta_{sc} = \theta_{c,sc} = \arcsin(1/n_{sc})$ (= 16.18° for InGaAs, e.g.), according to Snell's law. Whereas grazing incidence does not represent a realistic choice for LAEs, the Brewster angle $\theta^B_{vac} = \arctan\ n_{sc}$, with $\theta^B_{sc} = \arctan(1/n_{sc})$ (= 15.52° for InGaAs, e.g.), is even particularly favorable, as the incident laser light penetrates into the semiconductor without any reflection if polarized in the plane of incidence.

For the field in the z-direction, the intensity emitted by an accelerated elementary charge under the angle θ is given by Eq. (3.51) with θ_a replaced by θ, as θ_a coincides with the polar angle θ in this case. For periodic acceleration, $<a(t_r)>^2$ is again replaced by $a_0^2/2$. This yields

$$U_e^v(\theta, \varphi) = \frac{1}{2} \cdot \frac{\sqrt{\varepsilon_{sc}}}{16\pi^2 \varepsilon_0 c_0^3} (ea_0)^2 \sin^2\theta, \tag{3.95}$$

which does not depend on the azimuth angle φ. The superscript "v" has been added to account for the vertical orientation of the DC electric field. Compared with the in-plane, horizontal field present in typical photoconductors and vertical emission, the intensity for vertical fields with emission in the Brewster angle (critical angle) direction is reduced, by a factor of $[1/\sin(\arctan(1/n_{sc}))]^2$ (= 1/0.072 = 14 for InGaAs, e.g.) or $[n_{sc}]^2$ (= 13 for InGaAs, e.g.), respectively. In Section 3.6.2.5 we will see that the p-i-n diode offers many advantages over the photoconductive version, in spite of these reduction factors. An alternative option for the incident laser beam for THz generation in p-i-n-based emitters is the illumination from the backside, using a prism as depicted in Figure 3.57. Naively one would expect that angles close to $\pi/2$ would be ideal. However, we will learn in Section 3.6.2.4 that the critical angle for total reflection, θ_c, represents the optimum choice for this scenario.

In order to calculate the intensity and the power of the THz radiation from an LAE with vertical acceleration $U^v_{LAE}(\theta, \varphi)$ we follow the procedure from Section 3.6.2.3a adapted to the finite angle of incidence, θ^B_{vac}. The incidence under the finite angle θ^B_{vac} has mainly two consequences:

1. The shape of the area, where the photogeneration of charge carriers occurs, becomes elliptic, when illuminated by a Gaussian laser beam of radius ρ_0. For incidence in the xz-plane under the angle θ^B_{vac},

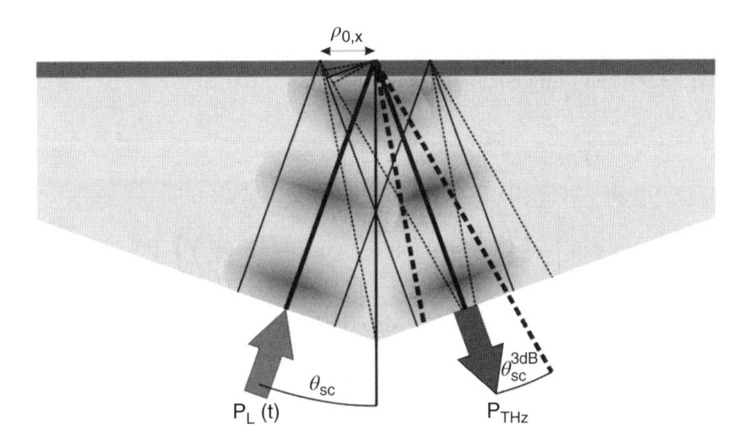

Figure 3.57 Excitation under the angle θ_{sc} of a vertical LAE using an attached prism.

the expressions for the carrier density distribution from Eqs. (3.63) and (3.64) are changed into

$$n(\vec{\rho}) = n_0 \exp[-(\rho_x/\rho_{0,x})^2 - (\rho_y/\rho_{0,y})^2] \tag{3.96}$$

with $\rho_{0,x} = \rho_0/\cos\theta_{\mathrm{vac}}^{\mathrm{B}} = \rho_0/\sin\theta_{\mathrm{sc}}^{\mathrm{B}} \approx \rho_0 n_{\mathrm{sc}}$ and $\rho_{0,y} = \rho_0$ and with

$$n_0 \propto \phi_{\mathrm{ph},0} = \cos\theta_{\mathrm{vac}}^{\mathrm{B}} P_{\mathrm{L}}/(\pi\rho_0)^2 \approx P_{\mathrm{L}}/(n_{\mathrm{sc}}\pi{\rho_0}^2). \tag{3.97}$$

2. The maxima of the array factor are found at the wave vectors of the refracted and reflected beams

$$\vec{k}_{\mathrm{B}}^{\mathrm{sc}} = n_{\mathrm{sc}}(k_{0,x}, 0, k_{0,z}) = n_{\mathrm{sc}}k_0(\sin\theta_{\mathrm{sc}}^{\mathrm{B}}, 0, \cos\theta_{\mathrm{sc}}^{\mathrm{B}}) \tag{3.98}$$

and

$$\vec{k}_{\mathrm{B}}^{\mathrm{vac}} = (k_{0,x}, 0, k_{0,z}) = k_0(\sin\theta_{\mathrm{vac}}^{\mathrm{B}}, 0, \cos\theta_{\mathrm{vac}}^{\mathrm{B}}), \tag{3.99}$$

and the corresponding angles $\theta_{\mathrm{sc}}^{\mathrm{B}}$ and $\theta_{\mathrm{vac}}^{\mathrm{B}}$,, respectively, as for these directions, the THz fields generated at different points x' in the r'-plane are in phase with the incident beam. As a consequence, the phase factor $\exp(-\mathrm{i}\vec{k}\cdot\vec{\rho})$ in Eq. (3.69) has to be replaced by $\exp(-\mathrm{i}(\vec{k}-\vec{k}_{\mathrm{B}})\cdot\vec{\rho})$.

With these modifications, the (continuous) array factor now becomes

$$\mathrm{AF}(\theta, \varphi; \theta_{\mathrm{B}}) = n_0^2 \left| \int \exp\left[-\left(\rho_x/\rho_{0,x}\right)^2 - (\rho_y/\rho_{0,y})^2\right] \exp(-\mathrm{i}[\vec{k}-\vec{k}_B]\cdot\vec{\rho})\mathrm{d}\rho_x \mathrm{d}\rho_y \right|^2 \tag{3.100}$$

Also the discretized version of this array factor, $\mathrm{AF}_{mn}(\theta, \varphi; \theta_{\mathrm{B}})$, is defined in analogy to the corresponding version, Eqs. (3.70) and (3.71). However, it should be noted that $\vec{\rho}_{ij} = (x_i, y_j, 0)$, now has to be discretized according to $x_i = (0, \pm 1, \pm 2 \ldots \pm m/2)b$ with $mb \approx 2\rho_{0x} \approx 2n_{\mathrm{sc}}\rho_0$, while the conditions $y_j = (0, \pm 1, \pm 2 \ldots \pm n/2)b$, with $nb \approx 2\rho_0$, and $b \ll \lambda_{\mathrm{sc}} = n_{\mathrm{sc}}\lambda_0$ remain unchanged.

As in the case of normal incidence, the expression for the array factor, Eq. (3.100), can be integrated. However, due to the finite value of the Brewster angle, the isotropy regarding the azimuth angle no longer persists. The result corresponding to Eq. (3.73) becomes

$$\mathrm{AF}(\theta, \varphi; \theta_{\mathrm{B}}) = \left(\frac{n_0\pi\rho_0^2}{\cos\theta_{\mathrm{vac}}^{\mathrm{B}}}\right)^2 \exp\left(-\frac{1}{2}\left(\left[k_x - k_{B,x}\right]\rho_{0,x}\right)^2\right) \exp\left(-\frac{1}{2}\left(k_y\rho_{0,y}\right)^2\right) \tag{3.101}$$

The emission into the air and into the semiconductor will be treated separately.

THz Emission into Air

With $\vec{k}_{\mathrm{vac}}^{\mathrm{B}}$ from Eq. (3.99) and using polar angles one obtains

$$\begin{aligned}\mathrm{AF}(\theta, \varphi; \theta_{\mathrm{vac}}^{\mathrm{B}}) = &\left(\frac{n_0\pi\rho_0^2}{\cos\theta_{\mathrm{vac}}^{\mathrm{B}}}\right)^2 \\ &\exp\left[-\frac{1}{2}\left(\frac{k_0\rho_0}{\cos\theta_{\mathrm{vac}}^{\mathrm{B}}}\left(\sin\theta_{\mathrm{vac}}\cos\varphi - \sin\theta_{\mathrm{vac}}^{\mathrm{B}}\right)\right)^2\right] \\ &\exp\left[-\frac{1}{2}\left(k_0\rho_0\sin\theta_{\mathrm{vac}}\sin\varphi\right)^2\right]\end{aligned} \tag{3.102}$$

As expected, the maximum of the array factor is found at $\theta_{\mathrm{vac}} = \theta_{\mathrm{vac}}^{\mathrm{B}}$ and $\varphi = 0$.

$$\mathrm{AF}(\theta, \varphi; \theta_{\mathrm{vac}}^{\mathrm{B}}) = \left(\frac{n_0\pi\rho_0^2}{\cos\theta_{\mathrm{vac}}^{\mathrm{B}}}\right)^2 \approx (n_0\pi\rho_0^2)^2 \tag{3.103}$$

Its value is increased by the factor $(1/\cos\theta^{\mathrm{B}}_{\mathrm{vac}})^2 \approx (n_{\mathrm{sc}})^2$, which reflects the elliptically increased illumination spot, due to the large incidence angle $\theta^{\mathrm{vac}}_{\mathrm{B}}$ of the laser beam.

The 3 dB angle for deviations $\Delta\theta^{3\mathrm{dB}}_{\mathrm{vac}}$ from $\theta^{\mathrm{B}}_{\mathrm{vac}}$ in the xz-plane ($\varphi = 0$) follows from the first exponential factor in Eq. (3.102)

$$\exp\left(-\frac{1}{2}\left(\frac{k_0\rho_0}{\cos\theta^{\mathrm{vac}}_{\mathrm{B}}}\left(\sin\theta^{3\mathrm{dB}}_{\mathrm{vac}} - \sin\theta^{\mathrm{B}}_{\mathrm{vac}}\right)\right)^2\right) = \frac{1}{2} \tag{3.104}$$

With $\sin\theta^{3\mathrm{dB}}_{\mathrm{vac}} - \sin\theta^{\mathrm{vac}}_{\mathrm{B}} = 2\cos[{}^1\!/\!{}_2(\theta^{3\mathrm{dB}}_{\mathrm{vac}} + \theta^{\mathrm{B}}_{\mathrm{vac}})]\sin[{}^1\!/\!{}_2(\sin\theta^{3\mathrm{dB}}_{\mathrm{vac}} - \sin\theta^{\mathrm{B}}_{\mathrm{vac}})] \approx 2\cos\theta^{\mathrm{B}}_{\mathrm{vac}}\sin[{}^1\!/\!{}_2(\theta^{3\mathrm{dB}}_{\mathrm{vac}} - \theta^{\mathrm{B}}_{\mathrm{vac}})] \approx \cos\theta^{\mathrm{B}}_{\mathrm{vac}}\Delta\,\theta^{3\mathrm{dB}}_{\mathrm{vac}}$, the increase factor $1/\cos\theta^{\mathrm{B}}_{\mathrm{vac}}$ and $\cos\theta^{\mathrm{B}}_{\mathrm{vac}}$ cancel and we obtain for large LAEs,

$$\Delta\theta^{3\mathrm{dB}}_{\mathrm{vac}} = \sqrt{2\ln 2}/(k_0\rho_0) \tag{3.105}$$

Thus, the 3 dB angle turns out to be the same as in the case of normal incidence (Eq. (3.79)).

The same applies for the 3 dB angle in the "air-Brewster-y"- (By-) plane, defined by the Brewster wave vector $\vec{k}^{\mathrm{B}}_{\mathrm{vac}}$ and the y-direction. In this plane, the 3 dB angle is defined by

$$\exp\left[-\frac{1}{2}\left(k_0\rho_0\sin\theta^{3\mathrm{dB}}_{\mathrm{vac}}\sin\varphi^{3\mathrm{dB}}_{\mathrm{vac}}\right)^2\right] \approx \exp\left[-\frac{1}{2}\left(k_0\rho_0\varphi^{\mathrm{B}y,3\mathrm{dB}}_{\mathrm{vac}}\right)^2\right] = \frac{1}{2} \tag{3.106}$$

When deriving Eq. (3.106) it has been taken into account that the angle in the By-plane is very similar to the angle in the xy-plane, as $\sin\theta^{3\mathrm{dB}}_{\mathrm{vac}} = 0.963 \approx 1$. Therefore, the 3 dB angle in the By-plane becomes practically the same as in the xz-plane

$$\varphi^{\mathrm{B}y,3\mathrm{dB}}_{\mathrm{vac}} = \sqrt{2\ln 2}/(k_0\rho_0) \tag{3.107}$$

In summary, the array factor for THz radiation, generated in *large* LAEs by a laser beam under Brewster angle incidence, emitted into air under the reflection angle, has the same far-field 3 dB angle as in the case of normal incidence (Eq. (3.79)). However, it should be noted that the area illuminated by a beam of the same width ρ_0 is increased by a factor $1/\cos\theta^{\mathrm{vac}}_{3\mathrm{dB}} \approx n_{\mathrm{sc}}$ (≈ 3.6 in GaAs and InGaAs, e.g.), which implies a maximum tolerable laser power higher by the same factor.

THz Emission into the Semiconductor

With $\vec{k}^{\mathrm{B}}_{\mathrm{sc}}$ from Eq. (3.98) the expression for the array factor, Eq. (3.101), becomes for the THz emission into the semiconductor

$$\mathrm{AF}(\theta,\varphi;\theta^{\mathrm{B}}_{\mathrm{sc}}) = \left(\frac{n_0\pi\rho_0^2}{\cos\theta^{\mathrm{B}}_{\mathrm{vac}}}\right)^2 \exp\left[-\frac{1}{2}\left(\frac{n_{\mathrm{sc}}\left(k_{0,x} - k^{B}_{0,x}\right)\rho_0}{\cos\theta^{\mathrm{B}}_{\mathrm{vac}}}\right)^2\right] \exp\left[-\frac{1}{2}\left(n_{\mathrm{sc}}k_{0,y}\rho_0\right)^2\right] \tag{3.108}$$

or, in polar coordinates,

$$\mathrm{AF}(\theta,\varphi;\theta^{\mathrm{B}}_{\mathrm{sc}}) = \left(\frac{n_0\pi\rho_0^2}{\cos\theta^{\mathrm{B}}_{\mathrm{vac}}}\right)^2 \exp\left[-\frac{1}{2}\left(\frac{n_{\mathrm{sc}}k_0\rho_0}{\cos\theta^{\mathrm{vac}}_{\mathrm{B}}}\left(\sin\theta_{\mathrm{sc}}\cos\varphi - \sin\theta^{\mathrm{B}}_{\mathrm{sc}}\right)\right)^2\right] \exp\left[-\frac{1}{2}\left(n_{\mathrm{sc}}k_0\rho_0\sin\theta_{\mathrm{sc}}\sin\varphi\right)^2\right] \tag{3.109}$$

As expected, the maximum of the array factor is now found at $\theta_{sc} = \theta_{sc}^{3dB}$ and $\varphi = 0$, whereas its value remains the same as for emission into air. The increase in the wave vector due to the refractive index n_{sc} results in reduced 3 dB angles compared to emission into air, as in the case of normal incidence of the laser beam discussed in the paragraph on *photoconductive scenario*. For incidence of the laser beam under the Brewster angle, however, the elliptic shape of the illuminated area now results in an elliptic angular dependence of the array factor, due to the rather small angle θ_{sc}^{3dB}, in contrast to the case of emission into air.

As before, the 3 dB angle in the xz-plane ($\varphi = 0$) follows from the first exponential factor in Eq. (3.109)

$$\exp\left[-\frac{1}{2}\left(\frac{n_{sc}k_0\rho_0}{\cos\theta_B^{vac}}\left(\sin\theta_{sc}^{3dB} - \sin\theta_{sc}^{B}\right)\right)^2\right] = \frac{1}{2} \tag{3.110}$$

With the same trigonometric relations for $\sin\theta_{sc}^{3dB} - \sin\theta_{sc}^{B}$ as before, noting that for the Brewster angle $\cos\theta_{sc}^{B} = \sin\theta_{vac}^{B}$ and using $\tan\theta_{vac}^{3dB} \approx n_{sc}$ Eq. (3.109) becomes

$$\exp\left[-\frac{1}{2}\left(n_{sc}k_0\rho_0\sin\left\{\theta_{sc}^{3dB} - \theta_{sc}^{B}\right\}\right)^2\right] = \frac{1}{2} \tag{3.111}$$

whence,

$$\Delta\theta_{sc}^{3dB} = \theta_{sc}^{3dB} - \theta_{sc}^{B} = \sqrt{2\ln 2}/(n_{sc}^2 k_0\rho_0) \tag{3.112}$$

for large LAEs. As expected, this angle is reduced by an additional factor $1/n_{sc}$, compared with Eq. (3.79) (with $k = k_{sc} = n_{sc}k_0$), since the illuminated area is elongated by a factor n_{sc}.

The 3 dB angle in the sc-Brewster-y- (By-) plane is determined by the second exponential factor in the expression for the array factor, Eq. (3.108)

$$\exp\left[-\frac{1}{2}\left(n_{sc}k_{0,y}^{3dB}\rho_0\right)^2\right] = \frac{1}{2} \tag{3.113}$$

Introducing the azimuth angle in the By-plane, Φ, as $k_{0,y}/k_0 = \sin\Phi \approx \Phi$, we obtain

$$\exp\left[-\frac{1}{2}\left(n_{sc}k_0\rho_0\sin\Phi_{3dB}\right)^2\right] = \frac{1}{2} \tag{3.114}$$

which yields, for *large* LAEs

$$\Phi_{3dB} \approx \sqrt{2\ln 2}/(n_{sc}k_0\rho_0) \tag{3.115}$$

Comparing this result for the azimuth plane with Eq. (3.79) for the case of normal incidence we see that the 3 dB angle of the array factor is the same. This was expected, as the width of the illuminated area in the y-direction is unchanged for laser beam incidence under the Brewster angle.

An array factor which yields strongly different THz beam profiles for the xz- and the By-plane is undesirable. In principle, this problem can be avoided, if an elliptic laser beam profile with $\rho_{0x} = \rho_0$ and $\rho_{0y} = \rho_0/\cos\theta_{vac}^{B}$ is used for compensation.

Another option is illumination through the semiconductor from the backside, as depicted in Figure 3.57. In this case the maximum of the array factor for THz emission into the semiconductor corresponds to the direction of reflection. The array factor for emission into air exhibits a maximum under the angle of refraction. For an angle of incidence $\theta_{sc} > \theta_c = \arcsin(1/n_{sc})$ no maximum of the array factor at finite angle θ_{vac} exists and destructive interference dominates its value at other angles $(\theta_{vac}, \varphi_{vac})$. The special case $\theta_{sc} = \theta_c$ will turn out particularly interesting. In the next Section 3.6.2.4 we will find that the radiation intensity of the vertical elementary quasi-dipoles $U_e^v(\theta, \varphi)$ in p-i-n structures exhibits a maximum at $\theta = \theta_c$, due to the modifications to $U_e^v(\theta, \varphi)$ introduced by the proximity to the interface to air.

In Figure 3.58a the array factors from Eqs. (3.102) to (3.109), in the xz-plane (i.e., $\varphi = 0$), normalized to the value at $\rho_0 = \lambda_0$ are shown for emission into air and into the semiconductor in a log-polar diagram for $\rho_0 = (0.03, 0.1, 0.3, 1.0, 3.0)\lambda_0$. Comparing Figure 3.58a with Figure 3.55 we note that the maxima of the array factors are not only shifted as expected from Snell's law, but, for the emission into the

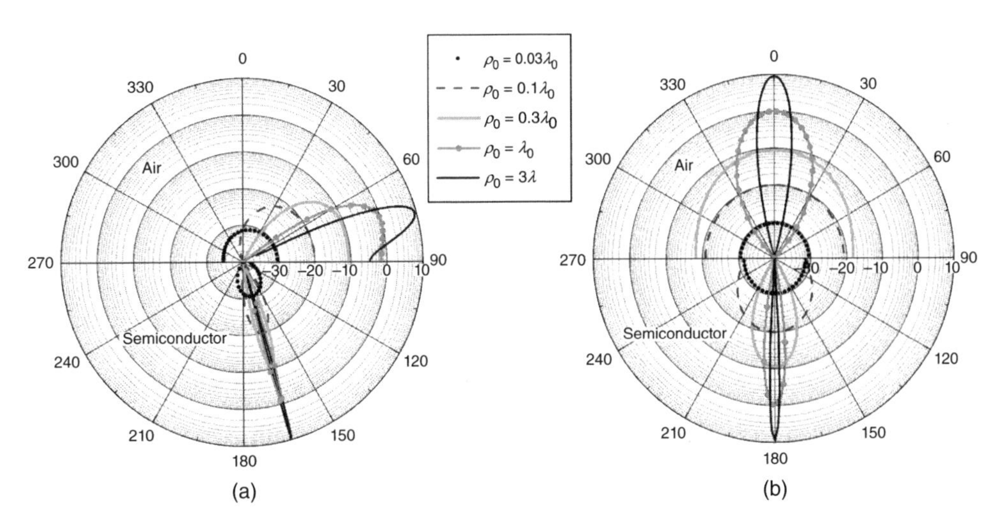

Figure 3.58 Array factors of a vertical LAE in the *xz*-plane. (a) and the By-plane (b) for various excitation spot radii according to Eq. (3.109).

semiconductor, we clearly see the further increased directivity due to the elongation of the illumination spot, as discussed above. In Figure 3.58b the corresponding results are shown for the "air-Brewster-*y*" and "semiconductor-Brewster-*y*" plane defined above. Here, the differences compared to Figure 3.55 are less pronounced.

3.6.2.4 Radiation Intensity Modifications with Respect to the Interface

According to our previous discussion in Section 3.6.2.3 the radiation intensity of LAEs for a given polar angle (θ, φ) can be evaluated in terms of a product of the radiation intensity of the individual elementary quasi-dipoles and a (continuous) array factor. In Section 3.6.2.3a the special cases of horizontal elementary dipoles in combination with normal incidence of the exciting laser beam (see Eqs. (3.67), (3.68), (3.69), and (3.73)) and in Section 3.6.2.3b vertical elementary dipoles in combination with incidence under the Brewster angle were evaluated (Eqs. (3.95), (3.100), and (3.102)). So far, however, the radiation intensities of horizontal and vertical elementary Hertzian dipoles embedded in a semiconductor, Eqs. (3.67) and (3.100) had been assumed. Only in Eq. (3.94) were horizontal Hertzian dipoles in air assumed for a comparison of the power emitted by LAEs into the air or into the semiconductor.

However, the radiation intensity $U_e(\theta, \varphi)$, emitted by a dipole in a semiconductor in the proximity of the interface to air is strongly modified, both for horizontal and vertical dipoles. This becomes obvious, if one considers the electromagnetic waves emitted by a dipole at a small distance from the interface in terms of simple optics. In Figure 3.59, the electromagnetic waves emitted by a dipole are classified into three groups:

1. For polar angles $0 < \theta_{sc} < \pi/2$ all the electromagnetic waves are emitted into the semiconductor.
2. For polar angles $\pi/2 < \theta_{sc} < \pi - \theta'_c$, or more convenient, $\theta'_c < \theta'_{sc} < \pi/2$, the electromagnetic waves are totally reflected at the interface semiconductor/air, suffering a phase shift, but continue traveling in the semiconductor.
3. For the small range $0 < \theta'_{sc} < \theta'_c$ the electromagnetic waves are refracted at the interface semiconductor/air. Only these waves originating from a rather narrow cone are contributing to the emission into air covering the angle range $0 < \theta'_{vac} < \pi$. A part of the intensity is also reflected at the interface.

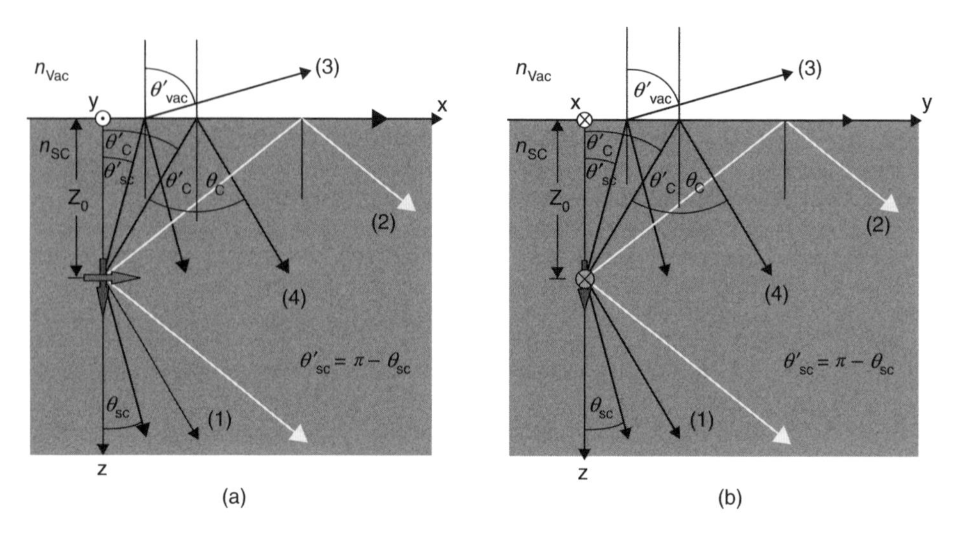

Figure 3.59 Definition of the variables used throughout the text. The angle θ is referenced relative to the z-axis, irrespective of the dipole orientation. For convenience, we also use the supplementary angle $\theta' = \pi - \theta$ with $0 < \theta' < \pi/2$ for the range of angles $\pi/2 < \theta < \pi$. (a) xz-plane and (b) yz plane. The numbered arrows indicate the three discussed cases: (1) emission into the substrate, (2) total internally reflected beams, (3) refracted beam into air. Case (4) represents emission under the critical angle for total internal reflection.

Therefore the radiation intensity and total power emitted into the upper hemisphere will be significantly less compared to the naive picture which assumes that $P_{e,vac} = P_{e,sc}/\sqrt{\varepsilon_{sc}}$ (as in Eqs. (3.93) and (3.94)).

From (1) and (2) it follows that the intensity of the radiation emitted into the semiconductor under angles $0 < \theta < \pi/2$ may even be enhanced by constructive interference with the waves emitted under the angle $\theta' = \pi - \theta$ and reflected under the angle θ. If the distance of the dipole from the interface z_0 is small compared to the THz wavelength, the interference depends only on the phase shift due to the reflection at the interface. It should be noted that total reflection occurs in the whole range $\theta'_c < \theta'_{sc} < \pi/2$. In this range, the amplitude of the electromagnetic wave is unchanged, but the phase shift Δ varies between 0 and π, depending on the angle θ'_{sc} and the polarization.

The problem of a Hertzian dipole in a material of refractive index n_1 $(= n_{sc})$ in proximity to the interface to another material of refractive index n_2 $(= n_{vac})$ was first solved for a radiating atom by Lukosz and Kunz [10–12]. Their results were adapted to the present situation in Ref. [77]. Only the components emitted into the far-field will be discussed in order to derive analytical expressions, and near-field effects will be neglected (they are considered small for a Hertzian dipole), following the argumentation of Lukosz and Kunz. In the following subsections these results are applied to calculate the radiation intensities $U_e^{if,v}(\theta, \varphi)$ and $U_e^{if,h}(\theta, \varphi)$ of vertical and horizontal elementary dipoles at the interface.

a) Radiation Intensity of a Vertical Elementary Quasi-Dipole in the Semiconductor Close to the Semiconductor/Air Interface

The radiation intensity for a vertical quasi-dipole embedded into the semiconductor given by Eq. (3.95) can be rewritten in terms of the total emitted power P_e, by adapting Eq. (3.53), as

$$U_e^v(\theta_{sc}) = \frac{1}{2} \cdot \frac{\sqrt{\varepsilon_{sc}}}{16\pi^2 \varepsilon_0 c_0^3} (ea_0)^2 \sin^2\theta_{sc} = \frac{3}{8\pi} P_e \sin^2\theta_{sc}, \quad \text{for } 0 \leq \theta_{sc} < \pi \tag{3.116}$$

The radiation intensity $U_{e,vac}^v(\theta'_{vac})$ emitted into the air by a vertical dipole in proximity to the semiconductor/air interface under an angle $0 \leq \theta'_{vac} \leq \pi/2$ is obtained [11, 12] by using the Fresnel equations in combination with the condition that the power emitted by the dipole into the solid angle $d\Omega_{sc} = 2\pi \sin\theta'_{sc} d\theta'_{sc}$ (with $0 \leq \theta'_{sc} < \theta'_c$) is partly reflected back into the semiconductor or transmitted into the air into the corresponding solid angle $d\Omega_{vac} = 2\pi \sin\theta'_{vac} d\theta'_{vac}$. The radiation intensity transmitted into air is then

$$U^{v}_{e,vac}(\theta'_{vac}) = U^{v}_{e,sc}(\theta'_{sc}) T(\theta'_{sc}) \frac{d\Omega_{sc}}{d\Omega_{vac}}, \tag{3.117}$$

where $T(\theta_{sc})$ is the Fresnel transmittivity. The angles θ'_{sc} and θ'_{vac} are related to each other by Snell's law, $\sin\theta'_{vac} = n_{sc} \sin\theta'_{sc}$. In this way, one obtains for the vertical dipole which only contains p-polarized components [11],

$$U^{if,v}_{e,vac}(\theta'_{vac}) = \frac{3}{8\pi} \cdot \frac{P_e}{n_{sc}} \cdot \left(\frac{\sin\theta'_{vac}}{n_{sc}} \right)^2 \cdot \left[\frac{2\cos\theta'_{vac}}{\cos\theta'_{sc} + n_{sc}\cos\theta'_{vac}} \right]^2 \quad \text{with } \theta'_{sc} < \theta'_c \tag{3.118}$$

Comparing Eq. (3.118) with Eq. (3.116) we note that (i) the factor P_e is reduced by the factor $1/n_{sc}$ (as naively expected), (ii) the factor $\sin^2\theta_{vac}$ is replaced by $\sin^2\theta_{vac}/n_{sc}{}^2$ describing the power redistribution from within the cone $0 < \theta'_{sc} < \theta'_c$ over the full range $0 < \theta'_{vac} < \pi/2$, and (iii) the interference term (square brackets in Eq. (3.118)) also is ≤ 1. In particular, we note that $U^{if,v}_{e,vac}(\theta'_{vac})$ is zero not only at $\theta'_{vac} = \pi$, but vanishes also at $\theta'_{vac} = \pi/2$.

The radiation intensity $U^{if,v}_{e,sc}(\theta_{sc})$ emitted into the semiconductor is determined by the superposition of the direct wave emitted under the angle $\theta_{sc} < \pi/2$ with the reflected component of the wave emitted under the angle $\theta'_{sc} = \pi - \theta_{sc}$. Since the solid angles are the same, the radiation intensity becomes

$$U^{v}_{e,sc}(\theta_{sc}) = U^{v}_{e,sc}(\theta_{sc})[1 + r(\theta'_{sc})]^2, \tag{3.119}$$

with the Fresnel field reflection coefficient $r(\theta'_{sc})$. For the range of partial transmission, that is, $\theta_{sc} < \theta_c$, one obtains

$$U^{if,v}_{e,sc}(\theta_{sc}) = \frac{3}{8\pi} \cdot P_e \sin^2\theta_{sc} \cdot \left[\frac{2\cos\theta_{sc}}{\cos\theta_{sc} + n_{sc}\cos\theta'_{vac}} \right]^2, \quad \text{for } 0 \leq \theta_{sc} \leq \theta_c \tag{3.120}$$

The difference in comparison with the fully embedded dipole, Eq. (3.116), is the squared term in brackets that reflects the superposition of the direct and the reflected electric field. For $\theta_{sc} = 0$, this term becomes $|1 + r(0)|^2 = [2/(1 + n_{sc})^2] = 0.19$, attributing for the losses due to partial transmission into air and destructive interference. For larger angles, the interference becomes constructive, reaching the maximum value of $|1 + r(\theta'_c)|^2 = 4$ at $\theta_{sc} = \theta_c$.

For the range of total reflection, that is, $\theta_c < \theta'_{sc} < \pi/2$, the constructive interference between direct and reflected wave turns into destructive with θ'_{sc} increasing from θ_c to $\pi/2$. This yields [11]

$$U^{if,v}_{e,sc}(\theta_{sc}) = \frac{3}{8\pi} \cdot P_e \sin^2\theta_{sc} \cdot \frac{(2\cos\theta_{sc})^2}{(n^2_{sc} - 1)[(n^2_{sc} + 1)\sin^2\theta_{sc} - 1]}, \quad \text{for } \theta_c < \theta_{sc} < \pi/2 \tag{3.121}$$

The interference term is again 4 at $\theta_{sc} = \theta_c$, ensuring the continuity of $U^{if,v}_{e,sc}(\theta_{sc})$. For $\theta_{sc} > \theta_c$, however, it decreases faster than $1/\sin^2\theta_{sc}$ and even becomes zero at $\theta_{sc} = \pi/2$.

Figure 3.60 depicts a log-scale polar diagram for both $U^{if,v}_{e,vac}(\theta'_{vac})$ and $U^{if,v}_{e,sc}(\theta_{sc})$. Also shown are the results for the dipole embedded into the semiconductor for $0 < \theta_{sc} < \pi/2$ and for the same dipole in air for $0 < \theta'_{vac} < \pi/2$ by dashed lines.

b) Radiation Intensity of a Horizontal Elementary Quasi-Dipole in the Semiconductor Close to the Semiconductor/Air Interface

The radiation intensity of a horizontal quasi-dipole oriented in the x-direction and embedded into the semiconductor given in Eq. (3.53), can be rewritten in terms of the total emitted power P_e as

$$U^{h}_{e}(\theta_{sc}, \varphi) = \frac{1}{2} \cdot \frac{\sqrt{\varepsilon_{sc}}}{16\pi^2\varepsilon_0 c_0^3}(ea_0)^2(1 - \sin^2\theta_{sc}\cos^2\varphi) = \frac{3}{8\pi} P_e (1 - \sin^2\theta_{sc}\cos^2\varphi) \tag{3.122}$$

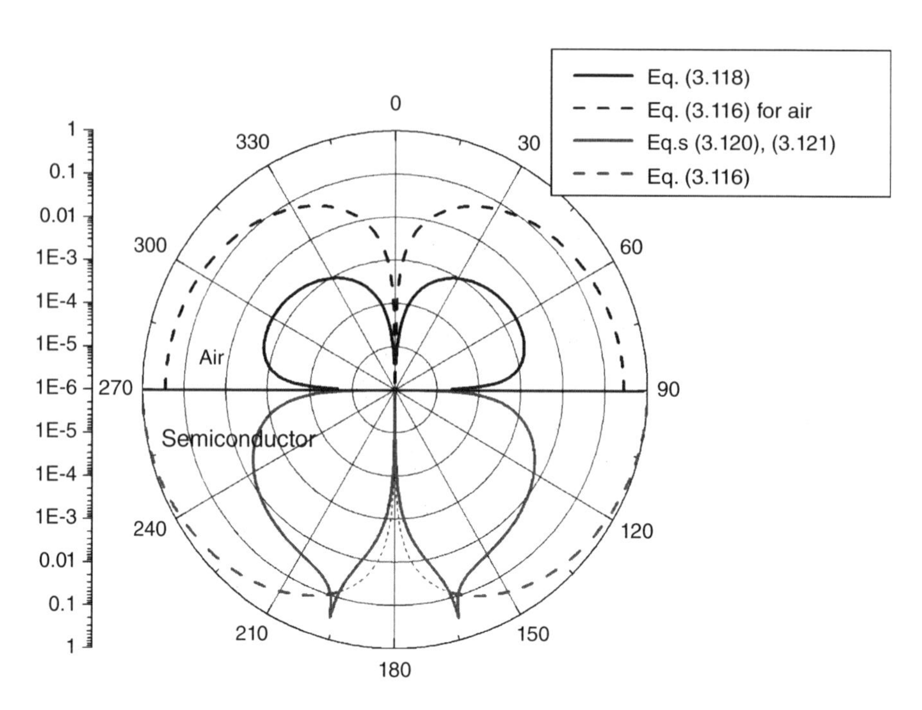

Figure 3.60 Dashed lines: Power emitted by a vertical Hertzian dipole embedded in air (top) or in a semiconductor with $n_{sc} = 3.6$. Solid lines: Radiation patterns of the dipole in proximity to the interface to air according to Eqs. (3.118), (3.120), and (3.121).

The radiation intensity now depends also on the polar angle φ. For $\varphi = 0$ the $\vec{k}$-vectors are in the xz-plane. As the polarization of the electromagnetic field of the THz waves is parallel to this plane (for this reason also called "E-plane"), this wave is referred to as "p-polarized" and the corresponding waves will be labeled by the superscript "p" in the following. This situation is depicted in Figure 3.59a. For $\varphi = \pi/2$ the $\vec{k}$-vectors are in the yz-plane (see Figure 3.59b). In this plane, the polarization of the electric field of the waves emitted by the dipole is perpendicular to this plane (also referred to as "H-plane," as the *magnetic* field is parallel to it). These waves are conventionally called "s-polarized" (from German "senkrecht" for "perpendicular") and they will be labeled by the superscript "s." For arbitrary values of the angle φ the waves are a superposition of p- and s-polarized waves. As the Fresnel boundary conditions at the interface are different for s- and p-polarized waves, this will also affect the modifications of the radiation intensity of the horizontal dipole in proximity to the interface. For the calculation of the corresponding radiation intensities $U_{\mathrm{e}}^{\mathrm{if,h,p}}(\theta_{\mathrm{sc}}, \varphi)$ and $U_{\mathrm{e}}^{\mathrm{if,h,s}}(\theta_{\mathrm{sc}}, \varphi)$ we proceed in a way analogous to the case of the vertical dipole. As before, only waves emitted under the angle $0 < \theta'_{\mathrm{sc}} < \theta'_{\mathrm{c}}$ are partially transmitted into air. For the p- and s-polarized components one finds

$$U_{\mathrm{e,vac}}^{\mathrm{if,h,p}}(\theta'_{\mathrm{vac}}, \varphi) = \frac{3}{8\pi} \cdot \frac{P_{\mathrm{e}}}{n_{\mathrm{sc}}} \cdot \left[\frac{2\cos\theta'_{\mathrm{vac}} \cos\theta'_{\mathrm{sc}} \cos\varphi}{\cos\theta'_{\mathrm{sc}} + n_{\mathrm{sc}}\cos\theta'_{\mathrm{vac}}} \right]^2, \text{ with } 0 < \theta'_{\mathrm{vac}} < \pi/2 \tag{3.123}$$

and

$$U_{\mathrm{e,vac}}^{\mathrm{if,h,s}}(\theta'_{\mathrm{vac}}, \varphi) = \frac{3}{8\pi} \cdot \frac{P_{\mathrm{e}}}{n_{\mathrm{sc}}} \cdot \left[\frac{2\cos\theta'_{\mathrm{sc}} \sin\varphi}{n_{\mathrm{sc}}\cos\theta'_{\mathrm{sc}} + \cos\theta'_{\mathrm{vac}}} \right]^2, \text{ with } 0 < \theta'_{\mathrm{vac}} < \pi/2, \tag{3.124}$$

respectively. In addition to the naively expected reduction by the factor $1/n_{sc}$, the radiation intensity is further reduced by the expression in squared brackets, which amounts to $4\cos^2\varphi/(1+n_{sc})^2$ (= 0.19, for $n_{sc}=3.6$ and $\varphi=0$) at $\theta'_{vac}=\theta'_{sc}=0$ and becomes zero at $\theta'_{vac}=\pi/2$.

The corresponding emission into the semiconductor for angles below the critical angle is again determined by the interference between the direct waves with $0<\theta_{sc}<\pi/2$ and the corresponding partially reflected ones with $0<\theta'_{sc}<\theta_c$. The radiation intensity becomes

$$U_{e,sc}^{if,h,p}(\theta_{sc},\varphi) = \frac{3}{8\pi}P_e\left[\frac{2n_{sc}\cos\theta'_{vac}\cos\theta_{sc}\cos\varphi}{\cos\theta_{sc}+n_{sc}\cos\theta'_{vac}}\right]^2, \text{ with } 0<\theta_{sc}<\theta_c \tag{3.125}$$

and

$$U_{e,sc}^{if,h,s}(\theta_{sc},\varphi) = \frac{3}{8\pi}P_e\left[\frac{2n_{sc}\cos\theta_{sc}\sin\varphi}{n_{sc}\cos\theta_{sc}+\cos\theta'_{vac}}\right]^2 \text{ with } 0<\theta_{sc}<\theta_c, \tag{3.126}$$

for the p- and s-polarized waves, respectively. Compared with the vertical dipole embedded into the semiconductor, the radiation intensity is increased by a factor $4(n_{sc}\cos\varphi/(1+n_{sc}))^2$ (= 2.45, for $n_{sc}=3.6$ and $\varphi=0$) at $\theta_{sc}=0$, due to constructive interference. For the p-polarized wave the radiation intensity decreases to zero with increasing θ_{sc} due to increasingly destructive interference, as the phase shift between the direct and the reflected wave increases. For the s-polarized wave the interference becomes fully constructive at $\theta_{sc}=\theta_c$, resulting in a fourfold increase of the radiation intensity at $(\theta_{sc},\varphi)=(\theta_c,0)$ compared to the embedded dipole.

In the range of total reflection of the THz waves, $\theta_c<\theta_{sc}\leq\pi/2$, the radiation intensities are again determined by the angular dependence of the phase shift. One obtains

$$U_{e,sc}^{if,h,p}(\theta_{sc},\varphi) = \frac{3}{8\pi}\cdot P_e\frac{(2n_{sc}\cos\theta_{sc}\cos\varphi)^2(n_{sc}^2\sin^2\theta_{sc}-1)}{(n_{sc}^2-1)[(n_{sc}^2+1)\sin^2\theta_{sc}-1]}, \text{ for } \theta_c<\theta_{sc}<\pi/2 \tag{3.127}$$

and

$$U_{e,sc}^{if,h,s}(\theta_{sc},\varphi) = \frac{3}{8\pi}\cdot P_e\frac{(2n_{sc}\cos\theta_{sc}\sin\varphi)^2}{(n_{sc}^2-1)} \text{ for } \theta_c<\theta_{sc}<\pi/2, \tag{3.128}$$

respectively.

In Figure 3.61, log-scale polar diagrams of $U_{e,vac}^{if,h}(\theta'_{vac},\varphi)$ and $U_{e,sc}^{if,h}(\theta_{sc},\varphi)$ are shown for the E-plane ($\varphi=0$) and the H-plane ($\varphi=\pi/2$). Also shown are the corresponding results for the dipole embedded into the semiconductor (for $0<\theta_{sc}<\pi/2$) and in air (for $\pi/2<\theta_{vac}<\pi$) by dashed lines.

3.6.2.5 Dependence of the Acceleration Term on the Transport Properties and on CW or Pulse Excitation

In Section 3.6.2.3 we assumed a fixed Gaussian distribution of photogenerated carriers, $n(\vec{\rho})$ according to Eq. (3.65) and a coherent quasi-acceleration $\vec{a}(t)=\vec{a}_0\cos\omega t$ for each of them. The assumption regarding the *spatial distribution* of the carriers is realistic, if a Gaussian laser beam is used for the carrier generation. This had allowed us to take into account correctly interference effects between carriers at different $\vec{\rho}$ via a "continuous array factor", AF(θ,φ). Together with the angular dependence of the radiation intensity of the accelerated carriers in proximity to the semiconductor/air interface, $U_e^{if,i}(\theta,\varphi)$ (with i = v or h for vertical or horizontal DC electric fields, respectively), calculated in the preceding section, we were able to obtain a realistic picture about the *angular distribution* of the THz radiation emitted at the THz frequency $f_{THz}=\omega_{THz}/2\pi$. For a realistic calculation of the *strength* of the emitted THz-fields,

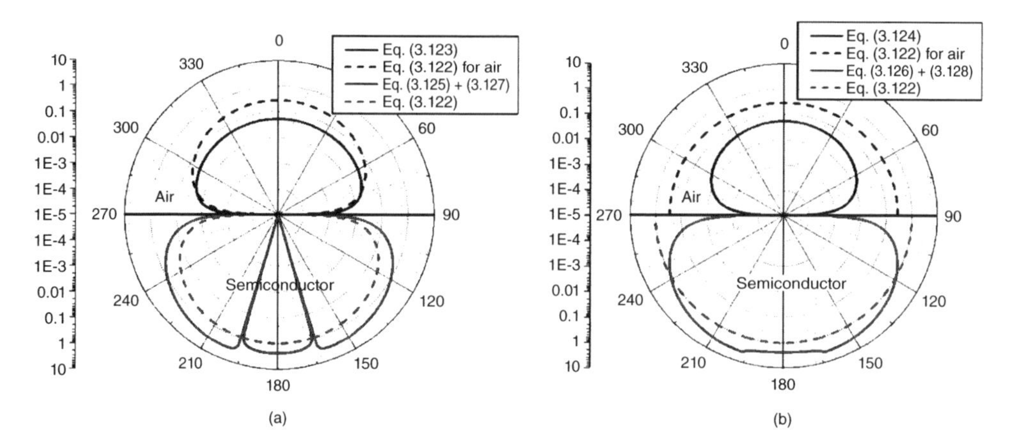

Figure 3.61 Radiation characteristics for the horizontal dipole. Dashed lines: dipole embedded in air (top) or in a semiconductor with $n_{sc} = 3.6$. Solid lines: dipole in proximity to the semiconductor to air interface (a) p-polarization ($\varphi = 0$). (b) s-polarization ($\varphi = 90°$). Dashed lines: horizontal Hertzian dipole embedded in air (top) or in a semiconductor with $n_{sc} = 3.6$. Solid lines: the dipole in proximity to the interface to air.

radiation intensity, and power it is necessary to consider in detail the temporal evolution of the number of carriers and their dynamics. In this way, it is no longer necessary to consider their *spatial distribution*. Therefore we can derive the required quantities in terms of generation rates for the *total number* of carriers, $dN(t')/dt'$. With a time-dependent laser power $P_L(t')$ of photon energy $h\upsilon$ and an external quantum efficiency η^{I}_{ext} for the generation of an electron–hole pair in the semiconductor per incident photon from Eq. (2.10) in Chapter 2 the total rate of electron–hole generation at the time t' becomes

$$\frac{\partial N\ (t')}{\partial t'} = \eta^{I}_{ext}\ \frac{P_L\,(t')}{h\upsilon} = (1 - R)[1 - \exp(-\alpha d)]\frac{P_L(t')}{h\upsilon} \tag{3.129}$$

Later, we will consider the two cases of particular interest for $P_L(t')$, that is, either pulsed excitation with a short Gaussian pulse as in Section 2.2.2.2

$$P_L(t') = P_0\ \exp[-(t'/\tau_{pls})^2], \tag{3.130}$$

or CW mixing with a beam obtained from two heterodyned lasers of same polarization and power P_0, differing in frequency by f_{THz}. This yields for Eq. (2.4) from Section 2.2.1

$$P_L(t') = 2P_0[1 + \cos\,\omega_{THz}t'] \tag{3.131}$$

Now we evaluate the acceleration dynamics of the photogenerated carriers. With a DC electric field E_0 in the semiconductor each charge carrier will experience a transient acceleration starting right at the time t' of its creation. For its contribution to THz generation at the time t its actual acceleration at that time is relevant. We now look at the individual carriers, indicated by the subscript, "e", with an acceleration of

$$a_e(t, t') = a_e\,(t - t') = \frac{\partial v_e(t - t')}{\partial t}, \tag{3.132}$$

not to be confused with the quasi-acceleration in Eq. (3.47). In Section 3.6.2.2, various scenarios for $a(t - t')$ were discussed. Depending on the device (photoconductor or p-i-n photodiode), the acceleration period can be extremely short at high fields, for example, resulting in acceleration times of a few tens of femtoseconds before reaching a constant drift velocity, v_{dr}, typically the high-field saturation velocity $v_{sat} \approx 10^7$ cm/s. But, in high-mobility semiconductors, the transient acceleration can extend over a few 100 fs at intermediate fields. In this case the transport may be ballistic. The ballistic transient $a_e(t-t')$ starts right after the generation at time t' with a constant acceleration $a_e = eE_0/m^*$. After reaching the maximum ballistic velocity $v_{bal}(\gg v_{sat})$ a deceleration down to v_{sat} immediately follows. Whether non-ballistic or ballistic, the acceleration transient terminates with a deceleration down to $v=0$, either due to capture by a deep trap, recombination with a charge carrier of opposite charge (e.g., in low temperature grown (LTG) GaAs photoconductors), recombination at a contact, or by reaching the field-free region of a p-i-n photodiode. Therefore, the integral of an individual photocarrier over its transient always yields

$$\int_{t'}^{\infty} a_e(t-t')\mathrm{d}t = 0, \tag{3.133}$$

independent of other details of the transport. This implies that all the positive contributions to the emitted THz field $E_{THz}(t)$ will be compensated by contributions of the opposite sign, as the THz field $E_{THz}(t)$ at any time is proportional to the acceleration $a_e(t-t')$

$$\int_{t'}^{\infty} E_{THz,e}(t-t')\mathrm{d}t = 0 \tag{3.134}$$

Of course, this is also true for a macroscopic THz pulse resulting from absorption of a short laser pulse. But, as already discussed in Chapter 2, the time dependence of $a_e(t-t')$ and $E_{THz,e}(t)$ may be highly asymmetric, depending on the experimental conditions. If, for instance, the range of negative acceleration and THz field covers a much longer time compared to that of positive fields, these fields will be much smaller and contribute only to the low frequency components in the Fourier spectrum of the THz pulse and their observation may be obscured.

It should be emphasized that both photogenerated electrons and holes contribute to the generation of the THz signal. As the contributions of the holes are generally smaller due to their heavier effective mass, and as holes do not exhibit ballistic transport, we restrict our evaluation to electrons. The contributions to the THz fields, originating from the holes, may simply be added.

We now calculate the macroscopic acceleration, $A(t)$, responsible for the THz emission at time t, and defined by the integral over all contributions resulting from all carriers generated at earlier times t',

$$A(t) = \int_{-\infty}^{t} a_e\ (t-t')\ \frac{\partial N(t')}{\partial t'}\mathrm{d}t'. \tag{3.135}$$

or, with Eq. (3.129),

$$A(t) = \int_{-\infty}^{t} a_e\ (t-t')\eta_{ext}^{I}\ \frac{P_L(t')}{h\upsilon}\mathrm{d}t' \tag{3.136}$$

For the evaluation of Eq. (3.136) it is instructive to consider the time dependence of $a_e(t-t')$ in comparison to that of $P_L(t')$. As, according to Eq. (3.133), the positive and negative contributions to $a_e(t-t')$ have similar weight, it is necessary for efficient THz generation to minimize destructive interference of positive $a_e(t-t_1')$ with negative $a_e(t-t_2')$ due to carriers generated earlier at $t_2' < t_1'$. This can be achieved, if, for instance, $P_L(t_1') \gg P_L(t_2')$, or vice versa. Our following evaluation of Eq. (3.136) for the cases of ballistic and non-ballistic transport under CW or pulse excitation will illustrate this in more detail.

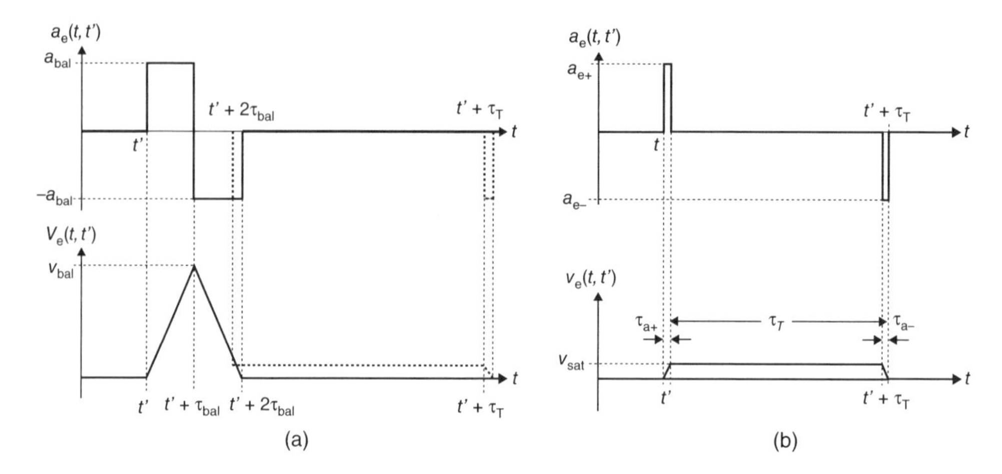

Figure 3.62 Acceleration and velocity for (a) ballistic transport and (b) transport at the saturation velocity.

a) Ballistic Transport

In Chapter 2 Monte Carlo (MC) simulations of ballistic transport in InGaAs were shown in the right panel of Figure 2.6b and some estimates were discussed later in connection with pulse excitation. The results of the MC simulations justify the following simplified picture (see Figure 3.62a). The maximum ballistic velocity v_{bal} is reached within the time τ_{bal} according to the equation

$$a\ \tau_{bal} = v_{bal}(\approx 10^8\text{cm/s, for GaAs, and } 2 \cdot 10^8 \text{ cm/s, for InGaAs, e.g.}) \tag{3.137}$$

As the acceleration scales linearly with the DC electric field, according to $a = eE_0/m^*$, the ballistic acceleration time $\tau_{bal} = \frac{v_{bal}m*}{eE_0}$ scales inversely with the field. After reaching v_{bal}, the electron is slowed down to v_{sat} ($\ll v_{bal}$) by a deceleration $\approx -a$ (due to scattering into the side valleys of the conduction band) within another period of time of τ_{bal}, that is,

$$a_e(t) = \begin{cases} v_{bal}/\tau_{bal} = a & t - \tau_{bal} < t' < t \\ -v_{bal}/\tau_{bal} = -a & t - 2\tau_{bal} < t' < t - \tau_{bal} \\ 0 & t' < t - 2\tau_{bal} \end{cases} \tag{3.138–3.140}$$

As $v_{sat} \ll v_{bal}$, the small contribution to $a(t - t')$ due to the final deceleration to $v = 0$ at the end of the lifetime of the ballistic carrier (see dashed line in Figure 3.62a) is (in general) negligible.

CW Photomixing: For CW mixing with periodic $P_L(t')$ from Eq. (3.131) and $a(t - t')$ from Eq. (3.140) and a device with a transport layer length adapted to the length covered by the electron during the acceleration–deceleration period, inserted into Eq. (3.138), yields

$$A(t) = \eta_{ext}^{I} \frac{2P_0}{h\upsilon} \cdot \frac{v_{bal}}{\tau_{bal}} \left(\int_{t-\tau_{bal}}^{t} \left[1 + \cos \omega_{THz} t'\right] dt' - \int_{t-2\tau_{bal}}^{t-\tau_{bal}} \left[1 + \cos \omega_{THz} t'\right] dt' \right) \tag{3.141}$$

After integration,

$$A(t) = -\eta_{ext}^{I} \frac{2P_0}{h\upsilon} \cdot \frac{2v_{bal}}{\tau_{bal}} \frac{(\cos \omega_{THz}\tau_{bal} - 1)}{\omega_{THz}} \sin \omega_{THz}(t - \tau_{bal}) \tag{3.142}$$

that is, the maximum amplitude of $A(t)$ occurs approximately for $\tau_{bal} = 1/(2f_{THz}) = ½\, T_{THz}$. For InGaAs, this condition is satisfied for $f_{THz} = 1$ THz at a (rather small) field of $E_0 \approx 10$ kV/cm (see Chapter 2, Figure 2.6b, right panel), and with $a = (v_{bal}/\tau_{bal}) = 4 \cdot 10^{20}$ cm/s^2, for example.

The macroscopic acceleration from Eq. (3.142) can now be expressed as a product of the number of electrons generated per acceleration–deceleration period,

$$N_0 = \eta_{ext}^{I} \frac{4\ P_0}{h\upsilon} \tau_{bal} \tag{3.143}$$

and a periodic quasi-acceleration with amplitude

$$a_0(\omega_{THz}\,\tau_{bal}) = \frac{v_{bal}}{\tau_{bal}} \left[\frac{1 - \cos\left(\omega_{THz}\,\tau_{bal}\right)}{\omega_{THz}\,\tau_{bal}} \right] \tag{3.144}$$

resulting in

$$A(t) = N_0 a(\omega_{THz}\,\tau_{bal}). \tag{3.145}$$

As the macroscopic acceleration $A(t)$ is the sum of the accelerations experienced at the time t by all the previously generated carriers, the meaning of the quasi-acceleration $a(t)$ is the *average contribution per carrier* at the time t. $A(t)$ is shifted in time by τ_{bal} because of the inversion symmetry of $a_e(t)$ with respect to τ_{bal}. This inversion symmetry also yields a phase shift by $\pi/2$ (cos → sin). The frequency dependence of the quasi-acceleration in Eq. (3.144) reflects the fact that at low frequencies ($f_{THz} \ll (2\tau_{bal})^{-1}$), the contributions of recently generated accelerating carriers interfere destructively with a similar number of earlier generated decelerating carriers resulting in a linear increase of $a_0\,(\omega_{THz}\,\tau_{bal})$. Obviously, destructive interference reaches a (first!) minimum and the quasi-acceleration its first maximum close to $f_{THz} = f_{THz}^{opt} = (2\tau_{bal})^{-1}$ where the THz period agrees with the acceleration-deceleration period of $2\tau_{bal}$.

N_0 and $a_0(\omega_{THz}\,\tau_{bal})$ from Eqs. (3.143) to (3.144) are the quantities to be used in the expressions for the array factors AF(θ, φ) (Eqs. (3.73) and (3.103) or (3.108) and (3.109)) and the radiation intensities $U_e^{if,i}(\theta, \varphi)$ (i = v or h) for vertical or in-plane electric fields, Eqs. (3.116)–(3.121) and (3.122)–(3.128), for CW-LAEs exhibiting ballistic transport. In particular, N_0 is related to the quantity n_0 entering into the array factors by

$$N_0 = n_0 \pi \rho_0{}^2. \tag{3.146}$$

Pulsed Excitation: For excitation with a Gaussian pulse with $P_L(t')$ according to Eq. (3.130), the macroscopic acceleration becomes

$$A(t) = \eta_{ext}^{I} \frac{P_0}{h\upsilon} \cdot \frac{v_{bal}}{\tau_{bal}} \left(\int_{t-\tau_{bal}}^{t} \exp\left[-t'^2/\tau_{pls}^2\right] dt' - \int_{t-2\tau_{bal}}^{t-\tau_{bal}} \exp\left[-t'^2/\tau_{pls}^2\right] dt' \right) \tag{3.147}$$

This expression can again be factorized in terms of the total number of carriers generated by the pulse

$$N_0 = \sqrt{\pi}\eta_{ext}^{I} \frac{P_0}{h\upsilon} \tau_{pls} \tag{3.148}$$

and a quasi-acceleration of the electrons

$$a(t;\ \tau_{pls}, \tau_{bal}) = \frac{v_{bal}}{\tau_{bal}} \cdot \frac{1}{\sqrt{\pi}\,\tau_{pls}} \left(\int_{t-\tau_{bal}}^{t} \exp\left[-t'^2/\tau_{pls}^2\right] dt' - \int_{t-2\tau_{bal}}^{t-\tau_{bal}} \exp\left[-t'^2/\tau_{pls}^2\right] dt' \right) \tag{3.149}$$

Nowadays it is easy to generate femtosecond-pulses with a width τ_{pls} smaller than τ_{bal} (i.e., $\tau_{pls} < \tau_{bal} = v_{bal}/a(E_0)$; see Chapter 2, Figure 2.6b, right panel) up to fields of about 100 kV/cm. For $\tau_{pls} \ll \tau_{bal}$ the integration of Eq. (3.147) leads to an error-function-broadened unity-step function with

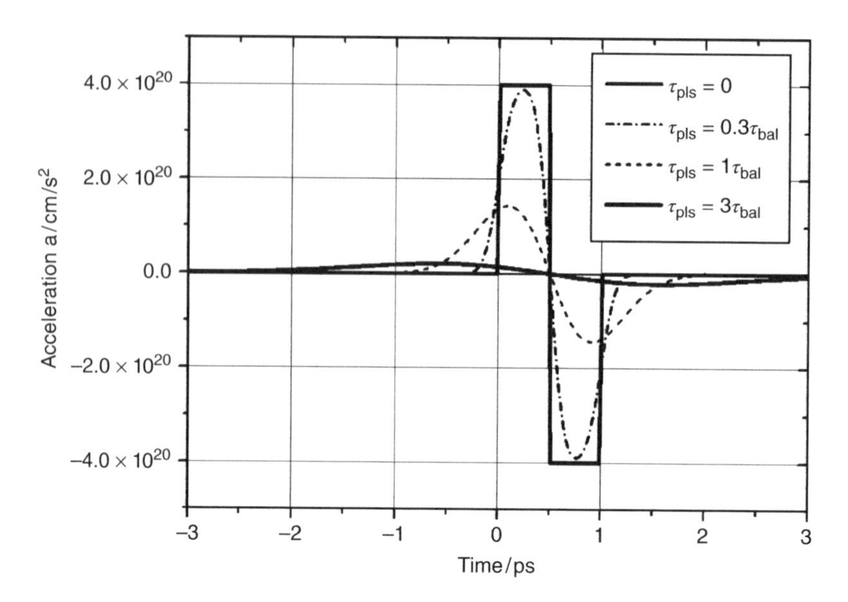

Figure 3.63 Quasi-acceleration for several pulse lengths for a ballistic flight time of $\tau_{bal} = 500$ fs.

a step to +1 at $t \approx 0$, a step down to −1 at $t \approx \tau_{bal,}$ and a step back to zero at $t \approx 2\tau_{bal}$. This reflects the nearly time-coherent acceleration and deceleration of the carriers which were generated within a short time interval compared with τ_{bal}. This acceleration results in a one-cycle quasi-rectangular THz pulse of fundamental frequency $f_{THz,1} = (2/\tau_{bal})$, and containing harmonics at $(3, 5, \dots (2n+1))f_{THz,1}$ (see also the maxima of the dashed line in Figure 2.9b of Chapter 2). For longer laser pulses the shape of $a(t;\, \tau_{pls,}\, \tau_{bal})$ and, hence, of the THz pulse $E_{THz}(t)$ approaches a quasi-harmonic single cycle pulse for $\tau_{pls} \approx \tau_{bal}$. For $\tau_{pls} > \tau_{bal}$ the amplitude decreases strongly due to destructive interference between accelerating and decelerating electrons. It should be noted that this discussion, based on the idealized model for ballistic transport is overestimating the higher harmonics. For illustration, $a(t; \tau_{pls,} \tau_{bal})$ is depicted for $\tau_{pls} = (0, 0.3, 1, 3)\tau_{bal}$ and $\tau_{bal} = 500$ fs in Figure 3.63. Obviously, the quasi-acceleration decreases quite quickly, if the laser pulse length τ_{pls} becomes similar to τ_{bal} or even exceeds τ_{bal}.

b) Non-Ballistic Transport in Semiconductors with Short Recombination Lifetimes

If the transport is non-ballistic, the following simplifying picture for the transport can be used. The velocity of a carrier generated by absorption of a photon at time t' increases monotonically from (close to) zero to its stationary value due to acceleration by the DC electric field E_0. It keeps this constant velocity during the whole transport time τ_T, until it is decelerated down to zero velocity by being captured by a recombination center (see Figure 3.62b). Here, we assume that the stationary velocity is of the order of the saturation velocity $v_{sat} \approx 10^7$ cm/s. We further replace the transient acceleration process by a constant acceleration $a_e^{\pm}$ acting during an acceleration period $\tau_a^{\pm}$, that is,

$$\int a_e^+(t-t')\mathrm{d}t' = a_e^+\,\tau_a^+ = v_{sat} \approx 10^7\ \mathrm{cm/s} \tag{3.150}$$

Below we shall see that the results for the macroscopic acceleration $A(t)$ are otherwise insensitive to the choice of a_e^+ and τ_a^+, as long as τ_a^+ is much smaller than the THz period $T_{THz} = 1/f_{THz}$ (for CW photomixing) or the pulse width τ_{pls} (for pulse excitation). Whereas all the electrons generated at the time t' experience the same (positive) acceleration pulse of duration τ_a^+ right after their generation, they experience the (negative) deceleration of the same strength *on the average* after the recombination time τ_{rec}. To take into account the stochastic character of the recombination processes we weight the deceleration pulse at any time $t > t'$ with the corresponding probability of recombining within the time interval $[t - \tau_a^-, t]$

$$p_{rec}\ (t - t') = \tau_a^- \frac{\partial}{\partial t} \exp[-(t - t')/\tau_{rec}] = \frac{\tau_a^-}{\tau_{rec}} \exp[-(t - t')/\tau_{rec}] \tag{3.151}$$

Thus, the positive contribution to the macroscopic acceleration $A^+(t)$ at the time t comes only from the short interval $t - \tau_a^+ < t' < t$

$$A^+(t) = \int_{t-\tau_a^+}^{t} a_e^+ \eta_{ext}^I \frac{P_L(t')}{h\upsilon} dt' \approx a_e^+ \tau_a^+ \eta_{ext}^I \frac{P_L(t')}{h\upsilon} = v_{sat} \eta_{ext}^I \frac{P_L(t')}{h\upsilon} \tag{3.152}$$

Obviously, Eq. (3.152) represents a good approximation, if $(dP_L(t)/dt)\tau_a^+ \ll P_L(t)$.

To obtain the negative contributions to the macroscopic acceleration at time t, one has to include the contributions of the carriers generated at previous times t' and recombine at time t with the appropriate statistical weight, according to Eq. (3.151),

$$A^-(t) = -\int_{-\infty}^{t} a_e^- \frac{\tau_a^-}{\tau_{rec}} \exp[-(t - t')/\tau_{rec}] \eta_{ext}^I \frac{P_L(t')}{h\upsilon} dt'. \tag{3.153}$$

Substituting $t' - t$ by the delay time $\tau = t' - t$ yields, with $a_e^- \tau_a^- = v_{sat}$

$$A^-(t) = -\frac{v_{sat}}{\tau_{rec}} \int_{-\infty}^{0} \exp(\tau/\tau_{rec}) \eta_{ext}^I \frac{P_L(t+\tau)}{h\upsilon} d\tau. \tag{3.154}$$

CW Photomixing: For CW photomixing with a laser signal $P_L(t')$ according to Eq. (3.131) we obtain

$$A^+(t) = v_{sat} \eta_{ext}^I \frac{2P_0}{h\upsilon}(1 + \cos \omega_{THz} t), \tag{3.155}$$

and

$$A^-(t) = -\frac{v_{sat}}{\tau_{rec}} \int_{-\infty}^{0} \eta_{ext}^I \frac{2P_0}{h\upsilon}[1 + \cos \omega_{THz}\ (t + \tau)] \exp(\tau/\tau_{rec}) d\tau, \tag{3.156}$$

resulting in

$$A^-(t) = -v_{sat} \eta_{ext}^I \frac{2P_0}{h\upsilon} \left[1 + \frac{1}{1 + (\omega_{THz} \tau_{rec})^2} (\cos \omega_{THz} t + \omega_{THz} \tau_{rec} \sin \omega_{THz} t) \right]. \tag{3.157}$$

Finally, we obtain for $A(t) = A^+(t) + A^-(t)$,

$$A(t) = v_{sat} \eta_{ext}^I \frac{2P_0}{h\upsilon} \left\{ \left[\frac{(\omega_{THz} \tau_{rec})^2}{1 + (\omega_{THz} \tau_{rec})^2} \right] \cos \omega_{THz} t - \frac{\omega_{THz} \tau_{rec}}{1 + (\omega_{THz} \tau_{rec})^2} \sin \omega_{THz} t \right\} \tag{3.158}$$

Here, again, we can express the macroscopic acceleration as a product of the average number of carriers in the LAE

$$N_0 = \frac{\partial N(t)}{\partial t} \tau_{rec} = \eta_{ext}^I \frac{2P_0}{h\upsilon} \tau_{rec} \tag{3.159}$$

and the amplitude of the quasi-acceleration of the individual charge carriers

$$a_0(\omega_{THz}) = \frac{v_{sat}}{\tau_{rec}} \cdot \frac{\omega_{THz}\tau_{rec}}{\sqrt{1+(\omega_{THz}\tau_{rec})^2}} \tag{3.160}$$

This is a remarkable result, since the quasi-acceleration and THz field of LAEs increase linearly with THz frequency before saturating for $\omega_{THz}\tau_{rec} > 1$. This is contrary to AEs, which exhibit a frequency-independent response for $\omega_{THz}\tau_{rec} < 1$ followed by a $1/\omega_{THz}$ roll-off for $\omega_{THz}\tau_{rec} > 1$. Again, the increase in the quasi-acceleration with THz frequency can be understood in terms of the positive and negative contributions to the total macroscopic acceleration. At frequencies $\omega_{THz} \ll \frac{1}{\tau_{rec}}$ the acceleration contributions to $A(t)$ are more or less compensated by the deceleration contributions, spread in time over τ_{rec}, if the generation rate does not change significantly within τ_{rec}, that is, if $(dP_L(t)/dt)\tau_{rec} \ll P_L(t)$. For increasing frequency the positive contributions, which result from the very short acceleration time τ_a^+ are unaffected, whereas the time dependence of the negative contributions is exhibiting a roll-off for $\omega_{THz}\tau_{rec} > 1$. Hence, increasingly, only the positive contributions survive in this frequency range.

Pulsed Excitation: For pulsed excitation with a Gaussian femtosecond pulse according to Eq. (3.130), the positive and negative contributions to the macroscopic acceleration, Eqs. (3.150) and (3.152), respectively, become

$$A^+(t) = \eta_{ext}^I \frac{P_0}{h\upsilon} \int_{t-\tau_a^+}^{t} \exp(-t'^2/\tau_{pls}^2)dt' \approx v_{sat}\eta_{ext}^I \frac{P_0}{h\upsilon} \exp(-t^2/\tau_{pls}^2) \tag{3.161}$$

we have used again $a_e^+\tau_a^+ = v_{sat}$ in the third term of Eq. (3.159) and

$$A^-(t) = -\frac{v_{sat}}{\tau_{rec}}\eta_{ext}^I \frac{P_0}{h\upsilon} \int_{-\infty}^{0} \exp(\tau/\tau_{rec}) \exp\left[-(t+\tau)^2/\tau_{pls}^2\right] d\tau \tag{3.162}$$

Although the pulse width is typically in the femtosecond range, we have assumed in Eq. (3.161) that $\tau_a^+ \ll \tau_{pls}$ still holds, which is a valid approximation for $\tau_{pls} > 50$ fs and large DC fields in the device. Under these conditions, the positive contributions to $A(t)$ and, hence the THz pulse practically replicate the shape of the laser pulse. For evaluation of the negative contribution, according to Eq. (3.162), both cases regarding $\tau_{rec} > \tau_{pls}$ and $\tau_{rec} < \tau_{pls}$ may be realized. If the recombination lifetime of the used material significantly exceeds "ultrashort" values of $\tau_{rec} \approx 100$ fs and the width of the laser pulse is short, say $\tau_{pls} < 100$ fs, the convolution in Eq. (3.162) yields a broadened exponential behavior

$$A^-(t) = -\frac{v_{sat}}{\tau_{rec}}\eta_{ext}^I \frac{P_0}{h\upsilon} \sqrt{\pi}\,\tau_{pls} \exp(-t/\tau_{rec}) \tag{3.163}$$

In this case, $A(t)$ and the THz pulse consist essentially of a positive (narrow) Gaussian pulse followed by an extended negative exponential contribution. The compensation of positive by negative contributions is small, as $A^+(t)$ has nearly no overlap with $A^-(t)$.

The situation is less favorable in terms of efficient THz generation, *if* τ_{rec} is significantly smaller than τ_{pls}, that is, if an "ultrashort τ_{rec}" semiconductor in combination with relatively slow pulsed laser is used. In this case, the shape of $A^-(t)$ is dominated by the (broad) Gaussian, convoluted with the (fast) exponential function. As a consequence, the macroscopic acceleration $A(t) = A^+(t) + A^-(t)$ at any time is strongly reduced as positive and negative contributions nearly compensate each other. Due to the temporal shift between $A^+(t)$ and $A^-(t)$, the shape of $A(t)$ approaches the time derivative of the Gaussian laser pulse, resulting in a nearly symmetric single-cycle THz pulse.

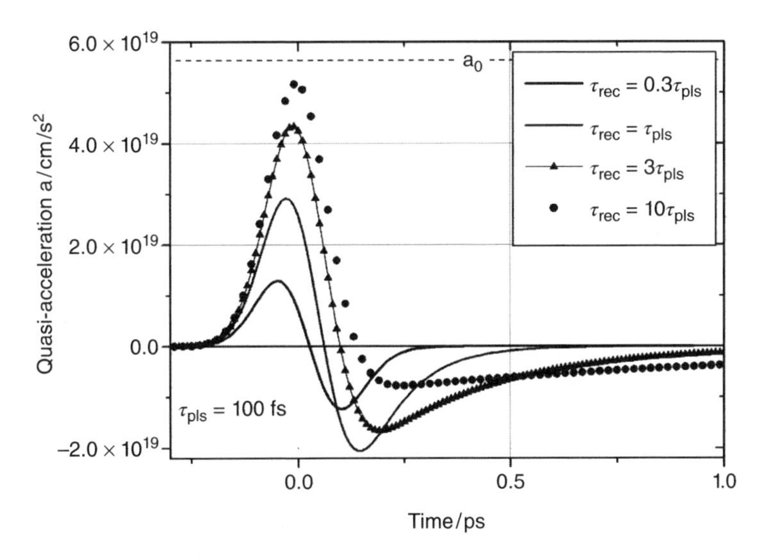

Figure 3.64 Quasi-acceleration for various recombination times, τ_{rec}. The optical pulse duration is $\tau_{pls} = 100\,\text{fs}$. The dashed line, marked a_0, corresponds to the maximum quasi-acceleration $a_0 = v_{sat}/(\pi^{1/2}\tau_{pls})$ from Eqs. (3.164) to (3.165).

If we express $A(t)$ again as a product of the number of carriers generated by the laser pulse and a quasi-acceleration of the individual carriers, $a(t)$, we get N_0 according to Eq. (3.147) and

$$a(t) \approx a_0 \left\{ \exp\left(-t^2/\tau_{pls}^2\right) - \frac{1}{\tau_{rec}} \int_{-\infty}^{0} \exp(\tau/\tau_{rec}) \exp\left[-(t+\tau)^2/\tau_{pls}^2\right] \mathrm{d}\tau \right\} \tag{3.164}$$

with

$$a_0 = \frac{v_{sat}}{\sqrt{\pi}\,\tau_{pls}}. \tag{3.165}$$

In Figure 3.64, the quasi-acceleration $a(t)$ from Eq. (3.164) is shown for $\tau_{rec} = (0.3, 1, 3, 10)\ \tau_{pls}$ and $\tau_{pls} = 100\,\text{fs}$.

3.6.2.6 Quantitative Predictions for LAEs and Experimental Results

In Sections 3.6.2.3–3.6.2.5 we have investigated the various features and processes determining the strength and angular dependence of the radiation intensity of LAEs as a function of an incident laser power with a Gaussian beam profile of radius ρ_0. This knowledge allows us now to make quantitative predictions for the expected THz radiation. The goal of this section is an application to some important scenarios and a comparison with experimental results from the literature.

We have learned that the intensity of the THz radiation emitted by a LAE under the space angle (θ, φ) can be expressed as a product of the (continuous) array factor, $\mathrm{AF}(\theta, \varphi)$, and the radiation intensity $U_e^{\mathrm{if},i}(\theta, \varphi)$, emitted by an elementary charge in proximity to the semiconductor/air interface (distance $\ll \lambda_0$) and accelerated by a vertical ($i = \mathrm{v}$) or horizontal ($i = h$) DC electric field,

$$U_{\mathrm{LAE}}^{\mathrm{if},i}\ (\theta, \varphi) = \mathrm{AF}\ (\theta, \varphi;\ \theta_0) U_e^{\mathrm{if},i}(\theta, \varphi) \qquad (i = \mathrm{v, h}) \tag{3.166}$$

The array factor takes into account the interference between THz waves originating from different spots of the LAE. Therefore, it depends not only on the radius of the laser beam, ρ_0, but also on the THz wavelength λ_0. In the case of non-vertical light incidence, the angle of incidence, θ_0, also has to be taken into account as discussed in Section 3.6.2.3b. The array factor also includes the square of the number of photogenerated charge carriers, $N_0 = n_0 \pi \rho_0^2$ and, therefore, increases with the square of the absorbed laser power (for a given value of ρ_0). In Section 3.6.2.3 we have calculated the array factors for the two cases of vertical light incidence and incidence under the Brewster angle (Figures 3.55 and 3.58, respectively) for constant n_0 and various values of ρ_0/λ_0.

In Section 3.6.2.4 we calculated the modifications of the THz radiation, introduced by the proximity of the charge carriers to the air/semiconductor interface (distance $\ll \lambda_0$) compared with that of an accelerated charge embedded into a semiconductor or in air (Figures 3.55 and 3.58) for a vertical or horizontal emitter. The modified radiation intensities $U_e^{\text{if},i}(\theta, \varphi)$, derived in Section 3.6.2.4, consist of a common prefactor $3/(8\pi)P_e$ and a dimensionless, angle-dependent factor which is different inside the semiconductor ($0 < \theta_{sc} < \pi/2$) and in air ($\pi/2 < \theta_{vac} < \pi$), for vertical and horizontal accelerating DC fields, and, in the case of the horizontal fields, for p- or s-polarized waves. The radiation intensity can generally be expressed as

$$U_{e,j}^{\text{if},i,l}(\theta, \varphi) = \frac{3}{8\pi} P_e f_j^{\text{if},i,l}(\theta, \varphi) \qquad (i = \text{v, h};\ j = \text{sc, vac};\ l = \text{p, s}), \tag{3.167}$$

where $f_j^{\text{if},i,l}(\theta, \varphi)$ takes the excitation geometry into account (vertical and horizontal dipoles, emission into vacuum or air, p- or s-polarized). The prefactor $3/(8\pi)P_e$, first introduced in Eq. (3.54), contains the square of the time-dependent quasi-acceleration $a(t)$

$$\frac{3}{8\pi} P_e = \frac{\sqrt{\varepsilon_{sc}}}{16\pi^2 \varepsilon_0 c_0^3}[ea(t)]^2 = \frac{\sqrt{\varepsilon_{sc}} Z_0}{16\pi^2 c_0^2}[ea(t)]^2 \tag{3.168}$$

for pulsed excitation and the square of the elementary quasi-acceleration amplitude $a(\omega_{\text{THz}})$

$$\frac{3}{8\pi} P_e = \frac{1}{2} \frac{\sqrt{\varepsilon_{sc}} Z_0}{16\pi^2 c_0^2}[ea(\omega_{\text{THz}})]^2, \tag{3.169}$$

for CW excitation. The factor ½ is due to the time average of $\langle a^2(t) \rangle$. In Eqs. (3.168) and (3.169) we have substituted the vacuum impedance $Z_0 = 1/(\varepsilon_0 c_0) = 377\,\Omega$.

Expressions for the elementary quasi-accelerations for CW and pulse excitation in LAEs have been derived for non-ballistic and ballistic transport in Section 3.6.2.5. In the following paragraphs we will use these results for an evaluation of photoconductive LAEs under CW and pulse excitation and subsequently to the corresponding scenarios in p-i-n-based LAEs.

Photoconducting LAEs

As already mentioned in the introductory section, in-plane electric fields and vertical laser incidence are usually used for photoconducting LAEs. With the corresponding array factor from Eq. (3.73), and the radiation intensities for carriers accelerated by horizontal fields, $U_{e,i}^{\text{if}}(\theta, \varphi)$ from Eqs. (3.123) to (3.128) we obtain for pulsed excitation

$$U_{\text{LAE},j}^{\text{if,h},l}(\theta_j, \varphi) = N_0^2 \exp\left[-\frac{1}{2}(k\rho_0 \sin\theta_j)^2\right] \frac{\sqrt{\varepsilon_{sc}} Z_0}{16\pi^2 c_0^2}[ea(t)]^2 f_j^{\text{if,h},l}(\theta_j, \varphi), \tag{3.170}$$

and for CW excitation

$$U_{LAE,j}^{if,h,l}(\theta_j, \varphi) = \frac{1}{2} N_0^2 \exp\left[-\frac{1}{2}(k\rho_0 \sin\theta_j)^2\right] \frac{\sqrt{\varepsilon_{sc}} Z_0}{16\pi^2 c_0^2}[ea(\omega_{\text{THz}})]^2 f_j^{if,h,l}(\theta_j, \varphi) \tag{3.171}$$

The prefactor ½ originates from the time average. The function $f_j^{\text{if,h},l}(\theta_j, \varphi)$ is:

$f_{\rm vac}^{\rm if,h,p}(\theta'_{\rm vac},\varphi)=\frac{1}{n_{\rm sc}}\cdot\left(\frac{2\cos\theta'_{\rm vac}\cos\theta_{\rm sc}\cos\varphi}{\cos\theta'_{\rm sc}+n_{\rm sc}\cos\theta'_{\rm vac}}\right)^2$ for p-polarization, emitted to the vacuum side and $0<\theta_{\rm vac}<\pi/2$, and $\theta'_{\rm vac}=\pi-\theta_{\rm vac}$ is defined in consistency with Figure 3.59 as the emission angle toward the air-side.

$f_{\rm vac}^{\rm if,h,s}(\theta'_{\rm vac},\varphi)=\frac{1}{n_{\rm sc}}\cdot\left(\frac{2\cos\theta'_{\rm vac}\sin\varphi}{n_{\rm sc}\cos\theta'_{\rm sc}+\cos\theta'_{\rm vac}}\right)^2$ for s-polarization, emission into vacuum, $0<\theta_{\rm vac}<\pi/2$.

$f_{\rm sc}^{\rm if,h,p}(\theta_{\rm sc},\varphi)=\left(\frac{2n_{\rm sc}\cos\theta_{\rm vac}\cos\theta_{\rm sc}\cos\varphi}{\cos\theta_{\rm sc}+n_{\rm sc}\cos\theta_{\rm vac}}\right)^2$ for p-polarization, emission into the semiconductor and $0<\theta_{\rm sc}<\theta_{\rm c}$.

$f_{\rm sc}^{\rm if,h,s}(\theta_{\rm sc},\varphi)=\left(\frac{2n_{\rm sc}\cos\theta_{\rm sc}\sin\varphi}{n_{\rm sc}\cos\theta_{\rm sc}+\cos\theta_{\rm vac}}\right)^2$ for s-polarization and emission into the substrate.

$f_{\rm sc}^{\rm if,h,p}(\theta_{\rm sc},\varphi)=\frac{4n_{\rm sc}^2\cos^2\theta_{\rm sc}\cos^2\varphi(n_{\rm sc}^2\sin^2\theta_{\rm sc}-1)}{(n_{\rm sc}^2-1)[(n_{\rm sc}^2+1)](\sin^2\theta_{\rm sc}-1)}$ for p-polarization, emission into the substrate and $\theta_{\rm c}<\theta_{\rm sc}<\pi/2$.

$f_{\rm sc}^{\rm if,h,s}(\theta_{\rm sc},\varphi)=\frac{4n_{\rm sc}^2\cos^2\theta_{\rm sc}\sin^2\varphi}{(n_{\rm sc}^2-1)}$ for s-polarization, emission into the semiconductor and $\theta_{\rm c}<\theta_{\rm sc}<\pi/2$.

CW Mixing

We consider first the case of CW mixing in a photoconductive LAE. In this case, the factor N_0 in Eq. (3.170) corresponds to the average number of photogenerated carriers in the photoconducting material with short recombination lifetime (compared to the transit time $\tau_{\rm tr}$ of the electrons from their point of generation to the contacts) $\tau_{\rm rec}<\tau_{\rm tr}$. According to Eq. (3.159), N_0 is given by the product of the photocarrier generation rate $dN/dt=\eta_{\rm ext}^{\rm I}2P_0/(h\upsilon)$ and the recombination lifetime. In order to establish the connection to photoconducting AEs, we introduce the *ideal* average photocurrent (if no carriers were lost by trapping or recombination) as

$$I_0=e\,\frac{dN}{dt}=e\,\eta_{\rm ext}^{\rm I}\,\frac{2P_0}{h\upsilon}\tag{3.172}$$

Its meaning is that each absorbed photon contributes one elementary charge to the photocurrent. The amplitude of the quasi-acceleration of the carriers for CW mixing was given in Eq. (3.158). Together with Eq. (3.172) inserted in Eq. (3.171), the radiation intensity can be written as

$$U_{{\rm LAE},j}^{{\rm if,h},l}(\theta_j,\varphi)=\frac{1}{2}I_0^2\exp\left[-\frac{1}{2}\left(k\rho_0\sin\theta_j\right)^2\right]\frac{\sqrt{\varepsilon_{\rm sc}}Z_0}{16\pi^2}\cdot\left(\frac{v_{\rm sat}}{c_0}\right)^2\frac{(\omega_{\rm THz}\tau_{\rm rec})^2}{1+(\omega_{\rm THz}\tau_{\rm rec})^2}f_j^{{\rm if,h},l}(\theta_j,\varphi)\tag{3.173}$$

This expression is fairly complex because of the angular dependences of $f_j^{{\rm if,h},l}(\theta_j,\varphi)$ for the general case. For sufficiently large LAEs with a beam radius ρ_0 significantly larger than the critical radius $\rho_{\rm c,vac}=0.195\lambda_0$, and $\rho_{\rm c,sc}=0.056\lambda_0$ defined by Eqs. (3.91) and (3.92), respectively, the angular dependence becomes dominated by the exponential factor $\exp[-½(k\rho_0\sin\theta_j)^2]$. In this case, the coefficients $f_j^{{\rm if,h},l}(\theta_j,\varphi)$ in Eq. (3.171) can be replaced by their values for $(\theta_{\rm sc},\varphi)=(\theta'_{\rm vac},\varphi)=(0,0)$, the latter corresponding to $(\theta_{\rm vac},\varphi)=(\pi,0)$, respectively. This yields

$$\begin{aligned}f_{\rm vac}^{\rm if,h,p}(\pi,0)&=0.05\\ f_{\rm vac}^{\rm if,h,s}(\pi,0)&=0\\ f_{\rm sc}^{\rm if,h,p}(0,0)&=2.5\\ f_{\rm sc}^{\rm if,h,s}(0,0)&=0\end{aligned}\tag{3.174}$$

A (nearly) plane wave is emitted and $U_{{\rm LAE},j}^{{\rm if,h},l}(\theta_j,\varphi)$ no longer depends on φ. The emitted wave is also p-polarized. The emitted power into the semiconductor $P_{\rm LAE,sc}^{\rm if,h,p}$ and into air $P_{\rm LAE,vac}^{\rm if,h,p}$ can be calculated analytically by integrating $U_{{\rm LAE},j}^{\rm if,h,p}(\theta_j,0)$ from Eq. (3.173) over the angles $\theta_{\rm sc}$ and $\theta_{\rm vac}$. This integration yields

$$P_{\rm LAE,sc}^{\rm h,p}\approx\frac{2.5}{(4\pi n_{\rm sc})}\cdot\left(\frac{\lambda_0}{\rho_0}\right)^2\frac{Z_0}{16\pi^2}\cdot\left(\frac{v_{\rm sat}}{c_0}\right)^2\frac{(\omega_{\rm THz}\tau_{\rm rec})^2}{1+(\omega_{\rm THz}\tau_{\rm rec})^2}I_0^2\tag{3.175}$$

for the semiconductor-hemisphere and

$$P^{\mathrm{h,p}}_{\mathrm{LAE,vac}} \approx \frac{0.05\,n_{\mathrm{sc}}}{4\pi} \cdot \left(\frac{\lambda_0}{\rho_0}\right)^2 \frac{Z_0}{16\pi^2} \cdot \left(\frac{v_{\mathrm{sat}}}{c_0}\right)^2 \frac{(\omega_{\mathrm{THz}}\tau_{\mathrm{rec}})^2}{1+(\omega_{\mathrm{THz}}\tau_{\mathrm{rec}})^2} I_0^2 \tag{3.176}$$

for the air-side. Now we can simplify the expressions for the emitted power by introducing the radiation resistances of LAEs

$$R^{\mathrm{h}}_{\mathrm{LAE}}(\rho_0, \omega_{\mathrm{THz}}) = 2\left[P^{\mathrm{h,p}}_{\mathrm{LAE,sc}}(\rho_0, \omega_{\mathrm{THz}}) + P^{\mathrm{h,p}}_{\mathrm{LAE,vac}}(\rho_0, \omega_{\mathrm{THz}})\right] / [I_{\mathrm{Ph}}(\omega_{\mathrm{THz}})]^2, \tag{3.177}$$

and the frequency-dependent (real) photocurrent

$$I_{\mathrm{Ph}}(\omega_{\mathrm{THz}}) = -\frac{\omega_{\mathrm{THz}}\,\tau_{\mathrm{rec}}}{\sqrt{1+(\omega_{\mathrm{THz}}\,\tau_{\mathrm{rec}})^2}} I_0. \tag{3.178}$$

Eq. (3.178) would predict a high photocurrent for long lifetimes. However, long lifetimes conflict with limitations by electrical power dissipation. Comparatively small lifetimes are still required.

We obtain an expression for the total power emitted by a large CW-LAE in terms of a product of a radiation resistance and the square of the photocurrent, analogous to that derived for AEs in Chapter 2, Eq. (2.7),

$$P^{\mathrm{if,h,p}}_{\mathrm{LAE}}(\rho_0, \omega_{\mathrm{THz}}) = \frac{1}{2}\, R^{\mathrm{h}}_{\mathrm{LAE}}\,(\rho_0, \omega_{\mathrm{THz}}) I^2_{\mathrm{Ph}}(\omega_{\mathrm{THz}}), \tag{3.179}$$

with the horizontal LAE radiation resistance of

$$\begin{aligned} R^{h}_{\mathrm{LAE}}(\rho_0, \lambda_0) &= \left(\frac{2.5}{n_{\mathrm{sc}}} + 0.05 n_{\mathrm{sc}}\right) \frac{Z_0}{32\pi^3} \left(\frac{\lambda_0}{\rho_0}\right)^2 \left(\frac{v_{\mathrm{sat}}}{c_0}\right)^2 \\ &= 0.333\ \Omega \cdot \left(\frac{\lambda_0}{\rho_0}\right)^2 \left(\frac{v_{\mathrm{sat}}}{c_0}\right)^2 \quad \text{for}\ \rho_0 \gg 0.056\ \lambda_0 \end{aligned} \tag{3.180}$$

or, alternatively, expressed in terms of THz (angular) frequency

$$R^{\mathrm{h}}_{\mathrm{LAE}}(\rho_0, \omega_{\mathrm{THz}}) = \left(\frac{2.5}{n_{\mathrm{sc}}} + 0.05 n_{\mathrm{sc}}\right) \frac{Z_0}{8\pi} \left(\frac{v_{\mathrm{sat}}}{\omega_{\mathrm{THz}}\,\rho_0}\right)^2 \approx 13.1\ \Omega \cdot \left(\frac{v_{\mathrm{sat}}}{\omega_{\mathrm{THz}}\,\rho_0}\right)^2, \tag{3.181}$$

where $n_{\mathrm{sc}} = 3.6$ was used. For $\rho_0 = \lambda_0$, for example, and $v_{\mathrm{sat}} = 10^7$ cm/s, the radiation resistance becomes $R^{\mathrm{h}}_{\mathrm{LAE}} \approx 37$ nΩ. This value is extremely small compared with the typical radiation resistances of AEs with $R_{\mathrm{A}} \approx 70\,\Omega$. Therefore, one might expect extremely poor performance for photoconducting CW-LAEs.

In spite of the formal similarity to photoconductive CW-AEs, there are significant qualitative and quantitative differences:

1. The expressions for the radiation resistance in Eqs. (3.180) and (3.181) were derived for the case of large LAEs with $\rho_0 \gg \rho_{\mathrm{c}}$. For a calculation of the power for the other limiting case, $\rho_0 \ll \rho_{\mathrm{c}}$, the exponential factor in Eq. (3.173) would have been ≈ 1. Therefore, it would have been necessary to perform the integration over the sc- and the air-hemisphere using the expressions for $f_j^{\mathrm{h},l}(\theta_j, \varphi)$. However, from Figure 3.61 we see that the modifications of the radiation intensities $U^{\mathrm{h},l}_{\mathrm{LAE},j}(\theta_j, 0)$ due to the proximity to the interface are quite significant, but not really dramatic. In particular, one can estimate that the rather strong reduction of radiation emitted into the air is compensated (or even overcompensated) by the enhanced emission into the semiconductor, which exceeds 3 dB for most angles (θ, φ). Therefore, we can assume, as a reasonable approximation, that the emitted power and the radiation resistance become independent of ρ_0 for $\rho_0 \ll \rho_{\mathrm{c,vac}}$ or $\rho_{\mathrm{c,sc}}$ from Eqs. (3.91) to (3.92), respectively. This yields approximations for the radiation resistance covering the whole range of laser beam radii, by replacing the factors $(\lambda_0/\rho_0)^2$ in Eq. (3.175) by $(\lambda_0/\rho_{\mathrm{c,sc}})^2/[1+(\rho_0/\rho_{\mathrm{c,sc}})^2] = (8\pi^2 n_{\mathrm{sc}}^2/3)/[1+(\rho_0/\rho_{\mathrm{c,sc}})^2]$

for the semiconductor side, and in Eq. (3.176) by $(\lambda_0 / \rho_{c,\,vac})^2 / [1 + (\rho_0/\rho_{c,\,vac})^2] = (8\pi^2/3)/[1 + (\rho_0/\rho_{c,\,vac})^2]$ for the air side, where $\rho_{c,\,vac} = n_{sc}\,\rho_{c,\,sc} = \lambda_0\sqrt{3/8\pi^2}$. This yields for the radiation resistance

$$R^{h}_{LAE}(\rho_0) \approx \left(\frac{2.5n_{sc}}{1 + \left(\rho_0 / \rho_{c,\,sc}\right)^2} + \frac{0.05n_{sc}}{1 + (\rho_0/\rho_{c,\,vac})^2} \right) \frac{Z_0}{12\pi} \left(\frac{v_{sat}}{c_0} \right)^2. \tag{3.182}$$

Even for $\rho_0 < \rho_{c,sc}$, the radiation resistance is in the range of $R^{h}_{LAE}(\rho_0) \approx 10\ \mu\Omega$ for $v_{sat} = 10^7$ cm/s. Again, this value is extremely small compared with the typical radiation resistances of AEs with $R_A \approx 70\,\Omega$.

2. Fortunately, the small values for the radiation resistances are, at least partly, compensated by the much larger values for current $I_{Ph}(\omega_{THz})$ compared with the corresponding values in photoconductive AEs. In our discussion following Eq. (3.158) the linear decrease of the frequency-dependent factor in Eq. (3.178) for $\omega_{THz} < 1/\tau_{rec}$ was attributed to the destructive interference between the positive and negative contributions to the macroscopic acceleration at lower frequencies. For $\omega_{THz} > 1/\tau_{rec}$ this destructive interference decreases and the saturation value I_0, corresponding to each absorbed photon contributing one elementary charge to the current, is obtained for $\omega_{THz}\tau_{rec} \gg 1$. In photoconductive AEs the term $\omega_{THz}\tau_{rec}$ in Eq. (3.178) is replaced by the "photoconductive gain," $g = \tau_{rec}/\tau_{tr} \approx 10^{-3}$ to 10^{-2} (where τ_{tr} stands for the transit time of an electron traveling with v_{sat} from one contact to the other; see frequency-dependent factor in Chapter 2, Eq. (2.19)). Hence, at $\omega_{THz}\tau_{rec} = 1$, the current contribution to the THz power is by a factor $g^2 \approx 10^{-6}$ to 10^{-4} smaller in photoconductive CW-AEs. Moreover it exhibits a $(\omega_{THz}\tau_{rec})^{-2}$ roll-off in the frequency range $\omega_{THz}\tau_{rec} \gg 1$, where the contribution to the power emitted in photoconductive CW-LAEs, in contrast, approaches its maximum value. The reason for the much better high frequency performance of the LAE is that radiation directly originates from the separation of the electron–hole pairs. This dipole current, represented by the numerator of Eq. (3.178), increases with increasing frequency and is independent of the transit-time to the contacts. For the photoconductive AEs, in contrast, only the displacement current at the contacts (which are the feeding points for the antenna) generates THz radiation. The displacement current is lowered by the photoconductive gain as compared to the ideal photocurrent.
3. The maximum tolerable absorbed laser power is limited by the heat flow. According to the considerations on thermal constraints in Section 2.2.3 the tolerable power for reasonably large devices scales with its cross-section as A if the heat conductivity of the substrate is sufficiently high in order to maintain a constant temperature across the LAE. The maximum tolerable power for photoconductive AEs with a typical cross-section of 50 μm² is about 50 mW. The cross-section of a LAE with diameter $\rho_0 = \rho_{c,sc}$ at 1 THz is $A = \pi\,\rho^2_{c,\,sc} = \pi\,(16.24\ \mu m)^2 = 829\ \mu m^2$. This yields a maximum tolerable (absorbed) laser power of nearly $P_L = 2P_0 = 1$ W. With these values inserted in Eq. (3.178) we expect for the maximum emitted THz power for an LAE at 1 W absorbed laser power at 800 nm a value of

$$\begin{aligned} P^{h}_{LAE}(\rho_0 = \rho_{c,\,sc}) &\approx \frac{1}{2}\left(\frac{2.5n_{sc}}{2} + \frac{0.05n_{sc}}{1 + \left(1/n_{sc}\right)^2} \right) \frac{Z_0}{12\pi}\left(\frac{v_{sat}}{c_0}\right)^2 \frac{(\omega_{THz}\,\tau_{rec})^2}{1 + (\omega_{THz}\,\tau_{rec})^2} \left(\frac{eP_L}{h\upsilon}\right)^2 \\ &\approx 2.3\ \mu W \cdot \frac{(\omega_{THz}\,\tau_{rec})^2}{1 + (\omega_{THz}\,\tau_{rec})^2}. \end{aligned} \tag{3.183}$$

This maximum value is approached for $\omega_{THz}\tau_{rec} > 1$. The corresponding value for an AE, driven at maximum tolerable power, that is, $P_L = 50$ mW with a fairly high gain of 10^{-2}, for comparison, is expected to be

$$\begin{aligned} P_{AE} &= \frac{1}{2} R_A g^2 \left(\frac{eP_{tot}}{h\upsilon}\right)^2 \frac{1}{1 + (\omega_{THz}\,\tau_{rec})^2} = \frac{1}{2}\ 70\,\Omega \cdot 10^{-2} \cdot (50/1.5\,mA)^2 \frac{1}{1 + (\omega_{THz}\,\tau_{rec})^2} \\ &\approx 4\,\mu W \frac{1}{1 + (\omega_{THz}\,\tau_{rec})^2} \end{aligned} \tag{3.184}$$

Although the AE provides higher power at lower frequencies, at $\omega_{THz}\tau_{rec} > 1$ the value for the LAE becomes larger. Of course, the optical-to-THz conversion efficiency for the LAE is only

$$\eta_{LAE} = P^{ph}_{LAE}/P_L = 2.5 \cdot 10^{-6}, \tag{3.185a}$$

whereas our estimate for the AE yields

$$\eta_{AE} = P_{AE}/P_L = 8 \cdot 10^{-5} \tag{3.185b}$$

that is, by a factor of about 30 better.

It should be added that the above-mentioned requirement of "sufficiently high" substrate heat conductivity is rather easy to fulfill for our example. As the substrate heating discussed in Chapter 2 is determined by Eq. (2.45) and the requirement that $\Delta T_{max} < 120$ K for a GaAs photomixer, the heat conductivity for the substrate λ_s has to satisfy the condition

$$\lambda_s > \frac{P_L}{\Delta T_{max}\ \sqrt{\pi A}} = \frac{1\ \mathrm{W}}{120\,\mathrm{K}\sqrt{\pi \cdot 829\,\mu\mathrm{m}^2}} \approx 1.6 \cdot \mathrm{W/cmK} \tag{3.186}$$

Thus, SiC ($\lambda_{SiC} = 3.5$ W/cmK) would be more than sufficient as a substrate, whereas Si ($\lambda_{Si} = 1.5$ W/cmK) just appears to be adequate. However, the active layer has to be thin (a few micrometers) and transferred to the substrate after epitaxial lift-off from its original substrate. With diamond as a substrate $\lambda_{dia} = 23$ W/cmK and with the same laser intensity of 1 mW/μm^2 the maximum area for the LAE could be increased by a factor of $(\lambda_{dia}/\lambda_s)^2 \approx 200$.

4. With the expressions from Eq. (3.182) for the radiation resistance we see that $R^{h}_{LAE}(\rho_0)$ *decreases* inversely with increasing area $\pi\rho_0{}^2$ of the illumination spot, that is, $\propto\ \rho_0{}^{-2}$ for $\rho_0 \gg (\rho_{c,sc}, \rho_{c,vac})$, *if* the current is kept constant. (Note that the THz *intensity* $U^{h}_{LAE,j}\ (\theta_j, \varphi)$ emitted under $\theta_{sc} = 0$ and $\theta_{vac} = \pi$, respectively, is not affected). As, at the same time, the maximum tolerable current increases linearly with increasing laser spot area, that is, $\propto\ \rho_0^2$, the decrease $\propto \rho_0^{-2}$ is overcompensated by the $\propto \rho_0^4$ increase of the square of the tolerable current Eq. (3.179), still resulting in a *linear* increase of THz *power* with laser spot area, that is, $\propto \rho_0^2$ (whereas the THz *intensity* $U^{h}_{LAE,j}\ (\theta_j, \varphi)$ emitted under $\theta_{sc} = 0$ and $\theta_{vac} = \pi$, even still increases quadratically, i.e., $\propto \rho_0^4$). Hence, for the example of the diamond substrate, the maximum THz power could be increased to about 400 μW.
5. One of the attractive features of LAEs is the highly directional radiation profile at sufficiently large values of ρ_0. If, due to large ρ_0, the 3-dB angle for emission into the substrate becomes significantly smaller than the critical angle for total reflection, arcsin $\theta_c = 1/n_{sc} = 16.13°$, the total reflection of the THz radiation transmitted through the substrate to the substrate/air interface is largely suppressed and the reflection losses, *R*, at this interface are not significantly larger than for vertical transmission. For $\rho_0 = 10\rho_{c,sc} \approx 0.56\lambda_0$, for example, $\theta^{3\,dB}_{sc} \approx 5.3°$ and $\theta^{3\,dB}_{vac} \approx 20°$. This implies that in this case nearly all the THz radiation transmitted through the planar substrate/air interface can be collimated by a simple THz lens or an off-axis parabolic mirror without using a hyper-hemispherical Si lens.
6. In order to take full advantage of both the increase in THz power and the highly directional radiation profile, very high laser power is required. According to (3) above, the laser power required for a LAE with $\rho_0 = \rho_{c,sc}$ to achieve about 60% of the maximum power emitted by an AE is about 20 times larger. A LAE with $\rho_0 = 10\rho_{c,sc}$ would provide about 60 times the THz power obtainable from an AE at 1 THz, but, at the same time the required laser power would be about 2000 times the maximum tolerable laser power of an AE, that is, on the order of 100 W. Therefore, high power lasers (or amplifiers) are required. Also, in order to avoid excessive substrate heating, the substrate should exhibit very high thermal conductivity. The best choice would be deposition of the semiconductor film on SiC or diamond after epitaxial lift-off from its substrate.
7. The maximum tolerable (absorbed) laser power density $P^{max}_L/A \sim 50\ \mathrm{mW}/50\ \mu\mathrm{m}^2 = 100\ \mathrm{kW/cm}^2$ yields a relatively small upper *thermal* limit for the average sheet carrier densities, $n^{th}_{0,s}$ in the absorbing layer

$$n^{th}_{0,s} = \frac{\partial n}{\partial t}\tau_{rec} = \frac{10^5\,\mathrm{W}}{\mathrm{cm}^2\ (h\nu)} 0.2\,\mathrm{ps} \approx 8.3 \cdot 10^{10}\ \mathrm{cm}^{-2} \tag{3.187}$$

for $h\upsilon = 1.5$ eV and $\tau_{rec} = 0.2$ ps, for example, and for the total number N_0^{th} in a laser spot of radius $\rho_0 = \rho_{c,sc} = 0.056\lambda_0 = 17\ \mu m$ we get $N_0^{th} = n_{0,s}^{th}\pi\rho_0^2 \approx 6.9 \cdot 10^5$ for $f_{THz} = 1$ THz. Below, we will see that extremely high average power can be obtained in LAEs under pulsed excitation, where $n_{0,s}^{th}$ will be replaced by a by several orders of magnitude larger upper limit, determined by electrical screening constraints.

We conclude that CW mixing with photoconducting LAEs without velocity overshoot is expected to work. However, it is not particularly appealing in terms of expected THz power and laser power to THz power efficiency. This is a consequence of the very low radiation resistance, which again is a consequence of the small factor v_{sat}/c_0 and the rather small value for the maximum tolerable average photo carrier (sheet) density $n_{0,s}^{th} < 10^{11}\ cm^{-2}$ from Eq. (3.187).

Pulsed Excitation

We will now consider photoconducting LAEs under pulse excitation. As already discussed for photoconducting AEs in Section 2.2.3, thermal heating normally does not limit the performance under pulse excitation. Typical repetition rates of pulsed lasers are of the order of $f_{rep} \approx 10^8$ Hz, and the laser-pulse and THz-pulse lengths, τ_{pls} and τ_{THz}, are in the sub-picosecond and picosecond-range. Therefore, the "duty factor" $f_{rep}\tau_{pls}$ is of the order of 10^{-4}. This implies that the density of carriers generated per pulse is allowed to exceed the maximum *average* density n_0^{th} of carriers participating in CW photomixing (Eq. (3.186)) by 4 orders of magnitude before thermal constraints become important. As the THz power scales quadratically with the number of carriers being accelerated, the peak power of the THz pulses could exceed the CW-power by 8 orders of magnitude. As a consequence, the *average power* emitted by a LAE under pulsed excitation could still be larger by 4 orders of magnitude. This corresponds to 4 orders of magnitude increase in the laser-to-THz conversion efficiency.

In Section 2.2.4 the electrical constraints were also examined. They may lead to field screening due to high electron–hole plasma densities. In Section 2.2.4.2 it was estimated that electron–hole plasma densities exceeding $n_0^{max} \approx 3 \cdot 10^{18}\ cm^{-3}$ (e.g., Eq. (2.64) and discussion of Eq. (2.68)) may screen the DC field within a few 100 fs. Plasma densities of $n_0^{max} \approx 3 \cdot 10^{18}\ cm^{-3}$ are generated within a layer thickness corresponding to an absorption length of $d_{abs} \approx 1\ \mu m$ by laser pulses with an energy of $\approx 80\ mJ/cm^2$ in GaAs ($h\upsilon \approx 1.5$ eV). This corresponds to a sheet density $n_{0,s}^{max} = n_0^{max} d_{abs} = 3 \cdot 10^{14}\ cm^{-2}$ for a 1 μm absorber layer and for a laser spot of radius $\rho_0 = \rho_{c,sc} = 17\ \mu m$ to a maximum number of carriers generated per pulse

$$N_0^{max} = n_{0,s}^{max}\pi\rho_0^2 \approx 2.7 \cdot 10^9 \text{ for } f_{THz} = 1 \text{ THz} \tag{3.188}$$

at an absorbed pulse energy of $E_{pls}^{max}(\rho_0) \sim 80\ mJ/cm^2 \cdot \pi\rho_0^2 = 0.73$ nJ.

Due to the relatively long delay time between the pulses of $1/f_{rep} > 10$ ns, the plasma decays (nearly) completely between successive pulses. It should be noted that the values for the pulsed sheet carrier density are by about 3 ½ orders of magnitude larger compared to the maximum average carrier densities by thermal constraints for CW in Eq. (3.187). At such high carrier densities, however, radiation field screening may affect the generated THz power.

Non-ballistic Transport

In Section 3.6.2.5b we considered a Gaussian laser pulse $P_L(t) = P_0 \exp(-t^2/\tau_{pls}^2)$ with pulse width τ_{pls}. The pulse generates $N_0 = \sqrt{\pi}\tau_{pls}\eta_{ext}^I (P_0/h\upsilon)$ electron–hole pairs per pulse, according to Eq. (3.148), and an ideal photocurrent $I_{Ph}(t) = I_0 \exp(-t^2/\tau_{pls}^2)$. The maximum I_0 is related to the laser pulse width and power by

$$I_0 = e\eta_{ext}^I (P_0/h\upsilon) = eN_0/(\sqrt{\pi}\tau_{pls}). \tag{3.189}$$

Following the same procedure as in the previous section for CW-mixing, the time-dependent power of the THz pulse can be written, formally, as

$$P_{LAE}^h(t) = \frac{1}{2} R_{LAE}^h(\rho_0) I_{Ph}^2(t). \tag{3.190}$$

The time dependence of $I_{Ph}(t)$ in Eq. (3.190) differs from that for the ideal photocurrent $I_{Ph}(t) \propto P_L(t)$, but it is related to it by the convolution with the time dependence of the acceleration kinetics. Therefore, it exhibits the same time dependence as the quasi-acceleration $a(t)$ from Eq. (3.164)

$$I_{Ph}(t) = I_0 \left(\exp\left(-t^2/\tau_{pls}^2\right) - \frac{1}{\tau_{pls}} \int_{-\infty}^{0} \exp(\tau/\tau_{rec}) \exp\left[-\frac{(t+\tau)^2}{\tau_{pls}^2}\right] \mathrm{d}\tau \right) \propto a(t) \tag{3.191}$$

Graphs of $a(t)$ are depicted in Figure 3.64 for $\tau_{pls} = 100$ fs and various values of $0.3\tau_{pls} < \tau_{rec} < 10\tau_{pls}$. The shape of $a(t)$ changes gradually as a function of τ_{rec}. At $\tau_{rec} = 0.3\tau_{pls}$ it corresponds to a nearly symmetric one-cycle THz pulse with an amplitude significantly smaller than the maximum amplitude $a_0 = \frac{v_{sat}}{\sqrt{\pi}\,\tau_{pls}}$ from Eq. (3.165). For $\tau_{rec} = 10\tau_{pls}$ the shape becomes very asymmetric exhibiting a positive one-half-cycle pulse, nearly reaching the value of a_0, and a much smaller and much longer negative contribution, decaying about exponentially $\propto \exp(-t/\tau_{rec})$ for large t. Also, we see that $\tau_{rec} > \tau_{pls}$ yields significantly larger THz pulses because of the larger amplitude of $a(t)$. Obviously, $a(t)$ exhibits a Fourier spectrum which changes drastically as a function of τ_{rec}/τ_{pls}.

The expressions for the radiation resistance R^h_{LAE} (ρ_0, ω_{THz}) turn out to be the same as for CW-mixing since the emitter geometry has not been changed. However, as $I_{Ph}(t)$ contains a more or less broad Fourier spectrum, one has to bear in mind that $\rho_{c,vac}(f_{THz}) = \lambda_0\sqrt{3/8\pi^2} = \sqrt{3/8\pi^2}\frac{c_0}{f_{THz}}$ and $\rho_{c,sc}(f_{THz}) = \rho_{c,vac}(f_{THz})/n_{sc}$ are actually frequency-dependent quantities. In order to take this into account, we substitute the factors $(\rho_0/\rho_{c,sc})^2$ and $(\rho_0/\rho_{c,vac})^2$ in the expression for the radiation resistance, Eq. (3.180) by

$$(\rho_0/\rho_{c,sc})^2 = f_{THz}/f_{c,sc}(\rho_0))^2, \text{ with } f_{c,sc}(\rho_0) = \frac{\sqrt{6}c_0}{4\pi n_{sc}\rho_0} \tag{3.192}$$

and

$$(\rho_0/\rho_{c,vac})^2 = (f_{THz}/f_{c,vac}(\rho_0))^2, \text{ with } f_{c,vac}(\rho_0) = \frac{\sqrt{6}c_0}{4\pi\rho_0} \tag{3.193}$$

The frequencies $f_{sc}(\rho_0)$ and $f_{vac}(\rho_0)$ are the 3 dB-frequencies for a f_{THz}^{-2} roll-off of the radiation resistance at a given laser beam radius ρ_0. For $\rho_0 = 100\ \mu$m, for example, we obtain $f_{c,sc}(100\ \mu\text{m}) = 0.1625$ THz and $f_{c,vac}(100\ \mu\text{m}) = 0.585$ THz, respectively. The radiation resistance is

$$R^h_{LAE}(f_{THz}) \approx \left(\frac{2.5n_{sc}}{1 + \left(f_{THz}/f_{c,sc}\right)^2} + \frac{0.05n_{sc}}{1 + (f_{THz}/f_{c,vac})^2} \right) \frac{Z_0}{12\pi} \left(\frac{v_{sat}}{c_0}\right)^2 \tag{3.194}$$

With a value in the range of $10^{-5}\ \Omega$ for $f_{THz} = f_{c,sc}$ as for the CW case. Because of the frequency dependence of the radiation resistance, a rigorous calculation of the radiation intensity and power emitted by LAEs under pulsed excitation requires an analysis in terms of the Fourier spectrum of $I_{Ph}(t)$ for the components at THz frequencies exceeding the 3-dB-frequencies $f_{vac}(\rho_0)$. In order to get a rough estimate also for large LAEs, we replace $I(t)$ in the expression for $P^{if,h}_{LAE}(t)$ by a one cycle sine-wave pulse corresponding to the peak frequency of its Fourier spectrum at $f^{peak}_{THz} \approx (4.5\tau_{pls})^{-1}$ (This value is deduced from Figure 3.64, where f^{peak}_{THz} ranges from $(6\tau_{pls})^{-1}$ to $(3\tau_{pls})^{-1}$ or $(1.7 \ldots 3.3)$ THz, depending on the ratio τ_{rec}/τ_{pls}). We get

$$P^h_{LAE}(t) \approx \frac{1}{2} 10^{-5}\Omega \cdot I_0^2 \sin^2(2\pi f^{peak}_{THz} t) \cdot \frac{1}{1 + (f^{peak}_{THz}/f_{c,sc}(\rho_0))^2} \text{ for } 0 < \text{t} < 1/f^{peak}_{THz} \tag{3.195}$$

So far, however, we have neglected radiation field screening which is to be expected for the given high photocarrier density of $n^{max}_{0,s} = 3 \cdot 10^{14}/\text{cm}^2$, according to Eq. (2.74) in Chapter 2. LTG GaAs features a typical mobility of $\mu = 400\ \text{cm}^2/\text{Vs}$ under dark conditions. Under illumination, the mobility gets even smaller due to enhanced electron–electron scattering with values in the 200 cm^2/Vs range [62]. The sheet conductance is $\sigma^{(2D)} = en_{0,s}^{\max}\mu = 96$ mS. The THz power is reduced by radiation field screening as $\eta_{RFS} = [1 + \sigma^{(2D)} Z_0/(n_{sc} + 1)]^{-2} = 0.31$.

We can now obtain an estimate for the maximum achievable power. Inserting the estimated upper tolerable limit for the number of carriers generated per pulse in a LAE of radius ρ_0 from Eq. (3.188) into the expression Eq. (3.189) for I_0 we obtain

$$P_{\mathrm{LAE}}^{\mathrm{h,max}}(t) \approx 36.2 \cdot 10^{-15}\,\mathrm{J\ s\ cm^{-4}} \cdot \left(\frac{\rho_0^4}{\tau_{\mathrm{pls}}^2}\right) \frac{1}{1 + (f_{\mathrm{THz}}^{\mathrm{peak}}/f_{\mathrm{c,sc}}(\rho_0))^2} \sin^2(2\pi f_{\mathrm{THz}}^{\mathrm{peak}} t) \tag{3.196}$$

for $0 < t < 1/f_{\mathrm{THz}}^{\mathrm{peak}}$. For a pulse length $\tau_{\mathrm{pls}} = 100$ fs and a laser beam radius of $\rho_0 = 100\,\mu$m this yields for the amplitude of the single cycle THz pulse

$$P_{\mathrm{LAE}}^{\mathrm{h,peak}}(t) = 36.2\,\mathrm{kW}\,\eta_{\mathrm{RFS}} \frac{1}{1 + (f_{\mathrm{THz}}^{\mathrm{peak}}/f_{\mathrm{c,sc}}(\rho_0))^2} = 60\ \mathrm{W} \tag{3.197}$$

for our value of $\rho_0 = 100\,\mu$m, $f_{\mathrm{c,sc}}(\rho_0) = 0.1625$ THz, $f_{\mathrm{THz}}^{\mathrm{peak}} \approx (4.5\tau_{\mathrm{pls}})^{-1} = 2.2$ THz, and $\tau_{\mathrm{pls}} = 100$ fs.

At the same time, this reduction of the spectral power at $f_{\mathrm{THz}}^{\mathrm{peak}}$ is associated with an increase in the directivity of the LAE, as $\rho_0 = 100\,\mu$m corresponds to $0.73\lambda_0$ (see Figure 3.55). From Eq. (3.73) we find a 3 dB angle of the array factor of $\theta_{\mathrm{sc,3dB}} = 2.91°$ for the emission of the 2.2 THz Fourier component into the semiconductor-substrate. This translates into $\theta_{\mathrm{sc,3dB}} = 10.52°$ after transmission into air. Therefore, the main Fourier components of the THz pulse can be collected by a detector with sufficiently large aperture without using any optical components. As an example, most of the power can be received with a Golay cell with a window of 1 cm diameter as detector, if the distance from the semiconductor is <2 cm.

For LAEs under pulse excitation it is interesting how much power will be contained within Fourier components corresponding to frequencies above and below $f_{\mathrm{THz}}^{\mathrm{peak}}$. The Fourier spectrum $I_{\mathrm{Ph}}(\omega)$ of $I_{\mathrm{Ph}}(t)$ in Eq. (3.191) is expected not to differ too much from that for the laser pulse, in particular for $\tau_{\mathrm{rec}} \gg \tau_{\mathrm{pls}}$ (see Figure 3.64). From the Fourier transform of the Laser pulse, $P_L(\omega) = P_0 \exp(-\omega^2 \tau_{\mathrm{pls}}{}^2/4)$ (see Section 2.2.2.2), we obtain a 3 dB frequency of $f_{\mathrm{3dB}} = \sqrt{\ln 2}/(\sqrt{2}\pi\tau_{\mathrm{pls}}) = 1.87$ THz for our example with $\tau_{\mathrm{pls}} = 100$ fs. Therefore, we expect $I_{\mathrm{Ph}}(\omega > 2\pi f_{\mathrm{THz}}^{\mathrm{peak}}) < I_{\mathrm{Ph}}(2\pi f_{\mathrm{THz}}^{\mathrm{peak}})$. Together with the fact, that, according to Eq. (3.197), we are deep in the f^{-2} power roll-off already at $f = f_{\mathrm{THz}}^{\mathrm{peak}}$, very little power for Fourier components above $f_{\mathrm{THz}}^{\mathrm{peak}}$ is expected. For $\omega < 2\pi f_{\mathrm{THz}}^{\mathrm{peak}}$, however, we expect $I_{\mathrm{Ph}}(\omega < 2\pi f_{\mathrm{THz}}^{\mathrm{peak}}) \approx I_{\mathrm{Ph}}(2\pi f_{\mathrm{THz}}^{\mathrm{peak}})$, or even larger. Therefore, the Fourier spectrum of the THz pulses is expected to increase quadratically toward lower frequencies in this frequency range, up to about $f = f_{\mathrm{c,sc}}(\rho_0)$, corresponding to 162 GHz for our example. For a comparison with experimental results, of course, it will be necessary to consider, in addition, also the details of the optics for the transmission of the THz radiation generated in the LAE to the detector.

Finally, we compare the maximum THz pulse energy with the corresponding laser pulse energy. With our estimates for the amplitude of the single cycle THz pulse from Eq. (3.197) and the THz pulse length of $T_{\mathrm{THz}} \approx 4.5\tau_{\mathrm{pls}} \approx 0.45$ ps we obtain for the maximum achievable THz pulse energy at $\rho_0 = 100\,\mu$m

$$E_{\mathrm{LAE}}^{\mathrm{h,max}}(\rho_0 = 100\,\mu\mathrm{m}) = \frac{1}{2} P_{\mathrm{LAE}}^{\mathrm{h,peak}}(\rho_0 = 100\,\mu\mathrm{m})\, T_{\mathrm{THz}} = 13.5\,\mathrm{pJ} \tag{3.198}$$

At a repetition rate of about $\upsilon_{\mathrm{rep}} = 10^8$ Hz this corresponds to a maximum average THz power of

$$\left\langle P_{\mathrm{LAE}}^{\mathrm{h,max}}(\rho_0 = 100\,\mu\mathrm{m}) \right\rangle = f_{\mathrm{rep}} E_{\mathrm{LAE}}^{\mathrm{h,max}}(\rho_0 = 100\,\mu\mathrm{m}) = 1.33\,\mathrm{mW} \tag{3.199}$$

The required laser pulse energy is

$$E_{\mathrm{pls}}^{\max}(\rho_0 = 100\,\mu\mathrm{m}) = (80\ \mathrm{mJ/cm^2})\ \pi(100\,\mu\mathrm{m})^2 = 25.1\,\mathrm{nJ} \tag{3.200}$$

corresponding to an average laser power

$$\left\langle P_{\mathrm{L}}^{\max}\left(\rho_0 = 100\,\mu\mathrm{m}\right)\right\rangle = f_{\mathrm{rep}} E_{\mathrm{pls}}^{\max}(\rho_0 = 100\,\mu\mathrm{m}) = 2.51\ \mathrm{W} \tag{3.201}$$

These values appear extremely appealing regarding both absolute values of THz power and laser-power to THz-power conversion efficiency. For the latter we find

$$\eta_{\mathrm{conv}} = E_{\mathrm{LAE}}^{\mathrm{h,max}}(\rho_0 = 100\,\mu\mathrm{m})/E_{\mathrm{pls}}^{\max}(\rho_0 = 100\,\mu\mathrm{m}) = 5.3 \cdot 10^{-4}. \tag{3.202}$$

Ballistic Transport

In Section 2.2.2.2 it was explained why photoconductive AEs require short-lifetime semiconductors for CW photomixing in order to avoid too high electron–hole plasma densities and the resulting field screening effects. The same arguments hold for LAEs. For pulsed operation, however, the recovery time between successive pulses is of the order of 10 ns, which drastically relaxes the requirements regarding recombination times (for both electron–hole recombination via recombination centers as well as recombination at the contacts due to transport). Therefore, photoconductive AEs, as well as LAEs, based on semi-insulating GaAs (SI GaAs), exhibiting higher mobilities and longer recombination lifetimes have been successfully demonstrated. SI GaAs is expected to show ballistic transport and velocity overshoot.

In the following we extend our considerations to photoconducting LAEs under pulsed laser operation which are using such materials as SI GaAs. The main difference compared with the previous case is the different quasi-acceleration for the electrons. Instead of Eq. (3.164) one has to use Eq. (3.149). A comparison of the corresponding graphs in Figures 3.63 and 3.64 indicates, first, that the maximum amplitude of the quasi-acceleration for the ballistic scenario, $a_0 = (v_{\mathrm{bal}}/\tau_{\mathrm{bal}})$, becomes much larger compared to the non-ballistic counterpart, $a_0 = v_{\mathrm{sat}}/(\pi^{1/2}\tau_{\mathrm{pls}})$, due to the higher achievable ballistic velocities. Moreover, we see that for a large amplitude of the quasi-acceleration it is necessary that the laser pulse width τ_{pls} is short compared to the acceleration time τ_{bal}, instead of the (easy to fulfill) requirement to be short compared to the recombination lifetime τ_{rec} in the previous case. The reasonable condition $\tau_{\mathrm{pls}} < 0.3\ \tau_{\mathrm{bal}}$, indeed, requires very short laser pulses, as τ_{bal} is typically very short in photoconductors, due to high electric fields between the contacts. For fields exceeding 40 kV/cm, for example, τ_{bal} becomes < 100 fs (see Chapter 2, Figure 2.6b, right panel; the situation for GaAs is similar to that for InGaAs).

With the quasi-acceleration for ballistic transport replacing the expression for the non-ballistic case in Eq. (3.191) we obtain for the THz power formally the same expression

$$P_{\mathrm{LAE}}^{\mathrm{if,\,h}}(t) = \frac{1}{2} R_{\mathrm{LAE}}^{\mathrm{h}}\ (\rho_0) I_{\mathrm{Ph}}^2\ (t) \tag{3.203}$$

but with $I_{\mathrm{Ph}}(t)$ exhibiting the time dependence of the ballistic quasi-acceleration from Eq. (3.149)

$$I_{\mathrm{Ph}}(t) = I_0 \left\{ \frac{1}{\tau_{\mathrm{bal}}} \left[\int_{t-\tau_{\mathrm{bal}}}^{t} \exp\left(-t'^2/\tau_{\mathrm{pls}}^2\right) \mathrm{d}t' - \int_{t-2\tau_{\mathrm{bal}}}^{t-\tau_{\mathrm{bal}}} \exp(-t'^2/\tau_{\mathrm{pls}}^2) \mathrm{d}t' \right] \right\}, \tag{3.204}$$

whereas the ideal photocurrent amplitude of the ideal photocurrent, I_0, remains the same. The radiation resistance

$$R_{\mathrm{LAE}}^{\mathrm{h}}(\rho_0, f_{\mathrm{THz}}) \approx \left(\frac{2.5 n_{\mathrm{sc}}}{1 + \left(f_{\mathrm{THz}}/f_{\mathrm{c,sc}}\left(\rho_0\right)\right)^2} + \frac{0.05 n_{\mathrm{sc}}}{1 + (f_{\mathrm{THz}}/f_{\mathrm{c,vac}}(\rho_0))^2} \right) \frac{Z_0}{12\pi} \left(\frac{v_{\mathrm{bal}}}{c_0} \right)^2 \tag{3.205}$$

is increased by the factor $(v_{\mathrm{bal}}/v_{\mathrm{sat}})^2$, which corresponds to a factor of 100 increase for GaAs. This yields $R_{\mathrm{LAE}}^{\mathrm{h}}(\rho_0, f_{\mathrm{THz}}) \approx 10^{-3}\Omega/(1 + [f_{\mathrm{THz}}/f_{\mathrm{c,sc}}(\rho_0)]^2)$ for $v_{\mathrm{bal}} = 10^8$ cm/s. The emitted power can be estimated similarly to that for the non-ballistic transport. At a reasonable electric field of 40 kV/cm we expect a single cycle THz-pulse with $f_{\mathrm{THz}} \approx 4$ THz and an amplitude of $I_{\mathrm{Ph}}(t)$ close to I_0, which becomes a factor of 2.5 times larger for $\tau_{\mathrm{pls}} \approx 40$ fs.

Assuming the same specifications of the LAE as before we expect a $100 \times (2.5)^2 = 625$ times larger maximum THz-power amplitude

$$P_{\mathrm{LAE}}^{\mathrm{h,peak}}(\rho_0 = 100\,\mu\mathrm{m}) = 37\,\mathrm{kW}, \tag{3.206}$$

Here, we also assumed the same average sheet conductivity (and *average* mobility under high optical excitation) for radiation field screening as for the non-ballistic case according to Ref. [62]. This leads to a maximum pulse energy $E_{\mathrm{LAE}}^{\mathrm{h,\,max}}(\rho_0 = 100\,\mu\mathrm{m}) = 9.3$ nJ for a THz pulse width of 1/(4 THz) = 0.25 ps, and a maximum average THz power at a repetition rate of about $f_{\mathrm{rep}} = 10^8$ Hz of $\left\langle P_{\mathrm{LAE}}^{\mathrm{h,max}}\left(\rho_0 = 100\,\mu\mathrm{m}\right)\right\rangle =$ 0.9 W with the same laser pulse energy, namely 25 nJ, and average laser power of 2.5 W as before. The laser-power/THz-power conversion efficiency is expected to become $\eta_{\mathrm{conv}} = 0.9/2.5 \approx 36\%$. These estimates are probably too optimistic. First, it is not clear, whether ballistic overshoot fully develops in SI GaAs. As the DC-fields have to be rather small to take advantage of ballistic transport (<40 kV/cm), the critical carrier density N_0^{max} may be significantly lower than the value given in Eq. (3.188), as screening of the applied field may become detrimental at higher laser fluence. Further, the fairly low average mobility under heavy optical illumination may be too optimistic. Larger average mobilities would lead to stronger radiation field screening. Finally, thermal constraints can also become relevant, as a significant fraction of the photogenerated carriers may reach the contacts due to the relatively long recombination lifetimes, which may yield an additional energy dissipation of several electronvolts per photogenerated charge carrier. The fact that pulsed AEs based on SI GaAs have proven at least competitive to LTG GaAs AEs in the pulsed regime, at least partly, supports these theoretical expectations.

Comparison with Experimental Results on Photoconducting LAEs

In the following section we will provide only a brief overview of the work on photoconducting LAEs already published.

In Chapter 2 we learned that it is important to use an interdigitated contact pattern for CW operation in order to combine an active area of optimum size with a reasonable photoconductive gain $g = \tau_{\mathrm{rec}}/\tau_{\mathrm{tr}}$ for the photocurrent. As the transit time $\tau_{\mathrm{tr}} = w_{\mathrm{G}}/v_{\mathrm{sat}}$ scales linearly with the width w_{G} of the gap between the fingers, a small value of w_{G} is favorable. Although the transit time is not relevant for the performance of photoconductive LAEs, interdigitated contact patterns represent a suitable choice to obtain horizontal DC electric fields of the required strength. For pulsed operation, an interdigitated structure also has to be used, however, with a larger spacing in the range of $w_{\mathrm{G}} \sim 7$–$10\,\mu$m [78]. However, the device structure has to be modified for LAEs. As the sign of the field between neighboring electrodes fingers alternates, it is necessary to make every second gap inactive for the generation and acceleration of charge carriers in order to avoid the THz fields emitted by the carriers in one stripe interfering destructively with those originating from the neighboring stripe. This can be achieved either by masking every second gap (Figure 3.65a; approach used in Refs. [67, 69–71, 74, 75]) or by removing the active material (Figure 3.65b; as used in Refs. [68, 73]).

Photoconductive CW LAEs

So far, only one group has reported on CW photomixing experiments with LAEs [75]. Eshaghi *et al.* present results obtained from a device as shown schematically in Figure 3.65a with a finger width and spacing of 10 μm fabricated on LTG GaAs. With a laser power of 900 mW ($\lambda_{\mathrm{L}} \approx 800$ nm) they achieved a maximum THz power of about 2 μW at $f_{\mathrm{THz}} = 1.2$ THz. This value agrees very well with our estimate of 2.5 μW at 1 W laser power, expected from Eq. (3.183) for $(\omega_{\mathrm{THz}}\tau_{\mathrm{rec}}) > 1$ and $\rho_0 \approx \rho_{\mathrm{c,sc}}$. The authors also report measurements of the far-field beam profile, confirming the expected relation between THz beam width and laser beam radius (with reference to [14]).

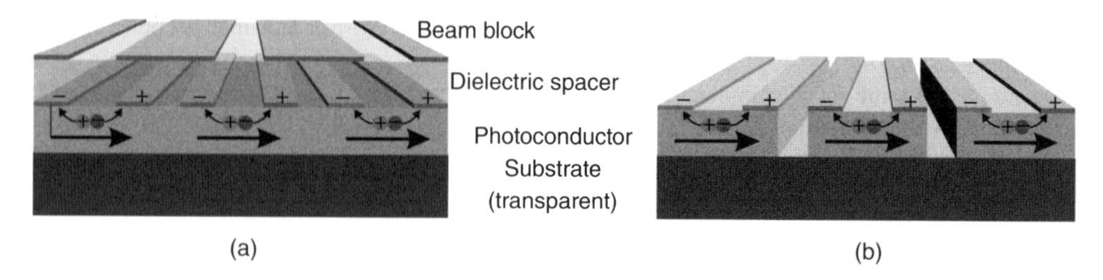

Figure 3.65 Device layouts for photoconducting LAEs. (a) Every second gap is shadowed by a metal mask (beam block) that is electrically isolated from the photoconductor and the electrodes by an insulating dielectric spacer. Therefore, only photocurrent dipoles of the same sign (indicated by the arrows) are generated, resulting in constructive interference in the far-field. In the covered gaps, no electron–hole pairs are generated such that no current flows. (b) Alternatively, the photoconductive material of every second gap can be removed. However, this is only possible if the substrate material does not absorb the laser signal. For GaAs and LTG GaAs that is usually grown on GaAs, a complicated flip-chip method has to be used, transferring the photoconductive material onto a new substrate and the original GaAs substrate has to be etched away. In contrast, InGaAs is typically grown on InP with a band gap energy of 1.344 eV whereas the exciting lasers have a photon energy in the range of 0.8 eV. The substrate is therefore transparent.

Photoconductive LAEs under Pulsed Excitation

According to the previous sections drastically increased average THz power and optical-to-THz conversion efficiencies are expected for LAEs operated under pulsed excitation with a low duty cycle (typically $< 10^{-4}$). A number of recent papers have confirmed these expectations for various types of photoconducting LAEs [67–74] Dreyhaupt *et al.* [67, 69] and Beck *et al.* [71] measured record THz fields and power in LAEs fabricated from SI GaAs with a sample design according to Figure 3.65a.

The best conversion efficiency of $2 \cdot 10^{-3}$ was achieved with $\tau_{\text{pls}} < 50$ fs laser pulses with $\lambda_{\text{L}} = 800$ nm at an excitation density of 20 mJ/cm^2 and DC fields of 70 kV/cm. The optimum fluence was determined as the upper limit for linear increase of the THz field amplitude at fixed laser spot radius. With the available pulse energy of 3 μJ a maximum total energy of the THz pulses of 6 nJ was achieved at a radius of the illumination spot of about $\rho_0 = 2.5$ mm.

Our estimates for the maximum THz pulse energy of ≈ 43 pJ (Eq. (3.198)) and the maximum tolerable laser pulse energy of ≈ 25 nJ (Eq. (3.200)) were made for $\rho_0 = 100$ μm, that is, for an area about a factor of $25^2 = 625$ smaller. Scaling up our values by the factor of 625 indicates that our estimate indeed underestimates the experimental value for the maximum tolerable fluence by $\sqrt{2}$ (as the pulse energy scales quadratically with the fluence!). It should be noted that the good agreement between theory and experiment was obtained for a model which assumes monotonic acceleration up to the saturation velocity without velocity overshoot. This indicates that velocity overshoot probably does not play a significant role in SI GaAs LAEs. Otherwise, much higher THz pulse energies and optical-to-THz conversion efficiencies should have been observed (see Eq. (3.206)).

Device structures with interdigitated contact stripes on the unstructured sample, as depicted in Figure 3.65a have the disadvantage that typically only about 25% of the device surface can be used. With a design according to Figure 3.65b one can improve the active area, in particular, if the active semiconductor layer is thin and the substrate does not exhibit a photoconductive response at the laser wavelength due to a higher band gap. This approach was first used by Awad *et al.* [68] with an LTG GaAs layer on a GaAs substrate with an AlAs sacrificial layer in between. In this way, the LTG GaAs layer could be removed from the substrate by an epitaxial lift-off process after deposition of the electrodes and etching of the gaps. Finally, the device structure was transferred to a sapphire substrate. The emphasis in this publication was mostly on the fabrication method, but there are no results which allow for a comparison with the results reported in Refs. [67, 69, 71]. Another version of the approach from Figure 3.65b was used in Refs. [73, 74]. In this case the goal was a photoconductive LAE operating at the telecom wavelength of 1550 nm. In Ref. [73] an $In_{0.52}Al_{0.48}As/In_{0.53}Ga_{0.47}As$ superlattice with a layer of ErAs clusters buried in the $In_{0.52}Al_{0.48}As$ layers and co-doped with p-doping was used. This approach was chosen to overcome the problem of fabricating a low-carrier-lifetime low-dark-conductivity material with the suitable bandgap for operation at 1550 nm. An example of a low-lifetime LTG InGaAs/LTG InAlAs superlattice is given in Ref. [74]. In both cases, no epitaxial lift-off process was required as the superlattice was grown lattice-matched on InP, which serves as an etch-stop layer and exhibits no absorption at 1550 nm because of the much larger bandgap of 1.344 eV. THz pulses with a power spectrum extending at least up to ~3.3 THz and 44 dB (using electro-optic sampling in ZnTe for detection) above the noise floor at 1 THz were measured in Ref. [73]. However, the materials still have to be improved before becoming competitive with LAEs based on 800 nm materials. This is also true for other attempts devoted to long-wavelength materials by Mittendorff *et al.* [74] using LTG InGaAs/LTG InAlAs heterostructures and by Peter *et al.* [70] using the quaternary semiconductor GaInAsN. Finally, we mention a very elegant approach for the generation of THz pulses using (quasi-) unidirectional horizontal Dember photo currents [72]. By using a saw tooth lateral modulation of the thickness of an absorption layer on top of the semiconductor surface extended regions of gradients of the generated electron–hole densities were achieved. Without any contacts THz fields comparable with those from interdigitated LAEs were detected.

p-i-n Diode-Based LAEs

We have learned in Section 3.6.2.4 that the intensity of the THz radiation emitted by carriers accelerated in semiconductors by a vertical field (i.e., oriented in the z-direction) exhibits a pronounced maximum at the critical angle for total reflection, θ_c, if the carriers are located in proximity to the semiconductor/air interface (distance $\ll \lambda_{THz}$). Therefore, the angle of incidence of the laser beam should be chosen such that the maximum of the array factor (at least nearly) coincides with θ_c, in order to optimize the achievable THz power of the LAE. We discussed as two good options either the use of the Brewster angle for illumination from the air (assuming $\theta_{vac,0} = \theta_{vac}^B$ in Figure 3.54b) or illumination from the backside through a bi-prism (assuming $\theta_{sc} = \theta_c$ in Figure 3.57). The former option has the advantage of zero reflection for the incident laser beam (for p-polarization), but the disadvantage of a strongly elliptical THz beam profile. Also, the maximum of the array factor in the semiconductor at $\theta_{sc} = \theta_{sc}^B = \arctan(1/n_{sc})$ (= 15.52°, for $n_{sc} = 3.6$), as discussed in Section 3.6.2.3 (see Figure 3.58), occurs at a slightly smaller angle compared with the maximum of the radiation intensity calculated in Section 3.6.2.4 (see Figure 3.60) at the angle $\theta_{sc} = \theta_c = \arcsin(1/n_{sc})$ (= 16.13°, for $n_{sc} = 3.6$). For large LAEs this disagreement may result in a significant loss of THz power. Also, one has to bear in mind that, due to the small difference between θ_{sc}^B and $\theta_{c,sc}$ and the finite 3 dB angle of the array factor, the beam profile of the THz radiation emitted into the semiconductor is strongly distorted and part of it experiences total reflection at the bottom interface to the air. All these disadvantages are avoided, if the configuration according to Figure 3.57 is used. However, there are reflection losses of about 32% at the air/semiconductor interface affecting both laser beam and THz beam.

In the following we will consider the, overall more appealing, case of illumination from the substrate side under the angle θ_c. The procedure for the calculation of the THz radiation intensity and power is basically the same as for the previous case of the photoconductors with horizontal fields. The array factor is nearly the same as for vertical light incidence, since the ellipticity of the illumination spot is negligibly small due to the small angle θ_c. The main difference is that the maximum of the array factor is found at $\theta_{sc} = \theta_c$. Therefore we introduce the polar angle $\Delta\theta_{sc} = \theta_{sc} - \theta_c$ for a coordinate system tilted by the angle θ_c in the xz-plane. The angular dependence of the array factor on the polar angle $\Delta\theta_{sc}$ is the same as on θ_{sc} for the case of vertical light incidence.

The counterpart to the expression Eqs. (3.170) and (3.171) for photoconductors with horizontal DC-fields becomes for vertical electric fields

$$U^{\mathrm{v}}_{\mathrm{LAE,sc}}(\theta_{\mathrm{sc}},\varphi;\theta_{\mathrm{c}}) = N_0^2 \exp\left[-\frac{1}{2}\left(k\rho_0 \sin \Delta\theta_{\mathrm{sc}}\right)^2\right] \frac{\sqrt{\varepsilon_{\mathrm{sc}}}Z_0}{16\pi^2 c_0^2}[ea(t)]^2 f^{\mathrm{if,v}}_{\mathrm{sc}}(\theta_{\mathrm{sc}},\varphi), \tag{3.207}$$

and

$$U^{\mathrm{v}}_{\mathrm{LAE,sc}}(\theta_{\mathrm{sc}},\varphi;\theta_{\mathrm{c}}) = \frac{1}{2}N_0^2 \exp\left[-\frac{1}{2}\left(k\rho_0 \sin \Delta\theta_{\mathrm{sc}}\right)^2\right] \frac{\sqrt{\varepsilon_{\mathrm{sc}}}Z_0}{16\pi^2 c_0^2}[ea(\omega_{\mathrm{THz}})]^2 f^{\mathrm{if,v}}_{\mathrm{sc}}(\theta_{sc},\varphi). \tag{3.208}$$

.

The emission into air (subscript j = vac in Eqs. (3.170) and (3.171)) becomes negligibly small in this case and will not be taken into account. Also, the superscript $l = (s, p)$ can be omitted, as only p-polarized radiation is emitted under acceleration in the z-direction. The dimensionless factor $f^{\mathrm{if,\,v}}_{\mathrm{sc}}(\theta_{\mathrm{sc}},\varphi)$ becomes, according to Eqs. (3.120) and (3.121),

$$f^{\mathrm{if,v}}_{\mathrm{sc}}(\theta_{\mathrm{sc}},\varphi) = \left[\frac{2\cos\theta_{\mathrm{sc}}\sin\theta_{\mathrm{sc}}}{\cos\theta_{\mathrm{sc}} + n_{\mathrm{sc}}\cos\theta'_{\mathrm{vac}}}\right]^2 \text{ for } 0 < \theta_{\mathrm{sc}} < \theta_{\mathrm{c}} \tag{3.209}$$

$$f^{\mathrm{if,v}}_{\mathrm{sc}}(\theta_{\mathrm{sc}},\varphi) = \frac{(2\cos\theta_{\mathrm{sc}}\sin\theta_{\mathrm{sc}})^2}{(n_{\mathrm{sc}}^2 - 1)[(n_{\mathrm{sc}}^2 + 1)\sin^2\theta_{\mathrm{sc}} - 1]} \text{ for } \theta_{\mathrm{c}} < \theta_{\mathrm{sc}} < \pi/2 \tag{3.210}$$

$f^{\mathrm{if,v}}_{\mathrm{sc}}(\theta_{\mathrm{sc}},\varphi)$ does not depend on φ because of the radial symmetry of the radiation intensity emitted by the vertical dipole (see Figure 3.60). The angle φ will therefore be omitted in the following. It exhibits a rather sharp cusp at $\theta_{\mathrm{sc}} = \theta_{\mathrm{c}}$ with $f^{\mathrm{if,v}}_{\mathrm{sc}}(\theta_{\mathrm{sc}}) = 4\sin^2\theta_{\mathrm{sc}} = 4/n_{\mathrm{sc}}^2 = 0.31$. The maximum of the array factor coincides with this cusp. For a calculation of the THz *power* $P^{\mathrm{v}}_{\mathrm{LAE}}$ by integration of $U^{\mathrm{v}}_{\mathrm{LAE,\,sc}}(\theta_{\mathrm{sc}},\varphi;\ \theta_{\mathrm{c}})$ from Eq. (3.208) over the space angle the tilted polar coordinate is recommended. Two points, however, should be noted:

1. For very large LAEs the angular dependence of $f^{\mathrm{if,v}}_{\mathrm{sc}}(\theta_{\mathrm{sc}})$ around its peak value of 0.31 becomes negligible if the 3 dB angle of the array factor becomes smaller than the 3 dB value of the cusp. In this case the space angle integration over $\Delta\theta_{\mathrm{sc}}$ yields the same factor $(1/2\pi n_{\mathrm{sc}}^2)/(\lambda_0/\rho_0)^2$ as previously for the case of normal incidence. This is expected to happen for about $\rho_0 > 3\lambda_0$.
2. For smaller LAEs the integration yields smaller contributions at polar angles $\Delta\theta_{\mathrm{sc}}$ in the xz-plane, due to the additional decrease of $f^{\mathrm{if,v}}_{\mathrm{sc}}(\theta_{\mathrm{sc}})$ for $\theta_{\mathrm{sc}} \neq \theta_{\mathrm{c}}$. The reduction is, however, less stringent, as there is no significant decrease for polar angles $\Delta\theta_{\mathrm{sc}}$ perpendicular to the xz-plane within a rather wide range of $\Delta\theta_{\mathrm{sc}}$.

CW Photomixing in p-i-n-Based LAEs Exhibiting Ballistic Transport

In p-i-n diode-based LAEs we are interested mostly in CW photomixing, taking advantage of ballistic transport, as discussed in Section 2.2.2 (see Figure 2.6). In these p-i-n photodiodes the absorbed photons generate an electron–hole plasma near the p-layer. While the holes are only moving into the neighboring p-layer, the electrons are ballistically accelerated through the intrinsic region toward the n-layer. For the calculation of the THz emission we have to insert the expressions from Eqs. (3.142) to (3.144) for N_0 and the quasi-acceleration a derived from the macroscopic quasi-acceleration $A(t)$ into Eq. (3.208). This yields again $P^{\mathrm{v}}_{\mathrm{LAE}}(\omega_{\mathrm{THz}}) = \frac{1}{2}R^{\mathrm{v}}_{\mathrm{LAE}}(\rho_0)I^2_{\mathrm{Ph}}(\omega_{\mathrm{THz}})$. The radiation resistance for very large LAEs ($\rho_0 \gg \rho_{\mathrm{c,sc}} = 0.05\lambda_0$) is

$$R^{\mathrm{v}}_{\mathrm{LAE}}(\rho_0) \approx \left(\frac{0.31\,Z_0}{32\pi^3 n_{\mathrm{sc}}}\right)\cdot\left(\frac{v_{\mathrm{bal}}}{c_0}\right)^2\cdot\left(\frac{\lambda_0}{\rho_0}\right)^2 \approx 1.45\,\mu\Omega\cdot\left(\frac{\lambda_0}{\rho_0}\right)^2 \tag{3.211}$$

It differs from the expression for photoconducting LAEs (with non-ballistic transport) by the prefactor $f_{\mathrm{sc}}^{\mathrm{if,\,v}}(\theta_{\mathrm{sc}}) f_{\mathrm{sc}}^{\mathrm{v}}(\theta_{\mathrm{sc}}) = 0.31$ replacing $f_{\mathrm{sc}}^{\mathrm{if,\,h}}(0,0) \approx 2.5$, and (v_{bal}/c_0) replacing (v_{sat}/c_0). For InGaAs with $v_{\mathrm{bal}} = 2 \cdot 10^8\,\mathrm{cm/s}$, the radiation resistance becomes by a factor of $\approx$50 larger, in spite of the unfavorable radiation intensity for vertical acceleration.

The expression for $I_{\mathrm{Ph}}(\omega_{\mathrm{THz}})$ becomes, according to Eqs. (3.144)–(3.146)

$$I_{\mathrm{Ph}}(\omega_{\mathrm{THz}}) = I_0 \frac{2}{\omega_{\mathrm{THz}} \tau_{\mathrm{bal}}} [1 - \cos(\omega_{\mathrm{THz}} \tau_{\mathrm{bal}})] \tag{3.212}$$

For an estimate of the maximum expected power we choose the maximum of $I_{\mathrm{Ph}}(\omega_{\mathrm{THz}})$ from Eq. (3.212), which occurs around $\omega_{\mathrm{THz}} \tau_{\mathrm{bal}} \approx 3\pi/4$ or $f_{\mathrm{THz}} \approx 3/8\,\tau_{\mathrm{bal}} \approx 3\,\mathrm{THz}$ with $\tau_{\mathrm{bal}} \approx 125\,\mathrm{fs}$ at an accelerating field of 40 kV/cm (see right panel of Figure 2.6b in Chapter 2). This maximum value then becomes $\approx 1.45\,I_0$. We choose $\rho_0 = \lambda_0 = 100\,\mu\mathrm{m}$, which fulfills the requirement $\rho_0 \gg \rho_{\mathrm{c,sc}} = 0.05\lambda_0$ for the validity of Eq. (3.211) and yields an illuminated area of $\approx 30\,000\,\mu\mathrm{m}^2$. Assuming again $100\,\mathrm{kW/cm^2} = 1\,\mathrm{mW/mm^2}$ as maximum tolerable absorbed power density, this would yield a maximum absorbed power of $\approx$30 W and a value for $I_0 \approx 40\,\mathrm{A}$, now for 1550 nm lasers. With these values we obtain $P_{\mathrm{LAE}}^{\mathrm{v,\,max}}(3\,\mathrm{THz}) = 2.5\,\mathrm{mW}$ 2.5 mW at $f_{\mathrm{THz}} = 3\,\mathrm{THz}$. The laser-power/THz-power conversion efficiency in this case becomes $\eta_{\mathrm{LAE}} = P_{\mathrm{LAE}}^{\mathrm{v,max}}(3\,\mathrm{THz})/P_{\mathrm{L}} = 8.5 \cdot 10^{-5}$. This value corresponds rather well to the maximum value estimated for photoconductive AEs (Eq. (3.185b)) and is 33 times the maximum value estimated for photoconductive LAEs (Eq. (3.185a)).

Such a high laser power, however, would require a rather demanding heat flow managing, using epitaxial lift-off technique for the p-i-n structure and deposition on a diamond substrate.

For the present example, the THz power is already emitted into a rather narrow cone. The 3 dB angle for the radiation intensity distribution is 10.8° in the semiconductor and after transmission into the air it is 42°. If one is interested rather in a more pronounced directivity while keeping the intensity at the maximum constant, the laser beam radius can simply be increased at the same laser power. The intensity at the spot center will remain the same, however, the total emitted power will decrease. As the changes in the 3 dB angle are reciprocal to the changes in the laser beam radius an extremely high directivity with a 3 dB angle of 3.7° in air could be achieved with a beam radius of 1 mm instead of 100 μm for our example. Under these conditions, the thermal managing of the LAE would no longer be any problem. At the same time, such a highly directive source would be particularly interesting for beam steering. For beam steering with LAEs one can take advantage of the fact that very small differences in the angles of incidence between the two laser beams used for the photomixing results in large changes of emission angle for the THz beam, magnified by a factor $\lambda_{\mathrm{THz}}/\lambda_{\mathrm{L}}$ [61, 79]. The viability of this approach has been demonstrated already in the CW photomixing work with photoconductive LAEs [75].

Pulsed Operation of p-i-n-Based LAEs Exhibiting Ballistic Transport

We now consider the same scenario but under pulsed laser excitation.

The expression for the radiation resistance remains the same (Eq. (3.211)) as before. For $I_{\mathrm{Ph}}(t)$ we adapt the results from our previous discussion of photoconductive LAEs exhibiting ballistic transport. There, we considered the THz power expected for the main Fourier component at the frequency $f_{\mathrm{THz}} = 1/2\tau_{\mathrm{bal}} \approx 4\,\mathrm{THz}$ at DC fields of about 40 kV/cm. The time dependence remains (basically) the same as in Eq. (3.204). However, we have to be aware that the performance of ballistic p-i-n diodes suffers from screening effects at much lower densities of the photogenerated plasma compared with photoconductors. For the photoconductors we took the maximum sheet density of the electron–hole plasma of $n_{0,\,\mathrm{s}}^{\mathrm{max}} \approx 3 \cdot 10^{14}\,\mathrm{cm^{-2}}$ generated per laser pulse, as estimated in Chapter 2. In a ballistic p-i-n diode, the sheet electron density $n_{\mathrm{s,pin}}$ forms a dipole layer together with the hole sheet density of opposite charge. When propagating in the *i*-layer toward the n-layer, this dipole layer screens the original field E_0 (40 kV/cm) by an amount

$$\Delta E = \frac{e n_{\mathrm{s,\,pin}}}{\varepsilon_0 \varepsilon_{\mathrm{sc}}} \tag{3.213}$$

In order to maintain ballistic transport for all the electrons in the ensemble, this screening field should be significantly smaller than E_0. This defines a maximum for the tolerable sheet electron density generated in the p-i-n diode per pulse.

$$n_{\mathrm{s,\,pin}}^{\max} < \frac{\varepsilon_0 \varepsilon_{\mathrm{sc}} E_0}{e} \tag{3.214}$$

($n_{0,s}^{\max} \approx 3 \cdot 10^{11}\ \mathrm{cm}^{-2}$, for our example) which turns out to be 3 orders of magnitude smaller than for photoconductors. For a laser pulse generating the same number N_0 of charge carriers as before, $N_0 = n_{0,\mathrm{s}}^{\max}\ \pi\rho_0^2 \approx 9.5 \cdot 10^{10}$ (from Eq. (3.188) with $\rho_0 = 100\ \mu\mathrm{m}$, and $n_{0,\mathrm{s}}^{\max} = 3 \cdot 10^{14}\ \mathrm{cm}^{-2}$) the beam radius ρ_0 has to be chosen as a factor of $\sqrt{1000} \approx 30$ larger, that is, $\rho_0 \approx 3\ \mathrm{mm}$.

The THz power for the single-cycle pulse of frequency $f_{\mathrm{THz}}^{\mathrm{peak}} = 1/2\tau_{\mathrm{bal}} \approx 4\ \mathrm{THz}$ now becomes

$$P_{\mathrm{LAE}}^{\mathrm{v}}(t) = \frac{1}{2} R_{\mathrm{LAE}}^{\mathrm{v}}(\rho_0) I_0^2 \sin^2(2\pi f_{\mathrm{THz}}^{\mathrm{peak}} t) \text{ for } 0 < \mathrm{t} < 2\tau_{\mathrm{bal}} = 1/f_{\mathrm{THz}}, \tag{3.215}$$

instead of the corresponding expression for the non-ballistic photoconducting LAE with pulse excitation, Eq. (3.195).

With the 4 THz-Fourier component ($\lambda_0 = 75\ \mu\mathrm{m}$) of the radiation resistance at $\rho_0 \approx 3\ \mathrm{mm}$ the radiation resistance according to Eq. (3.211) is $R_{\mathrm{LAE}}^{\mathrm{v}}\ (\rho_0) = 1.45\ \ \mu\Omega\ \ (\lambda_0/\rho_0)^2 = 0.9\ \mathrm{n}\Omega$. The peak current I_0 from Eq. (3.189) is $I_0 = eN_0/(\sqrt{\pi}\tau_{\mathrm{pls}}) = 2.14 \cdot 10^5\ \mathrm{A}$, for $\tau_{\mathrm{pls}} = 40\ \mathrm{fs}$. We obtain for the maximum power amplitude of the maximum THz power generated by our p-i-n-based LAE with $\rho_0 \approx 3\ \mathrm{mm}$ at 4 THz $P_{\mathrm{LAE}}^{\mathrm{v,peak}} = 1/2 \cdot 0.9 \cdot 10^{-9}\Omega\ (2.14 \cdot 10^5 \mathrm{A})^2 \approx 40\ \mathrm{W}$ and a pulse energy of $E_{\mathrm{LAE}}^{\mathrm{v,peak}} = P_{\mathrm{LAE}}^{\mathrm{v,peak}}/f_{\mathrm{THz}}^{\mathrm{peak}} = 10\ \mathrm{pJ}$. The laser pulse energy required to generate the 9.5×10^{10} carriers at 1550 nm ($h\upsilon_{\mathrm{L}} = 0.8\ \mathrm{eV}$) is $E_{\mathrm{L}} = N_0 h\upsilon_{\mathrm{L}} \approx 12\ \mathrm{nJ}$. It is smaller for the telecom wavelength material than for the GaAs material used in our estimates for photoconductive LAEs under pulsed operation, because of the photon energy of 0.8 eV instead of 1.4 eV. Comparing the performance of the p-i-n-based LAE with photoconducting LAEs we note that for the non-ballistic photoconductors the THz pulse energy (Eq. (3.198)) was only about a factor of 4 higher, in spite of the by 3 orders of magnitude higher tolerable carrier densities. As the pulse energy in the present case is a factor of nearly 2 lower, the conversion efficiency of $0.8 \cdot 10^{-3}$ is only about a factor of 2 smaller. At the same time, the THz beam profile becomes extremely narrow even for the low frequency components in the pulse spectrum, because of the extremely large beam radius of 3 mm. Of course, the comparison becomes much more favorable for the photoconducting LAE, if the photoconducting material (e.g.,SI GaAs) exhibits ballistic transport. How far this is the case is not clear. However, our comparison between theory and experimental results for such LAEs has not indicated significant contributions from ballistic transport for this case.

Comparison with Experiments

So far, there are no reports on CW photomixing in p-i-n-based LAEs. There are, however numerous reports on THz emission originating from laser pulse excitation of semiconductors with vertical fields. In many of them, the goal was the investigation of transport phenomena, such as Bloch oscillations in semiconductor superlattices [65], high-field transport like velocity overshoot [63], generation of THz plasma oscillations [66], or optical phonon excitation [64] due to the surface space charge field. In some investigations the goal was THz generation, as in Ref. [80, 81]. The reported experiments were primarily physics oriented and there are, to our knowledge, no reports about high THz pulse energies obtained by semiconductor structures with vertical fields. A particular problem concerning p-i-n based LAEs is the necessity of contact layers. If the doping concentrations of the top and bottom contacts (n, and p-contact, respectively) are chosen too high, the dipole generated by the electron-hole-pairs is quenched between two more or less conductive plates, preventing any far field emission. This problem can be mitigated by a sufficiently low doping concentration and thin contact layers. However, this will also increase the (DC) resistance of the contact layers. There needs to be a trade-off between THz absorbance of these layers and heating due to transport of the photocurrent within the resistive contacts. We are not aware of p-i-n diode-based LAEs, although, according to our estimates in the previous section, very attractive results are to be expected. However, depending on the doping density, especially of contact layers, free carrier effects (absorption and reflection) has to be considered as well. This might substantially influence the efficient extraction of THz-power from such devices.

References

[1] Balanis, C.A. (2012) *Antenna Theory: Analysis and Design*, John Wiley & Sons, Inc.
[2] Stutzman, W.L. and Thiele, G.A. (2012) *Antenna Theory and Design*, John Wiley & Sons, Inc.
[3] Collin, R.E. (1985) *Antennas and Radiowave Propagation*, McGraw-Hill, New York.
[4] Kraus, J.D. (1988) *Antennas*, McGraw-Hill Education.
[5] Oliner, A.A., Jackson, D.R., and Volakis, J. (2007) *Antenna Engineering Handbook*, McGraw-Hill.
[6] Johnson, R.C. and Jasik, H. (1984) *Antenna Engineering Handbook*, McGraw-Hill Book Company, New York.
[7] Kominami, M., Pozar, D., and Schaubert, D. (1985) Dipole and slot elements and arrays on semi-infinite substrates. *IEEE Transactions on Antennas and Propagation*, **33**, 600.
[8] Rutledge, D.B., Neikirk, D.P., and Kasilingam, D.P. (1983) Integrated circuit antennas. *Infrared and Millimeter Waves (Part II)*, **10**, 1.
[9] Brewitt-Taylor, C.R., Gunton, D.J., and Rees, H.D. (1981) *Planar antennas* on a dielectric surface. *Electronics Letters*, **17**, 729.
[10] Lukosz, W. and Kunz, R.E. (1977) Light emission by magnetic and electric dipoles close to a plane interface. I. Total radiated power. *Journal of the Optical Society of America*, **67**, 1607.
[11] Lukosz, W. and Kunz, R.E. (1977) Light emission by magnetic and electric dipoles close to a plane dielectric interface. II. Radiation patterns of perpendicular oriented dipoles. *Journal of the Optical Society of America*, **67**, 1615.
[12] Lukosz, W. (1979) Light emission by magnetic and electric dipoles close to a plane dielectric interface. III. Radiation patterns of dipoles with arbitrary orientation. *Journal of the Optical Society of America*, **69**, 1495.
[13] Brueck, S.R.J. (2000) Radiation from a dipole embedded in a dielectric slab. *IEEE Journal on Selected Topics in Quantum Electronics*, **6**, 899.
[14] Preu, S., Döhler, G.H., Malzer, S. *et al.* (2011) Tunable, continuous-wave terahertz photomixer sources and applications. *Journal of Applied Physics*, **109**, 061301.
[15] Ramo, S., Whinnery, J.R., and Van Duzer, T. (2007) *Fields and Waves in Communication Electronics*, John Wiley & Sons, Inc.
[16] Rutledge, D.B. (1980) Submillimeter integrated circuit antennas and detectors. PhD thesis. University of California, Berkeley, CA.
[17] Born, M. and Wolf, E. (1980) *Principles of Optics*, Pergamon, Oxford.
[18] Michalski, K.A. and Mosig, J.R. (1997) Multilayered media Green's functions in integral equation formulations. *IEEE Transactions on Antennas and Propagation*, **45**, 508.
[19] Sengupta, K. and Hajimiri, A. (2012) A 0.28 THz power-generation and beam-steering array in CMOS based on distributed active radiators. *IEEE Journal of Solid-State Circuits*, **47**, 3013.
[20] Drabowitch S., Papiernik A., Griffiths H. D., *et al.* (1998) *Modern Antennas*. Springer, 1998.
[21] Jackson, D.R., Burghignoli, P., Lovat, G. *et al.* (2011) The fundamental physics of directive beaming at microwave and optical frequencies and the role of leaky waves. *Proceedings of the IEEE*, **99**, 1780.
[22] Duffy, S.M., Verghese, S., McIntosh, K.A. *et al.* (2001) Accurate modeling of dual dipole and slot elements used with photomixers for coherent terahertz output power. *IEEE Transactions on Microwave Theory and Techniques*, **49**, 1032.
[23] Nguyen, T. and Park, I. (2012) Resonant antennas on semi-infinite and lens substrates at terahertz frequency, in *Convergence of Terahertz Sciences in Biomedical Systems*, Springer Netherlands, pp. 181–193.
[24] Filipovic, D.F., Ali-Ahmad, W.Y., and Rebeiz, G.M. (1992) Millimeter-wave douple-dipole antennas for high-gain integrated reflector illumination. *IEEE Transactions on Microwave Theory and Techniques*, **40**, 962–967.
[25] Filipovic, D.F., Gearhart, S.S., and Rebeiz, G.M. (1993) Double-slot antennas on extended hemispherical and elliptical silicon dielectric lenses. *IEEE Transactions on Microwave Theory and Techniques*, **41**, 1738.
[26] Endo, T., Sunahara, Y., Satoh, S., and Katagi, T. (2000) Resonant frequency and radiation efficiency of meander line antennas. *Electronic Communications in Japan (Part 2)*, **83**, 52.
[27] Woo, I., Nguyen, T.K., Han, H. *et al.* (2010) Four-leaf-clover-shaped antenna for a THz photomixer. *Optics Express*, **18**, 18532.

[28] Mushiake Y. (1992) Self-complementary antennas. *IEEE Antennas and Propagation Magazine*, **34**, 23.
[29] Nguyen, T.K., Ho, T.A., Han, H., and Park, I. (2012) Numerical sudy of self-complementary antenna characteristics on substrate lenses at terahertz frequency. *Journal of Infrared, Millimeter, and Terahertz Waves*, **33**, 1123.
[30] Rebeiz, G.M. (1992) Millimeter-wave and terahertz integrated circuit antennas. *Proceedings of the IEEE*, **80**, 1748.
[31] Computer Simulation Technology (CST) (2014) April https://www.cst.com/ (accessed 28 November 2014).
[32] ANSYS Simulation Driven Product Development (2014) HFSS, May 2014, http://www.ansys.com/ (accessed 28 November 2014).
[33] FEKO (2014) EM Simulation Software, May 2014, http://www.feko.info/ (accessed 28 November 2014).
[34] Chen, S.-Y. and Hsu, P. (2007) A simplified method to calculate far-field patterns of extended hemispherical dielectric lens. 2007 Asia-Pacific Microwave Conference (APMC 2007), IEEE, pp. 1–4.
[35] Neto, A., Maci, S., and De Maagt, P. (1998) Reflections inside an elliptical dielectric lens antenna. *IEE Proceedings-Microwaves, Antennas and Propagation*, **145**, 243.
[36] Balanis, C.A. (1989) *Advanced Engineering Electromagnetics*, John Wiley & Sons, Inc., New York.
[37] Collin, R.E. (2007) *Foundations for Microwave Engineering*, Wiley-IEEE Press.
[38] Chang, K., York, R.A., Hall, P.S., and Itoh, T. (2002) Active integrated antennas. *IEEE Transactions on Microwave Theory and Techniques*, **50**, 937.
[39] Segovia-Vargas, D., Castro-Galán, D., García-Muñoz, L.E., and González-Posadas, V. (2008) Broadband active receiving patch with resistive equalization. *IEEE Transactions on Microwave Theory and Techniques*, **56**, 56.
[40] Mayorga, I.C., Schmitz, A., Klein, T. *et al.* (2012) First in-field application of a full photonic local oscillator to terahertz astronomy. *IEEE Transactions on Terahertz Science and Technology*, **2**, 393.
[41] Gibson, P.J. (1979) The vivaldi aerial. 9th European Microwave Conference, 1979, p. 101.
[42] Yngvesson, K.S., Schaubert, D.H., Korzeniowski, T.L. *et al.* (1985) Endfire tapered slot antennas on dielectric substrates. *IEEE Transactions on Antennas and Propagation*, **33**, 1392.
[43] Yngvesson, K.S., Korzeniowski, T.L., Kim, Y.S. *et al.* (1989) The tapered slot antenna: a new integrated element for millimeter wave applications. *IEEE Transactions on Microwave Theory and Techniques*, **37**, 365.
[44] Maaskant, R., Ivashina, M.V., Iupikov, O. *et al.* (2011) Analysis of large microstrip-fed tapered slot antenna arrays by combining electrodynamic and quasi-static field modes. *IEEE Transactions on Antennas and Propagation*, **59**, 1798.
[45] Ivashina, M.V., Iupikov, O., Maaskant, R. *et al.* (2011) An optimal beamforming strategy for wide-field surveys with phased-array-fed reflector antennas. *IEEE Transactions on Antennas and Propagation*, **59**, 1864.
[46] Janaswamy, R., Schaubert, D.H., and Pozar, D.M. (1986) Analysis of the transverse electromagnetic mode linearly tapered slot antenna. *Radio Science*, **21**, 797.
[47] Janaswamy, R. and Schaubert, D.H. (1986) Characteristic impedance of a wide slotline on low-permittivity susbtrates. *IEEE Transactions on Microwave Theory and Techniques*, **34**, 900.
[48] Janaswamy, R. and Schaubert, D.H. (1987) Analysis of the tapered slot antenna. *IEEE Transactions on Antennas and Propagation*, **35**, 1058.
[49] de Lera Acedo, E., Garcia, E., Gonzalez-Posadas, V. *et al.* (2010) Study and design of a differentially-fed tapered slot antenna array. *IEEE Transactions on Antennas and Propagation*, **58**, 68.
[50] Holter, H., Chio, T.-H., and Schaubert, D.H. (2000) Elimination of impedance anomalies in single- and dual-polarized endfire tapered slot phased arrays. *IEEE Transactions on Antennas and Propagation*, **48**, 122.
[51] Schaubert, D. (1996) A class of E-plane scan blindnesses in single-polarized arrays of tapered-slot antennas with a ground plane. *IEEE Transactions on Antennas and Propagation*, **44**, 954.
[52] Andres-Garcia, B., Garcia-Munoz, L.E., Camara-Mayorga, I. *et al.* (2010) Antenna in the Terahertz band for radioastronomy applications. Proceedings of the Fourth European Conference on Antennas and Propagation (EuCAP), 2010, p. 1.
[53] Rizk, J.B. and Rebeiz, G.M. (2002) Millimeter-wave Fermi tapered slot antennas on micromachined silicon substrates. *IEEE Transactions on Antennas and Propagation*, **50**, 379.
[54] Munk, B.A. (2005) *Frequency Selective Surfaces: Theory and Design*, John Wiley & Sons, Inc.
[55] Yang, F. and Rahmat-Samii, Y. (2009) *Electromagnetic Band Gap Structures in Antenna Engineering*, Cambridge University Press.
[56] Mittra, R. (1988) Techniques for analyzing frequency selective surfaces – a review. *Proceedings of the IEEE*, **76**, 1539.
[57] Simons, R.N. (2001) *Coplanar Waveguide Circuits, Components, and Systems*, 1st edn, John Wiley & Sons, Inc., New York.

[58] Altshuler, E.E. (1989) Self- and mutual impedances on travelling- wave linear antennas. *IEEE Transactions on Antennas and Propagation*, **37**, 1312.
[59] Bauerschmidt, S., Preu, S., Malzer, S. *et al.* (2010) Continuous wave terahertz emitter arrays for spectroscopy and imaging applications. *Proceedings of SPIE*, **7671**, 76710D.
[60] Preu, S., Malzer, S., Döhler, G.H., and Wang, L.J. (2008) Coherent superposition of terahertz beams. *Proceedings of SPIE*, **7117**, 71170S.
[61] Preu, S., Malzer, S., Döhler, G.H. *et al.* (2006) Highly collimated and directional continous-wave terahertz emission by photomixing in semiconductor device arrays. *Proceedings of SPIE*, **6194**, 61940F.
[62] Darrow, J.T., Zhang, X.-C., Auston, D.H., and Morse, J.D. (1992) Saturation properties of large-aperture photoconducting antennas. *IEEE Journal of Quantum Electronics*, **28**, 1607.
[63] Leitenstorfer, A., Hunsche, S., Shah, J. *et al.* (2000) Femtosecond high-field transport in compound semiconductors. *Physical Review B*, **61**, 16642.
[64] Dekorsy, T., Auer, H., Bakker, H.J. *et al.* (1996) THz electromagnetic emission by coherent infrared-active phonons. *Physical Review B*, **53**, 4005.
[65] Waschke, C., Roskos, H.G., Schwedler, R. *et al.* (1993) Coherent submillimeter-wave emission from Bloch oscillations in a semiconductor superlattice. *Physical Review Letters*, **70**, 3319.
[66] Kersting, R., Unterrainer, K., Strasser, G. *et al.* (1997) Few-cycle THz emission from cold plasma oscillations. *Physical Review Letters*, **79**, 3038.
[67] Dreyhaupt, A., Winnerl, S., Decorsy, T., and Helm, M. (2005) High-intensity terahertz radiation from a microstructures large-area photoconductor. *Applied Physics Letters*, **86**, 121114.
[68] Awad, M., Nagel, M., Kurz, H. *et al.* (2007) Characterization of low temperature GaAs antenna array terahertz emitters. *Applied Physics Letters*, **91**, 181124.
[69] Dreyhaupt, A., Winnerl, S., Helm, M., and Dekorsy, T. (2006) Optimum excitation conditions for the generation of high-electric-field terahertz radiation from an oscillator-driven photoconductive device. *Optics Letters*, **31**, 1546.
[70] Peter, F., Winnerl, S., Schneider, H. *et al.* (2008) THz emission from a large-area GaInAsN emitter. *Applied Physics Letters*, **93**, 101102.
[71] Beck, M., Schäfer, H., Klatt, G. *et al.* (2010) Impulsive terahertz radiation with high electric fields from an amplifier-driven large-area photoconductive antenna. *Optics Express*, **18**, 9251.
[72] Klatt, G., Hilser, F., Qiao, W. *et al.* (2010) Terahertz emission from latera photo-Dember currents. *Optics Express*, **18**, 4939.
[73] Preu, S., Mittendorff, M., Lu, H. *et al.* (2012) 1550 nm ErAs:In(Al)GaAs large area photoconductive emitters. *Applied Physics Letters*, **101**, 101105.
[74] Mittendorff, M., Xu, M., Dietz, R.J.B. *et al.* (2013) Large area photoconductive terahertz emitter for 1.55 μm excitation based on an InGaAs heterostructure. *Nanotechnology*, **24**, 214007.
[75] Eshaghi, A., Shahabadi, M., and Chrostowski, L. (2012) Radiation characteristics of large-area photomixer used for generation of continuous-wave terahertz radiation. *Journal of the Optical Society of America B*, **29**, 813.
[76] Jackson, D.A. (1975) *Classical Electrodynamics*, 2nd edn, Chapter 9.2, John Wiley & Sons, Inc., New York.
[77] Döhler, G.H., Garcia-Muñoz, L.E., Preu, S. *et al.* (2013) From arrays of THz antennas to large-area emitters. *IEEE Transactions on Terahertz Science and Technology*, **3**, 532.
[78] Xu, M., Mittendorff, M., Dietz, R.J.B. *et al.* (2013) Terahertz generation and detection with InGaAs-based large-area photoconductive devices excited at 1.55 μm. *Applied Physics Letters*, **103**, 251114.
[79] Bauerschmidt, S.T., Döhler, G.H., Lu, H. *et al.* (2013) Arrayed free space continuous-wave THz photomixers. *Optics Letters*, **38**, 3673.
[80] Zhang, X.-C. and Auston, D.H. (1992) Optoelectronic measurement of semiconductor surfaces and interfaces with femtosecond optics. *Journal of Applied Physics*, **71**, 326.
[81] Johnston, M.B., Whittaker, D.M., Dowd, A. *et al.* (2002) Generation of high-power terahertz pulses in a prism. *Optics Letters*, **27**, 1935.

4

Propagation at THz Frequencies

Antti V. Räisänen[1], Dmitri Lioubtchenko[1], Andrey Generalov[1], J. Anthony Murphy[2], Créidhe O'Sullivan[2], Marcin L. Gradziel[2], Neil Trappe[2], Luis Enrique Garcia Muñoz[3], Alejandro Garcia-Lamperez[3], and Javier Montero-de-Paz[3]

[1]*Aalto University, School of Electrical Engineering, Department of Radio Science and Engineering, Espoo, Finland*

[2]*National University of Ireland Maynooth, Department of Experimental Physics, Maynooth, Co. Kildare, Ireland*

[3]*Universidad Carlos III de Madrid, Departamento de Teoria de la Señal y Communicaciones, Leganes, Madrid, Spain*

In this chapter we first briefly introduce Maxwell's equations and derive the Helmholtz equation, that is, a special case of the wave equation, and introduce its different solutions for fields that may propagate in THz waveguides. The second section describes different waveguides operating at THz frequencies. The third is devoted to the beam waveguide and quasi-optics. Material issues related to waveguides and quasi-optical components are discussed in the fourth section, and the chapter concludes with THz wave propagation in free space.

4.1 Helmholtz Equation and Electromagnetic Modes of Propagation

Maxwell's equations relate the electromagnetic fields, that is, the electric field strength E (V/m), electric flux density D (C/m^2 = As/m^2), magnetic field strength H (A/m), and magnetic flux density B (Wb/m^2 = Vs/m^2), and their sources, that is, the electric charge density ρ (C/m^3 = As/m^3), and electric current density J (A/m^2) to each other. The electric and magnetic properties of media bind the field strengths and flux densities; the constitutive relations are

$$\vec{D} = \varepsilon\vec{E} \tag{4.1}$$

$$\vec{B} = \mu\vec{H} \tag{4.2}$$

Semiconductor Terahertz Technology: Devices and Systems at Room Temperature Operation, First Edition.
Edited by Guillermo Carpintero, Luis Enrique García Muñoz, Hans L. Hartnagel, Sascha Preu and Antti V. Räisänen.

where ε is the permittivity (F/m = As/Vm) and μ the permeability (H/m = Vs/Am) of the medium, see, for example, Refs. [1–5].

Maxwell's equations in differential form are

$$\nabla \cdot \vec{D} = \rho \tag{4.3}$$

$$\nabla \cdot \vec{B} = 0 \tag{4.4}$$

$$\nabla \times \vec{E} = -\frac{\partial \vec{B}}{\partial t} \tag{4.5}$$

$$\nabla \times \vec{H} = \vec{J} + \frac{\partial \vec{D}}{\partial t} \tag{4.6}$$

Equation (4.3) is also known as Gauss' law, Eq. (4.5) as Faraday's law, and Eq. (4.6) as Ampère's law with the displacement current term added by Maxwell. The differential equations (4.3)–(4.6) describe the fields locally or in a given point. They allow us to obtain the change of field versus space or time.

Time harmonic fields, that is, fields having a sinusoidal time dependence at angular frequency of $\omega = 2\pi f$, may be presented as

$$\vec{A}(x, y, z, t) = \mathrm{Re}\{\vec{A}(x, y, z)\mathrm{e}^{\mathrm{j}\omega t}\} \tag{4.7}$$

At a given point (x, y, z), the field may be considered as a vector rotating on the complex plane and having a constant amplitude. The real part of the amplitude is the field value at a given instant. Most phenomena in terahertz techniques are time harmonic or can be considered as superpositions of several time harmonics. Assuming the $\mathrm{e}^{\mathrm{j}\omega t}$ time dependence, the time derivatives can be replaced by $\mathrm{j}\omega$. For such sinusoidal fields and sources, Maxwell's equations in differential form are

$$\nabla \cdot \vec{D} = \rho \tag{4.8}$$

$$\nabla \cdot \vec{B} = 0 \tag{4.9}$$

$$\nabla \times \vec{E} = -\mathrm{j}\omega\vec{B} = -\mathrm{j}\omega\mu\vec{H} \tag{4.10}$$

$$\nabla \times \vec{H} = \vec{J} + \mathrm{j}\omega\vec{D} = (\sigma + \mathrm{j}\omega\varepsilon)\vec{E} \tag{4.11}$$

In Eq. (4.11) we have taken into account that

$$\vec{J} = \sigma\vec{E} \tag{4.12}$$

where σ is the conductivity (S/m = A/Vm) of the medium.

A changing magnetic flux creates an electric field, and both a moving charge (current) and a changing electric flux create a magnetic field. These facts explain the creation of a propagating wave: a changing current, for example, in an antenna, creates a changing magnetic field, the changing magnetic field creates a changing electric field, the changing electric field creates a changing magnetic field, and so on.

In the above equations, the permittivity ε and permeability μ represent the properties of the medium. A medium is homogeneous if its properties are constant, independently of location. An isotropic medium has the same properties in all directions. The properties of a linear medium do not depend on the field strength.

In vacuum $\varepsilon = \varepsilon_0 = 8.8542 \times 10^{-12}\mathrm{F/m}$ and $\mu = \mu_0 = 4\pi \times 10^{-7}\mathrm{H/m}$. In other homogeneous media $\varepsilon = \varepsilon_\mathrm{r}\varepsilon_0$ and $\mu = \mu_\mathrm{r}\mu_0$, where the dielectric constant ε_r, that is, the relative permittivity, and the relative permeability μ_r depend on the structure of the material. For air we can take $\varepsilon_\mathrm{r} = \mu_\mathrm{r} = 1$. In general, in a lossy medium ε_r or μ_r are complex, as[1]

$$\varepsilon = \varepsilon_0\varepsilon_\mathrm{r} = \varepsilon_0(\varepsilon_\mathrm{r}' - \mathrm{j}\varepsilon_\mathrm{r}'') \tag{4.13}$$

and in an anisotropic medium they are furthermore tensors.

[1] The dielectric tensor in engineering notation is $\varepsilon_\mathrm{r} = \varepsilon_\mathrm{r}' - \mathrm{j}\varepsilon_\mathrm{r}''$, in physics notation it is, however, $\varepsilon_\mathrm{r} = \varepsilon_\mathrm{r}' + \mathrm{i}\varepsilon_\mathrm{r}''$. Conversion from one system to the other is simply achieved by replacing $-\mathrm{j}$ by i.

The loss in a medium is due to polarization of molecules but it may also be due to the conductivity of the material. In this case, in the material there are free charges that are moved by the electric field. When both loss mechanisms are introduced in Eq. (4.11), one obtains

$$\nabla \times \vec{H} = \left[\sigma + \mathrm{j}\omega\varepsilon_0\left(\varepsilon_\mathrm{r}' - \mathrm{j}\varepsilon_\mathrm{r}''\right)\right]\vec{E} = \mathrm{j}\omega\varepsilon_0\left(\varepsilon_\mathrm{r}' - \mathrm{j}\varepsilon_\mathrm{r}'' - \mathrm{j}\frac{\sigma}{\omega\varepsilon_0}\right)\vec{E} \tag{4.14}$$

which indicates that attenuation due to polarization and due to conduction are indistinguishable without a measurement at several frequencies. However, often $\sigma/(\omega\varepsilon_0)$ is included in ε_r''. Loss of a medium may be characterized by the loss tangent

$$\tan\delta = \frac{\varepsilon_\mathrm{r}'' + \sigma/(\omega\varepsilon_0)}{\varepsilon_\mathrm{r}'} \tag{4.15}$$

In the case of magnetic materials, the situation is analogous: the magnetic field polarizes the material magnetically. The permeability can be divided into a real part μ_r' and an imaginary part μ_r''; the latter causes magnetic loss.

In a source-free ($\rho = 0, \vec{J} = 0$), linear, and isotropic medium, Maxwell's equations are simplified into the following form:

$$\nabla \cdot \vec{E} = 0 \tag{4.16}$$

$$\nabla \cdot \vec{H} = 0 \tag{4.17}$$

$$\nabla \times \vec{E} = -\mathrm{j}\omega\mu\vec{H} \tag{4.18}$$

$$\nabla \times \vec{H} = \mathrm{j}\omega\varepsilon\vec{E} \tag{4.19}$$

When $\nabla\times$ operator is applied on both sides of Eq. (4.18), we obtain

$$\nabla \times \nabla \times \vec{E} = -\mathrm{j}\omega\mu\nabla \times \vec{H} \tag{4.20}$$

which leads, after utilizing vector identity $\nabla \times \nabla \times \vec{A} = \nabla(\nabla \cdot \vec{A}) - \nabla^2\vec{A}$ and Eq. (4.16), to

$$\nabla^2\vec{E} = -\omega^2\mu\varepsilon\vec{E} = -k^2\vec{E} \tag{4.21}$$

This equation is called the Helmholtz equation, which is a special case of the wave equation

$$\nabla^2\vec{E} - \mu\varepsilon\frac{\partial^2\vec{E}}{\partial t^2} = 0 \tag{4.22}$$

The constant $k = \omega\sqrt{\mu\varepsilon}$ is called the wavenumber (1/m).

Let us consider propagation of a wave in a lossless medium, where ε_r and μ_r are real. Then k is also real. Let us assume that the electric field has only the x component, the field is uniform in the x and y directions, and the wave propagates into the z direction. The Helmholtz equation reduces to

$$\frac{\partial^2 E_x}{\partial z^2} + k^2 E_x = 0 \tag{4.23}$$

which has a solution as

$$E_x(z) = E^+\mathrm{e}^{-\mathrm{j}kz} + E^-\mathrm{e}^{\mathrm{j}kz} \tag{4.24}$$

where E^+ and E^- are arbitrary amplitudes of waves propagating into the $+z$ and $-z$ directions, respectively. The values of E^+ and E^- may be determined by the boundary conditions. In the time domain, Eq. (4.24) can be expressed as

$$E_x(z,t) = E^+\cos(\omega t - kz) + E^-\cos(\omega t + kz) \tag{4.25}$$

where E^+ and E^- are assumed to be real constants.

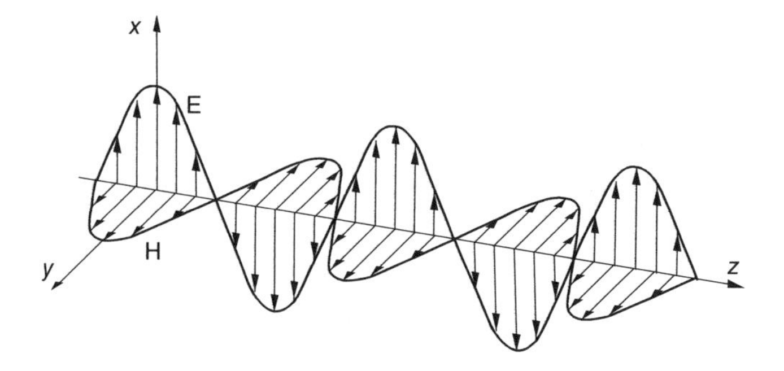

Figure 4.1 Plane wave propagating into the $+z$ direction.

The magnetic field of a plane wave can be solved from Eq. (4.19). The result is

$$H_y = \frac{1}{\eta}(E^+ e^{-jkz} - E^- e^{jkz}) \tag{4.26}$$

that is, the magnetic field has a component which is perpendicular to the electric field and to the direction of propagation. The ratio of the electric and magnetic fields is called the wave impedance, and it is $\eta = \sqrt{\mu/\varepsilon}$. In vacuum $\eta_0 = \sqrt{\mu_0/\varepsilon_0} \approx 120\pi\ \Omega \approx 377\ \Omega$. Figure 4.1 illustrates a plane wave propagating into the $+z$ direction. The fields of a plane wave repeat themselves periodically in the z direction; the wavelength is

$$\lambda = \frac{2\pi}{k} = \frac{2\pi}{\omega\sqrt{\mu\varepsilon}} = \frac{1}{f\sqrt{\mu\varepsilon}} \tag{4.27}$$

The propagation velocity of the wave is

$$v = f\lambda = \frac{1}{\sqrt{\mu\varepsilon}} \tag{4.28}$$

In vacuum, the propagation velocity is the speed of light, $v = c = 1/(\mu_0\varepsilon_0)^{1/2} = 2.998 \times 10^8$ m/s.

In a lossy medium having conductivity σ, the Helmholtz equation gets the following form:

$$\nabla^2\vec{E} + \omega^2\mu\varepsilon\left(1 - j\frac{\sigma}{\omega\varepsilon}\right)\vec{E} = 0 \tag{4.29}$$

Compared to Eq. (4.21), here jk is replaced by a complex propagation constant

$$\gamma = \alpha + j\beta = j\omega\sqrt{\mu\varepsilon}\sqrt{1 - j\frac{\sigma}{\omega\varepsilon}} \tag{4.30}$$

where α is the attenuation constant and β is the phase constant. In the case of a plane wave propagating into the z direction, we have

$$\frac{\partial^2 E_x}{\partial z^2} - \gamma^2 E_x = 0 \tag{4.31}$$

leading to

$$E_x(z) = E^+ e^{-\gamma z} + E^- e^{\gamma z} \tag{4.32}$$

and in the time domain

$$E_x(z,t) = E^+ e^{-\alpha z}\cos(\omega t - \beta z) + E^- e^{\alpha z}\cos(\omega t + \beta z) \tag{4.33}$$

In the case of a good conductor ($\sigma \gg \omega\varepsilon$), we obtain the propagation constant as

$$\gamma = \alpha + \mathrm{j}\beta \approx \mathrm{j}\omega\sqrt{\mu\varepsilon}\sqrt{\frac{\sigma}{\mathrm{j}\omega\varepsilon}} = (1+\mathrm{j})\sqrt{\frac{\omega\mu\sigma}{2}} \tag{4.34}$$

When a plane wave meets a surface of a lossy medium in a perpendicular direction and penetrates into it, its field is damped into 1/e part over a distance called a skin depth:

$$\delta_{\mathrm{s}} = \frac{1}{\alpha} = \sqrt{\frac{2}{\omega\mu\sigma}} \tag{4.35}$$

In a general case, each field component can be solved from the general Helmholtz equation

$$\frac{\partial^2 E_i}{\partial x^2} + \frac{\partial^2 E_i}{\partial y^2} + \frac{\partial^2 E_i}{\partial z^2} - \gamma^2 E_i = 0, \quad i = x, y, z \tag{4.36}$$

The solution is found by using the separation of variables. By assuming that $E_i = f(x)g(y)h(z)$

$$\frac{f''}{f} + \frac{g''}{g} + \frac{h''}{h} - \gamma^2 = 0 \tag{4.37}$$

where the double prime denotes the second derivative. The first three terms of this equation are each a function of one independent variable. As the sum of these terms is constant (γ^2), each term must also be constant. Therefore, Eq. (4.37) can be divided into three independent equations of form $f''/f - \gamma_x^2 = 0$ having solutions as shown above.

The Helmholtz equation (4.21) is valid in the source-free medium of a waveguide. Let us assume that a wave is propagating in a uniform line along the z direction. Now, the Helmholtz equation is separable so that a solution having a form $f(z)g(x, y)$ can be found. The z-dependence of the field has a form of $\mathrm{e}^{-\gamma z}$. Thus, the electric field may be written as

$$\vec{E} = \vec{E}(x, y, z) = \vec{g}(x, y)\mathrm{e}^{-\gamma z} \tag{4.38}$$

where $\vec{g}(x,y)$ is the field distribution in the transverse plane. By setting this to the Helmholtz equation, we get

$$\nabla^2\vec{E} = \nabla^2_{xy}\vec{E} + \frac{\partial^2\vec{E}}{\partial z^2} = \nabla^2_{xy}\vec{E} + \gamma^2\vec{E} = -\omega^2\mu\varepsilon\vec{E} \tag{4.39}$$

where ∇^2_{xy} includes the partial derivates of ∇^2 with respect to x and y. As similar equations can be derived for the magnetic field, the partial differential equations applicable for all uniform lines are

$$\nabla^2_{xy}\vec{E} = -(\gamma^2 + \omega^2\mu\varepsilon)\vec{E} \tag{4.40}$$

$$\nabla^2_{xy}\vec{H} = -(\gamma^2 + \omega^2\mu\varepsilon)\vec{H} \tag{4.41}$$

At first, the longitudinal or z-component of the electric or magnetic field is solved using the boundary conditions set by the structure of the line. When E_z or H_z is known, the x- and y-components of the fields may be solved from Eqs. (4.18) and (4.19). Equation $\nabla \times \vec{E} = -\mathrm{j}\omega\mu\vec{H}$ may be divided into three parts:

$$\partial E_z/\partial y + \gamma E_y = -\mathrm{j}\omega\mu H_x$$

$$-\gamma E_x - \partial E_z/\partial x = -\mathrm{j}\omega\mu H_y$$

$$\partial E_y/\partial x - \partial E_x/\partial y = -\mathrm{j}\omega\mu H_z$$

Correspondingly, $\nabla \times \vec{H} = \mathrm{j}\omega\varepsilon\vec{E}$ is divided into three parts:

$$\partial H_z/\partial y + \gamma H_y = \mathrm{j}\omega\varepsilon E_x$$
$$-\gamma H_x - \partial H_z/\partial x = \mathrm{j}\omega\varepsilon E_y$$
$$\partial H_y/\partial x - \partial H_x/\partial y = \mathrm{j}\omega\varepsilon E_z$$

From these, the transverse components E_x, E_y, H_x, and H_y are solved as functions of the longitudinal components E_z and H_z:

$$E_x = \frac{-1}{\gamma^2 + \omega^2\mu\varepsilon}\left(\gamma\frac{\partial E_z}{\partial x} + \mathrm{j}\omega\mu\frac{\partial H_z}{\partial y}\right) \tag{4.42a}$$

$$E_y = \frac{1}{\gamma^2 + \omega^2\mu\varepsilon}\left(-\gamma\frac{\partial E_z}{\partial y} + \mathrm{j}\omega\mu\frac{\partial H_z}{\partial x}\right) \tag{4.42b}$$

$$H_x = \frac{1}{\gamma^2 + \omega^2\mu\varepsilon}\left(\mathrm{j}\omega\varepsilon\frac{\partial E_z}{\partial y} - \gamma\frac{\partial H_z}{\partial x}\right) \tag{4.42c}$$

$$H_y = \frac{-1}{\gamma^2 + \omega^2\mu\varepsilon}\left(\mathrm{j}\omega\varepsilon\frac{\partial E_z}{\partial x} + \gamma\frac{\partial H_z}{\partial y}\right) \tag{4.42d}$$

In a cylindrical coordinate system, the solution of the Helmholtz equation has the form of $f(z)g(r,\varphi)$. As above, the transverse r- and φ-components are calculated from the longitudinal components:

$$E_r = \frac{-1}{\gamma^2 + \omega^2\mu\varepsilon}\left(\gamma\frac{\partial E_z}{\partial r} + \frac{\mathrm{j}\omega\mu}{r}\frac{\partial H_z}{\partial \varphi}\right) \tag{4.43a}$$

$$E_\varphi = \frac{1}{\gamma^2 + \omega^2\mu\varepsilon}\left(-\frac{\gamma}{r}\frac{\partial E_z}{\partial \varphi} + \mathrm{j}\omega\mu\frac{\partial H_z}{\partial r}\right) \tag{4.43b}$$

$$H_r = \frac{1}{\gamma^2 + \omega^2\mu\varepsilon}\left(\frac{\mathrm{j}\omega\varepsilon}{r}\frac{\partial E_z}{\partial \varphi} - \gamma\frac{\partial H_z}{\partial r}\right) \tag{4.43c}$$

$$H_\varphi = \frac{-1}{\gamma^2 + \omega^2\mu\varepsilon}\left(\mathrm{j}\omega\varepsilon\frac{\partial E_z}{\partial r} + \frac{\gamma}{r}\frac{\partial H_z}{\partial \varphi}\right) \tag{4.43d}$$

In a given waveguide at a given frequency, several field configurations may satisfy Maxwell's equations. These field configurations are called wave modes. Every mode has its own propagation characteristics: velocity, attenuation, and cut-off frequency. Because different modes propagate at different velocities, signals may be distorted due to this multi-mode propagation. Care must be taken to avoid the multi-mode distortion, for example, to use the waveguide at low enough frequencies, so that only one mode, the dominant or fundamental mode can propagate along the waveguide.

In lossless and, with a good approximation, in low-loss two-conductor waveguides, such as coaxial lines, fields can propagate as transverse electromagnetic (TEM) waves . TEM waves have no longitudinal field components. TEM waves may propagate at all frequencies, that is, the TEM mode has no cut-off frequency.

When $E_z = 0$ and $H_z = 0$, it follows from Eq. (4.42a–d) that also the x- and y-components of the fields are zero, unless

$$\gamma^2 + \omega^2\mu\varepsilon = 0 \tag{4.44}$$

Therefore, the propagation constant of a TEM wave is

$$\gamma = \pm\mathrm{j}\omega\sqrt{\mu\varepsilon} = \pm\mathrm{j}\frac{2\pi}{\lambda} = \pm\mathrm{j}\beta \tag{4.45}$$

The propagation velocity of a guided wave is described by two different quantities: the velocity of the constant phase points of the wave is called the phase velocity, v_p, and the velocity of the energy (of a narrow-band signal) is called the group velocity, v_g. The phase velocity can be obtained as the ratio of the angular frequency and the phase constant

$$v_p = \frac{\omega}{\beta} \tag{4.46}$$

and the group velocity is the derivative of the phase constant

$$v_g = \left(\frac{d\beta}{d\omega}\right)^{-1} \tag{4.47}$$

In case of a TEM wave propagating in a waveguide, there is no dispersion, and the wave propagates at the same velocity as a plane wave in free space having the same ε and μ as the insulator of the waveguide. The phase and group velocities are now equal: $v_p = v_g = (\mu\varepsilon)^{-1/2}$.

The wave equations for a TEM wave are

$$\nabla^2_{xy}\vec{E} = 0 \tag{4.48}$$

$$\nabla^2_{xy}\vec{H} = 0 \tag{4.49}$$

The fields of a wave propagating along the z direction satisfy the equation

$$\frac{E_x}{H_y} = -\frac{E_y}{H_x} = \eta \tag{4.50}$$

where $\eta = \sqrt{\mu/\varepsilon}$ is the wave impedance.

For static fields, the electric field may be presented as the gradient of the scalar transverse potential:

$$\vec{E}(x, y) = -\nabla_{xy}\Phi(x, y) \tag{4.51}$$

Because $\nabla \times \nabla f = 0$, the transverse curl of the electric field must vanish in order for Eq. (4.51) to be valid. This is the case for a TEM wave, because

$$\nabla_{xy} \times \vec{E} = -j\omega\mu H_z\vec{u}_z = 0 \tag{4.52}$$

where $\vec{u}_z$ is a unit vector pointing in the direction of the positive z-axis. Gauss' law in a sourceless space states that $\nabla \cdot \vec{D} = \varepsilon\nabla_{xy} \cdot \vec{E} = 0$. From this and Eq. (4.51) it follows also that $\Phi(x, y)$ is a solution of Laplace's equation or

$$\nabla^2_{xy}\Phi(x, y) = 0. \tag{4.53}$$

A wave mode may also have longitudinal field components in addition to the transverse components. Transverse electric (TE) modes have $E_z = 0$ but a nonzero longitudinal magnetic field H_z. Transverse magnetic (TM) modes have $H_z = 0$ and a non-zero E_z.

From Eq. (4.42a–d) it follows for a TE mode

$$\nabla^2_{xy}H_z = -(\gamma^2 + \omega^2\mu\varepsilon)H_z = -k_c^2 H_z \tag{4.54}$$

Correspondingly, for a TM mode

$$\nabla^2_{xy}E_z = -(\gamma^2 + \omega^2\mu\varepsilon)E_z = -k_c^2 E_z \tag{4.55}$$

The coefficient k_c is solved from Eqs. (4.54) or (4.55) using the boundary conditions set by the waveguide. For a given waveguide, usually an infinite number of k_c values exist. Each k_c corresponds to a propagating wave mode. It can be proved that in the case of a waveguide such as a rectangular or circular metal waveguide, in which a conductor surrounds an insulator, k_c is always a positive real number.

The propagation constant is

$$\gamma = \sqrt{k_c^2 - \omega^2 \mu \varepsilon} \tag{4.56}$$

If the insulating material is lossless, $\omega^2 \mu \varepsilon$ is real. The frequency at which $\omega^2 \mu \varepsilon = k_c^{\ 2}$ is called the cut-off frequency:

$$f_c = \frac{k_c}{2\pi \sqrt{\mu \varepsilon}} \tag{4.57}$$

The corresponding cut-off wavelength is

$$\lambda_c = \frac{2\pi}{k_c} \tag{4.58}$$

At frequencies below the cut-off frequency, $f < f_c$, there is no propagating wave. At frequencies higher than the cut-off frequency, $f > f_c$, waves can propagate and the propagation constant γ is complex. In a lossless line, γ is imaginary:

$$\gamma = \mathrm{j}\beta_g = \mathrm{j}\frac{2\pi}{\lambda}\sqrt{1 - \left(\frac{f_c}{f}\right)^2} \tag{4.59}$$

where λ is the wavelength in the free space composed of the same material as the insulator of the waveguide. The guided wavelength λ_g is

$$\lambda_g = \frac{2\pi}{\beta_g} = \frac{\lambda}{\sqrt{1 - (f_c/f)^2}} \tag{4.60}$$

and the phase velocity is

$$v_p = \frac{v}{\sqrt{1 - (f_c/f)^2}} \tag{4.61}$$

which is larger than the speed v of a plane wave in the same material as the insulator of the line. The propagation velocity of energy, that is, the group velocity

$$v_g = v\sqrt{1 - (f_c/f)^2} \tag{4.62}$$

is smaller than the speed of the plane wave in this material. Thus, the propagation velocity of TE and TM waves depends on frequency, and waveguides carrying these modes are dispersive.

TE and TM wave modes may propagate, except in closed metal waveguides, also, for example, in dielectric waveguides. However, typical wave modes in dielectric waveguides are hybrid wave modes, for which both electric and magnetic fields have longitudinal components.

4.2 THz Waveguides

Waveguides carry signals and power between different devices and within them. By connecting sections of waveguides together, we can form useful components for signal processing. Often, waveguides consisting of two or more conductors are called transmission lines, and wave-guiding structures having a single metal tube or no conductors at all are called waveguides. Here we call them all THz waveguides.

Several types of waveguides have been developed for various applications for millimeter and sub-millimeter wavelengths, see, for example, Ref. [6]. They are characterized by their attenuation, bandwidth, purity of wave mode, physical size, applicability for integration, and so on. Figure 4.2 shows different types of THz waveguides, some of which are briefly described and analyzed in the following.

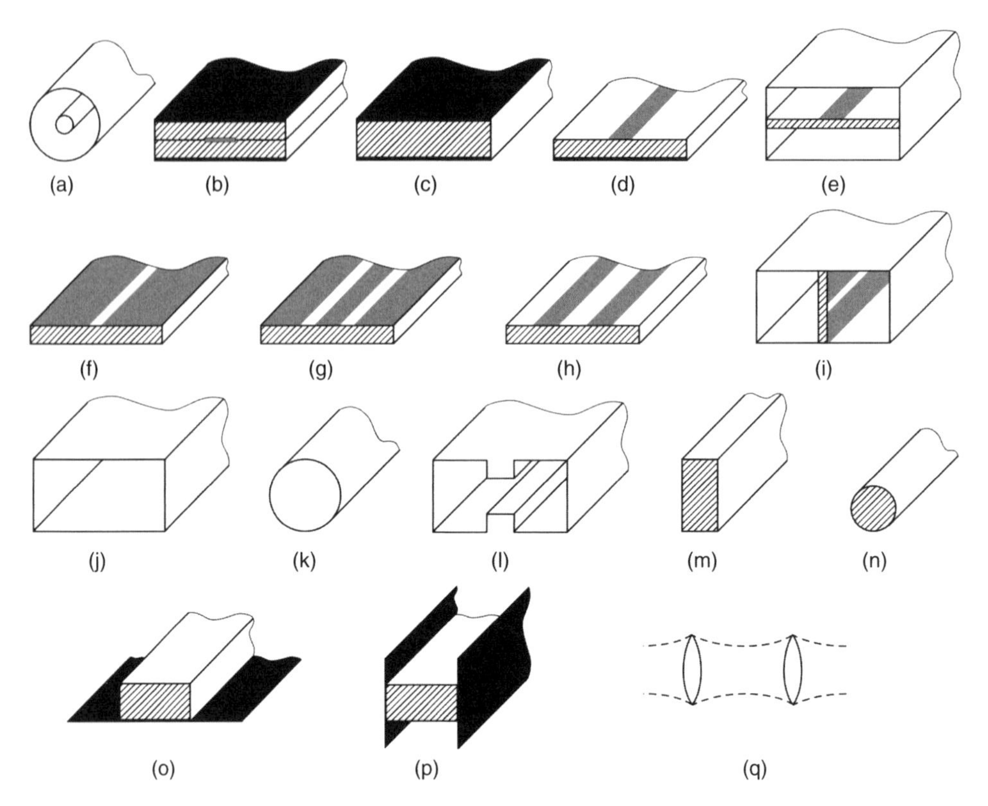

Figure 4.2 Waveguides used at frequencies 100 GHz to 1 THz: (a) coaxial line, (b) stripline, (c) parallel plate waveguide, (d) microstrip line, (e) suspended stripline, (f) slotline, (g) coplanar waveguide, (h) coplanar stripline, (i) fin line, (j) rectangular hollow metal waveguide, (k) circular hollow metal waveguide, (l) groove guide, (m) rectangular dielectric rod waveguide, (n) circular dielectric rod waveguide, (o) image guide, (p) H-guide, and (q) quasi-optical waveguide.

A quasi-optical waveguide or beam waveguide is made of focusing lenses or mirrors, which maintain the energy in a beam in the free space. Quasi-optical waveguides are used from about 100 GHz up to infrared waves. Section 4.3 is devoted to the beam waveguide and quasi-optics.

4.2.1 Waveguides with a Single Conductor: TE and TM Modes

4.2.1.1 Rectangular Metal Waveguide

A rectangular metal waveguide is a hollow metal pipe having a rectangular cross-section. It has relatively low losses and a high power-handling capability, although the latter capability is seldom needed in THz technology. Due to its closed structure, the fields are isolated well from the outside world. A large physical size and difficulty in integrating components within the waveguide are the main disadvantages. The usable bandwidth for the pure fundamental mode TE_{10} is less than one octave. Rectangular metal waveguides are used for various applications from microwave frequencies up to 1 THz and sometimes even at higher frequencies.

Both TE and TM wave modes may propagate in the waveguide shown in Figure 4.3. Because a rectangular metal waveguide is relatively easy to analyze theoretically, let us briefly go through its field analysis

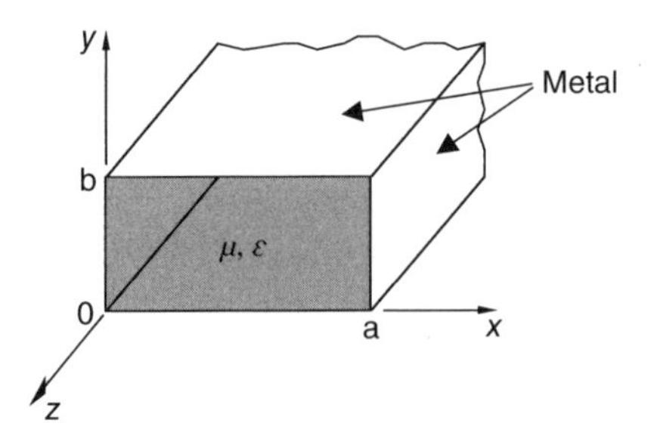

Figure 4.3 Rectangular metal waveguide.

for the TE modes and especially for the fundamental mode TE_{10} [1]. The longitudinal electric field of a TE mode is zero, $E_z = 0$, while for a TM mode the longitudinal magnetic field is zero, $H_z = 0$. We may assume that the solution of the longitudinal magnetic field has a form of $H_z(x, y) = A\ \cos(k_1 x)\cos(k_2 y)$, where A is an arbitrary amplitude constant. By introducing this field to the equation of the TE wave mode, Eq. (4.54), we note that the equation is fulfilled if

$$k_1^2 + k_2^2 = \gamma^2 + \omega^2\mu\varepsilon = k_c^2 \tag{4.63}$$

The transverse field components can be solved from Eqs. (4.42a–d):

$$E_x = \mathrm{j}\frac{\omega\mu k_2}{k_c^2}A\ \cos(k_1 x)\sin(k_2 y);\ E_y = -\mathrm{j}\frac{\omega\mu k_1}{k_c^2}A\ \sin(k_1 x)\cos(k_2 y) \tag{4.64}$$

$$H_x = -\frac{E_y}{Z_{TE}};\ H_y = \frac{E_x}{Z_{TE}}. \tag{4.65}$$

The wave impedance of the TE wave mode is

$$Z_{TE} = \frac{\eta}{\sqrt{1-(f_c/f)^2}}. \tag{4.66}$$

From boundary conditions for the fields in a metal waveguide it follows that the TM field cannot have a normal component at the boundary, that is, $\vec{n}\cdot\vec{H} = 0$. It follows from Eqs. (4.42c,d) that $\partial H_z/\partial x = 0$, when $x = 0$ or a, and $\partial H_z/\partial y = 0$, when $y = 0$ or b. Also the tangential component of the electric field must vanish at the boundary or $\vec{n}\times\vec{E} = 0$: $E_x(x,0) = 0$, $E_x(x,b) = 0$, $E_y(0,y) = 0$, $E_y(a,y) = 0$. It results from these boundary conditions that $k_1 a = n\pi$ and $k_2 b = m\pi$, where $n = 0, 1, 2, \ldots$ and $m = 0, 1, 2, \ldots$, and therefore,

$$k_c^2 = \left(\frac{n\pi}{a}\right)^2 + \left(\frac{m\pi}{b}\right)^2 \tag{4.67}$$

The cut-off wavelength and cut-off frequency are:

$$\lambda_{cnm} = \frac{2\pi}{k_c} = \frac{2}{\sqrt{(n/a)^2+(m/b)^2}} \tag{4.68}$$

$$f_{cnm} = \frac{1}{2\sqrt{\mu\varepsilon}}\sqrt{\left(\frac{n}{a}\right)^2+\left(\frac{m}{b}\right)^2} \tag{4.69}$$

The subscripts n and m refer to the number of field maxima in the x and y direction, respectively.

In most applications, we use the fundamental mode TE_{10} having the lowest cut-off frequency

$$f_{cTE10} = \frac{1}{2a\sqrt{\mu\varepsilon}} \tag{4.70}$$

In the case of an air-filled waveguide we obtain $f_{cTE10} = c/(2a)$. The cut-off wavelength is $\lambda_c = 2a$. If, as in the standard millimeter-wave guides, the waveguide width a is twice the height b, $a = 2b$, the TE_{20} and TE_{01} wave modes have a cut-off frequency of $2f_{cTE10}$.

The fields of the TE_{10} wave mode are solved from the Helmholtz equation and boundary conditions as

$$E_y = E_0 \sin\frac{\pi x}{a};\ E_x = H_y = 0;\ H_z = \mathrm{j}\frac{E_0}{\eta}\frac{\lambda}{2\pi}\cos\frac{\pi x}{a};\ H_x = -\frac{E_0}{\eta}\sqrt{1-\left(\frac{\lambda}{2a}\right)^2}\sin\frac{\pi x}{a} \tag{4.71}$$

E_0 is the maximum value of the electric field, which has only a y-component. All field components depend on time and the z-coordinate as $\exp(j\omega t - \gamma z)$.

The wave impedance of the TE_{10} mode is ($\eta = (\mu/\varepsilon)^{1/2}$):

$$Z_{TE10} = \left|\frac{E_y}{H_x}\right| = \frac{\eta}{\sqrt{1-[\lambda/(2a)]^2}} \tag{4.72}$$

For an air-filled waveguide at a frequency of $1.5f_c$, the wave impedance is 506 Ω. The wave impedance depends on frequency and is characteristic for each wave mode.

The power propagating in a waveguide is obtained by integrating Poynting's vector over the area of the cross-section S:

$$P_p = \frac{1}{2}\mathrm{Re}\int_S \vec{E}\times\vec{H}^*\cdot \mathrm{d}\vec{S} \tag{4.73}$$

The power propagating at the TE_{10} wave mode is

$$P_p = \frac{1}{2}\mathrm{Re}\int_S E_y H_x^* \mathrm{d}S = \frac{E_0^2}{Z_{TE10}}\frac{ab}{4} \tag{4.74}$$

The characteristic impedance Z_0 of a transmission line is the ratio of voltage and current in an infinitely long line or in a line terminated with a matched load having a wave propagating only into one direction. However, we cannot define uniquely the voltage and current of a waveguide as we can do for a two-conductor transmission line. The characteristic impedance, best suited for impedance matching issues, can be calculated from the power propagating in the guide P_p, and the voltage U_y obtained by integrating the electric field in the middle of the waveguide from the upper wall to the lower wall, as

$$Z_{0TE10} = \frac{U_y{}^2}{P_p} = \frac{(E_0 b)^2/2}{P_p} = \frac{2b}{a}Z_{TE10} \tag{4.75}$$

Note that the characteristic impedance depends on the height b whereas the wave impedance does not.

The finite conductivity of the metal, σ_m, causes loss in the surface of the waveguide. Also the insulating material may have dielectric loss due to ε_r'' and conduction loss due to σ_d. The attenuation constant can be given as

$$\alpha = \frac{P_l}{2P_p} \tag{4.76}$$

where P_l is the power loss per unit length. The power loss of conductors is calculated from the surface current density $\vec{J}_s = \vec{n}\times\vec{H}$ ($\vec{n}$ is a unit vector perpendicular to the surface) by integrating $\left|\vec{J}_s\right|^2 R_s/2$ over the surface of the conductor. The surface resistance of a conductor is $R_s = \sqrt{\omega\mu_0/(2\sigma_m)}$. The attenuation constant of conductor loss for the TE_{10} wave mode is obtained as

$$\alpha_{cTE10} = \frac{R_s}{\eta\sqrt{1-[\lambda/(2a)]^2}}\left(\frac{1}{b}+\frac{\lambda^2}{2a^3}\right) \tag{4.77}$$

In practice, the attenuation constant of conductor loss for THz waveguides is 1.2–1.5 times the value obtained from Eq. (4.77), mainly due to the surface roughness [7]. If the dielectric material filling the waveguide has loss, the attenuation constant of dielectric loss is for all wave modes

$$\alpha_d = \frac{\pi}{\lambda}\frac{\tan\delta}{\sqrt{1-(f_c/f)^2}} \tag{4.78}$$

The total attenuation constant is $\alpha = \alpha_c + \alpha_d$.

The recommended frequency range for waveguides operating at the TE_{10} wave mode is about 1.2–1.9 times the cut-off frequency f_{cTE10}. The increase of attenuation sets the lower limit, whereas the excitation of higher-order modes sets the higher limit. Hence, we need a large number of waveguides to cover the whole microwave and millimeter-wave range. Standard waveguides covering the 75–325 GHz range and commercially available semi-standard waveguides covering frequency range 325–1100 GHz are listed in Table 4.1.

In order to facilitate verifiable measurements, as well as for various other applications, a number of different components based on a standard hollow metal waveguide are available, both passive and active: terminations, attenuators, phase-shifters, directional couplers, mixers, frequency multipliers, amplifiers, and so on. It is essential that the waveguides of different parts of the system are aligned accurately. Therefore standardized waveguide flanges with alignment pins are also used, see for example, Ref. [8].

Besides the commercially available standard or semi-standard guides, a millimeter-wave or THz metallic waveguide is often made through milling a so-called split block [9], for example, when constructing mixers, frequency multipliers, and so on for THz frequencies. Other often used fabrication techniques are electroforming and micromachining. In the latter a channel is etched in a Si wafer (microelectromechanical systems (MEMS) techniques), and wafer bonding and sputtering are used to form the final hollow metal waveguide. Figure 4.4 shows the fabrication steps of a micromachined waveguide [10].

Another variation of the rectangular metal waveguide is the so-called substrate integrated waveguide (SIW), see, for example, Ref. [11]. A SIW is formed within a substrate utilizing the bottom and top metallization of the substrate as the broad walls while the sidewalls are formed with rows of thin metal vias. So far, SIW and its various applications for passive components have been demonstrated at frequencies up to about 100 GHz, but SIW may also be usable at higher frequencies.

Table 4.1 Standard and semi-standard waveguides at millimeter-wave and THz frequencies.

Abbreviation	a (mm)	b (mm)	f_c (GHz)	Range (GHz)
WR-10	2.54	1.27	59.0	75–110
WR-8	2.03	1.02	73.8	90–140
WR-6	1.65	0.83	90.8	110–170
WR-5	1.30	0.65	115.7	140–220
WR-4	1.09	0.55	137.2	170–260
WR-3	0.86	0.43	173.6	220–325
WR-2.2	0.57	0.28	263	325–500
WR-1.5	0.37	0.19	399	500–750
WR-1.0	0.25	0.13	590	750–1100

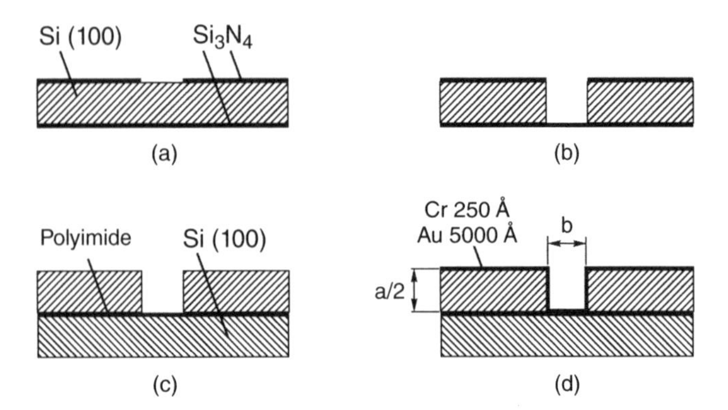

Figure 4.4 Fabrication of a hollow metal waveguide for THz frequencies on a Si wafer: (a) silicon nitride mask, (b) etching through the wafer, (c) bonding of wafers, and (d) metallization. (According to [10].)

4.2.1.2 Circular Metal Waveguide

A hollow metal waveguide may also be circular in cross-section. The analysis of a circular waveguide is best carried out using a cylindrical coordinate system. The principle of analysis is similar to that of the rectangular waveguide, see, for example, Ref. [3], and will not be presented here. The properties of a circular waveguide are generally the same as those of the rectangular metal waveguide. However, the usable bandwidth for single-mode operation is even narrower, only about 25%, see Figure 4.5. Therefore, many waveguide sizes are needed to cover a broad frequency range.

The fundamental mode in a circular metal waveguide is TE_{11}, the cut-off wavelength of which is $\lambda_c = 3.41a$, where a is the radius of the guide. In an oversized circular metal waveguide, a very low-loss TE_{01} mode can propagate. The cut-off wavelength of this mode is $\lambda_c = 1.64a$. The attenuation constant of the TE_{01} wave mode is

$$\alpha_{cTE01} = \frac{R_s}{a\eta} \frac{(f_c/f)^2}{\sqrt{1-(f_c/f)^2}} \tag{4.79}$$

The attenuation of the TE_{01} wave mode is very low if the operating frequency is much higher than the cut-off frequency f_c. At 100 GHz the theoretical attenuation for a guide with diameter of 40 mm is less

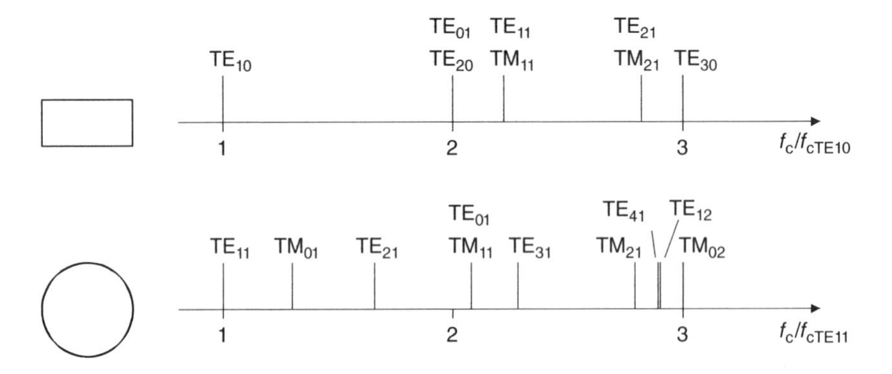

Figure 4.5 Cut-off frequencies of the lowest wave modes of rectangular and circular waveguides [1].

than 1 dB/km. However, many other modes can propagate in such an oversized waveguide. A low attenuation is achieved only if the excitation of unwanted modes is prevented.

4.2.1.3 Metal/Dielectric Interface

Electromagnetic waves can be guided also at the interface of a conductor and a dielectric, where the dielectric may be air or vacuum. Very well-known examples of such phenomena are the Zenneck wave [12] on a flat metal surface, the Sommerfeld wire [13], and the Goubau line [14]. These waves are inhomogenous plane waves which have a longitudinal electric field component, that is, they are TM waves.

Recently, such interfaces and waves have gained interest in connection with THz waveguiding, and are now called plasmonic interfaces and plasmonic waveguiding or plasmon polariton propagation. A plasmon is an elementary excitation of a charge density oscillation in plasma. Besides the smooth metal surface, the plasmonic structures also include surfaces patterned with corrugations, pits or holes. An instructive article bridging the plasmonics and Zenneck surface waves is presented in Ref. [15], and an overview of the THz plasmonic structures is presented, for example, in Ref. [16].

4.2.2 Waveguides with Two or More Conductors: TEM and Quasi-TEM Modes

Waveguides consisting of two or more conductors can carry a TEM mode. A parallel wire line is used in many applications at frequencies below 1 GHz. In principle it works also at THz frequencies, but due to radiation and losses of the needed supporting dielectric it has not found applications at THz. A parallel plate waveguide (Figure 4.2c) is another waveguide for the TEM mode; however, only its special case, a microstrip line carrying a quasi-TEM mode, is commonly used at frequencies above 100 GHz. A waveguide carrying a pure TEM mode also at THz frequencies is a coaxial line. Another such waveguide is a stripline (Figure 4.2b), where a metal strip is placed symmetrically between two substrates and their ground planes.

4.2.2.1 Coaxial Line

A coaxial line consists of two concentric conductors with circular cross-sections: an outer and an inner conductor. The space between the conductors is filled with low-loss insulating material, for example, air or Teflon. The coaxial line has a broad single-mode bandwidth from 0 Hz to an upper limit, which depends on the dimensions of the conductors. Commercially, coaxial line products are available up to 100 GHz. However, a coaxial line has been used at much higher frequencies in millimeter- and sub-millimeter-wave mixers as a filter structure.

Figure 4.6 illustrates the cross-section of a coaxial line. The inner radius of the outer conductor is r_o and the radius of the inner conductor is r_i; the relative permittivity of the insulator is ε_r. The fields are confined to the space between the conductors.

The fields of the TEM wave mode of such a cylindrically symmetric waveguide can be derived from Laplace's equation (4.53) using the scalar potential $\Phi(r,\varphi)$ (e.g., [1]). In the cylindrical coordinate system, Laplace's equation is written as

$$\frac{1}{r}\frac{\partial}{\partial r}\left(r\frac{\partial\Phi(r,\varphi)}{\partial r}\right)+\frac{1}{r^2}\frac{\partial^2\Phi(r,\varphi)}{\partial\varphi^2}=0 \tag{4.80}$$

Applying the boundary conditions $\Phi(r_\mathrm{o},\varphi)=0$ and $\Phi(r_\mathrm{i},\varphi)=V$, the potential is solved to be

$$\Phi(r,\varphi)=V\frac{\ln(r_\mathrm{o}/r)}{\ln(r_\mathrm{o}/r_\mathrm{i})} \tag{4.81}$$

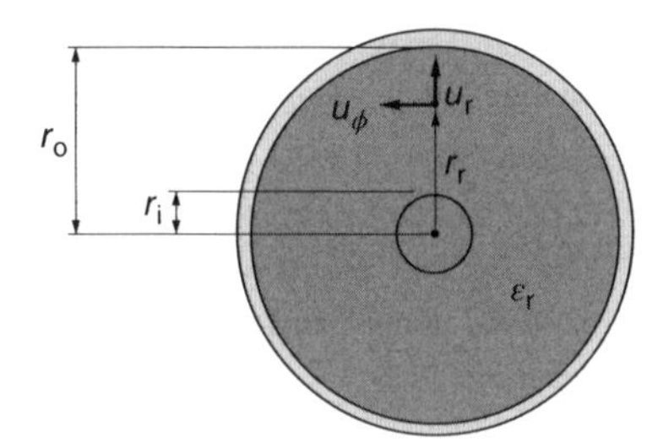

Figure 4.6 A cross-section of a coaxial line.

The electric field is the negative gradient of the potential:

$$\vec{E}(r,\varphi) = -\nabla\Phi(r,\varphi) = \vec{u}_r \frac{V}{\ln(r_o/r_i)}\frac{1}{r} \tag{4.82}$$

where $\vec{u}_r$ is the unit vector in the radial direction. This is also the electric field of a cylindrical capacitor. The magnetic field of the coaxial line is

$$\vec{H}(r,\varphi) = \frac{1}{\eta}\vec{u}_z \times \vec{E}(r,\varphi) = \vec{u}_\varphi \frac{V}{\eta r \ln(r_o/r_i)} = \vec{u}_\varphi \frac{I}{2\pi r} \tag{4.83}$$

where $\eta = \sqrt{\mu/\varepsilon}$ is the wave impedance, I is the current in the inner conductor, and u_φ is the unit vector perpendicular to the radial direction. The same current I also flows in the outer conductor but in the opposite direction.

The characteristic impedance of the coaxial line is

$$Z_0 = \frac{V}{I} = \frac{\eta}{2\pi}\ln(r_o/r_i) \tag{4.84}$$

The attenuation constant due to conductor loss is

$$\alpha_c = \frac{R_s}{4\pi Z_0}\left(\frac{1}{r_o}+\frac{1}{r_i}\right) \tag{4.85}$$

where the surface resistance of a conductor is $R_s = \sqrt{\omega\mu_0/(2\sigma_m)}$. In practice, the value of the surface resistance R_s is larger than the theoretical one, for example, due to the surface roughness. The attenuation constant due to dielectric loss is

$$\alpha_d = \frac{\pi}{\lambda}\tan\delta \tag{4.86}$$

TE and TM wave modes may also propagate in a coaxial line, if the operating frequency is larger than the cut-off frequency of these wave modes. To avoid these higher-order modes, the operating frequency should be chosen according to the dimensions. An approximate rule is that the circumference corresponding to the average radius should be smaller than the operating wavelength:

$$\lambda > \pi(r_o + r_i) \tag{4.87}$$

Consequently, the coaxial lines used at THz frequencies should be thin enough to make sure that only the TEM wave mode may propagate.

4.2.2.2 Microstrip Line

A microstrip line is a special case of a parallel plate waveguide, where one of the plates is made narrow (strip) and there is an insulating substrate between the plates. The metal layer on the opposite side of the strip operates as a ground plane. The advantages of the microstrip line are a broad bandwidth, small size, and a good applicability for integration and mass production. The disadvantages are fairly high losses and radiation due to the open structure. Microstrip lines are often used at 100 GHz and above.

Figure 4.7 shows a cross-section of a microstrip line. The substrate is made of a low-loss dielectric material such as, for example, quartz (fused, $\varepsilon_r = 3.8$, or crystalline, $\varepsilon_r = 4.2$), polytetrafluoroethylene (Teflon, $\varepsilon_r = 2.1$), or aluminum oxide (alumina, $\varepsilon_r = 9.7$).

Because the field in a microstrip line is in two different media, air and substrate, having different propagation constants, at higher frequencies the field also has longitudinal components and the wave mode is called quasi-TEM. The analytical solution of the quasi-TEM wave mode is so complicated that the practical design of microstrip lines is based on graphs, approximate equations, or a full wave simulator.

The phase constant of a TEM wave can be expressed as

$$\beta = \frac{2\pi}{\lambda} = \omega\sqrt{\mu\varepsilon_0\varepsilon_{\mathrm{r,eff}}} \tag{4.88}$$

where $\varepsilon_{\mathrm{r,eff}}$ is the effective relative permittivity, obtained, for example, by calculating the capacitance of the line per unit length, for example, with a full-wave simulator.

The sources of losses in a microstrip line are: conductor loss in the strip and ground plane, dielectric and conduction losses in the substrate, radiation loss, and surface wave loss. Assuming a constant current distribution over the strip width w, the attenuation constant due to metal loss is

$$\alpha_c = \frac{R_s}{Z_0 w} \tag{4.89}$$

The accuracy of this equation is best for a wide strip. The thickness of the conductors should be at least four times the skin depth $\delta_s = \sqrt{2/(\omega\mu\sigma_m)}$. A metallization thickness less than twice the skin depth

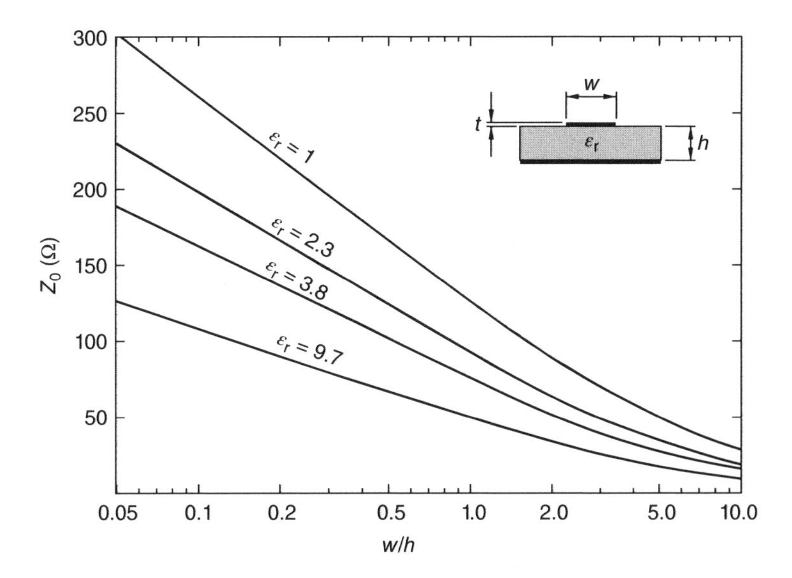

Figure 4.7 The cross-section of a microstrip line and the characteristic impedance Z_0 as a function of the ratio of strip width to substrate height w/h for different substrate materials [1].

yields excessive attenuation. Dielectric loss is usually much lower than conductor loss. The attenuation constant due to dielectric loss is

$$\alpha_{\mathrm{d}} = \pi \frac{\varepsilon_{\mathrm{r}}(\varepsilon_{\mathrm{r,eff}} - 1)}{\sqrt{\varepsilon_{\mathrm{r,eff}}}(\varepsilon_{\mathrm{r}} - 1)} \frac{\tan \delta}{\lambda_0} \tag{4.90}$$

where λ_0 is the free-space wavelength. Discontinuities of the microstrip line produce radiation to free space which increases rapidly as the frequency increases.

Surface waves are waves that are trapped by total reflection within the substrate. They may produce unwanted radiation from the edges of the substrate and spurious coupling between circuit elements. Therefore, it is important that the operational frequency of a microstrip is below frequencies f_{T} and f_{CT} [17]:

$$f_{\mathrm{T}} = \frac{c \arctan \varepsilon_{\mathrm{r}}}{\sqrt{2}\pi h \sqrt{\varepsilon_{\mathrm{r}} - 1}} \tag{4.91}$$

and

$$f_{\mathrm{CT}} = \frac{c}{2\sqrt{\varepsilon_{\mathrm{r}}}}(w + 0.4h)^{-1} \tag{4.92}$$

where c is the speed of light in vacuum, ε_{r} is the relative permittivity of the substrate, h is the substrate thickness, and w is the strip width. Equation (4.91) follows from the equal phase velocity of the quasi-TEM mode and the surface wave mode TM_0 in the case of a narrow strip, and Eq. (4.92) from the transversal resonance in the case of a wider strip.

4.2.2.3 Coplanar Waveguide

A coplanar waveguide (CPW) has ground-plane conductors on both sides of a metal strip, see Figure 4.2g. All conductors are on the same side of the substrate. However, there may be another ground-plane on the opposite side of the substrate. Both series and parallel components can easily be integrated with this line. CPWs are commonly used, for example, in monolithic integrated circuits operating at millimeter wavelengths.

The analytical solution of the wave mode in CPW is hard. The higher the frequency, the more the wave mode differs from TEM, and at THz frequencies it rather resembles TE. Therefore, practical design of CPW lines is mainly based on graphs [18, 19] or a full wave simulator.

4.2.2.4 Other Planar or Quasi-Planar Waveguides

The slotline, Figure 4.2f, and coplanar stripline, Figure 4.2h, are planar waveguides that carry at low frequencies a nearly pure TEM wave mode but at THz frequencies a large longitudinal magnetic field component appears, and the mode is far from TEM.

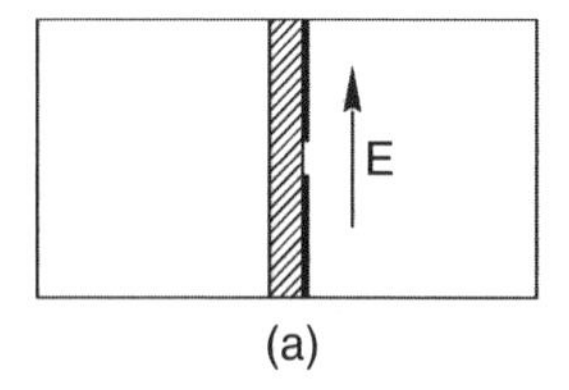

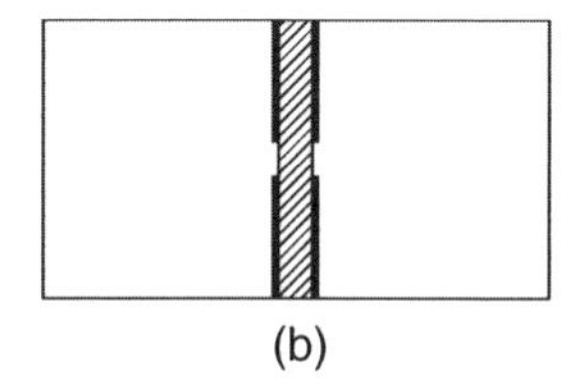

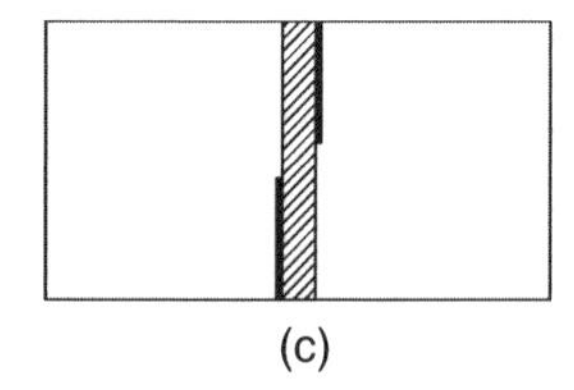

Figure 4.8 (a) Unilateral, (b) bilateral, and (c) antipodal finline.

A suspended stripline, see Figure 4.2e, has a substrate with a metal strip in the H-plane of a rectangular waveguide. A finline, Figures 4.2i and 4.8, is composed of a substrate with a slotline in the E-plane of a rectangular (metal) waveguide, see for example, Ref. [20, 21]. The field concentrates in the slot of the metallization of the substrate. Instead of TEM, the wave mode resembles a combination of TE and TM modes. The finline has a low radiation loss, and components can be integrated quite easily with this line. Finlines are used up to 200 GHz in millimeter-wave hybrid integrated circuits. It is an especially useful waveguide type for filters at millimeter-waves because, due to the longer wavelength, the filter structure is easier to fabricate in finline than, for example, in microstrip.

4.2.3 Waveguides with No Conductor: Hybrid Modes

A dielectric rod waveguide (see e.g., Ref. [22]), either rectangular (Figure 4.2m) or circular (Figure 4.2n), is made of a dielectric low-loss material. For example, an optical fiber is a circular dielectric rod waveguide. A rectangular dielectric rod waveguide is used, for example, in THz antennas [23].

However, as also partially shown in Figure 4.2, there are many different configurations of a dielectric waveguide: image guide (Figure 4.2o), insulated image guide, H guide (Figure 4.2p), non-radiative dielectric (NRD) [24] guide, and so on. These latter dielectric waveguides also involve one or two conductors. The difference between the H guide and the NRD guide is that in the latter the spacing between the conductor plates is less than half the wavelength in vacuum, while in the H guide the distance is larger. The conductor losses decrease as the spacing increases. On the other hand, in the NRD guide, the electromagnetic field with E-field parallel with the conductor plates is below cut-off outside the dielectric region, and no radiation occurs with this mode from discontinuities such as bends.

4.2.3.1 Rectangular Dielectric Rod Waveguide

Dielectric waveguides are promising transmission lines for THz frequencies due to their low propagation loss, low cost, compatibility with standard GaAs and/or Si technology, and so on (see, e.g., Ref. [25–27]). The electromagnetic problems related to dielectric waveguides for the circular cross-section can be solved in closed form in terms of the Bessel functions. But in the case of the rectangular cross-section no closed-form solution exists.

In the open rectangular dielectric waveguide the propagating modes are *hybrid modes,* and according to the standard nomenclature the propagating mode is called E^y_{mn}, if its electric field is polarized mainly along the y-direction, and E^x_{mn}, if the strongest electrical field component points along the x-direction according to Ref. [28, 29].

There are several different numerical methods that can be applied to the problem of dielectric waveguide structures, particularly to the open rectangular dielectric waveguide. The simplest but approximate approach is Marcatili's method.

Marcatili's Method

One of the existing approximate methods for the rectangular dielectric waveguide is the Marcatili method [28]. It was developed for low ratios between the core permittivity and that of the cladding region (slightly more than 1), but it has been shown that it works well even for high permittivity ratios (around 10) in a rather wide frequency region.

In this method, the complete cross-section area of the rectangular open dielectric waveguide, Figure 4.9, is divided into five regions with homogeneous but possibly different refractive indices n_υ.

The fields in the shadowed regions are not considered: these regions are much less essential for the waveguide properties than the other regions. In all the other regions, the fields are assumed to be

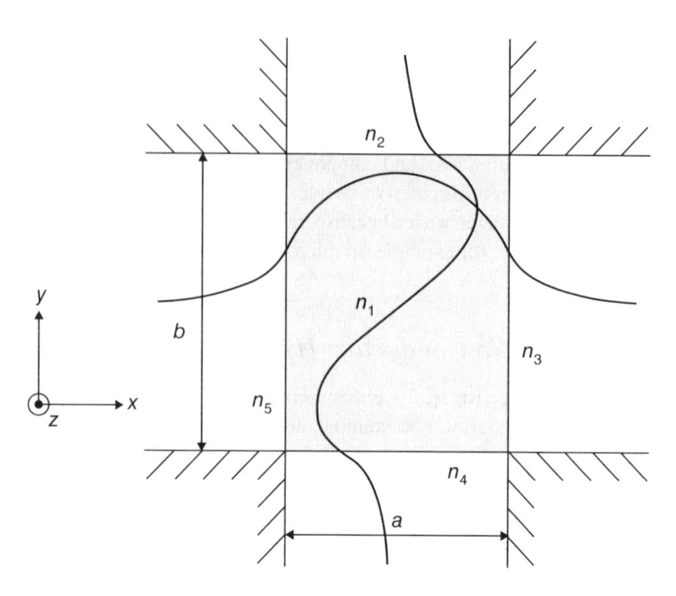

Figure 4.9 The geometry of a rectangular dielectric waveguide in Marcatili's method and the distributions of the H_x component for the E^y_{12} mode.

approximately (co)sinusoidally distributed inside the waveguide and decaying exponentially outside. That is, we express the field components, say, H_x, as follows:

$$H_{x\upsilon} = \exp(-\mathrm{j}k_z z + \mathrm{j}\omega t)\begin{cases} M_1\cos(k_x x + \alpha)\cos\left(k_y y + \beta\right) & \text{for} \quad \upsilon = 1 \\ M_2\cos(k_x x + \alpha)\exp(-k_{y2}y) & \text{for} \quad \upsilon = 2 \\ M_3\cos(k_y y + \beta)\exp(-k_{x3}x) & \text{for} \quad \upsilon = 3 \\ M_4\cos(k_x x + \alpha)\exp(k_{y4}y) & \text{for} \quad \upsilon = 4 \\ M_5\cos(k_y y + \beta)\exp(-k_{x5}x) & \text{for} \quad \upsilon = 5 \end{cases} \tag{4.93}$$

where $M_{1\dots5}$ are unknown amplitude coefficients, k_x and k_y are the propagation constants in region 1 (refractive index n_1) in the horizontal and vertical directions, respectively, k_{x3} and k_{y2} are the decay factors in the outer regions, and α and β are additional phase constants. If the system is symmetric, meaning that $n_2 = n_4$ and $n_3 = n_5$, the field distributions are given by cosine or sine functions with a null or a maximum at the rod center, and in this case constants α and β equal either 0 or $\frac{\pi}{2}$. This notation allows us to write both symmetric and antisymmetric field distributions in a compact form, as in Eq. (4.93). For the orthogonal polarization we can write similar expressions for H_y, but at this stage let us assume that $H_y = 0$. The other field components can be then expressed through H_x, using Eqs. (4.18) and (4.19) in component form. Finally, the field components are written as

$$H_{y\upsilon} = 0 \tag{4.94}$$

$$H_{z\upsilon} = -\frac{\mathrm{j}}{k_z}\frac{\partial H_{x\upsilon}}{\partial x} \tag{4.95}$$

$$E_{x\upsilon} = -\frac{1}{\omega\varepsilon_0 n_\upsilon^2 k_z}\frac{\partial^2 H_{x\upsilon}}{\partial x\partial y} \tag{4.96}$$

$$E_{y\upsilon} = \frac{k_0^2 n_\upsilon^2 - k_{y\upsilon}^2}{\omega\varepsilon_0 n_\upsilon^2 k_z} H_{x\upsilon} \tag{4.97}$$

$$E_{z\upsilon} = \frac{\mathrm{j}}{\omega\varepsilon_0 n_\upsilon^2} \frac{\partial H_{x\upsilon}}{\partial y} \tag{4.98}$$

where k_0 is the wavenumber in vacuum.

4.2.3.2 Rectangular Dielectric Rod Waveguide in Air

To analyze the open rectangular waveguide (Figure 4.9) we return to the main Eq. (4.93) and consider waveguides in air, with $n_{2,3,4,5} = 1$. Due to this symmetry α, $\beta = 0$ or $\pm\frac{\pi}{2}$ in Eq. (4.93). The symmetry also allows us to consider only the region $x > 0$, $y > 0$, see Figure 4.9. To find the eigensolutions we should match the components H_x, E_z, H_z, E_x at $y = b/2$ and E_y, E_z, H_z at $x = a/2$. In his original paper [28] Marcatili assumed that the refractive index of the dielectric waveguide was only slightly larger than that of air: $n_1 - 1 \ll 1$, therefore $k_{x,y} \ll k_z$, because the fields are more spread out from the rod when the dielectric material properties approach the properties of air. Thus, E_x can be neglected, as it is proportional to $k_x k_y$. Now, let us match the strongest field component, E_y, at $x = a/2$. Neglecting k_y^2 as compared with $k_0^2 n_1^2$, we come to

$$\frac{k_0^2}{\omega\varepsilon_0 k_z} M_1 \cos\left(k_x \frac{a}{2} + \alpha\right) = \frac{k_0^2}{\omega\varepsilon_0 k_z} M_3 \exp\left(-k_{x3}\frac{a}{2}\right) \tag{4.99}$$

From here,

$$M_3 = M_1 \cos\left(k_x \frac{a}{2} + \alpha\right) \exp\left(k_{x3}\frac{a}{2}\right) \tag{4.100}$$

When we write the boundary condition for E_z at $x = a/2$ and solve the resulting equation, we do not come to any useful result, because from

$$-\frac{\mathrm{j}}{\omega\varepsilon_0 n_1^2} M_1 k_y \cos\left(k_x \frac{a}{2} + \alpha\right) = -\frac{\mathrm{j}}{\omega\varepsilon_0 k_y} M_3 \exp\left(-k_{x3}\frac{a}{2}\right) \tag{4.101}$$

we arrive at $\frac{1}{n_1^2} = 1$, which leaves us with a trivial relation. For this reason we will not take into account this boundary condition. Fortunately, the boundary condition for H_z gives

$$\frac{\mathrm{j}}{k_z} M_1 k_x \sin\left(k_x \frac{a}{2} + \alpha\right) = \frac{\mathrm{j}}{k_z} M_3 k_{x3} \exp\left(-k_{x3}\frac{a}{2}\right) \tag{4.102}$$

which can be rewritten as

$$k_x \sin\left(k_x \frac{a}{2} + \alpha\right) = k_{x3} \cos\left(k_z \frac{a}{2} + \alpha\right) \tag{4.103}$$

or, after some transformations,

$$k_x a = \pi n - 2\arctan\left(\frac{k_x}{k_{x3}}\right) \tag{4.104}$$

where $k_{x3} = \sqrt{k_0^2(n_1^2 - 1) - k_x^2}$, and $n = 1, 2, 3, \ldots$ At $y = b/2$ we similarly match H_x and E_z and do not consider H_z. From the boundary condition for H_x we get

$$M_1 \cos\left(k_y \frac{b}{2} + \beta\right) = M_2 \exp\left(-k_{y2}\frac{b}{2}\right) \tag{4.105}$$

$$-\frac{\mathrm{j}}{\omega\varepsilon_0 n_1^2 k_y} M_1 \sin\left(k_y \frac{b}{2} + \beta\right) = -\frac{\mathrm{j}}{\omega\varepsilon_0 k_{y2}} M_2 \exp\left(-k_{y2}\frac{b}{2}\right) \tag{4.106}$$

or

$$\frac{1}{n_1^2} k_y \sin\left(k_y \frac{b}{2} + \beta\right) = k_{y2} \cos\left(k_y \frac{b}{2} + \beta\right) \tag{4.107}$$

From here we arrive at

$$k_y b = \pi m - 2\arctan\left(\frac{1}{n_1^2} \frac{k_y}{k_{y2}}\right) \tag{4.108}$$

where $k_{y2} = \sqrt{k_0^2(n_1^2 - 1) - k_y^2}$ and $m = 1, 2, 3, \ldots$ Thus, to calculate the propagation constant of a rectangular dielectric waveguide, we have to solve Eqs. (4.104) and (4.108) and then find

$$k_z = \sqrt{k_0^2 n_1^2 - k_x^2 - k_y^2} \tag{4.109}$$

Thus, we have to solve for two dielectric slabs (as in Section 4.3): one vertical and one horizontal, with thicknesses a and b, respectively. Another approach can be found in Ref. [29], where all the transversal field components are expressed through the longitudinal ones for the case of arbitrary values of $n_{2,3,4,5}$ (not necessarily symmetric cladding). In the case of symmetrical waveguides the resulting equations are the same as above.

Indices m and n in the equations for the eigenmodes (4.108) and (4.108) denote how many extrema the distribution of the main field component has in the horizontal and vertical directions, respectively.

In Figure 4.10 a,b, typical propagation and loss characteristics are shown. One can see that the dielectric waveguide has two fundamental modes without a low-frequency cut-off, which are degenerate if the cross-section is a square with $a = b$. At very high frequencies the loss factor tends to that for plane waves in the core because of the strong field concentration in the rod. At low frequencies losses are small because the fields are only weakly concentrated in the rod (for this reason, however, the usage of rod waveguides at very low frequencies is not practical).

Goell's Method

Instead of expanding the field distribution in sinusoidal functions, Goell [30] proposed to use the following approach: he expanded the longitudinal components of the electric and magnetic fields as

$$E_{z1} = \sum_{n=0}^{+\infty} a_n J_n(\beta_t r) \sin(n\theta + \varphi_n) \exp(-\mathrm{j}k_z z + \mathrm{j}\omega t) \tag{4.110}$$

$$H_{z1} = \sum_{n=0}^{+\infty} b_n J_n(\beta_t r) \sin(n\theta + \psi_n) \exp(-\mathrm{j}k_z z + \mathrm{j}\omega t) \tag{4.111}$$

inside the core of the dielectric waveguide and

$$E_{z0} = \sum_{n=0}^{+\infty} c_n K_n(qr) \sin(n\theta + \varphi_n) \exp(-\mathrm{j}k_z z + \mathrm{j}\omega t) \tag{4.112}$$

$$H_{z0} = \sum_{n=0}^{+\infty} d_n K_n(qr) \sin(n\theta + \psi_n) \exp(-\mathrm{j}k_z z + \mathrm{j}\omega t) \tag{4.113}$$

outside the core. Here $\beta_t = \sqrt{k_1^2 - k_z^2}$ and $q = \sqrt{k_z^2 - k_0^2}$ are the transverse propagation constants inside and outside the dielectric rod, $J(\beta_t r)$ and $K(qr)$ are the Bessel functions and the modified Bessel functions of the second kind, respectively. Variables a, b, c, and d are weighting coefficients, and φ and ψ are phase constants. Definitions of angle θ and distance r are shown in Figure 4.11.

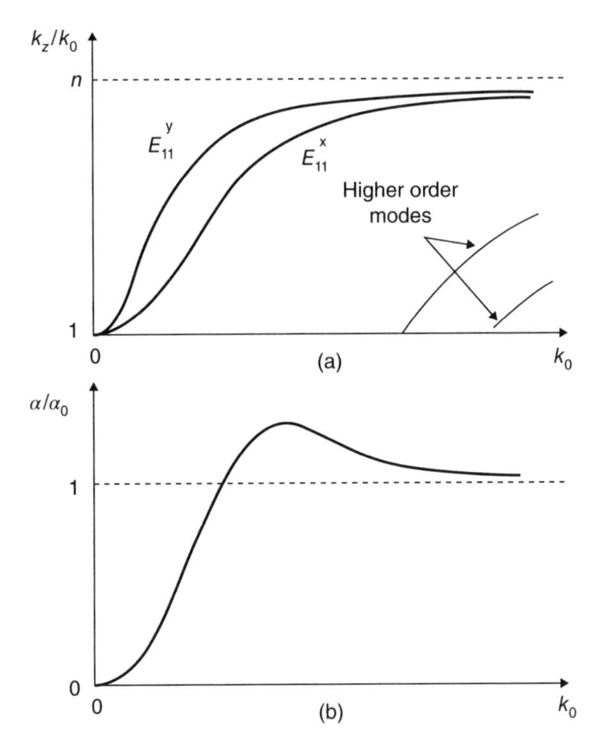

Figure 4.10 Typical dependence of (a) the normalized propagation factor k_z/k_0 and (b) the loss factor α/α_0 for the fundamental mode on the wavenumber in vacuum k_0. Here α_0 is the propagation loss of a plane wave in the infinite medium filled with the same material as the dielectric waveguide core.

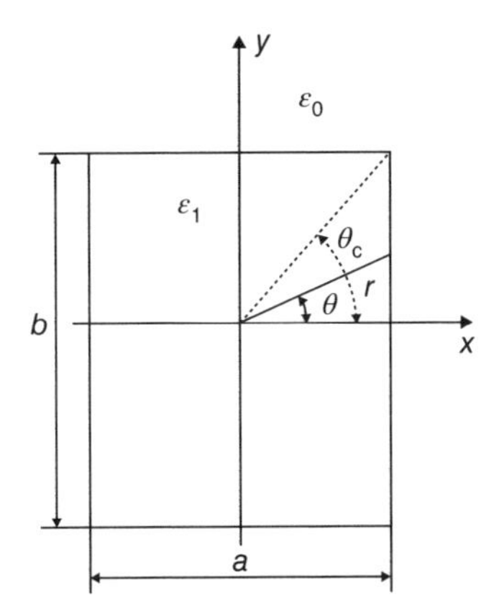

Figure 4.11 Dimensions of the dielectric waveguide cross-section and the corresponding coordinate system.

The sums are truncated at some finite number of terms N. Then, the tangential field components are expressed and "tailored" (requiring the boundary conditions to be satisfied) at several points of the interface. The symmetry simplifies the analysis [30], because one can consider only the first quadrant ($0 < \theta < \pi/2$) and only odd or even numbers of n and their corresponding harmonics. For example, for the odd harmonic case Goell chose the matching points as:

$$\theta_m = \left(m - \frac{1}{2}\right)\frac{\pi}{2N} \tag{4.114}$$

with respect to the angle $\theta = 0$, $m = 1, 2, \ldots, N$. As a result of this "tailoring", a system of equations is obtained, for which the determinant of its matrix should be equal to zero, otherwise no nonzero solution can exist. This method is rather fast, does not require many expansion terms (we used 10–12 terms), and is considered as classical.

A more rigorous solution without the limitation of matching only at particular points of the interface is proposed in Ref. [31]. The same expansion of the longitudinal field components is used except that instead of sine (cosine) functions complex exponents are taken. Then, the exact boundary conditions are written in terms of the longitudinal field components and their derivatives both in the tangential and normal directions. As a result, an infinite system of equations is obtained and the determinant of the matrix is set to zero. The solution in this case is much more complicated but more rigorous. In the next section more details on Goell's method will be explained, and the method will be extended to a more general anisotropic case.

Comparison of Marcatili's and Goell's Method

In the original Marcatili method it is assumed that the refractive index of the dielectric waveguide is only slightly larger than unity. Let us check what happens if we try to calculate the fundamental E^y_{11} mode for, say, a silicon dielectric waveguide ($n = 3.41$) with cross-section dimensions $0.5 \times 1.0\,\text{mm}^2$. Next we compare the results with the same characteristics calculated with a more accurate numerical Goell's method [30].

One can see in Figure 4.12 that in spite of its simplicity Marcatili's method works quite well for dielectric waveguides made of even high permittivity materials like silicon. However, it works well only when the wave is well guided, that is, at high enough frequencies. At lower frequencies accurate calculations become more complicated [32].

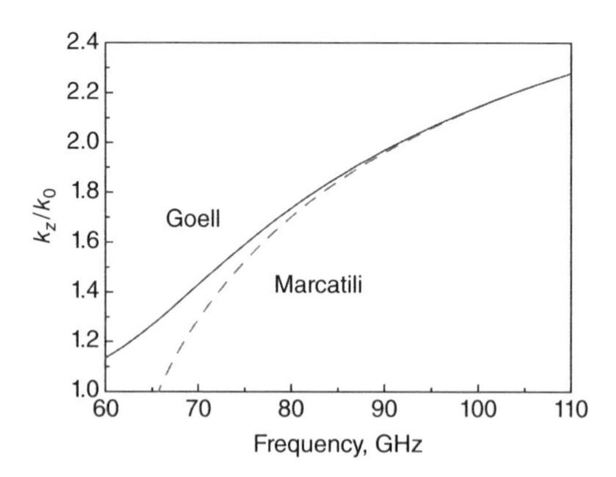

Figure 4.12 Comparison of the results of calculations of the propagation factor k_z/k_0 by Goell's and Marcatili's methods for a silicon dielectric waveguide with 0.5×1.0 mm^2 cross section, E^y_{11} mode.

4.3 Beam Waveguides

As losses in metallic and dielectric waveguides increase with frequency, it becomes advantageous to transmit electromagnetic radiation through free-space. In this section we describe beam waveguides: optical systems used to propagate radiation as free-space beams. Unlike traditional optics, at GHz and THz frequencies beams may be only moderately large when compared with the wavelength and so diffraction effects can become important. In this case we often talk about *quasi-optics*. Gaussian beam modes are a natural mode-set to describe quasi-optical beams with a well-defined direction of propagation, just as TE and TM modes were used to describe propagation in metallic waveguides (Section 4.2.1). In practice, the lowest-order Gaussian mode often suffices for a good description of a propagating beam. We begin this section by describing the behavior of these Gaussian beams.

Quasi-optical or beam waveguides are typically made of lenses or mirrors used to refocus the diffracting beam at regular intervals. Here we describe the operation and modeling of these components as well as the horn antennas often used to launch them. A more detailed discussion of quasi-optical beams and components can be found in, for example, Ref. [33–36].

4.3.1 Gaussian Beam

4.3.1.1 Gaussian Beam Propagation

Consider one component, $E(x, y, z)$, of the complex electric field of a monochromatic, spatially coherent beam propagating in the positive z direction in free space. It satisfies the time-independent, or Helmholtz (Eq. (4.22)), wave equation:

$$\nabla^2 E + k^2 E = 0 \text{ where } k = \omega\sqrt{\mu\varepsilon} = \omega/c = 2\pi/\lambda. \quad (4.115)$$

If we rewrite the electric field component as $E(x, y, z) \equiv u(x, y, z)\exp(-jkz)$ then the Helmholtz equation becomes $\nabla^2 u - 2jk\,{}^{\partial u}/_{\partial z} = 0$. If the beam is paraxial (largely collimated), then $u(x, y, z)$ varies only slowly with z over a distance comparable with a wavelength and $|(\partial^2 u)/(\partial z^2)| \ll |k\,\partial u/\partial z|$. In this case, the Helmholtz equation becomes the paraxial wave equation

$$\partial^2 u/\partial x^2 + \partial^2 u/\partial y^2 - 2jk\,\partial u/\partial z = 0. \quad (4.116)$$

A simple solution to this paraxial wave equation can be written in the form

$$u(x, y, z) = \frac{u_0}{(q_0 + z)} \exp\left(-j\frac{k\left(x^2 + y^2\right)}{2(q_0 + z)}\right) \quad (4.117)$$

where u_0 and q_0 are constants. If we choose $q_0 = jkw_0^2/2$, then at $z = 0$ we obtain a Gaussian-shaped beam:

$$u(x, y, 0) \propto \exp\left(-\frac{\left(x^2 + y^2\right)}{w_0^2}\right) \quad (4.118)$$

with a planar phase front, and 1/e in amplitude beam radius of w_0. On propagating away from $z = 0$, the exponent in the expression for $u(x, y, z)$ has both real and imaginary terms:

$$u(x, y, z) = \frac{u_0}{(q(z))} \exp\left[\left(-j\frac{k\left(x^2 + y^2\right)}{2q_r(z)}\right) - \left(\frac{k\left(x^2 + y^2\right)}{2q_i(z)}\right)\right] \quad (4.119)$$

where

$$q(z) = q_0 + z = j\frac{kw_0^2}{2} + z \text{ and } \frac{1}{q(z)} = \left(\frac{1}{q_r(z)}\right) - j\left(\frac{1}{q_i(z)}\right). \quad (4.120)$$

We can associate the real part of the exponent in Eq. (4.119) with a Gaussian amplitude profile of radius $w(z)$ and the imaginary part with the phase of a spherical wave front of radius of curvature $R(z)$ in the paraxial limit:

$$u(x, y, z) = \frac{u_0}{(q(z))} \exp\left[\left(-j\frac{k\left(x^2+y^2\right)}{2R(z)}\right) - \left(\frac{\left(x^2+y^2\right)}{w^2(z)}\right)\right] \tag{4.121}$$

where

$$\frac{1}{q(z)} = \frac{1}{R(z)} - j\frac{\lambda}{\pi w^2(z)}. \tag{4.122}$$

$q(z)$ is known as the complex radius of curvature, complex beam parameter, or Gaussian beam parameter, $R(z)$ is the phase radius of curvature and $w(z)$ is the beam radius. After normalization $\left(\iint_{\infty} E(x, y, z)\, E^*(x, y, z)\mathrm{d}x\mathrm{d}y = 1\right)$ we can write $E(x, y, z)$ in its usual form:

$$E(x, y, z) = \sqrt{\frac{2}{\pi w^2(z)}} \exp\left[-\left(\frac{\left(x^2+y^2\right)}{w^2(z)}\right)\right] \cdot \exp\left[-j\left(\frac{k\left(x^2+y^2\right)}{2R(z)} + kz - \varphi_0(z)\right)\right] \tag{4.123}$$

where $\varphi_0(z)$, the Gaussian beam phase slippage, or Guoy phase, is given by

$$\varphi_0(z) = \tan^{-1}\left(\frac{\lambda z}{\pi w_0^2}\right). \tag{4.124}$$

From Eqs. (4.120) and (4.122) we can derive the variation of beam radius and phase radius of curvature with propagation distance:

$$w(z) = w_0\left[1 + \left(\frac{\lambda z}{\pi w_0^2}\right)^2\right]^{1/2} \tag{4.125}$$

$$R(z) = z + \frac{1}{z}\left(\frac{\pi w_0^2}{\lambda}\right)^2. \tag{4.126}$$

The beam has a minimum in extent, the beam waist radius w_0, at $z = 0$ (Figure 4.13), where it also has an infinite radius of curvature. As the beam propagates away from the waist, its amplitude drops and it spreads out. The beam remains quasi-collimated (increases by a factor of $\sqrt{2}$ or less) for propagation distances $-z_c < z < z_c$ (Figure 4.14) where z_c, known as the confocal distance or Rayleigh range, is given by

$$z_c = \frac{\pi w_0^2}{\lambda}. \tag{4.127}$$

The beam is often divided into a near ($z \ll z_c$) and a far ($z \gg z_c$) field. In the far-field the beam radius grows linearly with z with an asymptotic beam growth angle

$$\theta_0 = \tan^{-1}\left(\frac{\lambda}{\pi w_0}\right). \tag{4.128}$$

Away from the beam waist the phase radius of curvature first drops and then increases. Eventually $R(z) \rightarrow z$. The phase slippage $\varphi_0(z)$ varies gradually with z and tends to $\pi/2$ in the far-field (Figure 4.15). The characteristic parameters $w(z)$, $R(z)$ and $\varphi_0(z)$ (Eqs. (4.124)–(4.126)) can be used to determine the properties of a Gaussian beam, the simplest solution to the paraxial wave equation, as it propagates in free space.

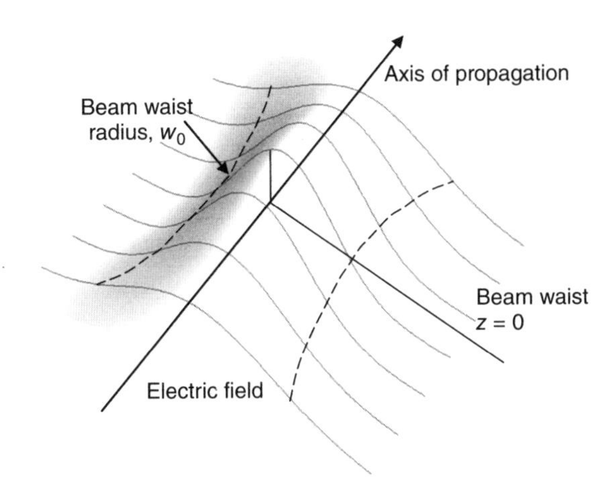

Figure 4.13 Propagating Gaussian beam showing how the beam radius increases with distance from the beam waist location.

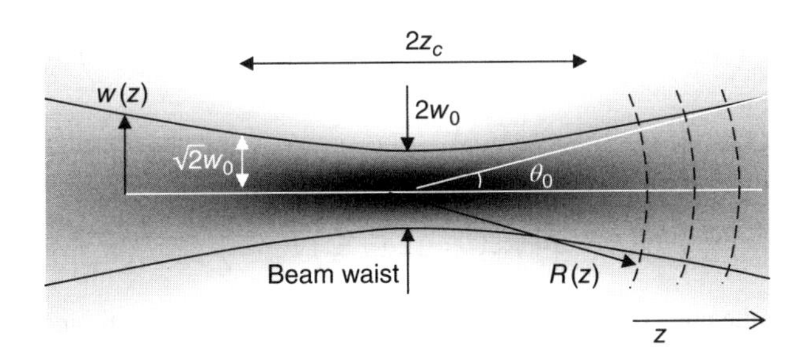

Figure 4.14 Characteristic parameters of a propagating Gaussian beam.

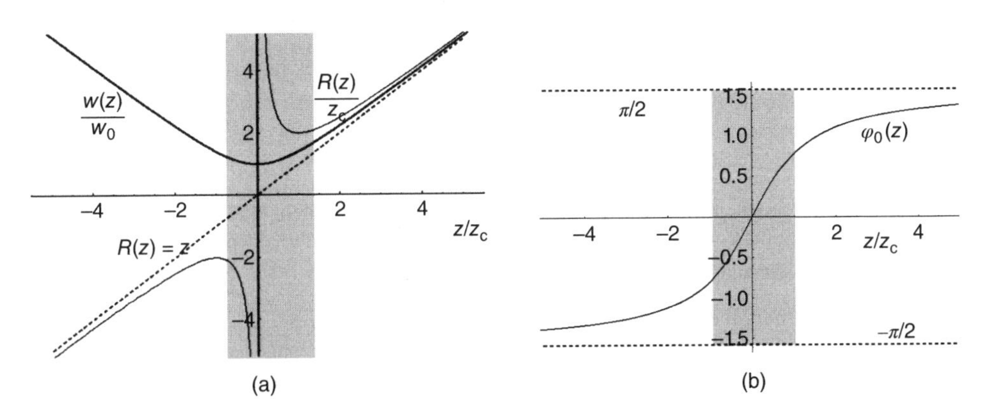

Figure 4.15 Variation of (a) beam radius and phase radius of curvature and (b) phase slippage as a function of distance from the beam waist.

4.3.1.2 Higher-Order Modes

A Gaussian beam is the simplest solution to the paraxial wave equation and suffices in many cases to describe beam propagation in an optical system. There are situations, however, when we need to model more complex field distributions, and in these cases we can use higher-order beam mode solutions. Higher-order modes are characterized by the same beam radius ($w(z)$) and phase radius of curvature ($R(z)$) as the fundamental mode above; only the phase-slippage differs. The most appropriate mode set to use depends on the symmetry of the system being modeled. In cylindrical coordinates we can use Gaussian–Laguerre modes:

$$\Psi_{pm}(r,\phi,z) = \left(\frac{2p!}{\pi\,(p+m)!}\right)^{1/2}\frac{1}{w(z)}\left(\frac{\sqrt{r}}{w\,(z)}\right)^{m} L_p^m\left(\frac{2r^2}{w^2\,(z)}\right)\exp\left(-\frac{r^2}{w^2\,(z)}\right)$$
$$.\exp(-\mathrm{j}kz)\exp\left(-\mathrm{j}\frac{\pi r^2}{\lambda R\,(z)}\right)\exp(-\mathrm{j}(2p+m+1)\varphi_0(z)).\exp(\mathrm{j}m\phi) \tag{4.129}$$

where L_p^m are generalized Laguerre polynomials. The phase slippage, $\varphi(z) = (2p + m + 1)\varphi_0(z)$, is now greater than that of the fundamental mode ($\varphi_0(z)$) by an amount that depends on the mode parameters p and m. In rectangular coordinates we use Gaussian–Hermite modes:

$$\Psi_{mn}(x,y,z) = \left(\frac{1}{\pi w_x\,(z)\,w_y(z)2^{m+n-1}m!n!}\right)^{1/2} H_m\left(\frac{\sqrt{2}x}{w_x\,(z)}\right)H_n\left(\frac{\sqrt{2}y}{w_y\,(z)}\right)$$
$$.\exp\left(-\frac{x^2}{w_x^2\,(z)}-\frac{y^2}{w_y^2(z)}\right).\exp(-\mathrm{j}kz)\exp\left(-\mathrm{j}\frac{\pi x^2}{\lambda R_x\,(z)}-\mathrm{j}\frac{\pi y^2}{\lambda R_y(z)}\right)$$
$$.\exp\left(-\mathrm{j}\left(m+\frac{1}{2}\right)\varphi_{0x}(z)-\mathrm{j}\left(n+\frac{1}{2}\right)\varphi_{0y}(z)\right) \tag{4.130}$$

where $H_{m/n}$ are Hermite polynomials, and the modes can have different widths and phase radii in the x and y directions. Again, there is a mode-dependent phase-slippage term.

The higher-order Gaussian–Hermite and Gaussian–Laguerre modes consist of polynomials superimposed on the fundamental Gaussian mode (Figure 4.16) and constitute complete orthonormal sets of modes that are each solutions to the paraxial wave equation. Any arbitrary solution of this wave equation (any paraxial beam) can therefore be expressed as a superposition of Gaussian modes, the particular mode set, Ψ_i, being chosen as appropriate for the symmetry of the problem:

$$E(r,z) = \sum_i A_i\Psi_i(r,z;w(z),R(z),\varphi_i(z)) \tag{4.131}$$

where $E(r, z)$ is the paraxial field at a location r in the plane given by z and A_i are the mode coefficients. $w(z)$, $R(z)$ and $\varphi_i(z)$ are the beam radius, phase radius, and mode-dependent phase slippage, as before.

Once the mode coefficients are known, the beam can be reconstructed at any plane along its propagation path by simply calculating new values of the beam parameters, $E(r,z') = \sum_i A_i\Psi_i(r,z;w(z'),R(z'),\varphi_i(z'))$. Since the form of the amplitude distribution of each mode remains unchanged as they propagate (they are simply scaled as $w(z)$) it is the relative change in the phase slippage term between modes that is responsible for the changing amplitude distribution of a diffracting beam. If the field is known over some surface S, then the mode coefficients can be evaluated from

$$A_i = \iint_S E^*(r,z)\Psi_i(r,z;w(z),R(z),\varphi_i(z))\mathrm{d}s. \tag{4.132}$$

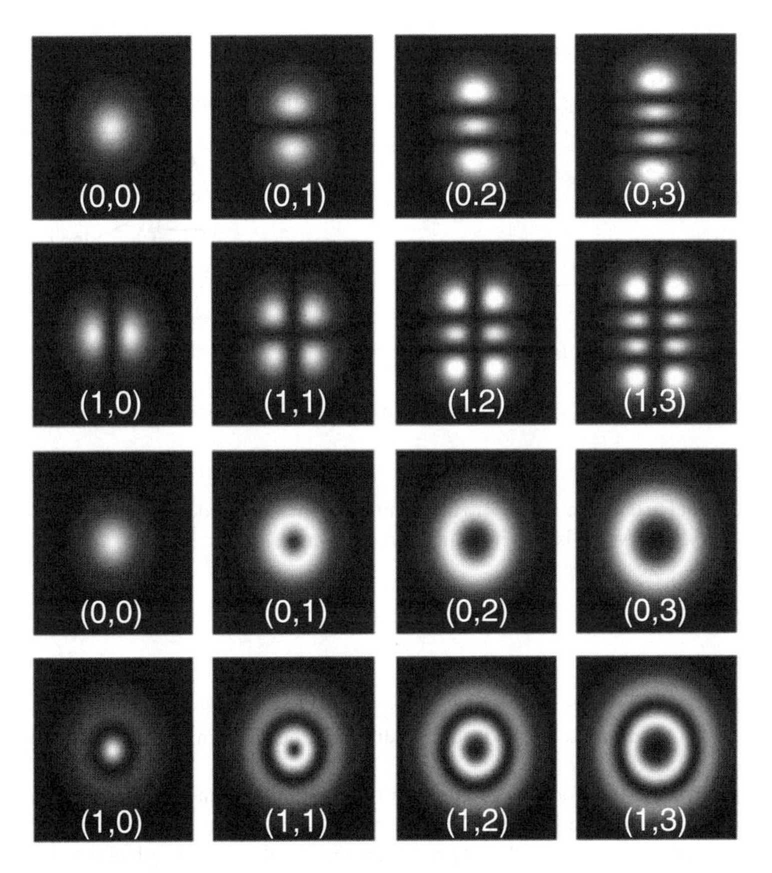

Figure 4.16 Intensity (Ψ^2) profiles of some Gaussian–Hermite (top two rows) and Gaussian–Laguerre (lower two rows) modes. The plots extend to $\pm 3w_0$.

This modal decomposition need only be carried out once, so long as there is no further scattering between modes (e.g., caused by truncation of the beam at an aperture). The choice of the optimum beam mode set (in terms of the parameters $w(z)$ and $R(z)$) is important for the efficiency of the Gaussian beam mode approach. Often a source field can be represented to a high accuracy by the sum of only a few modes and the mode set that maximizes the power in the fundamental is chosen. Other useful approaches take into account the ability of the highest-order mode used to model the edge of the field.

Equations (4.124)–(4.126) allow us to model the evolving shape and increasing lateral extent of a beam as it propagates in free space. Most practical systems must employ a series of reflecting or refracting elements to periodically re-focus such a beam and these are discussed next.

4.3.2 Launching and Focusing Components: Horns, Lenses, and Mirrors

A thin lens or curved mirror placed in the path of a propagating beam transforms the radius of curvature of the incident wavefront, R_{in}, into R_{out}, the radius of curvature of the transmitted beam, according to $1/R_{out} = 1/R_{in} - 1/f$ where f is the focal length of the transforming element. Using an appropriate combination of phase curvature transformations, a quasi-collimated beam can be maintained over long distances in what is known as a beam waveguide. In general, in long-wave optics, the phase radius of

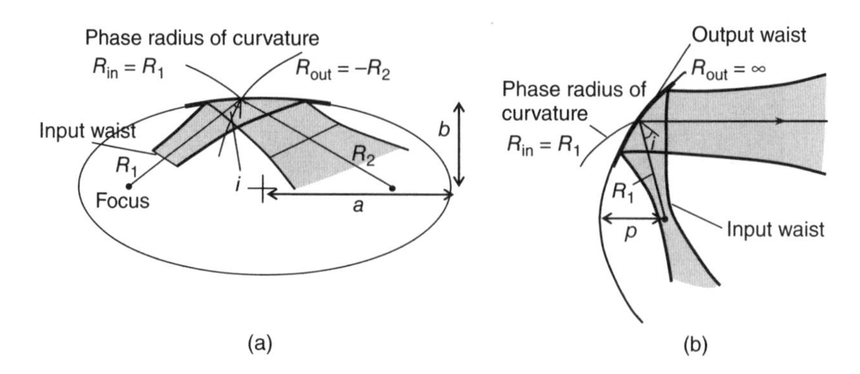

Figure 4.17 (a) Ellipsoidal and (b) paraboloidal refocusing elements.

curvature is not equal to the distance back to the beam waist and amplitude distributions are not re-imaged at waists. Typically, however, it is the reforming of beam waists, and not imaging, that is of interest when dealing with simple Gaussian beams.

4.3.2.1 Mirrors

Refocusing (curved) mirrors are commonly employed for re-collimation in quasi-optical beam guides (Figure 4.17). They are also useful for feeding a divergent beam from a source into a beam guide or focusing a beam onto a detector.

A reflecting surface in the form of an ellipsoid of revolution around its major axis transforms the phase radius of curvature of an incident Gaussian beam $R_{in} = R_1$ into a phase radius of curvature for the reflected beam of $R_{out} = -fR_{in}/(R_{in} - f) = -R_2$, where f is the focal length of the mirror. The major and minor axes and eccentricity of the ellipsoid required are given by

$$a = \frac{1}{2}(R_1 + R_2); \quad b = \sqrt{R_1 R_2}\cos(i) \tag{4.133}$$

$$e = \sqrt{1 - b^2/a^2} \tag{4.134}$$

and

$$f = R_1 R_2/(R_1 + R_2) \tag{4.135}$$

where $2i$ is the bending angle of the mirror. Usually, the beam waists do not coincide with the geometrical foci of the ellipsoid.

If either the incident or the reflected beam is required to have a waist at the mirror, then the surface to use is a paraboloid of revolution around the axis of symmetry, and the focal length of the mirror section should match the phase front radius of curvature $f = R_1$ of the non-collimated beam. If the distance from the vertex to the focus of the parent parabola is p, then $p = R_1\cos^2 i$, where i is the angle of incidence at the mirror. Parabolic mirrors can be used as substitutes for ellipsoidal mirrors if $R_{in} \geq R_{out}$ or vice versa.

Normally, mirrors are used in an off-axis configuration to prevent the beam from interfering with itself. To minimize beam distortion, the angle of incidence of the beam i should be less than 45° (bending angle $2i < 90°$). Because of projection effects, the reflected beam from an off-axis mirror suffers from amplitude distortion (beam squint). For a reflected Gaussian beam, the maximum is no longer on the propagation axis; however, the Gaussicity (fractional power coupling to a pure Gaussian) of such a beam is still high for long focal ratios and small angles of incidence. The Gaussicity is given by

$$K_G = 1 - \frac{1}{2}\left(\frac{w_m \tan i}{2f}\right)^2 \tag{4.136}$$

where w_m is the beam waist at the mirror [37]. In a two-mirror system, the distortion can be compensated. If the mirror does not have the correct ellipsoid of revolution shape, then phase distortion also occurs at the mirror, which leads to additional phase aberration effects. Another consequence of off-axis configurations is that a certain fraction of the power in an incident linearly polarized beam is scattered into the cross-polar direction:

$$P_{Xpol} = \left(\frac{w_m \tan i}{2f}\right)^2 \tag{4.137}$$

Again, the effect can be mitigated by special arrangements in multi-element optical systems [38, 39].

Many common metals (aluminum, copper, gold) behave as ideal conductors in the THz region and so metal or metal-coated mirrors are widely used. Small penetration depths mean that a thin metal layer is sufficient and reflectivity values close to 1 can be achieved. Manufacturing errors in these mirrors may give rise to more or less randomly distributed deviations from a nominal reflector surface. Ruze [40] has analyzed such errors in the case where they are much smaller than the wavelength and their effect is to scatter power from the main beam into a wider error beam, resulting in a loss of efficiency. For a reflector with rms surface error ε, and normal incidence, this surface efficiency is given by

$$\eta_{surf} = \exp\left(-\left(\frac{4\pi\varepsilon}{\lambda}\right)^2\right). \tag{4.138}$$

For beam guides with many reflecting components such effects must be taken into consideration.

4.3.2.2 Lenses

Similar to traditional optical lenses, THz lenses are used to focus or collimate radiation by refraction through curved-shaped dielectric materials. Refractive THz optical system designs are on-axis and so have many advantages, in terms of aberrations, over off-axis reflective designs. At optical wavelengths glass is an ideal, low absorptive material to use. At THz wavelengths dielectric plastic materials, such as polypropylene and polyethylene (PE, in particular ultra-high molecular weight (UHMW) PE) are commonly used. These all have low levels of absorption and a relatively low refractive index value, leading to the requirement of thick lens elements. Higher refractive index materials, such as quartz or silicon, suffer from larger reflections at the surface, described by the Fresnel equations, and require antireflection coatings which are not as well developed as in the optical regime. On the other hand the use of these coatings on higher refractive index material such as fused quartz ($n = 1.95$), silicon ($n = 3.4$), and germanium ($n = 4$) allows larger radii of curvature in the optical designs and can reduce associated optical aberrations and improve the broadband performance of optics. Currently many innovative spectrally broadband metamaterial-based gradient index (GRIN) lenses are being developed that focus to a spot size of approximately one wavelength, and rapid development in this field is leading to the evolution of "flat lenses" [41, 42]. Modeling the optical properties of lenses can be adequately achieved with the *ABCD* matrix formulation described next. More accurate models can be achieved using method of moments techniques or physical optics (Section 4.3.7) where an extra propagation through the lens is included to simulate internal reflections.

4.3.2.3 Gaussian Beam Transformation

To model the effect of an optical element on the waist, phase radius and slippage of a propagating Gaussian beam, we can use the convenient *ABCD* ray transfer matrix formalism that is commonly used in

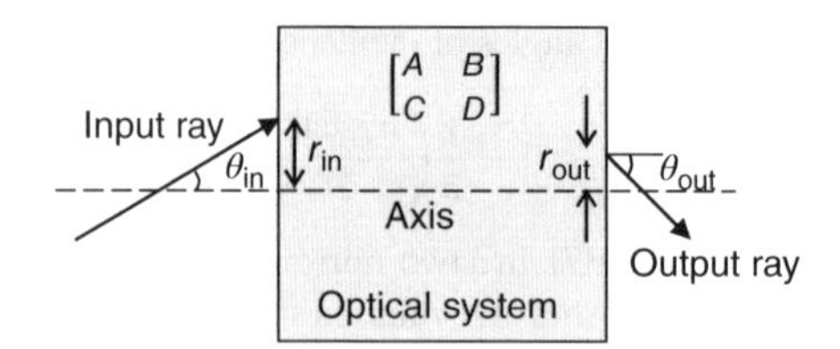

Figure 4.18 In the ray matrix formalism an optical element or combination of elements is represented by an *ABCD* matrix that modifies the input ray's displacement and slope. In quasi-optics the *ABCD* matrix elements are used to transform the input Gaussian beam parameter.

geometrical optics (see e.g., Ref. [36]). In this case the *ABCD* or *ray matrix* for the optical element operates on an input ray displacement and slope. It transforms (r_{in}, θ_{in}) at the input plane to (r_{out}, θ_{out}) at the output plane (Figure 4.18) according to the matrix equation

$$\begin{bmatrix} r_{out} \\ \theta_{out} \end{bmatrix} = \begin{bmatrix} A & B \\ C & D \end{bmatrix} \begin{bmatrix} r_{in} \\ \theta_{in} \end{bmatrix} \tag{4.139}$$

For Gaussian beams, the four parameters of the ray matrix operate on the complex radius of curvature q (which can be constructed from w and R), giving

$$q_{out} = \frac{Aq_{in} + B}{Cq_{in} + D} \tag{4.140}$$

and on the phase slippage giving

$$\varphi_{out} = \varphi_{in} - Arg\left[A + B\left(\frac{1}{q_{in}}\right)\right]. \tag{4.141}$$

The new beam radius and phase radius of curvature can be recovered from q (Eq. (4.122)):

$$R(z) = \left(Re\left[\frac{1}{q(z)}\right]\right)^{-1} \text{ and } w(z) = \sqrt{\frac{-\lambda}{\pi Im\left[{}^{1}/_{q(z)}\right]}} \tag{4.142}$$

Some common ray transformation matrices are:		
	$\begin{bmatrix} 1 & d \\ 0 & 1 \end{bmatrix}$	for propagation a distance d in free space or in a medium of constant refractive index
	$\begin{bmatrix} 1 & 0 \\ -{}^{1}/_{f} & 1 \end{bmatrix}$	for a thin lens of focal length f ($f > 0$ for converging)
	$\begin{bmatrix} 1 & 0 \\ -{}^{2}/_{R} & 1 \end{bmatrix}$	for reflection from a curved mirror (normal incidence, $R > 0$ for center of curvature before interface)

For a multi-element system made up of a series of optical components, the full *ABCD* matrix is given by the product of the corresponding *ABCD* matrices for the individual components, including any distance propagated in free space or in a medium. The matrices must be cascaded in reverse order so the first optical element is on the right-hand side of the product:

$$M_{total} = M_N \cdot M_{N-1} \cdot \cdots \cdot M_2 \cdot M_1 = \begin{bmatrix} A & B \\ C & D \end{bmatrix}_{total}. \tag{4.143}$$

4.3.2.4 Horns

Horn antennas are quite often used as feed systems for launching and detecting THz waves, particularly at longer wavelengths [43]. A horn antenna can be regarded as a flared section of waveguide that acts as a smooth transition to free space. If the flare is gentle, then the field structure at the horn aperture is readily predicted from waveguide theory. To a good approximation, the field at the aperture has a well-defined spherical phase front (expressed using the usual quadratic approximation for paraxial beams as $\exp(-jkr^2/2R)$, where r is the radial distance in the aperture and R is the phase radius of curvature). Horns typically produce beams that have low sidelobe levels and are well approximated by a Gaussian distribution. In terms of feeding quasi-optical systems, the most important parameters of a horn are the effective size and location of the best-fit Gaussian beam waist. These are given by

$$w_0 = \frac{w_{\text{ap}}}{\sqrt{1 + (\pi w_{\text{ap}}^2/\lambda R)^2}} \tag{4.144}$$

and

$$\Delta z = \frac{R}{1 + (\lambda R/\pi w_{\text{ap}}^2)^2} \tag{4.145}$$

where w_{ap} is the beam radius at the aperture of the horn. Δz refers to the distance of the phase center behind the horn aperture. Here we describe some of the most commonly used horn antenna types.

Smooth-Walled Horns

For the typical linearly-flared conical and pyramidal shapes encountered (Figure 4.19), the center of curvature of the phase front is located at the apex of the horn, that is, $R = L$ where L is the slant length of the horn. The sidelobe structure is determined by the high frequency structure of the horn aperture fields (such as sharp truncation of the field or its derivatives at the edge of the aperture).

A *pyramidal horn* antenna (of aperture width a and height b) produces a rectangular waveguide TE_{10} fundamental mode field at its aperture with a phase curvature given by the slant lengths (L_x and L_y) of the horn:

$$\vec{E}_{\text{ap}} = E_0 \cos\left(\frac{\pi x}{a}\right) \exp\left(-jk\left(\frac{x^2}{2L_x} + \frac{y^2}{2L_y}\right)\right)\hat{\mathbf{y}}; \quad |x| \le a/2,\ |y| \le b/2. \tag{4.146}$$

For optimum operation as a quasi-Gaussian feed, $L_x = L_y$ and $b/a = 0.7$, giving a symmetric best-fit Gaussian with a beam radius $w_{\text{ap}} = 0.35a = 0.5b$ at the horn aperture.

A *diagonal horn* is an improvement on a pyramidal smooth-walled horn as it produces a beam pattern with the same E- and H-plane patterns. A flared square horn section is rotated by 45° with respect to a rectangular waveguide feeding it with a single TE_{10} mode. Two orthogonal TE_{10} modes are thus excited

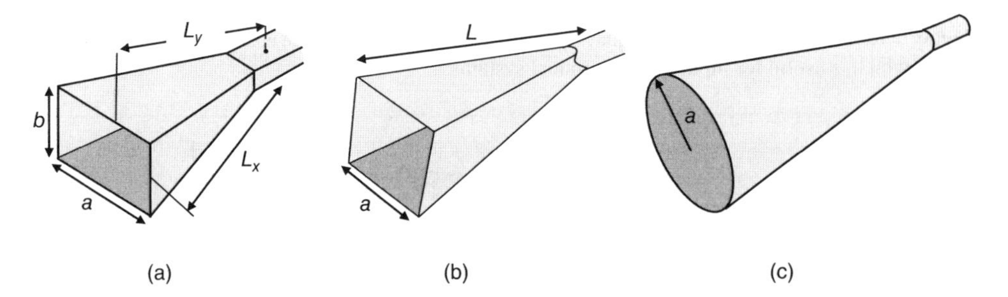

Figure 4.19 Three examples of smooth-walled horn antennas used for launching quasi-optical beams. (a) pyramidal horn, (b) diagonal horn, and (c) conical horn.

in the horn with equal phase and amplitude. For a horn of side length a and slant length L, the electric field at the aperture is given by

$$\vec{E}_{ap}(x,y) = E_0\left(\cos\left(\frac{\pi y}{a}\right)\hat{\mathbf{x}} + \cos\left(\frac{\pi x}{a}\right)\hat{\mathbf{y}}\right)\exp\left(-jk\left(\frac{x^2+y^2}{2L}\right)\right); |x|, |y| \leq \frac{a}{2}. \tag{4.147}$$

This yields a quasi-Gaussian field with beam radius $w_{ap} = 0.43a$ at the aperture.

A *conical horn* antenna of radius a and slant length L produces a cylindrical waveguide TE_{11} fundamental mode field of the form:

$$\vec{E}_{ap}(r,\phi) = E_0\left(\left(J_0\left(\frac{\chi r}{a}\right) + J_2\left(\frac{\chi r}{a}\right)\cos 2\phi\right)\hat{\mathbf{x}} + J_2\left(\frac{\chi r}{a}\right)\sin 2\phi\hat{\mathbf{y}}\right).\exp\left(-jk\left(\frac{r^2}{2L}\right)\right);$$
$$\chi = 1.841, |r| < a. \tag{4.148}$$

For operation as a quasi-Gaussian feed, the best-fit Gaussian has beam radius $w_{ap} = 0.76a$ at the horn aperture. A *dual-mode horn* is an improvement on a smooth-walled conical horn. It produces a nearly symmetric tapered field in a circular horn by also exciting the TM_{11} mode (with correct amplitude and phase relative to TE_{11}) with a carefully designed step plus phasing section in the throat of the horn. For a dual-mode horn $w_{ap} = 0.59a$.

Corrugated Horns

Corrugating the surface of a horn allows hybrid (combination of TE and TM) modes to propagate. One of the most widely used designs is the conical corrugated horn or "scalar feed" (Figure 4.20) [44, 45]. Such a horn, with $\lambda/4$-deep corrugations at the aperture, operates under the balanced hybrid condition and produces a HE_{11} mode. This gives an aperture field of the form:

$$\vec{E}_{ap}(r) = E_0\left(J_0\left(\frac{\chi r}{a}\right)\hat{\mathbf{y}}\right)\exp\left(-jk\left(\frac{r^2}{2L}\right)\right); \quad |r| < a, \tag{4.149}$$

where a is the radius of the horn aperture and L is its slant length. The beam radius of the best-fit Gaussian at the horn aperture is given by $w_{ap} = 0.64a$. Conical corrugated horns have the desirable properties of axial beam symmetry, low sidelobe levels and low cross-polarization. The beam from a conical corrugated horn has 98% coupling to a fundamental Gaussian beam.

Shaped horn designs have also been developed to fulfill criteria such as increased coupling to a fundamental Gaussian mode or reduced horn length.

4.3.2.5 Lens Antennas

Lens antennas tend to be used at higher frequencies in the THz band when the manufacture of horn antennas becomes problematic. A planar antenna typically acts as the basic radiating source, and this produces a highly divergent beam. The lens then quasi-collimates this beam to produce a far-less-divergent far-field pattern, which is useful for input to quasi-optical systems.

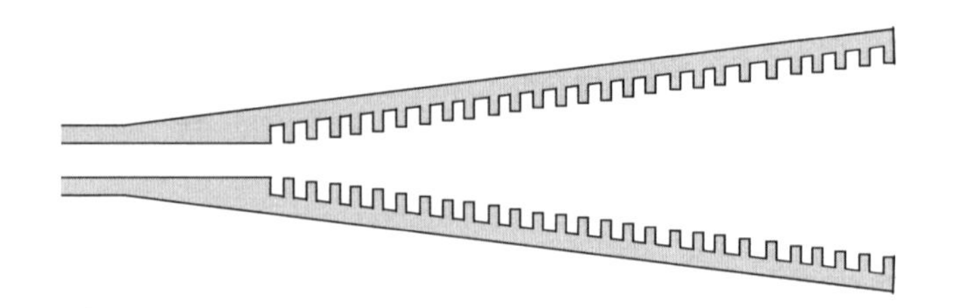

Figure 4.20 Cross-section of a corrugated conical horn.

4.3.3 *Other Components Needed in Beam Waveguides*

4.3.3.1 Beam Splitters, Polarizing Grids, Roof Mirrors, Interferometers, Filters, and Diffractive Optical Elements

Also useful in beam waveguides are a range of other quasi-optical components and subsystems that function in an analogous way to waveguide circuit devices in microwave systems. Such components may be inherently frequency-selective devices (such as wave plates) or may be frequency-independent devices (such as polarizing grids). A review of the operation of such components is presented in [33, 46], along with their application in both frequency-dependent and frequency-independent subsystems. For example, polarizing wire grids can be used to linearly polarize an unpolarized beam or change the direction of polarization of a beam already linearly polarized (Figure 4.21a). Since the part of the beam that is not transmitted ends up being reflected, polarizing grids are also useful as beam splitters and beam combiners. In the case of beam splitters the reflected and transmitted beams have orthogonal polarization directions that depend on the orientation of the wires of the grid with respect to the directions of propagation and polarization of the incident beams. An example of a diplexer based on such polarization splitting by a wire grid is illustrated in Figure 4.21b [47]. Alternatively thin sheets of transparent plastic, such as Mylar, are also often used as non-polarization sensitive beam splitters, when only a small fraction of power in the reflected beam is required.

Another very useful component is the roof mirror which rotates the direction of polarization of an incident wave by an angle of $2\theta_i$, if the angle between the incident polarization and the axis of the roof mirror is θ_i (see Figure 4.21c). The field component parallel to the axis of the roof mirror is not affected in the double reflection, while the component perpendicular to the axis has its direction reversed. Roof mirrors combined with polarizing grids can therefore be used as path length compensators, as in Figure 4.21d.

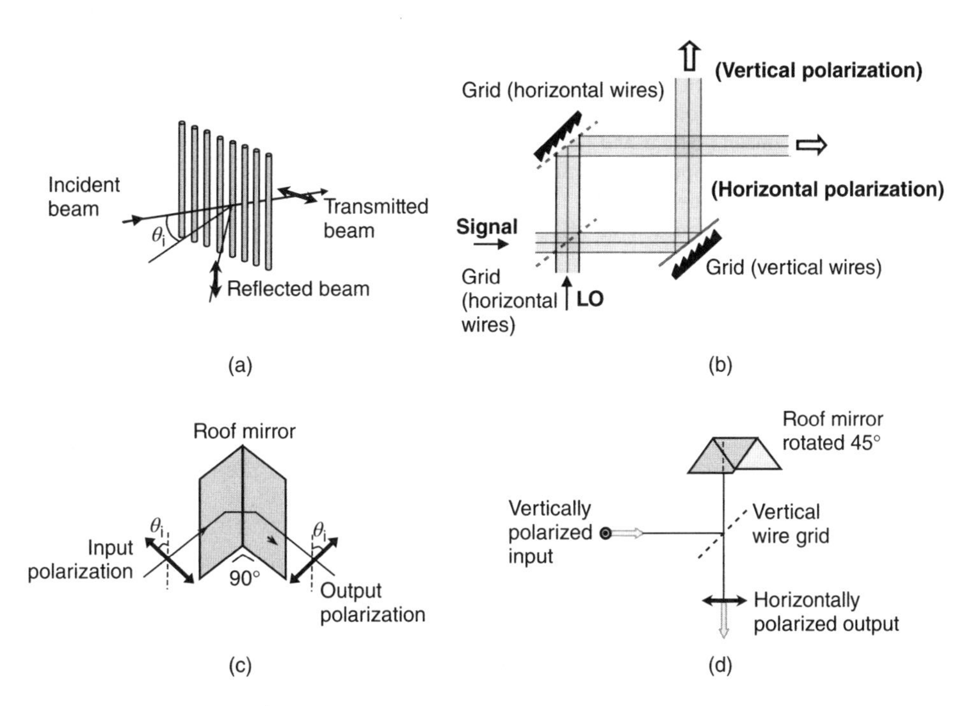

Figure 4.21 (a) Polarizing grid with orthogonally polarized reflected and transmitted beams, (b) diplexer based on polarization splitting [47] (c) roof mirror rotating the input polarization direction; and (d) roof mirror combined with a polarizing grid in a path length compensator.

A Michelson polarizing interferometer (Martin–Puplett Interferometer) is a combination of a polarizing grid acting as the beam splitter and two roof mirrors that reflect the orthogonally polarized beams back along the incident path from the beam splitter, but with their polarization directions rotated by 90°. They can then be recombined at the beam splitter before propagation to the output port, as shown in Figure 4.22. The difference in the path length between the two arms and the wavelength of the radiation determine the polarization properties of the recombined beams. Thus, further grids at the input and output ports of the system allow a four-port dual beam interferometer to be realized, which is useful in heterodyne systems for efficient injection of a local oscillator signal, and as a sideband filter for a signal beam with rejection of the unwanted band at the output port (i.e., the two beams of interest are separated in wavelength).

Waveplates which act as polarization transducers are useful quasi-optical devices that produce a differential phase shift between two orthogonal electric-field polarization directions (a half-wave plate providing a differential phase shift of 180° between two orthogonal electric field polarization directions while a quarter-wave plate transforms linear to circular polarization and vice versa.) Waveplates can be generated through mechanical milling or other techniques [48].

A quasi-optical polarizer based on arrays of self-complementary sub-wavelength holes is reported in Ref. [49]. A naturally isotropic material can be made to simulate an anisotropic birefringent dielectric by changing its geometry as, for example, in Ref. [50]. Dielectric strips of alternating permittivity can be combined to form composite dielectrics with the required dielectric constant, provided the wavelength is much longer than the strip thickness. Such strips can also be used as filters, mirrors, and so on [51].

Extremely useful high- and low-pass filters can also be made with inductive and capacitative grids, respectively [52]. Resonant grids can be used for band-pass filtering while a perforated plate filter acts as a high-pass filter with a sharp cut-off and even with profile shaping for lensing [48]. A transmission line matrix method can be used as a convenient way of calculating the frequency response of systems composed of cascaded elements.

Diffractive optic elements (when typically placed at a Fourier plane in a 4-f system) transform the field distribution (amplitude and/or phase) of an incident coherent beam, thereby producing the desired field distribution at a displaced second plane. Such components operate like spatial filters by redirecting segments of a wave front through the use of interference and/or phase control [53]. Fresnel phase and amplitude plates, Damann gratings and Fourier gratings [54–56] can, for example, be used to generate multiple images of a single input beam for multiplexing a local oscillator source quasi-optically with an array of detectors in a heterodyne system. There are a number of dielectric materials which have useful mechanical and optical properties in the THz part of the spectrum for use in transmission phase gratings in particular [33].

Generally, in designing diffractive optical elements it is not possible to use analytical techniques but rather approaches such as simulated annealing or genetic algorithms, and phase-retrieval techniques such as the Gerchberg–Saxton algorithm can be applied to find optimal solutions for the diffractive optical element [57–59]. Another useful example of beam-shaping is in the production of "diffraction-free beams", in which the amplitude does not change in form or scale while propagating [60]. These beams

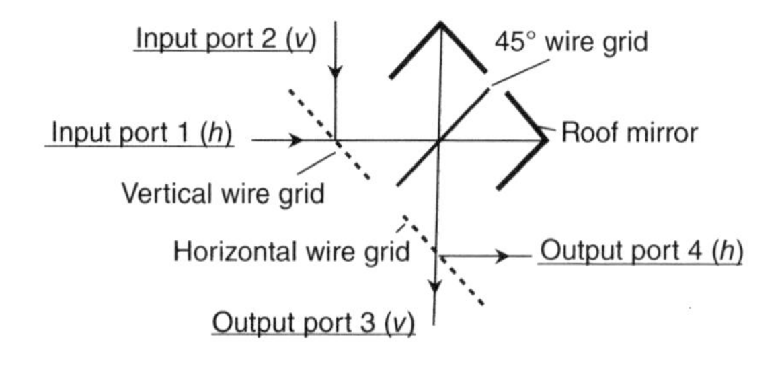

Figure 4.22 A four-port dual-beam interferometer (diplexer).

have the amplitude cross-section of a Bessel function and, ideally, are infinite in extent. Pseudo-Bessel beams (having finite extent and power) have been generated using dielectric conical-shaped lenses called axicons [61, 62]. A resonant Fresnel lens can also be designed which has lower optical loss characteristics than the alternative thin lens [63].

An issue for the optimal design of all quasi-optical devices is the consequences of Gaussian beam propagation through such devices in terms of beam spreading, truncation levels, and phase slippages, for example. The component should allow a clearance of $4w$ for a Gaussian beam, where w is the best fit Gaussian radius at the component. The consequence of beam truncation will be extra diffraction of the beam into the side-lobe structure and loss in transmitted power.

4.3.4 Absorbers

Absorbing materials are important in THz systems to reduce reflectivity from non-optical surfaces. Reflected power inevitably leads to a loss of sensitivity and undesirable effects such as standing waves, fringing, multiple images, and ghosting. These effects can be greatly reduced by efficient anti-reflection coatings on metal surfaces. Coatings can be tuned for given frequencies. In THz systems standing waves are troublesome for coherent detectors and can exist between all optical components and source and detector antennas. Many commercial products are available as foams and elastomers as well as rigid polymers with geometrical-shaped surfaces to minimize reflections at a design wavelength. This material pressed into cones or pyramids on polymer sheets is effective and is referred to as tessellating terahertz radar absorbing material. Conductive vinyl films can also be used within waveguides and the interior of cavities to reduce Q factors or to alter the flow of high frequency currents.

Microwave techniques can also be applied to shorter wavelengths using a Salisbury screen. This simple geometry is used to minimize back-scattering from a flat metal plate. A resistive sheet having a resistivity close to that of free space (377 Ω) is placed roughly one quarter of a wavelength in front of the conducting plate. This is a narrow-band technique. Many metamaterial-type structures are now realized also to minimize reflection. Metamaterials are constructed by layering material with subwavelength structures which are capable of manipulating electromagnetic radiation [64, 65] and have many THz applications. The first metamaterial THz absorber was reported in 2008 and used loss in a metal as a key mechanism to absorb THz waves [66]. This metamaterial absorber was made from three subwavelength layers including a frequency selective surface (FSS), a dielectric spacer layer, and a metallic back layer. This FSS determines the absorption frequency, the metallic back layer reflects the transmitted resonance frequency, and the spacer layer acts as a subwavelength cavity which makes the waves reflected from the metallic layer out of phase with respect to the reflected waves from the FSS [67].

Soon after the first demonstration of a THz metamaterial absorber, different structures were fabricated to enhance the functionality of the absorber, including polarization independence and operation over multiple frequencies [66, 68, 69]. Other promising artificial materials, such as graphene, have also been demonstrated [70]. Graphene as a material is optically transparent, it has high absorption per monolayer in the optical range and more so in the THz, with reported absorption up to 100% [71].

4.3.5 Modeling Horns Using Mode Matching

The details of the performance of a corrugated or smooth-walled horn, including any necessary waveguide transitions and couplers, can be obtained from a full electromagnetic simulation over the volume of the structure and some of its surroundings. This is increasingly possible because of the advances in the availability of computing resources, but it still remains a significant challenge because of the relative size of such structures in terms of the wavelength of the radiation (they are necessarily electrically large structures). Alternative methods, such as *mode-matching,* based on a combination of extra physical insight into the propagation of radiation in different parts of the structure and the use of computational resources, offer more efficient solutions in situations where there is a significant degree of symmetry in the structure (for example, axial symmetry in conical horns). This is particularly advantageous at the design stage.

In Section 4.2 the principles of propagation of electromagnetic waves in the THz domain, in various transmission lines, were discussed for the case of infinitely long guides, or more specifically with guide boundary conditions that are uniform along the propagation axis. In practical applications we are always dealing with finite sections of guides and therefore it is necessary also to consider the boundary conditions at the discontinuities at both ends of the guide section.

It will be assumed here that within the guide length we know the supported modes ($\vec{E}_i$ and $\vec{H}_i$) and corresponding complex propagation constants γ_i as a function of frequency ($i = 1 \ldots N$). The boundary conditions at the guide ends (ports 1 and 2) can therefore be expressed in terms of the complex mode amplitudes of all supported modes, both forward and backward propagating. Propagation delay and amplitude change between ports 1 and 2 are related to the guide length l and the propagation constant γ_i of the particular mode.

If we describe the amplitudes of the modes excited at port 1 and propagating toward port 2 using a column vector $[a_i^1]$ and the modes propagating from the direction of port 2 by $[b_i^1]$, and similarly at port 2 using $[a_i^2]$ and $[b_i^2]$ then for a section of uniform waveguide we get:

$$[b_i^2] = [S_{21}][a_i^1], \quad [b_i^1] = [S_{12}][a_i^2] \tag{4.150}$$

where $[S_{21}]$ and $[S_{12}]$ are the appropriate scattering matrices [43]. In this case of no actual change of the boundary conditions at the ports, the other two scattering matrices are zero, and therefore, assuming that at each port the reference frame is reset, so that $z = 0$ [43]

$$[S_{21}] = [S_{12}] = \begin{bmatrix} e^{-\gamma_1 l} & 0 & \cdots & 0 \\ 0 & e^{-\gamma_2 l} & & 0 \\ \vdots & \vdots & \ddots & \vdots \\ 0 & 0 & \cdots & e^{-\gamma_N l} \end{bmatrix}$$

$$[S_{11}] = [S_{22}] = [0] \tag{4.151}$$

If there is an actual discontinuity at one of the ports, due to the next, dissimilar guide section (with modes characterized by $\vec{E}'_j$, H'_j, and γ'_j and indexed by $j = 1 \ldots M$) then the full set of scattering matrices needs to be calculated for the junction itself (from the boundary conditions at the discontinuity):

$$\begin{aligned}
[S_{11}] &= ([R]^+ + [P]^+[Q]^{-1}[P])^{-1}([R]^+ - [P]^+[Q]^{-1}[P]) \\
[S_{12}] &= 2([R]^+ + [P]^+[Q]^{-1}[P])^{-1}[P]^+ \\
[S_{21}] &= 2([P]([R]^+)^{-1}[P]^+ + [Q])^{-1}[P] \\
[S_{21}] &= ([P]([R]^+)^{-1}[P]^+ + [Q])^{-1}([P]([R]^+)^{-1}[P]^+ - [Q])
\end{aligned} \tag{4.152}$$

where matrices $[P]$, $[Q]$, and $[R]$ characterize the coupling of modes across the junction and self-reactance either side of the junction, respectively:

$$\begin{aligned}
[P] = [P_{ij}] &= \int_\Sigma \vec{E}_j^{(t)} \times \vec{H}_i^{(t)} \cdot d\vec{S} \\
[R] = [R_{i'i}] &= \int_\Sigma \vec{E}_{i'}^{(t)} \times \vec{H}_i^{(t)} \cdot d\vec{S} \\
[Q] = [Q_{j'j}] &= \int_\Sigma \vec{E}_{j'}^{(t)} \times \vec{H}_j^{(t)} \cdot d\vec{S}
\end{aligned} \tag{4.153}$$

The integration is of the transverse components of the modal fields over the whole relevant cross-section of the junction (Σ), which might be finite (for example, metal waveguides) or infinite (dielectric guides). $[Q]$ and $[R]$ matrices are diagonal for orthogonal mode sets [43, 72].

It is possible to find analytical expressions for the elements of these matrices for some classes of junctions. For example, for a junction between two sections of circular waveguide with a step-wise *increase* in guide radius (from a to b) and a perfect electric conductor (PEC) wall closing the gap, it can be shown that:

$$[P] = \begin{bmatrix} [P_{EE}] & 0 \\ [P_{ME}] & [P_{MM}] \end{bmatrix}$$
$$[R] = [R_{i'i}]$$
$$[Q] = [Q_{j'j}] \tag{4.154}$$

In the expressions above the modes were grouped TE first, followed by TM modes, on both sides of the junction. The elements of the coupling matrices are indexed by the mode numbers of the circular waveguide modes n, l, and given by:

$$P_{\mathrm{TM}_{n'l'}-\mathrm{TM}_{nl}} = \frac{(1+\delta_{n0})\pi C_{nl} C_{nl'} a}{Z^{b*}_{\mathrm{TM}_{nl'}}\left[\left(\frac{p_{nl'}}{b}\right)^2 - \left(\frac{p_{nl}}{a}\right)^2\right]} \left(\frac{p_{nl'}}{b}\right) J'_n(p_{nl}) J_n\left(\frac{p_{nl'}a}{b}\right)\delta_{n'n}$$

$$P_{\mathrm{TE}_{n'l'}-\mathrm{TE}_{nl}} = \frac{-(1+\delta_{n0})\pi D_{nl} D_{nl'}}{Z^{b*}_{\mathrm{TE}_{nl'}}\left[\left(\frac{q_{nl'}}{b}\right)^2 - \left(\frac{q_{nl}}{a}\right)^2\right]} q_{nl} J_n(q_{nl}) J'_n\left(\frac{q_{nl'}a}{b}\right)\delta_{n'n}$$

$$P_{\mathrm{TE}_{n'l'}-\mathrm{TM}_{nl}} = 0$$

$$P_{\mathrm{TM}_{n'l'}-\mathrm{TE}_{nl}} = \frac{\pi n D_{nl'} C_{nl} ab}{2Z^{b*}_{\mathrm{TM}_{nl'}}[p_{nl'} \cdot q_{nl}]} J_n(q_{nl}) J_n\left(\frac{p_{nl'}a}{b}\right)\delta_{n'n}$$

$$Q_{\mathrm{TM}_{n'l'}-\mathrm{TM}_{nl}} = \frac{1}{Z^{b*}_{\mathrm{TM}_{nl}}}\delta_{n'n}\delta_{l'l}$$

$$Q_{\mathrm{TE}_{n'l'}-\mathrm{TE}_{nl}} = \frac{1}{Z^{b*}_{\mathrm{TE}_{nl}}}\delta_{n'n}\delta_{l'l}$$

$$R_{\mathrm{TM}_{n'l'}-\mathrm{TM}_{nl}} = \frac{1}{Z^{a*}_{\mathrm{TM}_{nl}}}\delta_{n'n}\delta_{l'l}$$

$$R_{\mathrm{TE}_{n'l'}-\mathrm{TE}_{nl'}} = \frac{1}{Z^{a*}_{\mathrm{TE}_{nl}}}\delta_{n'n}\delta_{l'l} \tag{4.155}$$

where C and D values are the normalization coefficients for the TM and TE waveguide modes, Z represents the impedance of the relevant mode while p_{nl} and q_{nl} represent the l-th root of the Bessel function J of order n and l-th root of its derivative, respectively [73]. In cases where a closed form solution cannot be found for the coupling matrices, their elements can be calculated numerically and tabulated for re-use, at a significantly higher computational cost.

Using the scattering matrices for a straight section (S; Eq. (4.151)) and a junction section (J; Eq. (4.152)) the scattering matrices characterizing a more complex passive 2 port (e.g., corrugated) waveguide structure can be determined. This process employs step-wise application of the cascading relation [43]:

$$[S_{11}^{AB}] = [S_{11}^{A}] + [S_{12}^{A}]([I] - [S_{11}^{B}][S_{22}^{A}])^{-1}[S_{11}^{B}][S_{21}^{A}]$$
$$[S_{12}^{AB}] = [S_{12}^{A}]([I] - [S_{11}^{B}][S_{22}^{A}])^{-1}[S_{12}^{B}]$$
$$[S_{21}^{AB}] = [S_{21}^{B}]([I] - [S_{22}^{A}][S_{11}^{B}])^{-1}[S_{21}^{A}]$$
$$[S_{22}^{AB}] = [S_{22}^{B}] + [S_{21}^{B}]([I] - [S_{22}^{A}][S_{11}^{B}])^{-1}[S_{22}^{A}][S_{12}^{B}] \tag{4.156}$$

where A and B are the two sections being cascaded, with port 2 of A exactly matching port 1 of B (uniform boundary conditions and the same number of modes being considered). $[I]$ is an identity matrix of the required size.

For example, to determine the scattering matrices of a structure consisting of two dissimilar waveguides joined together (*SJS*) the scattering matrix of the first straight section is cascaded with the junction section (matching the number of modes) and the result cascaded with the second straight section. This *mode-matching* process can be repeated an arbitrary number of times to generate the matrices for more complex elements, in particular corrugated horns. Smooth-walled horns can also be modeled using this approach by progressive approximation of the slanted wall with steps of decreasing size to verify convergence.

Often one end of the waveguide structure opens into free space. This situation can also be characterized using mode-matching, by treating the free space as a guide characterized by an infinite number of modes, for example, plane waves with varying directions of the wavevector and polarization (seen as TE and TM modes at the junction).

The expressions in Eq. (4.152) can only be exact (regardless of excitation) if all possible modes (propagating and evanescent, the total number being generally infinite) are included in the matrices. In practical applications the set of modes being considered is always finite, but convergence can be achieved by sufficiently extending the mode set. The fact that straight sections are of finite length makes it necessary to include a sufficient number of evanescent modes in their modal description, as they can still contribute to the transfer of power across the section [72, 73].

Horn analysis using the mode-matching approach produces results in excellent agreement with finite-difference time-domain (FDTD) methods, assuming that care is take to include a sufficient number of modes at each junction (see Figure 4.23).

In the case where the rank of the $[S_{12}]$ and $[S_{21}]$ matrices is 1 the structure is *single-moded,* as any excitation at one port produces the same combination of modes, and therefore the same field structure,

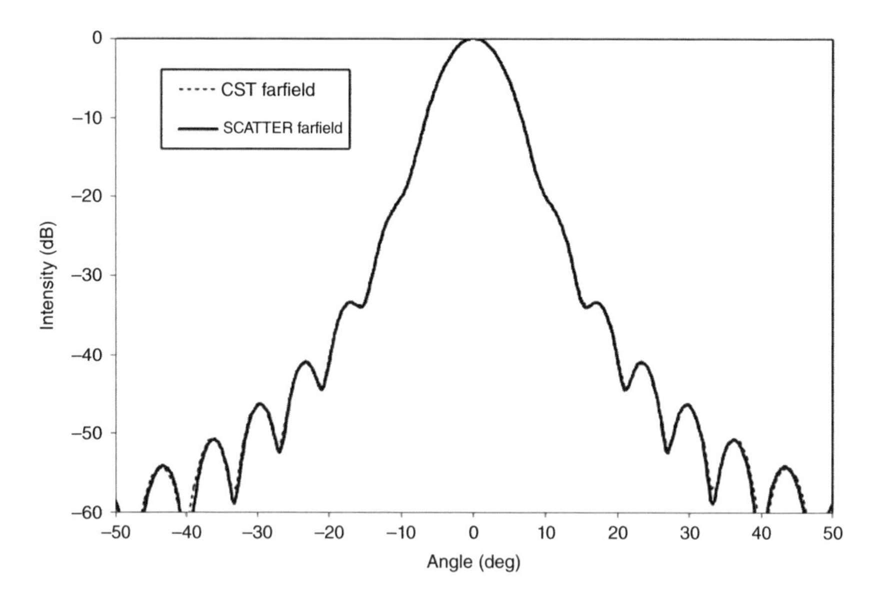

Figure 4.23 Comparison of the simulated performance of a corrugated horn designed to operate at a center frequency of 661 GHz, as predicted by a FDTD simulation (CST Microwave Studio) and mode-matching code (SCATTER).

at the other port, subject to a scalar amplitude factor. If at least one of the ranks is greater than 1 the structure is *multi-moded* in the respective direction.

4.3.6 Multi-mode Systems and Partially Coherent Propagation

If the waveguide system feeding a horn antenna is oversized, so that it can transmit more than the fundamental mode, then the horn is said to be multi-mode. In such cases the higher order modes in the waveguide can independently couple to an incoherent detector or source, with increased throughput over a single mode system [74, 75]. The on-axis gain and beam width of such horn antennas will be increased, which can be useful in systems where increased power coupling to a detector is an advantage and coherent illumination by the horn is not a vital requirement [73]. In such cases multi-mode horn antennas provide increased on-axis gain with low sidelobe levels (important for the off-axis rejection of noise, for example) [73, 76]. However, the relative coupling of the waveguide modes to the detector in the case of a receiver (or a source in the case of a transmitter) affects the details of the beam pattern, which thus becomes less easy to control and may require careful modeling.

For a multi-mode horn fed by an oversized waveguide with N independent modes propagating, the modified equation relating the beam solid angle Ω_A to the effective area of the horn aperture A_e is given by:

$$\Omega_A A_e = N\lambda^2 \tag{4.157}$$

where λ is the wavelength (Figure 4.24). The effective area A_e grows to be approximately equal to the physical area of the horn aperture A_{ph} as it becomes more highly over moded, while at the same time the beam solid angle area also grows approximately as the number of modes. This can have the effect of reducing the spatial resolution of the horn in imaging systems where the horn beam is matched to a reflecting antenna (or telescope) with low spillover, since A_e (which couples to the point spread function) will grow with N for the same Ω_A that fills the reflector. However, for a system with re-imaging optics a cold stop at the far field of a horn (or its image) could be used to terminate spillover to a low temperature, thus recovering spatial resolution while maintaining high on-axis aperture efficiency.

An example of the use of multi-mode horns (Figure 4.24) is in the two highest frequency channels of the high frequency instrument (HFI) on the European Space Agency Planck satellite (which has mapped the anisotropies of the cosmic microwave background radiation) [77]. There the requirement for high sensitivity and similar resolution to the lower frequency HFI bands (with single mode horns), while still requiring precise control of beam pattern characteristics, was an ideal application for the multi-mode horn

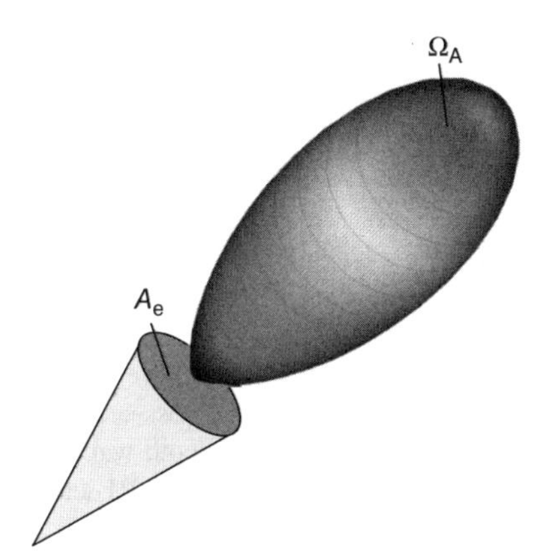

Figure 4.24 Antenna beam pattern.

systems [76]. Thus, since the Planck focal plane horns directly illuminated the Gregorian telescope secondary reflector, the requirement for very low edge taper and spillover levels meant that the horn aperture areas had to be increased, reducing the spatial resolution of the system as indicated above but increasing the on-axis gain and beam size on the sky by a large factor, thus increasing single pixel sensitivity to both point and extended sources.

Such multi-mode horn antennas are now also being proposed for other far-infrared systems, where increased throughput is also required but when also the coupling to the source is controlled via a cold absorbing stop in the far-field of the horn or its image (see e.g., Ref. [78]). In this case, for imaging applications, the horn aperture can be made significantly smaller than the point spread function, thus giving high aperture efficiency on-axis, but with good control of the power coupled off-axis into the beam.

In general, the modeling of a waveguide-fed multi-mode horn antenna system can be efficiently undertaken using a scattering matrix mode matching approach. For an oversized system a number of independent waveguide modes are excited at the entrance port (rather than one coherent field excited, as in a single mode system). These modes are scattered at any discontinuities encountered in the waveguide as well as in the horn antenna (which can be modeled as a set of short waveguide sections of varying radii) [43]. For cylindrically symmetric horns circular waveguide modes are used as the basis set for the scattering matrix formulism. In general, for circular waveguides these modes are degenerate in the sense that there are two orthogonal modes for each independent solution to the waveguide equation, where one mode can be transformed into the other degenerate partner by an appropriate rotation with respect to the axis of the guide (except for the $n = 0$ modes). The actual coupling to a source or the detector then depends on the polarization characteristics of the individual mode of the degenerate pair. As well as this, modes can only scatter for the case of circularly symmetric discontinuities into other modes of the same azimuthal order (same dependence on φ) and degeneracy type, which makes for efficient computation. Thus, when only the fundamental mode TE_{11} mode propagates in the waveguide feeding the horn (single mode case) then mode scattering in the horn can only occur to azimuthal order $n = 1$ modes, whereas, in a multi-mode horn, in general, modes of azimuthal order $n \neq 1$ also propagate, depending of course on their cut-off properties in the waveguide. An advantage of the scatter matrix approach is that it clearly allows shaped horns to be modeled as well as conical horns, since both types of horn are readily represented as a cascade of very short sections of waveguide of varying radius in a staircase-like profile that approximates the actual smooth conducting wall of the horn.

If the cylindrical waveguides and horns are corrugated (as was the case in the Planck example discussed above) then hybrid HE and EH modes (combinations of TE and TM fields) [43, 79] become the effective true modes of propagation and these can be derived from the scattering matrices using singular value decomposition (SVD) techniques. The actual propagation in the horn as well as in the waveguide can still be simulated as a cascade of short section circular waveguides representing the waveguide grooves and fins, with scattering between the corresponding TE and TM modes occurring at each corrugation. Other combinations are also clearly possible, such as an over-moded smooth-walled guide feeding a multi-mode corrugated horn but care must be taken in such a case to match the smooth waveguide to the corrugated horn such that there is not significant reflection at the throat of the horn and that non-ideal hybrid modes are not generated (e.g., slow wave modes).

In the case of pyramidal or shaped horns of rectangular cross-section then the appropriate basis set of modes are the TE and TM rectangular waveguide modes. These are less convenient than the circular waveguide modes in that modes at any symmetric discontinuity can generally scatter to a greater range of higher order modes, which leads to a slower computation. There is also no mode degeneracy as is the case for circular waveguides so more individual modes have to be considered for waveguides of the same cross-sectional area, although for symmetric square cross-section geometries clearly the symmetry would allow more efficient computation. Such shaped smooth-walled multi-mode square cross-section horns are being proposed and investigated as feeds for the SAFARI detector array on the proposed SPICA space telescope for far-infrared (THz) astronomy [80].

The prediction of the beam pattern for a multi-mode horn waveguide system can be simplified for the case where it feeds a bolometric detector mounted in a quasi-blackbody integrating cavity. In this case reciprocity can be applied and, to a reasonable approximation, we can assume all waveguide modes at the waveguide entrance port are excited independently of each other with no phase relationship between them. Thus, many modes will contribute incoherently to the beam on the sky produced by the horn and the transmission $[S_{21}]$ scattering matrix for the waveguide horn system contains the information about how each input independent mode at the entrance cavity port of the waveguide is scattered in being transmitted to the horn aperture. The far-field beam pattern of the horn is then given by summing in quadrature of the beam patterns of these individual scattered input modal fields at the horn aperture. An SVD analysis of the transmission $[S_{21}]$ matrix for the system will allow the number of independent modes of propagation to be determined, as this is given by the number of non-zero singular values. These Schmidt field independent channels for power transmission are given by the left and right singular vectors U and V [81] and act as a useful basis set for the system as a whole, from which, for example, the far-fields of the horn can be very efficiently calculated.

It is also possible to include an optical system feeding the multi-mode horn in the scattering matrix formalism, using, for example, Gaussian beam modes to represent propagation in free space and scattering at any stops and aberrating optical components [82]. If the detector is included in the description of the system then, viewing the source as the input port, the $[S_{11}]$ matrix now gives information about which modes are absorbed and which modes are reflected at the horn aperture (or the entrance port to a complete optical-horn system).

In a multi-mode infrared optical system the radiation field will be partially spatially coherent. It is possible to compute the scattering matrices for the complete multi-mode (i.e., partially coherent) optics–horn–waveguide system without reference to the source or detector, so that the waveguide (or an optical stop) acts as the mode filter that limits the throughput of the system. The power transmitted by the system can be expressed in terms of the independent Schmidt fields of this matrix which at one port couples to the source and at the other port to the detector. Each of these true component modes of propagation, although having no fixed phase relationship with other modes after propagation, will of course propagate in the system according to the laws of diffraction, and the total field intensity at the output plane will be given by the sum of the intensities of the component modes (fields add in quadrature). Geometrical optics can even be employed in highly over-moded systems as an efficient approach. However, in the long-wavelength limit systems tend to be at most few-moded and an approach incorporating diffraction techniques and scattering matrices is necessary. For partially coherent fields within the modal formulism propagation of any potential field can be very elegantly described in terms of coherence matrices. These track the evolution of the mutual coherence function [81, 83]. Over-moded horn antennas can then be used to couple to partially coherent source fields that have propagated through the optical system [75, 76, 79] and the power detected by the systems predicted. A number of papers by Withington have further developed the modal formulism in terms of partial coherence that is extremely powerful, especially for optical noise calculation in lossy systems, for example, see Ref. [84].

4.3.7 Modeling Techniques for THz Propagation in THz Systems

In this chapter we have discussed how quasi-optical beams can be modeled by decomposing an assumed source field into modes, each a solution of the paraxial wave equation. Propagation to the next optical surface then simply involved recombining scaled modes with an appropriate mode-dependent phase-slippage term included. We have used Gaussian beam modes but Gabor modes and plane waves are also commonly used. The approximations made with this technique are the same as those made when using Fresnel diffraction integrals to propagate a field. A plane-wave analysis has the significant advantage that it is does not have to be limited to paraxial fields.

Of course, electromagnetic fields are vector fields and here we have considered only one component. In practice, this scalar approach is sufficient when considering narrow-angle paraxial fields. To rigorously analyze wide-angle beams or effects such as scattering of power between co- and cross-polar components other more rigorous modeling techniques, such as physical optics, should be used. In general, optical modeling is concerned with calculating the electromagnetic field over a surface when the field, or currents, over some other surface is known. Techniques such as the method of moments attempt to calculate the current distribution over a surface precisely, but in practice approximations often have to be made. In the physical optics approximation, when a field is incident upon an aperture it is assumed that the field over the opaque region is zero, and the field over the transparent region is the same as if no aperture were there. This is reasonable when the radius of curvature of the reflector is many wavelengths, but is not valid at an edge. The geometrical theory of diffraction is often used in addition to estimate the effect of aperture edges. Propagation of the field onto the next optical component requires diffraction integrals to be calculated for each field point. This can be significantly more computationally demanding than the paraxial mode method discussed above.

At optical wavelengths, away from boundary shadows and abrupt changes in intensity, energy can be considered to be transported along curves, or rays, that obey geometrical laws. This ray-tracing (geometrical optics) can also be accurate for systems that are highly over-moded. In the THz regime, however, the wavelength is typically an appreciable fraction of component size, and systems tend to be at most few-moded, so diffraction cannot be neglected.

4.4 High Frequency Electric Characterization of Materials

As described by Eqs. (4.13) and (4.14), the classical approach for the electric characterization of materials is given by a frequency invariant, complex permittivity $\varepsilon = \varepsilon' - \mathrm{j}\varepsilon''$ and also a constant real conductivity σ. The permittivity accounts for the polarization phenomena inside the material, related to bound electrons in the molecules, while the conductivity models the behavior of free charges under the effect of electric fields. In this way, insulators ($\sigma = 0$) and conductors ($\sigma \to \infty$) are just the extreme cases where one of those phenomena is clearly dominating. For terahertz and higher frequencies this approach becomes invalid. In general, as the frequency is increased polarization and conduction phenomena start to depend strongly on the frequency, becoming simultaneously relevant and difficult to differentiate. In this situation the electrical properties of materials are better characterized by just a complex permittivity function in terms of the frequency or, equivalently, a complex conductivity one.

As an illustration of the limit of simple electric characterizations of materials, consider a metallic surface illuminated by a normally incident harmonic plane wave. At low frequencies the reflected electric field is always in phase opposition with respect to the incident one (additionally, for ideal conductors the amplitude is equal, actually canceling the total field on the surface). This corresponds to the metal electrons, that is, the induced currents, instantly following the electric field (this is just a statement of Ohm's law). However, at some point when the frequency is increased, the response time of the electrons becomes comparable to the period of the incident field, delaying the induced currents. As a result, at infrared or optical frequencies the reflection coefficient phase is always less than 180°, and clearly a real conductivity is not enough to model the metal.

A rigorous characterization of the complex permittivity should be based on the physics of the material electronic band structure. However, from a practical point of view such models are often too complicated, or restricted to a particular material. The alternative approach is the use of simpler phenomenological models based on classical mechanics, where the charges behave either as free particles, or bound particles that may oscillate as resonators. A description of the most usual models, as is the purpose of the following sections, starts with the free-electron Drude model, and continues to Lorentz–Drude and Brendel–Bormann models, that improve and extend the first one by introducing the interactions of bound charges as resonators. The main drawback of this type of model is the need for an experimental characterization of the model parameters for each material. Fortunately, an extensive knowledge base for many materials is already available [85].

4.4.1 Drude Model

The Drude model is based on the assumption that the material is composed of a lattice of fixed ions and a gas of free electrons. The electrons are accelerated by an applied electric field suffering a diffusive motion, while the only interactions that produce scattering are collisions between moving electrons and ions or other fixed particles. The behavior of the system is fully characterized by the lifetime of the free electrons between collisions τ, or the relaxation time [86]. It should be noted that this model neglects any other interactions, such as long-range interactions between free electrons and bound electrons in the ions, between free electrons themselves, or electron–phonon interactions. In spite of its simplicity, the Drude free-electron model provides a good approximation for metal conductivity frequencies for metals with a relaxation frequency $1/\tau$ below the far-infrared range of the spectrum, where other effects, such as interband transitions, are produced. Additionally, the model is not limited to metals: nonconductors show a free-electron type of behavior at sufficiently high frequencies. Finally, as stated before, it is the basis for other more refined models that overcome its limitations.

The derivation of the material electrical properties using the Drude model is rather straightforward. Consider the average momentum $\langle\vec{p}\rangle$ of the system electrons. Without excitation, the relaxation time τ is the relationship between the momentum itself and its variation, $\mathrm{d}\langle\vec{p}\rangle/\mathrm{d}t = \langle\vec{p}\rangle/\tau$, since it is the time constant of the exponential decay of $\langle\vec{p}\rangle$. However, when an electric field $\vec{E}$ is applied the equation of motion is non-homogeneous,

$$\frac{\mathrm{d}\langle\vec{p}\rangle}{\mathrm{d}t} = -\frac{\langle\vec{p}\rangle}{\tau} - e\vec{E} \tag{4.158}$$

where $-e$ is the electron charge. Of course, instead of $\langle\vec{p}\rangle$, it is more reasonable to formulate the problem in terms of the current density, which is directly proportional as their relationship shows,

$$\vec{J} = -\frac{n_\mathrm{e} e}{m}\langle\vec{p}\rangle \tag{4.159}$$

where n_e is the density of electrons (or just carriers in a more general case) and m is the electron mass.

For a DC problem, the derivative in Eq. (4.158) vanishes, and therefore the conductivity σ_0, that relates the applied field and current density through Ohm's law $\vec{J} = \sigma_0\vec{E}$, results in

$$\sigma_0 = \frac{n_\mathrm{e} e^2 \tau}{m}. \tag{4.160}$$

And in the case of an alternating sinusoidal field with frequency $f = \omega/2\pi$, the conductivity is a frequency-dependent complex function,

$$\sigma(\omega) = \frac{\sigma_0}{1 + \mathrm{j}\omega\tau} = \frac{\sigma_0}{1 + \omega^2\tau^2} - \mathrm{j}\omega\tau\frac{\sigma_0}{1 + \omega^2\tau^2} \tag{4.161}$$

An alternative to the complex conductivity is the dielectric function or complex electrical permittivity, $\varepsilon(\omega) = \varepsilon'(\omega) - \mathrm{j}\varepsilon''(\omega)$, defined in Eq. (4.13), related to the former as $\sigma(\omega) = \mathrm{j}\omega\varepsilon(\omega) = \mathrm{j}\omega\varepsilon_0\varepsilon_\mathrm{r}(\omega)$. Using this relationship, the dielectric function results as

$$\varepsilon_\mathrm{r}(\omega) = \varepsilon_\infty - \frac{\omega_\mathrm{p}^2}{\omega(\omega - \mathrm{j}\Gamma)} \tag{4.162}$$

where Γ is the damping frequency (that is, $\Gamma = 1/\tau$) and ω_p is the plasma frequency,

$$\omega_\mathrm{p}^2 = \frac{n_\mathrm{e} e^2}{\varepsilon_0 m}. \tag{4.163}$$

Note that an additional term ε_∞, representing the high frequency dielectric function, has been introduced. This term, not present in the basic free-electron model described above, models the electronic polarizability limits [87]: for very high frequency excitation fields there are no polarization mechanisms, since the time required for any polarization process is much larger than the excitation period. As a result, ε_∞ is expected to be similar to the vacuum permittivity, $\varepsilon_\infty \approx 1$ [85]. In fact, the term $\varepsilon_\infty = 1$ is required in order to verify the Kramers–Kronig relations [88, 89]. From a practical point of view, Eq. (4.163) is more conveniently expressed as its real and imaginary parts,

$$\varepsilon_r'(\omega) = \varepsilon_\infty - \frac{\omega_\mathrm{p}^2}{\omega^2 + \Gamma^2} \tag{4.164}$$

and

$$\varepsilon_r''(\omega) = \frac{\omega_\mathrm{p}^2 \Gamma}{\omega(\omega^2 + \Gamma^2)}. \tag{4.165}$$

In summary, the Drude model requires the characterization of two coefficients, namely the relaxation frequency Γ and the plasma frequency ω_p. These coefficients have been characterized experimentally and are readily available for many materials, such as noble metals (Cu, Au, Ag) or other metals (Ti, Pb, Cr, Al) [90, 91].

4.4.2 Lorentz–Drude Model

The validity of the Drude model ends at frequencies at which bound-electron interactions appear. This interband part of the spectrum can be modeled using the Lorentz model, again a classical model. Now, the bound charges behave as harmonic oscillators. Using the same formulation from Eq. (4.158), the equation of a bound charge is

$$\frac{\mathrm{d}\langle\vec{p}\rangle}{\mathrm{d}t} = -\frac{\langle\vec{p}\rangle}{\tau} - e\vec{E} - C\langle\vec{r}\rangle \tag{4.166}$$

where the new term $C\langle\vec{r}\rangle$ is the harmonic restoring force that is nonzero when the charges are displaced from its equilibrium state, C being a constant containing the total equivalent charge of the nucleus and other electrons, and $\vec{r}$ the displacement vector, so that $\vec{p} = m\mathrm{d}\vec{r}/\mathrm{d}t$. This constant should be substituted by a tensor in an anisotropic material. An approach similar to that followed for the Drude model produces the following contribution of the Lorentz oscillator to the dielectric function,

$$\varepsilon_k(\omega) = \frac{\omega_k^2}{(\omega_k^2 - \omega^2) - \mathrm{j}\omega\Gamma_k} \tag{4.167}$$

where ω_k is its resonant frequency and Γ_k its bandwidth. Of course, in general, with multiple electron layers and interactions, a single resonator may not be enough to model the interband behavior of the material. If a set of n oscillators is taken into account in order to extend the Drude model, the dielectric function results as

$$\varepsilon_r(\omega) = \varepsilon_\infty - \frac{\omega_\mathrm{p}^2}{\omega(\omega - \mathrm{j}\Gamma)} + \sum_{k=1}^{n} \frac{f_k \omega_k^2}{(\omega_k^2 - \omega^2) - \mathrm{j}\omega\Gamma_k}. \tag{4.168}$$

where each f_k is the oscillator strength or weight factor, that measures the fraction of oscillators having the corresponding resonance. The previous equation constitutes the Lorentz–Drude model for the dielectric function, where it is easy to differentiate the intraband or free-electron effects (the Drude model term) and the interband or bound-charge effects (the Lorentzian terms). Note that this function still verifies the Kramers–Kronig relations.

The price to pay for the improved model is that the Lorentz–Drude requires three coefficients to be characterized for each oscillator, apart from the Drude ones: ω_k, Γ_k, and f_k. It should be noted that the bound-charge part of the dielectric constant can be composed of oscillators that correspond to interband transition energies, but this is not necessarily true. Additional oscillators can be added in order to model other effects, particularly absorption ones [85, 87]. In this case it is not easy to assign a physical interpretation to the oscillator parameters that represent just frequencies, widths, and heights of spectral lines with Lorentzian shape. It should be noted that this shape may result in a poor modeling of certain interband absorption characteristics, as is the case for noble metals [85].

4.4.3 Brendel–Bormann Model

In the case of oscillators used to model absorption, it has been shown that Gaussian line shapes give, in general, better results than the described Lorentzian ones. For the same parameters (frequency, width at half-maximum and height), the Lorentzian line function produces more broadening that may result in excessive absorption far from its central frequency [85, 92]. The model proposed by Brendel and Bormann [87] replaces each Lorentzian broadening function by the superposition of an infinite number of oscillators with a Gaussian profile. After some mathematical manipulations, the Brendel–Bormann dielectric function results in [85, 93],

$$\varepsilon_r(\omega) = \varepsilon_\infty - \frac{\omega_\mathrm{p}^2}{\omega(\omega - \mathrm{j}\Gamma)} + \sum_{k=1}^{n} \chi_k(\omega) \tag{4.169}$$

where each function χ_k is

$$\chi_k(\omega) = \mathrm{j}\sqrt{\frac{\pi}{8}}\frac{f_k\omega_k^2}{a_k\sigma_k}\left[w\left(\frac{a_k-\omega_k}{\sqrt{2}\sigma_k}\right) + w\left(\frac{a_k+\omega_k}{\sqrt{2}\sigma_k}\right)\right] \tag{4.170}$$

where $a_k^2 = \omega(\omega - \mathrm{j}\Gamma_k)$, choosing the root a_k with positive imaginary part, and

$$w(z) = \mathrm{e}^{-z^2}\mathrm{erfc}(z). \tag{4.171}$$

Note that the parameters that define this dielectric function have been defined previously except for σ_k. Indeed, each shape function now is determined by two width parameters, Γ_k and σ_k, that in combination also determine its profile. The extreme cases are $\Gamma_k \approx 0$ for purely Gaussian profiles of width σ_k, and $\sigma_k \approx 0$ for Lorentzian ones. The intermediate values define a family of profiles that can model flexibly both oscillations due to interband transition energies, and absorption bands. Again, the Kramers–Kronig relations are fulfilled. Parametrizations of the Brendel–Bormann model of a variety of materials based on measurements are available [85, 93].

4.5 Propagation in Free Space

4.5.1 Link Budget

In most applications, we are interested in how well a THz signal is transmitted from an antenna to another (communications) or back to the same antenna after reflection or scattering (radar). Let us assume two antennas in free space and their separation r is large compared to the far-field distance $r = 2D^2/\lambda$, where D is the largest dimension of the antenna perpendicular to the direction of signal arrival. The main beams of the antennas are pointing to each other and their polarizations are matched.

If the power accepted by the transmitting antenna, P_t, were transmitted isotropically, the power density at a distance of r would be

$$S_\mathrm{isot} = \frac{P_t}{4\pi r^2} \tag{4.172}$$

The maximum power density produced by the transmitting antenna having a gain of G_t is

$$S = \frac{G_t P_t}{4\pi r^2} \tag{4.173}$$

The power available from the receiving antenna is its effective area A_r times the power density of the incoming wave:

$$P_r = A_r S \tag{4.174}$$

Taking into account that the antenna gain is related to its effective area by $G_r = 4\pi A_r/\lambda^2$, we obtain

$$P_r = G_t G_r \left(\frac{\lambda}{4\pi r}\right)^2 P_t \tag{4.175}$$

where G_r is the gain of the receiving antenna.

In practice, many factors may reduce the power received, for example, errors in the pointing of the antennas, polarization mismatch, and loss due to the atmosphere. Losses due to impedance mismatches also have to be taken into account: the power accepted by the transmitting antenna is smaller than the available power of the transmitter, and the power accepted by the receiver is smaller than the available power from the receiving antenna. If the total loss of signal besides the free space loss is L_p, then the received signal power is

$$P_r = G_t G_r \left(\frac{\lambda}{4\pi r}\right)^2 \frac{1}{L_p} P_t \tag{4.176}$$

4.5.2 Atmospheric Attenuation

All weather phenomena occur in the troposphere that is the lowest part of the atmosphere. The troposphere extends on the poles to about 9 km and on the equator to about 17 km. The troposphere is inhomogeneous and constantly changing. Temperature, pressure, humidity, precipitation, and so on, affect the propagation of radio waves, and especially strongly the THz waves. In the troposphere the propagating electromagnetic fields attenuate, scatter, refract, and reflect; and noise originating from the atmosphere is added to the signal.

THz waves need a line-of-sight (LOS) path, and as the antenna beams at THz are narrow, the propagation resembles that in free space. The radio link hops at THz are short because of high attenuation due to precipitation and gas molecules. This attenuation due to atmospheric absorption and scattering can be divided into two parts: attenuation due to clear air and attenuation due to precipitation (raindrops, hail, and snow flakes) and fog. Attenuation of the clear air is mainly due to resonance states of oxygen (O_2) and water vapor (H_2O) molecules. An energy quantum corresponding to the resonance frequency may change the rotational energy state of the gas molecule. When the molecule absorbs an energy quantum, the molecule is excited to a higher energy state. When it returns back to equilibrium, that is, drops back to the ground state, it radiates the energy difference but not necessarily in the same direction and at the same frequency because returning to equilibrium may happen in smaller energy steps. Under pressure the molecular emission lines have a wide spectrum. Therefore, the energy quantum is lost from the propagating wave, and, for the same reason the atmosphere is always noisy at all frequencies.

Resonance frequencies of oxygen are 57–63 and 119 GHz, and of water vapor, for example, 183, 325, 557, 752, and 988 GHz. The amount of oxygen is always nearly constant but that of water vapor is highly variable with time and location. The attenuation constant due to water vapor is directly proportional to the absolute amount of water vapor, which is a function of temperature and humidity. Figure 4.25 presents the clear-air attenuation versus frequency. Between the resonance frequencies there are so-called spectral windows centered at frequencies, for example, 140, 220, and 340 GHz. At resonance frequencies the attenuation may be tens to thousands of decibels per kilometer.

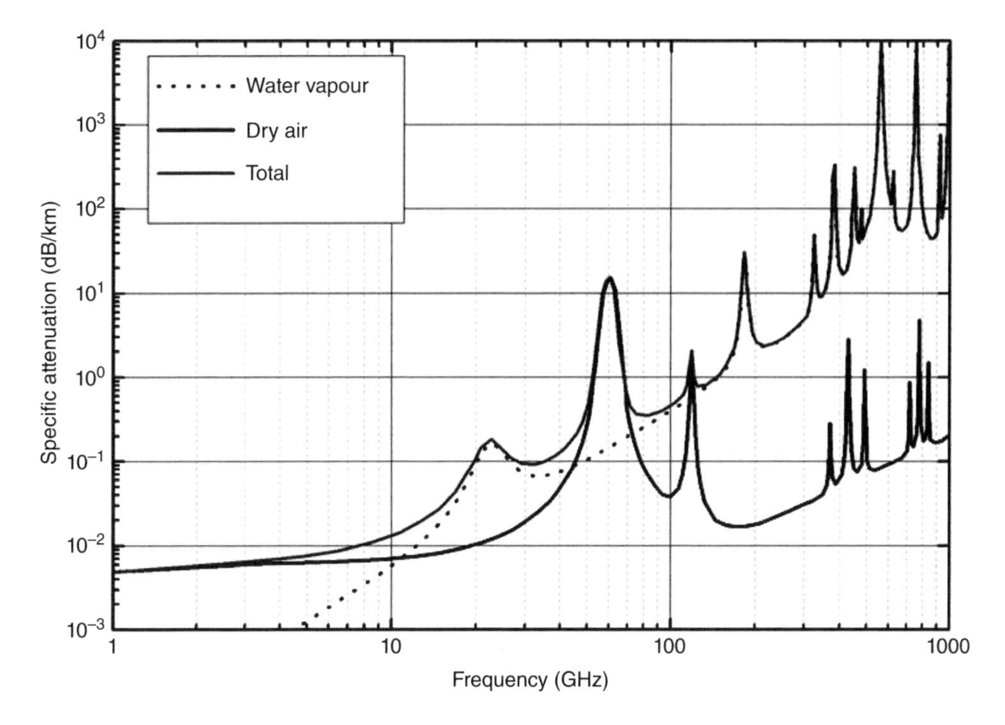

Figure 4.25 Attenuation in clear atmosphere versus frequency from 1 GHz to 1 THz ($T = 15\,°C$, pressure = 1013 hPa, water vapor density 7.5 g/m^3).

Rain and fog increase atmospheric attenuation. Attenuation of rain is mainly due to scattering: the electric field of the radio wave polarizes the water molecules of the raindrop, and then the raindrop acts like a small electric dipole radiating isotropically. At 100 GHz in moderate rain (5 mm/h) attenuation is 3 dB/km, but in pouring rain (150 mm/h) it is above 40 dB/km. At 1 THz the attenuation values are nearly the same. The attenuation constant due to fog and clouds is nearly directly proportional to the amount of water, and, for example, a fog of 0.1 g/m attenuates about 0.5 dB/km at 100 GHz and 6 dB/km at 1 THz.

Turbulence in the troposphere may also cause scintillation, that is, random changes in amplitude and phase of the wave as it propagates via different routes due to turbulence (the refractive index may vary strongly over short distances).

References

[1] Räisänen, A.V. and Lehto, A. (2003) *Radio Engineering for Wireless Communication and Sensor Applications*, Artech House, Boston, MA.

[2] Collin, R.E. (2001) *Foundations for Microwave Engineering*, 2nd edn, IEEE Press, New York.

[3] Collin, R.E. (1991) *Field Theory of Guided Waves*, IEEE Press, New York.

[4] Pozar, D.M. (1998) *Microwave Engineering*, 2nd edn, John Wiley & Sons, Inc., New York.

[5] Ramo, S., Whinnery, J., and van Duzer, T. (1965) *Fields and Waves in Communication Electronics*, John Wiley & Sons, Inc., New York.

[6] Benson, F.A. and Tischer, F.J. (1984) Some guiding structures for millimetre waves. *IEE Proceedings*, **131**, 429.

[7] Tischer, F.J. (1979) Experimental attenuation of rectangular waveguides at millimeter wavelengths. *IEEE Transactions on Microwave Theory and Techniques*, **27**, 31.

[8] Hesler, J.L., Kerr, A.R., Grammer, W., and Wollack, E. (2007) Recommendations for waveguide interfaces to 1 THz. Proceedings of the 18th International Symposium on Space Terahertz Technology, 2007, pp. 100–103.

[9] Räisänen, A.V., Choudhury, D., Dengler, R.J. *et al.* (1993) A novel split-waveguide mount design for millimeter and submillimeter wave frequency multipliers and harmonic mixers. *IEEE Microwave and Guided Wave Letters*, **3**, 369.

[10] McGrath, W.R., Walker, C., Yap, M., and Tai, Y.-C. (1993) Silicon micromachined waveguides for millimeter-wave and submillimeter-wave frequencies. *IEEE Microwave and Guided Wave Letters*, **3**, 61.

[11] Deslandes, D. and Wu, K. (2006) Accurate modeling, wave mechanisms, and design considerations of a substrate integrated waveguide. *IEEE Transactions on Microwave Theory and Techniques*, **54**, 2516.

[12] Zenneck, J. (1907) Über die Fortpflanzung ebener electromagnetischen Wellen längs einer ebenen Leiterfläche und ihre Beziehung zur drahtlosen Telegraphie. *Annalen der Physik*, **23**, 846.

[13] Sommerfeld, A. (1899) Über die Fortpflanzung elektrodynamischer Wellen an längs eines Drahtes. *Annalen der Physik und Chemie*, **67**, 233.

[14] Goubau, G. (1950) Surface waves and their application to transmission lines. *Journal of Applied Physics*, **21**, 1119.

[15] Sihvola, A., Qi, J., and Lindell, I.V. (2010) Bridging the gap between plasmonics and Zenneck waves. *IEEE Antennas and Propagation Magazine*, **52**, 124.

[16] Baragwanath, A.J., Gallant, A.J., and Chamberlain, J.M. (2013) Terahertz plasmonic structures, in *Terahertz Spectroscopy and Imaging*, Springer Series in Optical Sciences, vol. **171** (eds K.E. Peiponen, J.A. Zeitler, and M. Kuwata-Gonokami), Springer, pp. 539–568.

[17] Hoffmann, R.K. (1987) *Handbook of Microwave Integrated Circuits*, Artech House, Norwood, MA.

[18] Ghione, G. and Naldi, C. (1984) Analytical formulas for coplanar lines in hybrid and monolithic MICs. *Electronics Letters*, **20**, 179.

[19] Ghione, G. and Naldi, C.U. (1987) Coplanar waveguides for MMIC applications: effect of upper shielding, conductor backing, finite-extent ground planes, and line-to-line coupling. *IEEE Transactions on Microwave Theory and Techniques*, **35**, 260.

[20] Sharma, A.K. (ed) (1989) Special issue on quasi-planar millimeter-wave components and subsystems. *IEEE Transactions on Microwave Theory and Techniques*, **37** (2), pp. 273–437.

[21] Chang, K. (ed) (1989) *Handbook of Microwave and Optical Components, Microwave Passive and Antenna Components*, vol. **1**, John Wiley & Sons, Inc., New York.

[22] Lioubtchenko, D., Tretyakov, S., and Dudorov, S. (2003) *Millimeter-Wave Waveguides*, Kluwer Academic Publishers.

[23] Generalov, A.A., Lioubtchenko, D.V., and Räisänen, A.V. (2013) Dielectric rod waveguide antenna at 75 – 1100 GHz. Proceedings of the 7th European Conference on Antennas and Propagation, EuCAP2013, Gothenburg, Sweden, April 8–12, 2013, pp. 533–536.

[24] Yoneyama, T. and Nishida, S. (1981) Nonradiative dielectric waveguide for millimeter-wave integrated circuits. *IEEE Transactions on Microwave Theory and Techniques*, **29**, 1188.

[25] Lioubtchenko, D.V., Dudorov, S., Mallat, J. *et al.* (2001) Low loss sapphire waveguides for 75–110 GHz frequency range. *IEEE Microwave and Wireless Components Letters*, **11**, 252.

[26] Pousi, J.P., Lioubtchenko, D.V., Dudorov, S.N., and Räisänen, A.V. (2010) High permittivity dielectric rod waveguide as an antenna array element for millimeter waves. *IEEE Transactions on Antennas and Propagation*, **58**, 714.

[27] Pousi, P., Lioubtchenko, D., Dudorov, S., and Räisänen, A.V. (2008) Dielectric rod waveguide travelling wave amplifier based on AlGaAs/GaAs heterostructure. Proceedings of the 38th European Microwave Conference, Amsterdam, The Netherlands, October 28–30, 2008, pp. 1082–1085.

[28] Marcatili, E.A.J. (1969) Dielectric rectangular waveguide and directional coupler for integrated optics. *Bell System Technical Journal*, **48**, 2071.

[29] Marcuse, D. (1974) *Theory of Dielectric Optical Waveguides*, Academic Press, New York.

[30] Goell, J.E. (1969) A circular-harmonic computer analysis of rectangular dielectric waveguides. *Bell System Technical Journal*, **48**, 2133.

[31] Veselov, G.I. and Voronina, G.G. (1971) To the calculation of open dielectric waveguide with rectangular cross-section. *Radiofizika*, **14**, 1891 (in Russian).

[32] Sudbo, A.S. (1992) Why are accurate computations of mode fields in rectangular dielectric waveguides difficult? *Journal of Lightwave Technology*, **10**, 418.

[33] Goldsmith, P.F. (1998) *Quasioptical Systems: Gaussian Beam Quasioptical Propagation and Applications*, Wiley-IEEE Press, Hoboken, NJ.

[34] Lesurf, J.C.G. (1990) *Millimetre-Wave Optics, Devices and Systems*, Taylor & Francis, London.
[35] Martin, D.H. and Bowen, J.W. (1993) Long-wave optics. *IEEE Transactions on Microwave Theory and Techniques*, **41**, 1676.
[36] Siegman, A.E. (1986) *Lasers*, University Science Books, Sausalito, CA.
[37] Murphy, J.A. (1987) Distortion of a simple Gaussian beam on reflection from off-axis ellipsoidal mirrors. *International Journal of Infrared and Millimeter Waves*, **8**, 1165.
[38] Dragone, C. (1978) Offset multireflector antennas with perfect pattern symmetry and polarization discrimination. *AT&T Technical Journal*, **57**, 2663.
[39] Mizugutch, Y., Akagawa, M., and Yokoi, H. (1976) Offset dual reflector antenna. IEEE Antennas and Propagation Society International Symposium, 1976, pp. 2–5.
[40] Ruze, J. (1966) Antenna tolerance theory – a review. *Proceedings of the IEEE*, **54**, 633.
[41] Neu, J., Krolla, B., Paul, O. *et al.* (2010) Metamaterial-based gradient index lens with strong focusing in the THz frequency range. *Optics Express*, **18**, 27748.
[42] Savini, G., Ade, P., and Zhang, J. (2012) A new artificial material approach for flat THz frequency lenses. *Optics Express*, **20**, 25766.
[43] Olver, A.D., Clarricoats, P.J.B., Shafai, L., and Kishk, A.A. (1994) *Microwave Horns and Feeds*, IEE Electromagnetic Waves Series, vol. **39**, The Institution of Engineering and Technology.
[44] Clarricoats, P.J.B. and Olver, A.D. (1984) *Corrugated Horns for Microwave Antennas*, IEE Electromagnetic Waves Series, vol. **18**, Peter Peregrinus Ltd, London.
[45] Wylde, R.J. (1984) Millimetre-wave Gaussian beam-mode optics and corrugated feed horns. *Proceedings of IEE, Part H*, **131**, 258.
[46] O'Sullivan, C. and Murphy, J.A. (2012) *Field Guide to Terahertz Sources, Detectors, and Optics*, SPIE Press, Bellingham, WA.
[47] Herschel team (2011) HIFI Observers Manual, Chapter 2, HIFI instrument description, Section 2.3 Focal plane unit, http://herschel.esac.esa.int/Docs/HIFI/html/ch02s03.html (accessed 30 November 2014).
[48] Goldsmith, P.F. (1991) Perforated plate lens for millimeter quasi-optical systems. *IEEE Transactions on Antennas and Propagation*, **39**, 834.
[49] Beruete, M., Cia, M.N., Campillo, I. *et al.* (2007) Quasioptical polarizer based on self-complementary sub-wavelength hole arrays. *IEEE Microwave and Wireless Components Letters*, **17**, 834.
[50] Rulf, B. (1988) Transmission of microwaves through layered dielectrics—theory, experiment, and application. *American Journal of Physics*, **56**, 76.
[51] Finn, T., Trappe, N., and Murphy, J.A. (2008) Gaussian-beam mode analysis of reflection and transmission in multilayer dielectrics. *Journal of the Optical Society of America A*, **25**, 80.
[52] Ade, P.A.R., Pisano, G., Tucker, C., and Weaver, S. (2006) A review of metal mesh filters. *Proceedings of SPIE*, **6275**, 62750U-1-15.
[53] O'Shea, D.C., Suleski, T.J., Kathman, A.D., and Prather, D.W. (2004) *Diffractive Optics – Design, Fabrication, and Test*, Tutorial Texts in Optical Engineering, vol. **TT62**, SPIE Press, Bellingham, WA.
[54] Dammann, H. and Klotz, E. (1977) Coherent optical generation and inspection of two-dimensional periodic structures. *Optica Acta: International Journal of Optics*, **24**, 505.
[55] Murphy, J.A., O'Sullivan, C., Trappe, N. *et al.* (1999) Modal analysis of the quasi-optical performance of phase gratings. *International Journal of Infrared and Millimeter Waves*, **20**, 1469.
[56] May, R., Murphy, J.A., O'Sullivan, C. *et al.* (2008) Gaussian beam mode analysis of phase gratings. *Proceedings of SPIE*, **6893**, G8930.
[57] Johnson, E. and Abushagur, M. (1995) Microgenetic-algorithm optimization methods applied to dielectric gratings. *Journal of the Optical Society of America A*, **12**, 1152.
[58] Lavelle, J. and O'Sullivan, C. (2010) Beam shaping using Gaussian beam modes. *Journal of the Optical Society of America A*, **27**, 350.
[59] Kim, H., Yang, B., and Lee, B. (2004) Iterative Fourier transform algorithm with regularization for the optimal design of diffractive optical elements. *Journal of the Optical Society of America A*, **21**, 2353.
[60] Durnin, J. (1987) Exact solutions for non-diffracting beams. I. The scalar theory. *Journal of the Optical Society of America A*, **4**, 651.
[61] Monk, S., Arlt, J., Robertson, D. *et al.* (1999) The generation of Bessel beams at millimetre-wave frequencies by use of an axicon. *Optics Communication*, **170**, 213.
[62] Trappe, N., Mahon, R., Lanigan, W. *et al.* (2005) The quasi-optical analysis of Bessel beams in the far infrared. *Infrared Physics and Technology*, **46**, 233.

[63] Yurchenko, V.B., Ciydem, M., Gradziel, M. *et al.* (2014) Double-sided split-step mm-wave fresnel lenses: design, fabrication and focal field measurements. *Journal of the European Optical Society - Rapid Publications*, **9**, 14007.

[64] Zhang, J., MacDonald, K.F., and Zheludev, N.I. (2012) Controlling light-with-light without nonlinearity. *Light: Science and Applications*, **11**, 18.

[65] Iwanaga, M. (2012) Photonic metamaterials: a new class of material for manipulating light waves. *Science and Technology of Advanced Materials*, **13**, 053002.

[66] Ma, Y., Chen, Q., Grant, J. *et al.* (2011) A terahertz polarization insensitive dual band metamaterial absorber. *Optics Letters*, **36**, 945.

[67] Peng, X.Y., Wang, B., Lai, S. *et al.* (2012) Ultra-thin multi-band planar metamaterial absorber based on standing wave resonances. *Optics Express*, **20**, 27756.

[68] Shen, X., Cui, T.J., Zhao, J. *et al.* (2011) Polarization independent wide-angle triple-band metamaterial absorber. *Optics Express*, **19**, 9401.

[69] Tao, H., Bingham, C.M., Pilon, D. *et al.* (2010) A dual band terahertz metamaterial absorber. *Journal of Physics D: Applied Physics*, **43**, 225102.

[70] Alaee, R., Farhat, M., Rockstuhl, C., and Lederer, F. (2012) A perfect absorber made of a graphene micro-ribbon metamaterial. *Optics Express*, **20**, 28017.

[71] Fallahi, A. and Perruisseau-Carrier, J. (2012) Design of tunable biperiodic graphene metasurfaces. *Physical Review B*, **86**, 195408.

[72] Eleftheriades, G.V., Omar, A.S., Katehi, L.P.B., and Rebeiz, G.M. (1994) Some important properties of waveguide junction generalised scattering matrices in the context of the mode matching technique. *IEEE Transactions on Microwave Theory and Techniques*, **42**, 1896.

[73] Murphy, J.A., Colgan, R., O'Sullivan, C. *et al.* (2001) Radiation patterns of multi-moded corrugated horns for far-IR space applications. *Infrared Physics and Technology*, **42**, 515.

[74] Murphy, J.A. and Padman, R. (1991) Radiation patterns of few moded horns and condensing lightpipes. *Infrared Physics*, **31**, 291.

[75] Padman, R. and Murphy, J.A. (1991) Radiation patterns of scalar horns. *Infrared Physics*, **31**, 441.

[76] Murphy, J.A., Peacocke, T., Maffei, B. *et al.* (2010) Multi-mode horn design and beam characteristics for the Planck satellite. *Journal of Instrumentation*, **5**, T04001.

[77] Tauber, J., Mandolesi, J.-L., Puget, T. *et al.* (2010) Planck pre-launch status: the Planck mission. *Astronomy & Astrophysics*, **520**, A1.

[78] Trappe, N., Bracken, C., Doherty, S. *et al.* (2011) Modelling of horn antennas and detector cavities for the SAFARI instrument at THz frequencies. *Proceedings of SPIE*, 7938, Terahertz Technology and Applications IV, 79380J.

[79] Gleeson, E., Murphy, J.A., Maffei, B. *et al.* (2005) Corrugated waveguide band edge filters for CMB experiments in the far infrared. *Infrared Physics & Technology*, **46**, 493.

[80] Trappe, N., Bracken, C., Doherty, S. *et al.* (2012) Optical modeling of waveguide coupled TES detectors towards the SAFARI instrument for SPICA. *Proceedings of SPIE*, **8452**, 84520L.

[81] Withington, S. and Murphy, J.A. (1998) Modal analysis of partially coherent submillimeter-wave quasi-optical systems. *IEEE Transactions on Antennas and Propagation*, **46**, 1651.

[82] Trappe, N., Murphy, J.A., Withington, S., and Jellema, W. (2005) Gaussian beam mode analysis of standing waves between two coupled corrugated horns. *IEEE Transactions on Antennas and Propagation*, **53**, 1755.

[83] Withington, S., Yassin, G., and Murphy, J.A. (2001) Dyadic analysis of partially coherent submillimeter-wave antenna systems. *IEEE Transactions on Antennas and Propagation*, **49**, 1226.

[84] Withington, S., Hobson, M.P., and Campbell, E.S. (2004) Modal analysis of astronomical bolometric interferometers. *Journal of the Optical Society of America A*, **21**, 1988.

[85] Rakić, A.D., Djurišić, A.B., Elazar, J.M., and Majewski, M.L. (1998) Optical properties of metallic films for vertical-cavity optoelectronic devices. *Applied Optics*, **37**, 5271.

[86] Dressel, M. and Scheffler, M. (2006) Verifying the Drude response. *Annalen der Physik*, **15**, 535.

[87] Brendel, R. and Bormann, D. (1992) An infrared dielectric function model for amorphous solids. *Journal of Applied Physics*, **71**, 1.

[88] Fujiwara, H. (2007) Appendix 5: Kramers-Kronig relations, in *Spectroscopic Ellipsometry: Principles and Applications*, Wiley Online Library, pp. 357–360.

[89] Johnson, P.B. and Christy, R.W. (1972) Optical constants of the noble metals. *Physical Review B*, **6**, 4370.

[90] Ordal, M.A., Long, L.L., Bell, R.J. *et al.* (1983) Optical properties of the metals Al, Co, Cu, Au, Fe, Pb, Ni, Pd, Pt, Ag, Ti, and W in the infrared and far infrared. *Applied Optics*, **22**, 1099.

[91] Rakić, A.D. (1995) Algorithm for the determination of intrinsic optical constants of metal films: application to aluminum. *Applied Optics*, **34**, 4755.
[92] Kim, C., Garland, J., Abad, H., and Raccah, P. (1992) Modeling the optical dielectric function of semiconductors: extension of the critical-point parabolic-band approximation. *Physical Review B*, **45**, 11749.
[93] Djurišić, A.B., Fritz, T., and Leo, K. (2000) Modelling the optical constants of organic thin films: impact of the choice of objective function. *Journal of Optics A: Pure and Applied Optics*, **2**, 458.

5

Principles of THz Direct Detection

Elliott R. Brown[1] and Daniel Segovia-Vargas[2]
[1]*Wright State University, Dayton, OH, USA*
[2]*Universidad Carlos III de Madrid, Departamento Teoría de la Señal y Comunicaciones, Leganés-Madrid, Spain*

This chapter is a comprehensive review of the physical principles and engineering techniques associated with room-temperature THz direct detectors. It starts with the basic detection mechanisms, both rectifying and thermal. It then addresses the noise mechanisms using both classical and quantum principles, and the THz coupling using impedance-matching and antenna-feed considerations. Fundamental analyses of the noise mechanisms are provided because of lack of coverage in the popular literature. All THz detectors can then be described with a common performance formalism based on two metrics: noise-equivalent power (NEP) and noise-equivalent temperature difference (NETD). A summary is then provided of the typical, and best, detector performance reported to date, including the various complementary metal oxide semiconductor (CMOS)-based detectors. The chapter concludes with a comparison of the best THz detector results at room temperature, and demonstrates that none of the existing detector types are operating anywhere near fundamental theoretical limits, so there is room for significant performance advances.

5.1 Detection Mechanisms

The THz region is often considered as the transition between the radio frequency (RF) bands below 100 GHz and the popular infrared (IR) bands at wavelengths beyond ~12 μm. As such, it has a tendency to exhibit passive components and devices having both RF and IR characteristics. A good example is so-called "quasi-optical" components for which free-space radiation is coupled to passive components using mirrors, lenses, and other traditional optical components. Another good example is THz "direct" detectors which by themselves convert THz incoming radiation *directly* to a low-frequency baseband. In the THz region the direct detector is almost always a power-to-voltage or power-to-current converting device. That is, it is a device that puts out a voltage or current in proportion to the incoming power. There are many examples of such devices, but the most popular for room-temperature operation are field detectors and thermal detectors.

Semiconductor Terahertz Technology: Devices and Systems at Room Temperature Operation, First Edition.
Edited by Guillermo Carpintero, Luis Enrique García Muñoz, Hans L. Hartnagel, Sascha Preu and Antti V. Räisänen.

Field detectors, such as Schottky diodes, respond to the THz electric field and usually generate an output current or voltage through a quadratic term in their current–voltage characteristic. Thermal detectors are often composite devices consisting of a THz absorbing element coupled to some temperature-sensitive transducer. The THz absorber is generally isolated thermally from the environment so that the absorbed THz power raises the temperature of both the absorbing layer and the attached transducer. If the transducer is a thermistor, the detector is called a bolometer. If the transducer is a temperature-dependent capacitor based on pyroelectricity, it is called a pyroelectric detector. These are the two types that we will focus on in this chapter.

The thermistor in a bolometer is generally a device that displays a large change in resistance for a small change in temperature. Such behavior is displayed by certain transition metals and semi-metals. Furthermore, such materials can also be fabricated as THz resistors which effectively absorb THz radiation via the AC form of Ohm's law. Therefore, it is possible to construct bolometers such that the absorbing function and the thermistor function are carried out by the same element. Similarly, if the temperature-dependent material in a pyroelectric capacitor also absorbs THz radiation sufficiently, then the capacitor serves both functions as well. This type of bolometer and pyroelectric are referred to, henceforth, as "integrated" direct detectors.

A more recent THz detection mechanism is based on plasma-wave propagation and rectification in the 2D channel of high-electron-mobility transistors (HEMTs). The basic idea is that the 2DEG (two dimensional-electron-gas) of HEMTs or cryogenically cooled metal-oxide semiconductor field effect- transistors (MOSFETs) can support plasma wave propagation even at frequencies well above the unity-current cut-off frequency (f_τ), which is generally limited by the electron transit time between source and drain. For the typical sheet concentrations in these devices ($\sim 10^{12}/\text{cm}^2$), the plasma-wave group velocity is roughly one order of magnitude higher than that of the free electrons. So a change of gate-source voltage V_{GS} affects the drain-source voltage roughly 10-times faster than in conventional field effect transistors (FETs). And, as described below, the drain-source voltage is a nonlinear function of V_{GS}, so an AC gate-source voltage yields a DC term in the drain-source potential, which is the basis for THz detection.

5.1.1 E-Field Rectification

Simply stated, a square-law field rectifier detector is a device or circuit that takes an input signal and produces an output that is proportional to its square,

$$X_{\text{OUT}} = AX_{\text{IN}}^2, \tag{5.1}$$

where X_{OUT} could be a current or voltage and A is a proportionality constant. The utility of such a device in detection is twofold: (i) rectification, or producing a DC output from a purely AC input or (ii) frequency down-conversion (i.e., mixing). Rectification is most easily understood in the special case of a coherent signal

$$X_{\text{IN}}(t) = B\cos(\omega t + \varphi) \tag{5.2}$$

If put through a square-law detector, the output becomes

$$X_{\text{OUT}}(t) = \tfrac{1}{2} AB^2\{1 + \cos[2(\omega t + \varphi)]\}. \tag{5.3}$$

So if the square-law detector is followed up by a low pass filter (i.e., a time-domain integrator) with integration time $\tau \gg 2\pi/\omega$, then the second term will not contribute and the output of the filter will be

$$X_{\text{OUT}} = \tfrac{1}{2} AB^2 \equiv \Re P_{\text{S}} \tag{5.4}$$

where $\Re$ is a constant and P_{S} is the average absorbed input signal power. In the language of sensor theory, $\Re$ is usually called the *responsivity* and is measured either in A/W or V/W. Presumably, A is known, so that B can be determined from the X_{OUT} DC term. Square-law detectors are preferred in RF systems over cubic and other possible detectors partially for this reason. The proportionality constant A is dependent only on the detector characteristics and not on the power level, at least up to a point where saturation and higher-order effects begin to occur. A single calibration of A in the "small-signal" regime is all that is required to use the square-law detector over a wide range of input power.

The paradigm THz rectifier is the metal-semiconductor barrier (Schottky) diode. Originally this was fabricated as a sharpened metal wire, or "whisker", contacting a bare semiconductor such as galena (PbS). Later, with the advent of microelectronic fabrication, the whisker was replaced by a lithographically defined metal contact deposited on the virgin semiconductor surface and the resulting current–voltage characteristic was found to be well described by the highly asymmetric (with respect to bias voltage V), generic expression

$$I = I_S[\exp\left(eV/\alpha k_B T\right) - 1] \tag{5.5}$$

where I_S is the reverse-bias saturation current, k_B is the Boltzmann constant, T is the contact temperature, and α (typically $1.0 < \alpha < 2.0$) is a physically-rich parameter called the ideality factor. With the importance of rectifying detectors to radar receivers during World War II, detailed studies were made of the small-signal responsivity of such devices and the following general expression was found for the short-circuit current responsivity [1]:

$$\mathfrak{R}_I = \frac{1}{2}\frac{\mathrm{d}^2I/\mathrm{d}V^2}{\mathrm{d}I/\mathrm{d}V} \tag{5.6}$$

When applied to the ideal metal-semiconductor diode Eq. (5.5), Eq. (5.6) becomes

$$\mathfrak{R}_I = e/\left(2\alpha k_B T\right) \tag{5.7}$$

At 300 K and assuming $\alpha = 1$, this calculates to $\mathfrak{R}_I = 19.3$ A/W, which represents a maximal short-circuit responsivity. Notice that this result and Eq. (5.7) in general is independent of bias voltage and also the area of the diode to the extent that α is independent of these operating parameters. If the rectifier is terminated in a load impedance (e.g., high input-impedance amplifier) much greater than $\mathrm{d}V/\mathrm{d}I$, then the diode is self-loading and Eq. (5.7) can be written as an open circuit voltage responsivity. For the ideal metal-semiconductor diode, this becomes

$$\mathfrak{R}_V \approx \mathfrak{R}_I \cdot \mathrm{d}V/\mathrm{d}I = \frac{1}{2}\frac{\mathrm{d}^2I/\mathrm{d}V^2}{\left(\frac{\mathrm{d}I}{\mathrm{d}V}\right)^2} = eR_j/\left(2\alpha k_B T\right) \tag{5.8}$$

where $R_j \equiv \mathrm{d}V/\mathrm{d}I$ is the differential resistance. Since Schottky rectifiers are usually biased with constant voltage, Eq. (5.8) is re-written as:

$$\mathfrak{R}_V = [\exp\left(-eV/\alpha k_B T\right)]/\left(2I_S\right) \tag{5.9}$$

This clearly slows that the voltage responsivity does depend on bias voltage and device area (through I_S).

Figure 5.1 shows a family of I–V curves (a) and the corresponding $\mathfrak{R}_V$–V curves (b) for a relatively recent type of Schottky diode consisting of a single-crystal ErAs semimetallic contact on a quaternary semiconductor (InGaAlAs) having a tunable bandgap via the Al fraction [2]. This exhibits a number of interesting effects. For 0% Al, we have a metal contact on $In_{0.53}Ga_{0.47}As$, a semiconductor having a band gap of approximately 0.75 eV at room temperature. The I–V curve looks very linear (i.e., "ohmic"). As the Al fraction is increased, we begin to see more unilateral current flow and an increase in the peak electrical responsivity in Figure 5.1b. For an Al fraction of 0.6, the peak responsivity approaches the maximum theoretical value of 19.3 A/W at approximately 0.05 V bias. For a further increase in Al fraction, the maximum remains about the same, but the voltage at which the maximum occurs increases to approximately 0.1 V. Both maximum voltages are way lower than what one would measure in the

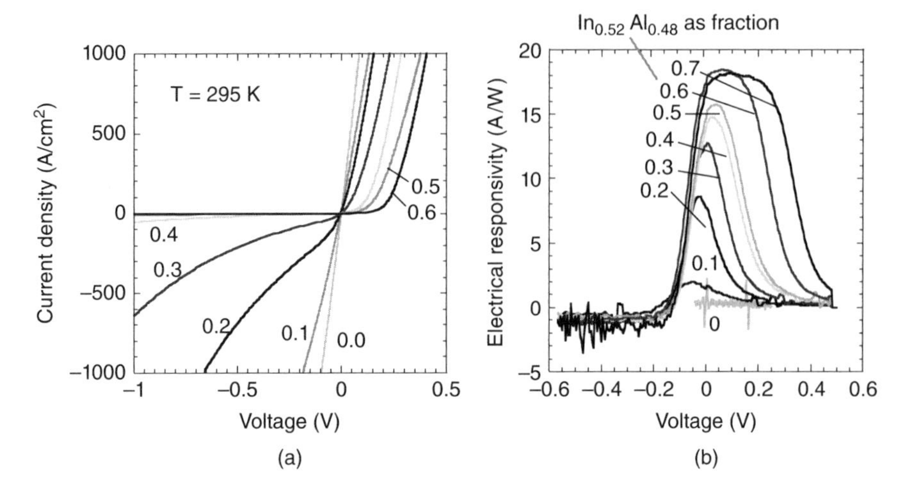

Figure 5.1 (a) Room-temperature *I–V* curves and (b) corresponding $\Re$–*V* curves for a family of Schottky diodes fabricated from ErAs (semimetal)–InGaAlAs (semiconductor) junctions and with various In, Ga, and Al fractions [2]. (©2007 IEEE. Reprinted, with permission, from IEEE Microwave Magazine vol. 8, p. 54 (2007).). See plate section for color representation of this figure.

more common metal–GaAs or metal–Si Schottky diodes, largely because the band gap is much less. In fact, the material combinations in Figure 5.1 were developed most for fabricating THz Schottky rectifiers operating with zero bias. This is easier to implement in practice than biased rectifiers, and because there is no bias current they display no 1/*f* noise of any type.

It would appear that Schottky rectifiers are relatively simple, and this is why historically they have usually been the first detector type to be applied in practically all emerging frequency bands, starting with microwaves during World War II. As interest grew in THz frequency-operation during the 1960s and beyond, it was realized that reliable, planar-fabricated Schottkys could be realized simply by reducing their area to sub-micron dimensions to reduce the junction capacitance commensurately. However, by so doing, the resistive part of the impedance also increases, such that the $R_D C$ time constant remains roughly the same, at least near zero bias. Hence, the Schottky THz impedance is still reactive, making it difficult to conjugately match to any realistic THz coupling element, such as a waveguide or planar antenna. This topic will be addressed later in the chapter.

5.1.2 *Thermal Detection*

Two of the three detection technologies explored in this chapter belong to the class of detectors known as thermal detectors. The basic idea behind all such detectors is that the incident radiation causes a temperature change in some detection element, similar to a thermometer, which can then be measured electronically. While this sounds like a relatively simple concept, its realization can range from relatively simple in the case of the bolometric detectors discussed below, to intricate as in the Golay cell.

5.1.2.1 Bolometers

A bolometer is a type of thermal detector that acts like a thermistor, that is, a temperature-dependent resistor, for which the temperature change is created by the incident THz radiation. A schematic diagram of

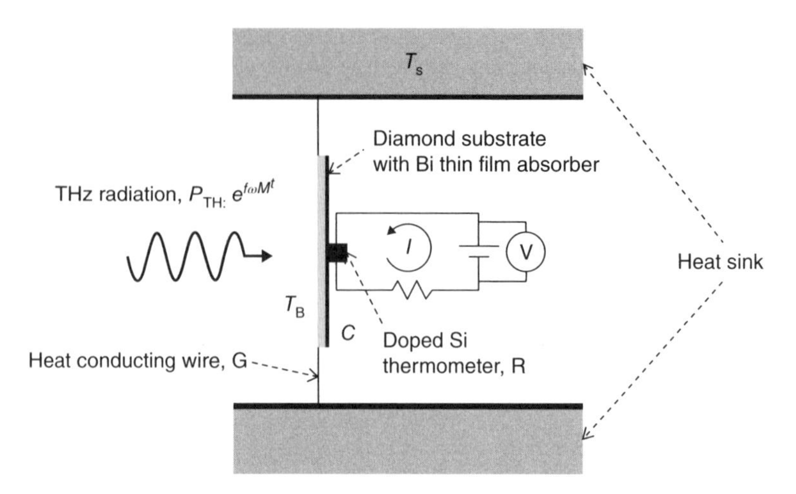

Figure 5.2 Schematic of a bolometer detector. (With kind permission of Springer Science+Business Media, Yun-Shik Lee, Principles of Terahertz Science and Technology. New York: Springer, 2009, pp. 151–154.)

bolometer functionality is shown in Figure 5.2. Bolometers typically include two functions: (i) a radiation coupling component, such as an integrating cavity or planar antenna on a substrate, and (ii) a thermistor element to transduce the change of temperature to a change of electrical resistance, which is then measured in a read-out circuit via an applied bias. For room-temperature operation, the thermometer is often made of bismuth or some semimetal having relatively low electrical conductivity compared to the transition or noble metals, but a similar thermal coefficient of resistance, α, defined by

$$R(T) = R_0(1 + \alpha \Delta T) \tag{5.10}$$

R_0 is the resistance at some reference temperature T_0 (usually 300 K) and $\Delta T = T - T_0$ with ΔT generally $\ll T_0$. If a two-terminal resistor is made of such material and then biased with a constant-current of I_B, a temperature change of the resistor will lead to a change of terminal voltage given by

$$\Delta V = V(R) - V\left(R_0\right) = I_B\left(R - R_0\right) = I_B R_0 \alpha \Delta T \tag{5.11}$$

Since ΔT is assumed much less than T_0, then linear thermal transport theory can be applied by which $\Delta T \approx \Delta P/G_0$, where ΔP is the change in average power dissipated in R that creates the temperature rise, and G_0 is the thermal conductance at T_0 that links the resistor to the environment (e.g., heat sink). Hence we can write

$$\Delta V \approx I_B R_0 \alpha \Delta P/G \tag{5.12}$$

And the electrical responsivity is given by

$$\Re_V \equiv \Delta V/\Delta P \approx I_B R_0 \alpha / G_0 \tag{5.13}$$

This is a much more complicated responsivity than for Schottky rectifiers in that all four parameters depend on the fundamental bolometer material characteristics, and three of them (I_B, R_0, and G_0) also depend on the size, and operational conditions. This gives bolometers a very broad design space, but also introduces trade-offs that limit the room-temperature sensitivity to values well below those of high-quality rectifiers once the requirements of efficient coupling of radiation and limited current bias are factored in.

According to Eq. (5.13) the current bias wants to be as high as possible to maximize the responsivity. But in practice it is limited by heating of the bolometer element to the point of thermal runaway. To see this we can rewrite Eq. (5.10) with R defined in terms of the bias conditions ($R = V/I$) and ΔT determined entirely by the electrical bias power, $P = V \cdot I$, such that

$$R(T) = V/I = R_0 \left(1 + \alpha V \cdot I/G_0\right) \tag{5.14}$$

Solving Eq. (5.14) for V, we get

$$V = R_0 I/[1 - R_0 \alpha I^2/G_0] \tag{5.15}$$

Notice that the denominator can display a zero, meaning that if α is positive the voltage diverges at a current given by

$$I_{max} = [G_0/R_0 \alpha]^{1/2} \tag{5.16}$$

Such a singular expression is reminiscent of regenerative feedback in circuit theory. However, if α is negative, the voltage is stable, reminiscent of degenerative feedback. In either case, the phenomenon leading to Eq. (5.15) is called *electrothermal feedback*, and is evident in all types of bolometers. It is important to realize, however, that even in the case of degenerative electrothermal feedback, there is still a maximum bias current dictated by the rise in temperature of the bolometer with respect to the ambient. Just as in an electrical fuse, Joule heating will eventually overwhelm the thermal dissipation and cause the bolometer element to be damaged or even melt.

Thermal transport also dictates another important aspect of bolometer behavior – the temporal response. Assuming the bolometer has a heat capacity C that is much greater than the heat capacity of the thermal link to the environment, then a step increase in temperature of the bolometer will create a temperature change having a characteristic (natural exponential) time constant of

$$\tau \approx C/G. \tag{5.17}$$

For macroscopically large bolometers (>1 mm in all dimensions), this time constant tends to be prohibitively large at room temperature, often approaching 1 s. But fortunately, C scales with volume while G tends to scale with peripheral area, so an isotropic reduction in volume tends to reduce the time constant accordingly to the millisecond level or even less.

Bolometer size is also important in the THz region because of radiative coupling. In the IR region the radiation coupling is relatively simple because the thermistor function and radiation absorption can occur in the same detector element. But in the THz region, this is practically impossible because of the much longer wavelengths. A much better approach is to separate the thermistor and radiation coupling functions by coupling the bolometer element to some type of waveguide or antenna, be it planar, horn, or otherwise. To obtain a good THz impedance match, this then constrains R_0 in Eq. (5.11) to typical antenna and waveguide source impedances of order 100 Ω. This constraint can be mitigated by coupling the bolometer to an integrating cavity, for example, but such an approach is not conducive to proliferating the bolometers into 1D or 2D arrays – a very useful exercise if the detectors are to be used for spatial imaging.

A fundamental issue with bolometers is that the room-temperature values of α are rather small, tend to have the same absolute value for all metallic solids including semimetals, but can differ in sign. The common absolute value turns out to be $\alpha \sim 0.003$. This follows from the fact that all metals tend to have a bulk resistivity that depends linearly on T around 300 K, so that $(1/R)dR/dT = (1/\rho)\, d\rho/dT \approx 1/T$. Naturally, one is then led to consider other possible bolometer materials, especially semiconductors. And indeed, they can be engineered to have values of α 10 or more times higher than $1/T$ around 300 K. The problem with semiconductors, as will be emphasized later, is that their resistivity tends to be too high to couple efficiently to THz antennas of any type. To achieve a good impedance match, the semiconductor has to be physically large compared to the antenna feed dimension. Again, this is not an issue if the semiconductor bolometer is coupled through an integrating cavity or similar coupling structure, which becomes particularly advantageous in cryogenic operation ($\ll$100 K) where the α values in semiconductors get much higher than at room temperature. The resulting responsivity of a typical Si or Ge bolometer can be of the order of 10^7 V/W at liquid helium temperatures (4.2 K). And the external coupling can be maintained usefully high because of the integrating cavity. But suffice it to say that at room temperature, semiconductor-based bolometers have not been competitive with metallic ones, especially semimetallic ones, as will be addressed later in the chapter.

5.1.2.2 Pyroelectric Detectors

Pyroelectric detectors are also thermal detectors but they rely on a completely different physical phenomenon. Figure 5.3 shows a schematic of how a typical pyroelectric detector works and is implemented [1]. The detector is based on a crystal in which each unit cell has a built-in dipole moment pointing along a particular axis of the crystal. As shown in Figure 5.3a this creates a net electric polarization along with surface charge densities on opposite ends of the crystal that are normally canceled by internal charges. The polarization, however, is temperature dependent so a change in the temperature of the crystal will create a transient macroscopic polarization and surface charge while the internal bound charges are moving to a new equilibrium. If the crystal is cut precisely across the axis of polarization and inserted into a circuit, a polarization current will flow when this transient surface charge is created. One practical effect of this type of detection mechanism is that the detector is inherently incapable of responding to a steady radiation source, so the THz radiation must be modulated [4].

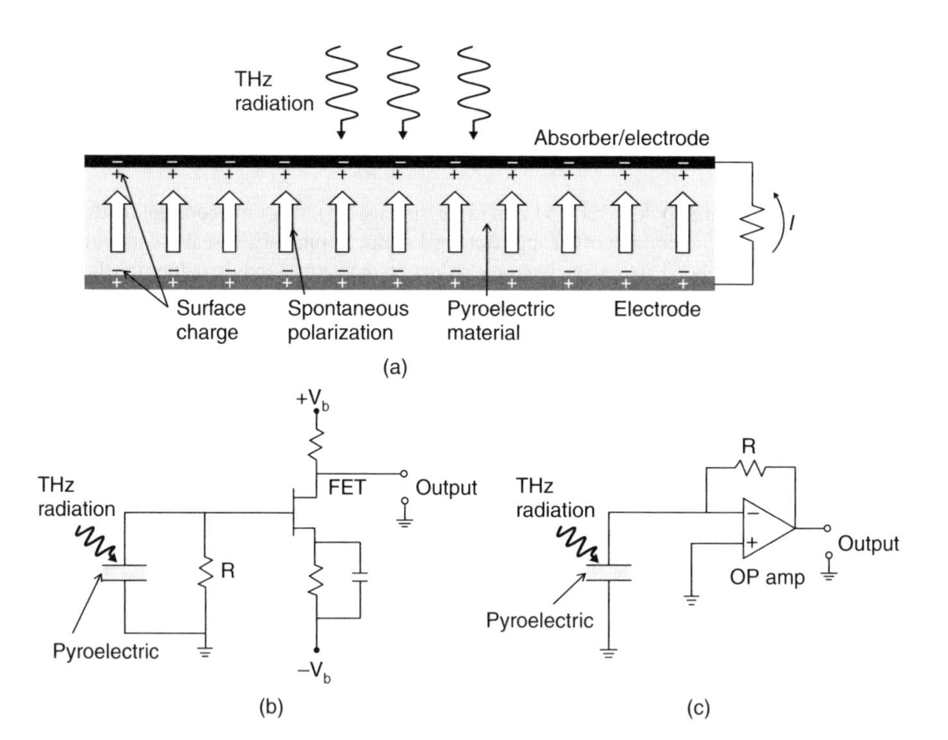

Figure 5.3 (a) Schematic diagram of a typical pyroelectric detector. Exemplary detection circuits for a pyroelectric detector coupled to, (b) voltage and (c) current amplifiers. (With kind permission of Springer Science+Business Media, Yun-Shik Lee, Principles of Terahertz Science and Technology. New York: Springer, 2009, pp. 151–154.)

A plot of the electric polarization of the crystal (P) versus temperature (T) shows that the polarization curves downward with increasing temperature, reaching zero at a point known as the Curie temperature. The slope of this graph at any given temperature, $p = \mathrm{d}P/\mathrm{d}T$, is known as the pyroelectric coefficient. The magnitude of the current created by a particular change in temperature is then

$$I = pA\frac{\mathrm{d}(\Delta T)}{\mathrm{d}t} \tag{5.18}$$

where ΔT is the change in temperature of the crystal and A is the area of the pyroelectric crystal.

In practice, the current generated is too small to be of much use by itself so some sort of amplification must be included as part of the detector circuitry. This circuitry is of two kinds depending on whether the pyroelectric crystal and the electrodes attached to it are coupled to a current amplifier or a voltage amplifier. Figure 5.3b,c shows exemplary circuits for both.

5.1.2.3 Golay (Pneumatic) Detectors

The Golay cell, invented as an IR detector [5], is sometimes known as a pneumatic detector because the radiation absorbing element heats a small amount of gas that expands as the temperature increases. This gas is contained in a small chamber that has an absorbing film on the front to efficiently transfer the thermal energy from the radiation to the gas, and a flexible mirror on the back that moves as the gas expands and contracts. In order to detect the change in volume of the chamber, an optical system consisting, in essence, of a light source, a line screen, a lens, and a photocell arranged in such a way that changes in the position of the mirror change the amount of light that is incident on the photocell and thus change the output voltage of the photocell. Like some other thermal detectors, the incident radiation must be modulated because of the AC readout circuit of the photocell.

Figure 5.4 shows the block diagram of a Golay cell. Since the Golay cell is broadband and operates by absorbing the incident radiation, it is very sensitive to thermal IR that peaks near 10 μm at room temperature. In order to detect THz, a window is usually used to block IR and pass THz. These windows can be made out of a variety of materials, but they are usually composed of high density polyethylene

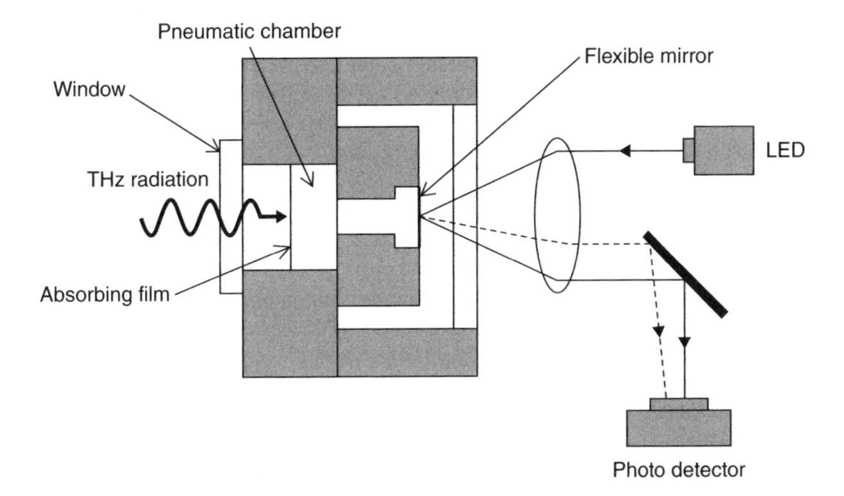

Figure 5.4 A schematic diagram of a Golay cell. (With kind permission of Springer Science+Business Media, Yun-Shik Lee, Principles of Terahertz Science and Technology. New York: Springer, 2009, pp. 151–154.)

(HDPE), or diamond [6] with some IR blocking material such as black polyethylene. Of course, the complexity of the Golay cell entails many sources of error and non-uniformity in the manufacturing process, Golay cells are fragile and their performance can be quickly damaged by one careless action, such as exposure to too much radiative power.

5.1.3 Plasma-Wave, HEMT, and MOS-Based Detection

As described first by Dyakonov and Shur [7], the high sheet concentration of free electrons in HEMTs and cryogenic MOSFETs can support plasma wave propagation between the source and drain, even at frequencies well above the free-electron transit-time cut-off of the device. This is because plasma waves are a collective excitation of the entire electron population, considered as a Fermi fluid. Such excitations generally propagate much faster than the free-electron velocity and do transfer energy at the appropriate group velocity. The remarkable prediction of Dyakonov and Shur was that the 2DEG in HEMTs and MOSFETs would create a superlinear response, and therefore a DC term, in the drain-source voltage in response to a purely AC voltage applied between the gate and source. This can be explained more precisely in the following way. The change in gate-source potential of a HEMT affects two parameters of the plasma-wave propagation: (i) the sheet charge density, through the relation $\rho = C_G \cdot V_{GS}$, where C_G is the gate capacitance and (ii) the plasma-wave velocity through the relation, $v = (eV_{GS}/m^*)^{1/2}$ (where e is the electron charge, m^* is the electron effective mass, and V_{GS} is the gate-source voltage relative to threshold). Hence, there is a superlinear dependence of the current density $J = \rho \cdot e \cdot v$ on V_{GS}. So if an incident THz signal is applied across the gate-source contacts, a non-zero DC term will occur in the drain-source current, which will manifest as a non-zero DC voltage if the drain-source termination has high-impedance. Experimental realization of this new mode of THz detection will be addressed below in Section 5.7.3.

The work of Dyakonov and Schur opened the possibility to use discrete FET transistors as terahertz detectors due to a non-resonant response of the 2DEG. The previous paragraph explained that the conventional low-frequency operation of a cold (non-biased source-drain junction) FET detection could be extended to frequencies well above the FET cut-off frequencies in the THz regime. The real large-scale interest in FET as THz detectors comes from the fact of using the well established Si technology of CMOS devices that can integrate in the same wafer the Si-CMOS devices for detection with the read-out electronics allowing room-temperature THz detectors to be absolutely competitive with the Schottky barrier diodes. The main advantages of Si-CMOS technology are their room-temperature operation, fast response times, and easy on-chip integration with read-out electronics leading to forward array fabrication. The detection mechanism can be explained either by plasma wave or by an alternative model of distributed resistive self-mixing that allows a more direct detector design.

Figure 5.5 shows two possible detection mechanisms for THz radiation. That in Figure 5.5a considers the antenna integration while that in Figure 5.5b only takes the FET resistive mixer consideration into account. It can be noted that both present an asymmetry between the source and drain needed to induce the photoresponse. In Figure 5.5a the asymmetry is achieved by special antenna connections in balanced antennas such as the bow-tie where the drain readsout the DC current once the balanced ports of the antenna are connected between the gate and source of the device. In Figure 5.5b the asymmetry is achieved by incorporating an external C_{gd} capacitor between the gate and drain that can take the DC current at the drain once the incident signal is applied between the gate and source. This model can be considered as a quasi-static approach since it is a simplified version of the detection mechanism that will be only valid for GHz frequencies. Unfortunately in "garden-variety" FETs having micron-scale gate length, the intrinsic gate-drain (and gate-source) capacitances are large enough to symmetrize the effect.

The simplified model presented in Figure 5.5b tries to explain the detection mechanism but is only valid at GHz frequencies. At THz frequencies, however, the capacitor C_{gd} no longer couples the signal between the gate and drain. Then, the approximation shown breaks down and has to be replaced by a non-quasi-static one, according to the theory presented by Dyakonov and Schur. This theory is based on

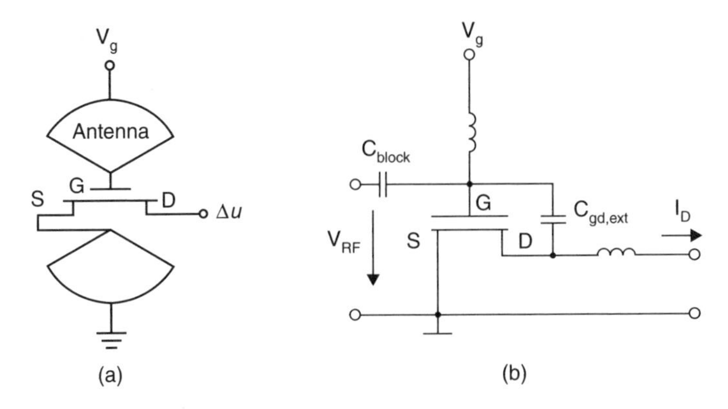

Figure 5.5 Configurations for plasma-wave detection in FETs. (a) THz configuration directly integrated with a bow-tie antenna (Knap 2013 [8]) and (b) lower-frequency equivalent circuit showing MOSFET detector coupled through an external gate-drain (C_{gd}) capacitor [9]. (Reproduced with permission from IEEE J. Solid-State circuits, vol. 44, p. 1968 . Copyright 2009, AIP Publishing LLC.)

the electron dynamics in the channel and on the continuity equation. For THz frequencies those equations can be simplified as follows

$$\frac{\partial n}{\partial t} + \frac{\partial (nv)}{\partial x} = 0 \tag{5.19}$$

$$v = -\mu \frac{\partial U}{\partial x} \tag{5.20}$$

where $n(x,t)$ is the electron density, μ the electron mobility, $v(x,t)$ the average drift velocity, and $U(x,t)$ the electrical potential along the channel. This reduced pair of differential equations (you can see Ref. [10] for the complete set) can be represented in the form of a RC transmission line that describes the non-resonant self-mixing. This can be seen in Figure 5.6a (taken from Ref. [10]) where the gray area of the waveguide indicates the channel of the transistor. For this non-resonant case, plasma oscillations are overdamped, as shown in Figure 5.6b [11].

Previous paragraphs state that THz radiation can be coupled between the gate and source of the FET; this THz wave simultaneously modulates the carrier density and its drift velocity, leading to nonlinearity and inducing a plasma wave that propagates in the channel. At cryogenic temperatures resonant plasma modes can be excited leading to narrowband tunable detection (for this case the product $\omega_0 \tau \gg 1$, where ω_0 is the fundamental plasma harmonic frequency and τ is the momentum relaxation time depending on the carrier mobility). At room temperature plasma waves are overdamped and the carrier density decays with a distance of the order of tens of nanometers. For this case the product $\omega_0 \tau \ll 1$ and the FET response is a smooth function of frequency and gate voltage and leads to non-resonant detection.

The applied voltage propagates through the channel from the source (left) to the drain (right) showing an exponential damping. It can be said that efficient resistive mixing takes place close to the source while the rest of the device acts as a distributed capacitance and parasitic series resistance. This makes this long-channel device be usable for direct power detection at THz frequencies. Then, the response signal can be given as

$$\Delta u = \frac{e \cdot u_a^2}{4m^* s^2} \left[1 - \frac{1}{\sinh^2 Q + \cos^2 Q} \right] \tag{5.21}$$

Where e is the electron charge, u_a is the THz voltage on the gate relative to the source, m^* is the effective mass of the electron, s is the plasma wave velocity, and Q is the ratio of the gate length to the characteristic length of the voltage decay from source to drain. After some algebraic operations, in Ref. [8], the

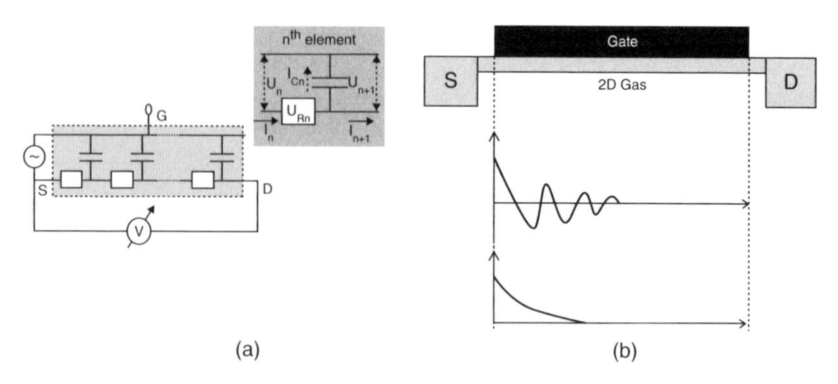

Figure 5.6 (a) Equivalent distributed circuit for the THz non-resonant self-mixing [10]. (Reproduced with permission from A. Lisauskas, U. Pfeiffer, E. Ojefors, P.H. Bolivar, D. Glaab and H. Roskos, "Rational design of high responsivity detectors of terahertz radiation based on distributed self-mixing in silicon field effect transistors," J. Appl. Phys. 105, 114511 (2009) . Copyright 2009, AIP Publishing LLC.) (b) Qualitative space dependence of electron oscillations excited by THz signal: upper part for $\omega\tau \gg 1$, bottom part for $\omega\tau \ll 1$ [11]. (Reproduced with permission from A. Gutin, V. Kachorovskii, A. Muraviev, and M. Shur "Plasmonic terahertz detector response at high intensities," J. Appl. Phys. Vol. 112, 014508 (2012) . Copyright 2012, AIP Publishing LLC.)

maximum value of the response signal is given as

$$\Delta u = \frac{e \cdot u_{\mathrm{a}}^{2}}{4\eta k_{\mathrm{B}} T} \tag{5.22}$$

Where η is a parameter depending on the sub-threshold slope of the channel conductivity and k_{B} is the Boltzmann constant. From the previous expression the responsivity that relates the detected signal to the THz power can be obtained. It can be seen that, ideally, the responsivity may become exponentially large below threshold. For actual conditions this is not really so since it depends on the antenna coupling and on the loading effects. Thus, as the input impedance of the read-out circuit becomes smaller, due to a simple dividing effect, the detected signal will decrease. A similar effect can be extracted from the imaginary part of the input impedance since a small capacitance can lead to large RC constants. In this way an increase in the modulation frequency can also lead to a decrease in the detected signal.

An alternate model for explaining the detection mechanism at lower frequency in CMOS technology can be given from Figure 5.5b [9]. This model has been chosen for simplicity since the results in the THz band will be qualitatively valid. Then, it can be seen that there is an extra gate to drain capacitor which serves to facilitate the self-mixing [12]. This capacitor allows simultaneous coupling of the input signal to the drain of the FET (acting as the "local oscillator" port) and to the gate (acting as the "RF port"). It can also be seen that a DC gate-bias V_{g} may be provided through an RF-choke. Thus, the voltages at each port can be written as

$$v_{\mathrm{gs}}(t) = v_{\mathrm{RF}}(t) + V_{\mathrm{g}} \tag{5.23}$$

$$v_{\mathrm{ds}}(t) = v_{\mathrm{RF}}(t) = V_{\mathrm{RF}} \sin \omega t \tag{5.24}$$

As the device is operated in the linear region, the drain current is obtained through multiplication of the voltage and the transconductance as

$$i_{\mathrm{ds}}(t) = v_{\mathrm{RF}}(t) \cdot g_{\mathrm{ds}}(t) \tag{5.25}$$

The transconductance, for the quasi-static case, can be written as

$$g_{\mathrm{ds}}(t) = \frac{W}{L} \cdot \mu \cdot C_{\mathrm{OX}} \cdot \left[\frac{v_{\mathrm{RF}}(t)}{2} + \left(V_{\mathrm{g}} - V_{\mathrm{thres}}\right) \right] \tag{5.26}$$

For the non-quasi-static case, the transconductance per unit length is written as

$$G_{ds}(v(x,t)) = \mu C_{\mathrm{OX}} W \left(v(x,t) - V_{\mathrm{th}}\right) \tag{5.27}$$

By taking the DC term in Eq. (5.26) it can be seen that the read-out current is given as

$$I_{\mathrm{ds}} = \frac{W}{L} \cdot \mu \cdot C_{\mathrm{OX}} \frac{V_{\mathrm{RF}}^2}{4} \tag{5.28}$$

The read-out voltage of the detector can be obtained by multiplying the DC current by the channel resistance. This can be written, whenever the drain-to-source voltage is in the linear region, as

$$V_{\mathrm{ds}} = \frac{I_{\mathrm{ds}}}{G_{\mathrm{ds}}} = \frac{V_{\mathrm{RF}}^2}{4\left(V_{\mathrm{g}} - V_{\mathrm{thres}}\right)} \tag{5.29}$$

From this expression the electrical (voltage) responsivity can be obtained as

$$\mathfrak{R}_V = \frac{V_{\mathrm{ds}}}{P_{\mathrm{in}}} = \frac{\dfrac{V_{\mathrm{RF}}^2}{4\left(V_{\mathrm{g}} - V_{\mathrm{thres}}\right)}}{\dfrac{V_{\mathrm{RF}}^2}{R_{\mathrm{in}}}} = \frac{R_{\mathrm{in}}}{4\left(V_{\mathrm{g}} - V_{\mathrm{thres}}\right)} \tag{5.30}$$

The maximum responsivity is obtained in the subthreshold region since the detection current is generated across a larger internal DC resistance. In Si CMOS technology, on-chip amplification close to the detector prevents external capacitive loading, avoiding both the reduction of the responsivity and the reduction of the bandwidth.

5.2 Noise Mechanisms

In statistical mechanics and then again in quantum mechanics, we learn that upon measurement every physical quantity, including radiation, is subject to fluctuations. The received power in sensor systems is generally very weak, typically orders-of-magnitude weaker than it is in communications systems. So an important issue is the "masking" of the signal by fluctuations in the power (i.e., the "noise") in the receiver. The "noise" is the totality of all the electronic and electromagnetic mechanisms, especially those in the detector itself. This section examines these noise mechanisms in the context of THz direct detectors operating at room temperature and in the ambient environment. A fundamental, physical viewpoint is adopted primarily because noise is one of the least understood topics in THz detectors and sensors as a whole. Therefore, this section could also serve for pedagogical purposes in academic courses or professional workshops.

5.2.1 Noise from Electronic Devices

Within every sensor system, particularly at the front end, are components that contribute significant noise to the detection process and therefore degrade the ultimate detectability of the signal. The majority of this noise usually comes from electronics, particularly the first device, which is often a mixer or direct detector. After this there is generally a low-level amplifier that contributes separate noise. The majority of noise from such devices falls into two classes: (i) thermal noise and (ii) shot noise. Thermal noise in semiconductors is caused by the inevitable fluctuations in voltage or current associated with the resistance in and around the active region of the device. This causes fluctuations in the voltage or current in the device by the same mechanism that causes resistance – the Joule dissipation that couples energy between charge carriers and electromagnetic fields. The form of the thermal noise is very similar to that for free-space blackbody radiation. And the Rayleigh–Jeans approximation is generally valid for THz radiation and room-temperature operation, so that the Johnson–Nyquist theorem applies. However, one must account for the fact that the device is coupled to a transmission-line circuit or similar, not to a free space mode, and the device may not be in equilibrium with the radiation as assumed by the blackbody model. These issues are addressed by Nyquist's generalized theorem [13]

$$V_{\mathrm{rms}} = [\langle \Delta V \rangle^2]^{1/2} = [4k_B T_D \mathrm{Re}\{Z_D\} \Delta f]^{1/2} \tag{5.31}$$

where T_D, Z_D, and Δf are the temperature, differential impedance, and bandwidth of the device. Even this generalized form has limitations since it is often difficult to define the temperature of the device if it is operating well away from thermal equilibrium.

Shot noise is a ramification of the device being well out of equilibrium, usually by virtue of electrical bias and some barrier blocking the flow of charge carriers. It is generally described as fluctuations in the current transmitted through the barrier caused by random changes in the rate of carriers incident on the opposite side. Because of the general importance of this phenomenon to THz detectors, and all radiative detectors for that matter, we will examine it in more detail than thermal noise with a similar physical basis as presented in Ref. [14]. From statistical kinetic theory, which generally applies well to charge carriers in non-degenerate semiconductors, we know that there is a mean time τ between collisions, and a mean rate of collisions $1/\tau$ for each carrier. We then assume that the probability of *having* a single collision between time t and $t + \delta t$ is exponential, $P_1(t) = \exp(-t/\tau) \cdot \delta t/\tau$ [15], which follows from the "relaxation time" approximation. This can be generalized to an ensemble of uncorrelated collisions to predict the probability of N collisions between 0 and time t,

$$P_N(t) = [(t/\tau)^N/N!] \cdot \exp(-t/\tau), \tag{5.32}$$

which is a form of the famous Poisson probability density function. To apply Eq. (5.32) to the shot-noise phenomenon, we think of $1/\tau$ – the collision rate R – also as the impinging rate on the barrier,

$$P_N(t) = [(Rt)^N/N!] \cdot \exp(-Rt). \tag{5.33}$$

Hence, $P_N(T)$ is to be thought of as the probability of N carriers incident on the barrier in time t. If we limit the ensemble to those carriers with enough kinetic energy to cross the barrier, Eq. (5.33) still applies with N constrained to be consistent with the ensemble-averaged current i through the barrier during measurement sampling time t_S:

$$\langle i \rangle = q \langle N \rangle / t_S \equiv I, \tag{5.34}$$

where q is the signed charge. And consistent with the ergodic hypothesis, this ensemble averaged current should equal the time-averaged current I over the same interval t_S.

The Poisson density function has the remarkable property that the variance is equal to the mean, that is,

$$\left\langle (N - \langle N \rangle)^2 \right\rangle \equiv \left\langle (\Delta N)^2 \right\rangle = \langle N \rangle \tag{5.35}$$

Applying this to Eq. (5.34) and given that t_S is deterministic, we can calculate the variance of the electrical current

$$\left\langle (i - \langle i \rangle)^2 \right\rangle \equiv \left\langle (\Delta i)^2 \right\rangle = q^2 \left\langle (\Delta N)^2 \right\rangle / t_S{}^2 = \left(q^2 / t_S{}^2 \right) \langle N \rangle = q \langle i \rangle / t_S, \tag{5.36}$$

which can also be thought of as the mean-square fluctuation of i during interval t_S. According to sampling theory, $1/t_S$ – the sampling rate – determines the effective bandwidth over which power can be measured unambiguously according to the Nyquist sampling relation, $B_{max} = 1/(2t_S)$, so that we can write

$$\langle (\Delta i)^2 \rangle = 2q\langle i \rangle \cdot B_{\max} \equiv 2qI \cdot B_{\max} \quad (5.37)$$

In real devices, t_S is usually dictated by an intrinsic time constant, such as the characteristic transit time of carriers across the barrier, or the device internal RC time in devices for which the transit time is relatively short. In this case, B is replaced by an effective bandwidth B_{eff} as determined by the longest of the internal characteristic times

$$\langle (\Delta i)^2 \rangle = 2qI \cdot B_{\text{eff}} \quad (5.38)$$

Equation (5.38) is the famous Schottky relation [16], discovered even before the Johnson–Nyquist form of thermal noise given above. While an approximation, it has withstood the test of time and been found to be accurate in many THz-detector devices, such as Schottky diodes. However, some quantum-effect devices, or devices that can store space charge in the barrier region, such as resonant tunneling diodes, have displayed a modified form of Eq. (5.38) given by [17]

$$\langle (\Delta i)^2 \rangle = 2q\Gamma \cdot I \cdot B_{\text{eff}} \quad (5.39)$$

where Γ is a numerical factor representing the degree to which the random Poissonian fluctuations of transmission times are modified by the transport or charge storage between the cathode (or emitter) and anode (or collector). If $\Gamma = 1$, there is no effect and the anode current has the "full" shot noise. When $\Gamma < 1$, the transport or charge storage reduces the fluctuations, often through some form of degenerative feedback mechanism, and the shot noise is said to be suppressed. When $\Gamma > 1$, the transport or charge storage increases the fluctuations, often through some form of regenerative feedback mechanism, and the shot noise is said to be enhanced.

5.2.2 *Phonon Noise*

By their physical nature, THz thermal detectors are inherently sensitive to fluctuations in the incident power, be it by radiative coupling (photons) or thermal conductance (phonons) to the environment. Such fluctuations get transduced into electrical noise in the same way that the average absorbed electromagnetic radiation gets transformed into a steady electrical output signal. It is therefore a fundamental matter to estimate what the power fluctuations are for a given thermal detector type in its environment. For THz thermal detectors, especially those operating at room temperature, the most important effect is power fluctuation via phonons.

To estimate the power spectrum of phonon fluctuations, we start with the same kinetic analysis as for carrier shot noise, but this time applied to a volume of phonons incident on a thermal link between the detector at temperature T and the bath. As with the charge carriers, we assume that the incident rate of phonons on the link fluctuates with Poissonian statistics. From fundamental solid-state physics, the mean energy of a phonon is $<U(\upsilon)> = h\upsilon/[\exp(h\upsilon/k_B T) - 1]$, where υ is the corresponding (atomic lattice) vibrational frequency and h and k_B are Planck's and Boltzmann's constants, respectively. A key part of the analysis is to assume the "classical limit" whereby $h\upsilon << k_B T$, such that $<U(\upsilon)> \approx k_B T$. While arguably valid at room temperature for solids having Debye temperatures <300 K, this assumption must certainly be corrected for thermal detectors operating at cryogenic temperatures (see below). Another part of the analysis is that the phonons that arrive at the link from the detector flow ballistically across the link, just like the charge carriers in the shot noise analysis. Even in the classical limit, we expect this number to be a function of υ and also of the detector volume V, both of which factor into the 3D acoustical-phonon density-of-states [18].

Given these assumptions, the average power in frequency interval $d\upsilon$ and volumetric element dV is given as mean energy per phonon times the total number of phonons per sampling time t_S

$$\langle P(\upsilon, V)\rangle = k_B T_D \langle N(\upsilon, V)\rangle / t_S \tag{5.40}$$

Given the Poissonian assumption about phonon transport rate, the mean-square fluctuation (i.e., variance) of power is given by

$$\left\langle [\Delta P(\upsilon, V)]^2 \right\rangle = \left(k_B T_D\right)^2 \left\langle [\Delta N(n, V)]^2 \right\rangle / t_S{}^2 = \left(k_B T_D\right)^2 \langle N(n, V)\rangle / t_S{}^2 \tag{5.41}$$

The total thermal power will be the integral over all frequencies and the specific detector volume:

$$\langle P\rangle = \int_0^{\upsilon_{max}}\int_0^{V} \langle P(\upsilon, V)\rangle \, d\upsilon dV = \left(k_B T_D / t_S\right) \int_0^{\upsilon_{max}}\int_0^{V} \langle N(\upsilon, V)\rangle \, d\upsilon dV \equiv G T_D \tag{5.42}$$

where G is the macroscopic thermal conductance between the detector and its environment. The corresponding total power fluctuation is,

$$\left\langle (\Delta P)^2 \right\rangle = \int_0^{\upsilon_{max}}\int_0^{V} \left\langle [\Delta P(\upsilon, V)]^2 \right\rangle d\upsilon dV = \left(k_B T_D / t_S\right)^2 \int_0^{\upsilon_{max}}\int_0^{V} \langle N(\upsilon, V)\rangle \, d\upsilon dV \tag{5.43}$$

Substitution of the last step of Eq. (5.42) into the last step of Eq. (5.43) yields

$$\left\langle (\Delta P)^2 \right\rangle = \left(k_B T_D / t_S\right)^2 G \cdot T_D \left(t_S / k_B T_D\right) = k_B T_D{}^2 G / t_S \tag{5.44}$$

Application of the same (Nyquist) sampling argument as for shot noise, namely $B_{max} = 1/(2t_S)$, yields the result

$$\left\langle (\Delta P)^2 \right\rangle = 2 k_B T_D{}^2 G \, B_{max} \tag{5.45}$$

Equation (5.45) requires some clarification and generalization to be practical. First, all thermal detectors have a characteristic thermal time constant, $\tau_\tau \approx C/G$, where C is the heat capacity. This generally limits the bandwidth to an effective value B_{eff}, a much lower value than in rectifying devices. Secondly, unlike the analysis of shot noise where the electrical current flow across the barrier was considered unilateral, thermal detectors often have a second significant source of power fluctuations associated with heat flowing from the bath to the detector. Hence, a second term must be added to Eq. (5.45) that is ostensibly uncorrelated to that from the detector to the bath. Hence, we can just add the variances of the two components:

$$\left\langle (\Delta P)^2 \right\rangle = 2 k_B G \, B_{eff} \left(T_D{}^2 + T_B{}^2\right) \approx 4 k_B G \, B_{eff} T_D{}^2 \tag{5.46}$$

where the last step follows when $T_D \approx T_B$, which is often true in room-temperature thermal detectors. Equation (5.46) is a famous result in thermal-detector theory, often cited in the literature as a general result from thermodynamics. However, from the above simple-minded derivation, it requires statistical transport theory, which is usually beyond the scope of thermodynamics [19].

Even more remarkably, Eq. (5.46) can be generalized to handle obvious shortcomings of the above analysis, such as the cryogenic operating regime where $k_B T < h\upsilon$ and $T_D > T_B$. Both of these cases were covered by the seminal analysis of Boyle and Rodgers using Planck (massless-Boson) statistics for the phonon population [20]. They showed that

$$\left\langle (\Delta P)^2 \right\rangle = 2 k_B G \, B_{eff} \left(T_D{}^5 + T_B{}^5\right) / T^2 \tag{5.47}$$

where T is the arithmetic mean between T_D and T_B. Note that it reduces to the classic formula when $T_D \approx T_B$, meaning that the physical effect of the Planck statistics is embedded in G, not the temperature dependence. To further see how much physical information is contained in G, we note that Eqs. (5.46) and (5.47) have also been applied successfully for calculating the power fluctuation of thermal detectors exchanging heat by thermal mechanisms other than phonons. Two good examples are thermal conduction by electromagnetic radiation (photons), and thermal convection by fluids, usually air. The challenge, then, is to calculate the physics-rich quantity, G.

5.2.3 Photon Noise with Direct Detection

No analysis of detector behavior anywhere in the electromagnetic spectrum is complete without a discussion of radiation fluctuations. Quantum mechanics teaches us that the act of measuring any physical observable is inherently probabilistic with a measure of uncertainty given by Planck's constant h. Therefore, we expect all radiative sources to fluctuate in power when measured by any detector, possibly at a very low level compared to the other noise sources. The theory of radiative power fluctuations is relatively simple in two extremes: (i) coherent radiation and (ii) thermal radiation. In the THz region, the latter is more important in the case of direct detectors having wide spectral bandwidth. To carry out the analysis, we start by considering n, the number of photons in spatial mode m, as a random variable to calculate the fluctuation in the incident power:

$$\Delta P_{\text{inc}} \equiv P_{\text{inc}} - \langle P_{\text{inc}} \rangle = \sum_{m}^{M} \int_{\upsilon_0}^{\upsilon_0+\Delta\upsilon} h\upsilon \cdot \{n_m(\upsilon) - \langle n_m(\upsilon) \rangle\} d\upsilon \equiv \sum_{m}^{M} \int_{\upsilon_0}^{\upsilon_0+\Delta\upsilon} h\upsilon \cdot \Delta n_m(\upsilon) \cdot d\upsilon \tag{5.48}$$

An important measure of noise is the variance, or mean-square fluctuation

$$\left\langle (\Delta P_{\text{inc}})^2 \right\rangle = \left\langle \left(\sum_{m}^{M} \int_{\upsilon_0}^{\upsilon_0+\Delta\upsilon} h\upsilon \cdot \Delta n_m(\upsilon) \cdot d\upsilon \right) \cdot \left(\sum_{m'}^{M'} \int_{\upsilon_0}^{\upsilon_0+\Delta\upsilon} h\upsilon \cdot \Delta n_{m'}(\upsilon) \cdot d\upsilon \right) \right\rangle \tag{5.49}$$

where m' is a dummy summation index. In evaluating this expression we utilize the fact that perfectly random fluctuations in different orthogonal modes are uncorrelated, so that only cross-products having the same mode index will survive the ensemble average

$$\left\langle (\Delta P_{\text{inc}})^2 \right\rangle = \left\langle \sum_{m}^{M} \left(\int_{\upsilon_0}^{\upsilon_0+\Delta\upsilon} h\upsilon \cdot \Delta n_m(\upsilon) \cdot d\upsilon^2 \right) \right\rangle \tag{5.50}$$

This is further simplified by the "band-limited" assumption: $\Delta\upsilon$ being narrow enough that $h\upsilon$ and Δn can be considered constant over the range of the integration:

$$\left\langle (\Delta P_{\text{inc}})^2 \right\rangle \approx \left\langle \sum_{m}^{M} (h\upsilon_0 \Delta n_m(\upsilon_0) \cdot \Delta\upsilon)^2 \right\rangle = \sum_{m}^{M} \cdot (h\upsilon_0\Delta\upsilon)^2 \left\langle [\Delta n_m(\upsilon_0)]^2 \right\rangle \tag{5.51}$$

In the special but common case in sensitive THz detectors of only one spatial mode, such as typically occurs with antenna or waveguide coupling, we find

$$\left\langle (\Delta P_{\text{inc}})^2 \right\rangle = (h\upsilon_0\Delta\upsilon)^2 \left\langle [\Delta n(\upsilon_0)]^2 \right\rangle \tag{5.52}$$

For direct detectors in the THz region, one should consider first the fluctuations of background thermal radiation, given how important these are in the higher-frequency IR bands. It is a basic exercise of quantum statistical mechanics to show:

$$\left\langle (\Delta n)^2 \right\rangle \equiv \left\langle (n - \langle n \rangle)^2 \right\rangle = \left\langle n^2 \right\rangle - (\langle n \rangle)^2 = \langle n \rangle (\langle n \rangle + 1) = f_p(1 + f_p) \tag{5.53}$$

where f_p is the Planck function. This expression has an interesting low-frequency, or Rayleigh–Jeans limit, where $h\upsilon \ll k_B T, f_P \gg 1$ and $<(\Delta n)^2>$ goes to $f_P{}^2$, which is approximately $(k_B T/h\upsilon_0)^2$. This limit is quite accurate in the THz region at terrestrial temperatures. Substitution into Eq. (5.52) then yields:

$$\left\langle \left(\Delta P_{\mathrm{inc}}\right)^2 \right\rangle = \left(k_B T \Delta\upsilon\right)^2 \tag{5.54}$$

And we can write

$$P_{\mathrm{rms}} = \left[\left\langle \left(\Delta P_{\mathrm{inc}}\right)^2 \right\rangle\right]^{1/2} = k_B T \Delta\upsilon \equiv \left\langle P_{\mathrm{abs}} \right\rangle \tag{5.55}$$

We observe the remarkable result that for perfect coupling ($\eta = 1$), the rms absorbed power fluctuation is exactly equal to the average power – a sign of exponential (Boltzmann) statistics! Since the power spectral density is uniform in frequency (i.e., "white"), it is then given by:

$$S_P(\upsilon) = P_{\mathrm{rms}}/\Delta\upsilon = k_B T, \tag{5.56}$$

a form of the famous Johnson–Nyquist theorem for one mode in free space.

Although fundamental and interesting, the above derivation serves just one purpose for the present chapter, which is predicting the maximum possible *output* signal fluctuations from a THz direct detector in the presence of a 300 K background having the power spectrum in Eq. (5.56). To do this, we take advantage of the fact that all of the common THz detectors are essentially "square-law" devices, meaning that they convert incident power into output electrical signal according to Eq. (5.1):

$$y = A\,x^2 = \Re \cdot P_{\mathrm{in}} \tag{5.57}$$

where y is the output voltage or current. This is true even for thermal detectors, for which x is related to the incident electric field magnitude, so that x^2 is proportional to the incident power. The key to understanding the resulting fluctuations in y is to realize that different frequency components of Gaussian incident thermal noise, say at υ and υ', are statistically uncorrelated and therefore produce a zero DC output from the detector after averaging. However, they can still contribute a fluctuation and therefore, output noise, if their difference frequency $\Delta\upsilon$ is within the detector electrical bandwidth. So we need to calculate the output power spectral density.

This is a classic calculation of nonlinear signal-processing and probability theory carried out elegantly, for example, by Davenport and Root [21]. They prove that a square-law detector changes the noise statistics from Gaussian at the input to "chi-squared" at the output. For our purposes, a more qualitative method can be applied that misses the statistics but approximates the output-signal mean-square fluctuations, which suffices to get the SNR (signal-to-noise ratio) and NEP [22]. We draw the input power spectrum as a THz-band-limited white spectrum as shown in Figure 5.7a. If we look at small difference frequencies, there are a maximal number of input frequency components available, proportional to the input THz bandwidth $\Delta\upsilon$. But with increasing frequency, the number of available input components drops until we reach the minimal condition depicted in Figure 5.7b where just the extremes of the input spectrum are effective. The resulting output spectrum is then given approximately by

$$S_y(f) \approx 2\Re^2\left(S_x\right)^2 \Delta\upsilon\,(1 - f/\Delta\upsilon) \tag{5.58}$$

where S_x is the input power spectral density at any frequency in the passband, f is the post-detection frequency, and the factor of 2 accounts for the fact that the input power spectrum has an identical interpretation at negative frequencies (not shown in Figure 5.7). As plotted in Figure 5.7c, Eq. (5.58) is a distinctly-non-white triangular spectrum centered at zero frequency – *a distinctly non-Gaussian result.* But since the input noise is "white," we can write the fluctuations as:

$$\left(S_x\right)^2 \approx \left\langle \left(\Delta P_{\mathrm{in}}\right)^2 \right\rangle / \left\langle (\Delta\upsilon)^2 \right\rangle \tag{5.59}$$

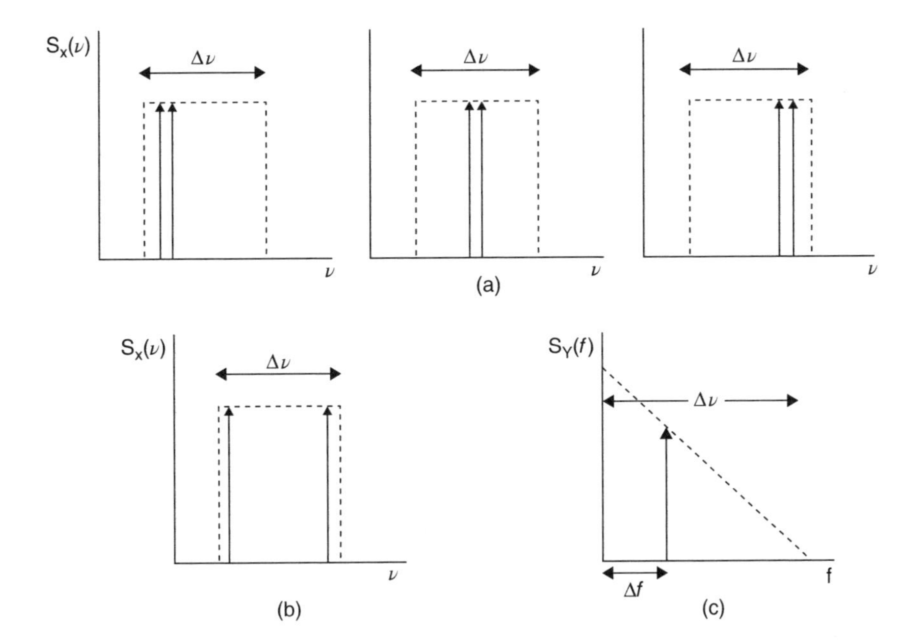

Figure 5.7 (a) Graphical representation of input power spectrum to a square-law detector created by thermal incident radiation over bandwidth $\Delta\upsilon$, and showing separate frequency components. (b) Same as (a) but with separate frequency components at edges of input band. (c) Output power spectrum from square-law detector as a function of difference frequency $\Delta f = |\upsilon_2 - \upsilon_1|$, and highlighting the specific difference frequency Δf.

Substitution into Eq. (5.58) then yields

$$S_y \approx 2\Re^2 \left\langle \left(\Delta P_{\text{in}}\right)^2 \right\rangle (1 - f/\Delta\upsilon)/\Delta\upsilon \tag{5.60}$$

The quantity we need is the mean-square fluctuation in output signal y, which is given by:

$$\left\langle (\Delta y)^2 \right\rangle = \int_0^\infty S_y(f)\,\mathrm{d}f \approx 2\Re^2 \left(\Delta P_{\text{in}}\right)^2 \int_0^{\Delta\upsilon} \frac{1 - f/\Delta\upsilon}{\Delta\upsilon}\mathrm{d}f \tag{5.61}$$

For practically all THz detectors, the output electrical bandwidth Δf is much less than the THz bandwidth, which entails limiting the above integral to a range 0 to $\Delta f \ll \Delta\upsilon$. Under this condition, the rather slowly varying triangular spectrum can be treated like a constant at the peak value. We then get the result,

$$\left\langle (\Delta y)^2 \right\rangle \approx 2\Re^2 \left(\Delta P_{\text{in}}\right)^2 \int_0^{\Delta f} \frac{\mathrm{d}f}{\Delta\upsilon} \approx 2\Re^2 \left(\Delta P_{\text{in}}\right)^2 \frac{\Delta f}{\Delta\upsilon} \tag{5.62}$$

Given terrestrial THz thermal radiation in the Rayleigh–Jeans limit, we can use Eq. (5.53) to re-write this as

$$\left\langle (\Delta y)^2 \right\rangle = 2\Re^2 \left(k_B T\right)^2 \Delta\upsilon \cdot \Delta f \tag{5.63}$$

As we should have expected, the mean-square output-signal fluctuation is proportional to the mean-square input power fluctuation, but also to the responsivity-squared and the product of the input and output bandwidths. So it is highly dependent on the detector type, and we will return to this later.

5.3 THz Coupling

Devices operating at terahertz frequencies become inseparable from the circuit. Therefore, issues relating to coupling the device to electromagnetic radiation, to antenna structures for electromagnetic radiation, and to device integration with THz circuits have to be addressed for practical consideration for the terahertz coupling. Since the device dimensions are often much smaller than the electromagnetic wavelength, the antenna structure needed for coupling the device with the electromagnetic radiation has to be designed as a unique entity inseparably comprising both elements the device and the antenna. In the microwave regime this concept has been studied and defined as the active integrated antenna (AIA) concept. Traditionally, antenna design for these receivers has been inefficient in terms of matching it with the diode. The antenna is often designed to be 50 matched while the capacitive part of the diode is compensated including a RF filter or RF matching network with an inductive part. The AIA concept was proposed to optimize the matching between the antenna and the active device. In addition to the matching optimization, a further improvement in the overall antenna efficiency should be considered by increasing the total efficiency which is essential within the THz-band. As a consequence, the responsivity of the receiver working within the THz-band will be optimized.

However the coupling design is done, some degree of impedance mismatch inevitably occurs and the power delivery from free space to the detector is imperfect. So a useful metric is an external power coupling factor, η, analogous to the external quantum efficiency in photonic devices. It is defined simply by

$$P_{\mathrm{abs}} = \eta P_{\mathrm{inc}} \tag{5.64}$$

where P_{inc} is the power incident on the device (or equivalently, the "available" power from free space).

5.3.1 *THz Impedance Matching*

The amount of power at THz frequencies is generally quite small so the coupling efficiency is a key factor that has to be taken into account. This analysis has to be undertaken in a twofold way. On the one hand the matching between the active device and the antenna itself has to be considered. On the other hand the matching and coupling between different media, guided and free-space, has to be taken into account. The first case will be developed in Section 5.3.2, while the second will be developed in this section.

The analysis of antennas embedded or just lying on the interface of two electrically different media has been carried out over a broad range of THz research [23]. The interest has increased in recent years due to, for example, the potential for new THz-generation devices based on the use of the substrate semiconductor properties, thereby avoiding the use of antennas. Antennas lying over a dielectric medium have some special features that distinguish them from antennas in free-space. In free-space, the power radiated from a planar antenna divides equally above and below the plane by symmetry, while an antenna on a dielectric mainly radiates primarily into the dielectric. This results both from the way in which elementary sources radiate and from the way in which waves propagate along metals at a dielectric interface. It can be explained by the impedance of both media. Let us assume that a dipole is placed on the interface between air and a dielectric. If the dipole is used as a receiving antenna it measures the electric field collinear to it. This electric field is the field transmitted through the interface. In the transmission-line model of wave propagation, a wave from a high impedance material (air) is incident on a low-impedance material (dielectric). The transmitted electric field is small, so the dipole response is small. The effect is the same as looking at the voltage across a coaxial cable terminated with a low resistance. When the wave is incident from the dielectric side, the transmitted electric field is large and the dipole response is large, since the material has a lower impedance than the air. The analogous situation in coaxial cable is the voltage seen at a high-impedance load. The result is that the response is larger when the wave is incident from the dielectric. By reciprocity, the power transmitted from the dipole is larger in the dielectric.

Another issue that has to be taken into account is the thickness of the substrates at THz frequencies which is on the order of a wavelength. Under this condition, the substrate is relatively thick and its effect has to be considered. From the ray point of view (Figure 5.8), the rays incident with an angle larger than the critical angle are completely reflected and trapped as substrate modes.

These substrate modes can be suppressed by the inclusion of a lens on the back side of the substrate with the same dielectric constant. This eliminates the problem, because, as shown in Figure 5.9, the rays are then incident nearly normal to the surface and do not suffer total internal reflection. The substrate lens takes advantage of the sensitivity of an antenna to radiation from the substrate side and eliminates the substrate modes. In addition, directivity is improved. The disadvantages of the substrate lens are those of any system using refractive optics: absorption and reflection loss. Among the available dielectric lenses, one is particularly attractive because it is aplanatic, adding no spherical aberration or coma. This is the extended (or hyper-) hemispherical lens with an extension of $d = r/n$.

5.3.2 Planar-Antenna Coupling

In the THz regime the detector often consists of a planar antenna integrated with a device (e.g., a zero-bias Schottky diode) together with a silicon lens [24] to suppress substrate modes due to the thick substrate and make the radiation pattern unidirectional. From the antenna point-of-view three parameters must be accounted for to determine the responsivity of the detector: the radiation propagation efficiency (ε_{rad}), impedance-matching efficiency (M), and the polarization matching efficiency (ε_{pol}):

$$\eta = \varepsilon_{rad} \cdot M \cdot \varepsilon_{pol} \tag{5.65}$$

This total radiative coupling efficiency expresses how close the performance is to ideal. Maximizing this efficiency will enhance the performance of the detector and it is a key factor that has been used to optimize the design of THz direct detectors integrated with planar antennas.

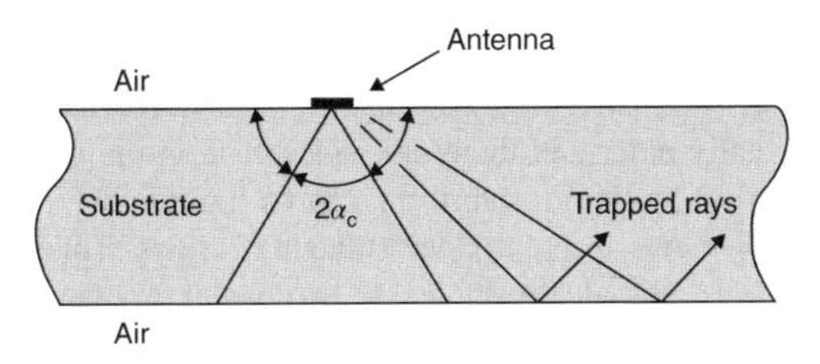

Figure 5.8 Antenna on a dielectric substrate showing the rays trapped as substrate modes. The critical angle is represented as α_c.

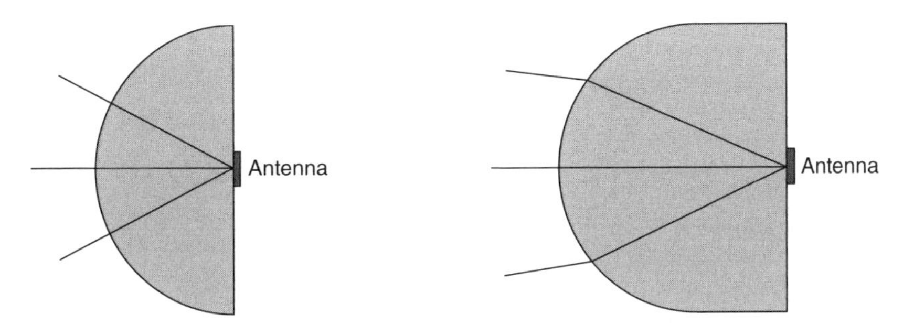

Figure 5.9 Hemispherical (a) and hyperhemispherical (b) lenses.

For lens-coupled planar antennas, the propagation efficiency is limited primarily by reflection-from and attenuation-in the lens. If it is made from high-resistivity silicon, as is common practice, the losses are usually negligible but the reflection can be substantial unless some form of anti-reflection coating is applied to the air/silicon interface. For incident radiation in the form of a Gaussian beam, the majority of power is incident on the Si lens near normal incidence so that the power reflectivity is given by $|\Gamma|^2 = [(n-1)/(n+1)]^2$, and $\varepsilon_{rad} \approx 1 - |\Gamma|^2$, where n is the THz refractive index of silicon. This is well-known in the THz region, $n \approx (11.65)^{1/2} = 3.41$ [25], which leads to $|\Gamma|^2 = 0.30$, and $\varepsilon_{rad} = 1 - |\Gamma|^2 = 0.70$.

The impedance matching efficiency at the THz signal frequency can be analyzed as the ratio between the power delivered to the detector acting as a passive load (P_L) and the power available from the antenna terminals acting as a generator (P_{avs}):

$$M = \frac{P_L}{P_{avs}} = \frac{4R_D R_A}{(R_D + R_A)^2 + (X_D + X_A)\,2} \tag{5.66}$$

where $Z_D = R_D + jX_D$ is the detector impedance and $Z_A = R_A + jX_A$ is the antenna impedance (see Figure 5.10). When $M = 1$, all the available power from the source is delivered to the load and the well-known conjugated matching condition can be obtained which states that the maximum power transferred to the load is obtained when $Z_A = Z_D{}^*$ (conjugate match, i.e., $\Gamma_A = \Gamma_D{}^*$). The M-factor is a key parameter to estimate the performance of the detector, especially for highly reactive detectors like Schottky rectifiers. As will be shown later, typical Schottky rectifiers have much higher values of R_D than the R_A of the antenna, and also have significant capacitance, making it very difficult to achieve the conjugate impedance matching condition. This tends to make M low at THz frequencies, of order 1% being typical.

5.3.3 Exemplary THz Coupling Structures

We have mentioned that planar antennas are now the preferred way of coupling room-temperature THz direct detectors to free space. Alternative methods such as metal-waveguide or integrating-cavity coupling also work and are actually older and therefore widespread in the literature. However, they also tend to be more difficult (and expensive) to fabricate, can be multimodal, and do not lend themselves to proliferation of detector elements in focal-plane (imaging) arrays. This is why we do not discuss them here. So in this section we graphically showcase two paradigm planar-antenna coupled detector structures, the first coupled to free space with a low-loss lens and the second coupled with a metal ground plane.

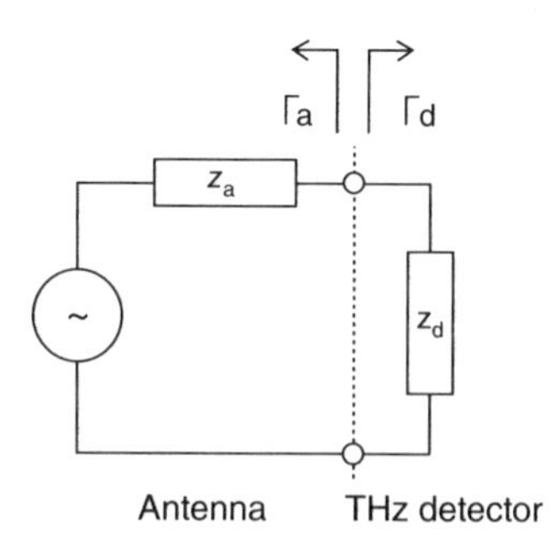

Figure 5.10 THz-frequency impedance-matching view of the Schottky rectifier.

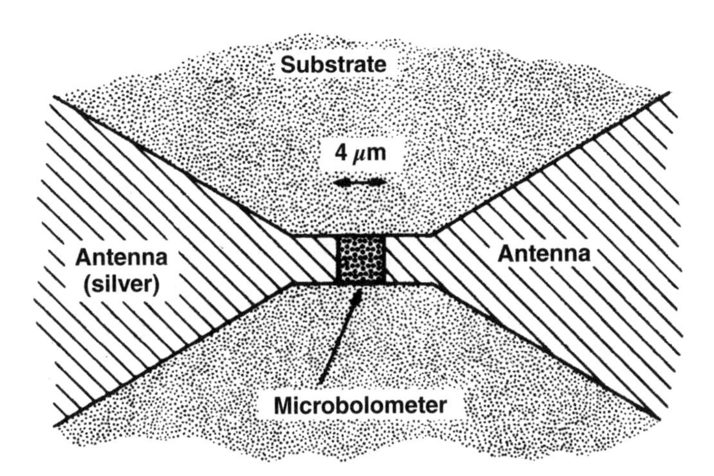

Figure 5.11 Schematic of a microbolometer mounted at the driving gap of a planar antenna [26]. (With kind permission of Springer Science+Business Media, D. P. Neikirk, D. B. Rutledge, and W. Lam, "Far-Infrared Microbolometer Detectors," Int. J. Infrared Millimeter Waves, 5, p. 245 (1984).)

Figure 5.11 shows perhaps the first successful THz planar-antenna structure: a microbolometer coupled to a self-complementary antenna. As will be emphasized later, in principle this structure can provide near-perfect impedance matching ($M \approx 1$) between the thin-film metallic or semimetallic microbolometer and the antenna on a dielectric substrate. A thin film of metal has very little reactance – just some kinetic inductance – at THz frequencies, so this structure should also have very wide bandwidth, approaching 1.0 THz or beyond. It integrates readily with hemispherical or hyperhemispherical lenses mounted on the back-side of the bolometer/antenna substrate, as shown in Figure 5.12a, which also exemplifies the antenna as a self-complementary log-spiral – one of the widest-band planar antennas ever demonstrated.

On inspection of Figure 5.12b, we also see another beneficial feature of the planar antenna coupling, which is electrical isolation of the THz antenna circuit from the read-out circuit. Notice that the read-out circuit is coupled to the outer extremity of the spiral antenna after one full turn of the spiral. This means that the THz radiation will be coupled primarily to the driving gap, the read-out circuit being isolated by the low-pass filtering properties of the spiral. So the rectified signal will flow through the spiral to the read-out arms relatively unimpeded. It also means that the read-out circuit can be tailored to a specific load impedance, independent of the THz antenna. Shown in Figure 5.12a is a standard 50 Ω coplanar waveguide (CPW) line, so the inclusion of an off-the-shelf, low-noise video output amplifier is straightforward. Shown in Figure 5.12c is the same coupling structure but with a Schottky rectifier mounted in the driving gap. This presents more impedance-matching challenge than the microbolometer, but still works well in the sub-THz region.

Figure 5.13 shows the planar-antenna coupled with a ground plane. The antenna is a resonant patch and the detector is a CMOS-type similar to those discussed in Section 5.1.3, both designed for 300 GHz operation [27]. Two differential feeds occur between the patch and the CMOS detector, and each feed is recessed into the patch – a common practice with patch antennas to provide for impedance transformation and better overall coupling efficiency [28]. Being Si-CMOS-based, this structure lends itself very well to proliferation into arrays via the super-successful technology of VLSI (very large scale integration). However, one drawback of this approach compared to Figure 5.12 is the antenna pattern. Patch antennas generally provide much lower gain than lens coupled antennas of all types, and suffer from substrate-mode and other loss mechanisms when fabricated on the large dielectric constant ($\varepsilon_r = 11.65$) of silicon substrates.

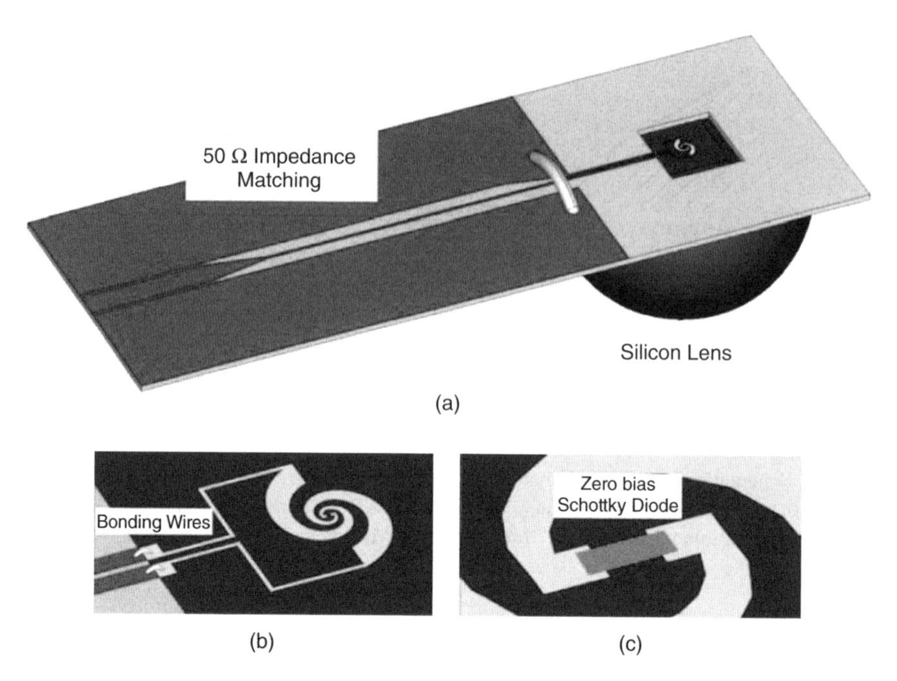

Figure 5.12 Schematic of the Schottky rectifier quasi-optical QO detector with log-spiral antenna. Materials: dark blue part is the silicon lens and silicon substrate where the gold (yellow part) planar antenna is grown, while the green part is FR-4 material and the brown part is copper. The bonding wires are included to connect gold lines over silicon to copper lines over FR-4 and a metallic wire is inserted to interconnect both ground planes of CPW line. The zero bias Schottky diode is epoxied in the middle of the antenna. (a) Complete SBD QO video detector 3D schematic. (b) Antenna zoom. (c) Schottky diode zoom.

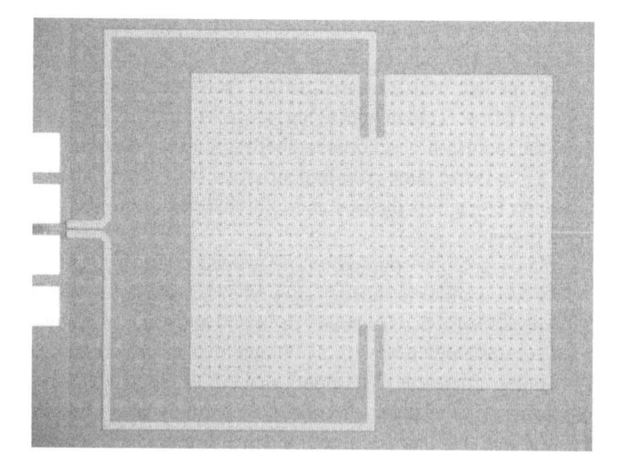

Figure 5.13 Differential patch antenna integrated with a CMOS detector.

5.3.4 Output-Circuit Coupling

Finally, there is the issue of output circuit coupling. While at first appearing to be a trivial issue, it can be relatively easy to lose signal strength (and SNR) at the output owing to voltage or current division between the differential resistance of the detector and the input resistance of the low-noise amplifier (LNA) connected to it. This is perhaps best shown by example, the Schottky rectifier once gain. Using Eq. (5.8) we can write the external responsivity of the rectifier as $\Re_{ext} = \eta\Re_V = \eta e R_D/(2\alpha k_B T)$, which directly shows the presence of the diode differential resistance, typically of order 10 kΩ. If we now connect the rectifier to a load resistance (no reactance), there will then be a voltage divider effect such that the fraction across the load relative to the (open-circuit) voltage across the detector will reduce the external responsivity to

$$\Re_{ext} = \eta \frac{eR_D}{2\alpha k_B T} \cdot \left[\frac{1}{1 + \frac{R_s}{R_D} + R_s R_D (\omega C)^2} \cdot \frac{R_L}{R_D + R_L} \right] \tag{5.67}$$

where ω is the angular output frequency and R_L is the load resistance. Unless high video bandwidth is required (say >10 MHz), the term in square brackets is close to unity, since a high-input-impedance operational amplifier (opamp) can be used having $R_L > 1\ \text{M}\Omega$ – a much higher value than practical Schottky rectifiers.

5.4 External Responsivity Examples

The combination of electrical responsivity introduced in Section 5.1 and external coupling efficiency forms a crucial combination called the external or optical, responsivity

$$\Re_{ext} = \eta \Re_X \tag{5.68}$$

where X is either current or voltage. In the small-signal operation that THz detectors usually experience, these two parameters are often separable, making the optical responsivity the most important metric with respect to signal processing (but not necessarily noise). We will now revisit each of the common detector types from Section 5.2 with simultaneous consideration of coupling efficiency and electrical responsivity. Unfortunately, much of the THz literature on direct detectors does not make this distinction, instead limiting the analysis to electrical responsivity. So our discussion will be limited to the detector types best understood at this point in time, which are the rectifiers and the bolometers. By so doing, we will see design trade-offs that ultimately must be made to optimize THz detector performance – a task that is generally more complicated than it first appears.

5.4.1 Rectifiers

As mentioned in Section 5.1, rectifiers generally are the simplest of all THz direct detectors, which is why they are so popular. The general analysis done for impedance matching in Section 5.3.3 can be made more specific by considering the equivalent circuit model shown in Figure 5.14 [29].

It is a bit more sophisticated than Eq. (5.66) through the explicit addition of a diode series resistance R_S which becomes increasingly important as the frequency increases from the microwave bands toward the THz regime.

$$M = \frac{4 \cdot R_D R_A}{\left(R_D + R_A + R_S - \omega^2 LCR_D\right)^2 + \left[\omega L + \omega R_D C \left(R_A + R_S\right)\right]^2} \tag{5.69}$$

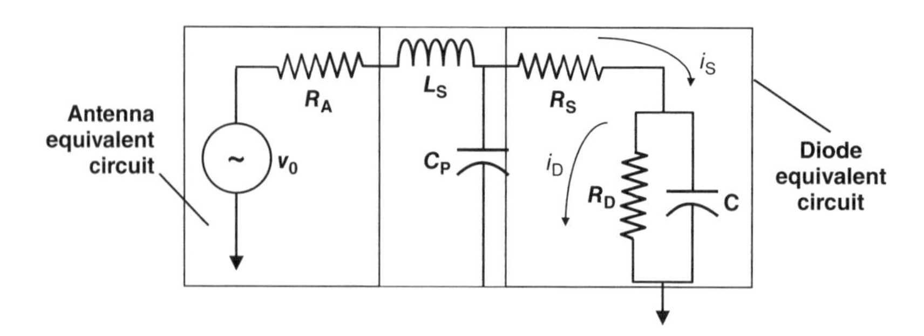

Figure 5.14 Equivalent circuit planar-antenna-coupled Schottky rectifier [2]. (©2007 IEEE. Reprinted, with permission, from IEEE Microwave Magazine vol. 8, p. 54 (2007).)

It is usually the sum of all dissipative components in the semiconductor outside the depletion layer of the Schottky diode itself. The model also includes parasitic elements L_S and C_P, which represent a series inductance and packaging capacitance, respectively. Generally L_S is more important than C_P in THz air-bridge type planar devices where the finger that contacts the anode is not contacting the high-permittivity GaAs (or InP) substrate. In that case we can assume $C_P \approx 0$ and can solve analytically for the impedance matching efficiency where R_A is the real part of the antenna driving-gap impedance. This ignores any reactive part of the antenna impedance, which is a fair assumption for self-complementary-type planar antennas such as bow-ties and log spirals.

It is informative to calculate Eq. (5.69) for typical values of all parameters at an interesting THz frequency – 650 GHz (which happens to be one of the highest-frequency useful atmospheric "windows"). We do this for a zero-bias rectifier as shown in Figure 5.1 for 20%-Al composition [29]. The diode anode area is 1 μm^2. The *I–V* and *R–V* curves are shown in Figure 5.15a. The corresponding zero-bias differential resistance and capacitance are approximately 7.5 kΩ and 2.0 fF, respectively. The antenna resistance is assumed to be 72 Ω, consistent with the theoretical value according to Booker's relation for self-complementary antennas on a GaAs substrate: $R_A = \eta_{eff}/2$ where $\eta_{eff} \approx [\mu_0/\varepsilon_{eff}]^{1/2}$, $\varepsilon_{eff} = (1 + \varepsilon_r)/2$, and $\varepsilon_r = 12.8$ for GaAs. For R_S we assume two values – 10 and 30 Ω – which roughly defines a range of possible values that depends on the specific fabrication technology. For L_S we consider 0 pH – an idealization, and 30 pH – a value found to resonate with the device (capacitance) at 650 GHz.

The resulting coupling efficiency is plotted in Figure 5.15b where we see an interesting behavior versus frequency. For both $L_S = 0$ and 30 pH, and practically independent of R_S, we see that the coupling is approximately 1.5% up to 100 GHz. For $L_S = 0$ pH, it drops at higher values down to ≈1.0% at 650 GHz. Hence, given the highest zero-bias electrical responsivity of ≈18 A/W from Figure 5.1, the maximum external responsivity is ≈0.18 A/W at 650 GHz. For $L_S = 30$ pH, the coupling rises above 100 GHz to a peak value of ≈10% at 650 GHz. Hence, the peak external responsivity at 650 GHz is 1.8 A/W. This is considered excellent coupling for Schottky rectifiers in spite of the limited bandwidth (~300 GHz) caused by the resonance. In contrast and calculated next, antenna-coupled bolometers can display coupling efficiencies approaching 100% and over bandwidths in excess of 1 THz. But because the zero-bias Schottky's are so "quiet" in physical noise relative to bolometers, the NEP of the Schottkys remains superior.

5.4.2 *Micro-Bolometers*

As described in Section 5.1.2.1, bolometers have a great deal of design latitude through their electrical responsivity $\Re_V \approx I_B\ R_0\ \alpha/G_0$. But what happens if we also consider the external coupling efficiency so that $\Re_{ext} = \eta \cdot I_B\ R_0\ \alpha/G_0$? This generally makes the optimization complicated except in the case of the antenna- or waveguide-coupled "microbolometer", as developed in the 1980s. By definition, a

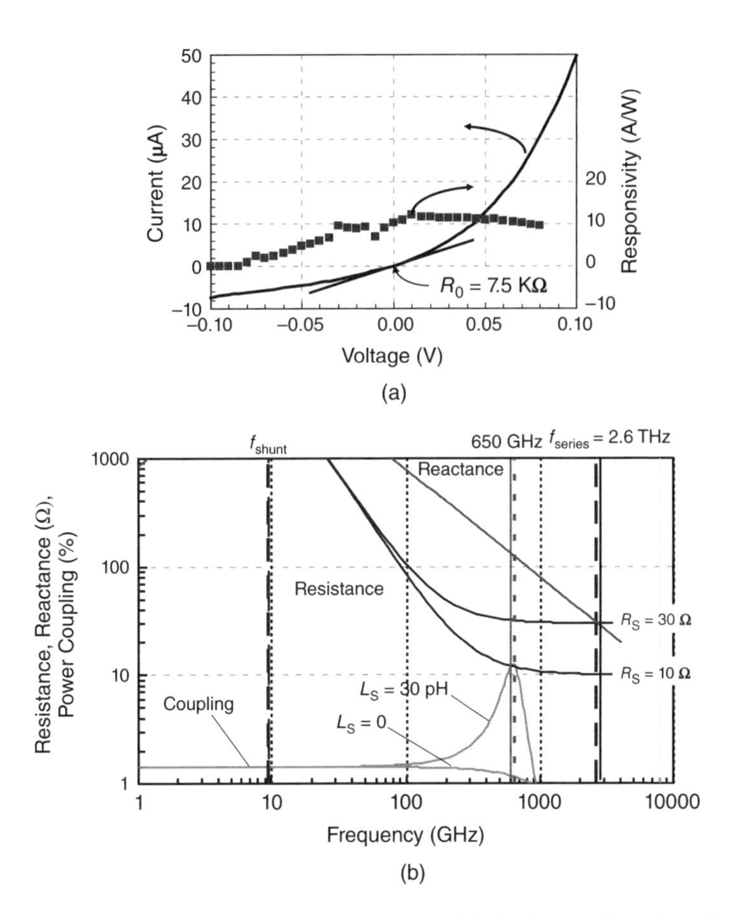

Figure 5.15 (a) *I–V* and current electrical responsivity of Schottky rectifier. (b) THz impedance-matching coupling efficiency of Schottky according to equivalent circuit in Figure 5.13.

microbolometer is physically much smaller than a wavelength but can still couple efficiently to THz radiation by integration with a waveguide or planar antenna. As described in Ref. [26], this leads to a simplification of the design if the bolometer is metallic because it should then display the same resistance R_0 both at THz and low (read-out) frequencies. Impedance matching then entails setting R_0 to the characteristic impedance Z_0 of the waveguide or antenna (a value generally less than 500 Ω), so that $\eta \approx 1$ and $\Re_{ext} \approx I_B Z_0 \alpha/G_0$. Aside from controlling I_B, the optimization then focuses simply on maximizing the ratio α/G_0. As discussed previously, α has approximately the same absolute value in all metallic solids, so the design emphasis focuses on the reduction of G_0 along with reduction in heat capacity C so that the time constant $\tau = C/G_0$ remains usefully small.

An important development in this vein occurred in the 1980s with the air-bridge microbolometer. The air-bridge entails lifting the thin-film bolometer element off the solid substrate by elegant microfabrication techniques, thereby eliminating the thermal conduction path to the substrate [30]. This constrains the thermal conduction to heat flow into the metal planar antenna to which the bolometer element is necessarily connected. Assuming operation in the ambient environment, there is also some convection to the surrounding air. Assuming that the conduction to the antenna metal dominates (an assumption that must be checked in each air-bridge microbolometer design), a simple but insightful analysis can be carried out taking advantage of the fact that the air-bridge bolometer metal usually has rectilinear dimensions l

(length), w (width), and t (thickness). In this case, the DC (and THz) resistance is $R_0 = \rho \cdot l/(w \cdot t)$ where ρ is the bulk electrical resistivity. The thermal conductance is a more complicated parameter since the temperature is not uniform across the bridge but rather peaks in the middle. Analysis has led to the approximate form $G_0 \approx 12 \cdot \kappa \cdot w \cdot t/l$ where K is the bulk thermal conductivity [26]. Substitution of the electrical expression into the thermal expression yields

$$G_0 = 12 \cdot K \cdot \rho / R_0 \tag{5.70}$$

Now setting $R_0 = Z_0$ for the impedance matching leads to a rule-of-thumb to minimize G_0 and thereby maximize the electrical responsivity without jeopardizing the external responsivity:

$$G_0 = 12 \cdot K / \left(s \cdot Z_0 \right) \tag{5.71}$$

where σ is the electrical conductivity. So minimizing G_0 entails minimizing the ratio κ/σ, which for bulk metallic solids is governed by the famous Wiedemann–Franz law of solid-state physics [14]:

$$K/\sigma = \left(\pi^2/3 \right) \left(k_B/e \right)^2 \cdot T \tag{5.72}$$

However, nm-scale thin films generally display a lower value of K/σ than the Wiedemann–Franz prediction because the electrical conductivity drops faster with decreasing thickness than the thermal conductivity [31]. So we write

$$K/\sigma = d \cdot \left(\pi^2/3 \right) \left(k_B/e \right)^2 \cdot T \tag{5.73}$$

where d is a metal- and dimension-dependent factor >1. Evaluation at 300 K yields $d \cdot 7.32 \times 10^{-6}$ W Ω K^{-1}. Substitution of $Z_0 = 72\,\Omega$ for a self-complementary antenna on GaAs then yields

$$G_0 = 12 \cdot d \cdot K / \left(\sigma \cdot Z_0 \right) = 1.22 \times 10^{-6} \cdot d[\mathrm{W/K}] \tag{5.74}$$

For the sake of estimation, we set $d = 2.0$, which yields

$$G_0 = 2.45 \times 10^{-6}[\mathrm{W/K}] \tag{5.75}$$

Substitution of $\alpha \approx 1/T$ then yields for the external responsivity from Eqs. (5.13) and (5.75)

$$\Re_{ext} \approx 12 \cdot I_B Z_0 \alpha / G_0 = 9.8 \times 10^4 \cdot I_B[\mathrm{V/W}] \tag{5.76}$$

where $\eta = 1.0$ (perfect coupling) has been assumed.

The external responsivity can now be estimated considering the limitations on I_B imposed by regenerative electrothermal feedback or overheating, as discussed in Section 5.1.2.1. We make a plausible assumption that the maximum temperature rise ΔT of the microbolometer with respect to ambient is 100 K, which should occur at the geometric center of the device. All of the heat is created by the bias power, $P_B = (I_B)^2 R(T)$ which we set to $(I_B)^2 R_0$ for the sake of estimation. Assuming further that G_0 is independent of temperature, we can write $\Delta T = P_B/G_0 = (I_B)^2 R_0/G_0 = (I_B)^2 Z_0/G_0$, where the last step follows from the antenna impedance matching condition. Solving for I_B using the nm-scale-corrected G_0 from above, we find, $I_B \approx 1.8$ mA – a typical value of microbolometer bias current in air-bridge microbolometers, but somewhat smaller than that in substrate-based devices which generally have higher G_0 so can support more current before overheating. Substitution of this into the external responsivity above yields $\Re_{ext} \approx 180$ V/W. By comparison, the original air-bridge bismuth bolometer displayed a DC responsivity of ≈100 V/W [30].

How then do THz researchers choose the metal for their microbolometers? The answer lies largely in practical considerations. To fabricate a metal air-bridge, the thickness must be large enough to obtain adequate strength and uniform electrical conductivity, but not so great that the microlithographic fabrication becomes too difficult. A ballpark minimum thickness is 20 nm (2×10^{-6} cm). For similar reasons, the geometry of the bridge is roughly square, such that $w = l$. Hence, the bolometer can be assumed to be one "square" of metal roughly 20 nm thick. The issue is then to match the antenna impedance Z_0 to such a square, that is, $\rho/t \approx Z_0$, or $\rho \approx 1.4 \times 10^{-4}\ \Omega\,\text{cm}$. Inspection of the metals commonly used in microfabrication, such as gold and aluminum, yields values of $\rho = 2.2 \times 10^{-6}$ and $2.7 \times 10^{-6}\ \Omega\,\text{cm}$, respectively [16]. Other transition metals have higher resistivity (e.g., $\rho = 12.6 \times 10^{-6}\ \Omega\,\text{cm}$ for chromium) but still nowhere near the target value. Semimetals get much closer (e.g., $\rho = 1.1 \times 10^{-4}\ \Omega\,\text{cm}$ for bismuth), so were the bolometer material of choice in early development of air-bridge microbolometers. However, many transition metals display significantly higher effective resistivity than the bulk values in such thin films, so some have been pursued (e.g., $\rho = 14.5 \times 10^{-6}\ \Omega\,\text{cm}$ in niobium but roughly twice this value in 20 nm films [32]).

5.5 System Metrics

5.5.1 Signal-to-Noise Ratio

A useful metric for all types of sensors is the power SNR.

$$\frac{S}{N} = \frac{\langle P \rangle}{\sqrt{\langle (\Delta P)^2 \rangle}} = \frac{\langle P \rangle}{S_{\text{P}} \cdot B_{\text{N}}} \tag{5.77}$$

where S_{P} is the power spectral density and B_{N} is the equivalent noise bandwidth at that point in the sensor

$$B_{\text{N}} = \left(G_{\text{max}}\right)^{-1} \int_0^\infty G(f)\,\text{d}f \tag{5.78}$$

where $G(f)$ is the sensor gain function versus frequency and G_{max} is the maximum value of this gain, B_{N} is generally dictated by sensor phenomenology, such as the resolution requirements and measurement time.

While at first appearing to add insurmountable complexity to sensor analysis, a great simplification results from the fact that radiation noise and two forms of physical noise discussed above are, in general, statistically Gaussian (the shot noise becomes Gaussian in the limit of large samples, consistent with the central limit theorem). A very important fact is that *any Gaussian noise passing through a linear component or network remains statistically Gaussian.* Hence, the output power spectrum S in terms of electrical variable X (current or voltage) will be white and will satisfy the important identity

$$\langle (\Delta X)^2 \rangle = \int_{f_0}^{f_0+\Delta f} S_X(f) \cdot \text{d}f \approx S_X(f) \cdot \Delta f, \tag{5.79}$$

where $\Delta f\,(\equiv B_{\text{N}})$ is the equivalent-noise bandwidth. Then one can do circuit and system analysis on noise added by that component at the output port by translating it back to the input port. In the language of linear system theory, the output and input ports are connected by the system transfer function $H_X(f)$, so the power spectrum referenced back to the input (reference) port becomes

$$S_X(\text{in}) = \frac{S_X(\text{out})}{|H_X(f)|^2}. \tag{5.80}$$

Because the different Gaussian mechanisms are statistically independent, the total noise at the reference point can be written as the uncorrelated sum

$$\left\langle \left(\Delta X_{\text{TOT}}\right)^2 \right\rangle = \sum_{j=1}^{N} \left\langle \left(\Delta X_i\right)^2 \right\rangle \text{ or } \left\langle \left(\Delta P_{\text{TOT}}\right)^2 \right\rangle = \sum_{j=1}^{N} \left\langle \left(\Delta P_i\right)^2 \right\rangle \tag{5.81}$$

5.5.2 *Sensitivity Metrics*

5.5.2.1 Noise Equivalent Power and Its Properties

Unfortunately, the noise figure (or factor) concept commonly applied in RF systems is not applicable or particularly useful to THz direct detectors since it presupposes linearity. THz direct detectors are inherently nonlinear elements in RF and THz systems, which is what allows them to extract information from incoming radiation in the form of a DC or low-frequency modulated signal. In this case, it is the SNR *at the point of detection or decision* that is generally the most important quantity in system performance. Because of reasons pertaining to statistical detection theory, generally this SNR must be greater than or equal to unity to have a reliable detection or decision. Therefore, a very useful metric for direct detection performance is to fix the "after detection" SNR, SNR_{AD}, at unity and then solve for the signal power *at the input to the system* sensor that achieves this. The resulting metric is the NEP which, most simply put, is the *input* signal power to the sensor required to achieve a SNR of unity at the *output.*

A simple example is helpful at this point. Suppose we have an ideal "square-law" detector, such as the Schottky-diode rectifier or a metal-film microbolometer. By definition, the square-law detector has a circuit transfer function of

$$X_{out} = \Re_X P_{in} \tag{5.82}$$

where X_{out} is the output signal (usually current or voltage), P_{in} is the input power, and $\Re_X$ is the "responsivity". The noise at the output of the detector is minimally given by the Nyquist generalized theorem, which for a Schottky rectifier is

$$P_N \propto \left\langle (\Delta i)^2 \right\rangle = 4k_B T_0 \mathrm{Re}\{Y_D\}\Delta f \approx 4k_B T_0 \Delta f / R_D \tag{5.83}$$

where Δf is the post-detection bandwidth. The corresponding signal power at the output is just $(X_{out})^2$, so that the output SNR (with X chosen as I) is

$$SNR_{AD} = \left(\Re_I P_{in}\right)^2 / \left(4k_B T_0 \Delta f / R_D\right) \tag{5.84}$$

Setting this SNR to unity and solving for P_{in}, we get

$$NEP_{AD} = \left(4k_B T_0 \Delta f / R_D\right)^{1/2} / \Re_I \tag{5.85}$$

For the purpose of comparing different sensor technologies, it is conventional to divide out the post-detection bandwidth effect (or equivalently, setting it equal to 1 Hz). This yields the normalized NEP_{AD} (in units of W $Hz^{-1/2}$), given by

$$NEP'_{AD} = NEP_{AD} \cdot \sqrt{\frac{1}{\Delta f}} = \left(4k_B T_0 / R_D\right)^{1/2} / \Re_I \tag{5.86}$$

We deduce the generic expression for a square-law rectifier including the external power coupling factor:

$$NEP'_{AD} = \left[\left\langle (\Delta X_{out})^2 \right\rangle / \Delta f\right]^{1/2} / \left(\eta \Re_X\right) \tag{5.87}$$

Although simple looking, there is a lot of interesting physics buried in this expression!

An equally fundamental effect is the NEP limited by incident radiation (photon) fluctuations in a single spatial mode according to the analysis conducted in Section 5.2.3. According to Eq. (5.63), $\langle(\Delta y)^2\rangle = 2\Re^2_{ext,y} \cdot (k_B T)^2 \Delta \upsilon \cdot \Delta f$, where y is the current or voltage. The average signal squared is simply $(y_s)^2 = (\Re_{ext,y} \cdot P_{inc})^2$, so that the $NEP = k_B T\,(2 \cdot \Delta\upsilon \cdot \Delta f)^{1/2}$ and the specific "background limited" NEP is

$$NEP'_{AD} = k_B T (2 \cdot \Delta \upsilon)^{1/2} \tag{5.88}$$

As a "worst case" scenario, we imagine a very broadband THz detector such as a microbolometer coupled to a self-complementary spiral antenna. In this case we can assume $\Delta\upsilon \approx 1.0\,\mathrm{THz}$, so that

$$\mathrm{NEP}'_{\mathrm{AD}} = 5.8 \times 10^{-15}\ \mathrm{WHz}^{-1/2} \tag{5.89}$$

This is so much lower than the typical NEP′ values in real THz detectors that it begs the question of why we bother deriving and analyzing it. The answer is physical completeness and demonstration of the richness of the NEP metric. If we were to do the same calculation for room-temperature detectors operating at higher frequencies into the IR bands, we would find that the background-limited $\mathrm{NEP}'_{\mathrm{AD}}$ increases dramatically since the blackbody spectral density increases so rapidly (as υ^2). By 30 THz or so (10 μm wavelength), the background-limited NEP′ begins to exceed common electrically- or thermally-limited NEPs under room-temperature operation, making it the primary concern in the design of IR sensors. Similarly, if one were to consider cryogenic operation at THz frequencies, then the electrically- or thermally-limited NEPs could drop down to values approaching $10^{-14}\,\mathrm{W\,Hz}^{-1/2}$ or lower (especially in bolometers), making the 300 K radiation noise comparable and requiring use of cold filters to reduce this noise component. For students or new researchers in the THz field, this trade-off can be quite mysterious because few if any books or review articles explain it as done here.

A useful system aspect of the NEP is its additivity. If there are N mechanisms contributing to the noise at a given node in the receiver, if the mechanisms are uncorrelated to the signal and to each other, if they obey Gaussian statistics, then the total NEP is the uncorrelated sum

$$\mathrm{NEP}^2_{\mathrm{TOT}} = \mathrm{NEP}^2_1 + \mathrm{NEP}^2_2 + \ldots\ \mathrm{NEP}^2_N \tag{5.90}$$

This property applies to any node, pre- or post-detection, and will be used explicitly when we discuss the contribution from electronic noise. In reality there are cases where a noise mechanism is correlated to the signal (e.g., radiation noise) or to another noise mechanism (current and voltage noise in transistors), so one must be careful in applying this addition formula. In such cases, one can always fall back on the SNR as a useful measure of overall system performance. A good example occurs when the rms power fluctuations associated with heat (phonon) transport are comparable to the electrical NEP of the detector. In this case, the system NEP after detection is given by:

$$\mathrm{NEP}_{\mathrm{AD}} \approx \eta^{-1}\left[\left\langle (\Delta P_{\mathrm{T}})^2 \right\rangle + (\mathrm{NEP}_{\mathrm{elect}})^2\right]^{1/2} \tag{5.91}$$

where P_{T} is the coupled thermal power.

5.5.2.2 Two Examples of THz Noise Equivalent Power

As a first practical example of the NEP metric, we consider a zero-bias, room-temperature Schottky rectifier coupled to a planar antenna. As described earlier, we know from Schottky-barrier physics that $\Re_I \approx 25$ A/W or less. Let us assume $\Re_I = 20$ A/W and $R_{\mathrm{D}} = 8\,\mathrm{k\Omega}$, and that the post-detection noise is dominated by Johnson–Nyquist fluctuations. We then can write $\mathrm{NEP}'_{\mathrm{AD}} = (4k_{\mathrm{B}}T_0/R_{\mathrm{D}})^{1/2}/(\eta\Re_I)$ to obtain $\mathrm{NEP}'_{\mathrm{AD}} \approx \eta^{-1}\cdot 7.0 \times 10^{-14}\,\mathrm{W\,Hz}^{-1/2}$ – a performance obviously dependent on power coupling. For the 8 kΩ Schottky diode operating at 650 GHz, we can apply the results shown in Section 5.4.1 whereby the resonant and non-resonant coupling efficiencies were approximately 10 and 1%, respectively. The corresponding $\mathrm{NEP}'_{\mathrm{AD}}$ values are 7.0×10^{-13} and $7.0 \times 10^{-12}\,\mathrm{W\,Hz}^{-1/2}$, respectively, which roughly straddle the best experimental results ever achieved in this frequency region for Schottky rectifiers.

As a second example of the NEP metric, we consider an air-bridge microbolometer designed according to the principles given above in Section 5.4.2, specifically with perfect THz impedance matching. This means that $\Re_{\mathrm{ext}} \approx 12\cdot I_{\mathrm{B}}\ Z_0\ \alpha/G_0 = 9.8 \times 10^4\cdot I_{\mathrm{B}}$ (V/W) = 180 V/W for $I_{\mathrm{B}} = 1.8$ mA, independent of THz frequency. We consider two separate cases: (i) a Johnson–Nyquist-noise limited NEP and (ii) a phonon-noise-limited NEP. For the first case, we have $\mathrm{NEP}'_{\mathrm{AD}} = (4k_{\mathrm{B}}T_0R_{\mathrm{D}})^{1/2}/\Re_{\mathrm{ext}}$

$= (4k_B T_0 Z_0)^{1/2}/(180\ \text{V/W}) = 6.0 \times 10^{-12}\ \text{W Hz}^{-1/2}$. For the phonon noise-limited case, we have $\text{NEP}'_{\text{AD}} = [4k_B (T_0)^2 G_0]^{1/2} = 3.5 \times 10^{-12}\ \text{W Hz}^{-1/2}$. This example is simplified in that it ignores some practical noise effects in bolometers, especially $1/f$ noise and ambient air fluctuations. So the best experimental NEP values are higher than these fundamental limits (see Section 5.7). Nevertheless, this example shows that the phonon noise can be the fundamental limiting mechanism when operating at room temperature, as opposed to Johnson–Nyquist noise in rectifiers.

5.5.2.3 Noise Equivalent Temperature Difference

For radiometric and thermal imaging systems, it is often convenient to express the sensitivity in terms of the change in temperature of a thermal source at the input that produces a post-detection SNR of unity. The resulting metric is the NETD, also called the noise equivalent change of temperature (NEΔT). Regardless of title, it is given mathematically by

$$\text{NETD} = \text{NEP}_{\text{AD}} / \left(\text{d}P_{\text{inc}} / \text{d}T|_{P_B} \right) \text{ or } \text{NETD}' = \text{NEP}'_{\text{AD}} / \left(\text{d}P_{\text{inc}} / \text{d}T|_{P_B} \right) \tag{5.92}$$

where P_{inc} is the incident power and P_B is the background power. If we assume that the receiver accepts just one spatial mode with power coupling η (valid for antenna-coupled detectors), and that the thermal source satisfies the Rayleigh–Jeans limit and fills the field-of-view of the sensor, then $P_{\text{inc}} = k_B T_B\ \Delta \upsilon$ and $P_{\text{abs}} = \eta k_B T_B\ \Delta \upsilon$ so the NETD′ becomes

$$\text{NETD}' = \text{NEP}'_{\text{AD}} / \left(k_B \cdot \Delta \upsilon \right) \tag{5.93}$$

where $\Delta \upsilon$ is the spectral equivalent noise bandwidth, not to be confused with the corresponding electrical quantity Δf given by Eq. (5.78). Clearly the NETD metric favors detectors having large spectral bandwidth, such as bolometers coupled to broadband self-complementary planar antennas.

The NETD metric brings some subtleties with respect to its accurate characterization. First is knowledge of $\Delta \upsilon$, which technically is given by

$$\Delta \upsilon \equiv \left(\eta_{\max} \right)^{-1} \int_0^{\infty} \eta(\upsilon)\, \text{d}\upsilon. \tag{5.94}$$

In most THz detectors, $\Delta \upsilon$ is not well known and very difficult to measure because $\eta(\upsilon)$ is a complicated function of frequency, especially for highly impedance-mismatched detectors like rectifiers. Furthermore, the external NEP'_{AD} may not be known, because of the lack of a powerful-enough THz coherent source to accurately measure it at any (spot) frequency υ_0, and therefore a lack of knowledge of η in Eq. (5.87). So researchers sometimes will resort to measurement of NEP'_{AD} and NETD by direct characterization with a THz blackbody. It is then important to be sure that the incident power from the blackbody is known accurately. This is particularly true for non-antenna-coupled detectors where the number of spatial modes may exceed unity and therefore the denominator of Eq. (5.92) may significantly exceed $k_B T \cdot \Delta \upsilon$ as assumed in Eq. (5.93). Because most THz thermal imaging is intended for the terrestrial environment, it is then best to do the NEP'_{AD} and NETD characterization with a room-temperature blackbody having brightness temperature as-close-as possible to 300 K.

Another subtlety of the NETD characterization is electrical bandwidth. For imaging applications, the output of the detector will be often be sampled at a time $t_{\text{samp}} = 1/30$ s to accommodate the frame-rate standard (30 frame/s) generally considered to be compatible with comfortable human viewing. This leads to the common usage of a *reference* NETD (Dr. Erich Grossman, personal communication), NETD″, given by the NEP'_{AD} of Eqs. (5.92) and (5.93) measured specifically at an electrical center frequency of 30 Hz, but multiplied by $[1/(2t_s)]^{1/2}$ – the equivalent electrical bandwidth. Clearly this will always lead to a value of NETD″ (units of K) significantly higher than the NETD′ (units of K Hz$^{-1/2}$).

5.6 Effect of Amplifier Noise

No discussion of THz detectors would be complete without addressing the effect of the essential LNA connected to the output. In fact this applies to direct detectors in general, independent of wavelength or operating temperature. It is a wonderful thing to have a high-responsivity, low-NEP detector, and quite another to have it integrated into a working sensor system and achieve comparable sensitivity. LNAs necessarily introduce their own noise which adds to the detector noise and degrades the overall NEP in a manner that depends on the specific amplifier noise mechanisms and the detector impedance. We will focus only on voltage-amplification LNAs, such as opamps, because of their relative simplicity of analysis compared to other types (e.g., transimpedance amplifiers). But our conclusions will be rather universal and hopefully convince the reader that the LNA performance should be considered early on when pursuing THz detectors in systems.

As well known in discrete-electronics, voltage amplifiers can often be represented by the simple equivalent circuit shown in Figure 5.16 [33] where the input impedance of the amplifier is assumed infinite and all the noise it creates is represented by two uncorrelated, white noise sources connected at the input. The voltage noise density generator $V_{N,A}$ has zero internal impedance and is quantified in units of V $Hz^{-1/2}$, uniform across the amplifier bandwidth. The current noise density generator $I_{N,A}$ has infinite impedance and is quantified in units of A $Hz^{-1/2}$. Assuming that the LNA bandwidth extends down to near DC, the THz detector can be represented by its low-frequency differential resistance.

This is generally valid for rectifiers, bolometers, and other resistive-type detectors, which minimally generate Johnson–Nyquist noise, as discussed above, and represented by the noise generator $V_{N,D} = (4k_B T_0 R_D)^{1/2}$, uncorrelated with the two amplifier noise sources. In this situation, the total voltage noise generated at the output of the detector (where the NEP_{AD} is referenced) is given by:

$$V_{N,tot} = \left[\left(4k_B T_0 R_D\right) + \left(V_{N,A}\right)^2 + \left(I_{N,A} \cdot R_D\right)^2 \right]^{1/2} \tag{5.95}$$

And the impact on the NEP′ is given by

$$NEP' = V_{N,tot} / \Re_{ext} \tag{5.96}$$

To proceed further we have to get specific. We choose a popular low-frequency, low-noise opamp that the author has used with low-noise THz rectifiers and bolometers for about 10 years. It has the following characteristics $V_N \approx 1.0$ nV Hz^{-1}, $I_N \approx 2.0$ pA Hz^{-1} (above ~100 Hz) [34]. We can then evaluate Eq. (5.95) using the detector dynamic resistance as a parameter. The resulting plot is shown in Figure 5.17a where we show the detector noise alone and the total noise including LNA contributions. The total noise is very close to the detector noise alone at a resistance $R_D = 500\ \Omega$, which is equal to V_{NA}/I_{NA}, and called the amplifier noise resistance R_N. This is the source resistance into the LNA that creates the best "noise match", thereby minimizing the effect of the LNA. At much lower R_D, the total noise is dominated by the LNA voltage noise, and at much higher R_D it is dominated by the LNA current noise. This behavior

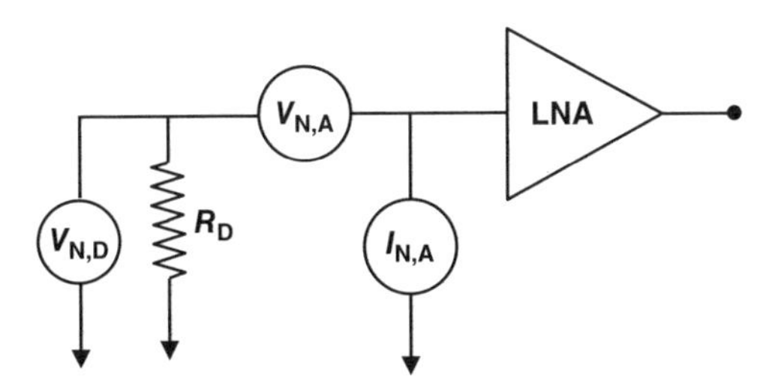

Figure 5.16 Noise equivalent circuit of THz detector connected to opamp-LNA.

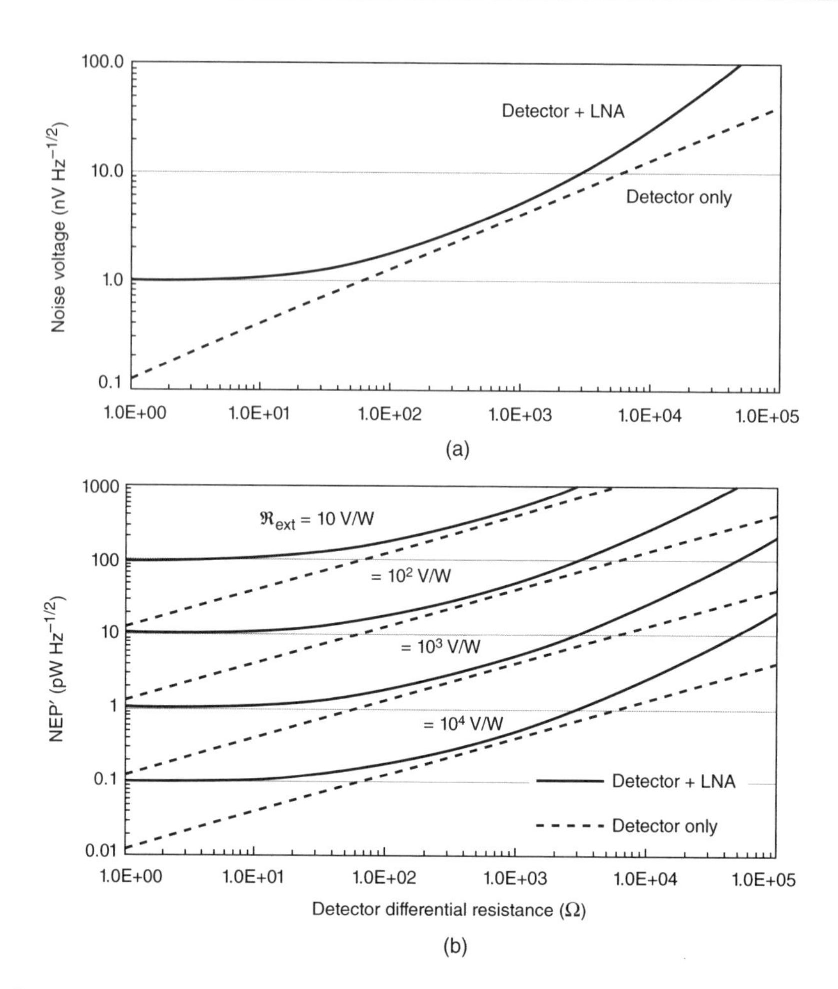

Figure 5.17 (a) Noise characteristics with THz detector connected to an opamp-LNA assuming noise characteristics of the AD797 (above ~100 Hz) and plotted against detector differential resistance. (b) Specific NEP values for THz detector versus its differential resistance given the AD797 noise characteristics and parameterized by detector external responsivity.

persists in the analysis of the NEP′, as calculated for Eq. (5.96) and plotted in Figure 5.17b, except now we need to include the detector external responsivity $\Re_{ext}$ as an additional parameter. We include four such values in the plot, 10, 100, 1000, and 10 000 V/W , which spans the values usually provided by THz room-temperature direct detectors. Again, the minimum impact of the amplifier noise occurs at $R_D = R_N$, but at much lower or higher values there is significant degradation of the NEP′ rather independent of $\Re_{ext}$. The best Schottky rectifiers are described reasonably well by the $\Re_{ext} = 1000$ V/W curves, and the best microbolometers by the 100 V/W curves.

5.7 A Survey of Experimental THz Detector Performance

In this section we tabulate experimental results on THz room-temperature direct detectors. Strong preference is given to results that include measured values of NEP or NETD. As this chapter has proposed, at THz frequencies, external or optical responsivity is far more telling than internal or electrical responsivity, and many publications only report the latter. Furthermore, responsivity is only part of the story in THz detectors, NEP being far more important from a sensor-system standpoint. As shown in the following tables, there is plenty of experimental NEP data for rectifier-type detectors, and even more for microbolometers, much of the microbolometer data being over 10 years old. The results tabulated for CMOS-based and plasma-wave detectors are recent and rapidly growing, testament to the fact that the CMOS technology is more generic and widely available. The data for pyroelectric detectors and Golay cells is the most sparse of all, which is mostly why we have not analyzed their performance in any detail. In fact, the 530 GHz data points at the end of each table were obtained by the author in cross-calibration experiments conducted several years ago.

The very best results to date for each detector type are compared in the bar-chart of Figure 5.18. In composing this, a decision had to be made about the test frequency. We chose to seek frequencies as close-as possible to 650 GHz as this is the center frequency of a well-known atmospheric window – perhaps the last one useful for terrestrial sensing before the atmosphere becomes prohibitively lossy. All values shown are external NEPs, meaning that they include the effects of the external coupling efficiency η. Interestingly, all the Si-CMOS- or SiGe-based devices are quite close to each other, hovering around 50 pW $Hz^{-1/2}$. In contrast the best pyroelectric detector and Golay cell lie just above and below 1 nW $Hz^{-1/2}$ which begs the question why they are so popular. The answer lies in their ease-of-use and their benchmarking utility. Both have physical wide apertures (5 or 6 mm) so it is quite easy to focus THz radiation into them and be assured that the majority of power is being collected (although not necessarily absorbed). Given this benefit and their approximately nanowatts sensitivity, they are useful devices to characterize THz sources of all types, which generally needs to be done with a SNR of at least 10 to be confident in the output power measurement. This means that the pyroelectric and Golay cell can establish whether or not a given source is putting out ~10 nW or more – an average power level thought to be minimally useful in THz sensor systems.

Also in the tables a strong preference is given to reported detector results in the conventional THz region, 300 GHz and above, since this frequency is the low-frequency edge of the older "submillimeter-wave" region. There are many excellent detector results in the literature that claim or allude to THz technology impact based on experimental results measured in the W-band (75–110 GHz) – one of the popular "millimeter-wave" bands for many decades. This approach is often dictated by the availability of source, control, and calibration components and instrumentation (usually in a rectangular metal waveguide) in the W band, but not at any higher frequency. The problem with

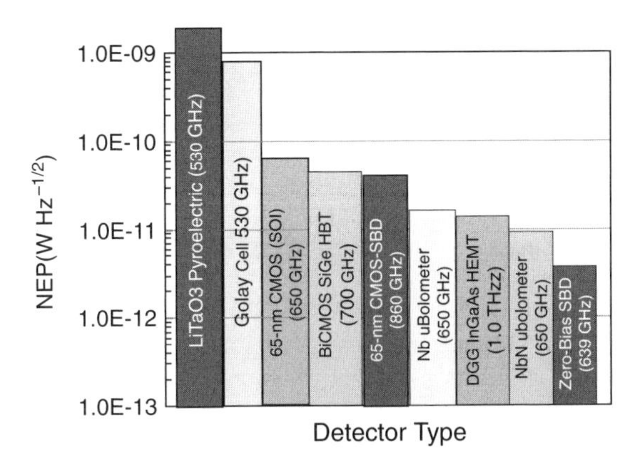

Figure 5.18 Comparison of the best room-temperature direct detectors operating >300 GHz and as close to 650 GHz as possible (SOI = silicon on insulator; HBT = heterojunction bipolar transistor; SBD = Schottky barrier diode); DGG = dual grating gate.

this is twofold. First, one of the great challenges with THz detectors is radiative coupling efficiency, which tends to drop rapidly as the frequency increases and for a variety of reasons. Some detector types (e.g., rectifiers) are more susceptible to this than others (e.g., bolometers), but there are always fundamental issues, such as skin-effect losses, planar-antenna feed losses, and so on, that come to bear as the frequency is increased from ~100 GHz to ~1.0 THz.

The second problem is that W-band direct detectors are of increasingly questionable importance because of technological competition. Integrated circuit technologies (both GaAs- and InP-based MMIC, Si RFIC, and SiGe) all work rather well up to 100 GHz already, with InP-based MMICs (monolithic microwave integrated circuits) operating well above 300 GHz. One of the key components from all of these is transistor-based LNAs with useful levels of gain (>10 dB) and noise figure (<6 dB). Given affordable and commercially available RF LNAs, the likelihood of engineers building sensor systems with direct-detection front-ends diminishes. This is particularly true for the Si-based radio frequency integrated circuits (RFICs) that can rather easily be proliferated into 1D- or 2D-receiver arrays.

Similarly, there are many publications and sometimes allusion to the THz impact of mid-IR measurements of direct detectors, especially thermal types in the long-wavelength infrared (LWIR) (8–12 μm band). Again the reason for many such measurements has to do largely with the availability of instrumentation, such as QCLs (quantum cascade lasers) and CO_2 lasers, sensitive power meters, and video-rate uncooled imaging arrays in the LWIR region. The problem in extending these results to THz wavelengths also includes fundamental issues such as radiative coupling. For example, microbolometers in the LWIR can be constructed in which the absorbing element and the thermistor element are one and the same. The dimensions of both can be made small enough for high sensitivity (low G) and high speed (small C/G), but large enough for diffraction-limited coupling to free-space optics in the LWIR. The diffraction-limited coupling then becomes a big problem in the THz region.

5.7.1 Rectifiers

Device type	Electrical responsivity (V/W)	Conditions	NEP (W Hz$^{-1/2}$); NETD (where provided)	References
ErAs/InGaAlAs Schottky in square spiral planar antenna	—	Zero bias, 639 GHz, 300 K	4.0E−12; NETD = 120 mK	[29]
Zero-bias Schottky diode, waveguide mounted	4000 at 100 GHz to 400 at 900 GHz	Zero bias	1.5E−12 near 150 GHz, 20E−12 at 800 GHz	[35]
Zero-bias InGaAs Schottky diode with log-spiral antenna	~200 for system, estimate 10^3 intrinsic for the diode	0.8 THz	5.0E−10	[36]
GaAs Schottky with traveling wave antenna	200 at 337 μm, 20 at 119 μm	—	0.1E−9 and 1.0E−9	[37]
Silicon diode thermal detector array	5.3	—	NETD = 1.7 °C	[38]
VDI model: WR2.8 ZBD	1500	260–400 GHz	2.7E−12	[39]
VDI Model: WR2.2 ZBD	1250	325–500 GHz	3.2E−12	[39]

Device type	Electrical responsivity (V/W)	Conditions	NEP ($W Hz^{-1/2}$); NETD (where provided)	References
VDI Model: WR1.9 ZBD	1000	400–600 GHz	4.1E−12	[39]
VDI Model: WR1.5 ZBD	750	500–750 GHz	5.1E−12	[39]
VDI Model: WR1.2 ZBD	250	600–900 GHz	16E−12	[39]
VDI Model: WR1.0 ZBD	200	750–1100 GHz	20E−12	[39]
VDI Model: WR0.8 ZBD	100	900–1400 GHz	40E−12	[39]
VDI Model: WR0.65 ZBD	100	1100–1700 GHz	40E−12	[39]
W-InSb point contact diode	25–30	313 μm, zero bias, room temperature	2.0E−09	[40]
GaAs Schottky diode	2000	214 μm	—	[41]
Ni-NiO-Ni MoM diode	—	3 kHz mod 98 GHz and 28.3 THz	50E−12 and 72E−12	[42]

5.7.2 *Thermal Detectors*

5.7.2.1 Bolometers

Device type	Electrical responsivity (V/W)	Conditions	NEP ($W Hz^{-1/2}$)	Citation
Bismuth, room temperature	1	70-Ω device; bias = 1 V; video mod of 300 Hz @ 185 GHz	3E−9	[43]
$Hg_{0.8}Cd_{0.2}Te$ HEB	0.30 at 17 mV bias, 36 GHz; 95 for 0.89 THz, 12 mV bias	Room temperature	2.4E−9 for 17 mV bias, 36 GHz; 7.4E−9 for 0.89 THz, 12 mV bias	[44]
Nichrome/tellurium composite microbolometer	120	Ambient, 30 kHz	6.7E−9	[45]
Si_xGe_y:H	170	0.934 THz, uncooled	0.2E−9	[46]
HgCdTe HEB	—	37 GHz, uncooled	4.00E−10	[47]

Device type	Electrical responsivity (V/W)	Conditions	NEP (W Hz$^{-1/2}$)	Citation
Vanadium oxide	—	Uncooled	320E−12 at 4.3 THz, 9E−13 @ 7.5–14 μm	[48]
Nb	—	Uncooled	4.00E−10	[49]
$Hg_{0.8}Cd_{0.2}Te$ HEB	95×10^{-3}	0.89 THz, 200 Hz mod, 300 K	7.40E−09	[50]
Niobium film	21	3.6 mA bias, 1 kHz mod, uncooled	1.10E−10	[32]
Ti; antenna-coupled microbolometer	—	10 kHz chop, 1.04 mA bias, room temperature	1.50E−11	[51]
Nb	8	0.25 V bias, room temperature, 3.6 mA, 1 kHz	4.5E−10	[52]
Nb_5N_6	400	0.4 mA bias, >10 kHz; electrical NEP	9.8E−12	[53]
Vanadium oxide array	1.5E4	1 V bias, 130 μm, uncooled	2.00E−10	[54]
Nb, polyimide; antenna coupled	450	<1 THz, no cryogenics mentioned	1.50E−11	[55]
Nb_5N_6	100.5	2 mA bias, 300 K, 1 kHz modulation	3.98E−10	[56]
Al/Nb; antenna coupled	85	1 kHz mod; 1.6 mA bias	2.50E−11	[57]
Free-standing Nb bridge; antenna coupled	163 (average over five devices)	—	17 E−12 (average over five devices); 650 GHz	[58] and Dr. Erich Grossman (personal communication)
Free-standing NbN bridge; antenna coupled	210 (average over 10 devices)	—	12.5 E−12 (average over 10 devices); 650 GHz	[58] and Dr. Erich Grossman (personal communication)
Nb	—	Uncooled, 3 kHz modulation	8.30E−11	[59]

5.7.2.2 Pyroelectrics

Device type	Electrical responsivity (V/W)	Conditions	NEP (W Hz$^{-1/2}$)	Citation
Philips P5219 deuterated L-alanine TGS pyroelectric	321	10 Hz modulation; amplifier with gain of 4.8, 91 GHz	3.1E−08	[60]
QMC Instruments	18 300	10 Hz modulation, room temperature	4.40E−10	[61]
	1 200	1.89 THz, <20 Hz modulation	—	[62]
LiTaO3 w/chrome	—	Room temperature, 530 GHz, Molectron Model SPH-45	2.0E−9	E.R. Brown and D. Dooley (unpublished)

5.7.2.3 Golay Cells

Device type	Electrical responsivity (V/W)	Conditions	NEP (W/Hz$^{-1/2}$)	Citation
Tydex GOLAY CELL GC-1X	100 000	21 Hz chopper	1.4E–10	[63]
Microtech Instruments Golay cell	10^4	12.5 Hz chopper	10E–8	[64]
Micro-array, LbL (layer by layer) polymer membranes over Si	—	30 Hz mod 105 GHz	300E–9	[65]
Tydex Golay cell, 6-mm-diameter Diamond Window	—	10 Hz modulation	7.0E–10	[62]

5.7.3 CMOS-Based and Plasma-Wave Detectors

Device type	Electrical responsivity	Conditions	NEP (W Hz$^{-1/2}$)	Citation
BiCMOS SiGe 0.25 µm, HBT	Current R_I 1 A/W at 0.7 THz	Array 3 × 5, chopper 125 kHz	50E–12 at 0.7 THz	[66]
BiCMOS SiGe 0.13 µm, HBT	Voltage R_V 11 kV/W at 0.17 THz	Chopper 0.16 kHz	10E–12 at 0.17 THz	[67]

Device type	Electrical responsivity	Conditions	NEP (W Hz$^{-1/2}$)	Citation
BiCMOS SiGe 0.25 μm, NMOS	Voltage R_V 80 kV/W at 0.6 THz	3 × 5 array, chopper 16 kHz	300E–12 at 0.6 THz	[68]
CMOS, 65 nm, NMOS	Voltage R_V 140 kV/W at 0.87 THz	32 × 32, chopper 5 kHz	100E–12 at 0.87 THz	[69]
CMOS, 65 nm, NMOS	Voltage R_V 0.8 kV/W at 1 THz	3 × 5, chopper 1 kHz	66E–12 at 1 THz	[70]
CMOS, 65 nm, SOI	Voltage R_V 1 kV/W at 0.65 THz	3 × 5, chopper 1 kHz	54E–12 at 0.65 THz	[71]
130-nm CMOS-SBD	Voltage R_V 0.323 kV/W at 0.28 THz	4 × 4, chopper 1000 kHz	29E–12 at 0.28 THz	[72]
65-nm, CMOS-SBD	—	1 element; 1 MHz mod	42E–12 at 0.86 THz	[73]
CMOS, 150 nm, NMOS	Voltage R_V 11 V/W at 4.1 THz	1 element	1330E–12 at 4.1 THz	[74]
InGaAs HEMT	R_V = 23 kV/W at 200 GHz	1 element	0.5E–12 at 200 GHz	[74, 75]
Asymmetric dual-grating gate InGaAs HEMT	R_V = 2.2 kV/W at 1.0 THz	1 element	15E–12 at 1.0 THz	[76] and Dr. Michael Shur (personal communication)
Asymmetric dual-grating gate InGaAs HEMT	R_C = 6.4 kV/W at 1.5 THz	1 element	50E–12 at 1.5 THz	[75] and Dr. Michael Shur (personal communication)

References

[1] Torrey, H.C. and Whitmer, C.A. (1948) *Crystal Rectifiers*, Chapter 11, McGraw-Hill, New York.
[2] Brown, E.R., Young, A.C., Zimmerman, J. *et al.* (2007) Advances in Schottky rectifier performance. *IEEE Microwave Magazine*, **8**, 54.
[3] Lee, Y.-S. (2009) *Principles of Terahertz Science and Technology*, Springer, New York, pp. 151–154.
[4] Kruse, P.W. (1997) Principles of uncooled infrared focal plane arrays, in *Uncooled Infrared Imaging Arrays and Systems*, vol. **3** (eds P.W. Kruse and D.D. Skatrud), Academic Press, San Diego, CA, pp. 25–29.
[5] Golay, M.J.E. (1947) A pneumatic infra-red detector. *Review of Scientific Instruments*, **18**, 357.
[6] Tydex http://www.tydexoptics.com/en/products/thz_optics/golay_cell/ (accessed 27 November 2014).
[7] Dyakonov, M. and Shur, M. (1996) Detection, mixing, and frequency multiplication of terahertz radiation by two-dimensional electronic fluid. *IEEE Transactions on Electron Devices*, **43**, 380.
[8] Knap, W., Rumyantsev, S., Vitiello, M.S. *et al.* (2013) Nanometer size field effect transistors for terahertz detectors. *Nanotechnology*, **24**, 214002.

[9] Ojefors, E., Pfeiffer, U.R., Lisauskas, A., and Roskos, H.G. (2009) A 0.65 THz focal plane array in a quarter-micron CMOS process technology. *IEEE Journal of Solid-State Circuits*, **44**, 1976–1968.
[10] Lisauskas, A., Pfeiffer, U., Ojefors, E. *et al.* (2009) Rational design of high responsivity detectors of terahertz radiation based on distributed self-mixing in silicon field effect transistors. *Journal of Applied Physics*, **105**, 114511.
[11] Gutin, A., Kachorovskii, V., Muraviev, A., and Schur, M. (2012) Plasmonic terahertz detector response at high intensities. *Journal of Applied Physics*, **112**, 014508.
[12] Maas, S. (1998) *The RF and Microwave Circuit Design Cookbook*, Artech House, Norwood, MA.
[13] Nyquist, H. (1928) Thermal agitation of electrical charge in conductors. *Physical Review*, **32**, 110.
[14] Kingston, R.H. (1978) *Detection of Optical and Infrared Radiation*, Springer, New York, Sect. 2.2.
[15] Ashcroft, N.W. and Mermin, N.D. (1976) *Solid State Physics*, Holt, Rinehart, and Winston, New York, p. 25.
[16] Schottky, W. (1918) Über spontane Stromschwankungen in verschiedenen Elektrizitätsleitern. *Annalen der Physik* (in German), **57**, 541.
[17] van der Ziel, A. (1986) *Noise in Solid State Devices and Circuits*, Chapter 3, John Wiley & Sons, Inc, NewYork.
[18] Kittel, C. (1996) *Introduction to Solid State Physics*, 7th edn, John Wiley & Sons, Inc, New York, Chapter 5, and Chapter 6 Table 3.
[19] Reif, F. (1965) *Fundamentals of Statistical and Thermal Physics*, McGraw-Hill, New York.
[20] Boyle, W.S. and Rodgers, K.F. Jr. (1959) Performance characteristics of a new low-temperature bolometer. *Journal of the Optical Society of America*, **49**, 66.
[21] Davenport, W.B. and Root, W.L. (1958) *An Introduction to the Theory of Random Signals and Noise*, McGraw-Hill, New York, Sec. 12-2.
[22] Kraus, J.D. (1986) *Radio Astronomy*, 2nd edn, Cygnus-Quasar, Powell, OH, Sect. 7-1.
[23] Chang, K., York, R.A., Hall, P.S., and Itoh, T. (2002) Active integrated antennas. *IEEE Transactions on Microwave Theory and Techniques*, **50**, 937.
[24] Rutledge, D.B., Neikirk, D.P., and Kasilingam, D. (1984) Integrated-circuit antennas, in *Infrared and Millimeter Waves: Millimeter Components and Techniques, Part II*, vol. **11** (ed K.J. Button), Academic Press, New York.
[25] Bolivar, P.H., Brucherseifer, M., Rivas, J.G. *et al.* (2003) Measurement of the dielectric constant and loss tangent of high dielectric-constant materials at terahertz frequencies. *IEEE Transactions on Microwave Theory and Techniques*, **51**, 1062.
[26] Neikirk, D.P., Rutledge, D.B., and Lam, W. (1984) Far-infrared microbolometer detectors. *International Journal of Infrared and Millimeter Waves*, **5**, 245.
[27] Segovia-Vargas, D., Montero-de-Paz, J., Crooks, J. *et al.* (2012) 300 GHz CMOS video detection using broadband and active planar antennas. European Conference on Antennas and Propagation 2012, EUCAP 2012.
[28] Balanis, C.A. (2005) *Antenna Theory*, 3rd edn, John Wiley & Sons, Inc, Hoboken, NJ, Sect. 14.2.1.
[29] Brown, E.R., Young, A.C., Bjarnason, J.E. *et al.* (2007) Millimeter and submillimeter wave performance of an ErAs:InAlGaAs Schottky diode coupled to a single-turn square spiral. *International Journal of High Speed Electronics and Systems*, **17**, 383.
[30] Neikirk, D.P. and Rutledge, D.B. (1984) Air-bridge microbolometer for far-infrared detection. *Applied Physics Letters*, **44**, 153.
[31] Stojanovic, N., Maithripala, D.H.S., Berg, J.M., and Holtz, M. (2010) Thermal conductivity in metallic nanostructures at high temperature: electrons, phonons, and the Wiedemann-Franz law. *Physical Review B*, **82**, 075418.
[32] MacDonald, M. E. and Grossman, E. N. Niobium microbolometers for far-infrared detection *IEEE Transactions on Microwave Theory and Techniques* **43**, p. 893 (1995).
[33] Horowitz P. and Hill W., *The Art of Electronics* 2nd edn, (Cambridge University Press, Cambridge, 1989), Figure 7.42.
[34] Analog Devices Ultralow Distortion Ultralow Noise Op Am, Data Sheet AD797, Analog Devices Corporation, Norwood, MA, http://www.analog.com/static/imported-files/data_sheets/AD797.pdf (accessed 28 November 2014).
[35] Hesler, J.L. and Crowe, T.W. (2007) NEP and responsivity of THz zero-bias Schottky diode detectors. International Conference on Infrared and Millimeter Waves (IRMMW-THz), 2007, p. 844.
[36] Semenov, A., Cojocari, O., Hubers, H.-W. *et al.* (2010) Application of zero-bias quasi-optical schottky-diode detectors for monitoring short-pulse and weak terahertz radiation. *IEEE Electron Device Letters*, **31**, 674.
[37] Kubarev, V.V., Kazakevitch, G.M., Uk Jeong, Y., and Lee, B.C. (2003) Quasi-optical highly sensitive Schottky-barrier detector for a wide-band FIR FEL. *Nuclear Instruments and Methods in Physics Research Section A: Accelerators, Spectrometers, Detectors and Associated Equipment*, **507**, 523.

[38] Xu, Y.-P., Huang, R.-S., and Rigby, G.A. (1993) A silicon-diode-based infrared thermal detector array. *Sensors and Actuators A*, **37-38**, 226.
[39] VDI http://vadiodes.com/index.php/en/products/detectors (accessed 28 November 2014).
[40] Bertolini, A., Carelli, G., Massa, C.A. *et al.* (1999) Detection and mixing properties of an InSb metal-semiconductor point contact diode. *International Journal of Infrared and Millimeter Waves*, **20**, 1121.
[41] Titz, R.U., Roser, H.P., Schwaab, G.W. *et al.* (1990) Investigation of GaAs Schottky barrier diodes in the THz range. *International Journal of Infrared and Millimeter Waves*, **11**, 809.
[42] Abdel-Rahman, M.R., Monacelli, B., Weeks, A.R. *et al.* (2005) Design, fabrication, and characterization of antenna-coupled metal-oxide-metal diodes for dual-band detection. *Optical Engineering*, **44**, 066401.
[43] Ling, C.C. and Rebeiz, G.M. (1990) A wideband monolithic submillimeter-wave quasi-optical power meter. IEEE Microwave Symposium Digest, 1990, p. 1315.
[44] Dobrovolsky, V.N., Sizov, F.F., Kamenev, Yu.E. *et al.* (2007) Millimeter and submillimeter semiconductor hot electron bolometer. Sixth International Kharkov Symposium on Physics and Engineering of Microwaves, Millimeter and Submillimeter Waves 2007, Vol. **1**, p. 198.
[45] Wentworth, S.M. and Neikirk, D.P. (1992) Composite microbolometers with tellurium detector elements. *IEEE Transactions on Microwave Theory and Techniques*, **40**, 196.
[46] Kosarev, A., Rumyantsev, S., Moreno, M. *et al.* (2010) SixGey:H-based micro-bolometers studied in the terahertz frequency range. *Solid-State Electronics*, **54**, 417.
[47] Dobrovolsky, V., Sizov, F., Zabudsky, V., and Momot, N. (2010) Mm/sub-mm bolometer based on electron heating in narrow-gap semiconductor. *Terahertz Science and Technology*, **3**, 33.
[48] Hu Q. and Lee A.W. (2010) Real-time, continuous-wave terahertz imaging using a microbolometer focal-plane array. US Patent 7692147 issued April 6, 2010.
[49] Hu, Q., de Lange, G., Verghese, S. *et al.* (1996) High-Frequency (>100 GHz) and High-Speed (<10 ps) Electronic Devices. Progress Report 139. Massachusetts Institute of Technology, Research Laboratory of Electronics.
[50] Dobrovolsky, V. and Sizov, F. (2007) A room temperature, or moderately cooled, fast THz semiconductor hot electron bolometer. *Semiconductor Science and Technology*, **22**, 103.
[51] Luukanen, A. (2003) High performance microbolometers and microcalorimeters: from 300 K TO 100 mK. PhD Dissertation. University of Jyvaskyla, Finland.
[52] Rahman, A., de Lange, G., and Hu, Q. (1996) Micromachined room-temperature microbolometers for millimeter-wave detection. *Applied Physics Letters*, **68**, 2020.
[53] Kang, L., Lu, X.H., Chen, J. *et al.* (2009) Room temperature Nb5N6 bolometer for THz detection. 34th International Conference on Infrared, Millimeter, and Terahertz Waves, pp. 1–2.
[54] Dem'yanenko, M.A., Esaev, D.G., Ovsyuk, V.N. *et al.* (2009) Microbolometer detector arrays for the infrared and terahertz ranges. *Journal of Optical Technology*, **76**, 739.
[55] Luukanen, A. and Viitanen, V.-P. (1998) Terahertz imaging system based on antenna-coupled microbolometers. *Proceedings of SPIE*, **3378**, 34.
[56] Lu, X.H., Kang, L., Chen, J. *et al.* (2008) A terahertz detector operating at room temperature. *Proceedings of SPIE*, **7277**, 72770N.
[57] Miller A. J., Luukanen A., Grossman E. N. Micromachined antenna-coupled uncooled microbolometers for terahertz imaging arrays *Proceedings of SPIE* **5411** (2004) 18–24.
[58] Dietlein C., Luukanen A., Penttila J.S., *et al.*, Performance comparison of Nb and NbN antenna-coupled microbolometers, *Proceedings of SPIE* **6549** (2007) 65490M.
[59] Rahman, A., Duerr, E., de Lange, G., and Hu, Q. (1997) Micromachined room-temperature microbolometer for mm-wave detectionand focal-plane imaging arrays. *Proceedings of SPIE*, **3064**, 122.
[60] Webb, M.R. (1991) A millimeter wave pyroelectric detector. *International Journal of Infrared and Millimeter Waves*, **12**, 1225.
[61] QMC Instruments http://www.qmcinstruments.co.uk/index.php?option=com_content&view=article&id=236&Itemid=530 (accessed 28 November 2014).
[62] Yang, J., Gong, X.-J., Zhang, Y.-D., and Jin, L. (2009) Research of an infrared pyroelectric sensor based THz detector and its application in CW THz imaging. *Proceedings of SPIE*, **7385**, 738521.
[63] Tydex Golay Cell GC-1X, http://tydexoptics.com/pdf/Golay_cell.pdf (accessed 28 November 2014).
[64] Microtech Instruments http://www.mtinstruments.com/downloads/Golay%20Cell%20Datasheet%20Revised.pdf (accessed 28 November 2014).
[65] Denison, D.R., Knotts, M.E., McConney, M.E., and Tsukruk, V.V. (2009) Experimental characterization of mm-wave detection by a micro-array of Golay cells. *Proceedings of SPIE*, **7309**, 7309J.

[66] Al Hadi, R., Grzyb, J., Heinemann, B., and Pfeiffer, U.R. (2013) A terahertz detector array in SiGe HBT technology. *IEEE Journal of Solid-State Circuits*, **48**, 2002–2010.
[67] Dacquay, E., Tomkins, A., Yau, K. *et al.* (2012) D-band total power radiometer performance optimization in an SiGe HBT technology. *IEEE Transactions on Microwave Theory and Techniques*, **60**, 813.
[68] Pfeiffer, U. and Ojefors, E. (2008) A 600-GHz CMOS focal-plane array for terahertz imaging applications. Proceedings of the European Solid-State Circuits Conference, p. 110.
[69] Al Hadi, R., Sherry, H., Grzyb, J. *et al.* (2012) A 1 k-pixel video camera for 0.7–1.1 terahertz imaging applications in 65-nm CMOS. *IEEE Journal of Solid-State Circuits*, **47**, 2999.
[70] Al Hadi, R., Sherry, H., Grzyb, J. *et al.* (2011) A broadband 0.6 to 1 THz CMOS imaging detector with an integrated lens. IEEE MTT-S International Microwave Symposium Digest, p. 1.
[71] Öjefors, E., Baktash, N., Zhao, Y. *et al.* (2010) Terahertz imaging detectors in a 65-nm CMOS SOI technology. Proceedings of the European Solid-State Circuits Conference, p. 486.
[72] Han, R., Zhang, Y., Kim, Y. *et al.* (2012) 280 GHz and 860 GHz image sensors using Schottky-barrier diodes in 0.13 m digital CMOS. IEEE International Solid-State Circuits Digest, p. 254.
[73] Boppel, S., Lisauskas, A., Seliuta, D. *et al.* (2012) CMOS-integrated antenna-coupled field-effect-transistors for the detection of 0.2 to 4.3 THz. IEEE SiRF Technical Digest, p. 77.
[74] Kurita, Y., Ducournau, G., Coquillat, D. *et al.* (2014) Ultrahigh sensitive sub-terahertz detection by InP-based asymmetric dual-grating-gate high-electron-mobility transistors and their broadband characteristics. *Applied Physics Letters*, **25**, 251114.
[75] Otsuji, T., Watanabe, S.A., Tombet, B. *et al.* (2013) Emission and detection of terahertz radiation using two dimensional plasmons in semiconductor nano-heterostructures for nondestructive evaluations. *Optical Engineering*, **53**, 031205.
[76] Watanabe, T., Boubanga Tombet, S., Tanimoto, Y. *et al.* (2013) InP- and GaAs-based plasmonic high-electron-mobility transistors for room-temperature ultrahigh-sensitive terahertz sensing and imaging. *IEEE Sensors Journal*, **13**, 89.

6

THz Electronics

Michael Feiginov[1,*], Ramón Gonzalo[2], Itziar Maestrojuán[3], Oleg Cojocari[4], Matthias Hoefle[4,5], and Ernesto Limiti[6]

[1] *Tokyo Institute of Technology, Department of Electronics and Applied Physics, Ookayama, Tokyo, Japan*
[2] *Universidad Pública de Navarra, Departamento de Ingeniería Eléctrica y Electrónica, Campus Arrosadia, Navarra, Spain*
[3] *Anteral S.L., edificio Jerónimo de Ayanz, Campus Arrosadía, Navarra, Spain*
[4] *ACST GmbH, Hanau, Germany*
[5] *Technische Universität Darmstadt, Institute for Microwave Engineering and Photonics, Darmstadt, Germany*
[6] *Università degli Studi di Roma Tor Vergata, Dipartimento di Ingegneria Elettronica, Rome, Italy*

6.1 Resonant-Tunneling Diodes

6.1.1 Historic Introduction

The concept of a resonant-tunneling diodes (RTDs) was put forward and experimentally demonstrated by Chang, Tsu, and Esaki in the beginning of the1970s [1, 2]. The fascinating concept that resonant tunneling can lead to a very particular *I–V* curve with a negative-differential-conductance (NDC) region attracted a lot of attention among physicists and engineers. Scientists all over the world started to study RTDs and related phenomena very intensively.

RTDs are tunnel devices and tunneling can be a very fast process, therefore it was expected that RTDs could operate at very high frequencies. Indeed, it was demonstrated in 1983 that RTDs can operate at frequencies up to 2.5 THz as passive devices [3]. Later, it was demonstrated that RTDs show response as nonlinear passive devices even up to 4 THz [4].

The most intriguing property of RTDs is that they are active two-terminal devices: due to NDC they could be used to build high-frequency oscillators (see also Section 2.3.1) and amplifiers. The fundamental frequencies of RTDs reached ~56 GHz in 1987 [5] and later, a frequency of 712 GHz was reported in 1991 [6]. In the subsequent almost 20 years nobody could achieve higher frequencies and, except for one work [7], nobody got close to it. Many attempts were made and all failed. Gradually, the

* Presently with Canon Inc., Frontier Research Centre, THz Imaging Division, Tokyo, Japan

Semiconductor Terahertz Technology: Devices and Systems at Room Temperature Operation, First Edition.
Edited by Guillermo Carpintero, Luis Enrique García Muñoz, Hans L. Hartnagel, Sascha Preu and Antti V. Räisänen.

opinion was established in the scientific community that RTDs had no future and were not suitable in practical applications.

However, a breakthrough came in 2010, when RTD oscillators with a fundamental-oscillation frequency slightly higher than 1 THz were finally reported [8]. Soon afterwards, in 2011, the frequency was increased to ∼1.1 THz [9], then to ∼1.3 THz in 2012 [10], and recently up to ∼1.4–1.5 THz [11, 12]. The output power at sub-THz frequencies is already in the milliwatt range: ∼0.6 mW at ∼600 GHz [13]. Sub-THz data communication links based on RTD oscillators were demonstrated recently [14] and RTD oscillators can be as small as a fraction of square millimeters [9]. The combination of these properties in one device makes RTD oscillators a unique technology. Now it is clear that RTDs have large perspectives at THz and sub-THz frequencies. RTDs could be the enabling technology to make THz applications mature enough finally to leave the premises of the research labs.

Back in 1991 and earlier, hollow waveguide resonators were used to make RTD oscillators operating up to 712 GHz [5, 6]. In 1997 another concept was suggested, RTDs were integrated with planar slot-antenna resonators, such oscillators reached a frequency of 650 GHz at that time [7]. Only in 2010 was the concept further improved to reach 1 THz [8]. This concept is often used nowadays in THz RTD oscillators.

6.1.2 Operating Principles of RTDs

The transmission properties of a double-barrier tunnel structure are illustrated in Figure 6.1. The barriers are highly transparent, when the energy of an electron incident on the barriers coincides with the energy of one of the resonant states in the quantum well (QW) between the barriers. The resonant tunnel transparency of the barriers can even be as high as 1, if the barriers are symmetric. The tunnel transparency of the barriers drops rapidly when the energy of the incident electron deviates slightly from the resonant energies. In a simplified picture, the barriers are transparent for the electrons with energy equal to that of the QW resonant states and are not transparent for all other electrons.

An RTD is usually a semiconductor structure with two tunnel barriers, as shown in Figure 6.1, and doped regions on both sides on the barriers, see Figure 6.2. At zero or low bias, the bottom of the two-dimensional (2D) QW subband is above the quasi-Fermi level in the emitter (left side of the structure), see Figure 6.2a. There are no electrons in the emitter which could tunnel through the QW

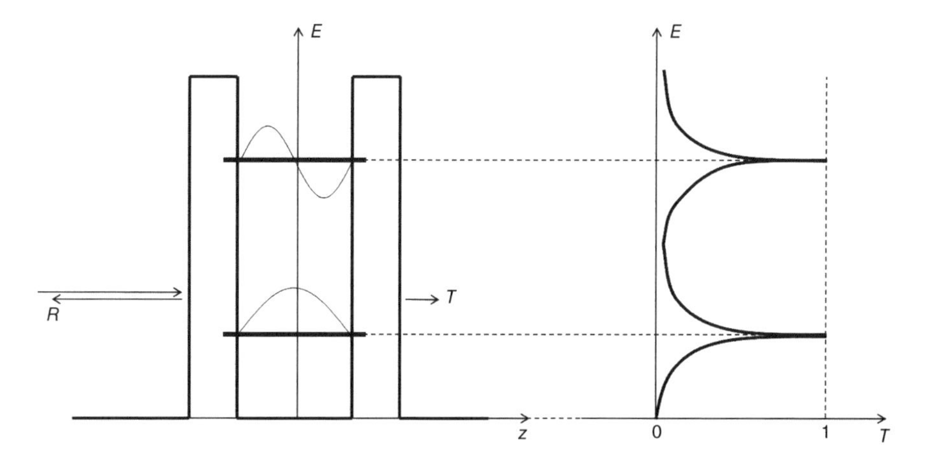

Figure 6.1 Electron transmission through two tunnel barriers. The tunnel coefficient (T) can be as high as 1, when the energy of the incident electrons is in resonance with one of the quantized states in the quantum well between the barriers. R is the reflection coefficient and E is the energy, the z axis is perpendicular to the barriers.

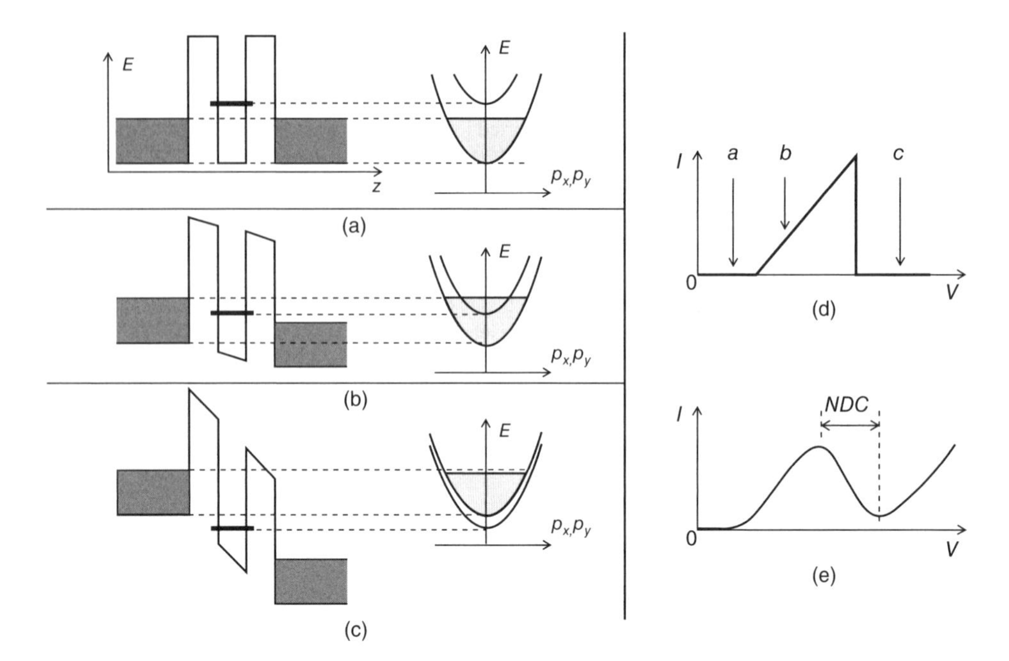

Figure 6.2 (a–c) Band diagram of an RTD at different biases. The right side of the pictures shows the electron spectrum of the QW subband (upper parabola in (a)) and the spectrum of the emitter electron states (the lower parabola in (a) filled by electrons up to the emitter quasi-Fermi level). The QW parabola shifts downwards with increase of bias (a–c). The electron quasi-momenta in the plain of the RTD barriers are denoted as p_x and p_y. (d) and (e) *I–V* curves of ideal and real RTDs, respectively.

resonant states in this situation. The current is close to zero at such biases, this is the region "a" in Figure 6.2d. At higher bias, the bottom of the resonant QW subband becomes lower than the emitter quasi-Fermi level, see Figure 6.2b, and the emitter electrons start to tunnel through the resonant states. The current starts to flow through the diode and it keeps growing with bias, since more and more QW resonant states are aligning with the filled emitter states. This is the region "b" with the positive differential conductance in Figure 6.2d. When we increase the bias even further, then, at some point, the QW subband shifts below the conduction-band bottom on the emitter side of the structure, see Figure 6.2c. Elastic tunneling from the emitter through the resonant states becomes impossible in this regime and the RTD current suddenly drops, region "c" in Figure 6.2d.

In an ideal case, the RTD *I–V* curve has a triangular form, shown in Figure 6.2d. The real RTD structures are not ideal, they have structural imperfections, the temperature is not zero, the electrons are subject to scattering, and so on, all these factors lead to broadening of the resonant features in the *I–V* curve and it becomes smoother. Sometimes, the resonant features get completely washed away, but good-quality RTDs show a smooth N-shaped *I–V* curve with well pronounced NDC region, see Figure 6.2e.

6.1.3 Charge-Relaxation Processes in RTDs

Tunnel electron lifetime (τ) is an obvious time constant in RTDs. If we put an electron in a subband in the QW, then τ is the time it will take for the electron to tunnel out of the QW through one of the barriers. Usually, it is assumed that the time τ imposes a fundamental limitation on the rate of the relaxation processes and on the reaction time of RTDs. It is also assumed that RTDs cannot operate at frequencies higher than $1/\tau$. This intuitive picture is not generally correct. The reason is Coulomb interaction effects.

Coulomb interaction can accelerate the relaxation processes in RTDs. It turns out that the processes are determined by a different time constant τ_{rel}, rather than τ [15, 16]. This can be illustrated with the help of a simpler structure with a single barrier and QW, as shown in Figure 6.3. Let us assume that we have a series of resonant states in the QW separated by the energy ΔE and each state is characterized by the same tunnel lifetime τ, see Figure 6.3a. In a stationary state, the contact (on the left side) and the QW are filled by electrons up to the same Fermi level, see Figure 6.3b. Let us introduce a perturbation in our system, we take an electron and move it from the contact into the QW, as shown in Figure 6.3c. The electron will occupy a state in the QW above the Fermi level, the states in the contact on the left side of it are empty and the electron can freely tunnel out of the QW. The relaxation time of such a process is τ. However, the electron is a charged particle and we have to take the Coulomb-interaction effects into account to describe the relaxation processes correctly. The additional electron will charge the QW and shift its bottom upwards, see Figure 6.3d. The Coulomb shift is equal to e^2/C, where e is the elementary electron charge and C is the capacitance between the QW and the contact. Due to the Coulomb shift, several additional filled QW states (their number is $\beta = e^2/C\Delta E$) will shift above the contact quasi-Fermi level. All these states (β plus the initial electron) will contribute to the relaxation process, but to bring the system back to equilibrium, only one electron out of $1 + \beta$ involved has to tunnel through the barrier. Therefore, τ_{rel} will be reduced by the factor $1 + \beta$ compared to τ. In practice, the parameter β can be large and that leads to strong acceleration of the relaxation processes due to Coulomb interaction.

The situation is very similar in double-barrier structures, see Figure 6.4. The stationary condition of the structure is sketched in Figure 6.4a. The electrons are tunneling from the emitter into empty states in the QW through a number of tunnel channels and from the filled states in the QW further to the collector. We treat the problem in a sequential-tunneling approximation [17]: the tunneling of electrons through the double-barrier structure is considered as a sequence of two independent tunneling events. First, an electron tunnels from the emitter into the QW (the process is characterized by the time constant τ_e)

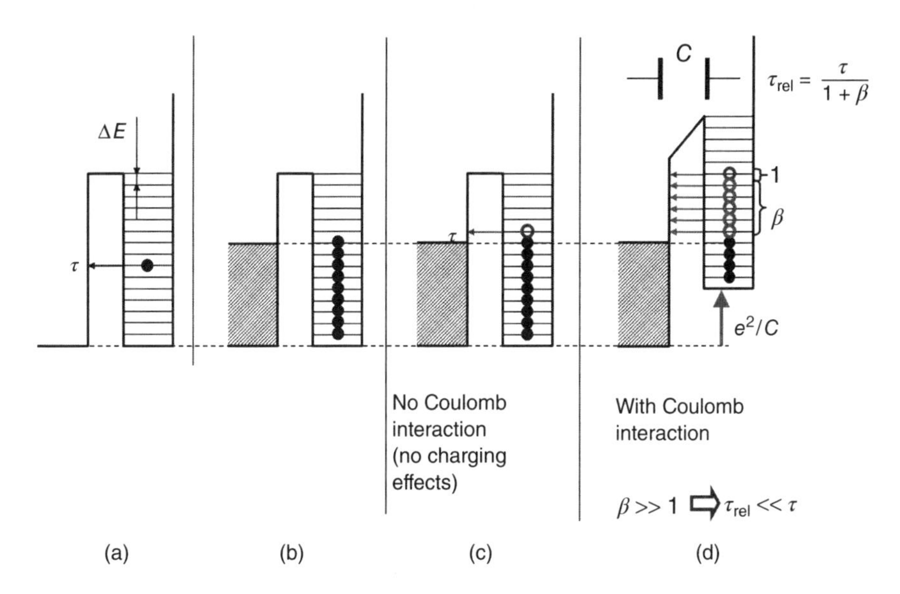

Figure 6.3 Charge-relaxation processes in a QW separated by a single barrier from the contact. See text for details of the processes in (a)–(d).

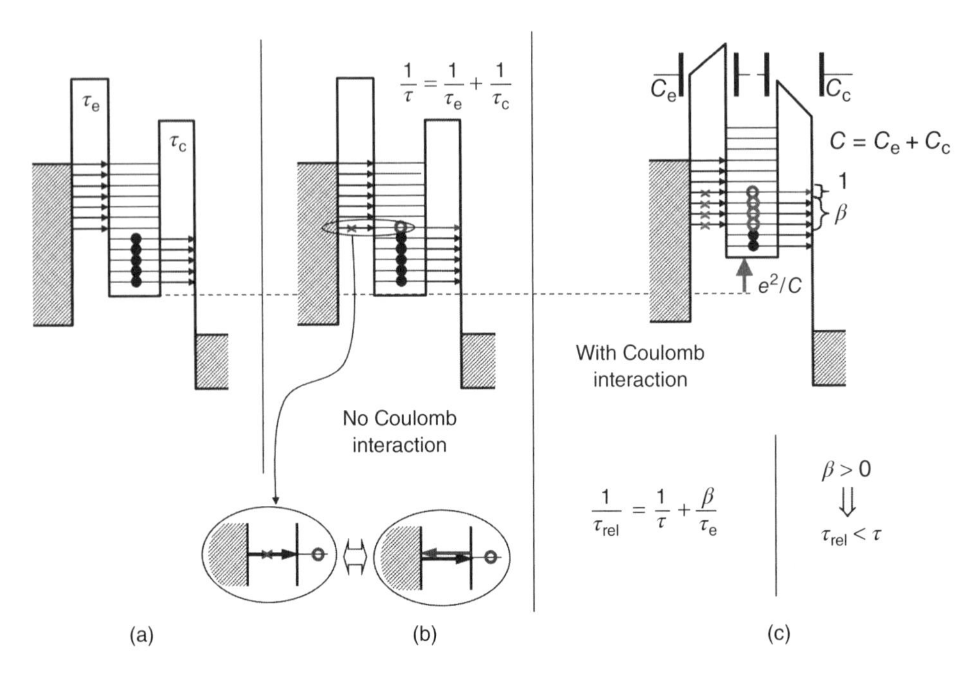

Figure 6.4 Charge relaxation processes in the double-barrier tunnel structures. See text for details of the processes in (a)–(c).

and then from the QW into the collector (the process is characterized by the time constant τ_c). As a next step, we introduce a perturbation into the system and move an electron from the contacts into the QW (Figure 6.4b). The additional electron in the QW will add an additional tunnel channel to the QW-to-collector current and will block a tunnel channel between the emitter and the QW. Blocking of a channel could be seen as if we have added a compensating tunnel channel from the QW to the emitter. Therefore, the additional electron in the QW will add two tunnel channels: one with the electron flow toward the emitter and the other with the flow toward the collector. The additional tunnel channels are responsible for the charge relaxation process with the time constant τ (Figure 6.4b). However, again, the electrons are charged particles and an additional electron in the QW should shift the QW bottom upwards by the same value e^2/C as before, see Figure 6.4c. As result, several ($\beta = e^2/C\Delta E$) additional tunnel channels between the emitter and the QW will be blocked or, equivalently, β tunnel channels will be added to the QW-to-emitter electron flow. The additional channels will accelerate the relaxation process and the relaxation time (τ_{rel}) will be given by:

$$\frac{1}{\tau_{rel}} = \frac{1+\beta}{\tau_e} + \frac{1}{\tau_c} = \frac{1}{\tau} + \frac{\beta}{\tau_e} \tag{6.1}$$

Again, the Coulomb interaction can significantly accelerate the relaxation process.

In the RTDs with 2D QW, we have to replace the series of the QW resonant states by a 2D subband with a 2D density of the electron states (ρ_{2D}). The qualitative picture described above does not change, but the parameter β will be [15, 16]:

$$\beta = \frac{e^2 S \rho_{2D}}{C}, \tag{6.2}$$

where C is the QW capacitance equal to the QW-emitter and QW-collector capacitances connected in parallel (see Figure 6.4c) and S is the RTD (mesa) area. The typical value of the parameter β in Eq. (6.2) in conventional RTDs is in the range 2–10. As a consequence, the Coulomb interaction can significantly, and sometimes dramatically, accelerate relaxation processes in RTDs [15, 16].

The above simple picture of Coulomb acceleration of the relaxation processes is applicable in the positive-differential-conductance region of the RTD *I–V* curve. The situation is more complicated in the NDC region. The 2D QW subband is becoming aligned with the emitter conduction-band bottom in the region. In this regime, small shifts of the QW subband can destroy or restore the resonant-tunneling conditions. That means that the tunneling rate through the emitter barrier (τ_e) is becoming very sensitive to the shift of the QW subband. We have to take this effect into account when we calculate the RTD relaxation time. If we do that properly [15, 16], then it turns out that the parameter β becomes negative in the NDC region and that means, from Eq. (6.1), that $\tau_{rel} > \tau$. If NDC is large, then the parameter β also becomes larger [15, 16]. As a consequence, $\tau_{rel} \gg \tau$ in RTDs with large NDC. In general, τ_{rel} is always longer, and often much longer, than τ in the NDC region of the RTD *I–V* curve, that is, RTDs are always relatively slow in the region [15, 16]. However, there is a way in which we can overcome the relaxation-time limit.

6.1.4 High-Frequency RTD Conductance

If we want to use RTDs in oscillators, then we should be concerned with the diode's differential conductance. It should be negative at the desired frequency, then we can use the RTDs to build oscillators. The analysis of the high-frequency behavior of RTDs can be done with the help of a Shockley–Ramo theorem [18, 19]. In the case of a double-barrier RTD, it can be written as:

$$j^{AC} = \frac{d}{l+d} j_e^{AC} + \frac{l}{l+d} j_c^{AC} + i\omega C_{ec} V^{AC} \tag{6.3}$$

where j is the total current flowing through the RTD emitter and collector (see Figure 6.5a), j_e and j_c are the emitter-barrier and collector-barrier tunnel currents, respectively; V is the voltage drop across the RTD barriers, the "AC" superscript denotes AC current and voltage components; d and l denote the emitter- and collector-barrier effective thicknesses, which include the screening and depletion lengths, C_{ec} is the geometrical emitter-collector capacitance. The tunnel currents through the emitter and collector barriers could be different in RTDs. Therefore we have two terms in Eq. (6.3) describing contributions of the barrier currents. The contribution of each of the currents is proportional to the factors describing which fraction of the total thickness of the RTD active region is occupied by the particular barrier (the "leverage" pre-factors in the first and second terms in Eq. (6.3)). The last term in Eq. (6.3) describes a purely imaginary capacitive contribution due to C_{ec}. Equation (6.3) includes both real- and displacement-current contributions.

Equation (6.3) allows us to analyze the behavior of RTD conductance at high frequencies [16, 20]. In the high-frequency regime, when $\omega\tau_{rel} \gg 1$, the electron concentration in the QW becomes a time-independent constant. The frequency is too high for the relaxation processes to follow the RTD-bias variation. In this regime, the collector-barrier current is changing solely due to variation of the collector-barrier transparency with bias. The mechanism leads to a positive contribution of the second term in Eq. (6.3) to the RTD conductance. The only place in RTD where we can get a negative contribution to the RTD conductance is the emitter barrier. The emitter-barrier current contribution is given by the first term in Eq. (6.3). Only there we are dealing with the resonant tunneling. In such a way, we see that only the emitter-barrier current gives the desirable negative contribution to the RTD conductance at high frequencies. The collector-barrier current plays a detrimental role at high frequencies: its contribution to the RTD conductance is positive.

The conventional RTDs often have weak doping or undoped spacer on their collector side. This is usually done to reduce RTD capacitance. In such RTDs, $d \ll l$. However, the negative contribution of the emitter-barrier current becomes strongly suppressed ($d/(d+l) \ll 1$) and the positive collector-barrier contribution plays a dominant role ($l/(d+l) \approx 1$) in such diodes at high frequencies. As a result, the RTD

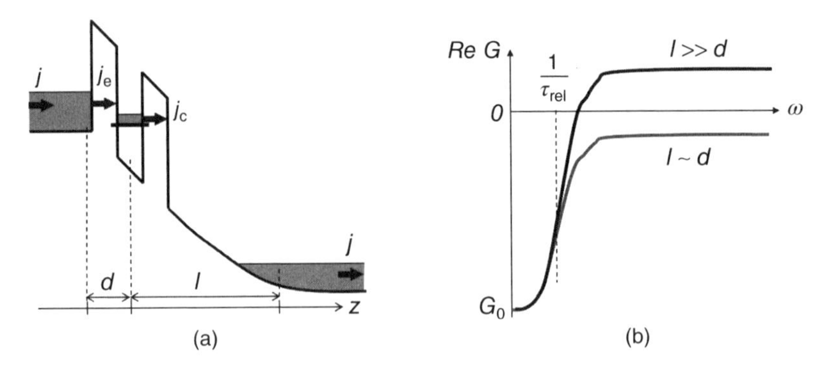

Figure 6.5 (a) Currents in different regions of RTDs. (b) Frequency dependence of conductance of conventional ($l \gg d$) RTDs and of RTDs with heavy collector doping ($l \sim d$).

conductance becomes positive at high frequencies, see Figure 6.5b. Such conventional diodes become passive lossy devices at high frequencies, they cannot be used to make oscillators working in the regime $\omega\tau_{rel} \gg 1$.

However, we can modify RTDs and make them work at high frequencies [16, 20]. We have to put heavy doping on their collector side. This eliminates or reduces the collector depletion region in such diodes. The effective thicknesses of both barriers will have similar values in this case ($l \sim d$). As a result, the emitter-barrier current contribution will be strongly enhanced in such diodes ($d/(d + l) \sim 0.5$) and the current will play a dominant role. In such diodes the high-frequency conductance should stay negative even in the high-frequency regime, see Figure 6.5b. In such a way we see that unconventionally-heavy doping of the collector allows one to extend the operating frequencies of RTDs and such diodes could be used to extend the operating frequencies of RTD oscillators [16, 20].

It has been proved experimentally that the above concept works. RTDs with a heavily doped collector have been studied in Ref. [21], where it was shown that the differential conductance of the diodes stays negative far beyond the relaxation-time and tunnel-lifetime limits, that is, when $\omega\tau_{rel} \gg 1$ and $\omega\tau \gg 1$, see Figure 6.6. It has been also shown in Ref. [21] that $\tau_{rel} > \tau$ in the NDC region of the I–V curve, although the opposite inequality ($\tau_{rel} < \tau$) is satisfied in the positive differential conductance region, as expected in theory [15, 16, 20].

6.1.5 Operating Principles of RTD Oscillators

There are several ways in which one can make THz and sub-THz oscillators with RTDs. For instance, one can use hollow-waveguide resonators, as shown in Figure 6.7a. An RTD is mounted inside a hollow waveguide in this case. Then a back short is put in the waveguide at some distance from the RTD. The piece of the waveguide between the RTD and the back short forms a hollow resonator. This approach was used in the 1980s and 1990s. Following this approach, a fundamental oscillation frequency of RTD oscillators of 712 GHz was achieved [6].

Later, another approach was suggested [7], where an RTD was mounted on a slot antenna. In principle, the slot antenna can work as a resonator for RTDs. However, RTDs usually represent a highly capacitive load and the length of the loaded antenna becomes much shorter than the length of an unloaded slot antenna with the same resonance frequency. The antenna works almost like a lumped-element inductor in this regime, since the antenna dimensions are small compared to the wavelength of the emitted radiation, see Figure 6.7c. The RTD connected to the antenna works similar to a lumped-element LC resonator. The emission efficiency of such resonators is quite low and additional elements could be connected to

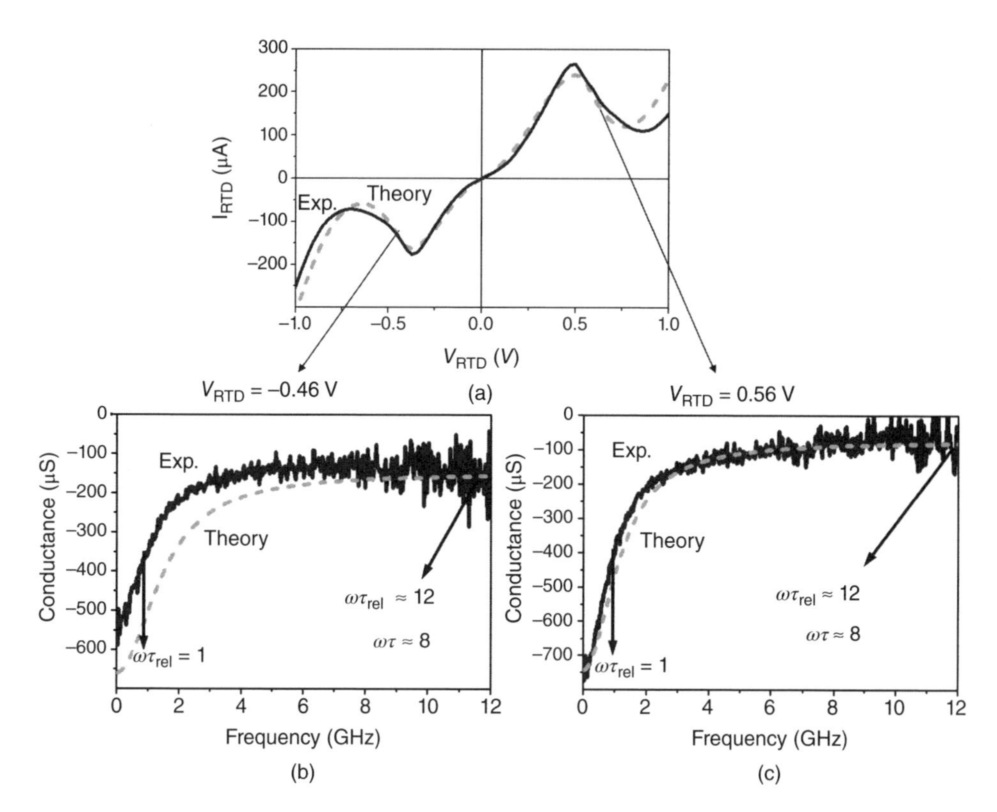

Figure 6.6 *I–V* curve (a) and frequency dependences of conductance (b,c) of an RTD with heavily doped collector studied in Ref. [21].

the antenna to improve its radiation efficiency. For example, an asymmetrically elongated slot antenna was studied in Ref. [13] to enhance the radiation efficiency and the slot antenna resonator was integrated with a planar Vivaldi antenna in another work [9], see Figure 6.8.

One more type of RTD resonator, which was suggested and demonstrated, is a patch antenna, see Figure 6.7b. The patch antennas are also resonant-type antennas. They have an advantage (compared to the slot antennas, which emit into the substrate) that they could be fabricated on top of a substrate and emit in the free space perpendicular to the substrate surface [22]. A Si lens is required in the case of a slot antenna to couple the radiation out of the substrate but the patch antenna oscillators do not need it. Therefore, the patch antennas allow one to get rid of additional lenses and that, in principle, simplifies construction of RTD THz sources and makes them easier to handle.

6.1.6 Limitations of RTD Oscillators

The oscillation conditions of an RTD oscillator are determined by a set of parameters. The oscillation frequency is defined by the resonance frequency of the resonator loaded by an RTD. To achieve oscillations, the total conductance of an RTD together with a resonator/antenna should be negative at the resonance frequency. The total conductance has several major positive contributions due to antenna/resonator radiation, ohmic loss in the antenna, RTD contact resistance, and so on. These positive contributions have to be compensated by the NDC of the RTD. Among the positive contributions, the RTD contact resistance

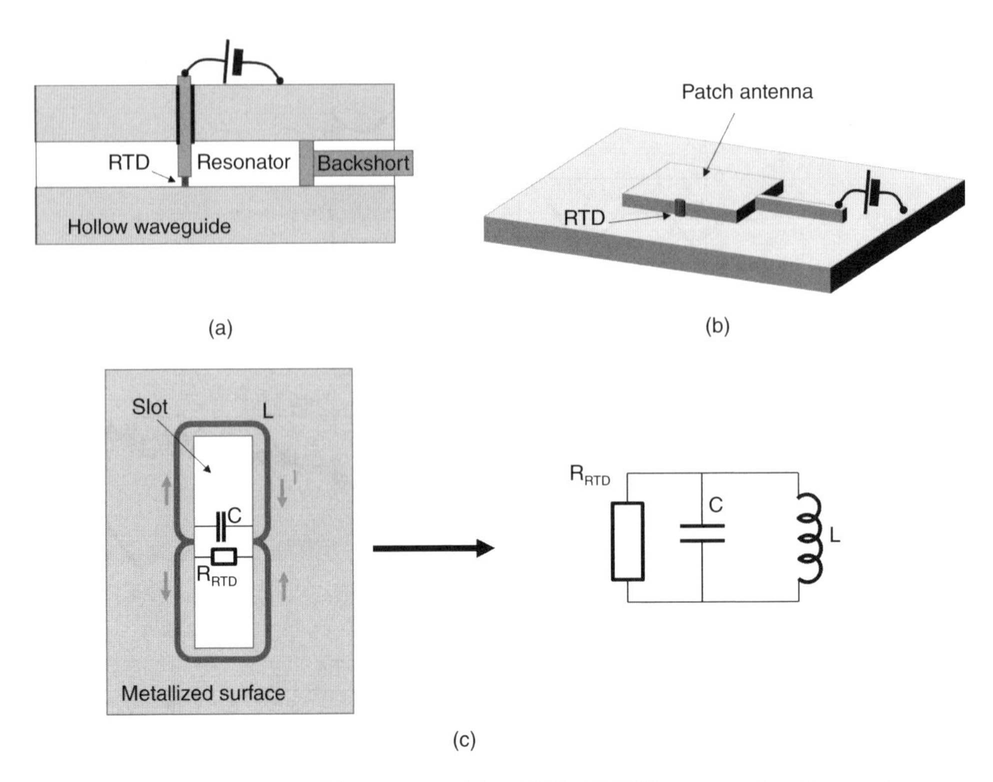

Figure 6.7 (a) Hollow-waveguide resonator with an RTD. (b) RTD resonator based on patch antenna. (c) Schematic of a slot-antenna resonator with an RTD mounted at the driving point of the antenna. If the RTD capacitance is large, the slot length becomes small compared to the emission wavelength. Then the current (I) loop around the edges of the slot creates a nearly lumped-element inductor.

is often the most critical parameter. In THz oscillators, the contact resistance is on the order of or below $10\,\Omega\,\mu\text{m}^2$ and it can be at the level of a few $\Omega\,\mu\text{m}^2$ in very good structures.

RTD oscillators should oscillate at the desired frequency and the oscillations at all other frequencies have to be suppressed. Often, it is problematic to satisfy this condition. RTDs exhibit NDC over a wide frequency range and NDC is usually higher at lower frequencies. This leads to low-frequency instability (parasitic oscillations) in the external RTD circuit. The external circuit is needed for the DC-bias supply to the diode. If special measures are not taken, RTDs almost always have parasitic oscillations at low frequencies. An easy way to suppress such parasitic oscillations is to include a parallel shunt resistor (with conductance G_{par}), which should be small enough to satisfy the condition $G_{\text{RTD}} + G_{\text{par}} > 0$, where G_{RTD} is the RTD conductance. For simplicity, we ignore here the contribution of contacts and other parasitic elements. The RTD together with the parallel resistor will have a positive conductance and that will keep the external circuitry stable. However, the electromagnetic field of the RTD resonator is usually well localized close to the resonator and the parallel shunt resistor (see an example in Figure 6.8) should have little influence on the oscillation conditions of the resonator at the desired THz or sub-THz frequency. This approach allows one to suppress the parasitic low-frequency oscillations.

To determine and analyze the oscillation conditions of an RTD oscillator, one needs to know the value of the RTD NDC at the desired oscillation frequency. One usually tries to make the RTD barriers as thin as possible, reducing τ_{rel} and the condition $\omega\tau_{\text{rel}} \ll 1$ is often satisfied at the oscillation frequency.

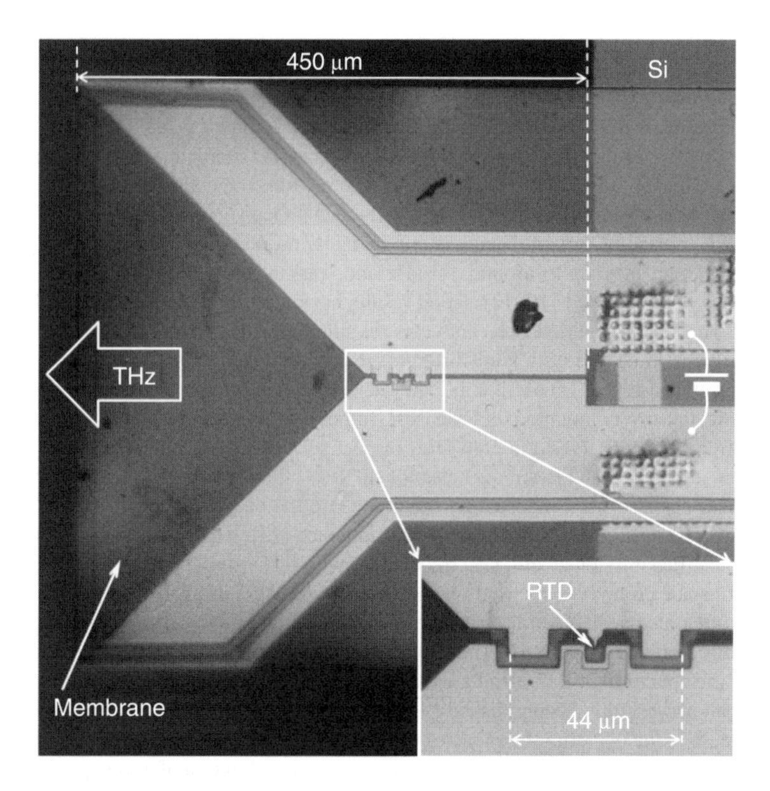

Figure 6.8 Slot-antenna RTD resonator integrated with planar Vivaldi antenna to enhance radiation efficiency. The oscillator with antenna is fabricated on a thin dielectric membrane. Reprinted with permission from Ref. [9] © 2011, AIP Publishing LLC.

This means that such RTDs are operating in a quasi-static regime and the AC NDC of the RTD at the operating frequency coincides with the DC NDC. The DC measurements of the RTD *I–V* curve provide information on the NDC at the desired oscillation frequency in such RTDs. The design of RTD oscillators is straightforward in this case.

However, when we move to higher frequencies, the condition $\omega\tau_{rel} \ll 1$ will be violated and the analysis of oscillation conditions becomes more complicated. RTD should have NDC at the target frequency, this is a necessary condition to achieve oscillations. In the conventional RTDs with a weak doping on the collector side and a large depletion layer there, the RTD NDC region is usually limited by the condition $\omega\tau_{rel} \sim 1$. This condition automatically limits the oscillation frequencies achievable with such RTDs. However, as we discussed in Section 6.1.4, the NDC region of RTDs could be extended beyond this limit, if high doping is applied on the collector side of the diode. The differential conductance of such RTDs stays negative at frequencies $\omega\tau_{rel} \gg 1$ and that allows one to increase the operating frequencies of RTD oscillators. The first experimental demonstration of the operation of RTD oscillators in the regime $\omega\tau_{rel} > 1$ and $\omega\tau > 1$ was described in Ref. [23], where RTD oscillators were working up to the frequency of 150 GHz and the parameters $\omega\tau_{rel} \approx 10$ and $\omega\tau \approx 3$ were achieved. That proved that RTD oscillators can operate beyond the tunnel and relaxation-time limits. Further, it has been shown experimentally that RTD oscillators can operate beyond those limits at higher frequencies, around ~560 GHz [24]. The later analysis of the RTD oscillators working at 1.1 THz in Ref. [9] has shown that the regime $\omega\tau_{rel} \geq 1$ should be achievable in the THz range and that should lead to realization of RTD oscillators working at 2–3 THz.

6.1.7 *Overview of the State of the Art Results*

There was no progress in the development of sub-THz oscillators for more than a decade since the 1990s However, RTD oscillators have been improving very rapidly in the last 5 years. Presently, there are three groups worldwide who have demonstrated THz and sub-THz RTD oscillators. The first was the group at Tokyo Institute of Technology (Tokyo Tech, Japan), where a fundamental oscillation frequency slightly higher than 1 THz was achieved for the first time in 2010 [8]. One year later, the group at the Technical University (TU) Darmstadt (Germany) demonstrated 1.1 THz oscillators [9]. Later, the Tokyo Tech group has increased the frequency up to around 1.3 THz and, very recently, oscillation frequencies around 1.4–1.5 THz have been reported [10–12]. Good results have also been reported by the Canon (Japan) group [22, 25]. The results reported in recent years are summarized in Figure 6.9.

The three different groups were studying different types of RTD oscillators. The Tokyo Tech group was focused on slot-antenna-type RTD oscillators. They have tried out different geometries of oscillators and antennas. The antennas were mounted on Si-lenses. Output powers around 1 and 0.36 μW were reported at the highest frequencies of 1.42 and 1.46 THz, respectively, and tens and several tens of microwatts were reported at slightly lower THz frequencies for single oscillators [11, 12]. The power combining of several oscillators has also been reported by the group and the output powers ~600 μW at ~600 GHz and ~200 μW at ~800 GHz were achieved with this approach [13]. Additionally, the group has recently demonstrated data transmission at sub-THz frequencies with RTD oscillators [14].

The TU Darmstadt group was focused on membrane-type RTD oscillators [9], where slot-antenna resonators are integrated with Vivaldi antennas for enhanced emission, see Figure 6.8. The estimated output power was around 1 μW at 1.1 THz and around 40 μW at ~0.5 THz. The focus of this group was to demonstrate the operation of RTD oscillators and RTDs beyond the tunnel and relaxation-time limits, and this has been successfully accomplished [9, 15, 16, 20–24]. The series of works lay down a basis for further increase in the operating frequencies of RTD oscillators. The RTD oscillators reported by the group are probably the smallest THz sources so far, they are just a fraction of a square millimeter in size, see Figure 6.8 [9].

The Canon group focused on RTD oscillators with patch antennas, see Figure 6.7b [22, 25]. They reported emission up to a frequency of 1.4 THz. However, the emission spectra show additional side lines and it is not yet clear if the emission was at the fundamental oscillation frequency [25]. The other particular property of the studied oscillators is that they are based on three-barrier RTDs, where NDC is achieved

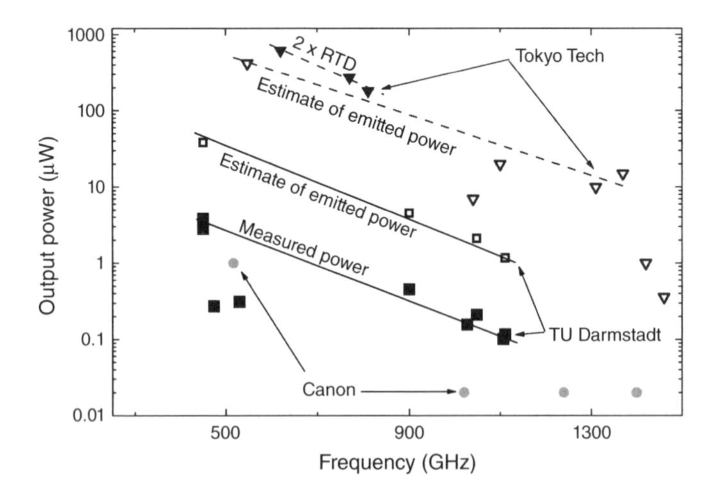

Figure 6.9 Summary of the recent results on sub-THz and THz oscillators.

due to tunneling between two resonant subbands in the adjacent QWs. The advantages and disadvantages of conventional (two-barrier) and three-barrier RTDs still need to be investigated and clarified.

The experimental results reported in recent years and their theoretical analysis suggest that RTD oscillators will probably reach the operation frequency of several THz and their THz output power will grow beyond 1 mW level in the next years.

6.1.8 RTD Oscillators versus Other Types of THz Sources

We have to compare RTD THz oscillators with other types of existing THz sources. In general, THz sources (as a part of THz systems) have until now been used exclusively as scientific instrumentation. Their use in practical real-world applications is prohibited due to essential deficiencies of the existing sources. For example, Schottky-diode multipliers [26] represent a well-developed mature technology extensively used nowadays in different THz instruments. However, the multipliers are relatively complex, expensive, and not very compact. Their output power is relatively low, below ~100 μW above 1 THz. THz quantum-cascade lasers (QCLs) [27] can provide high output power at the level of ~1 mW at 1–2 THz and above ~10 mW at higher frequencies, but they require cryogenic cooling, which makes such sources bulky and expensive. Additionally, THz QCLs operate only at the upper end of the THz frequency range. Monolithic microwave integrated circuit (MMIC) THz oscillators have exhibited rapid progress in recent years [28] and they have reached ~0.67 THz, but it seems that they can hardly operate at frequencies significantly higher than 1 THz in the foreseeable future.

THz photomixers (see Chapter 2) have a special position among THz sources. THz photomixing systems (both pulsed and continuous-wave) are now the most widely used THz measurement systems. They are commercially available and used in nearly all THz research laboratories. They have wide spectral coverage, are easily tunable and versatile THz measurement systems. They are basically ideal multipurpose scientific instruments for THz measurements. They are also attractive for some specific applications, for example, when a THz transmitter needs to be coupled to the optical signal transmitted over optical fibers (radio over fiber). However, photomixing systems could hardly be used in real-world THz applications. The photomixing systems are too complex and expensive for this purpose and they can provide only low output power. These factors prevent their use in affordable portable and handheld THz systems, like THz Wi-Fi, compact spectrometers, compact THz-cameras, and so on.

Compared to the existing THz sources, RTD oscillators seem to be a unique technology. The oscillators have the potential to satisfy all the necessary requirements for practical applications in the near future.

6.1.9 Future Perspectives

RTD oscillators are versatile sub-THz and THz technology, which can cover a wide frequency range from almost 0 up to 1.46 THz. RTD oscillators can already provide almost ~1 mW of output power at sub-THz frequencies. The analysis of the available experimental data indicates that RTD oscillators should work at several THz and their output power should surpass 1 mW very soon. Applications like high-speed data transmission have already been demonstrated with RTD oscillators. RTD oscillators are perhaps the most compact THz sources. They work at room temperature and require a low level of input DC power. All these factors open a vast and unique perspective for RTD oscillators in THz technology and real-world applications.

6.2 Schottky Diode Mixers: Fundamental and Harmonic Approaches

Despite the progress of THz low noise amplifiers (LNAs) [29–31], Schottky diode mixers are still used as the first element of receiver front-ends to down-convert the signal collected by the antenna to microwave frequencies where it can be amplified, demodulated, and analyzed easily. Schottky diode mixers have the advantage over other technologies of working at room temperature as well as cryogenic temperatures for improved noise performance, which makes them the technology of choice for many different applications; such as THz imaging cameras, biochemical material identification, medical imaging, wideband wireless communication systems, and so on.

A Schottky barrier diode uses a rectifying metal–semiconductor junction formed by plating, evaporating, or sputtering one of a variety of metals onto n-type or p-type semiconductor material. Generally, n-type silicon and n-type gallium arsenide (GaAs) are used in commercially available Schottky diodes. The principle of this kind of diode resides in the use of a metal contact deposited on a semiconductor n-type substrate, which provides high electron mobility. Then a very high-conductivity layer is grown on top of the substrate in order to assure low series resistance. The buffer and the substrate are heavily doped and an n-type epitaxial layer is grown on top of them. The contact between the metal anode and the epitaxial layer creates the Schottky contact [32].

When used as THz receivers, the main aim of the Schottky diode mixer devices is to perform frequency mixing, which consists in the conversion of a low power level signal (commonly called the radio-frequency (RF) signal) from one frequency to another by combining it with a higher power (local oscillator, LO) signal in a device with a nonlinear impedance (such as a diode or a transistor), see Figure 6.10. Mixing produces a large number of new frequencies which are the sums and differences of the RF and LO signals and their respective harmonics. In a down-converter mixer, the intermediate frequency (IF) corresponds with the desired output signal. In most applications this signal is the difference between the RF and LO frequencies.

The configuration most commonly employed is a heterodyne receiver. In this case, the Schottky diode acts as a nonlinear impedance to perform the mixing phenomenon. Heterodyne receivers are harmonic mixers in which the output signal is proportional to the product of the input voltages. It is defined as follows, Figure 6.10, [33]:

$$v_{\text{IF}} \cong K v_{\text{RF}}(t) v_{\text{LO}}(t) \tag{6.4}$$

where $v_{\text{RF}}(t)$ can be defined as a modulated carrier whose frequency is f_{RF}, $v_{\text{RF}}(t) = V_{\text{RF}} x(t)\cos(\omega_{\text{RF}} t + \phi(t))$, and $v_{\text{LO}}(t)$ is a sinusoidal signal with frequency f_{LO}, $v_{\text{LO}}(t) = V_{\text{LO}} \cos(\omega_{\text{LO}} t)$. The mixing of those signals will give v_{IF} which will consist of two modulated carriers of frequencies $f_{\text{RF}} + f_{\text{LO}}$ and $f_{\text{RF}} - f_{\text{LO}}$ with the same modulation as $v_{\text{RF}}(t)$. The output signal $v_{\text{IF}}(t)$ will be proportional to:

$$v_{\text{IF}}(t) \cong \frac{K}{2} V_{\text{LO}} V_{\text{RF}} x(t)[\cos((\omega_{\text{RF}} + \omega_{\text{LO}})t + \phi(t)) + \cos((\omega_{\text{RF}} - \omega_{\text{LO}})t + \phi(t))] \tag{6.5}$$

Therefore, the mixer output signal consists of two replicas of the input signal with a frequency shift of $\pm f_{\text{LO}}$, Figure 6.11. With this conversion, the input signal is translated to a much lower ($f_{\text{RF}} - f_{\text{LO}}$) frequency or to a much higher ($f_{\text{RF}} + f_{\text{LO}}$) frequency. From a practical point of view, the lower frequency signal will be selected (after a filtering process) for the receiver.

In a more general way, and taking into account the response of a nonlinear quadrupole, the output signal can be defined by a polynomial function:

$$v_{\text{IF}}(t) = K_1 v(t) + K_2 v(t)^2 + K_3 v(t)^3 + \ldots \tag{6.6}$$

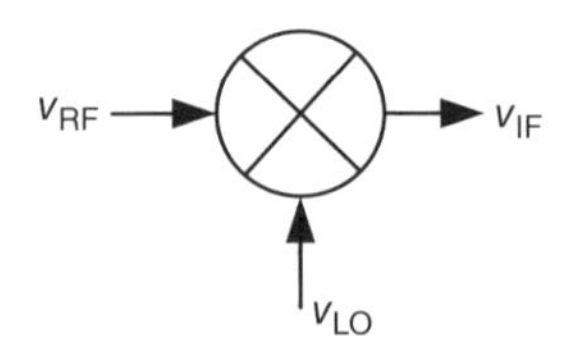

Figure 6.10 Mixer schematic.

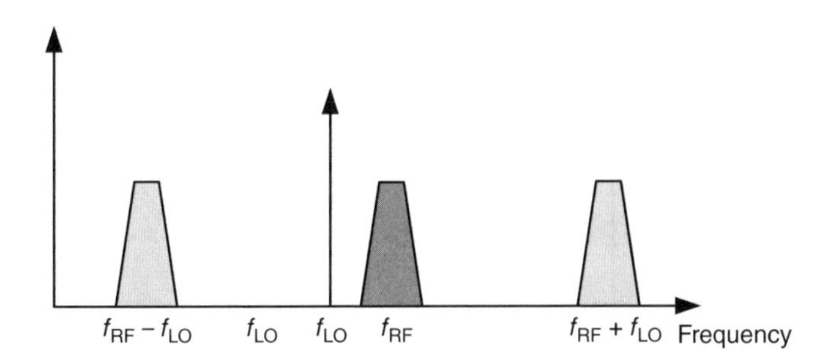

Figure 6.11 Input and output signals of a mixer.

Using Eq. (6.6) for an input signal is equal to the addition of $v_s(t)$ and $v_0(t)$, as defined previously, the corresponding output signal of the mixer scheme presented in Figure 6.10 is as follows:

$$\begin{aligned} v_{IF}(t) &= K_1(v_{RF}(t) + v_{LO}(t)) + K_2(v_{RF}(t) + v_{LO}(t))^2 + K_3(v_{RF}(t) + v_{LO}(t))^3 + \dots \\ &= \dots\ 2K_2 v_{RF}(t) v_{LO}(t)\ \dots \end{aligned} \tag{6.7}$$

Equation (6.7) is composed of a group of terms of the following form $K_{m,n}v_{RF}(t)^m v_{LO}(t)^n$. If a basic harmonic mixer is desired, the term that will give the product of the input voltages will depend only on the coefficient of second grade. The rest of the terms that are part of Eq. (6.7) will become spurious mixing products which generate signals of frequencies $mf_{RF} \pm nf_{LO}$. These spurious mixing products are known as intermodulation products of order $m+n$ which power decreases when the order increases. It is possible to reduce the intermodulation products influence by different methods: using filters, reducing the input power, or combining the output signal of several mixers.

The typical schema followed by a THz heterodyne receiver is presented in Figure 6.12.

6.2.1 Sub-Harmonic Mixers

Sub-harmonic mixers are of great interest as a first stage in the design of receivers at the THz frequency range. These mixers are a type of harmonic mixer in which the desired mixing product is given by $K_{1,2}v_{RF}(t)v_{LO}(t)^2$, leading to the following mixing signal, $f_{IF} = |2f_{LO} - f_{RF}|$.Clearly, the IF signal is composed of the second harmonic of the LO together with the RF signal (see Figure 6.13). This configuration is referred to as a sub-harmonic mixer.

This kind of mixer offers several advantages over basic harmonic mixers since it makes it possible to reduce the LO frequency to half and so increase the available power at the required LO frequency. Furthermore, if $(m+n)$ is even, the frequency $(mf_{RF} \pm nf_{LO})$ is removed at the output (anti-parallel diodes

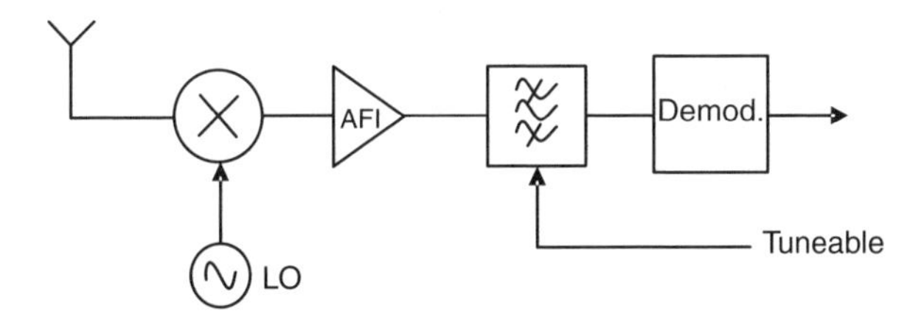

Figure 6.12 THz heterodyne receiver topology.

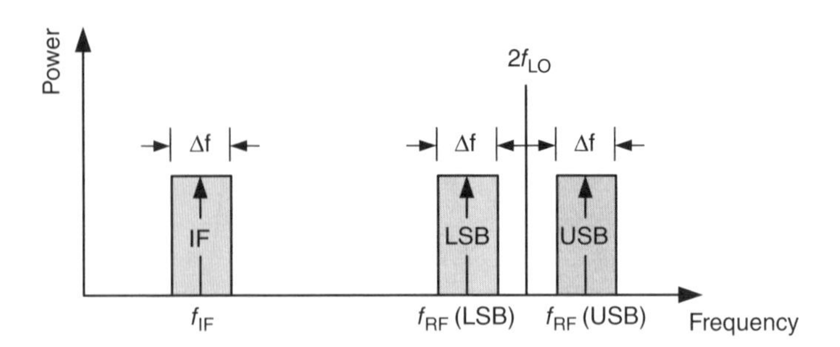

Figure 6.13 Sub-harmonic mixer principle of operation.

configuration). Moreover, there is also an increased IF bandwidth (lower impedances of the diodes at the IF frequency) and the f_{LO} is easily filtered and coupled. At the same time, this configuration presents some disadvantages, such as the conversion loss is higher, the power at LO must be higher as the pair of diodes is usually unbiased, and the noise of two diodes is generally higher than one.

The third order components, $2f_{RF} - f_{LO}$ or $2f_{LO} - f_{RF}$, are not produced by an ideally quadratic component and sub-harmonic mixers are bound to use a couple of quadratic components in order to obtain these mixing products.

In the sub-harmonic configuration, the suppression of some of the harmonics or mixing products is achieved by connecting the Schottky diodes in parallel or in series. There are basically two important interconnections, the anti-parallel and the anti-series. Sub-harmonic mixers usually make use of an anti-parallel configuration (see Figure 6.14).

The current through each diode A and B can be written as follows, together with the total current I:

$$I_A = f(V) = K_1 V + K_2 V^2 + K_3 V^3 + K_4 V^4 + K_5 V^5 + \ldots \tag{6.8}$$

$$I_B = -f(-V) = K_1 V - K_2 V^2 + K_3 V^3 - K_4 V^4 + K_5 V^5 + \ldots \tag{6.9}$$

$$I = I_A + I_B = f(V) - f(-V) = 2K_1 V + 2K_3 V^3 + 2K_5 V^5 + \ldots \tag{6.10}$$

Note that when a heterodyne mixer is analyzed V can be replaced by v_{IF}, Eq. (6.4). So that, the value of I for a sub-harmonic mixer with an anti-parallel diode configuration can be defined by:

$$\begin{aligned} I &= 2[K_1(v_{RF} + v_{LO}) + K_3(v_{RF} + v_{LO})^3 + K_5(v_{RF} + v_{LO})^5 + \ldots] \\ &= 2\begin{bmatrix} K_1\left(v_{RF} + v_{LO}\right) + K_3(v_{RF}{}^3 + 3v_{RF}{}^2 v_{LO} + 3v_{RF} v_{LO}{}^2 + v_{LO}{}^3) \\ + K_5(v_{RF}{}^5 + 5v_{RF}{}^4 v_{LO} + 10v_{RF}{}^3 v_{LO}{}^2 + 10v_{RF}{}^2 v_{LO}{}^3 + 5v_{RF} v_{LO}{}^4 + v_{LO}{}^5) + \ldots \end{bmatrix} \end{aligned} \tag{6.11}$$

Equation (6.11) confirms that the odd harmonics are the only ones produced by the pair of diodes while the even harmonics remain trapped inside the pair of diodes due to their anti-symmetry. Note that this statement if true if and only if both diodes are equals.

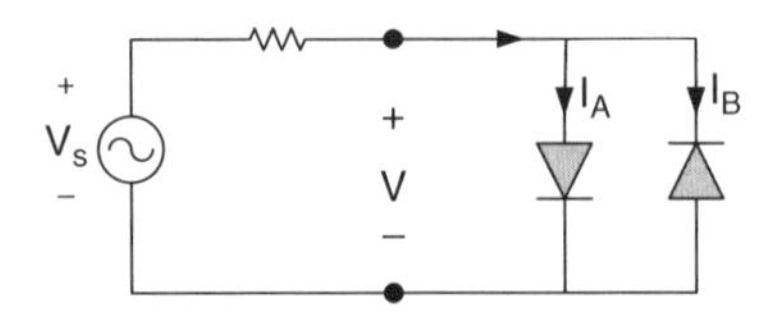

Figure 6.14 Anti-parallel diode configuration.

In order to characterize the sensitivity of a receiver, in general, and a mixer, in particular, two important parameters are used: the conversion loss and the equivalent noise temperature.

The mixer conversion loss, L_{mix}, is defined by the ratio of the input power of the RF signal to the IF output power signal. It is usually expressed in dB and can be calculated as follows [33]:

$$L_{mix}(\text{dB}) = -10\log\left(\frac{P_{IF}}{P_{RF}}\right) \tag{6.12}$$

On the other hand, mixers, as any other electronic components, are noise sources. Noise arising from the Schottky diode mixers is a combination of several sources, some of which are dominating at THz frequencies. The main noise sources are the thermal noise (arising from the random motion of thermally agitated electrons) and the shot noise (arising from the actuations of the number of electrons crossing the Schottky barrier). It is also important to take into account the hot electron noise which is proportional to the square of the current that goes through the diodes, increasing their series resistance and causing a variation in the mixer performance [34].

Other types of noises, such as flicker noise (1/*f*), are usually negligible at these frequencies.

Due to the difficulty in quantifying the contribution of each kind of noise to the overall noise of the mixer, the different noises presented in a mixer can be grouped and considered as a white noise source. Moreover, the white noise source can be modeled on a thermal noise equivalent source, and it is characterized by an equivalent noise temperature, T_e, which is proportional to noise power and expressed in Kelvin (K) [35].

$$T_e = \frac{N_0}{Bk_B} \tag{6.13}$$

where N_0 is the output noise power expressed in W, k_B represents Boltzmann's constant, 1.38×10^{-23} J/K, and B is the bandwidth in Hz.

The mixer noise temperature is typically measured by means of the Y-factor method [35]. It is important to take into consideration that when measuring the mixer noise temperature two cases can be contemplated; that is, noise coming from the imaging band or not. In the case where both signal and imaging band frequencies affect the mixer noise temperature, the mixer noise would be modeled through the use of the double side-band temperature, TDSB (see Figure 6.15b). If the model assumes that the noise is only coming from the signal frequency, the noise would be characterized through the single side-band noise temperature, TSSB (see Figure 6.15a). Following this approach, TDSB can generally be considered half TSSB. Independently of the selected model, the overall noise produced by the mixer is always the same.

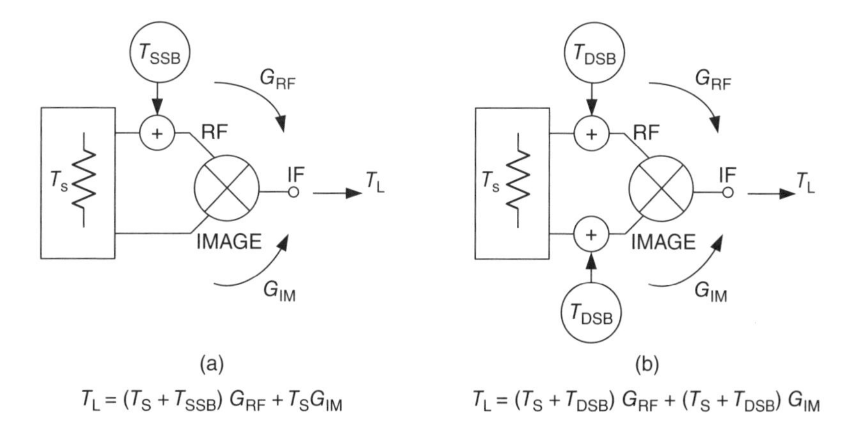

Figure 6.15 (a) TSSB and (b) TDSB mixer noise temperature models.

Table 6.1 State of the art of sub-harmonic mixers.

Mixer Reference	RF Frequency (GHz)	Substrate	Diodes	CL (dB)	NT (K)
[36]	92.1	RF/duroid 5880	Flip-chip	11.5	—
[37]	118	RF/duroid 5880	Flip-chip	12.6	—
[38]	150	Quartz (127 μm)	Flip-chip	6.7	—
[39]	180	GaAs	Integrated	4.9	608
[40]	183	Quartz	Flip-chip	5.1	530
[41]	224	Quartz	Flip-chip	6.4	—
[42]	220.5	Quartz (127 μm)	Flip-chip	7.1 simulated	—
[43]	240	Quartz	Integrated	5.4	510
[44]	330	Quartz (50 μm)	Flip-chip	6.3	700
[39]	366	GaAs	Integrated	6.9	1220
[45]	380	Quartz	Flip-chip	8.5	850
[46]	520–590	GaAs (5 μm)	Integrated	—	2200
[47]	585	Quartz	Integrated	8	1200
[48]	865.8	GaAs	Integrated	8.02	2660

Table 6.2 ESA sub-harmonic mixer requirements.

RF Frequency (GHz)	NT (K)	CL (dB)	RF bandwidth (GHz)
183	<500	<5	19.8
243	<700	<6.5	8
325	<900	<5	22
448	<170	<7	17.4
664	<2500	<9	13.4

Finally, for reader information, some sub-harmonic mixers parameters presented in the literature are gathered in Table 6.1.

As complementary information, the European Space Agency, [49], has also launched some projects related to the development of THz receivers for space applications, based on Schottky diode mixers. The specifications demanded for such THz receivers are summarized in Table 6.2.

6.2.2 *Circuit Fabrication Technologies*

When talking about Schottky diode mixers, two different implementations can be found; that is, integrated diodes [50–52] or discrete diodes [36, 44, 52]. In 1993 NASA's Jet Propulsion Laboratory (JPL) further integrated the planar diode on quartz structure by incorporating the mixer RF/LO and IF microstrip filter circuitry with the diode using their QUID process [53]. In general, if integrated diodes are employed, a better performance can be achieved than with discrete diodes [42, 54–56]. When the operational frequency increases (>400 GHz) the use of discrete diodes becomes more complicated since their functionality is related to the shape and thickness of the substrate. This is why integrated diodes are more frequently used at THz frequencies since the fabrication and tolerances required need to be better. Furthermore, it is not easy to have access to integrated diodes but it is relatively straightforward to obtain discrete diodes, also known as flip-chip diodes. One of the reasons why discrete diodes show worse mixer performance is the mixer fabrication process where these diodes need to be welded to the microstrip circuit by means of silver epoxy which increments the diode's series resistance, degrading the mixer performance.

Focusing on the use of flip-chip diode configurations, different technologies can be used to fabricate the corresponding microstrip THz mixer circuits. Within this section, two techniques will be commented; that is, the use of a laser milling machine and a photolithography process.

Using a laser milling machine, the fabrication process consists of depositing a layer of gold on a low loss THz substrate using a sputtering system. Then, the microstrip circuit, where the Schottky diode will be soldered, is printed into the golden substrate using a laser milling machine. A typical circuit is shown in Figure 6.16. In the figure, the gap where the diodes will be welded (in this case 80 μm wide), together with the RF and LO filters can be observed.

The photolithography process needs more sophisticated equipment; this includes the use of a spin coater for the resist coating process, a mask-aligner system to expose the circuits to UV light, together with the corresponding mask of the microstrip circuit. Furthermore, chemical products, such as a photoresist, a developer, and a stripper are also needed. This technique normally leads to better resolution in the manufactured circuits, which is very important at THz frequencies.

Some pictures of a microstrip mixer circuit fabricated using this process are shown in Figure 6.17. The circuit presents the same configuration as that in Figure 6.16.

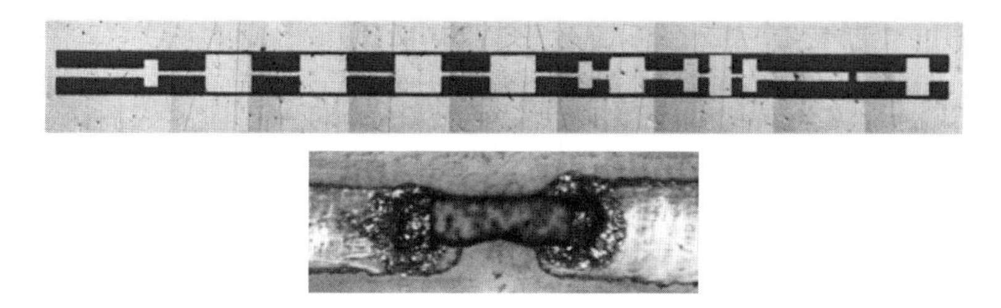

Figure 6.16 Photo of a complete microstrip mixer circuit fabricated with a UV laser milling machine. Detail of the flip-chip diode soldered into the gap.

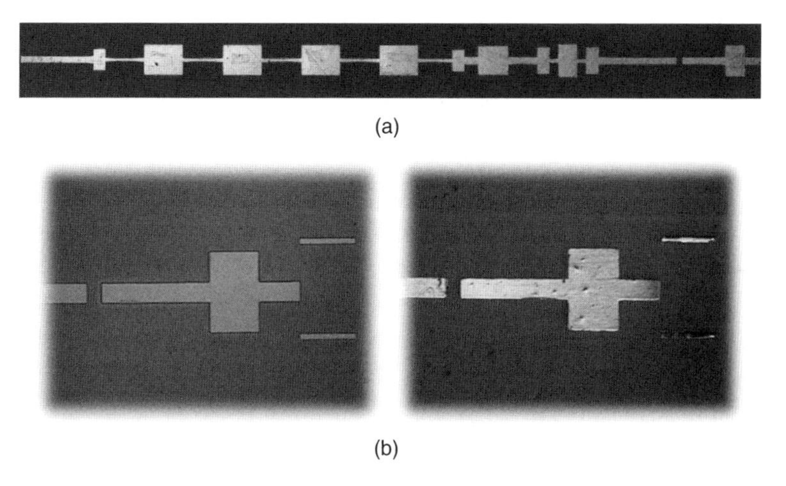

Figure 6.17 (a) Photo of the complete microstrip mixer circuit. (b) Detail of the microstrip circuit mask together with the corresponding fabricated section.

6.2.3 Characterization Technologies

There are several procedures commonly employed for measuring the performance of Schottky diodes THz mixers [57, 58]. In general they all require the de-embedding of the mixer performance from Y-factor measurements. The performance of a mixer can vary considerably, depending on the measurement technique employed.

Here, the three most commonly employed procedures are described. First, it is important to define the Y-factor method. This method is widely used to measure the noise temperature of a receiver and it is based on the Johnson–Nyquist noise of a resistor at two different temperatures in which the output power for each one is defined by

$$P_{\mathrm{rx}} = (T_{\mathrm{rx}} + T_R)k_{\mathrm{B}}G_{\mathrm{rx}}B, \tag{6.14}$$

where T_R are two known different temperatures (in K); that is, T_{COLD} and T_{HOT}, where T_{COLD} is obtained from a blackbody load of liquid nitrogen ($T = 77$ K) and as T_{HOT} the blackbody load is the room temperature ($T = 290$ K), for example, T_{rx} is the equivalent noise temperature of the receiver, G_{rx} represents its gain, B the bandwidth, and k_{B} the Boltzmann's constant. The Y-factor is obtained from:

$$Y_{\mathrm{rx}} = \frac{P_{\mathrm{HOT}}}{P_{\mathrm{COLD}}}. \tag{6.15}$$

6.2.3.1 Attenuator Procedure

This measurement procedure requires the receiver and IF Y-factor measurements to be performed with and without a coaxial attenuator connected to the input of the IF chain (between mixer and pre-amplifier). The measurements result in two simultaneous equations from which mixer noise temperature and conversion loss can be de-embedded. Figure 6.18 presents the two receiver measurement scenarios. The IF Y-factor measurements are performed by terminating the input of the IF chain with a 50 Ω coaxial load via a short length of semi-rigid cable. The coaxial load acts as a hot load at room temperature and a cold load when immersed in liquid nitrogen.

With the attenuator omitted the receiver Y-factor is

$$Y_{\mathrm{rx}} = \frac{P_{\mathrm{HOT}}}{P_{\mathrm{COLD}}} = \frac{T_{\mathrm{HOT}} + T_{\mathrm{rx}}}{T_{\mathrm{COLD}} + T_{\mathrm{rx}}}. \tag{6.16}$$

Rearranging, the noise temperature of the receiver is

$$T_{\mathrm{rx}} = \frac{T_{\mathrm{HOT}} - Y_{\mathrm{rx}}T_{\mathrm{COLD}}}{Y_{\mathrm{rx}} - 1}. \tag{6.17}$$

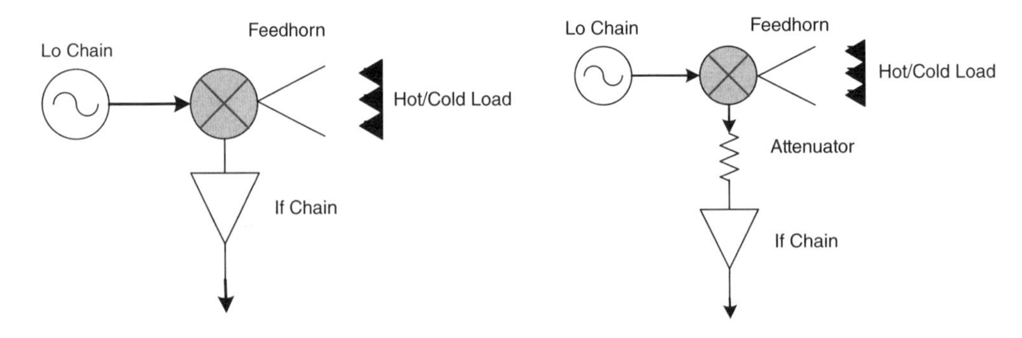

Figure 6.18 Attenuator method set-up.

With the attenuator in position the receiver Y-factor is

$$Y_{\text{rx,att}} = \frac{P_{\text{HOT}}}{P_{\text{COLD}}} = \frac{T_{\text{HOT}} + T_{\text{rx,att}}}{T_{\text{COLD}} + T_{\text{rx,att}}}. \tag{6.18}$$

Rearranging, the noise temperature of the receiver is

$$T_{\text{rx,att}} = \frac{T_{\text{HOT}} - Y_{\text{rx,att}} T_{\text{COLD}}}{Y_{\text{rx,att}} - 1}. \tag{6.19}$$

Two simultaneous equations result:

$$T_{\text{rx}} = T_{\text{mix}} + L_{\text{mix}} T_{\text{IF}} \tag{6.20}$$

$$T_{\text{rx,att}} = T_{\text{mix}} + L_{\text{mix}} T_{\text{IF,att}} \tag{6.21}$$

T_{IF} and $T_{\text{IF,att}}$ are calculated following the same process described above.

Rearranging Eqs. (6.20) and (6.21) the conversion loss and the noise temperature of the mixer are calculated from

$$L_{\text{mix}} = \frac{T_{\text{rx}} - T_{\text{rx,att}}}{T_{\text{IF}} - T_{\text{IF,att}}} \tag{6.22}$$

and

$$T_{\text{mix}} = T_{\text{rx}} - L_{\text{mix}} T_{\text{IF}} \tag{6.23}$$

6.2.3.2 Noise-Injection Procedure

This procedure for measuring mixer performance is similar to the previous method. The only difference is the equipment used to generate the two different IF noise temperatures. Here, noise is injected into the input of the IF chain using a noise source and directional coupler. Figure 6.19 presents the two receiver measurement scenarios with the noise source switched ON or OFF. The receiver and IF Y-factor measurements are performed as described previously.

With the noise source OFF, the receiver Y-factor is

$$Y_{\text{OFF}} = \frac{P_{\text{HOT}}}{P_{\text{COLD}}} = \frac{T_{\text{HOT}} + T_{\text{rx,OFF}}}{T_{\text{COLD}} + T_{\text{rx,OFF}}} \tag{6.24}$$

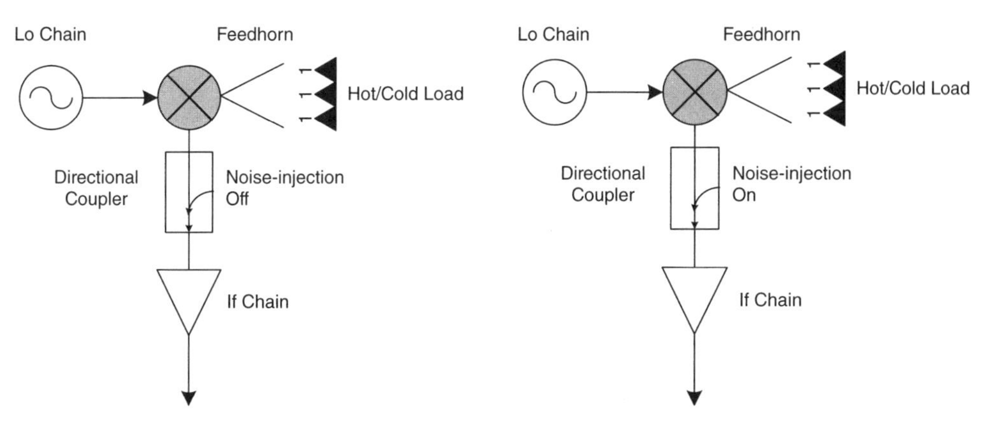

Figure 6.19 Noise-injection method set-up.

Rearranging, the noise temperature of the receiver yields

$$T_{\text{rx,OFF}} = \frac{T_{\text{HOT}} - Y_{\text{OFF}} T_{\text{COLD}}}{Y_{\text{OFF}} - 1} \tag{6.25}$$

With the noise source ON, the receiver Y-factor is

$$Y_{\text{ON}} = \frac{P_{\text{HOT}}}{P_{\text{COLD}}} = \frac{T_{\text{HOT}} + T_{\text{rx,ON}}}{T_{\text{COLD}} + T_{\text{rx,ON}}} \tag{6.26}$$

Rearranging, the noise temperature of the receiver yields

$$T_{\text{rx,ON}} = \frac{T_{\text{HOT}} - Y_{\text{ON}} T_{\text{COLD}}}{Y_{\text{ON}} - 1} \tag{6.27}$$

As before, two simultaneous equations result:

$$T_{\text{rx,OFF}} = T_{\text{mix}} + L_{\text{mix}} T_{\text{IF,OFF}}, \tag{6.28}$$

$$T_{\text{rx,ON}} = T_{\text{mix}} + L_{\text{mix}} T_{\text{IF,ON}}. \tag{6.29}$$

With knowledge of $T_{\text{IF,OFF}}$ and $T_{\text{IF,ON}}$ the mixer performance can be calculated.

Rearranging Eqs. (6.28) and (6.29) the conversion loss and the noise temperature of the mixer are calculated from:

$$L_{\text{mix}} = \frac{T_{\text{rx,OFF}} - T_{\text{rx,ON}}}{T_{\text{IF,OFF}} - T_{\text{IF,ON}}}, \tag{6.30}$$

and

$$T_{\text{mix}} = T_{\text{rx,OFF}} - L_{\text{mix}} T_{\text{IF,OFF}}. \tag{6.31}$$

6.2.3.3 Gain Procedure

This measurement procedure is different from those previously described requiring Y-factor measurements of a single receiver and IF chain. These measurements are used to directly calculate the gain of the mixer. The Y-factor measurements are performed in an identical manner to that described previously.

From the Y-factor measurement of the receiver the following equations result:

$$P_{\text{HOT,rx}} = (T_{\text{rx}} + T_{\text{HOT}}) k_{\text{B}} G_{\text{rx}} B, \tag{6.32}$$

$$P_{\text{COLD,rx}} = (T_{\text{rx}} + T_{\text{COLD}}) k_{\text{B}} G_{\text{rx}} B. \tag{6.33}$$

Rearranging, the receiver gain is obtained:

$$G_{\text{rx}} = \frac{P_{\text{HOT,rx}} - P_{\text{COLD,rx}}}{k_{\text{B}} B (T_{\text{HOT}} - T_{\text{COLD}})}. \tag{6.34}$$

From the Y-factor measurement of the IF chain the following equations result:

$$P_{\text{HOT,IF}} = (T_{\text{IF}} + T_{\text{HOT}}) k_{\text{B}} G_{\text{IF}} B, \tag{6.35}$$

$$P_{\text{COLD,IF}} = (T_{\text{IF}} + T_{\text{COLD}}) k_{\text{B}} G_{\text{IF}} B. \tag{6.36}$$

Rearranging, the gain of the IF chain is obtained:

$$G_{\text{IF}} = \frac{P_{\text{HOT,IF}} - P_{\text{COLD,IF}}}{k_{\text{B}} B (T_{\text{HOT}} - T_{\text{COLD}})} \tag{6.37}$$

The mixer gain is then calculated as:

$$G_{\mathrm{mix}} = \frac{G_{\mathrm{rx}}}{G_{\mathrm{IF}}} = \frac{P_{\mathrm{HOT,rx}} - P_{\mathrm{COLD,rx}}}{P_{\mathrm{HOT,IF}} - P_{\mathrm{COLD,IF}}} \tag{6.38}$$

and

$$L_{\mathrm{mix}} = \frac{1}{G_{\mathrm{mix}}}. \tag{6.39}$$

Once the conversion loss is known, the noise temperature of the mixer can be calculated using Eq. (6.23).

6.2.3.4 Results

In order to show the differences among the different measurement methods, a sub-harmonic mixer working at an RF frequency of 183 GHz is tested in this section. The configuration of the heterodyne THz receiver (including LO and RF chains and Schottky diode mixer) and the IF chain used for the mixer test are depicted in Figures 6.20 and 6.21.

The LO chain consists of a frequency synthesized sweeper that generates a 15.1667 GHz signal; a sextupler that obtains the LO frequency of 91 GHz; an isolator that prevents reflections from reaching the LO source where they could reflect again and combine with the incident signal; and a variable attenuator that attenuates the LO signal in order to obtain the desired output power. The maximum available power characterized at the mixer input is 5 mW.

The RF chain consists of a RF Feed-horn, TRG Alpha ind. 861G/387, that receives the radio frequency signal. Furthermore, the hot and cold loads that will be used to measure the Y-factor of the receiver are also represented.

The IF chain operates over the bandwidth 2.5–3.5 GHz. A bandpass filter is used to reduce noise and reject spurious signals outside the operational frequency range. A 3 dB resistive attenuator pad is used to

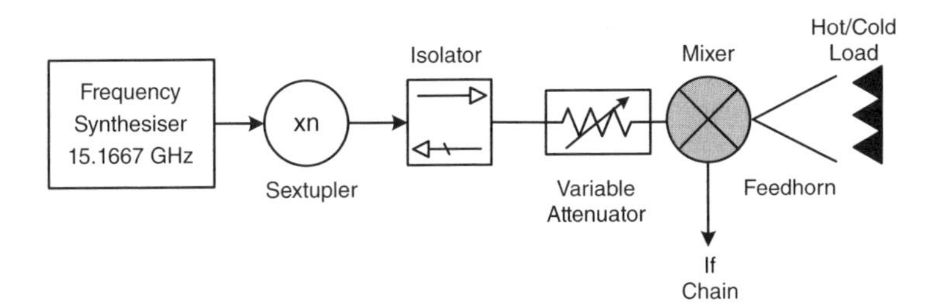

Figure 6.20 Receiver test set-up excluding the IF chain.

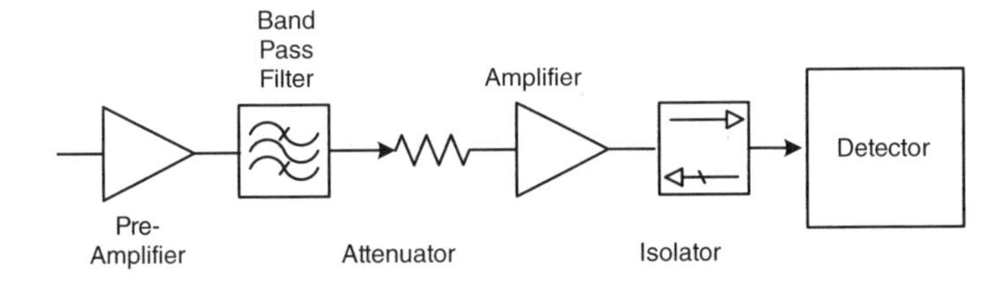

Figure 6.21 IF chain used for the mixer test.

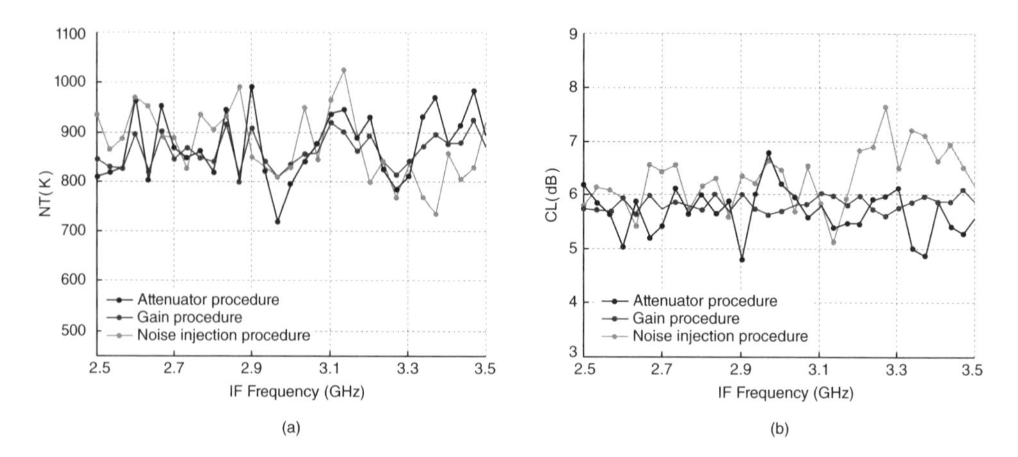

Figure 6.22 Mixer noise temperature (a) and conversion loss (b).

attenuate reflections and avoid oscillations. The isolator is used to prevent reflections that might combine with the incident signal. After the isolator the corresponding detector is placed. The measured NT of this chain is around 78 K.

The mixer under test is a sub-harmonic Schottky diode mixer; model number FDP-044 MTT C, provided by the Rutherford Appleton Laboratory (RAL), with a 160–200 GHz RF bandwidth. This device comprises an integrated anti-parallel diode circuit fabricated using the UMS foundry process [59].

The measured mixer noise temperature and conversion loss using the attenuator, gain and noise injection methods are depicted in Figure 6.22.

6.2.4 *Advanced Configuration Approach*

Usually, THz mixer designs are focused on sub-harmonic mixers, in which the RF frequency is twice the LO frequency. The actual technology available at the THz range suffers in most cases from insufficient power to feed the available components due to the high operational frequencies. This can result in an important problem when arrays of sub-harmonic mixers need to be pumped at these frequencies. However, a different approach to sort out this issue is presented here; this is the use of a fourth-harmonic mixer in which the RF frequency is four times the LO frequency. This configuration is a promising solution since the LO source can operate to a much lower frequency than in sub-harmonic mixer designs and, as a consequence, a wide range of sources are available at those frequencies [60].

A fourth-harmonic mixer configuration consists of a harmonic mixer that selects the fourth mixing product of the LO frequency and the RF. For instance, a LO frequency of 110 GHz has been chosen, so that, the RF frequency will be 440 GHz and the IF signal will be calculated by $f_{IF} = 4f_{LO} - f_{RF}$. It is well-known that the fourth harmonic mixing product has a lower level of power compared to the second harmonic mixing product ($f_{IF} = 2f_{LO} - f_{RF}$), for this reason a higher pumped LO signal will be demanded. This can be seen as a disadvantage but, if a sub-harmonic mixer is used instead, a LO frequency of 220 GHz should be used and, therefore, a more complicated and expensive LO power source would be required. In this case, fourth-harmonic mixers based on Schottky diodes are connected in antiparallel configuration following the same scheme as for sub-harmonic mixers (see Figure 6.14). This kind of configuration presents worse performances: higher noise temperature and conversion losses. Some measured results of the performance of a fourth harmonic mixer (see Figure 6.23) are presented in Figure 6.24.

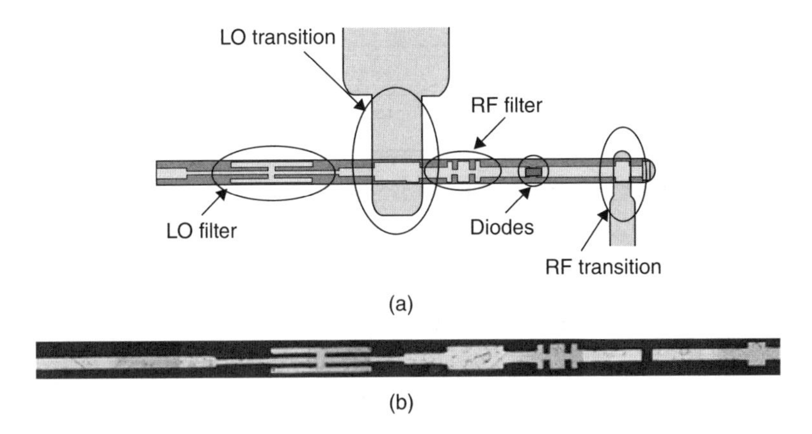

Figure 6.23 (a) Schematic of a fourth harmonic microstrip design. (b) Microstrip circuit fabricated with a lithographic process.

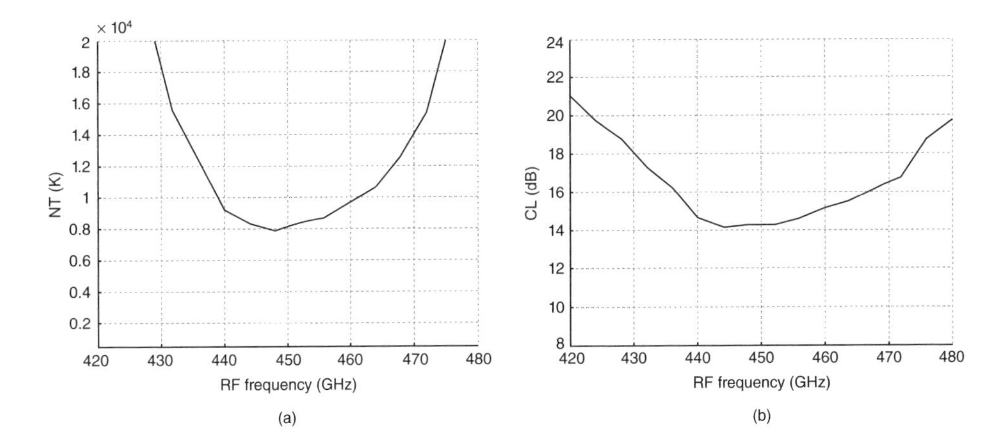

Figure 6.24 Measured noise temperature (NT) (a) and conversion loss (CL) (b) of a fourth harmonic mixer.

6.2.5 *Imaging Applications of Schottky Mixers*

The THz frequency range is slowly becoming more technologically relevant due to the increasing number of attractive applications which can be potentially developed, that is, medicine (skin cancer detection, caries detection, etc.), security and surveillance (detection of hidden weapons or explosives, detection of gases, etc.), viticulture (control of the vine state), food sector, space and aeronautics, industrial, passive tomography imaging, and investigation on proteomics in the pharmaceutical industry. Furthermore, THz waves are becoming of great interest for new and emerging wireless communications.

Within these applications, the implementation of passive or active imaging cameras has become of great interest due to the major advantages this frequency range offers compared to the visible range or infrared, such as, the availability of taking images under low visibility conditions (fog, clouds, smoke, sandstorms, etc.) or even at night.

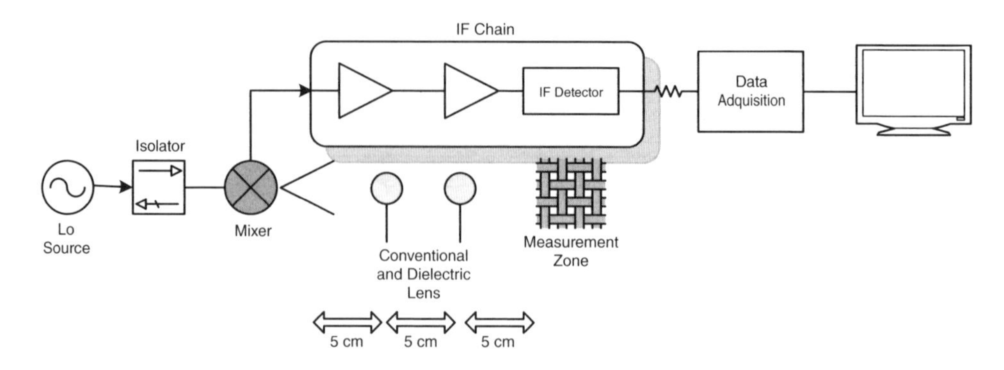

Figure 6.25 Imaging set-up at 220 GHz.

Usually two kinds of receiver configurations are used for these applications: direct or heterodyne receivers. Direct detection is typically obtained by means of a single diode detector or an array of diodes, a receiving antenna and a LNA. Since millimeter wave direct detectors do not provide sufficient sensitivity by themselves, LNAs are required to boost the receiver signal above the detector noise floor. Current technology allows the use, for frequencies around 90 and up to 250 GHz, of direct detection configurations, thanks to the availability of LNAs. However, when the operational frequency increases in order to allow the achievement of better resolution due to the shorter operational wavelength, having smaller systems, and higher contrast in the images, it is difficult to find commercially available LNAs with low noise values to be competitive with conventional mixers.

Only Research Institutes for investigation purposes have developed LNAs working at sub-millimeter wave frequencies, [30, 31]. This is one of the main reasons why heterodyne receivers are still the most typically used receiver configuration for imaging systems. Heterodyne receivers are based on fundamental and sub-harmonic mixers. As commented in previous sections, the sub-harmonic ones present a clear advantage over fundamental receivers; that is, they reduce the value of the LO frequency in order to have more power accessible to feed both a single or a group of mixers, with the same LO device [50, 51, 61–63].

A typical configuration used for a single detector imaging system is presented in Figure 6.25. This set-up consists of a feedhorn attached to the sub-harmonic mixer RF port and two lenses placed in front of them to improve the imaging resolution. At the output of the IF port two amplifiers (GAMP0100.0600SM10) with 35 dB gain are placed before the IF detector (ACSP-2656NZC3). The results are recorded by a computer and presented without any post-processing in Figure 6.26 for the case of a LO frequency of 118 GHz and compared with its corresponding image in the visible range. In this image it is possible to identify clearly a hand and a silver ring on one of her fingers. Note that when taking the image, the hand was introduced inside a plastic glove, Figure 6.26a. A good resolution has been obtained since the edge of both the hand and the ring are perfectly distinguished. Moreover, the passive sub-millimeter wave receiver is able to see through the plastic glove.

6.3 Solid-State THz Low Noise Amplifiers

The new millennium brought an increasing interest in THz electronics. The traditional applications in astronomy [64], including the search for spectral signatures in the interstellar dust via submillimeter wave detectors, leading to the construction of Space- and Earth-based radio telescopes and interferometers [65–68], are progressively being complemented by a growing number of different applications (e.g., see Ref. [69] for a pioneering list). The latter include not only THz imaging systems, both active and passive

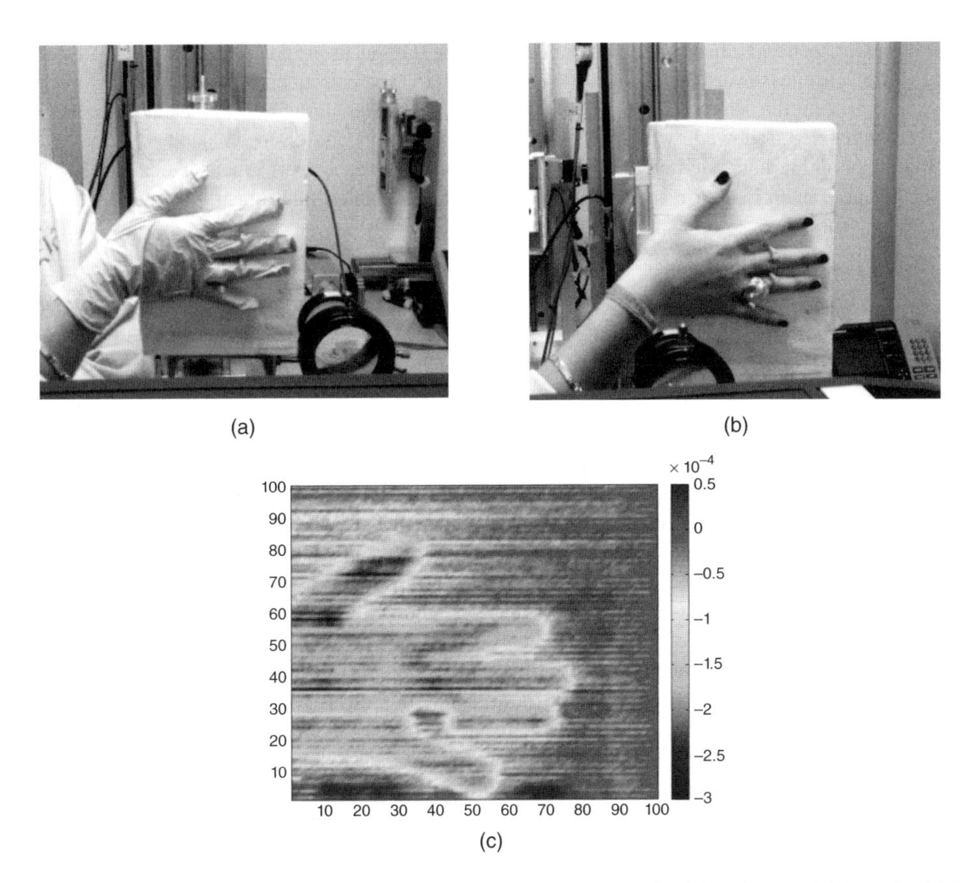

Figure 6.26 (a) Photo of a hand within a plastic glove, (b) the hand with a ring, and (c) passive THz image at 220 GHz.

[70, 71], for security and safety application, but also medical imaging [72], benefitting from the higher penetration depth of THz radiation. Ultra-wideband frequency ranges are available for next-generation secure broadband communications [73] and will open up the possibility for a series of new telecom services [74].

Solid-state terahertz electronics is rapidly moving from detector technology, mostly cryogenically cooled, toward the replication of system schemes already adopted at microwave and millimeter waves. In fact, heterodyne schemes, adopted widely both in radar and communication applications, would allow a greater flexibility and new potentials to the Transmit/Receive subsystems, in turn permitting the flourishing of exciting new capabilities. In addition, the availability of a flexible solid-state technology, providing not only mixing and/or detecting functionalities, but also low noise signal amplification and power transmission, actually allows a higher degree of integration and the subsequent realization of multi-pixel modules and arrays, together with implementation of MIMO (multiple-input multiple-output) schemes.

To this end, a fundamental step is the development of active device technologies with operating frequencies in excess of 1 THz. Such an extremely challenging goal is now becoming close to reality, given the recent demonstrations of LNAs operating at 850 GHz [75]. Indeed, such huge technological development is possible thanks to the increasing interest in security and defence application of terahertz electronics, as demonstrated by the US-funded DARPA program in THz Electronics [76].

The present section is therefore an attempt to describe the recent progress in the field of solid-state integrated electronics toward the realization of LNAs in the terahertz frequency range, therefore updating previous works on the same topic (as for instance [77]). To this goal, the relevant technologies will be briefly summarized, followed by the peculiar features of the transmission media adopted in the case of terahertz monolithic integrated circuits (TMICs). The state of the art performances obtained are then presented and framed in their respective technology, trying to sketch subsequently the perspectives for the coming years.

6.3.1 *Solid-State Active Devices and Technologies for Low Noise Amplification*

A critical issue for reaching the terahertz range for an integrated circuit is indeed represented by the development of transistors exhibiting maximum oscillation frequencies (f_{MAX}) above 1 THz, allowing in turn the design and realization of active functionalities with operating frequencies well above the 300 GHz limit, conventionally representing the lower boundary for the TMIC. To this goal, the natural candidate is the high electron mobility transistor (HEMT) realized onto InP material systems, being the device type to break the 100 GHz barrier in the 1990s (see Ref. [78] as an example). This property is due to its excellent electronic properties, related both to the electron saturated velocity (together with a bulk mobility of 5400 cm^2/Vs at 300 K) and to the HEMT structure, allowing the actual separation between the resulting two-dimensional electron gas and the scattering impurities. As is well known, HEMTs are heterostructure-based devices, exploiting the different bandgap of two semiconductor material systems. In the case of InP HEMTs, the heterostructure is formed by $In_xGa_{1-x}As/In_yAl_{1-y}As$ (where $x = 0.53$ and $y = 0.52$ for a lattice-matched structure), with the resulting formation of a potential well in the InGaAs, spatially confining the conduction-band electrons in a two-dimensional electron gas. In this way, electrons may fly freely with much higher resulting mobility. To increase further the latter parameter, the In content may be increased in the InGaAs layer (up to InAs), progressively creating more strain and therefore limiting the growth of the material to extremely thin layers. An increased mobility leads to higher device transconductance and hence higher gain, thus contributing to the increase in operating frequency. Advanced structures [79] include composite channels to increase further electron transport properties, a Si-δ doping in the Schottky layer to supply electrons and proper overall epitaxial growth using molecular beam epitaxy (MBE). A schematic view of the InP HEMT epitaxy is shown in Figure 6.27a.

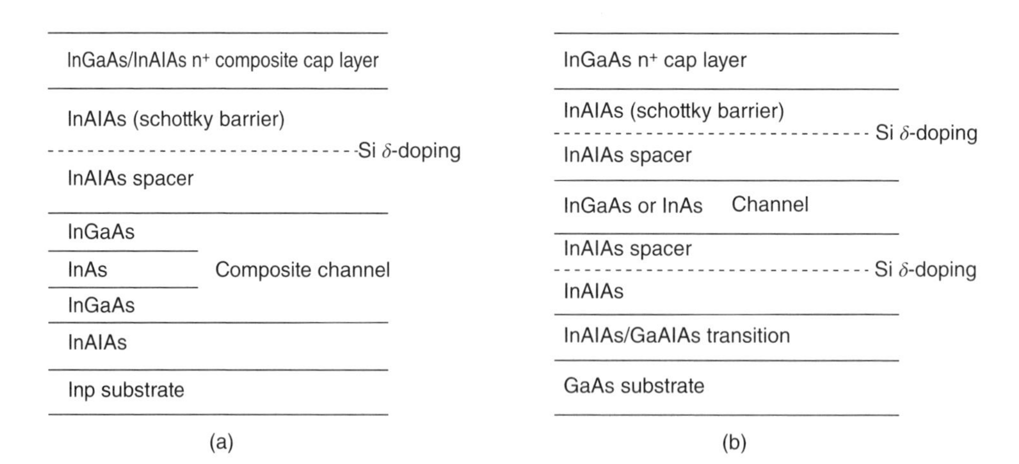

Figure 6.27 (a) Advanced InP HEMT [79] and (b) InP mHEMT epitaxial structures [80].

The resulting measured electron mobility exceeds 15 000 cm^2/Vs, with a sheet charge density of 3.3×10^{12} cm^2.

The second major modification in the basic device structure applies to the gate length, implying an aggressive scaling of the latter. Huge efforts have been made to progressively decrease it from 0.1 μm (that still represents an advanced technology for many commercial foundries) down to 35 to 20 nm. The only possibility to achieve such a small device feature consists in using electron beam lithography (EBL) techniques, appropriately adapted to reach sub-100 nm dimensions: at least two EBL layers are typically used, the first to achieve the smaller gate foot defining the gate length, the second to fabricate a 100 nm T-gate on top of the first (see, e.g., Ref. [81] for details). As already noted, the resulting gate structure is a T-shaped one (or mushroom gate), allowing simultaneous minimization of the gate length and parasitic gate resistance: clearly, great care must be devoted to the evaluation of device parasitic effects, both resistive and capacitive, not to frustrate the intrinsic device improvements. The widely accepted figure-of-merit for device frequency performance is its current gain cut-off frequency f_T, expressed as $f_T = g_m/(2\pi C_{gs})$ where g_m is the device transconductance and C_{gs} the gate-source intrinsic capacitance. Increasing the indium content leads ultimately to increased device transconductance, while decreasing the gate length results in a smaller gate-source capacitance: both effects imply an increase in device cut-off frequency. Similarly, f_{MAX}, the device maximum oscillation frequency, typically measured or extrapolated from measurements from the device power gain, is also adopted in conjunction with f_T to give a complete picture of device frequency performance. f_{MAX}, whose value is typically much higher than f_T, represents the highest operating frequency for the device acting as an active element but the use is normally restricted to frequencies lower than f_T. Given the aggressive scaling of the device gate length, a continuous improvement in f_T/f_{MAX} is being evidenced: the already impressive results in Ref. [82], consisting in $f_T = 586$ GHz and f_{MAX} ranging from 0.9 to 1.3 THz, associated with a $g_m = 2300$ mS/mm (for a 2×20 μm device) have been recently updated by a 1.4 THz f_{MAX} technology [75] obtained via 25 nm gate devices by the same team at Northrop Grumman demonstrating circuit operation at 850 GHz. The record-breaking device technology is, however, still under development, and, under the push of the cited DARPA (Defense Advanced Research Projects Agency) program [76], it is expected to be further improved shortly.

A different approach is possible if one attempts to decrease the inherent high costs of the InP technology by using a metamorphic structure [83]: in this case, InP HEMTs are fabricated onto an epitaxy that is grown on GaAs substrates, with larger wafer size, better robustness, and lower costs compared to InP technology. 3-in. InP substrates are then replaced by 4-in. GaAs wafers (or even potentially 6-in. ones), thus leading to higher uniformity and lower unit costs. In this case clearly the biggest issue resides in the transitional layer accommodating the strain from the bulk GaAs substrate to the epitaxial layer. In the case of Ref. [81], a 4-in. GaAs diameter wafer is adopted, and a 1.1 μm graded InAlGaAs buffer layer transforms the GaAs lattice constant to the InP one (Figure 6.27b). A high In content channel layer (80%) is adopted for the InGaAs, and, with EBL-defined gate length down to 35 nm, resulting in a cut-off frequency $f_T = 515$ GHz and f_{MAX} in excess of 1 THz (on 2×10 μm devices). A record 660 GHz f_T has also been demonstrated by the same Fraunhofer IAF research group [80], scaling down the gate length to 20 nm using two separate EBL layers for the T-gate fabrication, from an epitaxy similar to the one in Figure 6.27b: electron mobility of 9800 cm^2/Vs, with a sheet charge density of 6.1×10^{12} cm^2 has been achieved. A maximum transconductance $g_m = 2500$ mS/mm (for a 2×10 μm device) has been obtained. Such results clearly demonstrate that the metamorphic approach is a viable and effective alternative to the competing pure-InP technology.

Clearly also, device geometries other than the gate length have to be scaled accordingly. As an example, devices in Ref. [81] are fabricated with a source-drain spacing scaled down to 200 nm, to be compared with microwave devices exhibiting one-order-of-magnitude-higher spacing.

The HEMT structure, both in lattice-matched and metamorphic variants, indeed represents the state-of-the-art device in performance for low-noise applications, as will be illustrated later. Nevertheless, different active device structures are achieving record-breaking frequency performance and may represent an important alternative for future THz subsystems in functionalities different from the receiving one. This is the case for heterojunction bipolar transistors (HBTs) realized in InP substrates: InP HBTs demonstrated maximum oscillating frequencies over 1 THz [84] and demonstrator circuits do operate at frequencies as high as 600 GHz [85, 86], even if not evidencing state-of-the-art noise performance. An interesting roadmap for terahertz HBT technology can be found in Ref. [87]. Intrinsically the bipolar devices are well-suited for differential amplification, and therefore are especially tailored for balanced circuit design, matching specific antenna systems and therefore partially overcoming their noise performance limitations. In addition, their application to integrated signal generation and low-phase noise oscillators is indeed to be preferred to the HEMT counterpart.

Moving from III–V compound semiconductors to SiGe systems [88], the improvement in frequency performance has been even more pronounced, considering that half-terahertz devices have been demonstrated [89]: using a 0.13 μm BiCMOS, f_T/f_{MAX} of 300 GHz/500 GHz has been achieved, acting both on an epitaxial modification of the existing HBT structure and on further refinements aiming at reducing lateral device parasitic effects. Further improvements are expected, however, arising from physical limitations in the structure performance [90], as already evidenced from the results of ongoing FP7 European projects on the topic, such as [91], that recorded a 798 GHz f_{MAX} performance. Combining SiGe with the cost, yield, size, integration, and manufacturing advantages of silicon makes the device competitive in the marketplace, above all if compared with the costly approaches necessary for III–V based circuit fabrication. However, if record noise performance is considered, the gap to be filled remains quite remarkable. Similar considerations, even if with lower performance [92], apply to pure Si CMOS (complementary metal oxide semiconductor) technologies, where the scaling is accompanied by high k dielectrics and extremely small device features, reaching the 28 nm node and beyond.

6.3.2 Circuit and Propagation Issues for TMIC

The increase in operating frequencies to reach sub-millimeter wave frequencies necessarily implies major modifications in the traditional circuit design flow, well assessed in microwave electronics up to the K_a Band. In fact, MMICs rely on a well-defined flow, almost regardless of the functionality to be realized: a circuit-oriented commercial CAD tool is adopted in which, making use of extracted or already available device models, passive elements are appropriately mixed with passive elements taken from a foundry library, following a selected topology to achieve the desired functionality. Active device and passive elements models are mostly microstrip ones, that is, extracted from measured microstrip devices. The resulting circuit will probably be a microstrip one, eventually contacted, for circuit verification issues, via coplanar probes, and therefore incorporating coplanar probe pads (typically ground–signal–ground). The resulting fabricated circuit, once successfully probed with acceptable performance, is then mounted using a typical chip-and-wire process, that is, it is glued (in different ways depending on the functionality and thermal issues) onto a carrier and bonded via micro-wires to the rest of the system. The result is then packaged and the package, depending on the frequency of operation, may be connectorized or accessible via an appropriate waveguide flange. Such a design flow is simply not applicable to TMIC circuits, for a series of reasons that will be briefly recalled in the following.

As a first consideration, microstrip transmission media may not be adequate for TMIC circuit design. The reasons are twofold. On the one hand, the use of active devices in an amplifying stage actually forces one to connect one of the device terminals (typically the source for FET-based circuits) to ground to accommodate a proper biasing/operating scheme. In a microstrip, such ground connection is provided by via holes connecting the front and back sides of the circuit. At millimeter-wave frequencies the via hole parasitic effects (mainly resistive and inductive) are already quite high, therefore introducing detrimental effects on the device performance. Such effects become tremendously high as the frequency increases, therefore not guaranteeing an effective grounding and decreasing the transistor gain as a minimum. To mitigate the effect (and to eliminate substrate modes), substrates are thinned down from the usual ≈100 μm in MMICs to 50 μm or even 20–25 μm [68–70] for TMICs. Such a decrease also helps in reducing the via front-side area, thus reaching diameters on the order of substrate thickness, but, on the

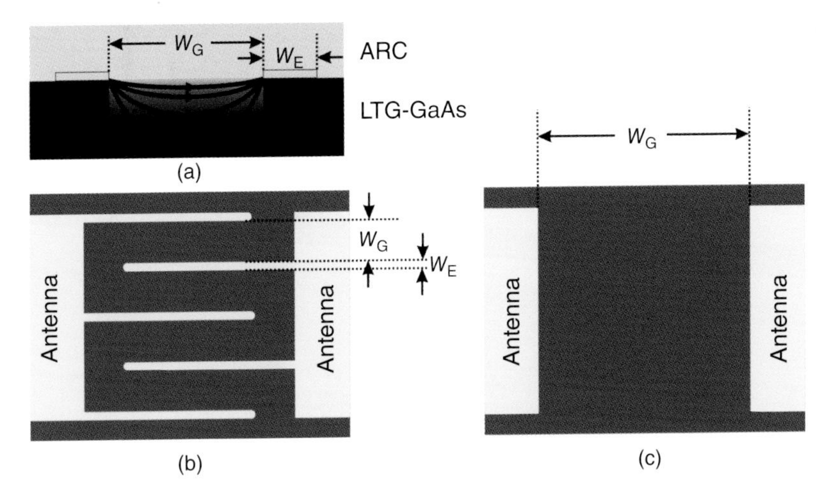

Figure 2.8 (a) Cross-section of a photoconductor. The electric DC fields, due to biasing the electrodes, are indicated by the bowed lines. The gradient illustrates the amount of photo-generated carriers in the gap (logarithmic scale; bright: many carriers, dark: few carriers). (b) Top view of a photoconductive mixer with fingers for CW operation. (c) Fingerless gap for pulsed operation.

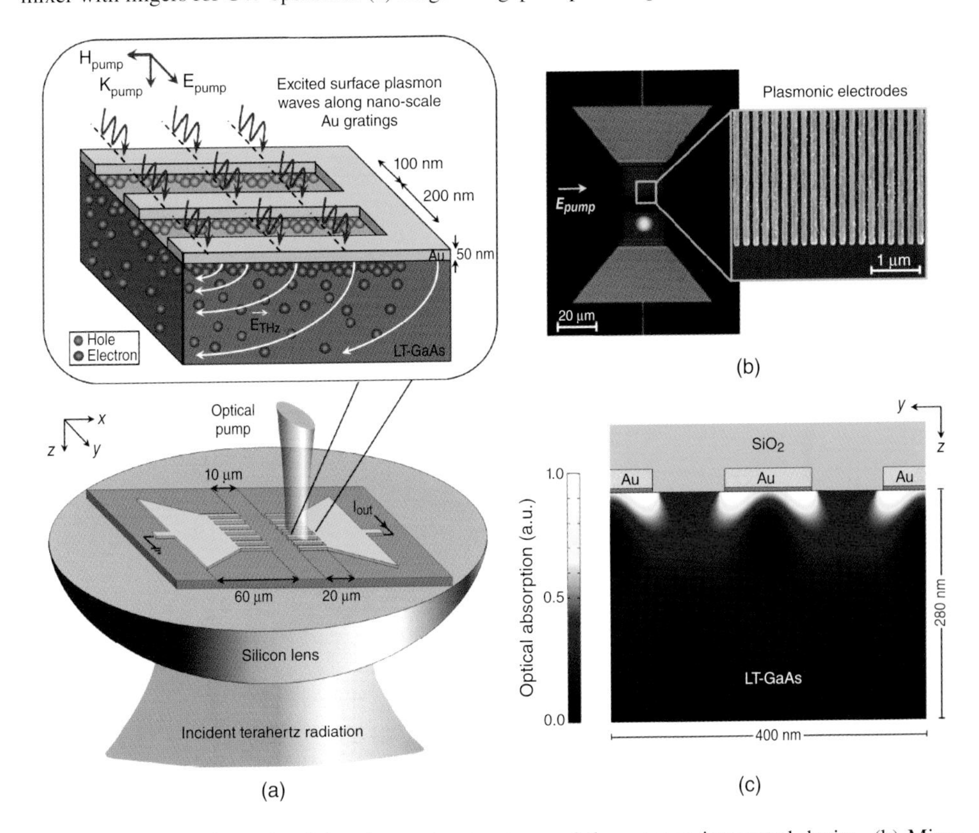

Figure 2.25 (a) Schematic of the plasmonic structure and the antenna-integrated device. (b) Micrograph of the structure. (c) Simulated optical absorption. Figures reproduced from Ref. [92] with permission from the Optical Society of America, © 2013 Optical Society of America.

Semiconductor Terahertz Technology: Devices and Systems at Room Temperature Operation, First Edition.
Edited by Guillermo Carpintero, Luis Enrique García Muñoz, Hans L. Hartnagel, Sascha Preu and Antti V. Räisänen.

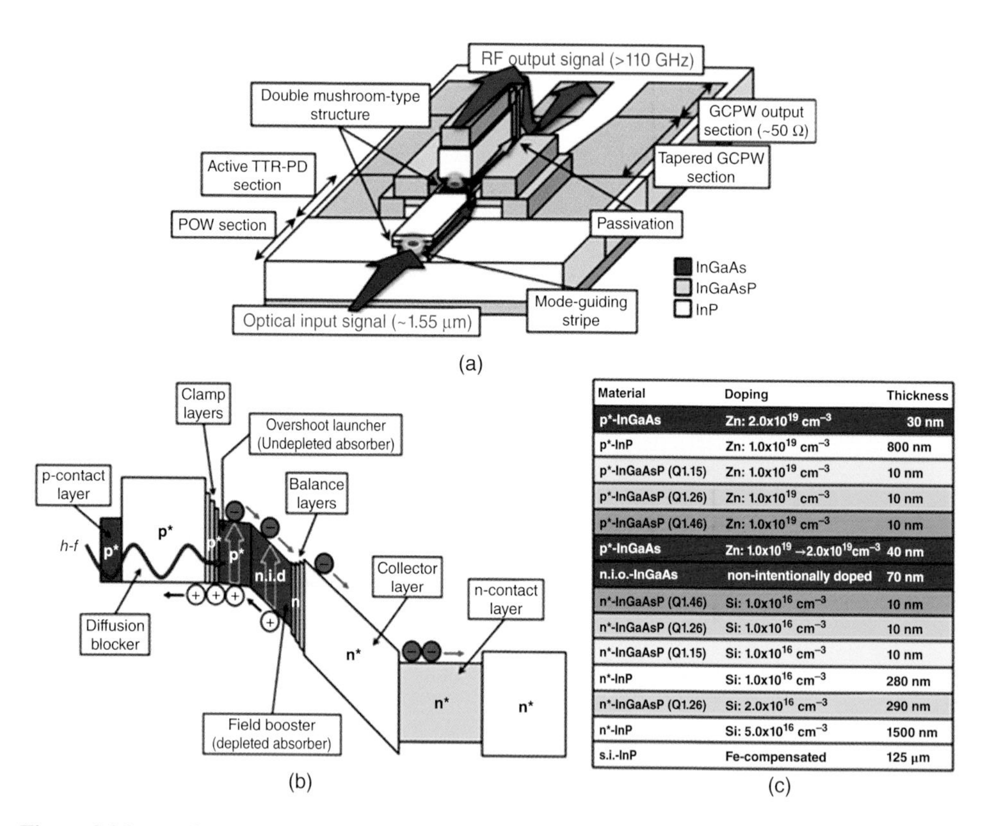

Material	Doping	Thickness
p*-InGaAs	Zn: 2.0x10^{19} cm^{-3}	30 nm
p*-InP	Zn: 1.0x10^{19} cm^{-3}	800 nm
p*-InGaAsP (Q1.15)	Zn: 1.0x10^{19} cm^{-3}	10 nm
p*-InGaAsP (Q1.26)	Zn: 1.0x10^{19} cm^{-3}	10 nm
p*-InGaAsP (Q1.46)	Zn: 1.0x10^{19} cm^{-3}	10 nm
p*-InGaAs	Zn: 1.0x10^{19} →2.0x10^{19}cm^{-3}	40 nm
n.i.o.-InGaAs	non-intentionally doped	70 nm
n*-InGaAsP (Q1.46)	Si: 1.0x10^{16} cm^{-3}	10 nm
n*-InGaAsP (Q1.26)	Si: 1.0x10^{16} cm^{-3}	10 nm
n*-InGaAsP (Q1.15)	Si: 1.0x10^{16} cm^{-3}	10 nm
n*-InP	Si: 1.0x10^{16} cm^{-3}	280 nm
n*-InGaAsP (Q1.26)	Si: 2.0x10^{16} cm^{-3}	290 nm
n*-InP	Si: 5.0x10^{16} cm^{-3}	1500 nm
s.i.-InP	Fe-compensated	125 μm

Figure 2.26 (a) Schematic view of the developed TTR-PD (b) with band diagram and (c) layer structure.

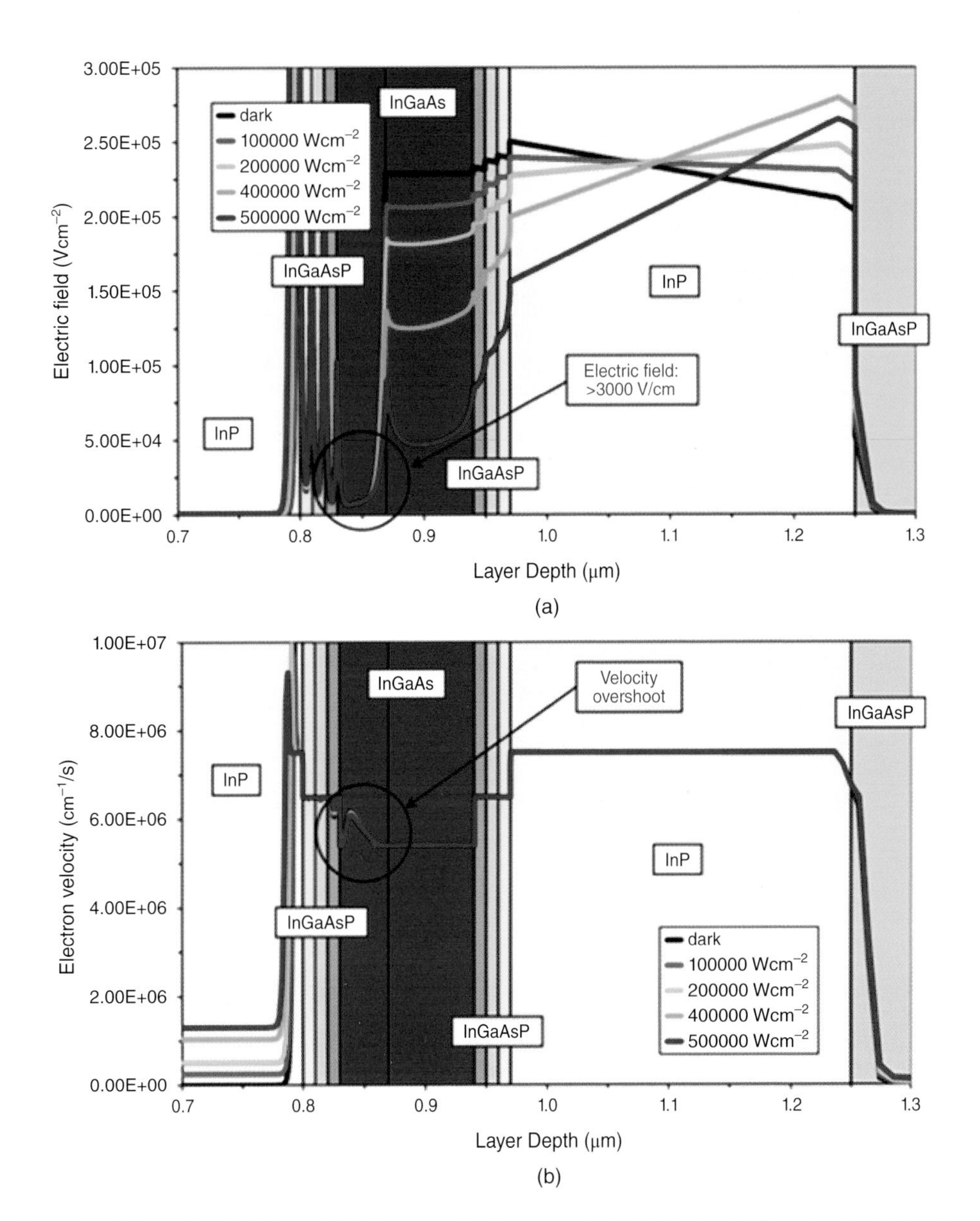

Figure 2.27 (a) Electric field and (b) electron velocity across the TTR diode at different optical intensity levels for a reverse bias of 8 V. Figure reproduced from ref. [98] with permission from the Optical Society of America, © 2014 Optical Society of America.

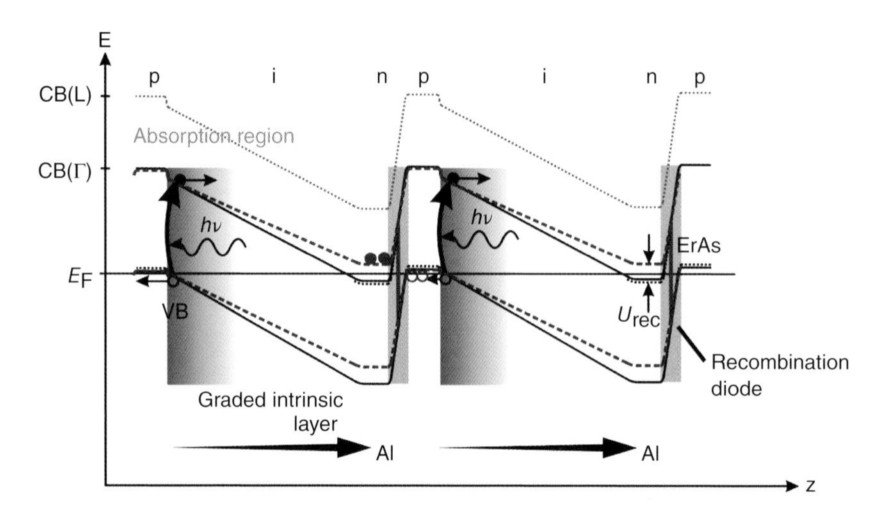

Figure 2.30 Band diagram of a n-i-pn-i-p superlattice photomixer ($N = 2$). Conduction band, valence band and L-sidevalley (dotted) are shown. The dashed line depicts the band diagram under illumination. A small forward bias, U_{rec}, evolves at the recombination diodes due to charge accumulation. The bias is only a fraction of the bandgap voltage. Reproduced and adapted with permission from ref. [18] S.Preu *et al.*, J. Appl. Phys. 109, 061301 (2011) © 2011, AIP Publishing LLC.

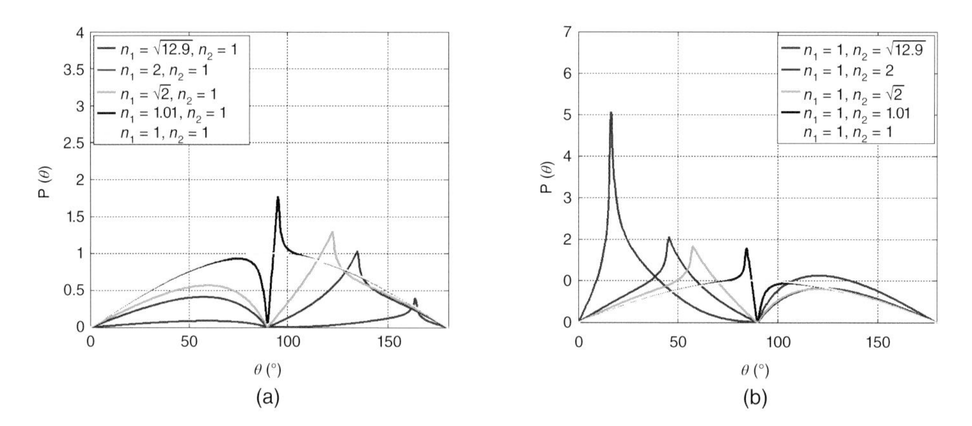

Figure 3.7 Angular distribution $P(\theta)$ of emitted power normalized to $n_1 = 1$ and $n_2 = 1$ case versus angle θ for vertical electric dipoles located on the interface ($z_0 = 0$). (a) $n_1 > n_2$ and (b) $n_1 < n_2$.

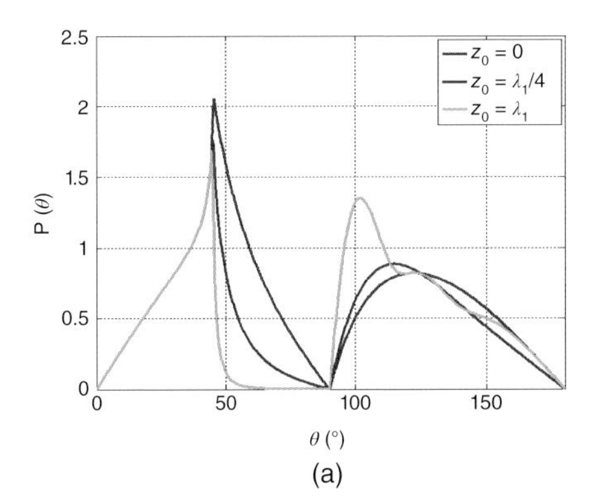

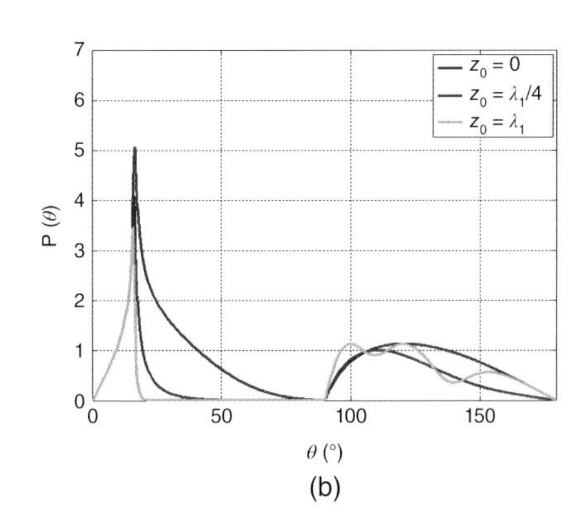

Figure 3.8 Angular distribution $P(\theta)$ of emitted power normalized to $n_1 = 1$ and $n_2 = 1$ case versus angle θ for vertical electric dipoles located at different distances z_0. (a) $n_1 = 1$ and $n_2 = \sqrt{2}$ and (b) $n_1 = 1$ and $n_2 = \sqrt{12.9}$.

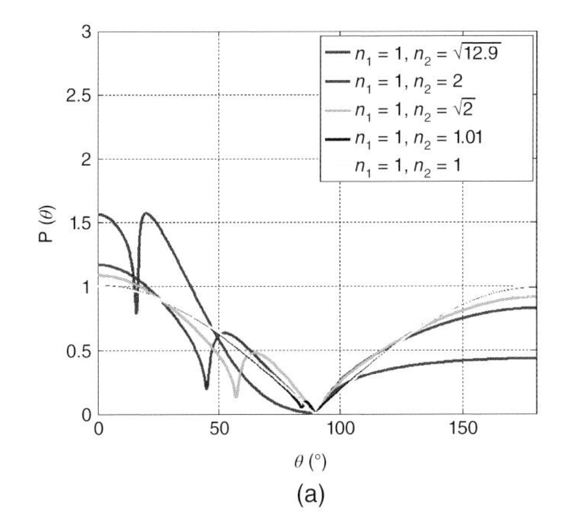

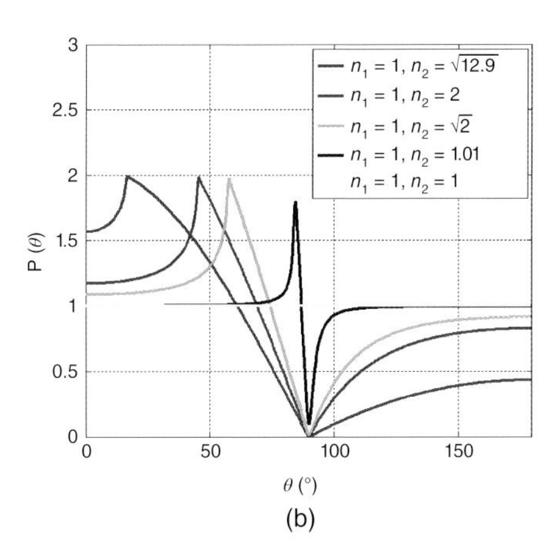

Figure 3.10 Angular distribution $P(\theta)$ of emitted power normalized to the free-space ($n_1 = 1$ and $n_2 = 1$) case versus angle θ for horizontal electric dipoles located on the interface ($z_0 = 0$). (a) E-plane and (b) H-plane.

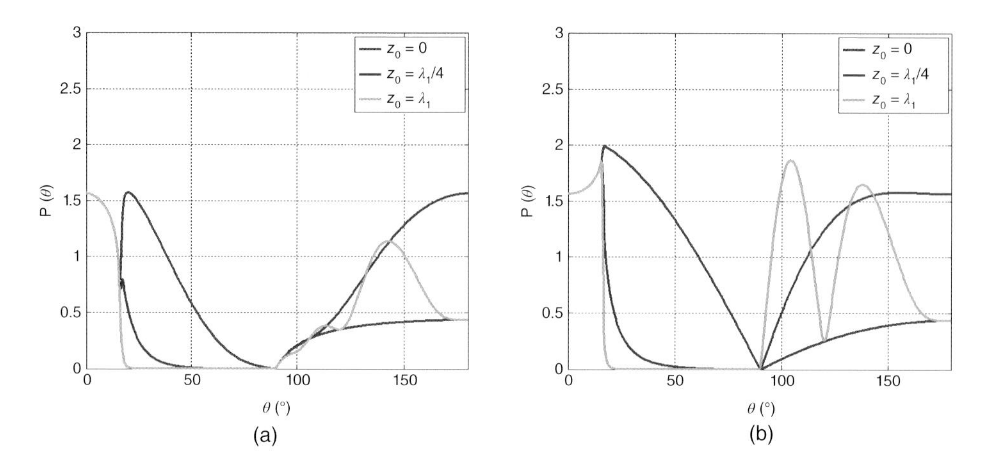

Figure 3.11 Angular distribution $P(\theta)$ of emitted power normalized to the free-space ($n_1 = 1$ and $n_2 = 1$) case versus angle θ for horizontal electric dipoles located at different distances z_0 with $n_1 = 1$ $n_2 = \sqrt{12.9}$. (a) E-plane and (b) H-plane.

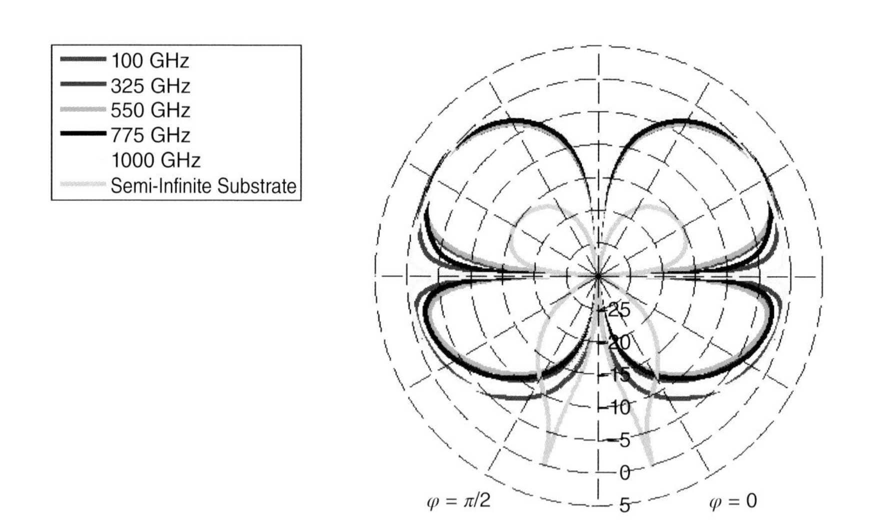

Figure 3.12 Radiation patterns of a vertical Hertzian dipole lying on a 350 μm GaAs substrate obtained at different frequencies. The case of a dipole on a semi-infinite GaAs substrate is shown as a reference.

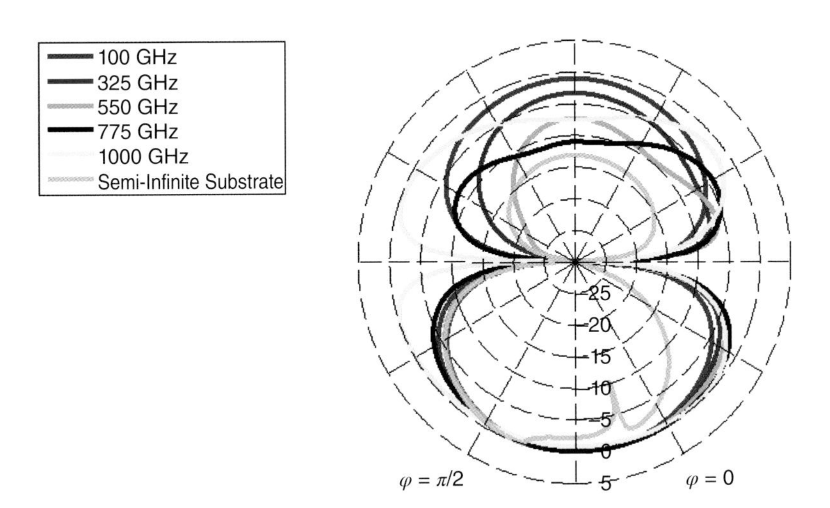

Figure 3.13 Radiation patterns of a horizontal Hertzian dipole lying on a 350 μm GaAs substrate obtained at different frequencies. The case of a dipole on a semi-infinite GaAs substrate is shown as a reference.

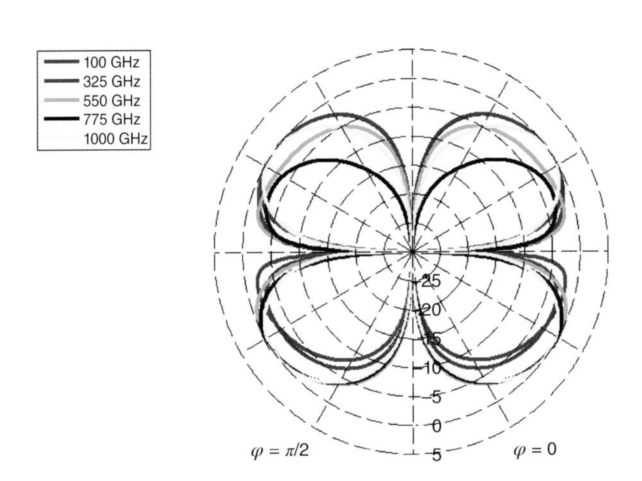

Figure 3.14 Radiation patterns of a vertical Hertzian dipole inside a 350 μm GaAs substrate obtained at different frequencies. The dipole is placed 25 μm away from the upper part of the wafer.

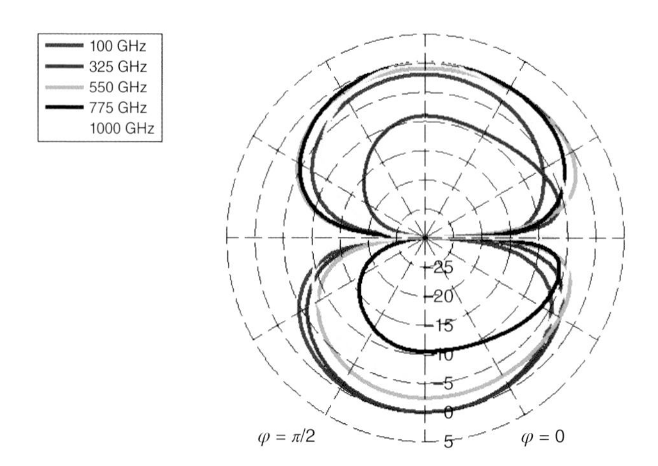

Figure 3.15 Radiation patterns of a horizontal Hertzian dipole inside a 350 μm GaAs substrate obtained at different frequencies. The dipole is placed 25 μm away from the upper part of the wafer.

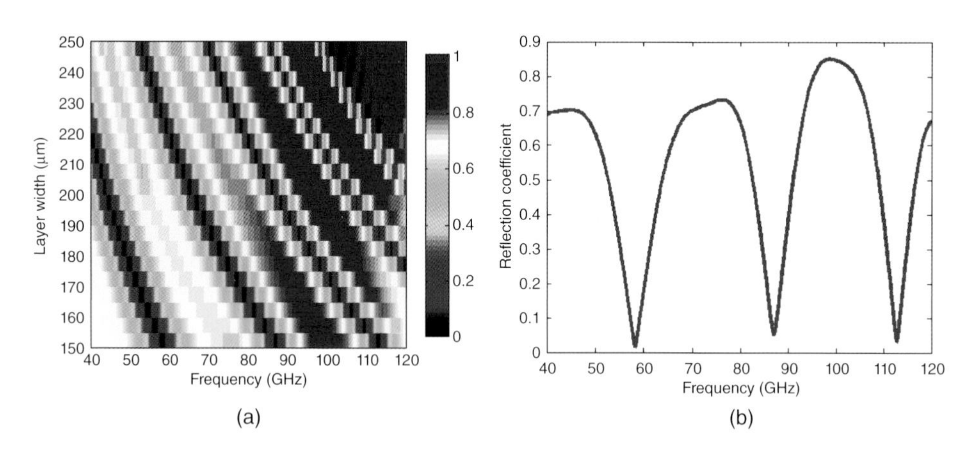

Figure 3.38 Reflection coefficients for six periods of stacking layers with $\varepsilon_{r1} = 12.9$ and $\varepsilon_{r2} = 2.2$ in terms of frequency and substrate width.

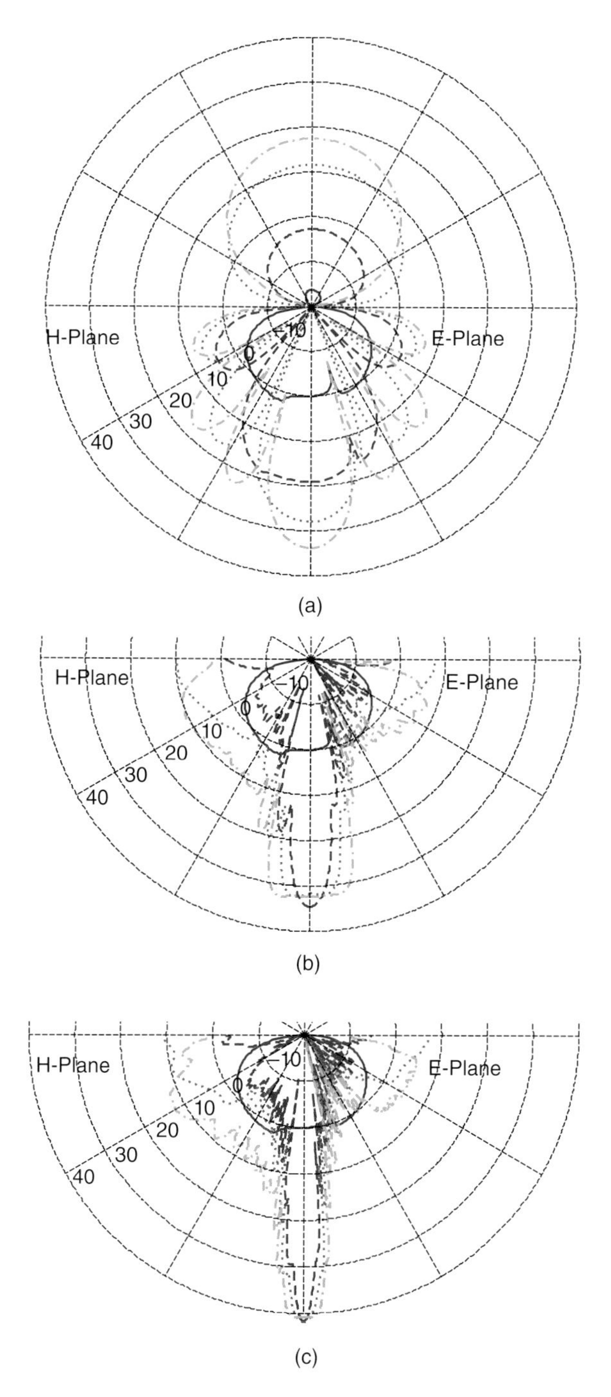

Figure 3.50 Radiation intensity patterns of a 3×3 array of isolated dipoles (- -), a 5×5 array of isolated dipoles (··), and a 7×7 array of isolated dipoles (-·-) with pitch $a = \lambda_0/(2n_{sc})$. (a) Arrays without lens, (b) with a 2 mm radius and 0.69 mm slab hyperhemispherical lens, and (c) with a 5 mm radius and 1.725 mm slab hyperhemispherical lens. For the sake of comparison also the radiation intensity pattern for the single AE is depicted (-).

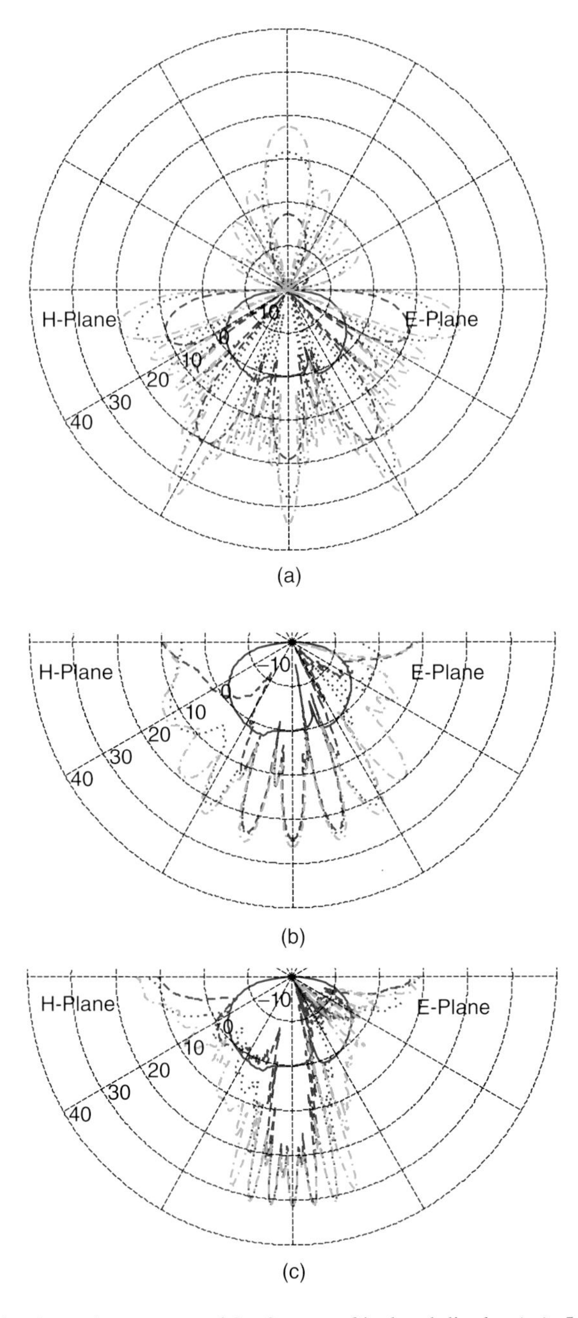

Figure 3.51 Radiation intensity patterns of 3×3 array of isolated dipoles (- -), 5×5 array of isolated dipoles (··), and 7×7 array of isolated dipoles (-··) with pitch $a = 4\lambda_0/(2n_{sc})$. (a) Arrays without lens, (b) with 2 mm radius and 0.69 mm slab hyperhemispherical lens, and (c) with 5 mm radius and 1.725 mm slab lens. For comparison also the radiation intensity pattern for the single AE is depicted (-). The lobes in (b) and (c) are directly related to the off-center positions of the pixels (three lobes for the 3×3, five lobes for the 5×5, and seven lobes for the 7×7 array). The angle between neighboring lobes closely corresponds to the values obtained from Eq. (3.43).

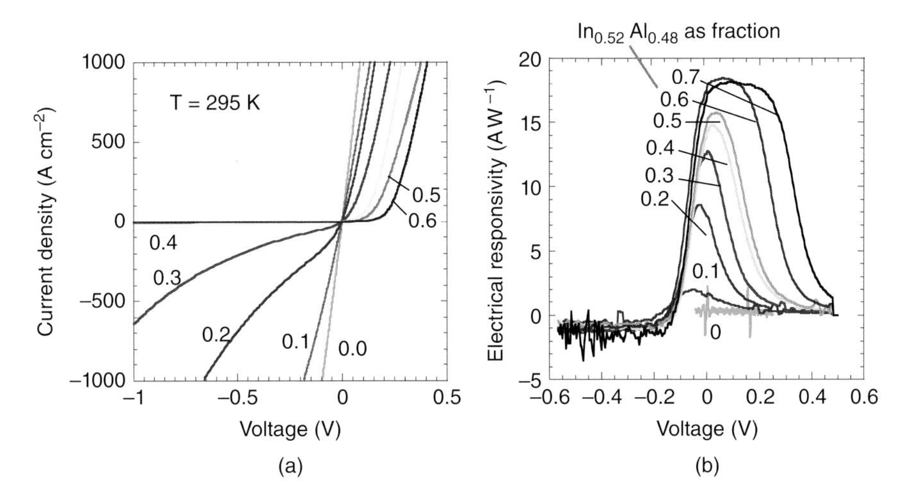

Figure 5.1 (a) Room-temperature *I*–*V* curves and (b) corresponding $\mathfrak{R} - V$ curves for a family of Schottky diodes fabricated from ErAs (semimetal)–InGaAlAs (semiconductor) junctions and with various In, Ga, and Al fractions [2]. (©2007 IEEE. Reprinted, with permission, from IEEE Microwave Magazine vol. 8, p. 54 (2007).).

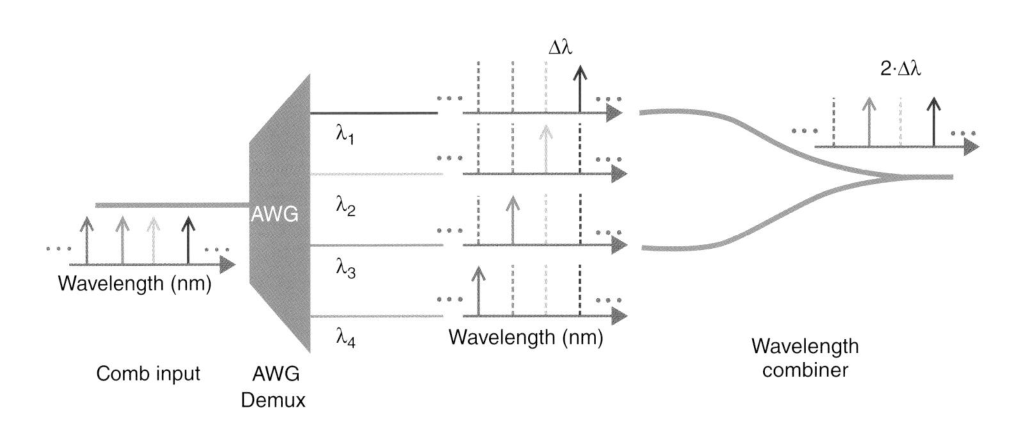

Figure 7.13 Block diagram of a planar lightwave circuit to generate a dual wavelength source.

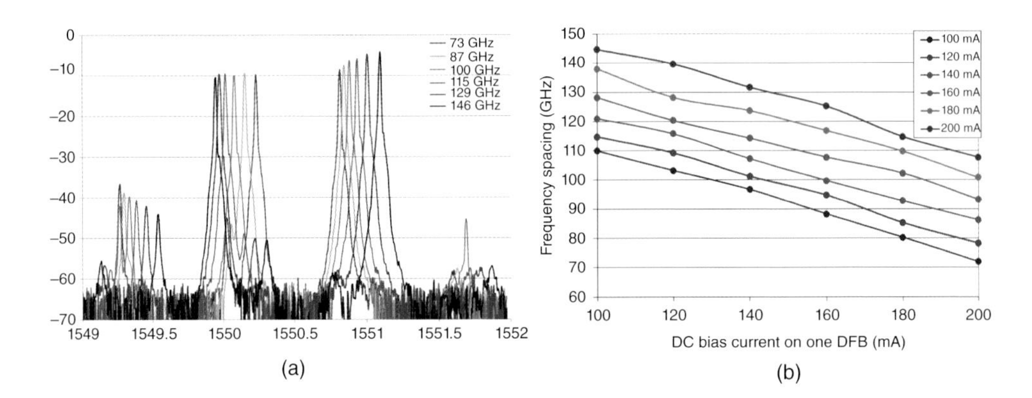

Figure 7.15 Dual wavelength output from two DFB lasers combined on a Y-junction. (a) observed on an optical spectrum analyzer and (b) wavelength spacing in terms of the current injected on one DFB (x-axis) for fixed values on the other (inset). (Fice 2012 [46]. Reproduced with permission of Optical Society of America.)

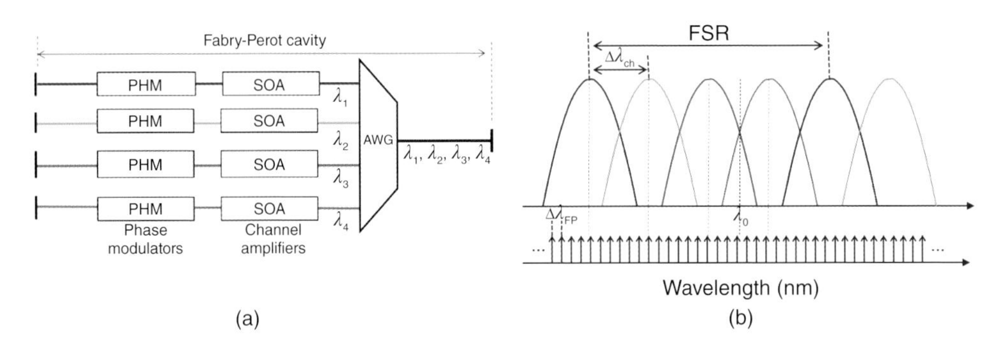

Figure 7.16 Dual wavelength output from an AWG laser. (a) schematic of the device, showing the AWG, its channel and common waveguides and the Fabry–Perot cavity, and (b) transfer function of the AWG, selecting the Fabry–Perot modes allowed to lase within each channel.

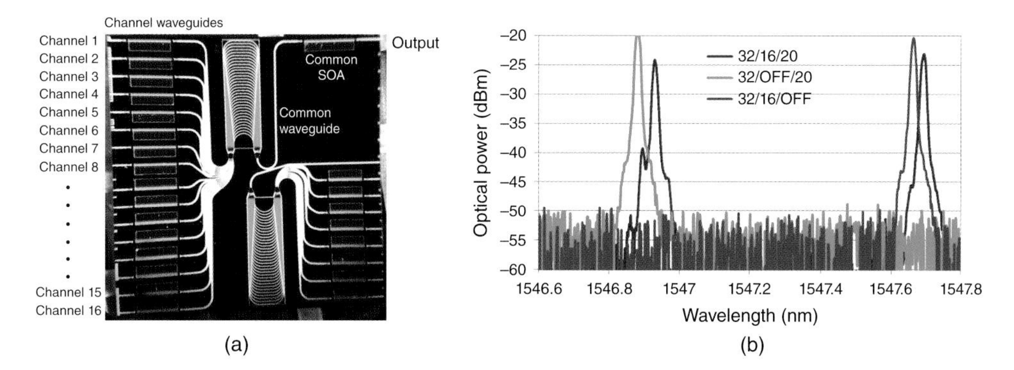

Figure 7.17 (a) Multi-wavelength AWG laser chip with two devices and (b) Dual wavelength output from an AWG laser, with bias in one channel, bias on the adjacent channel, and simultaneously biasing both.

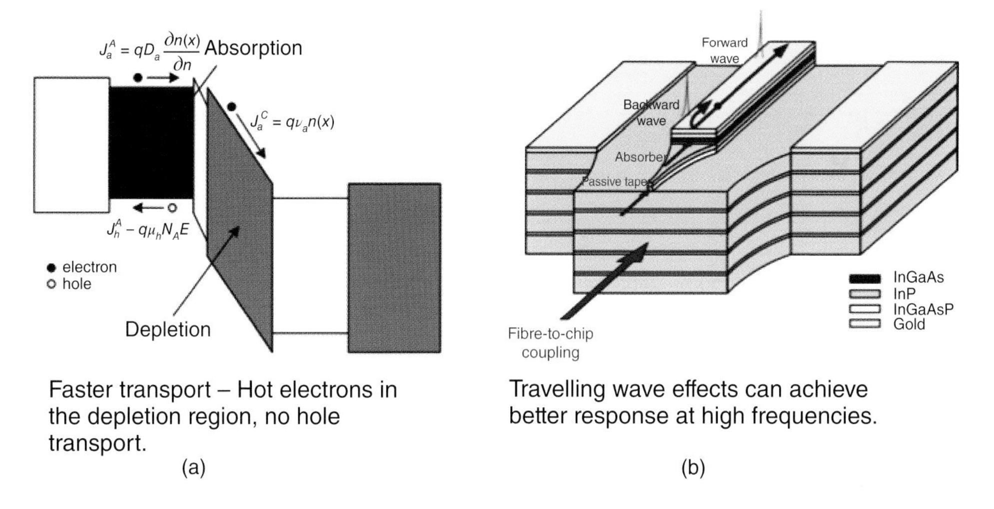

Figure 7.20 Schematic diagram of (a) a UTC-PD band structure and (b) aTW PD structure.

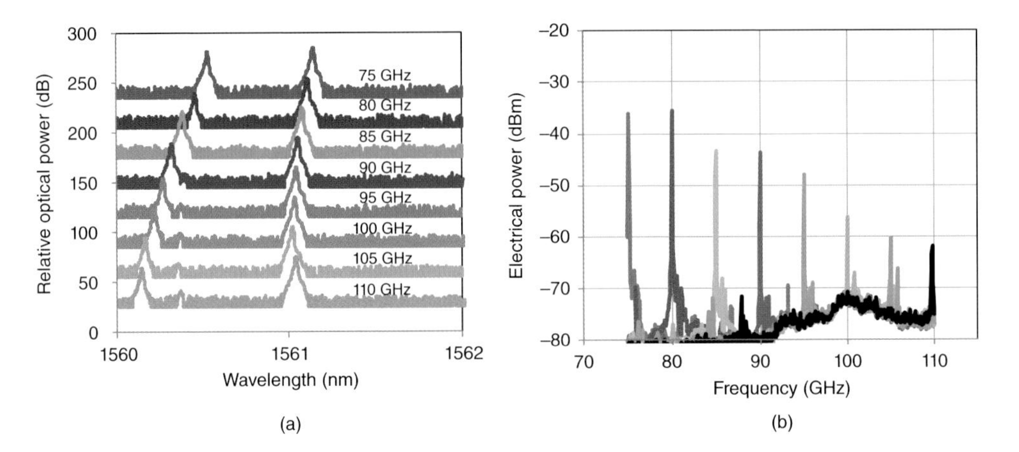

Figure 7.35 (a) Optical spectra of light generated by the chip for different bias currents applied to the DFB lasers and (b) corresponding electrical tone measured from one of the two UTC photodiodes.

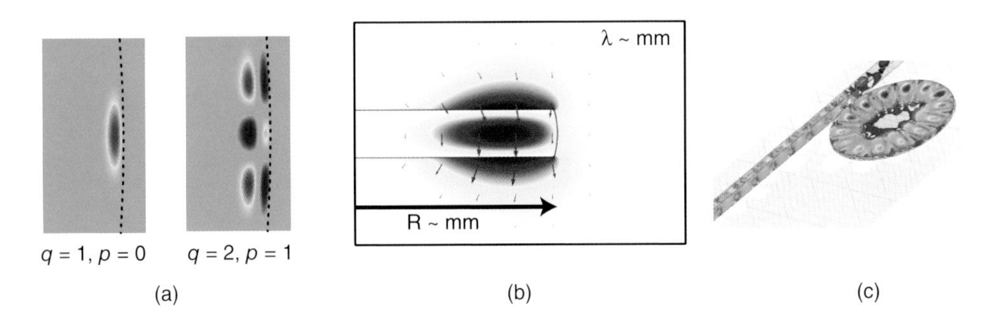

Figure 8.3 Modal distributions in a whispering gallery mode resonator. (a) Analytical solutions based on an expansion in spherical harmonics for a spherical geometry. (Adapted and reproduced from ref. [23] with permission of the Optical Society of America © 2013.) (b) Numerical solutions based on finite element methods. (c) Solutions based on a full 3D finite difference time domain which also include the coupling waveguide.

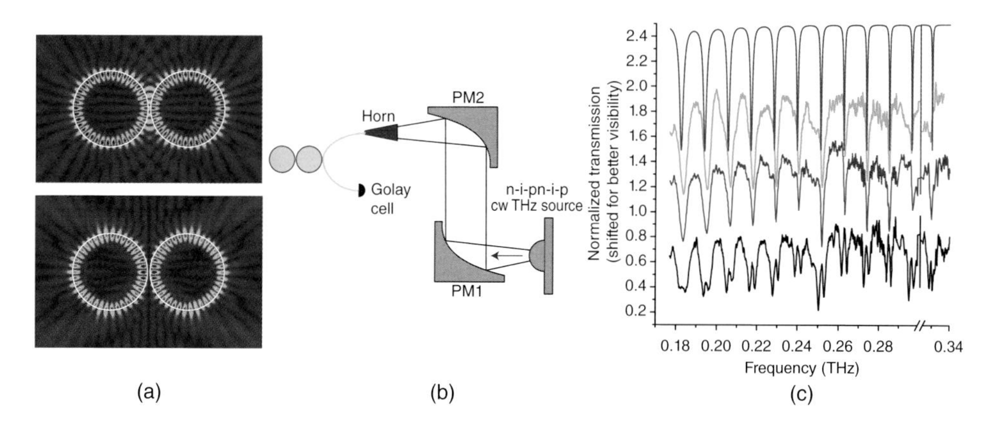

Figure 8.6 (a) When two spectrally identical disk resonators come within close proximity, they couple strongly and show a mode splitting similar to that of a hydrogen molecule. The field intensity distribution in two such coupled dielectric resonators is shown, calculated via a fast multipole method in 2D [30]. (b) Schematic set-up: THz light generated by beating two detuned lasers onto a n-i-pn-i-p photomixer and focused through two parabolic mirrors onto a horn and into a Teflon waveguide. The transmittance is measured. (c) Transmission spectra of two spectrally identical polyethylene WGM resonators (middle two traces). The bottom curve shows the transmission of both resonators in close proximity, showing clearly the mode splitting [31]. In the top trace data of a numerical simulation are shown. The individual traces are vertically shifted for clarity. (Reproduced from ref. [33] with permission of the Optical Society of America © 2008.)

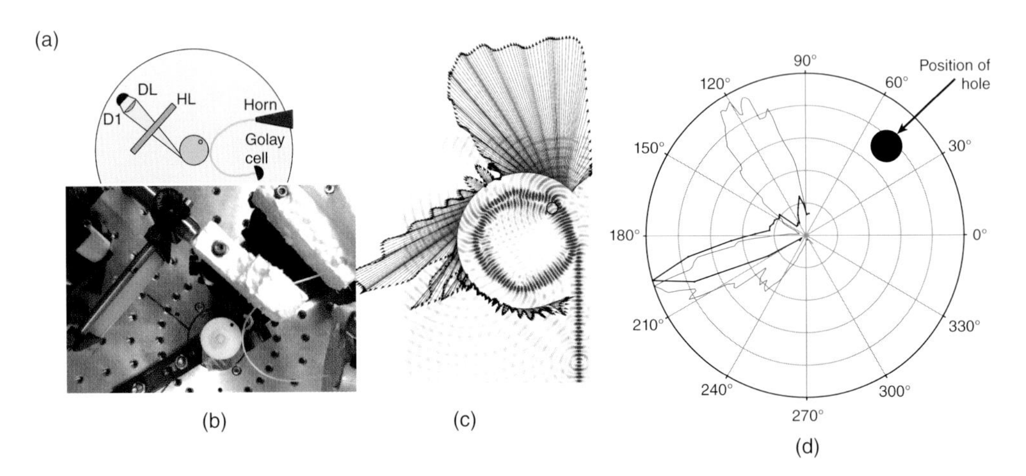

Figure 8.7 (a) A resonator is mounted at a fixed position in the center of a rotation stage. The detector (Golay cell, D1) scans the angular power distribution of the far-field of the resonator with an angular resolution of about 10°. (b) Photograph of resonator and waveguide. The waveguide is bent to improve coupling. (c) Result of the Poynting vector analysis of the resonator with hole. (d) Experimental data (black dotted) as well as far-field radiation pattern predicted from the theoretical Poynting-analysis on a linear scale. While the first main lobe predicted from the theoretical calculation around 120° is missing in the experimental data, experiment and theory agree well for the second lobe at 200°, most likely due to the slightly curved waveguide in the experiment. (Adapted and reproduced from ref [20] with permission of the Optical Society of America © 2013.)

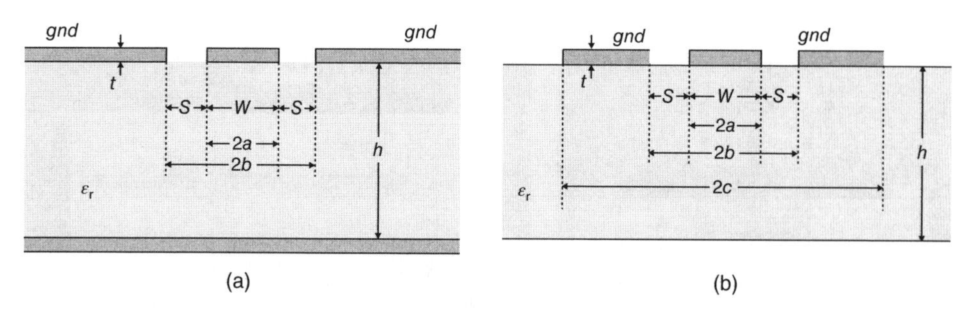

Figure 6.28 (a) Grounded coplanar waveguide structure and (b) coplanar waveguide structure with finite ground planes.

other hand, makes wafer handling quite tricky. In any case, for operating frequencies approaching the 1 THz frontier, this is not enough to reduce via parasitics. As a result, TMICs are not designed using a microstrip as a transmission media, but using a coplanar waveguide approach, typically with a back-side ground plane (grounded coplanar waveguide, GCPW), as sketched in Figure 6.28a. Such a ground plane is in any case to be accounted for if the chip is going to be mounted onto a metal carrier, acting as a metal back-side. In addition, size constraints on the circuit transverse dimensions lead to finite lateral conductors (grounds, Figure 6.28b), that in turn introduce further modifications to the electromagnetic modeling of the signal propagation in the circuit. Back-side processing and via formation is maintained and adopted to suppress the onset of eventual substrate and microstrip modes. Further, as in low-frequency CPWs, air bridges are used, further complicating the technology, to eliminate odd propagating modes close to major discontinuities.

The resulting circuits are therefore mostly in GCPW, leaving a few realizations in microstrip for the "lower" portion around 300 GHz. This modification, however, is not painless, since nowadays neither circuit simulators nor the foundries provide reliable models for components realized in such transmission media, and the electromagnetic simulator in turn may be particularly tricky to be used: the designer's expertise and a trial-and-error approach play an important role.

A noteworthy exception regards the possibility of using a microstrip together with benzocyclobutene (BCB) triple-layer interconnects [93]: in this case, a thin film microstrip wiring is used, and the ground-plane is formed in the first interconnect metal layer, shielding the RF signal formed in the topmost interconnect layer from unwanted coupling into the substrate, and thus preventing the onset of parasitic modes formation. Again, the approach has been demonstrated in the lower portion of the TMIC range only.

A further notable difference introduced by TMICs concerns the access and related mounting procedures. In fact, most of the MMICs are typically glued onto a carrier and wire-bonded (or ribbon-bonded) to external microstrip transmission media; for the TMIC, bonding actually introduces losses and unwanted effects that inevitably damage the circuit performance dramatically: as a consequence the "chip-and-wire" approach is very seldom used and ultimately only in the lower portion of the TMICs. In this case the coupling with the input and output waveguides is realized through waveguide probes, of alumina or quartz, and inserted typically in the E-plane of the appropriate waveguide. The microstrip end of the probe is then wired to the MMIC (as in Ref. [94], reported in Figure 6.29a). For higher frequencies, in order to minimize the losses associated with probe connection, the waveguide probes are directly realized via the same monolithic technology onto the chip, with electromagnetic coupling depending on the probe type. The same concept of E-plane probe, introduced for W-band circuits in the late 1990s, can be applied, but taking into account that the small waveguide dimensions and the high dielectric constant of the semiconductor material may not allow use of the full waveguide bandwidth, given the resulting dielectric loading. Several countermeasures can be adopted, among

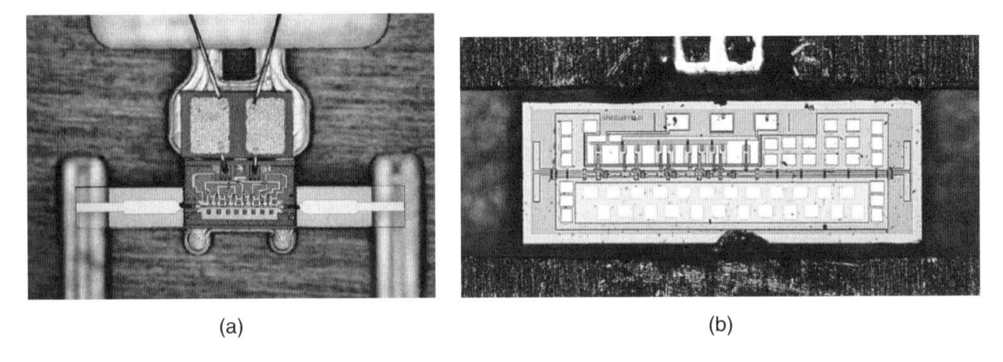

(a) (b)

Figure 6.29 (a) Quartz transitions to waveguide [94] and (b) integrated coplanar stripline dipoles [95].

which are the etching of part of the substrate around the probe or modifying the standard cavity dimensions. In addition, a different and interesting approach can be adopted if the waveguide probe is replaced by a dipole, integrated in the TMIC and realized in coplanar stripline transmission media ([95], Figure 6.29a,b): in this case the dipole is oriented in the E-plane, and the coplanar slot of the probe is transformed into the chip coplanar waveguide via air-bridge-assisted transition. Multiple chips can be mounted in cascade, thus eliminating the need for multiple flange connections.

As a result of this integrated arrangement for the amplifier chip coupling to the input/output waveguides, measured performance includes the effects of the transitions and cannot be effectively de-embedded, even if a rough estimate via back-to-back transition measurement is often attempted. For the sole purpose of the amplifier chip performance verification, a replica of the amplifier can be realized not containing the waveguide transition system but equipped with coplanar probe pads. In this case the chip performance may be directly assessed and eventual interactions with the transition may be isolated and counteracted.

Available results of TMIC LNAs are therefore influenced by such dichotomy, since many results do include integrated transitions insertion losses, affecting small-signal gain, noise figures, and matching performance; a few of them are measured directly on chip with coplanar probes: in this case an evaluation of the technological advances is directly available.

6.3.3 Low Noise Amplifier Design and Realizations

The race to break the 1 THz barrier with active circuits has to cope with several issues, adding to the inherent difficulties residing in the development of the appropriate active device(s) (e.g., InP HEMTs, mHEMTs, or HBTs). The first is the lack of accurate active device models at such extremely high frequencies. In most cases, device models, even if simply linear ones, are obtained from *S*-parameter measurements performed typically up to W-band (110 GHz) with on-wafer probe coplanar arrangements, provided a suitable calibration approach has been adopted and calibration kit is available. Almost standard (lower frequency) extraction procedures are adopted, and the resulting model is then extrapolated and tuned to match the measurements (*S*-parameter) obtained at higher frequencies with different set-ups (e.g., WR 3, WR 2.2, or WR 1.5 VNA extension modules). The result, given the inherent uncertainties, is far from being accurate but represents a suitable starting point to evaluate technological capabilities and to draft a design. The situation is even more complex for active device noise models that represent a key issue in LNA design. In this case the difficulty is twofold: on the one hand, direct extraction of device noise parameters based on tuner-based systems is simply not possible, even at W-band. Therefore, a possible approach resides in the use of noise-temperature models, based on the prior knowledge of the device equivalent-circuit model and a few noise measurements. In this way, the resulting noise model is intrinsically valid up to the range of the equivalent-circuit it is based upon. Further noise performance extrapolations within the THz range are, therefore, even less accurate than the linear ones. Attempts at verifying device noise performance, even at moderately high frequencies, suffer from a huge uncertainty arising from the output device mismatch and interactions with the noise receiver (power sensor). As a result, no device noise model data are available, to the author's knowledge, for the THz range. Furthermore, in the THz range, even at moderately high frequencies, the commonly adopted assumptions of white power spectral densities begin to fail, given the onset of quantum effects: the simple extrapolation of models extracted at lower frequencies may not represent the real noise behavior of the device.

A further nasty problem resides in the characterization and modeling of the passive structures (transmission lines, discontinuities, lumped elements, etc.): in most cases the approach is to build and measure, thus skipping the step of model extraction and trusting as much as possible the simulation capabilities of existing circuit-oriented and electromagnetic software. The latter are indeed growing in accuracy and potential, but are actually based on the structure parameters, that have to be carefully evaluated to yield a reliable simulation. They may also be derived from appropriate calibration kit measurements.

Finally, a technological difficulty does exist in tailoring the passive elements' values to the reduced dimensions of a TMIC. As an example, MIM (metal–insulator–metal) capacitor values necessary on chip to filter out the DC bias would result too large if their capacitance per unit area is not increased accordingly, together with the resistors' sheet resistance. Similarly, dry-etched through vias, fundamental to eliminate parallel plate waveguide modes, should be scaled accordingly to reach a higher density (lower spacing) and smaller front-side dimensions together with air bridges and metal lines features (down to 1 μm).

Given the above considerations, TMIC amplifiers are featured by a pretty standard configuration, composed of the cascade of 3–4 amplifying stages in the lower THz region, while up to 10 stages are cascaded when the gain perstage becomes impractically low approaching 1 THz. Such a simple configuration is clearly dependent on the need to keep circuit complexity to the lowest possible level, given the design uncertainties sketched above. Single stages are either of the common-source (CS) or cascode (CS-CG) type. For the former and simpler approach, a double bias scheme is adopted, with separate gate and drain bias lines for each stage, to allow maximum versatility in circuit tuning. On-chip simple filtering/stabilizing networks are provided for each bias line, basically composed by series R/shunt C elements toward the bias pads. Overall input and output matching networks are kept to minimum complexity and designed to get 50 Ω matching also at the amplifier's input, due mainly to the lack of reliable noise models (no optimum noise match is attempted): a combination of series L (line)/open stub/series L (line) is adopted for matching [75]. Interstage matching is again left to series lines of appropriate lengths and widths, together with DC-blocking series capacitors, that is, kept to the minimum complexity. The latter configuration, that is, the cascode connection in Figure 6.30, is a bit more complex, requiring the

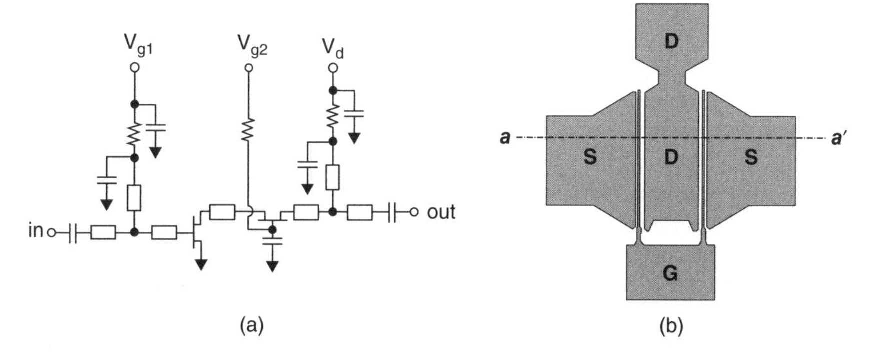

Figure 6.30 (a) Cascode connection and (b) typical interdigitated layout.

cascade of CS and CG (common-gate) transistors, therefore adopting two bias lines for the gates and a single drain bias (thus biasing the two transistors at the same current level [94]).

Regardless of the adopted stage configuration, selected devices for TMIC LNAs always exhibit a simple interdigitated layout (Figure 6.30b), featured typically by two fingers (in some cases a four fingers layout is adopted if allowed by the available air-bridge technology [96]) whose length decreases with increasing target operating frequencies, from 25 down to 4–5 μm; moreover, the two finger topology eases the drain and source direct connection, thus minimizing discontinuities in propagation. The resulting gate periphery is therefore from 50 to 10 μm, giving rise to a current consumption per stage on the order of 5–20 mA. Given the typically low drain biases (ranging from $V_{ds} = 1.2$ down to 0.8 V), this brings the power consumption of typical TMICs stages to 20 mW maximum down to 4 mW per stage.

To briefly summarize the attained performance to date, it should be noted that the commonly adopted figure cannot be just the gain perstage of the given amplifier to compare gain performance with a common metrics. In this respect it should be noted that cascade realizations inherently guarantee a higher gain perstage, due to their CS-CG topology as compared to simple CS ones. Further, both gain and noise performance are strongly influenced by the amplifier realization, for example, TMIC (on-wafer probed) or packaged (module). In the latter case, transitions from waveguide media, both at module input and output, increasingly worsen the module performance if compared with that of the bare chip. Moreover, for packaging constraints, the TMIC to be mounted inside the module is typically modified by the addition of transmission lines connecting the waveguide transition to the amplifier chip input. Both effects may degrade chip performance by several dBs [96].

Available data on realized TMIC LNAs are mostly restricted to the frequency range 300–500 GHz, where a clear trend can be evidenced: 5 dB perstage is attained on average with CS topologies, reaching 6.7 dB for CS-CG stages. The noise figure for packaged amplifiers is slightly higher than 8 dB in the lower frequency edge, while it increases to almost 12 dB at 480 GHz. Such a figure has clearly to be corrected by the transition and mounting contribution, totaling at least 2 dB. The most commonly adopted technology for such demonstrators is a 35 nm HEMT, either of the lattice-matched InP or metamorphic type. A few examples of InP DHBTs (double-heterojunction bipolar transistors) do exist [97], demonstrating quite good gain performance, even if the measured noise performance is typically worse than HEMT counterparts (by at least 3 dB), however, exhibiting the technology potentials. A quite complete listing of submillimeter wave realizations can be found in Table 1 of [77].

If frequency is further increased over 500 GHz, a few demonstrators actually exist, either realized in InP HEMT or mHEMT technologies, with increasingly smaller gate features. In particular, if lattice-matched InP HEMT devices are considered, the first demonstrator dates back to 2010 with cascode stages [98], while a remarkable upgrade was presented in 2011 [96], for operation at 670 GHz. In this case a TMIC amplifier composed of 10 CS stages, each 30 nm 2 μm × 7 μm devices, demonstrates a peak 30 dB gain at 670 GHz with a >25 dB gain from 635 to 680 GHz. A second TMIC, composed of 5 CS stages in the same technology and devices, mounted onto a module, exhibited a 13 dB noise figure at the same frequency. In a later contribution [79], a 10-stage amplifier was inserted in a module exhibiting a 12.5 dB noise figure at 670 GHz, competitive with GaAs Schottky diode receivers at the same frequency.

Very recently, the InP technology has been further improved to reach the record 850 GHz operating frequency [75]: in this case, 25 nm transistors were used to form a 9 CS-stage TMIC (see Figure 6.31a,b), demonstrating, for the first time, a peak gain of 6 dB at 850 GHz. Such a figure is clearly not practical for LNA packaging. However, with the same technology, a 4-stage frequency-scaled version has also been fabricated to operate at 425 GHz, exhibiting a record >20 dB gain over more than 150 GHz bandwidth, with associated 7.5 dB noise figure at the module level, again a state-of-the-art performance.

If metamorphic technology is now considered, the first result at frequencies higher than 500 GHz is presented in Ref. [99], while a 600 GHz amplifier in 35 nm technology is presented in Ref. [100]: six stages are adopted in GCPW, each adopting a 2 × 4 μm mHEMT, resulting in a maximum 20.3 dB gain at 610 GHz. With the same technology, a 230–300 GHz packaged amplifier has been realized, resulting in >20 dB gain and an associated noise figure of 6.1 dB. Also, in this case, a further recent update

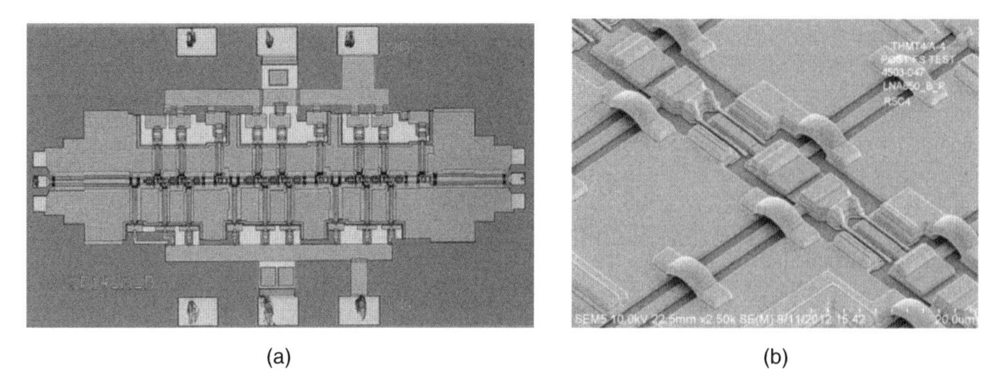

Figure 6.31 (a) Microphotograph of the nine-stage TMIC in Ref. [75] and (b) SEM picture of the front-side of (a) [75].

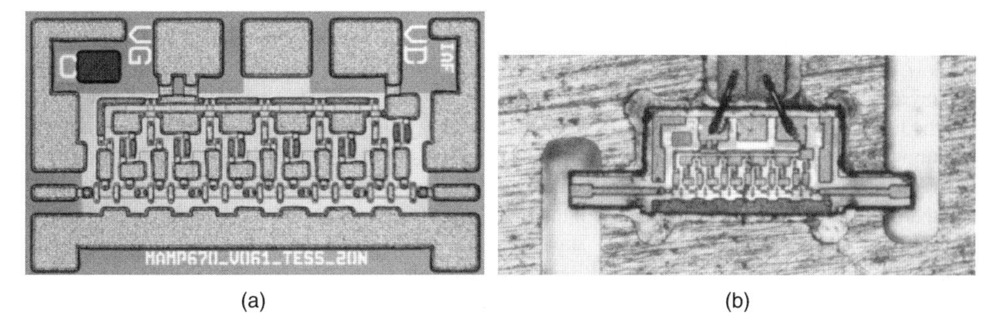

Figure 6.32 (a) Microphotograph of 600 GHz TMIC in Ref. [101] and (b) the same amplifier with integrated microstrip E-plane transitions [101].

[101] contributed with the measured noise performance of a packaged version of the same amplifier (see Figure 6.32), resulting in an average noise figure of 15.1 dB at center frequencies.

Also in the case of mHEMT technology, transistor gate features continue to shrink, leading to 20 nm gate length, as detailed in Ref. [102]: 8 CS stages were used, each composed of 2×5 μm transistors, to realize an amplifier achieving more than 15 dB gain from 500 to 635 GHz, with a peak gain of 23 dB at 570 GHz. The amplifier smoothly degrades the gain performance, exhibiting gain up to 700 GHz. No noise data are available at the moment. A peculiar feature of this demonstrator resides in the use of microstrip transmission media (as contrasted with the previously described realizations, all in GCPW), realized making use of BCB (as a low-*k* dielectric), representing therefore the state of the art in microstrip active circuits.

6.3.4 Perspectives

As it should be clear from the discussion above, the race to 1 THz amplifier operating frequency is going to be completed very soon: very likely the milestone indicated in the DARPA THz Electronics program ([76], i.e., February 2015) will be reached. However, as demonstrated by the many steps forward to date, the game is far from being finished: state-of-the-art demonstrators represent in fact simply statements of technological capabilities and potentials. To fully exploit the real performance of a technological node, further advances are needed: on the one hand on the device characterization and modeling side, and on the other one hand specific design methodologies have to be adopted. For the former point, characterization set-up costs and complexity have to be reduced, and complemented with ad hoc calibration procedures. Noise characterization is a problem if on-wafer measurements are required at THz frequencies, given the lack of appropriate set-ups and de-embedding methods. The problem is even more cumbersome if modeling is considered, taking into account that, as already mentioned, the use of device noise models borrowed from lower frequencies is questionable. A quantum noise approach is necessary to fully understand and dominate the noise generation in THz devices. Once the above-mentioned tools are developed, even the actual gate features may suffice to improve noise (and gain) performance drastically, making use of ad hoc design strategies, as in the case of the microwave and millimeter wave ranges.

6.4 Square-Law Detectors

Summarized with the term square-law detector are all devices that allow high frequency signal rectification with a quadratic dependence between the applied voltage and current. With a proper detector design, this nonlinear behavior leads to a defined dependence between the RF and DC or video signals. For low RF input power levels, the output voltage is proportional to the quadratic RF input voltage, or directly proportional to the RF input power. If the input power is enhanced beyond specific power levels, the linear diode region is reached. Here the output voltage is proportional to the square root of the input power. The otherwise constant detector voltage responsivity degrades with increased input power. For further increased input power levels, the detector saturates. For an exceeded maximum input power, and therefore power density at anode level, the diode might even be destroyed. In order to assure high detector linearity, the applicable dynamic range is defined within the quadratic region of the diode, being considered for most millimeter and submillimeter wave applications.

Schottky diodes with reduced barrier height are very efficient in signal rectification and have already been applied in many kinds of applications. The avoidance of bias currents reduces design complexity, power consumption, and particularly noise contribution. This technology evolution has opened the door for competitive diode-based devices to frequency ranges even up to the THz. Most commercially available Schottky detectors, waveguide coupled and quasi-optical, are broadband devices, providing only minor impedance matching with only acceptable signal to noise ratio (SNR). These performances cannot compete within most emerging detector application concepts:

1. Radiometry demands a maximum of SNR and linearity (deviation from square law) from low-barrier detectors to compete with mixer module performances.
2. Ultra broadband detectors require a fast response time and reasonable sensitivity over a multi decade bandwidth, for frequencies between 50 GHz and several THz.
3. Wireless communication networks require compact devices with large video bandwidth and maximum SNR.

Detector circuits for demodulation or power measurement with low-barrier Schottky diodes are all based on the same principle. A nonlinear device, the implemented diode, realizes the rectification of an RF signal to a video frequency or DC signal. All other circuit elements separate the relevant frequencies, and assure closed circuits for the RF and video frequencies.

The focus in this section is on radiometric detectors, with low-barrier Schottky diode characterization in particular. In contrast to radar, radiometry is a fully passive technique, detecting spectral parts of the blackbody radiation of objects. The type of the emitted or reflected signal is thermal noise at low power levels. Therefore, extremely sensitive receivers are required, providing very low noise contribution by the square-law detector itself.

The rectifying element in the exemplified detector design is a low-barrier Schottky diode from ACST GmbH [103]. Due to a strongly reduced barrier height, this diode type features a low differential resistance R_{diff} of down to a few hundred ohms only. Benefits are an enhanced power transfer to the diode and a low thermal noise level. The applied low-barrier Schottky diode of type 3DSF is fabricated by the film diode process (FD-process), suitable for fabrication of discrete diodes and integrated circuits for millimeter and submillimeter wave applications. This technology is further discussed in Section 6.5.

6.4.1 Characterization and Modeling of Low-Barrier Schottky Diodes

Square-law detector designs with maximum performance require detailed knowledge of the detector diode. Diode manufacturers typically provide relevant diode models. In case of incompleteness or inaccuracy, RF designers are required to set up their own diode models, adapted to the specific application and final device specifications. Figure 6.33 illustrates a typical parameter extraction procedure for low-barrier Schottky diodes. Besides the common *I–V* measurement, to extract intrinsic diode parameters, low-barrier Schottky diodes for radiometry have to be characterized for white noise and $1/f$ noise, which is essential for radiometer integration time as part of the signal processing.

For most Schottky diodes, the extraction of the junction capacitance C_j parameter is achieved with LCR meters. For low-barrier Schottky diodes, the values to be extracted from phase angle measurements of the complex impedance is in the range of the calibration accuracy of state of the art measurement equipment. Therefore, it is not possible to extract this parameter directly for low-barrier Schottky diodes. Hence, *S*-parameter measurements are performed within the frequency band of application, or as close as possible. The diode junction capacitance at zero bias, C_{j0}, can then be obtained by curve fitting of measured and simulated diode data. The diode model is further expanded with temperature and thermal noise measurements of the *I–V* data and the noise floor. Expanding frequency ranges also require inclusion of full 3D diode models for RF and thermal simulations.

A common first step in diode modeling is the DC characterization [103, 104]. From *I–V* data, applying first and second derivatives, R_{diff}, (Eq. 6.40) and DC current responsivity, β_0, (Eq. 6.41) are extracted, respectively. Low-barrier diodes are often compared by their R_{diff} at 0 V. This parameter is dependent on the anode size and the Schottky interface. The second derivative describes the diode's curvature in A/W. Half this value is also referred to as the diode's intrinsic current responsivity, β_0. For low-barrier devices, the current responsivity maximum is very close to 0 V. The absolute value is limited by technology,

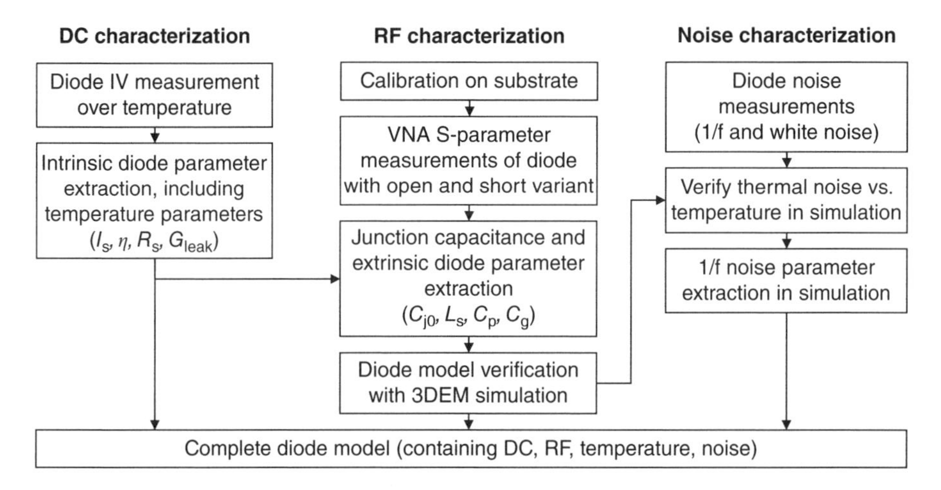

Figure 6.33 Low-barrier Schottky diode characterization chart.

providing quality criteria for the Schottky interface of the diode. This parameter describes the short circuit current responsivity of the diode, and therefore the maximal achievable current responsivity of the detector.

$$R_{\mathrm{diff}} = \frac{1}{\frac{\mathrm{d}I}{\mathrm{d}V}} = \frac{\mathrm{d}V}{\mathrm{d}I}. \tag{6.40}$$

$$\beta_0 = \frac{\frac{\mathrm{d}^2 I}{\mathrm{d}V^2}}{2\frac{\mathrm{d}I}{\mathrm{d}V}} = -\frac{\frac{\mathrm{d}^2 V}{\mathrm{d}I^2}}{2\left(\frac{\mathrm{d}V}{\mathrm{d}I}\right)^2}. \tag{6.41}$$

Figure 6.34 presents typical measurement data. Results are plotted for four different anode diameter sizes from one single semiconductor wafer. With curve fitting methods, the three data sets per diode, *I–V* curve, R_{diff}, and responsivity, are compared to simulation data, to gain an intrinsic diode model.

Most simulation tools provide standard diode models, which can be adapted to a specific diode. Physical intrinsic parameters are directly revealed, such as saturation current I_s, ideality factor η, series resistance R_s, and leakage conductance G_{leak}. If these physical parameters are of no interest for the designer, the *I–V* data might also be modeled with equation-based nonlinear components in simulation tools. The applied polynomial order depends on the required fitting accuracy and voltage or RF power range of the final application.

Temperature measurements are directly performed with the diode on wafer, or on single diodes, already separated. The recorded series of curves with *I–V* data over temperature is represented and fitted by the temperature coefficient in the standard diode model, or included in the polynomial equation. In the latter case, special attention is demanded for the modeling of the nonlinear temperature dependence of the diode.

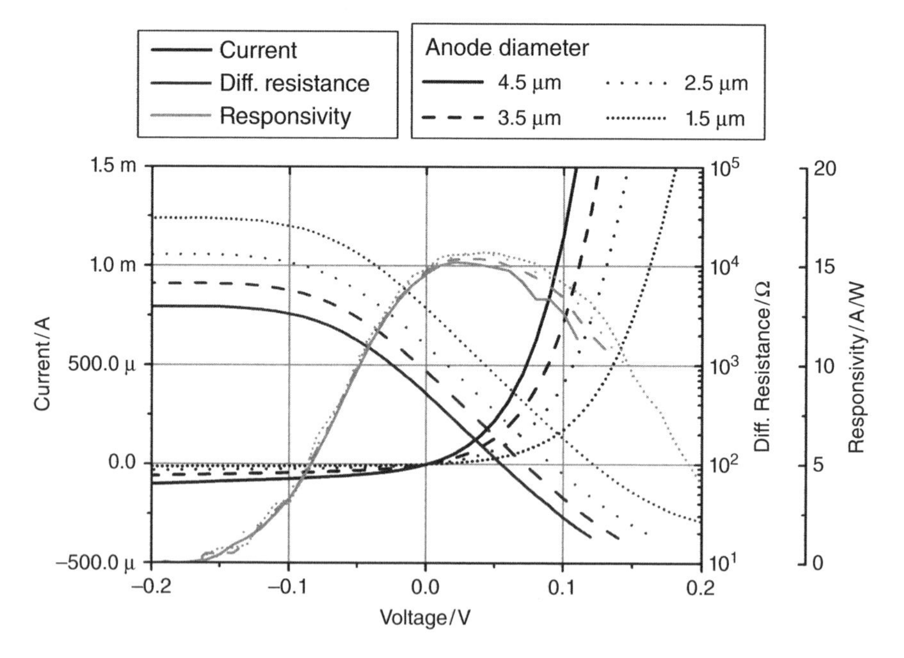

Figure 6.34 DC measurement result, with *I–V* curve, differential resistance R_{diff}, and DC current responsivity. Four different anode diameters are presented.

S-parameter measurements of Schottky diodes are performed either to verify the directly measured capacitance [104], or, in the case of low-barrier Schottky diodes, it is the only way to extract this diode parameter. The probe calibration for vector network analyzer measurements of the mounted diode is achieved with individually designed calibration kits, to de-embed the diode RF characteristic only. A common line structure is the grounded CPW line, which suits the GSG wafer probes for millimeter and submillimeter wave frequencies. In addition to the diode itself, open and short variants of the diode serve to increase the accuracy of the diode model. Knowledge of the housing or packaging diode parameters is gained, as the series inductance L_s and parasitic capacitances (Figure 6.35). Similar to the physical intrinsic diode parameters, this information is not essential for a detector design. However, at higher millimeter frequencies, these parameters can limit the overall detector performances, especially for smaller diodes, and should be considered as a design parameter.

Most often not only the RF responsivity, but rather the SNR of a detector is optimized. This is one major design parameter also for radiometry. Then, an accurate noise model is to be created. White noise, dominated by thermal noise, is directly modeled with knowledge of R_{diff}, but the $1/f$ noise contribution requires biased measurements on the single diode [105].

3D diode models are necessary for simulation and modeling at millimeter frequencies and above. Either the diode manufacturer provides the 3DEM models directly (Figure 6.35b), or they are generated by the detector designer with knowledge of the main diode dimensions and material properties. At frequencies below 100 GHz, equivalent circuits (Figure 6.35a) may provide sufficient accuracy. This depends on the individual design and relevant circuitry surrounding the diode.

After diode parameter extraction, the maximum achievable diode RF voltage responsivity at room temperature is calculated according to Eq. (6.42), derived from Ref. [106].

$$\gamma_{RF} = \frac{\beta_0 R_{diff}}{(1+\rho)\{1+\rho[1+(\omega C_j R_{diff})^2]\}} \tag{6.42}$$

The parameter ω is the angular frequency, the variables $\beta_0 = q_e/2\eta k_B T$ and $\rho = R_s/R_{diff}$, with the temperature T and the Boltzmann constant k_B. With an applied single diode with 4.5 μm anode diameter from Figure 6.35, a maximum value of 3600 mV/mW is possible with ideal impedance matching.

6.4.2 Design of Millimeter-Wave Square-Law Detectors

A general detection principle introduction has already been given in Section 5.1. State of the art detectors in the millimeter wave range are all realized on planar structures with transitions from a rectangular

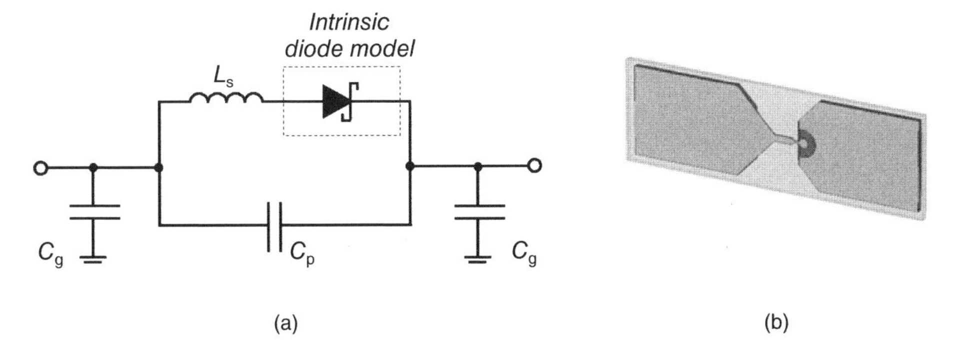

Figure 6.35 (a) Equivalent circuit for Schottky diodes, including the intrinsic diode model and parasitic elements. (b) 3DEM model of a low-barrier Schottky diode based on FD-process by ACST GmbH.

waveguide, or are quasi-optically coupled to lenses. The most reliable and accurate way to realize planar RF circuitry in waveguide coupled detectors is the application of a microstrip line or GCPW sections. The main part of the electrical field is inside the substrate or surrounding air. However, suspended striplines have also been reported. Since for a stripline the functionality is dependent on the field distribution between the line and the surrounding metal block, a highly accurate assembly process has to be ensured.

The presented diode types are suitable for detector applications for up to several hundreds of GHz, and even up to the THz range, due to their small series resistance and junction capacitance values. Even though diodes with smaller anodes do provide better responsivity results than those with larger anodes, the overall performance might decrease. R_{diff}, indirectly proportional to the anode size for equivalent technology, is the dominant white noise contributor. Therefore, the required SNR for detectors is fundamentally limited by the diode performance. By proper junction size and resistance choice, the SNR is maximized, potentially meeting required radiometer SNR specifications.

Further design steps commonly include the video connections in front and behind the diode. Additional impedance matching structures optimize the detector's performance, specifically responsivity and bandwidth. The real part of the complex RF impedance is typically below 50 Ω for low-barrier Schottky diodes. This low impedance is either matched to an intermediate reference impedance of 50 Ω on substrate, or directly with the design of the waveguide to microstrip transition. The second impedance matching aspect is the video amplifier section. For DC, the diode's R_{diff} is relevant for the voltage read-out. Hence, for plain DC detection, or low video bandwidth as for a radiometric detector, presented in Ref. [107], a large input impedance of the read-out equipment is the best choice in order to avoid additional loading of the diode. For larger video bandwidth, for example, telecommunication, the impedance has to be matched to a low impedance amplifier [108]. Based on the same diode type, quasi-optical detectors for communication within the 71–76 GHz band are developed at Universidad Carlos III de Madrid, in Spain. This communication system demonstrates a data rate of 1 Gbps with a bit error rate (BER) below 2×10^{-10}.

Within the last decade, low-barrier Schottky technology has appeared in various detector designs for direct detection radiometry. Although already established up to 40 GHz [109], recent ESA studies pushed the direct detection even above 200 GHz, with promising results for future radiometric space missions. Based on the ACST FD-technology, low-barrier Schottky diodes are applied in detector designs, suitable for direct detection receiver channels, at 89 GHz by ACST in Germany [107], at 166 GHz by Airbus Defence and Space in France, and at 243 GHz by OMNISYS in Sweden [110].

The particular design focus for radiometric detectors is on the best possible SNR and linearity for several decibels RF input power dynamic. Only by optimization on the diode, detector, and video amplifier level, can the required radiometric specifications be met. The diode is selected according to an optimum ratio between maximum RF responsivity, linearity, and minimum overall noise. The detector design provides the diode integration without limiting the major diode performance. The first video amplifier stage should minimize excess noise to the detector. This is an essential design aspect, since state-of the art detectors provide noise levels in the range of the noise levels of high performance operational amplifiers.

Figure 6.36 presents photographs of the recently developed detectors for 89 GHz by ACST and 166 GHz by Airbus Defence and Space, with waveguide interfaces for W-Band (WR10) and G-Band (WR5), respectively. The actual detector is placed close to the waveguide flange. The video amplifier section in the back-side cavity determines the overall size of the detector module. SNR values for the 89 GHz detector above 35 dB with 0.2 ms integration time are reported for RF input power levels above −30 dBm [107]. Measured NEP (noise-equivalent power) levels are below $6\,\mathrm{pW}/\sqrt{\mathrm{HZ}}$.

6.5 Fabrication Technologies

Since the beginnings of wireless electromagnetic experimentation and RF technology, the Schottky diode has been a cornerstone for RF system applications, such as small-signal square-law and envelope detectors, or large-signal mixers and frequency multipliers. No cooling requirements, large bandwidth and

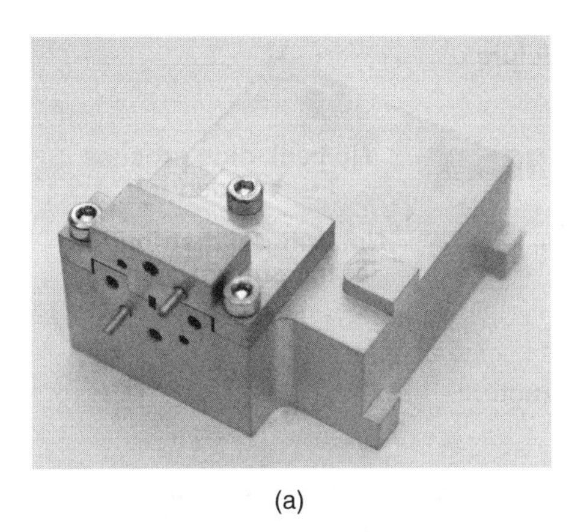

(a)

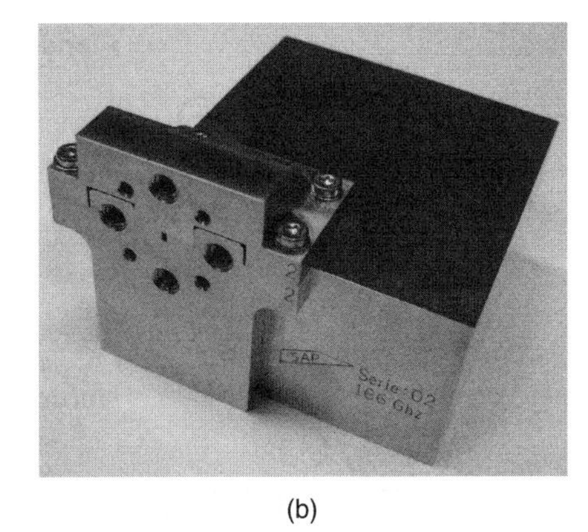

(b)

Figure 6.36 Photographs of the detector modules with low-barrier Schottky diodes for (a) 89 GHz by ACST and (b) 166 GHz by Airbus Defence & Space.

proven reliability and repeatability of Schottky diodes weigh heavily in their favor in comparison to other square-law elements.

6.5.1 Overview of Fabrication Approaches of Schottky Structures for Millimeter-Wave Applications

Whereas the theoretical model of the Schottky contact has been well understood since 1938, its practical employment for high frequency applications was delayed for several decades. Since existing fabrication technologies for microwave structures were not well-suited for millimeter-/submillimeter wave devices, development of dedicated processes was needed.

Different dedicated processes have been developed during recent decades by several companies and R&D groups around the world. Two structural concepts are usually considered for Schottky devices, namely, planar and vertical devices. Figure 6.37 illustrates these.

The planar concept implies both Schottky and ohmic contacts situated on the same side of the epitaxial layer. Such a technology is relatively simple because basically only one side of the semiconductor wafer is processed, whereas the bulk of the semiconductor wafer serves as a carrier substrate during processing. An undesirable effect of the planarization of Schottky and ohmic contacts is non-uniform distribution of the current density across the contact area. Local current overloading can occur on the anode region closest to the ohmic contact. High current density may heat electrons considerably above the lattice thermal energy, causing excess noise in the device and/or limiting the power capability of a planar structure.

In contrast, the advantage of vertical structures is the uniform distribution of current density across the entire contact area at the cost of more complex fabrication process. This involves substantial thinning or complete removal of the bulk semiconductor substrate and processing of the epitaxial layer from its back-side.

A whisker-contacted diode (WCD) was probably the first prototype of a vertical Schottky structure for millimeter-wave applications. A schematic diagram and an optical microscope photo of a WCD are shown in Figure 6.38. Although WCD have proven high performance at THz-frequencies [111, 112], there was great interest in developing technologies that are more mechanically robust and suitable for hybrid and monolithic integration. Since the initial development and demonstration of the surface-channel-etched (SCE) diode structure by Bishop *et al.* [113], the planar GaAs Schottky diode chip has been widely deployed in millimeter and submillimeter wave mixers and multipliers. A 3D model of a SCE-diode is shown in Figure 6.39.

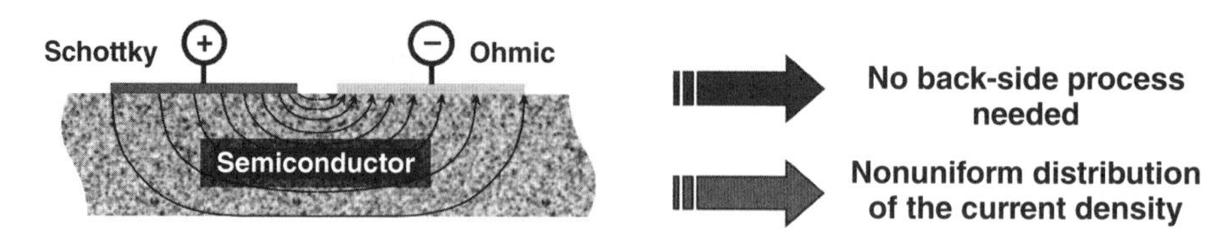

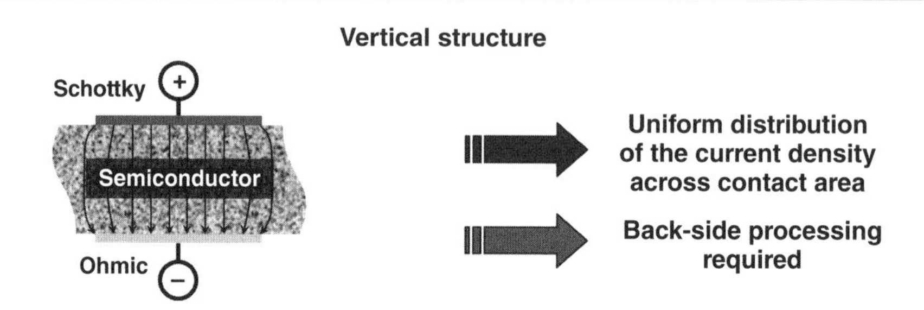

Figure 6.37 Two structural concepts for Schottky devices.

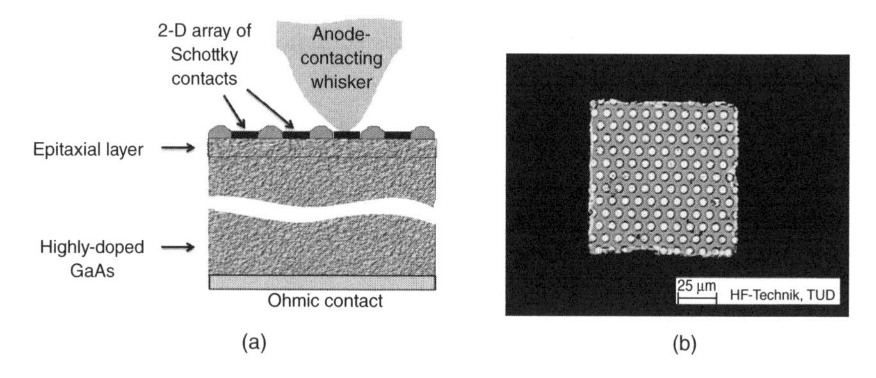

Figure 6.38 Schematic and an optical microscope photo of a WCD.

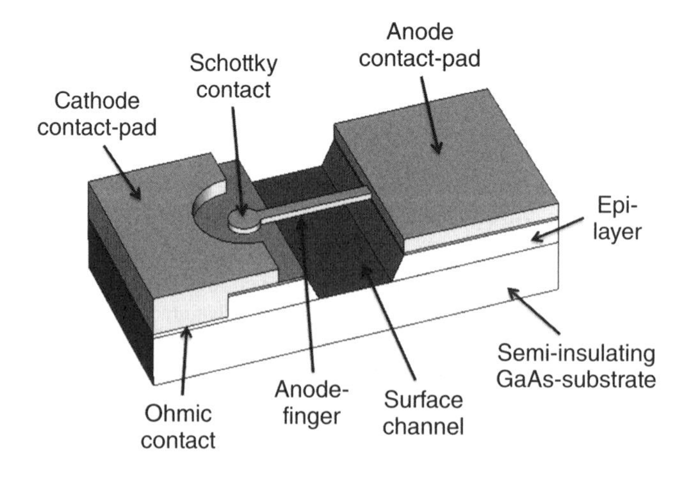

Figure 6.39 3D model of a SCE-diode.

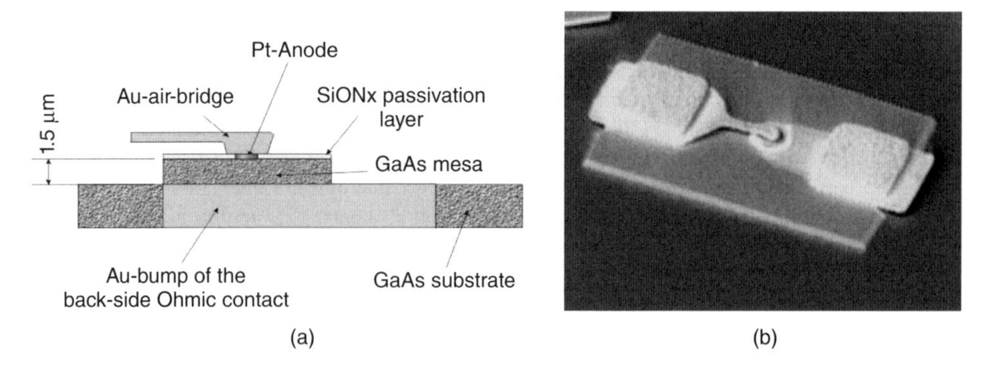

Figure 6.40 A sketch and a SEM-picture of a QVD.

The quasi-vertical diode (QVD), proposed by IHF, Technical University Darmstadt (TUD) [114], is an alternative to the SCE structure. Similar to the planar structures, QVD is suitable for hybrid integration and demonstrated good mixing performance at millimeter waves [115]. QVD structures are also suitable for monolithic integration with passive elements on a quartz substrate [116].

A sketch of a single-diode QVD structure is shown in Figure 6.40, together with a SEM-picture of a fabricated QVD diode. The active part of the device is the GaAs-mesa. It has a disk-like shape and is enclosed between the air-bridge-foot and an Au-bump, embedded in the substrate. The air-bridge connects the anode to the anode contact-pad, whereas the cathode contact-pad is electrically connected to the back-side ohmic contact through the Au-bump. All elements are attached to a membrane-substrate, in this case a 4 µm thin semi-insulating AlGaAs.

The QVD structure combines some advantages of both WCD and a planar structure. Its particular features are as follows:

- In contrast to the SCE-structure, the current flow is kept vertical, as in the case of the WCD-structure. The Schottky and ohmic contacts are parallel and the current flows from the anode on the top of the epitaxial layer to the back-side ohmic-contact. Therefore, the field distribution and current density are kept uniform across the entire anode area.
- The GaAs-mesa of a QVD-structure is usually as thin as 1.5 µm and is enclosed between the Schottky and the ohmic contacts. Therefore, the current path through the semiconductor is very short. This minimizes the impact of GaAs on the series resistance.
- Since the GaAs mesa is very thin, the heat generated at the Schottky interface can comparatively easily sink in the back-side Au-bump and other massive metallic elements of the circuitry. This suggests an enhanced power capability of QVD structures, in comparison to WCD structures.
- For technological reasons, Schottky contacts of any GaAs-structure have to be fabricated after ohmic contacts. In contrast to planar structures, ohmic contacts of QVD-structures are defined on the back-side of the wafer. This represents an advantage when high planarity of the front side is required for definition of very fine anodes.
- Because of a one-side processing, contact-pads of a SCE-structure can only be fabricated on the top of the diode chip. Therefore, the most convenient mounting approach of a discrete planar structure is the flip-chip approach. This means upside-down mounting of the diode chip on a substrate with pre-defined surrounding circuitry. In this case the anode/finger area, which is usually most sensitive to any parasitics, is placed close to the substrate material. Any substrate has worse dielectric characteristics in comparison to air and this can affect the overall performance. In contrast, a specific feature of QVD structures is a back-side gold-bump of the ohmic-contact and via-holes. This offers the opportunity for up-side-up mounting, which diminishes the influence of the substrate in the anode/finger area.

6.5.2 *Film-Diode Process*

The FD process represents a novel technology platform, recently developed at ACST GmbH. This technology is suitable for fabrication of both discrete diodes and integrated circuits, so-called terahertz monolithically-integrated circuits (THz-MICs). Figure 6.41 shows a single FD-structure mounted onto a quartz-substrate with pre-defined impedance matching circuitry.

The main particularity of FD-structures consists in a few micrometers thick transferred insulator membrane-substrate, which maintains integrity of all circuit elements. This approach aims at ultimate performance at millimeter- and submillimeter-waves and is scalable up to terahertz frequencies.

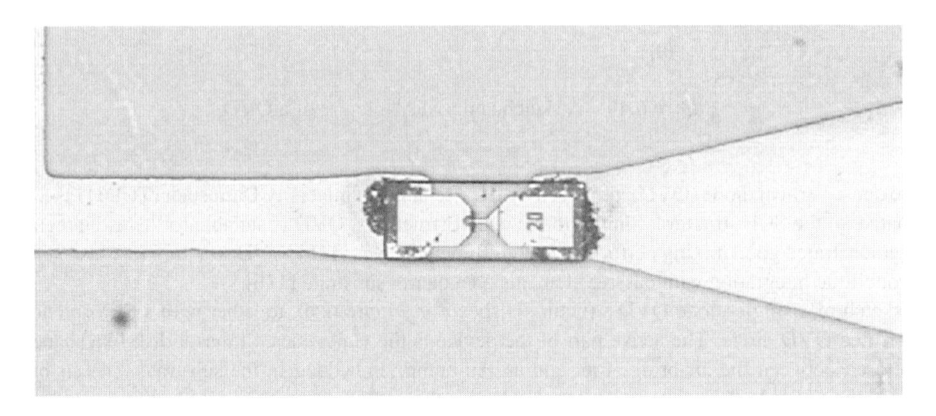

Figure 6.41 A single FD-structure mounted onto a quartz-substrate with pre-defined impedance matching circuitry. The diode is mounted by gluing using silver-epoxy. High mounting accuracy is facilitated by optical control of the mounting procedure via transferred insulator membrane-substrate. Lateral chip dimensions are about $60 \times 157\ \mu m^2$.

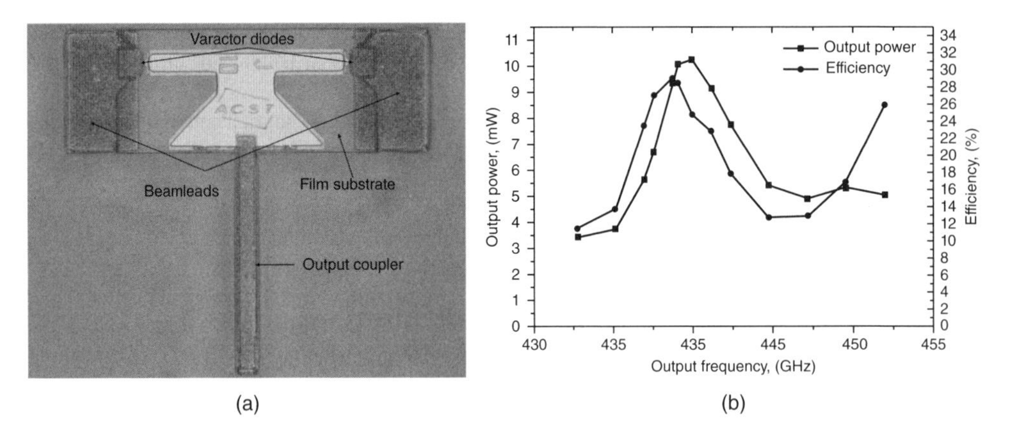

Figure 6.42 Integrated circuit of balanced 440 GHz frequency doubler. This consists on one varactor diode on each arm, input and output couplers, and beamleads for suspending the structure directly into the input waveguide of a metallic WG split-block. All the elements are monolithically integrated on transferred insulator membrane-substrate, which has overall dimensions of about $100 \times 250\ \mu m^2$. The doubler provides about 10 mW of output power at peak efficiency of more than 25%. This performance compares well with state-of-the-art performance at this frequency and is achieved in collaboration with GML and RPG in framework of an ESA/ESTEC project.

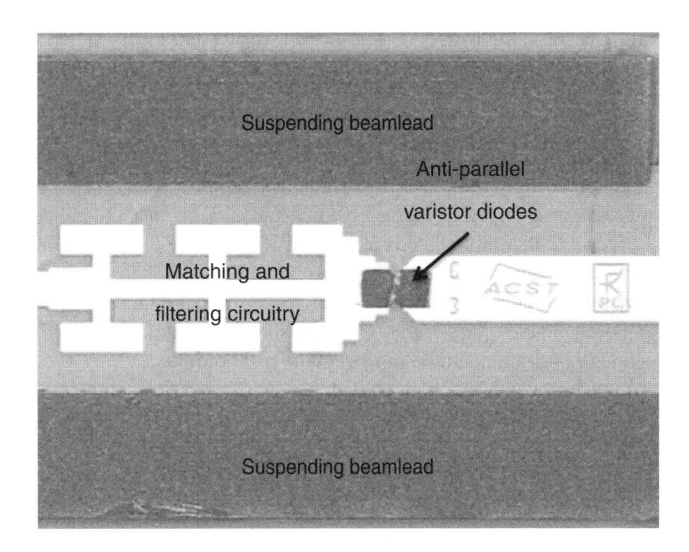

Figure 6.43 Photo of a 664 GHz subharmonic mixer IC in the diode region. The chip contains two anti-parallel varistor diodes integrated with matching and filtering circuitry and suspending beamleads on a transferred insulator substrate. DSB conversion loss and mixer temperature of the fabricated mixer were about 8 dB and 1600 K, respectively, which is an excellent performance at submillimeter waves. This has been achieved in collaboration with RPG in the framework of an ESA/ESTEC project.

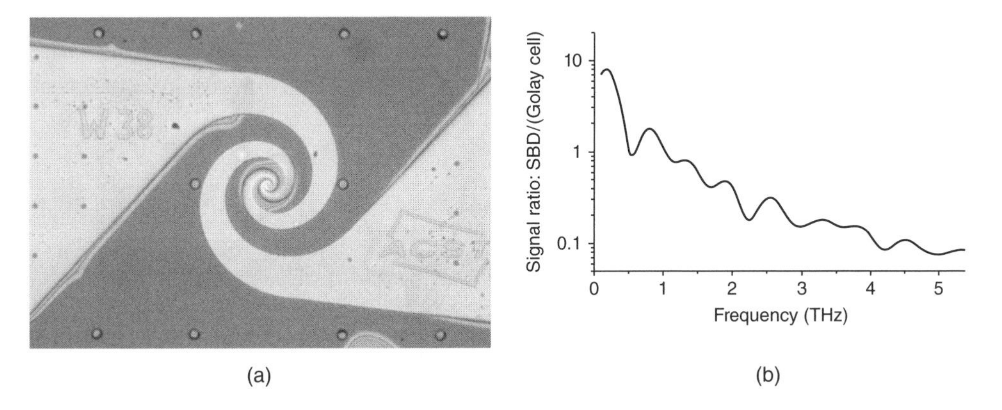

Figure 6.44 A log-spiral antenna feed by a zero-bias diode on a transferred insulator membrane-substrate. The structure is attached to a Si-lens to form a quasi-optical ultra-wide band detection. FTIR measurements were performed at the German national space agency/(DLR) to assess frequency response of this detector. These measurements show detector responsivity from about 50 GHz through well beyond 2 THz, which is the largest operation bandwidth to date for a Schottky detector. When compared to the Golay Cell the responsivity of the Schottky detector was higher below about 1 THz and lower beyond. This detector finds many applications because it simultaneously combines such qualities as ultra-wide frequency band, high responsivity, extremely short response time, and room-temperature operation. (Courtesy of DLR department THz-Instrumentierung, Institut für Planetenforschung.)

Similarly to QVD structures, the FD-process implies two-side processing of the semiconductor wafer. Schottky-contacts are formed on the front-side of the epi-layer whereas back-side processing offers the opportunity for ohmic metallization directly under the corresponding Schottky contacts. The main advantages of such an approach are the opportunity for improved thermal dissipation and uniform distribution of current density across the anode area. Moreover, two-side processing offers the opportunity for 2.5 D integration, that is, electrical wiring on both sides of the substrate. This implies more flexibility for device miniaturization, which is especially useful for high frequency applications.

The membrane-substrate is just a few micrometers thick and is made on a low-dielectric insulator. This allows for drastically reduced substrate-related circuit parasitics and RF losses. The membrane-substrate is very flexible, mechanically robust, and optically transparent. This significantly facilitates the assembly process by optical control during chip positioning. FD-structures can be assembled either by soldering or gluing on corresponding contact-pairs or they can be suspended on gold-beamleads, which is the best approach for a suspended-substrate mounted in waveguide technology.

Various circuit architectures can be accommodated by the FD process, including fabrication of single diodes, various diode arrays, and passive circuit elements, including MIM-capacitors and the surrounding circuitry. This technology can be extended to fabrication of active devices, like RTDs and transistors.

A few examples of realized FD structures at ACST GmbH are shown in the following figures (Figures 6.42–6.45). A short explanation is given in the corresponding captions.

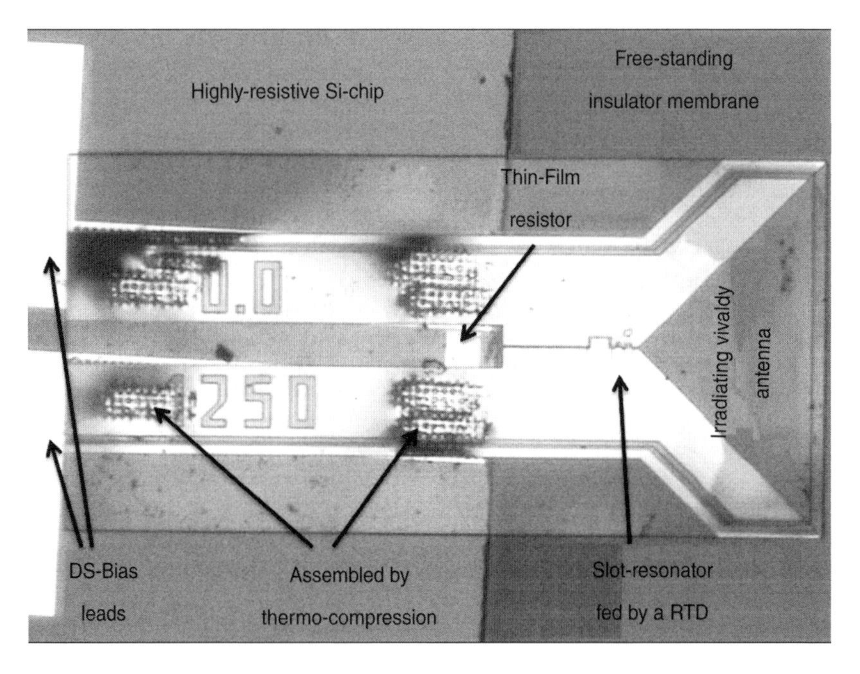

Figure 6.45 Microscope photo of an assembled planar RTD oscillator operating at about 1 THz. The RTD is enclosed in a slot-resonator, which is defined by two MIM capacitors. DC-bias is applied to the RTD via DC-leads. Generated THz-power is irradiated in free-space by a Vivaldi antenna. The IC has overall dimensions of about $500 \times 1000\,\mu m^2$ and is mounted on a high-resistive Si-chip for mechanical robustness and heat sinking. The active part of the circuit, namely the slot-resonator and free-space coupling antenna, are supported by an insulator membrane, which is free-standing in air. The thin-film resistor suppresses unwanted parasitic resonances. A series of such oscillators for various frequencies were developed in collaboration with TUD in the framework of an R&D activity supported by the Federal Ministry of Education (DFG). The highest operation frequency was observed at more than 1.1 THz. At the moment of experimental demonstration this was the highest frequency generated by a solid-state oscillator worldwide.

References

[1] Tsu, R. and Esaki, L. (1973) Tunneling in a finite superlattice. *Applied Physics Letters*, **22**, 562.

[2] Chang, L.L., Esaki, L., and Tsu, R. (1974) Resonant tunneling in semiconductor double barriers. *Applied Physics Letters*, **24**, 593.

[3] Sollner, T.C.L.G., Goodhue, W.D., Tannenwald, P.E. *et al.* (1983) Resonant tunneling through quantum wells at frequencies up to 2.5 THz. *Applied Physics Letters*, **43**, 588.

[4] Scott, J.S., Kaminski, J.P., Wanke, M. *et al.* (1994) Terahertz frequency response of an $In_{0.53}Ga_{0.47}As/AlAs$ resonant-tunneling diode. *Applied Physics Letters*, **64**, 1995.

[5] Brown, E.R., Sollner, T.C.L.G., Goodhue, W.D., and Parker, C.D. (1987) Millimeter-band oscillations based on resonant tunneling in a double-barrier diode at room temperature. *Applied Physics Letters*, **50**, 83.

[6] Brown, E.R., Söderström, J.R., Parker, C.D. *et al.* (1991) Oscillations up to 712 GHz in InAs/AlSb resonant-tunneling diodes. *Applied Physics Letters*, **58**, 2291.

[7] Reddy, M., Martin, S.C., Molnar, A.C. *et al.* (1997) Monolithic Schottky-collector resonant tunnel diode oscillator arrays to 650 GHz. *IEEE Electron Device Letters*, **18**, 218.

[8] Suzuki, S., Asada, M., Teranishi, A. *et al.* (2010) Fundamental oscillation of resonant tunneling diodes above 1 THz at room temperature. *Applied Physics Letters*, **97**, 242102.

[9] Feiginov, M., Sydlo, C., Cojocari, O., and Meissner, P. (2011) Resonant-tunnelling-diode oscillators operating at frequencies above 1.1 THz. *Applied Physics Letters*, **99**, 233506.

[10] Kanaya, H., Shibayama, H., Sogabe, R. *et al.* (2012) Fundamental oscillation up to 1.31 THz in resonant tunneling diodes with thin well and barriers. *Applied Physics Express*, **5**, 124101.

[11] Feiginov, M., Kanaya, H., Suzuki, S., and Asada, M. (2014) Operation of resonant-tunneling diodes with strong back injection from the collector at frequencies up to 1.46 THz. *Applied Physics Letters*, **104**, 243509.

[12] Kanaya, H., Sogabe, R., Maekawa, T. *et al.* (2014) Fundamental oscillation up to 1.42 THz in resonant tunneling diodes by optimized collector spacer thickness. *Journal of Infrared, Millimeter, and Terahertz Waves*, **35**, 425.

[13] Suzuki, S., Shiraishi, M., Shibayama, H., and Asada, M. (2013) High-power operation of terahertz oscillators with resonant tunneling diodes using impedance-matched antennas and array configuration. *IEEE Journal of Selected Topics in Quantum Electronics*, **19**, 8500108.

[14] Ishigaki, K., Shiraishi, M., Suzuki, S. *et al.* (2012) Direct intensity modulation and wireless data transmission characteristics of terahertz-oscillating resonant tunnelling diodes. *Electronics Letters*, **48**, 582.

[15] Feiginov, M.N. (2000) Effect of the coulomb interaction on the response time and impedance of the resonant-tunneling diodes. *Applied Physics Letters*, **76**, 2904.

[16] Feiginov, M.N. (2000) Does the quasibound-state lifetime restrict the high-frequency operation of resonant-tunneling diodes? *Nanotechnology*, **11**, 359.

[17] Luryi, S. (1985) Frequency limit of double-barrier resonant-tunneling oscillators. *Applied Physics Letters*, **47**, 490.

[18] Shockley, W. (1938) Currents to conductors induced by a moving point charge. *Journal of Applied Physics*, **9**, 635.

[19] Ramo, S. (1939) Currents induced by electron motion. *Proceedings of IRE*, **27**, 584.

[20] Feiginov, M.N. (2001) Displacement currents and the real part of high-frequency conductance of the resonant-tunneling diode. *Applied Physics Letters*, **78**, 3301.

[21] Feiginov, M.N. and Roy Chowdhury, D. (2007) Operation of resonant-tunneling diodes beyond resonant-state-lifetime limit. *Applied Physics Letters*, **91**, 203501.

[22] Sekiguchi, R., Koyama, Y., and Ouchi, T. (2010) Subterahertz oscillations from triple-barrier resonant tunneling diodes with integrated patch antennas. *Applied Physics Letters*, **96**, 062115.

[23] Feiginov, M., Sydlo, C., Cojocari, O., and Meissner, P. (2011) Operation of resonant-tunnelling oscillators beyond tunnel-lifetime limit. *EPL*, **94**, 48007.

[24] Feiginov, M., Sydlo, C., Cojocari, O., and Meissner, P. (2012) Operation of resonant-tunnelling- diode oscillators beyond tunnel-lifetime limit at 564 GHz. *EPL*, **97**, 58006.

[25] Koyama, Y., Sekiguchi, R., and Ouchi, T. (2013) Oscillations up to 1.40 THz from resonant-tunneling-diode-based oscillators with integrated patch antennas. *Applied Physics Express*, **6**, 064102.

[26] Maestrini, A., Thomas, B., Wang, H. *et al.* (2010) Schottky diode-based terahertz frequency multipliers and mixers. *Comptes Rendus Physique*, **11**, 480.

[27] Sirtori, C., Barbieri, S., and Colombelli, R. (2013) Wave engineering with THz quantum cascade lasers. *Nature Photonics*, **7**, 691.

[28] Deal, W.R. (2013) THz sources using indium phosphide high electron mobility transistors. *Proceedings of SPIE*, Terahertz Emitters, Receivers, and Applications IV, **8846**, 88460D.

[29] Gaier, T., Samoska, L., Fung, A. *et al.* (2007) Measurement of a 270 GHz low noise amplifier with 7.5 dB noise figure. *Microwave and Optical Technology Letters*, **17**, 7.

[30] Tessmann, A., Leuther, A., Massler, H. *et al.* (2007) 220 GHz low-noise amplifier MMICs and modules based on a high performance 50 nm metamorphic HEMT technology. *Physics of the Solid State C*, **4**, 5.

[31] Tessmann, A., Leuther, A, Massler, H. *et al.* (2008) A metamorphic 220–320 GHz HEMT amplifier MMIC. IEEE Compound Semiconductor Integrated Circuit Symposium.

[32] Maas, S.A. (2003) *Nonlinear Microwave and RF Circuits*, 2nd edn, Artech House Microwave Library.

[33] Sierra Pérez, M., Galocha Iragüen, B., Fernández Jambrina, J.L., and Sierra Castañer, M. (2003) *Electrónica de Comunicaciones*, Pearson Educación, S. A., Madrid.

[34] Maestrini, A., Thomas, B., Wang, H. *et al.* (2010) Schottky diode based terahertz frequency multipliers and mixers. *Comptes Rendus Physique*, **11**, 7–8.

[35] Räisänen, A. (1980) Experimental studies on cooled millimeter wave mixers, *Acta Polytechnica Scandinavica* **46**, pp. 45–55.

[36] Bo, X., Wenbin, D., and Minmin, H. (2012) A subharmonic mixer at W band. International Conference on Microwave and Millimeter Wave Technology (ICMMT), Vol. 3.

[37] Chen, Z., Zhang, B., Fan, Y. *et al.* (2010) Design of a 118-GHz sub-harmonic mixer using foundry diodes. International Symposium on Signals Systems and Electronics (ISSSE), Vol. 1.

[38] Liu, W., Zhang, Y., and Lu, Q. (2012) Design of a compact 150 GHz subharmonic mixer with high isolation. International Conference on Microwave and Millimeter Wave Technology (ICMMT), Vol. 2.

[39] Waliwander, T., Crowley, M., Fehilly, M., and Lederer, D. (2011) Sub-millimeter Wave 183 GHz and 366 GHz MMIC membrane sub-harmonic mixers. IEEE MTT-S International Microwave Symposium.

[40] Porterfield, D., Hesler, J., Crowe, T. *et al.* (2003) Integrated terahertz transmit/receive modules. 33rd European Microwave Conference.

[41] Zhang, B., Yong, F., Yang, X.F. *et al.* (2010) Optimization and design of a suspended 215–235-GHz subharmonically pumped mixer for space-borne radiometers. International Conference on Microwave and Millimeter Wave Technology (ICMMT).

[42] Wang, H., Sanghera, H., Alderman, B. *et al.* (2012) A performance comparison of discrete and integrated sub-harmonic Schottky diode mixers at 664 GHz. 23rd International Symposium on Space Terahertz Technology.

[43] Mehdi, I., Marazita, S.M., Humphrey, D.A. *et al.* (1998) Improved 240-GHz subharmonically pumped planar Schottky diode mixers for space-borne applications. *Microwave Theory and Techniques*, **46**, 12.

[44] Thomas, B., Maestrini, A., and Beaudin, G. (2005) A low-noise fixed-tuned 300–360-GHz sub-harmonic mixer using planar schottky diodes. *Microwave and Wireless Components Letters*, **15**, 12.

[45] Hesler, J.L. (2000) Broadband fixed-tuned subharmonic receivers to 640 GHz. Proceeding to 11th International Symposium on Space and Terahertz Technology.

[46] Schlecht, E.T., Gill, J.J., Lin, R.H. *et al.* (2010) A 520–590 GHz crossbar balanced fundamental Schottky mixer. *Microwave and Wireless Components Letters*, **20**, 7.

[47] Hui, K., Hesler, J.L., Kurtz, D.S. *et al.* (2000) A micromachined 585 GHz Schottky mixer. *Microwave and Guided Wave Letters*, **10**, 9.

[48] Thomas, B., Maestrini, A., Gill, J. *et al.* (2010) An 874 GHz fundamental balanced mixer based on MMIC membrane planar Schottky diodes. 21st International Symposium on Space Terahertz Technology.

[49] European Space Agency (2011) Reliability Assessment of Mixers and Multipliers for Microwave Imaging Radiometers. Statement of Work, Appendix 1 to AO 1-6900/11/NL/CT, IPD-SOW-189.

[50] Wang, H., Rollin, J., Thomas, B. *et al.* (2006) Design of a low noise integrated sub-harmonic mixer at 183 GHz using European schottky diode technology. 4th ESA Workshop on Millimetre Wave Technology and Applications.

[51] Thomas, B., Maestrini, A., Matheson, D. *et al.* (2008) Design of an 874 GHz biasable sub-harmonic mixer based on MMIC membrane planar schottky diodes. Infrared, 33rd International Conference on Millimeter and Terahertz Waves.

[52] Zhong, F., Bo, Z., Yong, F. *et al.* (2012) A broadband W-band subharmonic mixers circuit based on planar schottky diodes. International Conference on Industrial Control and Electronics Engineering (ICICEE).

[53] Medhi, I., Siegel, P., and Mazed, M. (1993) Fabrication and characterization of planar integrated Schottky devices for very high frequency mixers. Proceedings of the IEEE/Cornell Conference on Advanced Concepts in High Speed Semiconductor Devices and Circuits.

[54] Mehdi, I., Mazed, M., Dengler, R. *et al.* (1994) Planar GaAs Schottky diodes integrated with quartz substrate circuitry for waveguide subharmonic mixers at 215 GHz. IEEE MTT-S International Microwave Symposium Digest.

[55] Stake, J., Bryllert, T., Sobis, P. *et al.* (2010) Development of integrated submillimeter wave diodes for sources and detectors. European Microwave Integrated Circuits Conference (EuMIC).
[56] Alderman, B., Henry, M., Sanghera, H. *et al.* (2011) Schottky diode technology at Rutherford Appleton Laboratory. IEEE International Conference on Microwave Technology & Computational Electromagnetics (ICMTCE).
[57] Faber, M.T. and Archer, J.W. (1985) Computed aided testing of mixers between 90 and 350 GHz. *IEEE Transactions on Microwave Theory and Techniques*, **33**, 11.
[58] Kang, T.W., Kim, J.H., Lee, J.G. *et al.* (2011) Determining noise temperature of a noise source using calibrated noise sources and RF attenuator. *IEEE Transactions on Instrumentation and Measurement*, **60**, 7.
[59] UMS http://www.ums-gaas.com/ (accessed 29 November 2014).
[60] Schür, J., Ruf, M., and Schmidt, L.P. (2008) A 4th harmonic schottky diode mixer – facilitated access to THz frequencies. 33rd International Conference on Infrared, Millimeter and Terahertz Waves (IRMMW-THz).
[61] Wang, H., Maestrini, A., Thomas, B. *et al.* (2008) Development of a two-pixel integrated heterodyne Schottky diode receiver at 183 GHz. 19th International Symposium on Space Terahertz Technology.
[62] Gibson, H.J.E., Walber, A., Zimmermann, R. *et al.* (2010) Improvements in Schottky harmonic and subharmonic mixers for use up to 900 GHz. Proceedings of the European Integrated Circuits Conference.
[63] Alderman, B., Sanghera, H., Thomas, B. *et al.* (2008) Integrated Schottky structures for applications above 100 GHz. Microwave Integrated Circuit Conference.
[64] Phillips, T.G. and Keene, J. (1992) Submillimeter astronomy. *Proceedings of the IEEE*, **80**, 1662.
[65] Wyborn, N.D. (1997) The HIFI heterodyne instrument for FIRST: capabilities and performance. Proceedings of the European Space Agency Symposium, p. 19.
[66] Melnick, G.J., Stauffer, J. R., Ashby M. L. N. *et al.* (2000) The submillimeter wave astronomy satellite: science objectives and instrument description, *Astrophysical Journal* **539**, p.L77.
[67] Brown, R.L. (1998) Technical specification of the millimeter array. *Proceedings of SPIE*, **3357**, 231.
[68] Butler, B. and Wootten, A. (1998) ALMA Sensitivity, Supra-THz Windows and 20km Baselines. ALMA memo 276, http://legacy.nrao.edu/alma/memos/html-memos/alma276/memo276.pdf (accessed 29 November 2014).
[69] Siegel, P.H. (2002) Terahertz technology. *IEEE Transactions on Microwave Theory and Techniques*, **50**, 910.
[70] Grossman, E.N., Gordon, J., Novotny, D., and Chamberlin, R. (2014) Terahertz active and passive imaging. Proceedings of the 8th European Conference on Antennas and Propagation, Physical Communication (EuCAP).
[71] Grossman, E.N., Leong, K., Mei, X.B., and Deal, W.R. (2014) Passive 670 GHz imaging with uncooled low-noise HEMT amplifiers coupled to zero-bias diodes. *Proceedings of SPIE*, Passive and Active Millimeter-Wave Imaging XVII, **9078**.
[72] Arnone, D.D., Ciesla, C.M., Corchia, A. *et al.* (1999) Applications of terahertz (THz) technology to medical imaging. *Proceedings of SPIE*, Terahertz Spectroscopy and Applications II, **3828**, 209.
[73] Akyildiz, I.F., Jornet, J.M., and Han, C. (2014) Terahertz band: next frontier for wireless communications. *Physical Communication*, **12**, 16.
[74] Rodwell, M.J.W. (2014) 50–500 GHz wireless: transistors, ICs, and system design. 2014 German Microwave Conference (GeMIC), p. 1.
[75] Deal, W.R., Leong, K., Zamora, A. *et al.* (2014) Recent progress in scaling InP HEMT TMIC technology to 850 GHz. 2014 IEEE MTT-S International Microwave Symposium Digest (IMS), p. 1.
[76] Albrecht, J.D. (2011) THz electronics: transistors, TMICs, and high power amplifiers. Proceedings of the Compound Semiconductor MANTECH Conference.
[77] Samoska, L.A. (2011) An overview of solid-state integrated circuit amplifiers in the submillimeter-wave and THz regime. *IEEE Transactions on Terahertz Science and Technology*, **1**, 9.
[78] Weinreb, S., Lai, R., Erickson, N. *et al.* (1999) W-band InP wideband MMIC LNA with 30 K noise temperature. 1999 IEEE MTT-S International Microwave Symposium Digest (IMS), p. 101.
[79] Leong, K., Mei, G., Radisic, V. *et al.* (2012) THz integrated circuits using InP HEMT transistors. 2012 International Conference on Indium Phosphide and Related Materials (IPRM), p. 1.
[80] Leuther, A., Koch, S., Tessmann, A. *et al.* (2011) 20 nm metamorphic HEMT with 660 GHz f_T. 2011 Compound Semiconductor Week and 23rd International Conference on Indium Phosphide and Related Materials (CSW/IPRM), p. 1.
[81] Mikulla, M., Leuther, A., Bruckner, P. *et al.* (2013) High-speed technologies based on III-V compound semiconductors at Fraunhofer IAF. 2013 European Microwave Integrated Circuits Conference (EuMIC), p. 169.
[82] Lai, R., Mei, X., Sarkozy, S. *et al.* (2010) Sub 50 nm InP HEMT with f_T 586 GHz and amplifier circuit gain at 390 GHz for sub-millimeter wave applications. 2010 International Conference on Indium Phosphide and Related Materials (IPRM), p. 1.

[83] Ng, G.I., Radhakrishnan, K., and Wang, H. (2005) Are we there yet?-a metamorphic HEMT and HBT perspective. 2005 European Gallium Arsenide and Other Semiconductor Application Symposium, p. 13.
[84] Jain, V., Rode, J.C., Chiang, H.W. *et al.* (2011) 1.0 THz fmax InP DHBTs in a refractory emitter and self-aligned base process for reduced base access resistance. Proceedings of the 69th Annual Device Research Conference (DRC), p. 271.
[85] Seo, M., Urteaga, M., Hacker, J. *et al.* (2013) A 600 GHz InP HBT amplifier using cross-coupled feedback stabilization and dual-Differential Power Combining. 2013 IEEE MTT-S International Microwave Symposium Digest (IMS), p.1.
[86] Hacker, J., Urteaga, M., Munkyo, S. *et al.* (2013) InP HBT amplifier MMICs operating to 0.67 THz. 2013 IEEE MTT-S International Microwave Symposium Digest (IMS), p. 1.
[87] Rodwell, M., Lobisser, E., Wistey, M. *et al.* (2008) THz bipolar transistor circuits: technical feasibility, technology development, integrated circuit results. 2008 IEEE Compound Semiconductor Integrated Circuits Symposium (CSICS), p. 1.
[88] Yuan, J., Cressler, J.D., Krithivasan, R. *et al.* (2009) On the performance limits of cryogenically operated SiGe HBTs and its relation to scaling for terahertz speeds. *IEEE Transactions on Electron Devices*, **56**, 1007.
[89] Rucker, H., Heinemann, B., and Fox, A. (2012) Half-terahertz SiGe BiCMOS technology. Proceedings of the 2012 IEEE 12th Topical Meeting on Silicon Monolithic Integrated Circuits in RF Systems (SiRF), p. 1.
[90] (a) Schroter, M., Wedel, G., Heinemann, B. *et al.* (2011) Physical and electrical performance limits of high-speed SiGeC HBTs -Part I: vertical scaling. *IEEE Transactions on Electron Devices*, **58**, 3687; (b) Schroter, M., Krause, J., Rinaldi, N. *et al.* (2011) Physical and electrical performance limits of high-speed SiGeC HBTs – Part II: lateral scaling. *IEEE Transactions on Electron Devices*, **58**, 3697.
[91] Dot Seven FP7 Project DOTSEVEN: Towards 0.7 Terahertz Silicon Germanium Heterojunction Bipolar Technology, http://www.dotseven.eu/ (accessed 29 November 2014).
[92] ITRS (2013) ITRS 2013 Report, http://www.itrs.net/ITRS%201999-2014%20Mtgs,%20Presentations%20&%20Links/2013ITRS/Summary2013.htm(accessed 29 November 2014).
[93] Griffith, Z., Ha, W., Chen, P. *et al.* (2010) A 206-294 GHz 3-stage amplifier in 35 nm InP mHEMT, using a thin film microstrip environment. 2010 IEEE MTT-S International Microwave Symposium Digest (IMS), p. 57.
[94] Tessmann, A., Leuther, A., Hurm, V. *et al.* (2009) A 300 GHz mHEMT amplifier module. 2009 International Conference on Indium Phosphide and Related Materials (IPRM), p. 196.
[95] Deal, W., Mei, X., Radisic, V. *et al.*. (2010) Demonstration of a 0.48 THz amplifier module using InP HEMT transistors, *IEEE Microwave and Wireless Components Letters*, **20**, 289.
[96] Deal, W., Mei, X., Leong, K. *et al.* (2011) THz monolithic integrated circuits using InP high electron mobility transistors. *IEEE Transactions on Terahertz Science and Technology*, **1**, 25.
[97] Hacker, J., Seo, M., Young, A. *et al.* (2010) THz MMICs based on InP HBT technology. 2010 IEEE MTT-S International Microwave Symposium Digest (IMS), p. 1126.
[98] Deal, W., Leong, K., Mei, X. *et al.* (2010) Scaling of InP HEMT cascode integrated circuits to THz frequencies. 2010 IEEE Compound Semiconductor Integrated Circuit Symposium (CSICS), p. 1.
[99] Tessmann, A., Leuther, A., Lösch, R., and Massler, H. (2010) A metamorphic HEMT S-MMIC amplifier with 16.1 dB gain at 460 GHz. 2010 IEEE Compound Semiconductor Integrated Circuit Symposium (CSICS), p. 1.
[100] Leuther, A., Tessmann, A., Dammann, M. *et al.* (2013) 35 nm mHEMT technology for THz and ultra low noise applications. 2013 International Conference on Indium Phosphide and Related Materials (IPRM), p. 1.
[101] Tessmann, A., Leuther, A., Massler, H. *et al.* (2014) A 600 GHz low-noise amplifier module. 2014 IEEE MTT-S International Microwave Symposium Digest (IMS), p. 1.
[102] Leuther, A., Tessmann, A., Doria, P. *et al.* (2014) 20 nm metamorphic HEMT technology for terahertz monolithic integrated circuits. 2014 European Microwave Integrated Circuit Conference (EuMIC), p. 1.
[103] Hoefle, M., Penirschke, A., Cojocari, O., and Jakoby, R. (2011) Advanced RF characterization of new planar high sensitive zero-bias Schottky diodes. European Microwave Integrated Circuits Conference (EuMIC), p. 89.
[104] Möttönen, V., Mallat, J., and Räisänen, A. (2004) Characterisation of European millimetre-wave planar diodes. 34th European Microwave Conference, p. 921.
[105] Hoefle, M., Penirschke, A., Cojocari, O. *et al.* (2013) 1/f-noise prediction in millimeter-wave detectors based on quasi vertical Schottky diodes. 38th International Conference on Infrared, Millimeter and TerahertzWaves (IRMMW-THz).
[106] Bahl, I. and Bhartia, P. (2003) *Microwave Solid State Circuit Design*, 2nd edn, John Wiley & Sons, Inc.
[107] Hoefle, M., Penirschke, A., Cojocari, O. *et al.* (2014) 89 GHz zero-bias Schottky detector for direct detection radiometry in European satellite programme MetOp-SG. *Electronics Letters*, **50**, 606.

[108] Montero-de-Paz, J., Oprea, I., Rymanov, V. *et al.* (2013) Compact modules for wireless communication systems in the E-band (71–76 GHz). *Journal of Infrared, Millimeter, and Terahertz Waves*, **34**, 251.
[109] Goldstein, C., Trier, M., Maestrini, A., and Orlhac, J.C. (2006) Present and future R&T development in CNES for microwave radiometer. IEEE MicroRad, San Juan, Puerto Rico, p. 60.
[110] Sobis, P., Pellikka, T., Dejanovic, S. *et al.* (2013) A 243 GHz direct detection radiometer. 24th International Symposium on Space Terahertz Technology (ISSTT).
[111] Huber, K., Hillermeier, R., Brand, H. *et al.* (1997) 2.5 THz corner cube mixer with substrateless Schottky diodes. Proceedings of the 5th International Workshop on Terahertz Electronics, Grenoble, France.
[112] Kurtz, D.S., Hesler, J.L., Growe, W., and Weikle, R.M. (2002) Submillimeter-wave side-band generation using varactor Schottky diodes. *IEEE Transactions on Microwave Theory and Techniques*, **50**, 2610.
[113] Bishop, W.L., Mckinney, K., Mattauch, R.L. *et al.* (1987) A novel whiskerless Schottky diode for millimeter and submillimeter wave applications. IEEE MTT-S International Microwave Symposium Digest, p. 607.
[114] Simon, A., Grueb, A., Krozer, V. *et al.* (1993) Proceedings of the 4th International Symposium on Space Terahertz Technology, Los Angeles, CA, p. 392.
[115] Möttönen, V., Piironen, P., Zhang, J. *et al.* (2000) Subharmonic waveguide mixer at 215 GHz utilizing quasivertical Schottky diodes. *Microwave and Optical Technology Letters*, **27**, 93.
[116] Cojocari, O. (2007) Schottky technology for THz-electronics. PhD thesis. Shaker Verlag, ISBN: 978-3-8322-6260-0, pp. 144–146.

7

Selected Photonic THz Technologies

Cyril C. Renaud[1], Andreas Stöhr[2], Thorsten Goebel[3], Frédéric Van Dijk[4], and Guillermo Carpintero[5]

[1] *University College London, Department of Electronic & Electrical Engineering, London, UK*
[2] *Universität Duisburg-Essen, ZHO/Optoelektronik, Duisburg, Germany*
[3] *Fraunhofer Heinrich Hertz Institute, Terahertz Group/Photonic Components, Berlin, Germany*
[4] *Alcatel-Thales III-V Lab, Palaiseau Cedex, France*
[5] *Universidad Carlos III de Madrid, Departamento de Tecnología Electronica, Madrid, Spain*

7.1 Photonic Techniques for THz Emission and Detection

The THz spectral range is not exploited to its full potential due to the current limitations in sources and detectors. To open the frequency range for applications, photonic solutions have been at the technological forefront. For example, the advances of time domain spectroscopy (TDS) techniques using short pulse lasers have enabled the measurement of full spectroscopy data across the range [1, 2]. There are different types of systems and their development should be driven by the requirements of the potential application. Photonic techniques are desirable solutions for millimeter and THz generation in terms of their energy efficiency, bandwidth, and above all, their tuning range. Recent developments in this area target the improvement of optical-to-THz converters as well as pushing the level of integration of semiconductor laser sources in order to address the main drawbacks, cost, lack of power and spectral purity.

The purpose of this chapter is to describe the main types of photonically enabled THz systems and the expected performances from their components. Then a description of the key elements in designing each of the components and their limitation will be given. The final part of the chapter is a discussion on potential future development and the importance of photonic integration.

7.1.1 Overall Photonic System

Currently the most used THz systems, especially for spectroscopic applications, are based on time domain techniques, that is terahertz TDS [1, 2] (cf. Figure 7.1).

Semiconductor Terahertz Technology: Devices and Systems at Room Temperature Operation, First Edition.
Edited by Guillermo Carpintero, Luis Enrique García Muñoz, Hans L. Hartnagel, Sascha Preu and Antti V. Räisänen.

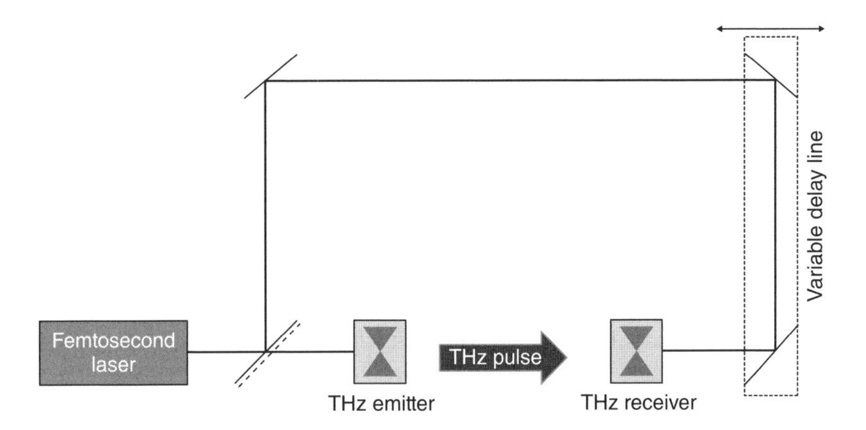

Figure 7.1 Simplified schematic of a TDS system.

A typical system includes a narrow optical pulse source – with a pulse-width typically of the order of 100 fs – mode locked laser. That source is used to excite either a nonlinear crystal or a photoconductive antenna as both emitter and receiver through a variable delay Michelson interferometer. From the emitter side the excitation creates a short electrical pulse that has a spectrum of a few THz. That pulse can be coherently detected and sampled using varying delay by utilizing electro-optic or photoconductive techniques [3, 4]. These systems typically rely on expensive lasers, such as Ti:Al_2O_3 lasers, to generate the optical pulses since most operational photoconductive switches are based on the GaAs material system.

On the other hand, one can develop terahertz continuous wave (CW) systems by heterodyning two lasers in a photomixer [5], whereas similar detection could be obtained in a photonically driven switch (cf. Figure 7.2). The advantage of such systems is that they can be based on optical communication components that are more cost effective and offer agility, high scanning speed, and superior frequency resolution with reduced requirements for sampling and data processing. However, the source power and the detection sensitivity are both relatively low [6]. One of the main obstacles for the development of this technology has been the limited availability of CW emitters and detectors that can operate efficiently at room temperature [7].

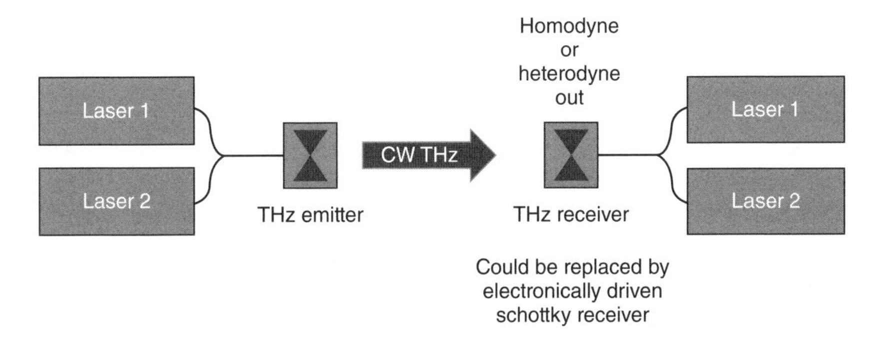

Figure 7.2 Simplified schematic of a heterodyne CW system.

7.1.2 *Basic Components Description*

In this section we summarize the main components of a THz photonic system and what are their key characteristics for THz application. This will be used as a reference when discussing the components in detail.

7.1.2.1 Laser Sources

We can distinguish three different types of laser sources for THz systems: mode locked lasers (as a short pulse source or a comb generator), optical frequency comb generators (OFCGs) (as a reference source for a set of frequencies that are multiples of the comb spacing), and single/dual frequency laser sources (photomixing solutions).

For a mode locked, the following parameters are key to the system design: pulse width (should be short enough to generate sufficient frequency components), pulse peak power (high enough to switch the photoconductive antenna or drive the nonlinear effects), and repetition rate (key in designing the Michelson interferometer and the sampling-triggering mechanism). Since in a TDS system the sampling of the transmitted pulse is made by the same original pulse sent to the receiver, the jitter of the source is mostly not relevant.

For the comb generators the key parameters are the line spacing, the span (hopefully a multiple of THz), the power per line, and the linewidth (phase noise) of a single line as it will determine the resulting frequency obtained from these sources. It is important to note that there are many ways of obtaining THz spanning comb sources that would output either pulsed (mode locked lasers) or CW signals [8–12], which could be important when put in combination with the THz photomixer or line filters.

CW lasers used for heterodyning can be divided into two kinds: single frequency and dual frequency. In both cases the key characteristics are the output power, the relative intensity noise (RIN), the linewidth, and the tuning range. Typically, we will focus on telecommunication technology based solutions as they offer the most prospects for application, potential integration, and tuning range (several THz).

7.1.2.2 Photonic Emitters and Detectors

One of the main components in TDS systems or in photomixing, used as both an emitter or a detector, is the photoconductive antenna that is typically made from a low-temperature grown gallium arsenide crystal (LT-GaAs) [13], though other technical solutions exist [14]. Here, the key parameters to assess are the recovery time, the mobility of the carrier, and the dark resistivity of the switch, all of which will influence the shape of the THz pulse and its intensity.

Nonlinear crystals are also used [15] through a different set of effects. In this case the main parameters are the phase matching conditions, the nonlinearity coefficient and the polarization dependence, as they will affect the choice of the mode lock laser source and the power of the THz pulse.

Finally, photodetectors could be used, mostly for photomixing [16] or less efficiently as detectors [17]. The key parameters are the responsivity (key in converting optical power into THz power efficiently), frequency response (3 dB point and roll-off), that is driven by both the carrier transit time, and the electrical characteristics, the saturation output power to assess the maximum power achievable. For such devices solutions to address those parameters exist in terms of both material and structure (waveguide or vertical).

Some other elements also need to be envisaged for the overall system design, such as passive waveguides to transit between the different active components which are mostly described in term of losses and become key in integrated designs, as discussed later in this chapter. A designer should also consider amplifiers in integrated and CW systems as they might be required to generate enough power at the cost of a decreased signal to noise ratio. For certain applications modulators might also be used. For these last two it is more of a general optical system design problem, as described in [18], and will not be discussed in this chapter.

7.1.3 Systems Parameters, Pulsed versus CW

In order to compare the two systems we are going to use spectrometry as an example of an application field and in that case the test parameters should be as follows:

- Spectral range
- Signal-to-noise ratio (SNR) and dynamic range
- Spectral resolution.

Considering that both systems could be used for spectroscopy, and that commercial solutions are available for both, it is a fair comparison of their performances [19]. However, one should bear in mind that the choice of system remains highly dependent on the application domain.

In comparison to the TDS system shown in Figure 7.1, we can schematize a CW spectroscopy system (homodyne) in the form shown in Figure 7.3.

It can be seen that in that case we have an interferometric type of signal, generating a signal which should follow a sinusoidal evolution as a function of path length:

$$I_{\mathrm{detect}} \propto E_{\mathrm{THz}} \cos(\varphi_{\mathrm{THz}} + \varphi_{\mathrm{opt}}) \tag{7.1}$$

where E_{THz} is the amplitude of the THz electrical field, φ_{THz} is the accumulated phase on the THz path, which depends on the THz frequency ω_{THz} and the path length for the THz beam, and φ_{opt} is the phase difference between the two optical paths.

Such a system operating at 1500 nm [20] has enabled the measurement of the data shown in Figure 7.4, which are to date the best results obtain from a THz CW spectroscopy system. While a typical TDS spectrum from one of the commercial systems [21] is shown in Figure 7.5.

On pure performances we then have the results shown in Table 7.1.

As one can see from the straightforward comparison of the operating characteristics of the two systems, the TDS system is slightly advantageous in term of spectral range, and SNR, with shorter measurements. On the other hand the CW system demonstrates very similar SNR and spectral range with better resolution, which can even go down to 100 MHz or lower. However the increase in resolution is obtained through more sampling points, which each need 200 ms integration time to reach the required SNR.

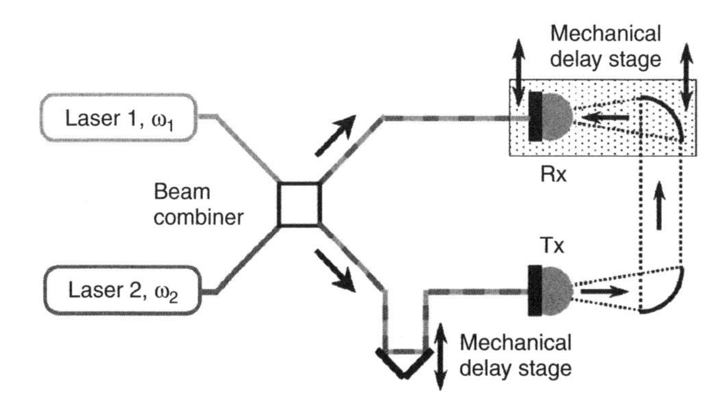

Figure 7.3 Typical schematic for a THz CW homodyne spectroscopy system.

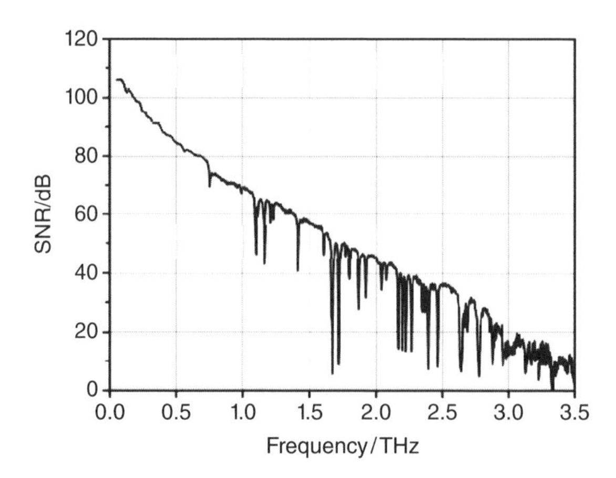

Figure 7.4 Spectrum from a THz CW system with 200 ms integration time and lasers at 1550 nm. (Reproduced and adapted with permission from Ref. [20], © 2013 Optical Society of America.)

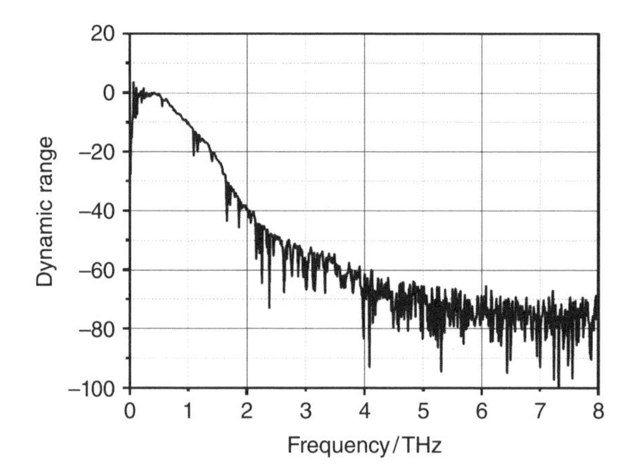

Figure 7.5 Spectrum from a THz TDS system [21].

Table 7.1 Performance parameters of THz systems reported in [20] and [21].

	TDS system	THz CW system
Spectral range	4 THz (5 THz)	3.5 THz
Spectral resolution	5 GHz	1 GHz
SNR @ 3 THz	20 dB (up to 40 dB)	16 dB
Measurement time	50 ms (25 s for higher SNR)	>10 min

To obtain these operating conditions each system requires the following:

TDS system:

Pulsed laser with 120 fs pulses or less and about 30 mW average output power for each arm and a typical electrical power to drive the system (including cooling) of the order of 5 W minimum for a fiber laser based system (higher for titanium sapphire based systems). The THz pulse emitted is about 1 μW in power.

THz CW system:

Two extended cavity lasers with 30 mW output each and a typical electrical power to drive the system of the order of 2 W. The emitted signal ranges from hundreds of microwatts at 100 GHz to tens of nanowatts at 3 THz.

Overall we can see that while a CW system requires less energy to drive for a higher resolution and uses off-the-shelf components from the optical communication industry, it can only match a THz TDS system performance in term of SNR and spectral range at the expense of a longer measurement time. It is to be noted that the laser average power required for both systems is relatively similar, while the bias voltage required for a photodiode emitter would be much lower than that required for a photoconductive antenna (by a factor of 5).

7.2 Laser Sources for THz Generation

7.2.1 Pulsed Laser Sources

Time domain (TD) has already been mentioned as a common technique to produce THz waves, requiring femtosecond optical pulses. Semiconductor lasers have been pursued as a compact source for femtosecond pulses as the gain bandwidth of their active region sets the lower limit for the duration of mode locked pulses on the order of 50–100 fs. These figures are by far shorter than any measured pulse width achieved from a semiconductor laser to date [22]. Mode locked pulses are generated from phase locking the longitudinal modes, resulting in pulsed optical output with a repetition rate defined by the cavity length and the number of pulses in the cavity. TD generally relies on passive mode locking and requires an intracavity saturable absorber (SA) to lock the cavity modes in phase. Mode locking is a least energy situation in which light can exist in the cavity, using the gain of the media in the most efficient way to turn the SA into transparency. Energy builds up by adding the individual modes of the cavity in phase, creating a short pulse of light that concentrates the optical energy. This pulse is then capable of saturating the SA for the short time interval that the pulse propagates through the SA. As a main advantage, we must point out that it does not require any electrical modulation and therefore does not have an electrical limitation. The fundamental frequency of the pulses, known as the fundamental repetition rate (f_{RP0}), is determined by the cavity length as

$$f_{RP0} = \frac{c}{n2L_{cav}} \quad (7.2)$$

where c is the speed of light, n the refractive index in the medium, and $2L_{cav}$ is the length in the cavity that light must propagate to return to the initial start point. Figure 7.6 shows the basic scheme of a mode locked device, generating a pulse train over time with the corresponding optical frequency spectrum.

One of the main problems with this technique is that as the appearance of the pulses comes from the interplay between the gain and absorber sections, this happens for a given set of conditions that allow establishing a self-stabilized pulse that travels back and forth through the cavity. Usually, this interplay also allows other regimes of operation, such as Q-switching or Q-switched mode locking, with a negative impact for the device. A deep understanding of the physics of the device is needed to suppress these undesirable operating regimes [23].

As THz sources, passive mode locked lasers can be used in different forms. One approach is based on the fact that the shorter the pulse (decreasing its pulse duration, τ_P), the greater the frequency

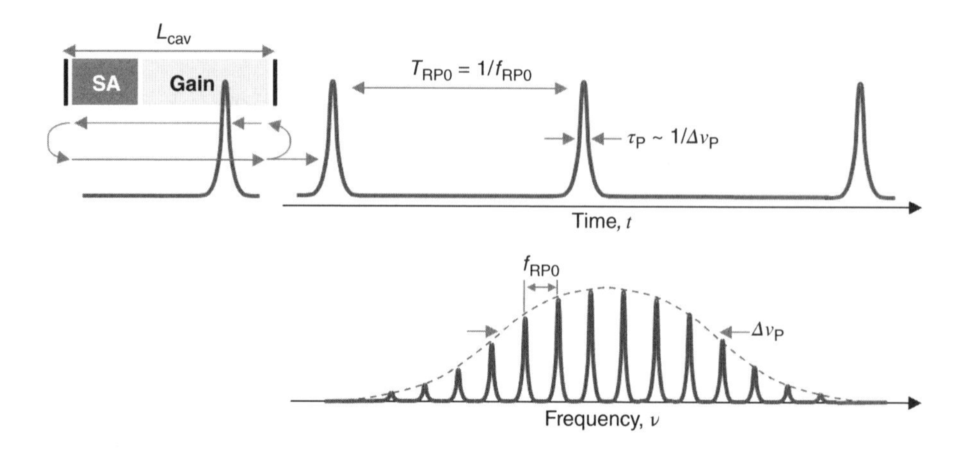

Figure 7.6 Mode-locked laser output in the time domain and the frequency domain.

bandwidth ($\Delta\nu_P$). As shown in Figure 7.6, there exists an inverse relationship between pulse duration and spectral bandwidth. In order to generate frequency components in the THz range, femtosecond pulse durations must be generated. THz generation therefore requires ultrafast laser pulses. Combined with a photoconductor made from low-temperature-grown GaAs (LT-GaAs), operating at 850 nm, this is one of the most mature THz technologies in commercial use [24]. However, the ultrafast laser pulse sources are complex, bulky, and power-hungry titanium-sapphire lasers, which highlights the lack of compact sources for THz.

Another competitive photonic pulse source is mode locked fiber lasers, capable of producing short pulses down to a few femtoseconds with high peak power [25]. The large gain bandwidth of the fiber, which can be pumped by means of diode lasers, allows these ultra-short pulses to be compact and low cost. However, the gain media used, ytterbium-doped and erbium-doped fibers operate within the wavelengths around 1050 and 1560 nm. Thus a critical step forward for their use has been the development of optoelectronic converters that operate in the C-band telecommunication wavelength range (1530–1565 nm), which opened the door to telecom components.

A current alternative to these short pulse sources are mode-locked laser diodes (MLLDs). Their main advantages are being small in size and easy to operate just by injecting an electrical current. Short pulses have been reported in the far-infrared region, between 820 and 840 nm, with duration down to 158 fs and peak power (including a tapered laser amplifier) up to 6.5 kW [26]. The problems that semiconductor pulsed sources experience to generate short pulses come from the strong dependence of the refractive index of the medium on the carrier density, leading to self-phase modulation (SPM) effects that cause chirp of the generated pulses.

A second approach using mode locked laser diodes is to push upwards the repetition rate frequency of the pulse train, locking the device at higher cavity harmonics without modifying the fundamental cavity length. This approach, known as harmonic mode locking (HML), was very promising to increase the data rate of time division multiplexed (TDM) data transmission systems in the 1990s before the telecom bubble burst in the early 2000s. Several approaches were demonstrated, which were facilitated by the early monolithic integration efforts to place the gain and SA onto the same chip. Integration allowed reducing the cavity length increasing the repetition rate. However, it was soon noticed that in order to generate frequencies within the THz range, from 300 GHz to 1 THz, the cavity length should be decreased below 100 μm. With such a cavity length it is not possible to introduce the gain section and the SA, and still have the gain section overcome the losses. Figure 7.7 shows different routes that were developed in order to achieve HML. The simplest procedure is to increase the bias current on the gain section. Higher order harmonics are produced when the gain section current goes above the value for stable operation

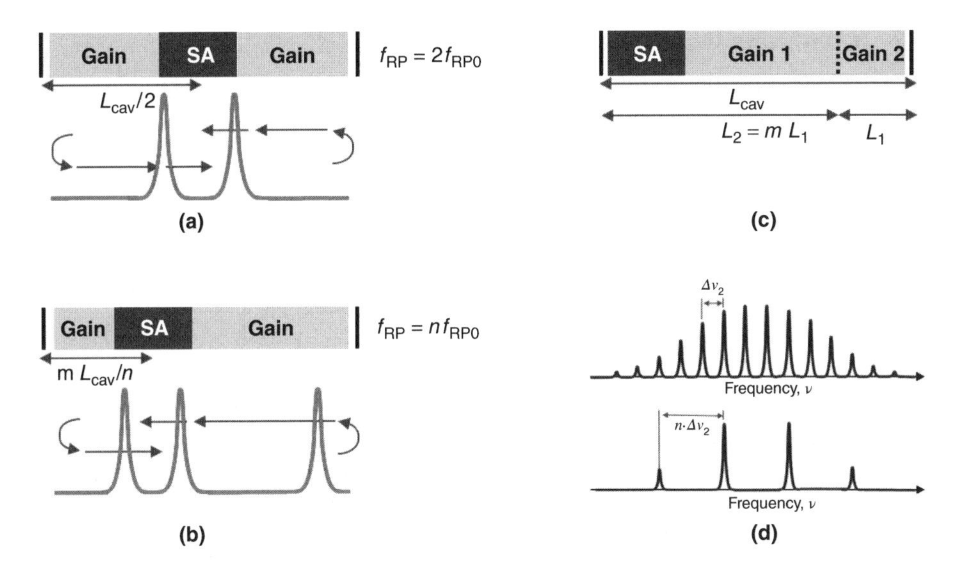

Figure 7.7 Approaches to increase the repetition rate frequency of mode-locked lasers. (a) Colliding passive mode locking (CPM), (b) asymmetric colliding passive mode locking (ACPM) and (c) compound cavity mode locking (CCML) and (d) CCML optical mode filtering scheme.

of the fundamental repetition rate. With such an approach, it was demonstrated that the repetition rate could be set at 400 GHz, 800 GHz, and 1.54 THz just by fixing the current at 137, 154, and 179 mA [27].

A radical new approach developed when it was realized that the SA could be moved into the center of the cavity. This simple change resulted in two counter-propagating pulses that meet at the SA every round trip, doubling the repetition rate [28]. This technique is known as colliding passive mode locking (CPM)(cf. Figure 7.7a). It was soon realized that further increases in the repetition rate could be achieved by locating the SA at an asymmetric location, shifting it from the midpoint toward one facet, developing a new structure, known as asymmetric colliding passive mode locking (ACPM) (cf. Figure 7.7b). The usual approach is to locate the saturable absorber section at a prime multiple (m) of an integer fraction (n) of the cavity length, $m \cdot L_{cav}/n$, to achieve a repetition rate at the nth-harmonic of the fundamental, demonstrating repetition rates up to 860 GHz [29].

Another important HML technique is compound cavity mode locking (CCML) (cf. Figure 7.7c). In CCML, two cavities with defined ratios of their lengths, $L_1/L_2 = 1/m$, are coupled together. Since the compound cavity modes must fulfill the phase condition for both cavities simultaneously as shown in Figure 7.7d, the ratio of their lengths determines when their modes align in frequency, becoming a solution for the compound structure. The repetition rate frequency, proportional to the mode spacing, is therefore increased to the nth-multiple, with $n = m + 1$. With such an approach, devices with cavity lengths from 900 to 615 μm produced repetition rates extending from 131 GHz to 2.1 THz for $m = 3$ to $m = 33$, respectively, delivering around 2 mW of optical power at the facets [30].

It has been reported that short pulse widths produce higher THz power when used with Uni-Travelling Carrier Photodiodes (UTC-PD) optoelectronic converters. When compared with heterodyne schemes, after polarization alignment, the pulsed system outputs ~7 dB more THz power for the same level of incident optical power on the UTC-PD, suggesting that peak optical power matters more than mean optical power in the conversion process [31].

However, TD techniques have two major drawbacks: limited frequency resolution, and the fixed relation between the repetition rate and the cavity length, which cannot be tuned over a wide range without moving parts. These weaknesses are precisely addressed by the frequency domain techniques.

7.2.2 Continous Wave (CW) Sources

In contrast to TD techniques, based on short optical pulses, frequency domain techniques – also known as continuous wave – aim to generate a continuous wave having constant amplitude over time. Optical heterodyne generation (OHG) is a CW technique in which, as represented in Figure 7.8a, two optical modes having slightly different frequencies are combined in a single output. Assuming two waves of equal amplitude (cf. Figure 7.8b), and using the principle of superposition, we find that the total field in the common output is the sum of the two modes, resulting in a signal given by:

$$E_1 \sin(2\pi f_1 t + \varphi_1) + E_2 \sin(2\pi f_2 t + \varphi_2) = 2A \sin\left[\frac{2\pi\,(f_2 + f_1)\,t}{2}\right] \cos\left[\frac{2\pi\,(f_2 - f_1)\,t}{2}\right] \tag{7.3}$$

The resulting wave, shown in Figure 7.8c, is the product of a sine wave oscillating at the average frequency $(f_1 + f_2)/2$ and a cosine wave oscillating at the difference frequency $(f_2 - f_1)/2$. When this wave is feed into an optoelectronic converter (a photoconductor or photodiode), which behaves as a square law detector, the high frequency is filtered out and the resulting photocurrent is a signal that results from the down-conversion of the two optical frequencies, with its frequency determined by the beat at twice the cosine, $f_{\text{beat}} = (f_2 - f_1) = \Delta\upsilon$ [32]. For this reason, OHG is also known as photomixing, first reported at

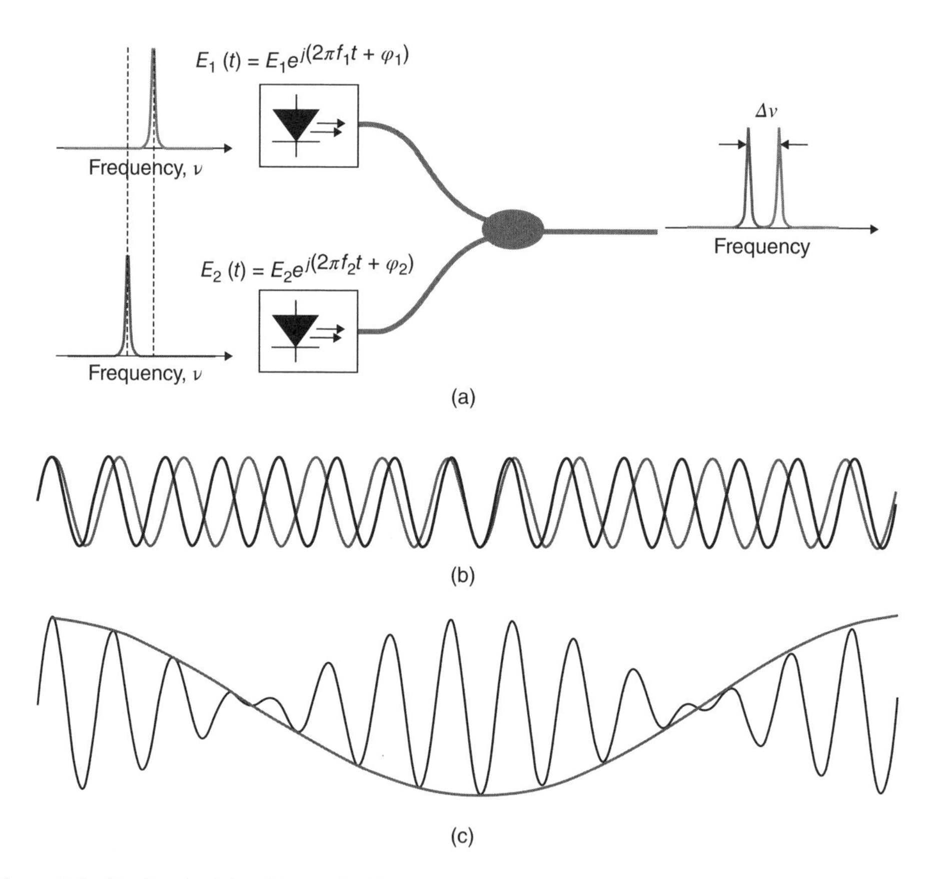

Figure 7.8 Basic principle of the optical heterodyne generation: (a) dual wavelength source combining two slightly different optical frequencies, (b) time evolution of the two signals separately, and (c) time evolution of the combined signal.

terahertz frequencies using a low-temperaturegrown GaAs (LTG GaAs) photomixer at 780 nm reaching up to 3.8 THz [33].

Compared to TD techniques, the OHG technique offers a wide range of advantages. The most straightforward is that it concentrates the energy at a single terahertz frequency, improving the spectral density and the SNR. Another advantage is that OHG can produce broadly tunable terahertz sources just by using any wavelength tuning mechanism on one (or both) optical sources. More importantly, OHG enables the modulation and distribution of the THz signal in the optical domain, making use of the existing optical fibers using techniques similar to radio-over-fiber (RoF). Since THz waveguides do not yet exist, and the propagation through air does not allow covering distances beyond a few meters, the distribution of the optical signal through fiber to the optoelectronic converter is a very attractive option.

A key factor reducing the cost has been the availability of optoelectronic converters with sufficient electrical bandwidth operating in the telecom-based wavelengths for long-haul optical fiber communications. Turning to the C-band (1530–1565 nm) or L-band (1565–1625 nm) allows one to use the mature InP semiconductor laser technology to develop the required dual wavelength sources and to use off-the-shelf components, such as semiconductor lasers, semiconductor optical amplifiers (SOAs), or erbium-doped fiber amplifiers (EDFAs) – with wide optical bandwidths of several THz-, as well as high-speed modulators. The most straightforward solution to get a dual wavelength source is to combine the output from two discrete and independent single frequency semiconductor lasers [34]. Common single frequency semiconductor laser structures are based on frequency selective feedback schemes, such as distributed feedback (DFB) lasers, in which a grating layer is incorporated along the active gain section, and distributed Bragg reflector (DBR) lasers, in which the grating is processed in a low-loss passive section [35]. With this approach, the main concern is the poor frequency stability and large phase noise, as measurements have shown a frequency stability of about 10 MHz/h and a phase noise level of −75 dBc/Hz at a 100 MHz offset [36]. These issues have an impact on the linewidth of the generated terahertz signal, related to the sum of the optical linewidths from each of the modes that are mixed.

When the two independent lasers are combined, the noise from each one is uncorrelated to the other, which has a negative impact on the beat frequency. Unfortunately, there are several mechanisms governing the change in lasing wavelength of a semiconductor laser, the most important being linked to the dependence of the active region refractive index on injected current (about 0.005 nm/mA) and temperature (around 0.13 nm/°C) [37]. To make matters worse, on single frequency lasers, such as DFB and DBR structures, we can vary the wavelength through changes in the refractive index affecting the effective grating pitch. Thus, frequency drift appears from the independent temperature fluctuation of each laser. Also, the optical beams need to be precisely controlled, in order to have the same relative intensity and polarization in order to maximize the amplitude excursion of the beat signal. A positive aspect of the temperature dependence of wavelength is that it can be used to vary the wavelength of one (or both) lasers, allowing tuning of the output frequency. Of course, the most straightforward solution usually does not provide the best performance, although it usually provides good lessons. The problems observed using discrete and independent lasers can be addressed in different ways to "build a better mouse trap", or in this case, a better dual wavelength source.

Since using two independent sources to generate each of the mixed wavelengths produces poor frequency stability and large phase noise, establishing a common wavelength reference should improve the situation. The most straightforward method is shown in Figure 7.9, using a CW single mode laser followed by an external modulator, such as an electro-absorption (EA) modulator or a Mach–Zehnder modulator (MZM), driven by an electrical signal of frequency f_{RF}. The modulation generates two optical sidebands, a lower sideband (LSB), and an upper sideband (USB), with their phases locked together. Thus the phase noise of the generated photonic signal is directly related to the phase noise of the electrical signal driving the external modulator. This technique is mainly intended for RoF applications, to translate an electrical wireless carrier signal into the optical domain for routing it through fiber to a remote antenna unit (RAU), not for signal generation at the THz range, since it already requires an electrical signal at the desired frequency. However, the fact that the signal quality characteristics are directly inherited from the electrical driving source makes it attractive for high frequency generation.

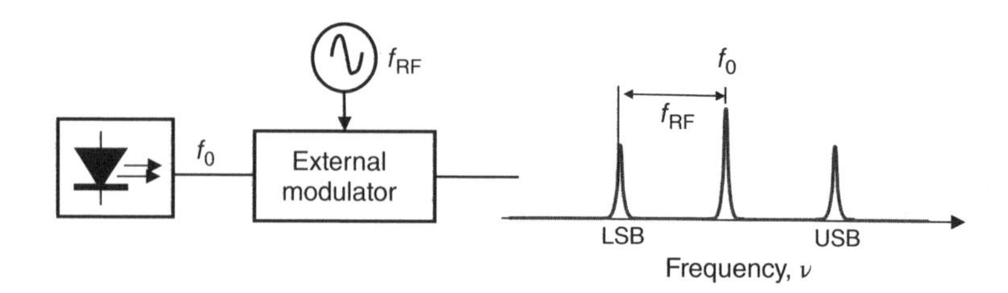

Figure 7.9 Photonic signal generation technique based on external modulation.

An approach to multiply the frequency of an electrical signal in the optical domain is by using the nonlinear transfer function of an external MZM to create a carrier-suppressed modulation, shown in Figure 7.10. This technique allows doubling of the frequency of the electrical signal f_{RF} by appropriately selecting the bias point of the MZM. Setting the DC bias voltage so that the insertion loss becomes a maximum, suppresses the optical center frequency at f_0. Every semi-period of the electrical modulation drives the MZM out of the maximum loss into the maximum transmission, given that the amplitude of the electrical signal is V_π. The output signal consists now of an optical suppressed carrier and dual sideband (SC-DSB), achieving a dual wavelength source. The great advantage is that it provides the stability and tuning range of the electrical source, with the drawback that it requires an electrical signal with frequency of at least half the target, having successfully been applied to generate the carrier waves for transmission systems in the millimeter wave range [38]. For the THz regime, it still remains impractical given the frequency limitation of the available sources and the modulators.

7.2.3 *Noise Reduction Techniques*

We have already discussed how using two independent sources to generate each of the wavelengths mixed in the OHG technique produces poor frequency stability and large phase noise. However, using two independent single wavelength-tunable laser diodes offers the best solution to a CW frequency-tunable source capable of reaching the THz frequency range. Several techniques have therefore been developed to lock the optical wavelengths from the two independent sources, aiming to reduce the noise and drift.

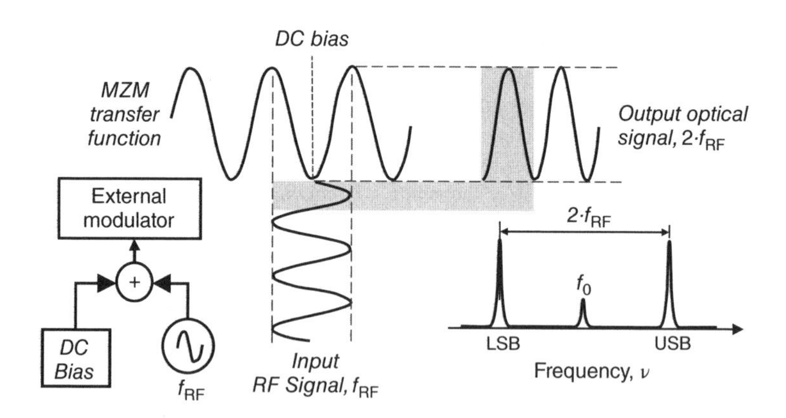

Figure 7.10 Photonic signal generation technique based on carrier suppressed double side band modulation on a MZM.

In the end, we need a stable frequency reference and a method to compare the frequency of the generated signal to this reference.

The two wavelengths can be lock to each other by means of an optical phase-locked loop (OPLL). An OPLL is a feedback system that enables electronic control of the phase of the output of a laser source, locking the two wavelengths to a microwave reference electrical signal. The block diagram of the feedback loop is shown in Figure 7.11. It consists of two semiconductor lasers (one free-running), a high speed photodiode, a phase detector (microwave mixer), a low pass loop filter, and a microwave reference oscillator. The operation principle is based on comparing the signal generated by heterodyning the two lasers on the photodiode to the reference in the phase detector. The resulting phase-error signal is fed back to one of the lasers, changing its wavelength acting on the bias current, to force it into tracking the free-running laser at a frequency offset corresponding to the frequency of the microwave reference oscillator. This causes a significant reduction in the phase noise on the optically generated microwave signal, and yet allows the OPLL to tune the generated frequency. The wide linewidth of DFB and DBR semiconductor lasers make OPLLs difficult to design and implement, requiring wideband feedback electronics and small loop delays [39]. Although phase-locked, we are still limited to two wavelengths separated by the frequency offset corresponding to the frequency of the microwave reference oscillator.

In order to have the noise reduction advantages offered by an OPLL, while freeing ourselves from the limitation imposed by the microwave reference to access THz wavelength separation, a double arm OPLL THz photonic oscillator has been proposed [40], following the schematic in Figure 7.12. The system allows the two semiconductor lasers to be mutually phase locked, locking each one of them to a different wavelength from an external optical reference, usually an OFCG. An ideal optical frequency comb is an optical spectrum made up by multiple optical wavelengths over a wide optical wavelength range (comb span), all equidistant from each other by a fixed value (comb spacing), and all phase locked to each other.

7.2.4 *Photonic Integrated Laser Sources*

In the previous sections we have described different laser diode structures that can be used as light sources for millimeter- and terahertz-wave signal generation using either pulsed or CW photonic techniques. These structures require different active and passive components to be combined, including laser sources emitting at different wavelength and wavelength combiners. Photonic integration allows placing different photonic components with various functions into a single chip, not just to reduce system complexity, size, and cost, but also to improve their performance [41]. One early demonstration showed

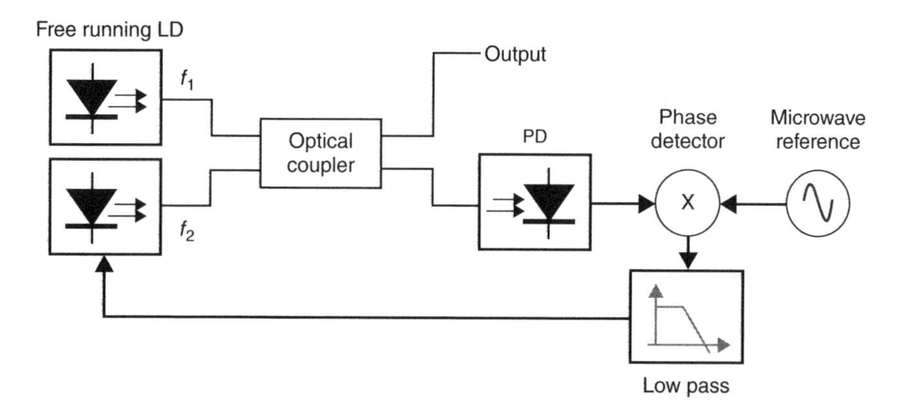

Figure 7.11 Block diagram of a single arm optical phase-locked loop (OPLL).

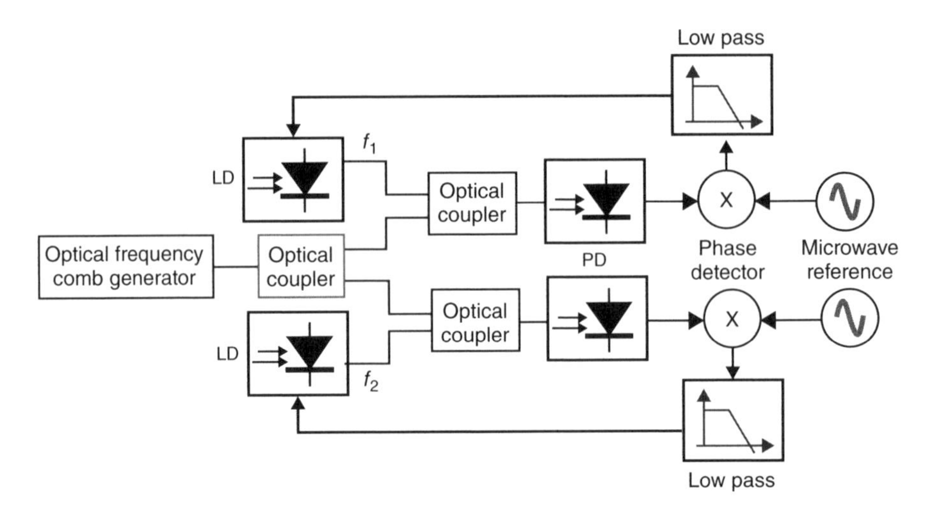

Figure 7.12 Block diagram of a double arm optical phase-locked loop (OPLL).

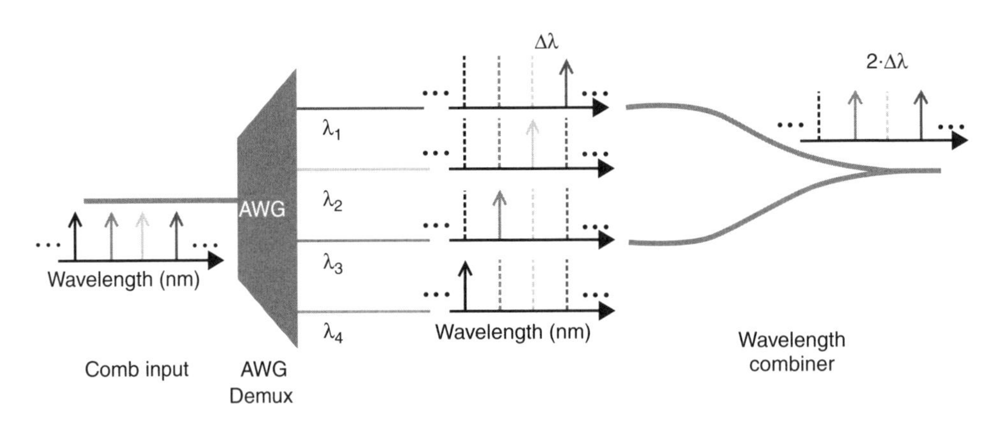

Figure 7.13 Block diagram of a planar lightwave circuit to generate a dual wavelength source. See plate section for color representation of this figure.

the advantages that photonic integration could bring in terms of improving the noise characteristics of the generated signal [42]. The scheme shown in Figure 7.13, consisted in an arrayed-waveguide grating (AWG) to separate the wavelengths of an OFCG into different output channel waveguides. The AWG was followed by a 3-dB combiner to select the outputs from two of these channels, achieving a dual wavelength source for optical heterodyning. When these elements were integrated together into the same planar lightwave circuit (PLC), the phase noise was reduced by the suppression of optical path length differences, achieving a phase noise of less than −75 dBc/Hz at an offset frequency of 100 Hz.

To date, several semiconductor laser structures specifically designed to produce two longitudinal modes on a single output waveguide have been reported. Most of these structures exploit in a different manner the potential of integrated frequency selective reflectors, DFB or DBR. An early solution consisted of a two-section DBR laser with an active gain section and a passive DBR, included in the latter was a periodic-phase-shift to produce a transmission "fringe" into the grating reflection band. This in turn created the two-wavelength emission [43]. At present, we can distinguish two main approaches, the parallel and the axial, to achieve a dual wavelength source with a single output from a chip. In the parallel approach, multiple discrete lasers are integrated adjacent to each other on the chip, and use a wavelength combiner element to place the two wavelengths into a single optical output waveguide. In the second approach, which we might call the axial approach, the two optical modes coexist along the same cavity in which the selection filters are located. The main advantage of this approach is that the two modes will experience the same electrical and thermal fluctuations. The noise of the two wavelengths will be correlated, reducing its impact by the common-mode noise rejection as the beat note is a frequency difference.

Considering first the parallel structures, these commonly use two single frequency DFB or DBR semiconductor laser structures, each with its own grating selecting a different emission wavelength, and a followed on-chip multiplexer to combine the output of each laser. These structures can then be organized in terms of the type of multiplexer used to combine the wavelengths.

The simplest combiner that can be used for a dual wavelength source is a Y-junction, integrated on the same active material [44] or in a butt-coupled passive epilayer [45], as shown in Figure 7.14a. An effective technique to reduce the linewidth is to increase the length of the optical cavity, both having an inverse relation. It has been shown that for a 2500 μm long cavity, the beat note linewidth was 600 kHz without any additional control [45]. This level of quality for the beat note was sufficient to use it as a carrier frequency generator in a wireless link operating at 146 GHz [46]. As observed in Figure 7.15(a), the shift in the grating pitch for each DFB laser allows dual wavelength emission. One important issue of

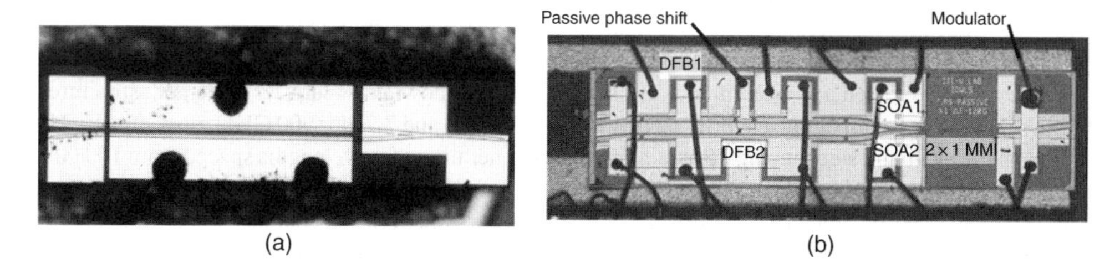

Figure 7.14 Dual wavelength source based on two DFB lasers: (a) combined with a Y-junction and (b) combined with an MMI.

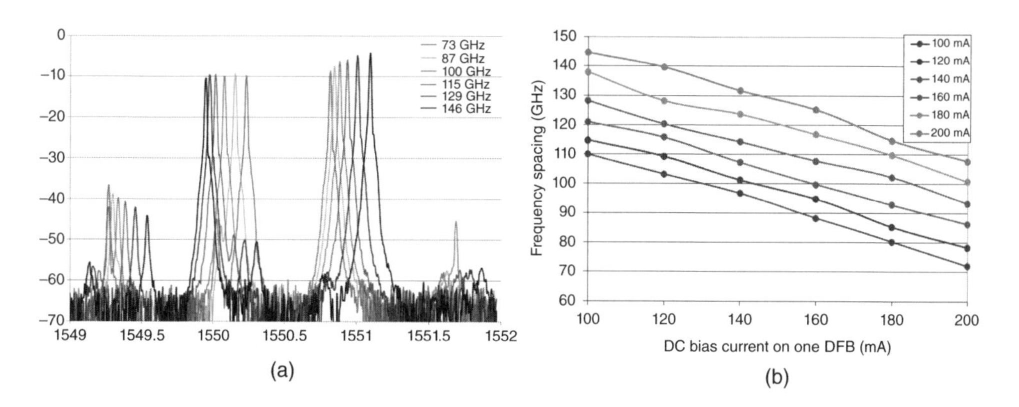

Figure 7.15 Dual wavelength output from two DFB lasers combined on a Y-junction. (a) observed on an optical spectrum analyzer and (b) wavelength spacing in terms of the current injected on one DFB (x-axis) for fixed values on the other (inset). (Reproduced and adapted with permission from Ref. [46], © 2012 Optical Society of America.) See plate section for color representation of this figure.

these structures is that the wavelength tuning that is observed is mainly due to thermal heating dependent on the injection current level. In addition, there is some thermal crosstalk between the two lasers, as shown in Figure 7.15(b), as the wavelength for a given input current in one laser changes depending on the current level on the other. The major drawback is that Y-junction combiners have issues with repeatability as well as introducing undesirable back-reflections to the laser. The next step forward is to substitute the multiplexer by a multimode interference (MMI) combiner, as shown in Figure 7.14b. An MMI is a combiner that operates on a self-imaging property that divides the power at each one of the inputs evenly among its outputs. When a 2×2 MMI combiner is used, with two inputs and two outputs, half of the light from each DFB is combined at the outputs. The back reflections at the MMI inputs are reduced by more than 10 dB by replacing the straight ends by tapered waveguides [47].

Another type of structure where multiple discrete lasers are integrated adjacent to each other are multi-wavelength lasers based on AWGs. This element is another type of intra-cavity optical filter, with particular properties that allow it to have a multiplexing/de-multiplexing action [48]. The basic schematic of the device is in Figure 7.16a, showing the AWG function combining the fixed wavelengths of the channel waveguides (on the left-hand side of the AWG) onto a common waveguide. The operation principle, shown in Figure 7.16b, is that the AWG selects the lasing wavelength within each channel by filtering the Fabry–Perot modes among those under the AWG passband. The Fabry–Perot modes are formed in the cavity between the end mirrors. Each channel has a frequency selective response, around the AWG central wavelength (usually $\lambda_0 = 1550$ nm). The channel wavelengths are spaced by design at a fixed value, known as the channel spacing ($\Delta\lambda$). In addition, the response of the AWG is periodic, and the frequency response of each channel repeats every free spectral range (FSR). Injecting current into a channel SOA turns on the lasing wavelength of the corresponding channel. When multiple SOA channels are activated, the channel wavelengths are combined in the common waveguide.

Two examples of such a device, fitted into a single chip, are shown in Figure 7.17a. In both structures, the AWG have channel spacing and FSR set at 0.8 nm (100 GHz) and 7.2 nm (900 GHz), respectively. The device on the left-hand side, having 16 channels, allows generation of wavelengths spaced from 100 GHz to 1.5 THz, in steps of 100 GHz. As shown in Figure 7.17(b), AWG lasers are able to deliver these wavelengths independently, delivering simultaneously as many wavelengths as channels are activated, with their spacing being determined by the fixed AWG channel spacing. Arrayed-waveguide grating lasers (AWGLs) are able to deliver many wavelengths, spaced by the fixed AWG channel spacing. When a narrow channel bandwidth is used, these devices have been shown to emit in a single mode despite the length of the cavity, producing very stable and reproducible devices [49]. Recently, the optical linewidth of each of the channel wavelengths has been evaluated using a self-heterodyne technique to find a full width half maximum lower that 130 kHz [50]. There are several factors that may account for such extremely

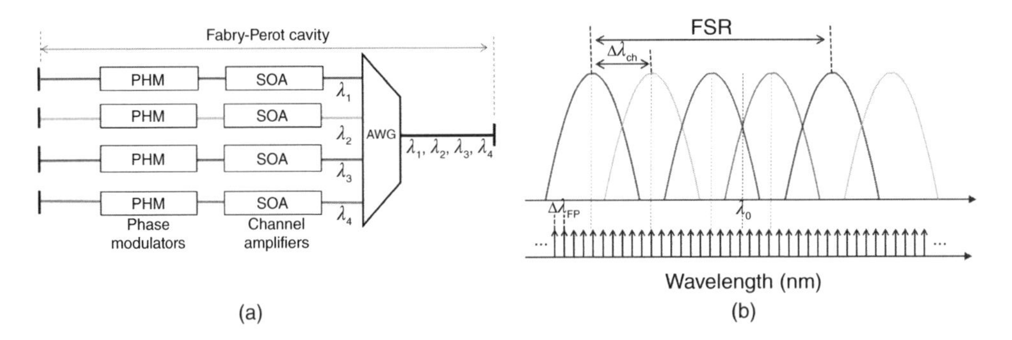

Figure 7.16 Dual wavelength output from an AWG laser. (a) schematic of the device, showing the AWG, its channel and common waveguides and the Fabry–Perot cavity, and (b) transfer function of the AWG, selecting the Fabry–Perot modes allowed to lase within each channel. See plate section for color representation of this figure.

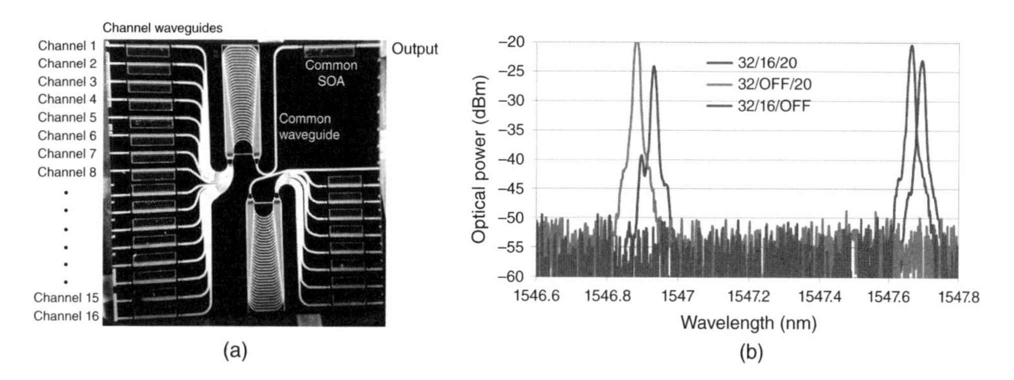

Figure 7.17 (a) Multi-wavelength AWG laser chip with two devices and (b) Dual wavelength output from an AWG laser, with bias in one channel, bias on the adjacent channel, and simultaneously biasing both. See plate section for color representation of this figure.

good performance. First, we must note that the device includes an intra-cavity filter, with a narrow pass-band to eliminate Fabry–Perot side-modes. The AWG therefore reduces the amplified spontaneous emission (ASE) noise from the amplifiers. In addition, the fabrication of these devices requires active/passive integration technology, fabricated using an extended cavity laser configuration. Due to the AWG, the devices have a long cavity (>2 mm), most of which is a low loss passive waveguide. Since the cavity length is one of the factors involved in the determination of the optical linewidth, we expect these devices to be of interest in the generation of high purity carrier wave signals. Shown in Figure 7.18 is the electrical beat note generated by optical heterodyning the optical modes from adjacent channels of an AWGL.

The axial approach to implementing a dual wavelength source has a wide range of different structures, in which both modes propagate through a structure in which at least an optical filter is located. Single filter dual wavelength structures need to have at least three sections within the cavity, as shown in Figure 7.19a,

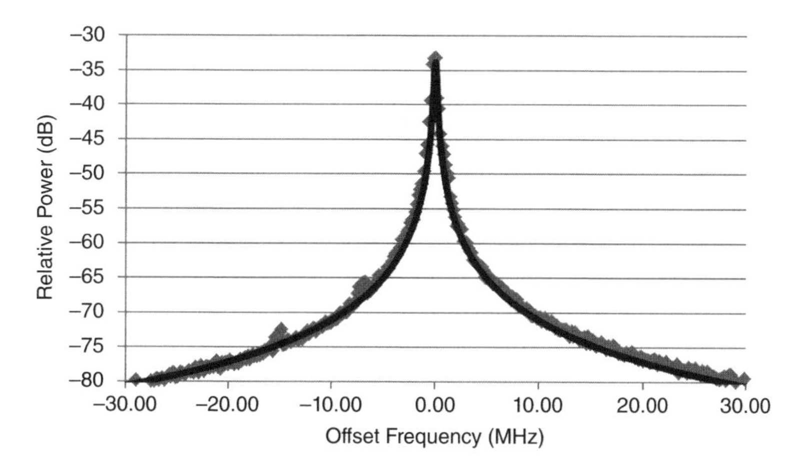

Figure 7.18 Beat note at 95.15 GHz obtained by photomixing on a UTC-PD the two modes on adjacent channels of an AWG laser (resolution bandwidth (RBW) = 50 kHz, video bandwidth (VBW) = 300 Hz). (Reproduced and adapted with permission from Ref. [50], © 2012 Optical Society of America.)

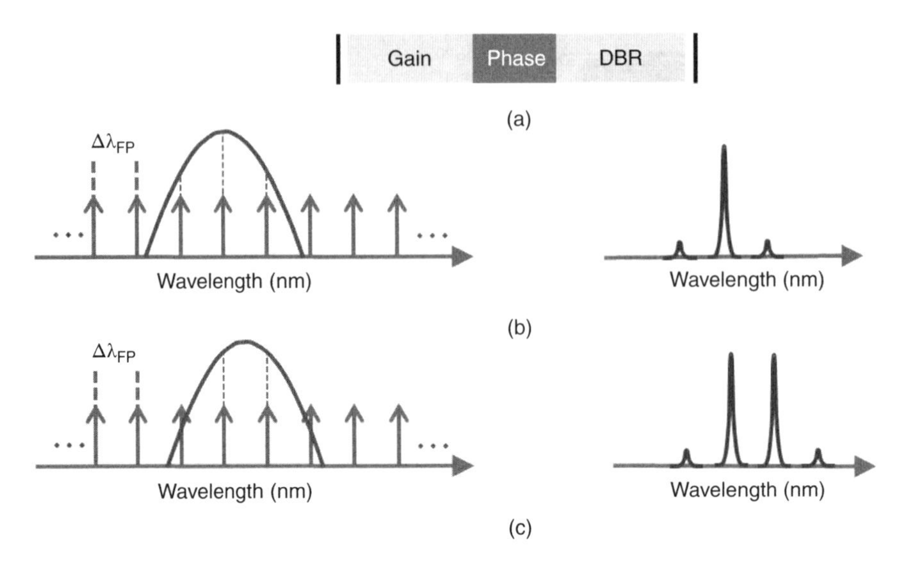

Figure 7.19 (a) Dual mode source based on an axial structure with a single DBR section, (b) DBR bias set for single mode operation, and (c) DBR bias for dual mode operation.

namely a gain section, a phase section, and a DBR grating. The current injected into the DBR section allows tuning of the center wavelength of the grating filter, moving from situations in which a single Fabry–Perot mode is selected, Figure 7.19b, into the most desirable ones in which two modes have similar losses in the grating, achieving a dual mode lasing, Figure 7.19c. The main drawback again, as in pulsed schemes, is that the wavelength spacing among the modes depends greatly on the cavity length, thus the frequencies are restricted to around 100 GHz [51]. This scheme has also been used introducing into the cavity a SA so that the modes filtered by the DBR section are locked, helping to reduce the phase noise of the generated electrical signal [52].

Another alternative for axial structures is to introduce two different grating structures to select the two lasing modes in the cavity. The main problem here is that the modes must propagate through the other mode selection filter. One approach to this has been to use surface lateral gratings along the guiding layer, defining two different Bragg grating periods, one for each side of the ridge [53]. This approach generates two wavelengths but lacks tuning. In order to enable frequency tuning, a different approach needs to be followed. Recently, a structure in which two DFB lasers are placed in-line, separated by a passive phase modulator section, has been reported. Frequency tuning of each of the DFB lasers is achieved by a titanium wire heater [54].

7.3 Photodiode for THz Emission

7.3.1 PD Limitations and Key Parameters

As discussed in Section 7.1.2 for photodetectors (PDs) in THz photonic systems we need to consider four different parameters: responsivity, bandwidth, roll-off, and saturation power. These parameters can be summarized by a figure of merit demonstrating the efficiency of the device in converting light into THz signal ($\eta_{THZ} = P_{max\text{-}THz}/P^2_{opt}$). Within this section we will look at the ultimate performances of each current PD technology looking into these parameters with the use of the figure of merit as an ultimate comparison tool.

To start with bandwidth, it is limited by two factors, electrical and transit time-related, as shown in Eq. (7.4) describing the normalized output power of the device, $P(f)$ as a function of frequency, f.

$$P(f) = \frac{1}{1+\left(\frac{2\pi W_{\mathrm{d}} f}{3.5\overline{\upsilon}}\right)^2} \frac{1}{1+(2\pi RCf)^2} \tag{7.4}$$

where W_{d} is the thickness of the depletion layer, R and C are the total resistance and capacitance of the device, and $\overline{\upsilon}$ is the average carrier drift velocity, which in a heterojunction photodiode is given by Eq. (7.5) [55].

$$\overline{\upsilon} = \sqrt[4]{\frac{2}{\frac{1}{v_{\mathrm{e}}^4}+\frac{1}{v_{\mathrm{h}}^4}}} \tag{7.5}$$

where v_{e} and v_{h} are the electron and hole velocities, respectively.

It is clear that both the electrical and transit time limit need to be addressed in THz PDs. Indeed work on PDs has been principally looking at two avenues for solving that problem. Hot carrier type PDs (cf. Figure 7.20a) [57] allow achieving shorter transit times while traveling wave PDs (cf. Figure 7.20b) [56] address the electrical bandwidth limitation. It is also important to note that traveling wave design helps in improving the roll-off beyond the 3 dB bandwidth point thus enhancing the output power at higher frequency.

Saturation power is mostly limited by space charge and power distribution, so again in this case the use of single carrier fast transport and waveguide distributed design will help performances at high frequencies.

A good example is to simulate a device at fixed capacitance and fixed absorbing area with different structures [58]. In the case of Figure 7.21, the devices are a vertically illuminated UTC-PD, a traveling wave PIN and a traveling wave UTC-PD. This simple simulation shows the clear advantage of optimizing both the transit time and the electrical limit, in particular at frequencies beyond the 3 dB limit. It is also interesting to note that this does not take into account the enhancement in term of saturation power that a UTC-PD can offer.

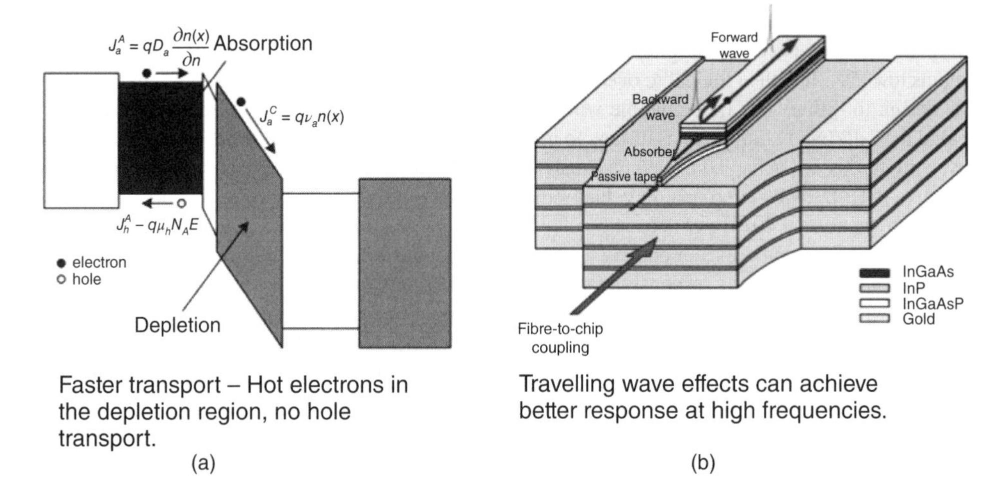

Figure 7.20 Schematic diagram of (a) a UTC-PD band structure and (b) aTW PD structure. See plate section for color representation of this figure.

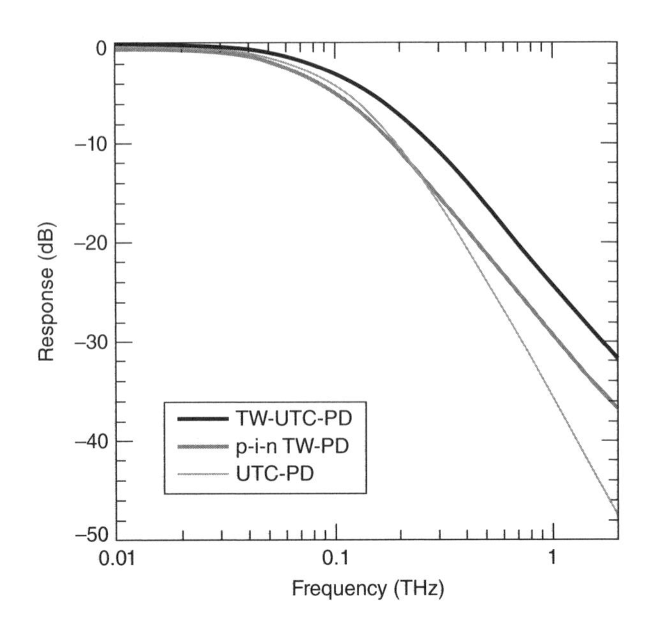

Figure 7.21 Modeling results for different PD structures with identical capacitance and area [58].

Finally, it is important to note that optical coupling plays an important role in achieving the required high responsivity in a small area detector (required to have a low capacitance).

While much of the theory of the operation of fast PDs is described in Chapter 3, we will here concentrate on showing the best performances shown by these different devices. We will only describe waveguide detectors as, expected from the discussion in this introduction, they have demonstrated the best performances to date.

7.3.2 Traveling Wave UTC-PD Solution

It is understood that to achieve fast operation the absorption layer of a UTC-PD will generally be thinner than that of a normal PIN PD, resulting in a degraded responsivity in a vertical structure illumination configuration. In these cases despite a relatively high emitted power the figure of merit of the device remains low [59]. It is then logical to overcome that issue with the design of an edge illuminated waveguide design [60] that is then compatible with traveling wave design techniques to enhance the roll-off beyond the 3 dB bandwidth point as discussed in Section 7.3.1. However for such devices a true traveling wave structure is not achieved as velocity matching could only happen with practical PD in periodical structures [61]. Designs approaching velocity matching are possible and will give roll-off significantly better than the normal 40 dB per decade.

A typical device would use an epitaxy as described in Table 7.2, and would have a 2×25 μm ridge waveguide absorber enabling responsivity of the order of 0.5 A/W for 3 dB bandwidth of 108 GHz when integrated with a mode converter [62]. This device, depending on the application, can then be coupled to an antenna either through a coplanar interface or an integrated planar antenna (cf. Figure 7.22).

Such devices can then be characterized across frequencies up to several THz using heterodyne excitation. The best published results are summarized in Figure 7.23. It is important to note that all measurements are highly dependent on the calibration of the power measurement systems and assessment of the noise sources. However, coplanar measurements use a calibrated millimeter wave system and are more reliable, thus giving a good reference in the lower frequency range. Importantly, one can note that in the range of interest the roll-off is of the order of −30 dB/decade; a significant improvement from the standard 40 dB/decade.

Table 7.2 Typical epitaxial structure for a TW-UTC-PD [56].

Doping (cm^{-3})	Material	Function	Thickness (nm)
Zn ($>2 \cdot 10^{18}$)	Q1.3	P contact and barrier	200
Zn ($1.5 \cdot 10^{18}$)	InGaAs	Absorber	120
S (uid)	Q1.3	Spacer	30
S (uid)	InP	Carrier collection	300
S ($2 \cdot 10^{18}$)	Q1.3	Waveguide	300
S ($4 \cdot 10^{18}$)	InP	n-contact	600
	Q1.2	Diluted waveguide	40
	Repeated		
Fe	SI-InP	Substrate	350 000

uid = unintentionally doped

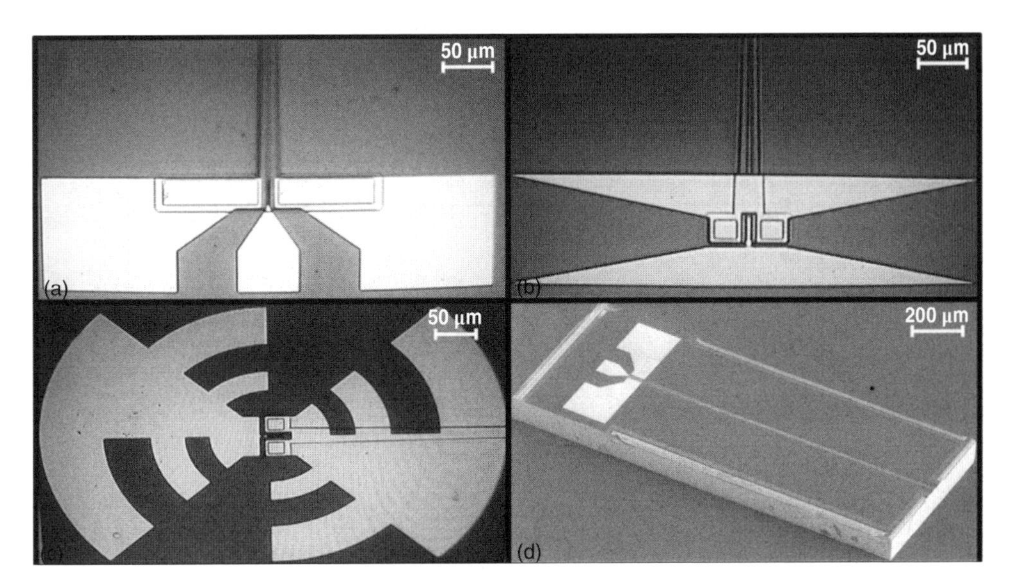

Figure 7.22 Images of the fabricated devices. (a) Coplanar waveguide, (b) bow-tie, (c) log-periodic, and (d) cleaved chip with coplanar waveguide.

Although the achieved powers are at the top of the range from such devices, it is far more interesting to look at the figure of merit. This allows to fully assess the performances of the device as it combines all the parameters, including the responsivity and compares it to other known photomixing solutions. In Figure 7.24, we show a summary of the figure of merit for the device described in this chapter compared to other photomixing technologies [58]. It clearly demonstrates the enhancement in terms of efficiency of the traveling wave uni traveling carrier design. However, looking into the details of the design requirements for the thicknesses of the absorber and depletion region, one can see that these affect both carrier transport and traveling wave design performances [58]. In particular, further optimization of the carrier transport would give better transport performances combined with a better junction capacitance, which in turn enhance traveling wave results. For that reason we discuss further design mechanisms and performances from optimized devices for carrier transport in the next section.

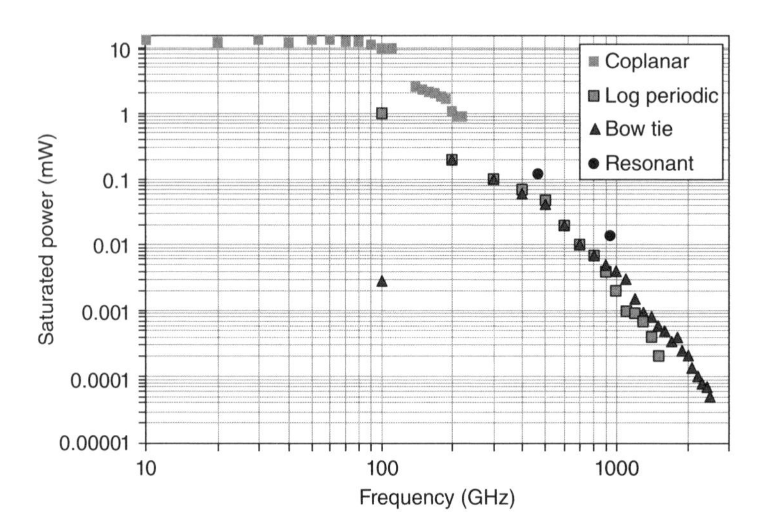

Figure 7.23 Measured saturated power out of TW-UTC PDs in different configurations (coplanar probing, resonant antenna, log periodic antenna, and bow tie antenna) [62–64].

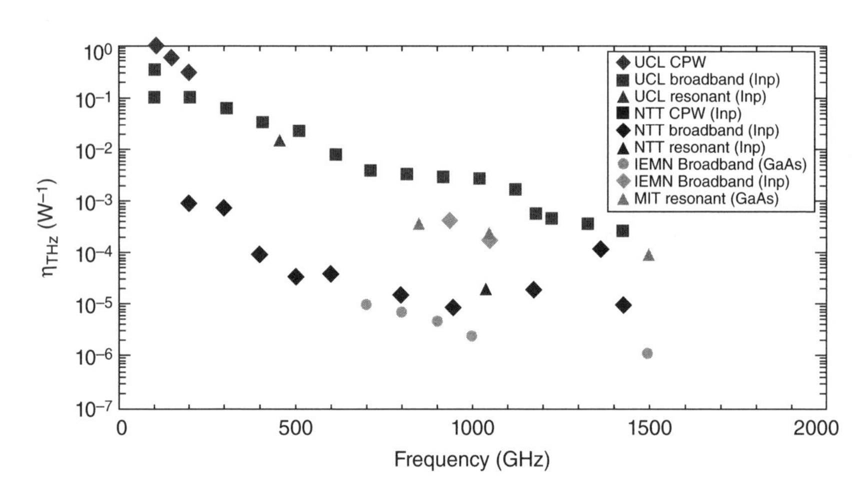

Figure 7.24 Comparative figure of different photomixers [58].

7.4 Photonically Enabled THz Detection

Despite the high availability of power detectors, such as Golay cells and pyroelectric detectors (cf. Chapter 5), a key issue which has been addressed from the beginning of photonic-based THz techniques is concepts for the coherent detection of the generated THz radiation. Whether for spectroscopy or THz imaging, for full analysis of the device under test (DUT), the real and the imaginary part of the refractive index and permittivity, respectively, have to be identified. Photonically enabled THz detection offers the advantage of precise sampling of the electrical THz pulses in time-domain systems via femtosecond lasers and the possibility to realize local oscillators at THz frequencies in CW THz systems. The most common methods, including all commercial systems, are (i) photoconductive antennas and

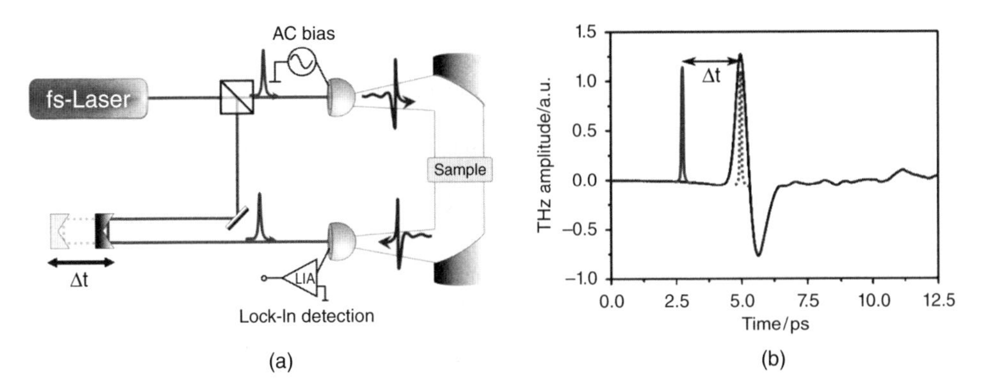

Figure 7.25 Basic structure of a TDS system (a) and photoconductive sampling of the THz pulse via a delayed optical femtosecond-pulse (b). To adjust arrival time of the optical sampling pulse at the receiver, a variable delay stage is applied.

(ii) electro-optical sampling via birefringent crystals. In addition, fast photodiodes or optically pumped Schottky mixers are demonstrated concepts. The operation principle of these techniques and basic design considerations are addressed in the subsequent sections.

7.4.1 Pulsed Terahertz Systems

In pulsed THz systems, the receiver performance is a crucial parameter which has to be carefully optimized if one wants to profit from the wide bandwidth of these systems. Since the width of the detected pulse (and therefore the obtainable THz bandwidth) is determined by the convolution of the incident THz pulse with the transfer function of the receiver, the receiver contributes significantly to the system performance.

A widespread technique for coherent THz detection in pulsed THz systems is photoconductive sampling of THz pulses by means of optical femtosecond pulses in semiconductors. As illustrated in Figure 7.25, the THz pulse and an optical sampling pulse are both incident onto the receiver. The optical pulse induces a time-variant photoconductivity $\sigma(t)$, which can be described by the following convolution, which takes the temporal function of the optical pulse into account [65]

$$\sigma(t) \propto e \int_{-\infty}^{+\infty} P_{\text{opt}}(t')\mu(t-t')n(t-t')\mathrm{d}t'. \tag{7.6}$$

The semiconductor properties are described by the mobility μ and the carrier density n. For further analysis, the optical pulse is considered short as compared to the trapping time of the carriers. This allows approximation of the optical pulse as delta function. Thus, the carrier density after excitation is given by

$$n(t) = \begin{cases} N(0)\,\mathrm{e}^{-t/\tau_c} & t > 0 \\ 0 & t < 0 \end{cases}, \tag{7.7}$$

where τ_c is the average trapping time. Similarly it follows for the mobility:

$$\mu(t) = \begin{cases} \mu_e\left[1-\mathrm{e}^{-t/\tau_s}\right] & t > 0 \\ 0 & t < 0 \end{cases} \tag{7.8}$$

where τ_s is the average scattering time due to defects or other carriers. Since the repetition rate of femtosecond lasers is in the 100 MHz range, the repetition time (~10 ns) is much shorter than typical integration times. Thus, the detected signal is a time-averaged photocurrent which only depends on the difference in the arrival time Δt of the THz and the optical pulse at the receiver. Therefore, the current density can be expressed by the convolution of the time-dependent conductivity $\sigma(t)$ and the incident THz field:

$$J(\Delta t) = \int_{-\infty}^{+\infty} \sigma(t - \Delta t)\, E_{\mathrm{THz}}(t)\mathrm{d}t \tag{7.9}$$

If not only the laser pulse but also the carrier lifetime is much shorter than the THz pulse (i.e., it can also be approximated by a delta function), the induced photocurrent density is directly proportional to the electric THz field ($J(\omega) \propto E_{\mathrm{THz}}(\omega)$). In the long term limit, the detected current is proportional to the integral of the electric field. In the intermediate regime, which is the most realistic case, the photocurrent response does depend significantly on the time constants τ_c and τ_s. If the time domain function of $n(t)$ and $\mu(t)$ is Fourier transformed according to

$$F(\omega) = \frac{1}{\sqrt{2\pi}} \int_{-\infty}^{+\infty} f(t)\mathrm{e}^{-\mathrm{i}\omega t}\mathrm{d}t \tag{7.10}$$

both carrier density and mobility feature a frequency dependence proportional to

$$F(\omega) \propto \frac{1}{1 + \omega^2 t_{\mathrm{s,c}}^2}. \tag{7.11}$$

Thus, a photoconductive receiver features a low pass characteristic because the photoconductivity decreases with increasing frequency. This signifies that short carrier lifetimes are essential for the realization of wideband THz systems. The concepts (e.g., low-temperature-growth) to achieve fast trapping of photoexcited carriers in semiconductors are identical to those for photoconductive emitters, which is why Chapter 2 is referred to at this point.

Nevertheless, the design criteria for photoconductive switches operated as a receiver are different from the ideal parameters in the emitting regime. The difficulty is the existing trade-off between high mobility and fast carrier trapping, where the ideal semiconductor should feature both. But since the trapping of photoexcited carriers is a scattering process, semiconductors with short carrier lifetimes feature a lower mobility, and vice versa. For the realization of efficient THz emitters, a high mobility is preferable to fast trapping [66]. On the receiver side, the dynamic range of the detector is limited by the carrier lifetime which causes an increased frequency roll-off and Nyquist noise [67] (cf. Figure 7.26).

The advantage of photoconductive sampling is the convenient handling and the simple system geometry. However, if very wide bandwidths (>8 THz) are required, for example, for the analysis of carrier dynamics in semiconductors [68, 69], different detection methods have to be applied due to the low-pass characteristics of the photoconductive switches.

An alternative technique, which features an almost flat frequency response, is electro-optical sampling. This method is based on birefringence in crystals, where the incident THz field alters the refractive index of the material, as illustrated in Figure 7.27. In general, the difference in the refractive index due to an external field is given by [70, 71]

$$\Delta n = n^3 r_{ij} E_{\mathrm{THz}}, \tag{7.12}$$

where n is the refractive index at the optical probe wavelength and r_{ij} is the relevant electro-optical or Pockels coefficient, respectively. For the orientation of electric field, probe beam and material axes in Figure 7.27, the induced phase retardation between the optical polarization components is given by

$$\Delta\varphi = \frac{2\pi}{\lambda} n^3 r_{ij} E_{\mathrm{THz}} d, \tag{7.13}$$

where d denotes the thickness of the crystal in the direction of propagation.

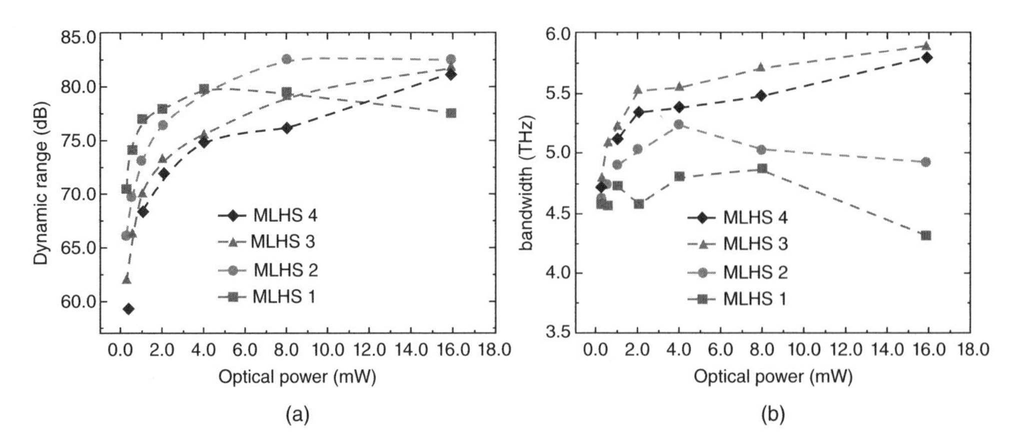

Figure 7.26 Dynamic range (a) and bandwidth (b) of a 1.5 μm TDS system which was equipped with four different receivers, based on InGaAs/InAlAs multilayer-heterostructures (MLHSs) with different carrier lifetimes. For the identical emitter, the two samples with the shortest carrier lifetime (0.3 ps for MHLS 3 and 0.2 ps for MHLS 4) lead to the highest bandwidth. (Reproduced and adapted with permission from Ref. [67], © 2013 Optical Society of America.)

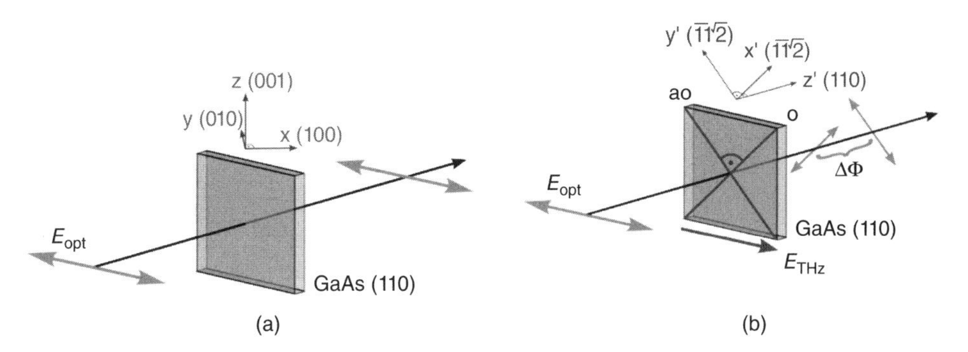

Figure 7.27 Induced birefringence in a GaAs crystal by an external electric field. In the absence of the electric field (a), the GaAs is isotropic and the optical probe beam passes the crystal unchanged. In the presence of an electric field (b), the refractive index becomes different for the two polarization components of the optical probe beam, which causes a phase shift.

The frequency response of the EO detector depends on the group velocity dispersion of the probe beam and the refractive index dispersion of the THz beam propagating through the crystal. Thus, there is a trade-off between the sensitivity and the bandwidth of the detector. A thick crystal allows longer interaction of the optical probe pulse and the THz-signal, which increases the phase difference of the polarization components and therefore the detectable signal. At the same time, the detector bandwidth is decreased due to the group-velocity mismatch (cf. Figure 7.28).

Suitable materials depend on the optical wavelength. Classical detectors for 800 and 1060 nm are based on ZnTe [72] or GaP [73], respectively, where for 1.5 μm GaAs is preferable due to the long coherence length [74] at this optical wavelength. Within recent years, organic crystals such as DAST,

[1] Rainbow Photonics DSTMS.

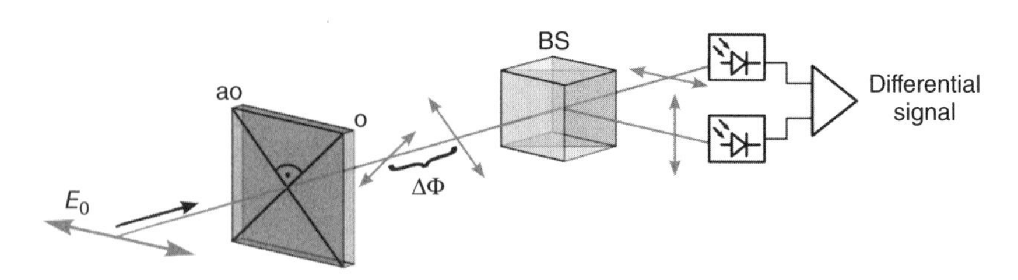

Figure 7.28 Detection of the optical phase shift: behind the detector crystal, a polarization beam splitter separates the horizontal and the vertical polarization components, where each is detected by a photodiode. In the absence of a THz field, the differential signal of the two photodiodes is set to zero. In the presence of an incident THz field, the optical power in the two separated polarization components will become unbalanced, causing the differential signal of the photodiodes to deviate from zero.

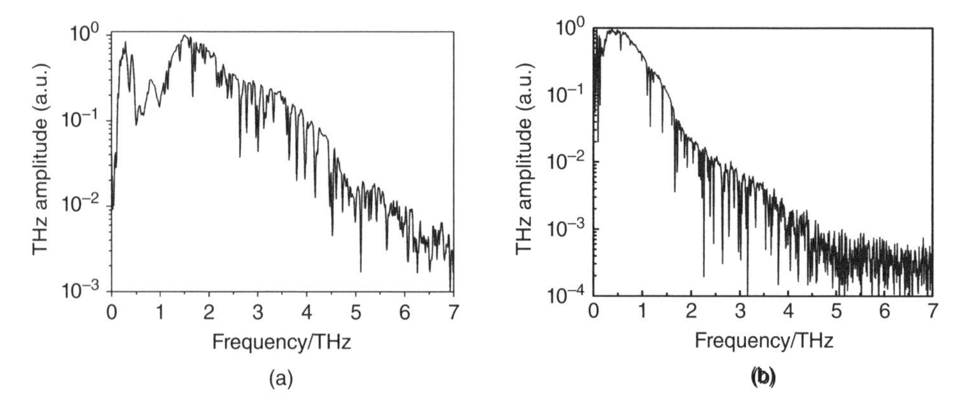

Figure 7.29 Comparison of two different detection techniques in a pulsed THz system for an identical emitter. Electro-optical sampling (a) features a larger bandwidth whereas the photoconductive antenna (b) is superior in terms of dynamic range. The EO crystal used in (a) was a 300 μm thick DSTMS crystal which also caused the dip between 500 GHz and 1.5 THz due to absorption.[1]

OH1, or DSTMS [75] have been introduced, which feature larger electro-optic coefficients. A comparison between the two different detection techniques is shown in Figure 7.29, where the identical 1.5 μm TDS system was equipped with a photoconductive receiver (LTG-InGaAs/InAlAs) and an organic crystal (DSTMS).

7.4.2 Optically Pumped Mixers

For the detection of CW THz radiation, optoelectronic down-conversion, also called photomixing, is the most common approach. This heterodyne (or, in special cases homodyne) technique has the advantage over direct detection, that the THz frequency is mixed to lower frequencies, which facilitates the measurement due to the availability of the required components. In principle, photoconductors [4] and photodiodes [17] can be applied as a mixer, where photoconductors are more widespread due to the simple fabrication process and the availability of a suitable semiconductor (LTG GaAs). With the increasing interest in photomixing systems operating at the optical telecommunication wavelength of 1.5 μm, photodiodes have become more popular in recent years.

In general, the operation of a photoconductor-based mixer can be modeled as follows. A THz wave

$$E_{\mathrm{THz}} = E_0 \cdot \cos(\omega_{\mathrm{THz}} \cdot t + \varphi_{\mathrm{THz}}) \tag{7.14}$$

is incident on a photosensitive semiconductor. This electric field will generate a proportional voltage, which takes the role of the bias voltage for THz emission. In parallel, the detector is illuminated with the optical local oscillator signal (cf. Figure 7.30), composed of two optical modes with a slight wavelength difference, which can be written as

$$E(t) = E_1 \cdot \mathrm{e}^{\mathrm{i}\omega_1 t} + E_2 \cdot \mathrm{e}^{\mathrm{i}\omega_2 t + \mathrm{i}\varphi}. \tag{7.15}$$

If the relative phase φ between the two signals is set to zero, the total optical power incident on the detector can be written as:

$$\begin{aligned} P_{\mathrm{opt}}(t) &\sim E_1(t) \cdot E_2^*(t) \\ &= E_1^2 + E_2^2 + E_1 E_2 \mathrm{e}^{\mathrm{i}(\omega_1 - \omega_2)t} + E_1 E_2 \mathrm{e}^{\mathrm{i}(\omega_1 + \omega_2)t} \\ &= E_1^2 + E_2^2 + 2E_1 E_2 cos((\omega_1 - \omega_2)t). \end{aligned} \tag{7.16}$$

The optical signal modulates the resistance of the photoconductor, which is transformed into a current via the voltage induced by the electric THz field. For most photomixing systems, THz frequency and local oscillator frequency are identical, since both are generated from the same pair of laser. In this homodyne regime, the detected (DC) current follows [76]

$$I_{\mathrm{det}} = \frac{1}{2} E_0 e \mu_e \eta \frac{\tau}{hf} \cdot \frac{\sqrt{P_{\mathrm{opt1}} \cdot P_{\mathrm{opt2}}}}{\sqrt{1 + \tau^2 \omega_{\mathrm{THz}}^2}} \cos(\varphi_{\mathrm{THz}} - \varphi) \tag{7.17}$$

As expected, the trade-off between mobility and trapping time is also present in the case of CW THz systems. Thus, also for CW THz detection, the material parameters have to be carefully analyzed and optimized to find the right balance between the different material properties. Concerning the design of the contact metallization structure and the antennae, the same considerations are valid as for the design of photoconductive emitters, which is described in detail in Chapter 2.

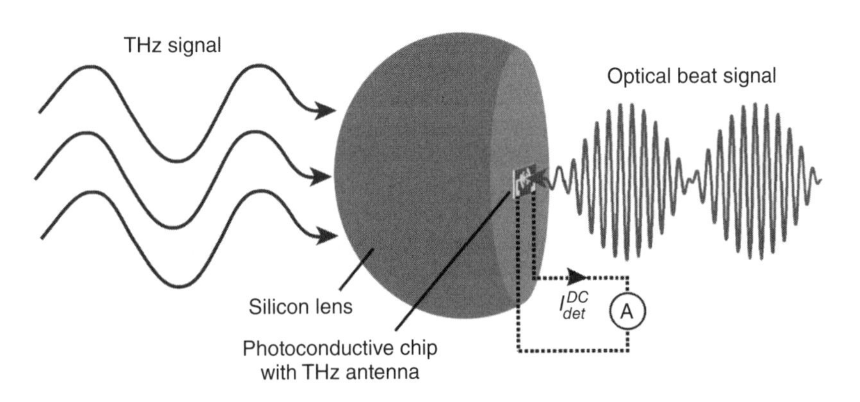

Figure 7.30 Coherent detection of THz radiation using a photomixer.

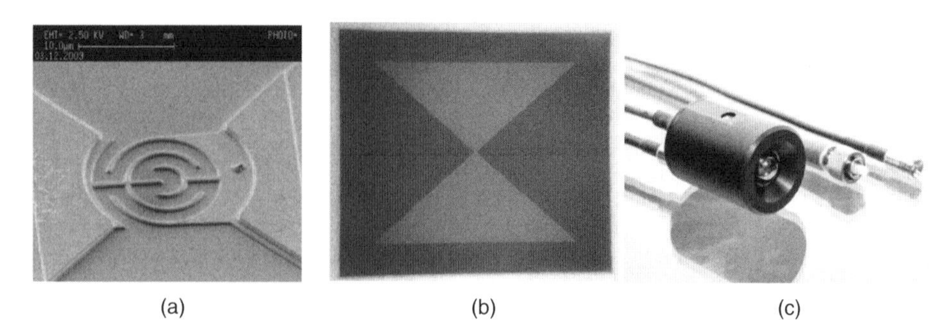

Figure 7.31 Commercially available photomixer for 1.5 μm optical wavelength. The mixer itself (a) features a radial-symmetric interdigital finger metallization with a diameter of 10 μm. The photomixer is located in the feedpoint of a bow-tie antenna, with 800 μm diameter (b). For convenient handling, the receiver chip is mounted into a fiber-coupled module (c), with a footprint of 25 mm.

The choice of semiconductor certainly depends on the optical wavelength. Since the first suitable material, which offers the required balance between fast trapping and high mobility was low-temperature-grown (LTG) GaAs, the first coherent CW THz systems were pumped at 800 nm optical wavelength (cf. Figure 7.31).

With increasing interest in THz technologies, THz systems at 1.5 μm optical wavelengths have become intensively requested, since the required optical components are available off-the-shelf in high quality at low price. Further, optical integration technologies, which have been established for telecommunications, can be applied for the realization of dual-color laser sources or a fully integrated THz emitter. For the realization of photoconductive detectors operating at 1.5 μm optical wavelength, several material systems have been explored, including GaAsSb [77] or Br-doped InGaAs [78]. Another material system, where a detailed description can be found in Chapter 2, is $In_{0.53}Ga_{0.47}As/In_{0.52}A_{l0.48}As$ multilayer structure, as shown in Figure 7.32. In combination with a photodiode-based emitter, generating ~10μW of THz power at 100 GHz, a signal-to-noise-ratio of 105 dB and 3.5 THz bandwidth have been achieved.

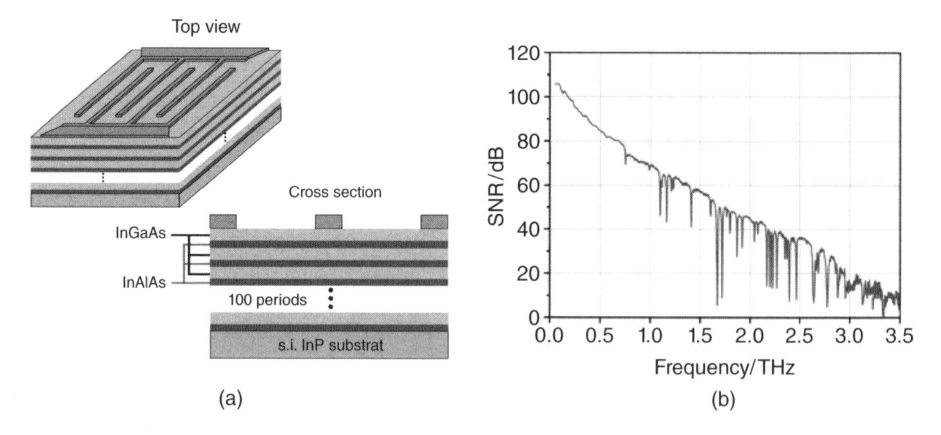

Figure 7.32 Photomixer structure (a) and SNR performance (b) of a 1.5 μm photomixing system for an integration time of 200 ms. (Reproduced and adapted with permission from Ref. [20], © 2013 Optical Society of America.)

An alternative, less frequently used approach for the coherent detection of CW THz radiation is photodiodes. The advantage of using photodiodes as the detector is that they can be operated in two detection modes, depending on the bias condition [79]. In forward bias, they work as a square-law (i.e., power) detector, in reverse bias, they can be applied for heterodyne/homodyne detection. The fundamental difference between the photoconductive and the photodiode approach is that, in an ideal case, no photocurrent is generated in a photoconductor if no THz signal is incident. In contrast, a photodiode will generate a photocurrent (and therefore a photovoltage) if illuminated with light, due to the build-in potential of the pn-structure. This DC part is inevitable and has to be separated from the modulated signal via a bias-tee. Up to now, photoconductive detectors have been preferred due to their higher signal-to-noise-ratio.

7.5 Photonic Integration for THz Systems

Integration is the technology by which a large number of individual components can be fabricated on a single common substrate, arranged to perform a given function. Planar technology was developed in the 1960s by Jack Kilby and Robert Noyce to fabricate silicon semiconductor devices and integrated circuits (ICs). It brought such advantages in both cost and performance that silicon has been the primary host material for electronics since then. These advantages fueled the continuous effort to increase the scale of integration, which as foreseen by Gordon Moore, has experienced an exponential growth rate, doubling the number of transistors on an IC every two years. The dimension for the success of silicon comes from both its technological and economic advantages.

The impact that planar technologies had in the development of electronics motivated the efforts to develop photonic integration techniques. The integration of a large number of optical components on a single substrate in order to develop complex functions started in the 1970s, known as "active integrated optics", concentrating on the InP material system to develop components in the fiber optics communication spectral region 1.1–1.6 μm. This technology evolved into optoelectronic integrated circuits (OEICs), which refers to the monolithic integration of optical devices and electronics [80], and photonic integrated circuits (PICs), which aim at the integration of a large number of optical devices on a common substrate [41]. Despite these efforts, the growth of PIC technology is occurring at a slower pace, PICs being several generations behind ICs, as shown by some key figures of merit, such as the cost per unit component [81].

The generation of signals with frequencies ranging from the millimeter to the THz range is one key application for photonic technologies. The photonic approach provides important advantages, among which we can highlight the low-phase noise, the wide tuning range (several GHz) and the ultra-wide modulation bandwidth (up to 100 Gbit/s). In addition, we can make use of optical fibers and RoF technologies to provide a low-loss media for the distribution of the signals. Such characteristics have been demonstrated to be key in the development of ultra-broadband wireless data transmission systems to generate high frequency carrier waves [82]. These systems usually require additional functional building blocks to introduce some level of signal processing, such as data modulation. In the following sections we tackle the issues that using PIC technology raises for the integration of a photonic-based wireless communications transmitter operating in the millimeter to THz frequency range. The building blocks that compose such a system, shown in Figure 7.33, include photonic carrier signal generator, an optical modulator and an optoelectronic converter coupled to an antenna. The photonic carrier signal generation will provide the optical signal that when shined onto the optoelectronic conversion module (a photodiode or photoconductor) produces the millimeter/THz frequency wave, which is emitted into the medium using an antenna. The optical data modulation takes advantage of the wide bandwidth offered by optical components to modulate the carrier signal in the optical domain, as the modulation speeds that are required lie beyond 10 Gbit/s. If the main objective is to integrate all of these elements onto a single substrate, which reduces significantly the optical coupling losses between the elements and the packaging costs, it becomes of major importance to define the primary host material.

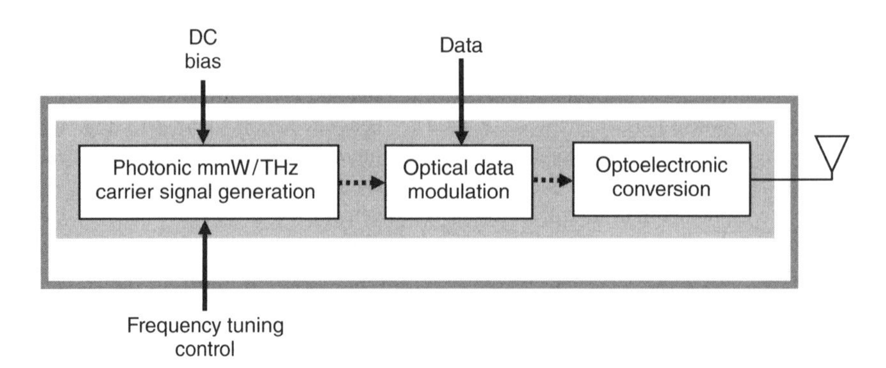

Figure 7.33 Building blocks of a photonic-based wireless transmitter, with optical carrier signal generation in the millimeter/THz wave range.

7.5.1 Hybrid or Monolithic Integrations

There are two main host materials (known also as material systems) which are available for monolithic integration of PICs, each having its own advantages and disadvantages [83]. One is based on the group of III–V materials (mainly In, P, Ga, As, and Al), which, having a direct band gap, allow efficient optical gain. The other, based on group IV materials (mainly Si, Ge, and N), does not have a direct band gap, so it can only be used for passive components. However, it has a very low propagation loss and has a high quality oxide (SiO_2) that allows one to achieve a high index contrast, which are extremely useful in reducing the size of the components.

Indium phosphide (InP) is a III–V group substrate, which is the most widely used platform for monolithic integration of photonic components. InP-based technologies are the most advanced from the huge demand for components to deploy the current optical fiber communication networks, operating at the wavelength windows for minimum-dispersion (1.3 μm) and minimum attenuation (1.5 μm). InP substrate is a good host for monolithic integration due to the wide range of different materials that can be grown. On InP substrate it is possible to grow easily materials for optical emission, guiding, modulation, amplification, and detection. Reliability of devices made on InP is also known to be good. The main drawbacks of InP substrates are that the material is very fragile and very expensive.

Gallium arsenide (GaAs) is another III–V group substrate that can be used in order to integrate THz systems. As for InP substrate, it can also be used to integrate optical emission, guiding, modulation, amplification, and detection. GaAs substrate is the best host for optical elements working in the 0.6–1 μm range and can, for this reason, be interesting for THz systems exploiting LTG GaAs. But maturity of integration technologies on GaAs, even if their development started earlier, is less than for InP due to less effort in this field. Devices developed on GaAs also face some reliability issues that require particular care for passivation steps and reduce the integration flexibility.

Silicon substrate is nowadays encountering a big success for photonic circuit integration. Fabrication of silicon PICs gains from technical development made for silicon-based microelectronics and by the large fabrication capacity. Even if silicon is not adapted for emission and detection, integration schemes of passive optical elements on silicon, associates with silica and silicon nitride, and active elements based on silicon germanium (SiGe) or on different III–V elements have been developed. These integration schemes use either hetero-epitaxy or wafer bonding techniques. The maturity of silicon-based integration is not sufficient to make a THz transceiver now but this could change rapidly as developments on this platform are very fast.

All these three material systems accept the integration of different components. All allow the processing of different passive components, such as optical waveguides, optical multiplexers and demultiplexers, phase modulators, or Bragg mirrors, or active ones such as PDs. To date, InP is the most common material in order to generate light, providing SOA, and lasers.

When an optical system composed of different elements is grown on a single chip, on any of the above material systems, a monolithic PIC is developed. However, when we aim to utilize the advantages from each material simultaneously, to develop the most convenient material for its realization, we require integrating different materials. The methods that allow us to integrate different materials are broadly known as hybrid integration techniques. Currently the main driver is to integrate InP, the main substrate material of photonics, with silicon, the main substrate material for electronics [84].

7.5.2 Monolithic Integration of Subsystems

As presented before, photonic-based THz generation requires at least a laser source and a photomixer. The laser source can be a pulse source or a dual wavelength source. The photomixer can be a photodiode, a photoconductor, or even a nonlinear frequency converter. This results in a multi-element system that can gain a lot from integration in a single integration platform through monolithic integration in a single semiconductor chip.

If the different elements of a THz emitter or a THz receiver are integrated in a single chip, the resulting system will be more compact. As all optical couplings between the elements of the system will be done in the chip, the optical losses will be lower and the packaging complexity significantly reduced. As all active elements, that often require a thermal control, are integrated in the same chip, only a single thermoelectric temperature control will have to be used which will result in a reduced packaging complexity and a lower power consumption of the system.

Monolithic integration is also a way to reduce the sensitivity to environmental fluctuation. For example, in the case of a dual wavelength source, if the two laser sources are integrated on the same chip, they will encounter the same environment-related temperature fluctuations that will result in the same optical frequency shift. As a result the frequency difference between the two optical tones will be less affected. Integration is also a way to improve phase stability between different elements: all free-space optical interfaces or optical fiber interfaces that are used in non-integrated systems induce phase fluctuation due to mechanical and thermal phenomena that will not occur in monolithic systems.

The first challenge encountered to integrate these devices is the need to develop the laser resonator without relying on the cleave facets. Several approaches have been demonstrated to integrate laser diodes on-chip:

Ring resonators: This early solution uses closed resonator structures, which offer single mode operation and lithographic control of the cavity length [85]. Unfortunately ring structures were soon demonstrated to support two counter-propagating fields which give rise to complex phenomena due to their mutual interaction. These structures require S-shape waveguides or asymmetries to be included to suppress one propagation direction [86].

Distributed feedback lasers (DFB): This is one of the most common approaches to achieve a single-mode monolithic laser structure. This laser relies on a grating structure close to the active layer of the laser, periodically varying the index of refraction, building a distributed reflector along the cavity that selects the lasing mode [45]. Therefore, to operate this structure actually requires anti-reflection coating on the facets to suppress the Fabry–Perot cavity modes.

Distributed Bragg reflectors (DBR): This is another approach that uses a grating mirror structure which, in contrast to the DFB, places the grating mirror structures in a passive material at the extremes of the active region. Some structures use one cleaved facet on one end of the cavity and a DBR on the other [87]. This technique has recently been improved demonstrating mode-locking when a surface-etched grating is used [88], with a simpler fabrication process than a DBR.

On-chip interference reflectors: This is a relatively recent approach to achieve on-chip laser sources. Here, the optical resonator is defined by multimode interference reflectors (MIRs), which have a simple fabrication technique requiring only the use of a deep etch fabrication step [89], and which have already been demonstrated to implement multiwavelength sources.

DFB and DBR laser sources for THz emission based on PICs have already been demonstrated on InP [45, 54] and on GaAs [90]. The target of these demonstrations was mainly to integrate at least two lasers for heterodyne generation or to add functionality to the device. There have also been demonstrations of photonic integrated devices for improved photodiode-based photomixers. In this case, the main advantage of integration is to combine more than one photodiode in order to increase the total power that can be emitted. Indeed a single photodiode, as the size is reduced to keep a large bandwidth, is also limited in terms of optical power handling due to heating effects. Increasing the number of photodiodes that can be coherently combined has been proven to be a good way to improve the total generated THz power [91].

There has been less demonstration of fully integrated photonic-based THz emitters, integrating both laser emitters and photomixers in the same chip. Even if this would be the ultimate solution to get cost effective power efficient emitters based on these technologies, their fabrication complexity is high, requiring mastering fabrication of high performance lasers and photomixers in a single fabrication run. There is one example of this type of fabrication that was demonstrated to be applied to high data rate wireless transmissions [92]. The chip that was realized is presented in Figure 7.34. It integrates two DFB lasers to generate two optical tones, SOA to adjust the optical power in the circuit, optical couplers, electro-optical modulators in order to modulate data on the two optical tones and high speed photodiodes to do the mixing process to generate the data-modulated THz signal. An optical output on the opposite side of the chip is used to monitor the optical spectrum. All these elements were integrated in a single InP chip.

The resulting optical spectra and the corresponding beat notes are presented in Figure 7.35. Even if only generation from 75 to 110 GHz is presented in the graphs, generation from 3 to 120 GHz was demonstrated with these chips. Output power at 120 GHz of up to −10 dBm has been obtained. This definitely demonstrates that photonic integration is a very promising solution for THz systems.

7.5.3 *Foundry Model for Integrated Systems*

One of the main problems that needs to be addressed in the field of photonics is reduction of the development costs [41]. It is very expensive to set up and run a clean-room facility to fabricate these chips. This issue was already encountered in the electronics field, where several industries with fabrication facilities, willing to make them more profitable, have given access to their fabrication platform to designers sharing the chip space among different users, giving rise to the concept of multi-project wafer (MPW) runs. This has been facilitated by the fact that there are few design tools for which the manufacturers disclose their component libraries and design rules contained within a kit, usually know as a design manual.

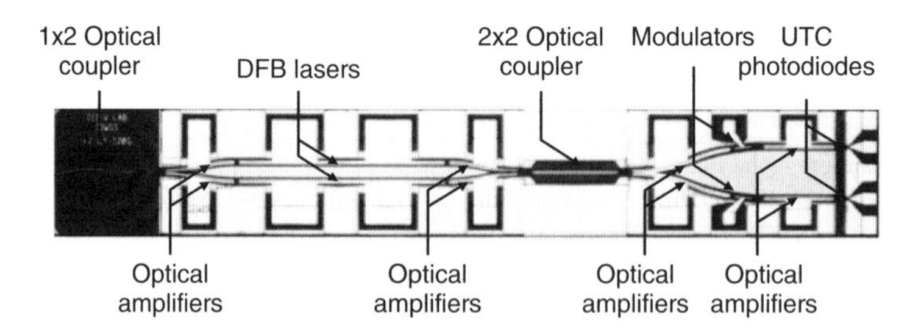

Figure 7.34 InP photonic integrated heterodyne THz generation circuit. The chip is 4.2 mm long and 750 μm wide.

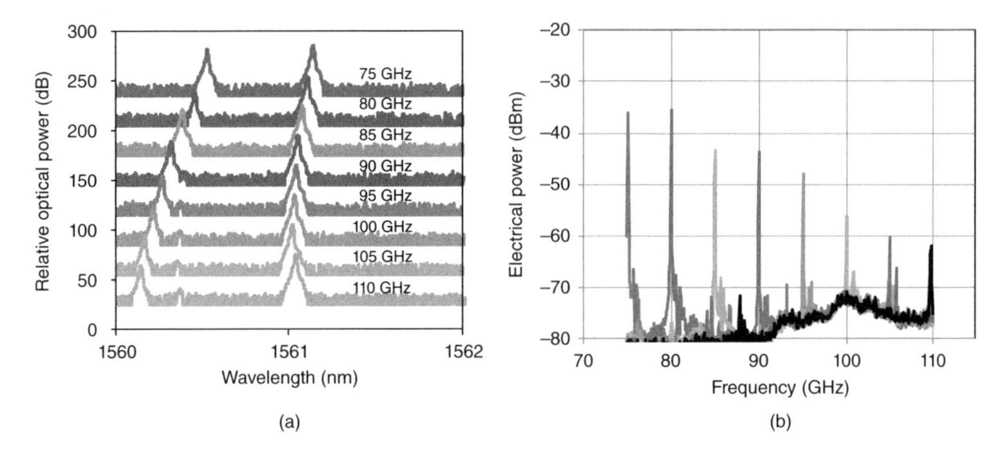

Figure 7.35 (a) Optical spectra of light generated by the chip for different bias currents applied to the DFB lasers and (b) corresponding electrical tone measured from one of the two UTC photodiodes. See plate section for color representation of this figure.

A design manual contains the description of the different building blocks available to the designer, their characteristics, as well as any constraints imposed by the fabrication process. These building blocks, which can be used for a broad range of different applications, are the first step toward a generic integration. Each industry platform provides a set of building blocks into a standard design tool. This strategy causes a dramatic reduction in the entry costs to developing custom photonic ICs into novel or improved products, bringing them within reach even of SMEs. Also, new types of SMEs have appeared whose business model is to act as design houses that give support to other industries that want to include photonic ICs in their products.

This process was started under the European Network of Excellence on Photonic Integrated Components and Circuits (ePIXnet) [93]. Within this project, InP and silicon technology platforms were created (JePPIX and ePIXfab, respectively), which are currently being transferred from a University foundry model into industrial platforms (Oclaro, UK and Fraunhofer HHI, DE), developing software design kits and standardized packaging solutions.

References

[1] Auston, D.H. and Cheung, K.P. (1985) Coherent time-domain far-infrared spectroscopy. *Journal of the Optical Society of America B*, **2**, 606.

[2] Grischkowsky, D., Keiding, S., van Exter, M., and Fattinger, C. (1990) Far -infrared time-domain spectroscopy with terahertz beams of dielectrics and semiconductors. *Journal of the Optical Society of America B*, **7**, 2006.

[3] Wu, Q. and Zhang, X.C. (1995) Free-space electro-optic sampling of terahertz beams. *Applied Physics Letters*, **67**, 3523.

[4] Verghese, S., McIntosh, K.A., Calawa, S. *et al.* (1998) Generation and detection of coherent terahertz waves using two photomixers. *Applied Physics Letters*, **73**, 3824.

[5] McIntosh, K.A., Brown, E.R., Nichols, K.B. *et al.* (1995) Terahertz photomixing with diode lasers in low temperature grown GaAs. *Applied Physics Letters*, **67**, 3844.

[6] Seeds, A.J., Fice, M.J., Balakier, K. *et al.* (2013) Coherent THz photonics. *Optics Express*, **21**, 22988.

[7] Miles, R.E., Zhang, X.C., Eisele, H., and Krotkus, A. (2007) *Terahertz Frequency Detection and Identification of Materials and Objects*, Springer, p. 341.

[8] Kourogi, M., Nakagawa, K., and Ohtsu, M. (1993) Wide-span optical frequency comb generator for accurate optical frequency difference measurement. *IEEE Journal of Quantum Electronics*, **29**, 2693.

[9] Shen, P., Gomes, N.J., Davies, P.A. *et al.* (2005) Fibre ring based optical frequency comb generator with comb line spacing tunability. LEOS Summer Topical Meetings 2005, San Diego, CA, Paper MB 2.2.

[10] Gee, S., Quinlan, F., Ozharar, S. *et al.* (2005) Optical frequency comb generation from modelocked diode lasers – Techniques and applications. LEOS Summer Topical Meetings 2005, San Diego, CA, Paper MB 2.3.

[11] Kuizenga, D.J. and Siegman, A. (1970) FM-laser operation of the Nd:YAG laser. *IEEE Journal of Quantum Electronics*, **6**, 673.

[12] Renaud, C.C., Pantouvaki, M., Gregoire, S. *et al.* (2007) A monolithic MQW InP/InGaAsP-based comb generator. *IEEE Journal of Quantum Electronics*, **43**, 998.

[13] Chou, S.Y. and Liu, M.Y. (1992) Nanoscale tera-hertz metal-semiconductor-metal photodetectors. *IEEE Journal of Quantum Electronics*, **28**, 2358.

[14] Dekorsy, T., Auer, H., Waschke, C. *et al.* (1995) Emission of submillimeter electromagnetic waves by coherent phonons. *Physical Review Letters*, **74**, 738.

[15] Zhang, X.C., Jim, Y., and Ma, X.F. (1992) Coherent measurement of THz optical rectification from electro-optic crystals. *Applied Physics Letters*, **61**, 2764.

[16] Teich, M.C. (1969) Field- theoretical treatment of photomixing. *Applied Physics Letters*, **14**, 201.

[17] Rouvalis, E., Fice, M.J., Renaud, C.C., and Seeds, A.J. (2011) Optoelectronic detection of millimetre-wave signals with travelling-wave uni-travelling carrier photodiodes. *Optics Express*, **19**, 2079.

[18] Agrawal, G.P. (2002) *Fiber-Optic Communication Systems*, 3rd edn, Wiley-Interscience, New York.

[19] TOPTICA Photonics AG http://www.toptica.com/products/terahertz_generation.html (accessed 27 November 2014).

[20] Göbel, T., Stanze, D., Globisch, B., Dietz, R.J.B., Roehle, H., and Schell, M. (2013) Telecom technology based continuous wave terahertz photomixing system with 105 decibel signal-to-noise ratio and 3.5 terahertz bandwidth. *Optics Letters*, **38**, 4197.

[21] TOPTICA Photonics AG http://www.toptica.com/products/terahertz_generation/fs_packages/teraflash.html (accessed 27 November 2014).

[22] Williams, K.A., Thompson, M.G., and White, I.H. (2004) Long-wavelength monolithic mode-locked diode lasers. *New Journal of Physics*, **6**, 179.

[23] Keller U., Herziger G., Weber H., Poprawe R.. (2007) Ultrafast solid-state lasers, *Springer Materials – The Landolt-Börnstein Database*. Springer-Verlag, pp. 33–167.

[24] Siegel, P.H. (2002) Terahertz technology. *IEEE Transactions on Microwave Theory and Techniques*, **50**, 910.

[25] Richardson, D.J., Nilsson, J., and Clarkson–, W.A. (2010) High power fiber lasers: current status and future perspectives. *Journal of the Optical Society of America B*, **27**, B63.

[26] Balzer, J.C., Schlauch, T., Klehr, A. *et al.* (2013) High peak power pulses from dispersion optimized modelocked semiconductor laser. *Electronics Letters*, **49**, 838.

[27] Arahira, S., Oshiba, S., Matsui, Y. *et al.* (1994) Terahertz-rate optical pulse generation from a passively mode-locked semiconductor laser diode. *Optics Letters*, **19**, 834.

[28] Chen, Y.K. and Wu, M.C. (1992) Monolithic colliding-pulse mode locked quantum-well lasers. *IEEE Journal of Quantum Electronics*, **28**, 2176.

[29] Shimizu, T., Ogura, I., and Yokoyama, H. (1997) 860 GHz rate asymmetric colliding pulse modelocked diode lasers. *Electronics Letters*, **33**, 1868.

[30] Yanson, D.A., Street, M.W., McDougall, S.D. *et al.* (2002) Ultrafast harmonic mode-locking of monolithic compound-cavity laser diodes incorporating photonic-bandgap reflectors. *IEEE Journal of Quantum Electronics*, **38**, 1.

[31] Moeller, L., Shen, A., Caillaud, C., and Achouche, M. (2013) Enhanced THz generation for wireless communications using short optical pulses. 38th International Conference on Infrared, Millimeter, and Terahertz Waves (IRMMW-THz), September 1–6, 2013.

[32] Nagatsuma, T., Hirata, A., Shimizu, N. *et al.* (2007) Photonic generation of millimeter and terahertz waves and its applications. Proceedings of the 19th International Conference on Applied Electromagnetics and Communications, ICECom, pp. 145–148.

[33] Brown, E. and McIntosh, K. (1995) Photomixing up to 3.8 THz in low-temperature-grown GaAs. *Applied Physics Letters*, **66**, 228.

[34] Matsuura, S., Tani, M., and Sakai, K. (1997) Generation of coherent terahertz radiation by photomixing in dipole photoconductive antennas. *Applied Physics Letters*, **70**, 559.

[35] Buus, J. (1991) *Single Frequency Semiconductor Lasers*, SPIE Optical Engineering Press.

[36] Hirata, A., Harada, M., Sato, K., and Nagatsuma, T. (2003) Low-cost millimeter-wave photonic techniques for Gigabit/s wireless link. *IEICE Transactions on Electronics*, **86** (7), 1123.

[37] Fukuda, M., Mishima, T., Nakayama, N., and Masuda, T. (2010) Temperature and current coefficients of lasing wavelength in tunable diode laser spectroscopy. *Applied Physics B: Lasers and Optics*, **100**, 377.
[38] Weiss, M., Huchard, M., Stohr, A. *et al.* (2008) 60-GHz photonic millimeter-wave link for short- to medium-range wireless transmission upto 12.5 Gb/s. *Journal of Lightwave Technology*, **26** (15), 2424.
[39] Johansson, L.A. and Seeds, A.J. (2000) Millimeter-wave modulated optical signal generation with high spectral purity and wide-locking bandwidth using a fiber-integrated optical injection phase-lock loop. *IEEE Photonics Technology Letters*, **12**, 690.
[40] Balakier, K., Fice, M.J., Ponnampalam, L. *et al.* (2014) Monolithically integrated optical phase lock loop for microwave photonics. *IEEE Journal of Lightwave Technology*, **32** (20), 3893.
[41] Smit, M., Leijtens, X., Bente, E. *et al.* (2011) Generic foundry model for InP-based photonics. *IET Optoelectronics*, **5** (5), 187.
[42] Hirata, A., Togo, H., Shimizu, N. *et al.* (2005) Low-phase noise photonic millimeter-wave generator using an AWG integrated with a 3-dB combiner. *IEICE Transactions on Electronics*, **E88-C**, 1458.
[43] Io, S., Suehiro, M., Hirata, T., and Hidaka, T. (1995) Two-longitudinal-mode laser diodes. *IEEE Photonics Technology Letters*, **7**, 959.
[44] Huang, J., Sun, C., Xiong, B., and Luo, Y. (2009) Y-branch integrated dual wavelength laser diode for microwave generation by sideband injection locking. *Optics Express*, **17**, 20727.
[45] Van Dijk, F., Accard, A., Enard, A. *et al.* (2011) Monolithic dual wavelength DFB for narrow linewidth heterodyne beat-note generation. Proceedings of the IEEE International Topical Meeting on Microwave Photonics (MWP 2011), p. 73.
[46] Fice, M.J., Rouvalis, E., van Dijk, F. *et al.* (2012) 146-GHz millimeter-wave radio-over-fiber photonic wireless transmission system. *Optics Express*, **20**, 1769.
[47] Hanfoug, R., Augustin, L.M., Barbarin, Y. *et al.* (2006) Reduced reflections from multimode interference couplers. *Electronics Letters*, **42**, 465.
[48] Smit, M.K. and van Dam, C. (1996) PHASAR-based WDM-devices: principles, design and applications. *IEEE Journal on Selected Topics in Quantum Electronics*, **2**, 236.
[49] Doerr, C.R., Zirngibl, M., and Joyner, C.H. (1995) Single longitudinal-mode stability via wave mixing in long-cavity semiconductor lasers. *IEEE Photonics Technology Letters*, **7**, 962.
[50] Carpintero, G., Rouvalis, E., Ławniczuk, K. *et al.* (2012) 95 GHz millimeter wave signal generation using an arrayed waveguide grating dual wavelength semiconductor laser. *Optics Letters*, **37**, 3657.
[51] Ondo, A.A., Torres, J., Palermo, C. *et al.* (2006) Two-mode operation of 4-sections semiconductor laser for THz generation by photomixing. *Proceedings of SPIE*, Millimeter-Wave and Terahertz Photonics, **6194**, 61940L.
[52] Acedo, P., Lamela, H., Garidel, S. *et al.* (2006) Spectral characterisation of monolithic modelocked lasers for mm-wave generation and signal processing. *Electronics Letters*, **42**, 928.
[53] Pozzi, F., De La Rue, R.M., and Sorel, M. (2006) Dual-wavelength InAlGaAs–InP laterally coupled distributed feedback laser. *IEEE Photonics Technology Letters*, **18**, 2563.
[54] Gobel, T., Stanze, D., Troppenz, U. *et al.* (2012) Integrated continuous-wave THz control unit with 1 THz tuning range. 37th International Conference on Infrared, Millimeter, and Terahertz Waves (IRMMW-THz), p. 23.
[55] Kato, K. (1999) Ultrawide-band/high- frequency photodetectors. *IEEE Transactions on Microwave Theory and Techniques*, **47**, 1265.
[56] Ito, H., Kodama, S., Muramoto, Y. *et al.* (2004) High-speed and high-output InP-InGaAs unitraveling-carrier photodiodes. *IEEE Journal on Selected Topics in Quantum Electronics*, **10**, 709.
[57] Giboney, K.S., Rodwell, M.J.W., and Bowers, J.E. (1996) Traveling-Wave photodetector design and measurements. *IEEE Journal on Selected Topics in Quantum Electronics*, **2**, 622.
[58] Rouvalis, E. (2011) Indium phosphide based photodiodes for continuous wave terahertz generation. PhD thesis. University College London.
[59] Ishibashi, T., Kodama, S., Shimizu, N., and Furuta, T. (1997) High-speed response of uni-traveling-carrier photodiodes. *Japanese Journal of Applied Physics*, **36**, 6263.
[60] Demiguel, S., Li, N., Li, X. *et al.* (2003) Very high-responsivity evanescently coupled photodiodes integrating a short planar multimode waveguide for high-speed applications. *IEEE Photonics Technology Letters*, **15**, 1761.
[61] Beling, A., Bach, H.-G., Mekonnen, G.G. *et al.* (2007) High-speed miniaturized photodiode and parallel-fed traveling-wave photodetectors based on InP. *IEEE Journal on Selected Topics in Quantum Electronics*, **13**, 15.
[62] Rouvalis, E., Renaud, C.C., Moodie, D. *et al.* (2012) Continuous wave terahertz generation from ultra-fast InP-based photodiodes. *IEEE Transactions on Microwave Theory and Techniques*, **60**, 509.

[63] Renaud, C.C., Moodie, D., Robertson, M., and Seeds, A.J. (2007) High output power at 110 GHz with a waveguide uni-travelling carrier photodiode. LEOS 2007. 20th Annual Meeting of the IEEE Lasers and Electro-Optics Society, p. 782.

[64] Renaud, C.C., Robertson, M., Rogers, D. *et al.* (2006) A high responsivity, broadband waveguide uni-travelling carrier photodiode. Proceedings of the SPIE, Photonics Europe 2006, Strasbourg.

[65] Lee, Y.-S. (2009) *Principles of Terahertz Science and Technology*, Springer, New York.

[66] Dietz, R.J.B., Globisch, B., Gerhard, M. *et al.* (2013) 64 μW pulsed terahertz emission from growth optimized InGaAs/InAlAs heterostructures with separated photoconductive and trapping regions. *Applied Physics Letters*, **103**, 061103.

[67] Dietz, R.J.B., Globisch, B., Roehle, H. *et al.* (2014) Influence and adjustment of carrier lifetimes in InGaAs/InAlAs photoconductive pulsed terahertz detectors based on: towards 6 THz bandwidth and 90dB dynamic range. *Optics Express*, **22** (16), 19411.

[68] Cooke, D.G., Jepsen, P.U., Lek, J.Y. *et al.* (2013) Picosecond dynamics of internal exciton transitions in CdSe nanorods. *Physical Review B*, **88**, 241307.

[69] Ulbricht, R., Hendry, E., Shan, J. *et al.* (2011) Carrier dynamics in semiconductors studied with time-resolved terahertz spectroscopy. *Reviews of Modern Physics*, **83**, 543.

[70] Boyd, S. (1992) *Nonlinear Optics*, Academic Press, San Diego, CA.

[71] Chen, Q., Tani, M., Jiang, Z., and Zhang, X.-C. (2001) Electro-optic transceivers for terahertz-wave applications. *Journal of the Optical Society of America B*, **18**, 823.

[72] Wu, Q. and Zhang, X.-C. (1997) Free-space electro-optics sampling of mid-infrared pulses. *Applied Physics Letters*, **71**, 1285.

[73] Chang, G., Divin, C.J., Liu, C.-H. *et al.* (2006) Power scalable compact THz system based on an ultrafast Yb-doped fiber amplifier. *Optics Express*, **14**, 7907.

[74] Nagai, M., Tanaka, K., Ohtake, H. *et al.* (2004) Generation and detection of terahertz radiation by electro-optical process in GaAs using 1.56 μm fiber laser pulses. *Applied Physics Letters*, **85**, 3974.

[75] Vicario, C., Ruchert, C., and Hauri, C.P. (2013) High field broadband THz generation in organic materials. *Journal of Modern Optics*. doi: 10.1080/09500340.2013.800242

[76] Siebert, K. (2002) Optoelectronic generation and detection of coherent cw THz radiation for imaging applications. PhD thesis. University of Frankfurt, Germany.

[77] Sigmund, J., Sydlo, C., Hartnagel, H.L. *et al.* (2005) Structure investigation of low-temperature-grown GaAsSb, a material for photoconductive terahertz antennas. *Applied Physics Letters*, **87**, 252103.

[78] Fekete, L., Němec, H., Mics, Z. *et al.* (2012) Ultrafast carrier response of Br1-irradiated In0.53Ga0.47As excited at telecommunication wavelengths. *Journal of Applied Physics*, **111**, 093721.

[79] Nagatsuma, T., Kaino, A., Hisatake, S. *et al.* (2010) Continuous-wave terahertz spectroscopy system based on photodiodes. *PIERS Online*, **6**, 390.

[80] Yariv, A. (2008) The beginnings of optoelectronic integrated circuits – A personal perspective. *Journal of Lightwave Technology*, **26** (9), 1172.

[81] Liang, D. and Bowers, J.E. (2009) Photonic integration: Si or InP substrate? *Electronics Letters*, **45**(12), 578.

[82] Nagatsuma, T., Horiguchi, S., Minamikata, Y. *et al.* (2013) Terahertz wireless communications based on photonics technologies. *Optics Express*, **21** (20), 23736.

[83] Doerr, C.R. (2013) Integrated photonic platforms for telecommunications: InP and Si. *IEICE Transactions on Electronics*, **E96-C** (7), 950.

[84] Liang, D., Roelkens, G., Baets, R., and Bowers, J.E. (2010) Hybrid integrated platforms for silicon photonics. *Materials*, **3**, 1782.

[85] Hohimer, J.P., Craft, D.C., Hadley, G.R. *et al.* (1991) Single-frequency continuous-wave operation of ring resonator laser diodes. *Applied Physics Letters*, **59** (26), 3360.

[86] Tahvili, M., Barbarin, Y., Leijtens, X. *et al.* (2011) Directional control of optical power in integrated InP/InGaAsP extended cavity mode-locked ring lasers. *Optics Letters*, **36**, 2462.

[87] Joshi, S., Chimot, N., Rosales, R. *et al.* (2013) Mode locked InAs/InP Quantum dash based DBR Laser monolithically integrated with a semiconductor optical amplifier. International Conference on Indium Phosphide and Related Materials (IPRM).

[88] Hou, L., Haji, M., and Marsh, J.H. (2013) Monolithic mode-locked laser with an integrated optical amplifier for low-noise and high-power operation. *IEEE Journal on Selected Topics in Quantum Electronics*, **19**, 1100808.

[89] Kleijn, E., Smit, M.K., and Leijtens, X.J.M. (2013) Multimode interference reflectors: a new class of components for photonic integrated circuits. *Journal of Lightwave Technology*, **31**, 3055.

[90] Zimmerman, J.W., Price, R.K., Reddy, U. *et al.* (2013) Narrow linewidth surface-etched DBR lasers: fundamental design aspects and applications. *IEEE Journal of Selected Topics in Quantum Electronics*, **19**, 1503712.
[91] Song, H.-J., Ajito, K., Wakatsuki, Y.A. *et al.* (2012) Uni-travelling-carrier photodiode module generating 300 GHz power greater than 1 mW. *IEEE Microwave and Wireless Components Letters*, **22**, 363.
[92] van Dijk, F., Kervella, G., Chtioui, M. *et al.* (2014) Integrated InP heterodyne millimeter wave transmitter. *IEEE Photonics Technology Letters*, **26**, 965.
[93] JePPIX ROADMAP (2013) The Road to a Multi-Billion Euro Market in InP-Based Integrated Photonics, http://www.jeppix.eu/document_store/JePPIXRoadmap2013.pdf (accessed 27 November 2014).

8

Selected Emerging THz Technologies

Christian Damm[1], Harald G. L. Schwefel[2,3], Florian Sedlmeir[2],
Hans Hartnagel[1], Sascha Preu[1], and Christian Weickhmann[1]
[1]*Technische Universität Darmstadt, Dept. of Electrical Engineering and Information Technology, Darmstadt, Germany*
[2]*Max Planck Institute for the Science of Light, Erlangen, Germany*
[3]*University of Otago, Department of Physics, Dunedin, New Zealand*

8.1 THz Resonators

Resonators are key components for a number of applications. Their key properties are their resonance frequencies, their size, and their losses. These losses, such as coupling, radiation, and absorption losses, determine the frequency range in which electromagnetic waves can interact with a certain resonance and determine how much power of the incident wave can be coupled into the resonator. This frequency range, also named linewidth, is inversely proportional to the photon lifetime within the cavity. The lifetime determines how much electromagnetic energy the resonator can store if driven resonantly. A long photon lifetime combined with a small footprint of the resonant system can lead to significant field enhancements of the injected fields. Absolute resonance frequencies are directly determined by the size and the dielectric properties of the resonator.

Compared to optics or electronics, resonators are by far less common in the THz domain up to now. As in optics, THz resonators mainly find their application in coherent THz sources such as gas [1, 2], solid state [3], and quantum cascade lasers [4, 5]. Depending on the particular system, one can find different resonant structures, such as external mirror based cavities, integrated Bragg mirrors, or ring resonators. Here, the resonators enhance the coherence time of the laser, decrease its lasing threshold and can enable single modal operation by mode selection.

A particularly interesting resonator type is the dielectric resonator, first described by Richtmyer in 1939 [6]. It consists of a dielectric cylinder and became of crucial importance in microwave technology in the following decades due to its small size and high mechanical and thermal stability. In the optical domain, light scattering on spherical particles was first described at the beginning of the last century by Mie [7] and Debye [8]. These resonators have in common especially high quality factors if the size of the resonator is significantly larger than the wavelength, such that the light confinement

Semiconductor Terahertz Technology: Devices and Systems at Room Temperature Operation, First Edition.
Edited by Guillermo Carpintero, Luis Enrique García Muñoz, Hans L. Hartnagel, Sascha Preu and Antti V. Räisänen.

can be described by continued total internal reflection (TIR). Already in 1961 such a dielectric resonator was used as a feedback device in one of the first solid state lasers and termed a whispering gallery mode (WGM) resonator [9]. Its huge success in the optical domain came in 1989 after Braginsky showed ultra-high quality (Q) factors and therefore strong nonlinear effects in fused silica microspheres. Due to their high Q factors in both the microwave and optical regimes, these resonators are used as very narrow band frequency filters and, still mainly in the optical domain, as highly efficient nonlinear devices.

A few years ago, people started to exploit the huge bandwidth of these resonators to combine microwaves with the optical regime to build highly efficient electro-optic modulators in materials showing high electro-optic coefficients, such as lithium niobate [10, 11] and lithium tantalate [12]. These schemes are based on coupling a microwave cavity with an optical one. Recently, there have been efforts to extend the frequency range of such a process from the microwave into the THz regime [13], see Section 8.1.5.

The main application for resonators in the microwave and THz frequency range so far is material characterization and sensing [14, 15]. As we will see in the next section, the resonance frequency and the linewidth depend strongly on the resonator's geometry and its dielectric properties. If the geometry is known with a sufficient accuracy compared to the used wavelength, it is possible to extract very precise information about the real and imaginary parts of the dielectric constant of the used material. This is particularly easy in the THz compared to the optical domain.

In all dielectric resonators the electromagnetic wave is guided in a transparent material with a high dielectric constant. Contrary to a cavity with conducting boundaries, the electric field can leak out of the cavity and – in the case of TIR – show an exponentially decaying (evanescent) field component. Such resonators are therefore called "open cavities" as they can interact with their environment. This interaction can be detrimental but it can also be ideal for sensing the environment. In all of these resonant systems a simple measurement of resonance positions and their linewidth is sufficient to determine all of the resonator's properties. These measurements are highly accurate due to the precision with which frequency measurements can be taken.

8.1.1 Principles of Resonators

If two electromagnetic waves overlap, they can interfere constructively or destructively, depending on the relative phase between them. A simple set-up where two such waves continuously interfere is a ring resonator, consisting of three perfectly reflecting mirrors placed at the edges of an equilateral triangle, each at a distance $L/3$ from each other, see Figure 8.1a. A plane wave traveling between these three mirrors would interfere constructively with itself after a round-trip if an integer number m of wavelengths fit into the optical path length $nL = m\lambda_0 = mc_0 / \upsilon_0$, where n is the index of refraction, λ_0 the vacuum wavelength, υ_0 the frequency, and c_0 the speed of light in free space. At such a wavelength, the system is said to be in resonance. In order to couple to such a resonator one of the mirrors would need to be not perfectly reflecting but semi-transparent. Then coupling can be modeled as a black box that has a complex incident (outgoing) field a_1 (b_1), respectively, which is coupled to the inside fields a_2 and b_2. Schematically, such a system can be described by a transfer matrix [16], which relates the fields by the complex coupling parameters t and r. The coupling between the field amplitudes is given by a linear system of equations:

$$\begin{pmatrix} b_1 \\ b_2 \end{pmatrix} = \begin{pmatrix} r & t^* \\ -t & r^* \end{pmatrix} \begin{pmatrix} a_1 \\ a_2 \end{pmatrix}, \tag{8.1}$$

Due to energy conservation the matrix has to be unitary, which leads to $|t|^2 + |r|^2 = 1$. If one considers a phase change θ during one round-trip and that the internal field can acquire some losses ($\alpha \in \mathbb{R}$) we can write the field a_2 as $a_2 = \alpha \exp(i\theta) b_2$. With a normalized input field $a_1 = 1$ we can solve for the intensities

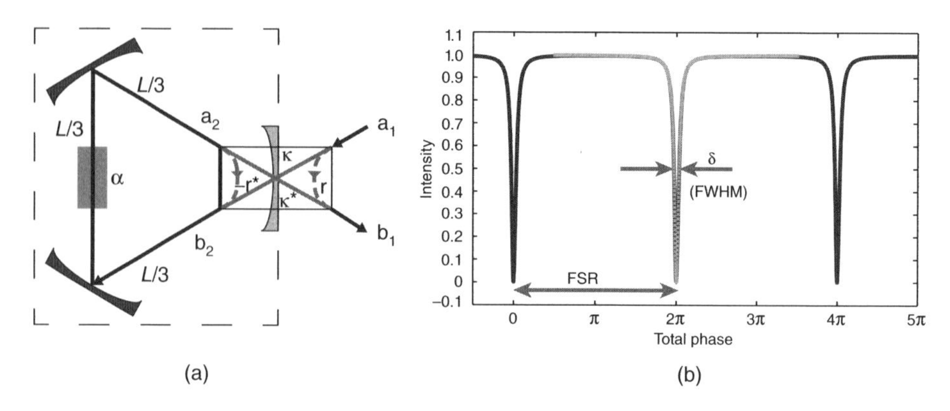

Figure 8.1 (a) Schematic of the coupling. The incoming field a_1 is related to the reflected field b_1 and the internal cavity field b_2. The coupling coefficients are r and κ. This coupling scheme can be applied to the coupling to a ring resonator of length L and some loss α. (b) Reflected intensity for a critically coupled resonator as a function of the total phase $\theta + \phi$, from Eq. (8.3). The shape for low coupling and low internal losses around a resonance is that of a Lorentzian, given by Eq. (8.4).

within the resonator

$$|b_1|^2 = \frac{\alpha^2 + |r|^2 - 2\alpha|r|\cos(\theta)}{1 + \alpha^2|r|^2 - 2\alpha|r|\cos(\theta)}, \quad \text{and} \quad |a_2|^2 = \frac{\alpha^2(1 - |r|^2)}{1 + 2\alpha^2|r|^2 - 2\alpha|r|\cos(\theta)}. \tag{8.2}$$

The system is said to be in resonance if the total phase $\theta = 2\pi l$ where l is an integer. In this case the equations simplify to

$$|b_1|^2 = \frac{(\alpha - |r|)^2}{(1 - \alpha|r|)^2} \quad \text{and} \quad |a_2|^2 = \frac{\alpha^2(1 - |r|^2)}{(1 - \alpha|r|)^2}. \tag{8.3}$$

In the picture of the ring cavity (Figure 8.1a) the phase incurred by the light in one round-trip is $\theta = 2\pi nL/\lambda_0$ which is an integer multiple of 2π if $nL = m\lambda_0$. This corresponds directly to the intuitive picture of resonances given above. If the internal losses α equal the coupling losses $\alpha = |r|$ the system in resonance is *critically coupled* and no field exits the cavity $|b_1|^2 = 0$. Every multiple of 2π phase build-up corresponds to a resonance. The distance between two such resonances in frequency space is the free spectral range (FSR), yielding $\mathrm{FSR}_\upsilon = c_0/nL$, with its inverse being the round-trip time T. The resonant frequency can therefore also be given by $\upsilon = m\Delta\upsilon_{\mathrm{FSR}}$. For low internal and coupling losses $\alpha \approx |r| \approx 1$ the reflected intensity around a resonance ω_0 is very well approximated by a Lorentzian

$$I(\omega) \sim \frac{\frac{1}{2}\delta\omega}{(\omega - \omega_0)^2 + \left(\frac{1}{2}\delta\omega\right)^2}, \tag{8.4}$$

whose linewidth, or full width at half maximum (FWHM) is given by $\delta\omega$, see Figure 8.1b. In terms of the coupling coefficients it is given by

$$\delta\omega = \frac{2|\kappa|^2 c_0}{nL}. \tag{8.5}$$

This linewidth is the inverse of the lifetime τ of a photon stored in the resonator. An important figure of merit is the quality factor Q or loss tangent $\tan\delta$, which is defined as

$$Q = \omega\tau = \frac{\omega}{\delta\omega} = \frac{1}{\tan\delta}. \tag{8.6}$$

Another important factor is the finesse, F. The finesse quantifies the loss per round-trip $F = \pi/(1 - |r|^2) = \pi/|t|^2$ and can also be considered as the average number of cavity round-trips before a photon leaves the cavity. For a critically coupled resonator the internal and external losses are equal, thus the finesse yields directly the power inside the cavity P_R to the power coupled into the cavity P_I

$$P_R = \frac{F}{2\pi} P_I. \tag{8.7}$$

Correspondingly the quality factor represents the inverse of the loss per optical cycle

$$Q = \frac{P_R}{P_I/(\omega T)} = m \cdot F. \tag{8.8}$$

where T is the round-trip time, and $P_I/(\omega T)$ is the power lost per optical cycle. We finally can relate the finesse to experimentally easily accessible quantities

$$F = \frac{Q}{m} = \frac{\Delta\upsilon_{FSR}}{\upsilon} Q = \frac{\Delta\upsilon_{FSR}}{\delta\upsilon}. \tag{8.9}$$

namely the ratio of the FSR_υ and the linewidth $\delta\upsilon$.

8.1.2 Introduction to WGM Resonators

Monolithic dielectric resonators feature a different concept of confining light. If a plane electromagnetic wave is impinging from a dielectric with refractive index n_{in} onto a dielectric with smaller refractive index $n_{out} < n_{in}$ the light can be partially refracted and reflected, see Figure 8.2a. The angles are governed by Snell's law, where the incidence angle χ with respect to the normal is related to the refraction angle β by $n_{in} \sin \chi = n_{out} \sin \beta$. It can be seen that at an angle $\chi > \chi_{crit} = \arcsin(n_{out}/n_{in})$ the refraction angle becomes complex, which means that there is no refraction possible. Starting at this point, all the light is said to undergo total internal reflection (TIR). The complex angle, and therefore the complex wavevector, corresponds to an evanescently decaying field into the optically less dense medium. As TIR only depends on the angle of incidence and the refractive indices, it is a very broadband process.

The basic concept behind a WGM resonator is that of multiple consecutive TIRs, see Figure 8.2b. In order to take these multiple consecutive reflections to the extreme we can imagine a perfectly circular dielectric. There the angle of incidence would always be conserved between consecutive reflections and thus the light would always be totally internally reflected. Introducing a curved boundary does have a small effect on the TIR. If the radius of curvature of the boundary becomes comparable to the wavelength within the resonator there is a certain amount of radiative emission [9, 17]; such loss is considered as internal loss. In a WGM resonator there are many different loss channels. They can all be lumped into the quality factor by assuming the Lambert–Beer law $I(x) = I_0 \exp(-\alpha x)$ yielding

$$Q_{max} = \frac{2\pi n}{\lambda_0 \alpha}. \tag{8.10}$$

where α is the total optical loss coefficient, with the most significant individual components given by material, radiative, surface scattering, and coupling loss, respectively

$$\alpha = \underbrace{\alpha_{material} + \alpha_{radiative} + \alpha_{surface\ scattering}}_{internal} + \underbrace{\alpha_{coupling}}_{external}. \tag{8.11}$$

In the THz domain the material losses dominate, still allowing quality factors up to 10 000, with the current materials available [18].

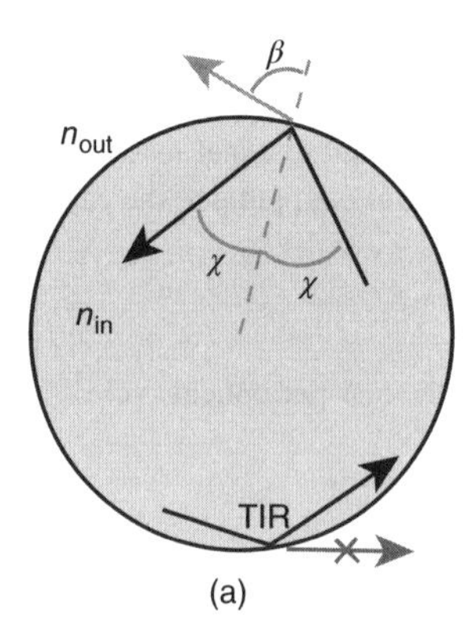

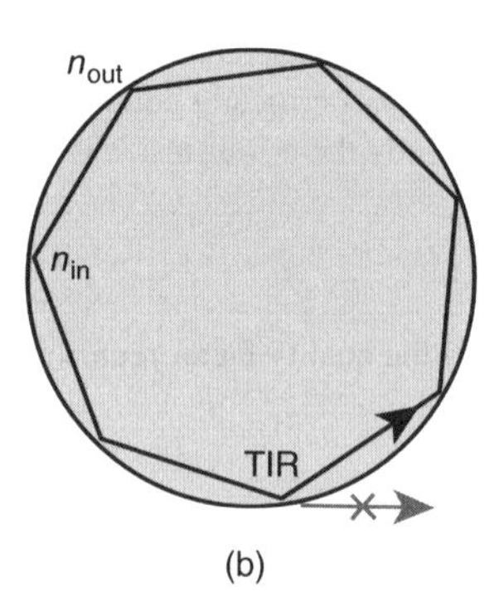

Figure 8.2 Schematic of a dielectric whispering gallery mode resonator. (a) In the ray-dynamical description, light incident at an angle χ can be either partially refracted out ($\chi < \chi_{crit}$) or totally internally reflected ($\chi > \chi_{crit}$). (b) If the phase matching conditions after a round-trip are fulfilled, a resonance can build up as described in Section 8.1.1. For low loss dielectrics, Q factors of up to 10 000 have been achieved in the THz domain, limited mainly by material losses; in the optical domain $Q \sim 10^{11}$ is possible, limited predominantly by material absorption and surface roughness.

Contrary to simple linear resonators, the modal structure of WGM resonators is more complex. In order to find an expression for the field distribution inside a WGM resonator we need to turn to Maxwell's equations and, in particular, the Helmholtz equation

$$[\nabla^2 - (nk_0)^2]\vec{E} = 0 \tag{8.12}$$

which needs to be solved with the appropriate boundary conditions. WGM resonators are azimuthally symmetric. The solutions to the Helmholtz equation in spherical coordinates with imposed Sommerfeld radiation conditions are well known [19] and result in a product of spherical Bessel functions j_m and h_m and spherical Harmonics, if a fully spherical geometry is assumed:

$$\begin{aligned} E^{<}_{lmq}(r,\theta,\varphi) &\sim j_m(n_{in}k_q r) \times P^m_l(\cos\theta)e^{im\varphi}, \\ E^{>}_{lmq}(r,\theta,\varphi) &\sim h_m(n_{out}k_q r) \times P^m_l(\cos\theta)e^{im\varphi} \end{aligned} \tag{8.13}$$

where the associate Legendre polynomials $P^m_l(\cos\theta)$ yield the spherical harmonics $Y^m_l(\theta,\varphi) = P^m_l(\cos\theta)e^{im\varphi}$. Here the integer m in front of the azimuthal angle φ gives us the number of oscillations that fit into the WGM along the equator. The associate Legendre polynomials provide with $p = l - |m|$ the number of nodes in the polar direction and the Bessel-function $j_m(n_{in}k_q r)$ provides the number of q field maxima within the radial direction of the WGM (for an illustration see Figure 8.3a). Differences arising between a spherical and a disk geometry can be considered by imposing a different local coordinate system [20]. For THz WGM resonators, where the size of the resonator is close to that of the wavelength, fully analytic methods provide useful understanding but are not sufficient as the field interacts strongly with the three-dimensional geometry of the set-up. In order to investigate disk or ring structures, fully numerical methods, such as finite element methods, can be utilized [21, 22]. They become particularly efficient if azimuthal geometries are imposed [23] (see Figure 8.3b). In the next section we will consider coupling into such a resonator. Especially in the THz domain where the coupler is often similar in size to the wavelength, the coupler perturbs the system significantly such that these effects can only be considered in a full 3D calculation (see Figure 8.3c).

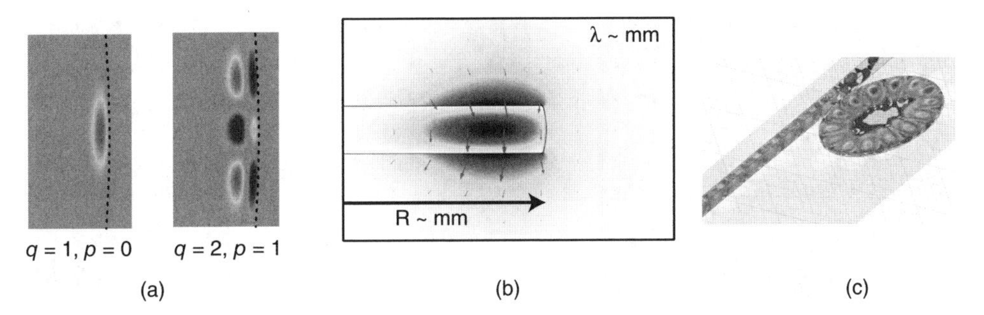

Figure 8.3 Modal distributions in a whispering gallery mode resonator. (a) Analytical solutions based on an expansion in spherical harmonics for a spherical geometry. (Adapted and reproduced from Ref. [20] with permission of the Optical Society of America © 2013.) (b) Numerical solutions based on finite element methods. (c) Solutions based on a full 3D finite difference time domain which also include the coupling waveguide. See plate section for color representation of this figure.

8.1.3 Evanescent Waveguide Coupling to WGMs

Coupling light into a dielectric resonator can be achieved by a number of different means. Two factors are of importance, the ideality of the coupling and the criticality. The former specifies the ratio of the power coupled into the desired mode, the latter the energy exchange rate between the coupled light and the resonator mode at resonance (see Section 8.1.1). The simplest approach is free-beam coupling. If the main loss mechanism of the resonator is radiative, then by time reversing the field one can couple light into the resonator directly from the far-field. This method is efficient only for rather leaky cavities or for specially designed asymmetric microcavities using the reverse directional emission direction [24, 25], see also Section 8.1.4.2. In general, ideality and criticality are low for free-space coupling. Therefore, we focus on near-field coupling methods, as they are the method of choice for high Q WGM resonators.

Two basic near-field approaches can be employed to couple radiation into high Q resonators. The basic idea is to introduce an evanescent field close to the WGM resonator, which is comparable to the evanescent field leaking out of the WGM resonator. We have seen in Section 8.1.1 that if the fields a_1 and a_2 are appropriately out of phase the emitted field b_1 can vanish. This condition can be achieved if the phase velocity of the radiation within the coupling device corresponds to that of the resonant mode in the WGM. Usually such phase-matching is considered in terms of a matching of the effective refractive index n_{eff}.

Prism coupling with frustrated TIR is among the oldest approaches used in the optical domain [26]. An incident beam of radiation is focused inside the high-index coupling prism and the angle of incidence is adjusted in such a way that phase matching between the WGM and the evanescent TIR light from the prism is achieved. The gap between the prism and the resonator is optimized to obtain critical coupling [26–28]. In the THz domain this method would also work but is effectively limited by the required focusing optics and the prism size.

Another more promising near-field coupling approach regularly used in coupling THz WGM resonators is waveguide coupling. In the optical domain this corresponds usually to optical fibers, which are tapered down close-to and below the wavelength of the light, such that an evanescent field manifests around the vicinity of the taper. Such structures are very fragile but in the THz domain they can be readily implemented as tapered polymer or fused silica rods (see Figure 8.4a). These low index amorphous waveguides have been successfully used to couple to low index Teflon $n = 1.43$, polyethylene $n = 1.53$, and silica $n = 1.95$ resonators. For high refractive index materials, such as birefringent sapphire $n_o = 3.0$, $n_e = 3.4$ (at 100 GHz), waveguides with higher bulk refractive index are necessary. Gallium arsenide

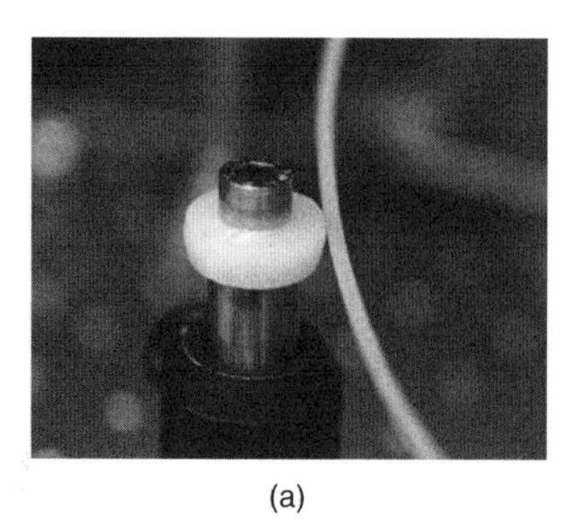

(a)

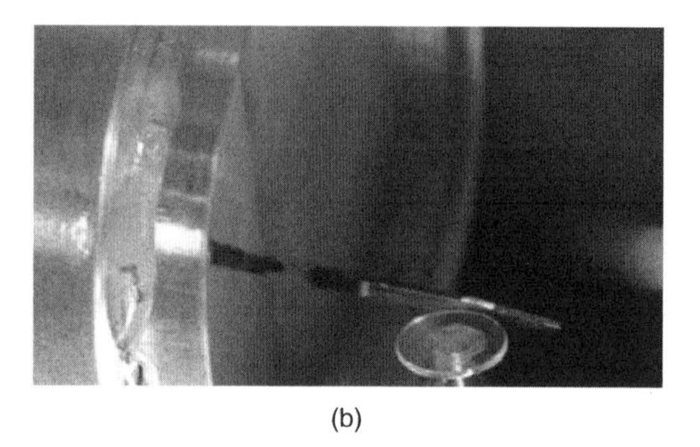

(b)

Figure 8.4 Two different dielectric waveguides for THz radiation. (a) Tapered Teflon waveguide for coupling to low refractive dielectric disk resonators. (b) Gallium arsenide waveguide coupling to a high index crystalline WGM resonator.

dielectric rod waveguides (DRWs) (see Figure 8.4b) have been used to efficiently couple low frequency THz radiation into a sapphire WGM resonator [18] (see Figure 8.5a for the set-up). A typical spectrum of such a high-Q ($> 10^4$) resonator is shown in Figure 8.5b, where two different non-degenerate mode families are excited.

8.1.4 Resonant Scattering in WGM Resonators

Up to now we have only considered the interaction of a single resonator with a non-resonant waveguide, and considered scatterers as sources for loss within the resonator. If the interaction with such a scatterer is, however, strong, both systems can interact coherently and form a combined resonance. The interaction

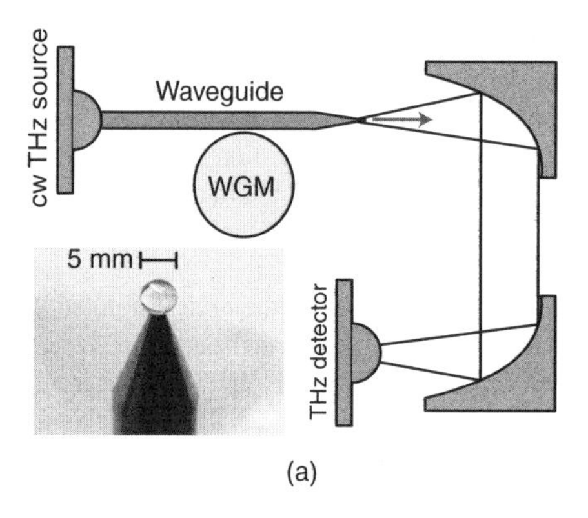

(a)

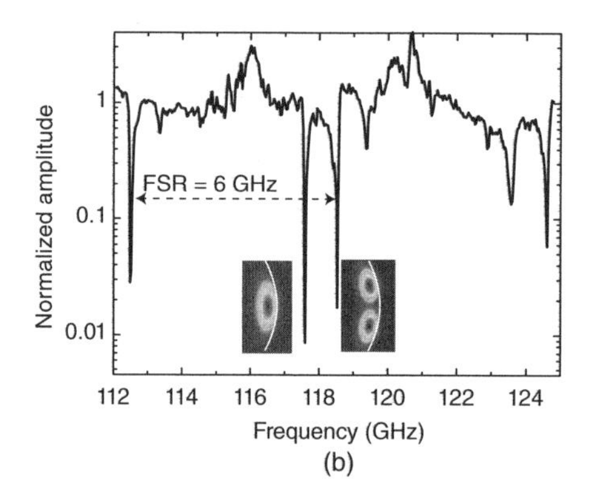

(b)

Figure 8.5 Waveguide coupling of a crystalline DRW to a WGM resonator. (a) The schematic of the coupling set-up shows the GaAs waveguide directly attached to a photomixer chip. Via the waveguide, the radiation can efficiently couple to the sapphire resonator, which is shown in the inset photograph. These crystalline waveguides can also act as antennas. (b) A transmission spectrum of the waveguide coupled to a sapphire sphere. Resonance dips can be observed when the power is resonantly coupled to the sphere and dissipated in it. Two different, non-degenerated, mode families are excited, their simulated intensity distributions are shown in the insets.

of two identical resonators with each other can be easily modeled and is similar to the hydrogen molecule and thus they are often called *photonic molecules* [29]. In the optical domain, it is very challenging to fabricate two resonators that show exactly the same spectral response as their geometries have to be identical on a sub-wavelength scale. In the THz frequency domain this challenge is feasible.

8.1.4.1 Coupled Spectrally Identical WGM Resonators

In a fundamental picture, resonances of a WGM resonator and atoms can be compared with each other. Both systems are rotationally symmetric and feature a non-equidistant spectrum. If two atoms are brought into proximity they can form a molecule if their combined wavefunction is of a binding type. Two interacting identical WGM resonators will also form a combined resonance and their degenerate modes are split, see Figure 8.6a. Experimentally it was shown that it is possible to fabricate multiple dielectric disks made out of high density polyethylene (HDPE) that show identical spectral response in the THz domain [31], see Figure 8.6b,c.

The coupled disk spectrum shows mode splitting as expected (see Figure 8.6c). However, there is a small asymmetry between the bonding and antibonding modes with respect to the Q-factor [32, 33]. Most resonances show antibonding modes with a higher Q-factor. A similar effect was found in the microwave range [34] and was described as an interplay between mode splitting and absorption. Recently, similar effects were shown in coupled semiconductor resonators [35].

8.1.4.2 WGM Resonators with Finite Scatterer

Unperturbed WGM resonators are fully symmetric and therefore do not allow efficient directional free-space coupling. Thus, several different schemes have been developed to induce directional emission,

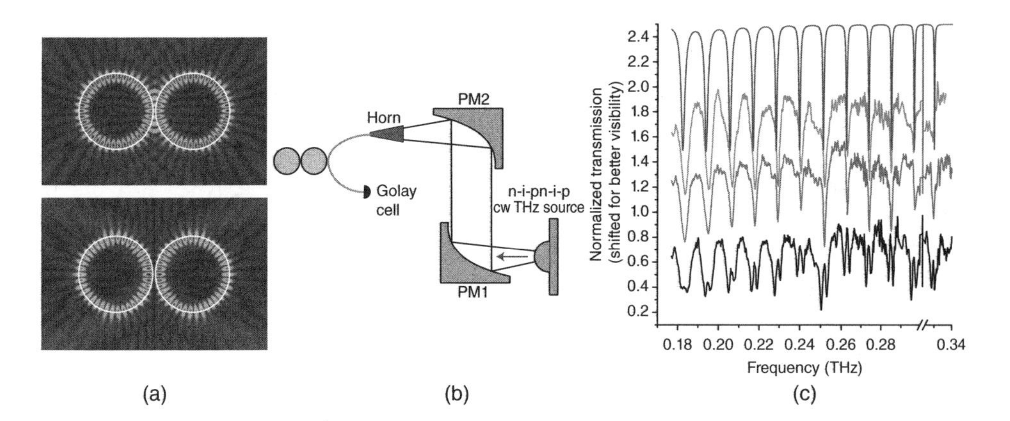

Figure 8.6 (a) When two spectrally identical disk resonators come within close proximity, they couple strongly and show a mode splitting similar to that of a hydrogen molecule. The field intensity distribution in two such coupled dielectric resonators is shown, calculated via a fast multipole method in 2D [30]. (b) Schematic set-up: THz light generated by beating two detuned lasers onto a n-i-pn-i-p photomixer and focused through two parabolic mirrors onto a horn and into a Teflon waveguide. The transmittance is measured. (c) Transmission spectra of two spectrally identical polyethylene WGM resonators (middle two traces). The bottom curve shows the transmission of both resonators in close proximity, showing clearly the mode splitting [31]. In the top trace data of a numerical simulation are shown. The individual traces are vertically shifted for clarity. (Reproduced from Ref. [31] with permission of the Optical Society of America © 2008.) See plate section for color representation of this figure.

as this would allow effective free-space coupling due to time inversion symmetry. Smoothly deforming the boundary can change the optical mode from a regular to a wave-chaotic behavior. This can keep the Q factors high, but at the same time allows directional emission [36–42].

Another way to achieve directional emission is to introduce a finite scatterer within the resonator's mode volume. Theoretical calculations initiated by Wiersig *et al.* [43–46] already described efficient outcoupling of a high Q mode via interaction with a low Q mode without ruining the Q factor. Besides directional emission at high Q factors, finite scatterers may further assist to improve the threshold behavior of WGM-based lasers [47]: they can be used to perturb non-lasing modes which suppresses spontaneous emission. Symmetrically spaced perturbations on the rim or above the disk can act as gratings and efficiently couple light out of, for example, THz quantum cascade laser resonators [5], and also provide a novel way to generate optical angular momentum beams [48]. Scatterers therefore present an excellent alternative to improve both the directionality and lasing performance of WGM-based lasers.

A recent experiment with polymer THz WGM resonators showed that inducing a small hole within the mode volume can drastically change the emission direction [17] while keeping the Q-factor high, see Figure 8.7.

8.1.4.3 Sensing Applications

The properties of dielectric resonators make them powerful tools for material characterization and sensing applications. As the major part of the electromagnetic field travels inside the dielectric, the spectral position and the linewidth of a particular resonance depend strongly on the real and imaginary part of the dielectric constant of that material. In the microwave regime, a dielectric sandwiched between

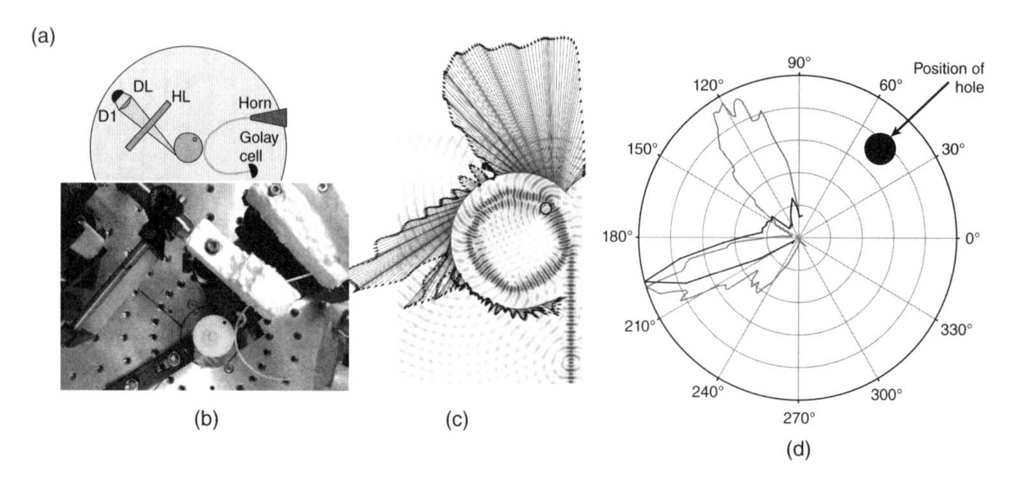

Figure 8.7 (a) A resonator is mounted at a fixed position in the center of a rotation stage. The detector (Golay cell, D1) scans the angular power distribution of the far-field of the resonator with an angular resolution of about 10°. (b) Photograph of resonator and waveguide. The waveguide is bent to improve coupling. (c) Result of the Poynting vector analysis of the resonator with hole. (d) Experimental data (black dotted) as well as far-field radiation pattern predicted from the theoretical Poynting-analysis on a linear scale. While the first main lobe predicted from the theoretical calculation around 120° is missing in the experimental data, experiment and theory agree well for the second lobe at 200°, most likely due to the slightly curved waveguide in the experiment. (Adapted and reproduced from Ref [17] with permission of the Optical Society of America © 2013.) See plate section for color representation of this figure.

two conductive metal plates, a so-called "post resonator," is a common device for material characterization. If one of the loss constants, either that of the metal or the dielectric, is known with certain accuracy, the other can be extracted from the linewidth of the resonances [49–51]. While this technique is mainly used in the microwave domain, first attempts have been made to extend it into the THz regime [14].

In optics, people are more interested in actual sensing applications, exploiting the small field portion leaking out of the WGM resonator evanescently. A typical measurement scheme consists of a fused silica microsphere, which is immersed in water. As soon as the refractive index of the water changes, the resonance spectrum of the resonator will change and thus give a measure of the refractive index of the surrounding [52].

8.1.5 Nonlinear Interactions in WGM

If the amplitude of an electromagnetic wave in a dielectric is sufficiently strong, the dielectric response of the medium can no longer be considered linear and the dielectric constant becomes a function of the electric field strength. Such a nonlinear response allows interaction of fields of different frequencies with each other. Nonlinear frequency conversion of far-infrared or microwave signals into the optical domain has been actively used for detection of such signals [13, 53–60]. The relative ease of optical detection compared to, for example, detection in the THz or sub-THz range, in combination with an intrinsically noiseless character of nonlinear frequency conversion, explains the close attention this method has been receiving. Its main drawback, however, is its low conversion efficiency. The highest power conversion efficiency known to us is about 0.5%, which has been recently achieved [13] for a 100 GHz signal using 16 mW of continuous-wave (CW) optical pump at 1560 nm. This number corresponds to the photon-number conversion efficiency of approximately 2.6×10^{-6}, following the Manley–Rowe relation. One of the main reasons for the low efficiency is that most transparent materials show only a very weak nonlinear response and therefore such interactions usually require very high field intensities. Especially in the CW THz regime, these high field intensities are difficult to achieve. Furthermore, in such experiments a good field overlap between microwave and optical modes is difficult to achieve due to the very different wavelengths.

WGM resonators are the ideal platform to address this issue. They feature very small modal volumes and therefore enhance the local field strength significantly. Furthermore, they can be fabricated out of the best available nonlinear materials and provide long interaction length within the nonlinear medium as the fields are almost fully confined within. Second order nonlinear processes are orders of magnitude stronger than higher order processes, however, they can only occur in certain materials. Only crystals which do not possess inversion symmetry can show such nonlinear $\chi^{(2)}$ response. Examples are crystals such as lithium niobate ($LiNbO_3$) and single crystal quartz. It is possible to fabricate high quality WGM resonators from these crystals [11, 61]. Both materials are also (semi-)transparent in the THz domain. Due to the dispersionless nature of TIR, such a resonator can support WGMs of both frequency domains simultaneously and hence allows nonlinear interaction between them.

All resonant, highly efficient and noiseless upconversion of microwave and especially THz radiation into the optical domain would open up numerous possibilities in microwave imaging and communications. Reaching the photon-counting regime in the sub-THz or THz range would be an important achievement for quantum information processing and computing, sub-millimeter spectroscopy and astronomy, security, and for other areas where highly sensitive detection of microwave radiation is desired. It has been proposed [13, 62] that using a lithium niobate ring resonator supporting WGM type modes in the low THz regime can allow unity conversion efficiency into the optical regime with moderate pump powers. The basic idea of such a scheme is shown in Figure 8.8.

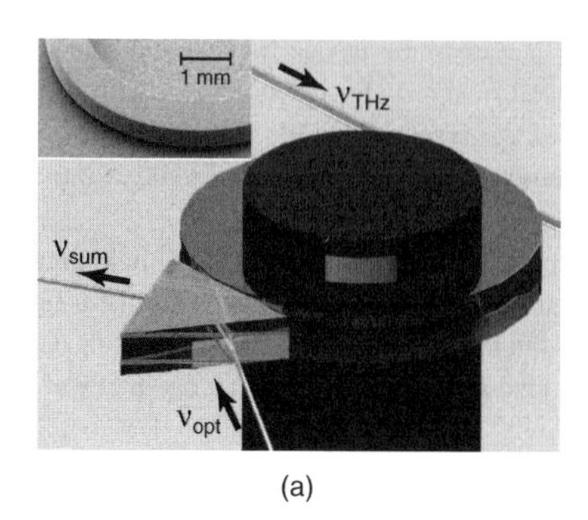

(a)

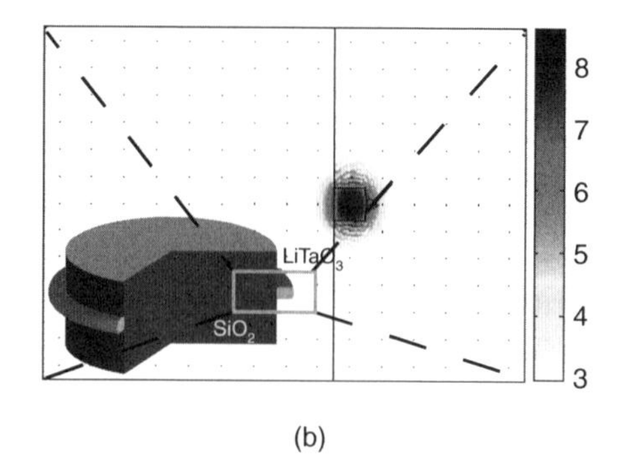

(b)

Figure 8.8 (a) Schematic of the proposed THz detector set-up. A ring resonator fabricated out of an optical $\chi^{(2)}$ material, in which the THz υ_{THz}, optical pump υ_{opt}, as well as the optical signal υ_{sum}, are resonant. If the phase-matching conditions are fulfilled single THz photon sensitivity in the υ_{sum} signal should be achievable at moderate pump powers and Q value requirements. The inset shows an SEM picture of a ring fabricated from $LiNbO_3$, polished to optical precision. (b) Good spatial modal overlap of the THz and optical modes is required in order to achieve strong nonlinear interactions. Contrary to a sphere resonator, a ring resonator that "pushes" the THz mode toward the outside cannot be analytically solved. Numerical simulation of the THz field distribution, here in a $LiTaO_3$ ring resonator, is necessary and shows that good modal overlap with the optical modes, which travel within 1 μm of the rim is possible. The simulations also provide the dispersion relation required in order to find the phase-matching condition.

8.2 Liquid Crystals

8.2.1 Introduction

Liquid crystals (LCs) have been used successfully in optics and, recently, in the microwave domain. The remarkable combination of liquidity and anisotropy is unique to this class of materials. It has since been discovered that a number of LC blends tend to conserve their good properties – in particular low loss and high anisotropy – far into the terahertz range [63–66].

In the following section the concepts and the technical application of LCs in microwave, millimeter-wave and THz devices will be introduced. A short discussion will provide the reader with the basic concepts of this unique material and literature for further reading. Nevertheless, this section will focus on the particular results in the field needed to understand both characterization and application of LC in devices and, in particular, toward the terahertz regime.

8.2.1.1 History

LCs were first described in 1888 by Friedrich Reinitzer while experimenting with benzoic acid and its cholesterol derivates. The phenomenon was then first described theoretically by Otto Lehmann shortly after Reinitzer's discovery.

For another 80 years, with the exception of the time during the wars, the phenomenon was studied continuously by various groups but remained rather a niche field. After 1960, research intensified with the first liquid crystal displays (LCDs) discussed in the early 1970s [67].

With the commercialization and broad availability of LCs and LCDs in the 1990s, interest in other fields of application grew. In the late 1990s and early 2000s, several groups started to study LCs in the field of microwaves and millimeter-waves. They discovered that it was possible to design molecules (singlets) and blends (mixtures thereof) with low losses and high anisotropy in the microwave range. Finally, in the mid-2000s Merck KGaA commercialized the first LCs for microwave and millimeter-wave applications.

8.2.1.2 Typical Liquid Crystals

Today, an enormous number of LC compounds are known, but only a few are commonly used. Typical LCs available off the shelf are blends of several singlet compounds. A recurring compound is 4-cyano-4′-pentylbiphenyl (5CB) which belongs to a wider family of compounds, the cyanobiphenyls (n-CB and n-OCB). Data for 5CB is widely available, see for instance [64, 68].

A commercially available blend is for instance K15 which has been used in many publications and referred to as the "fruit fly" of LCs. It is actually an LC for display applications, offers good tuning speed and a fair tunability, but the losses in the microwave and millimeter-wave range are very high.

There is a great variety of molecules and blends exhibiting numerous phases. For the sake of clarity we focus on calamitic nematic LCs in this section, as they are the most widely used group of LCs for our purpose. Most of the methods and equations shown, however, do generally also describe other types of nematic LCs.

8.2.1.3 Physical Properties of LCs

The physical properties of LCs in general are a rather complex topic for which the reader is advised to refer to Ref. [69]. We will therefore introduce a number of simplifications. First, LCs will be treated as ordinary homogeneous dielectrics (anisotropic nonetheless, although we will see that even this can be simplified) which holds for a limited temperature range for a given LC mixture.

In order to understand the limitations of this approach, one has to keep in mind the general properties of dielectrics over a wide frequency range.

The dielectric permittivity ε describes the macroscopic interaction of matter and electromagnetic waves, more specifically the interaction of dipoles (i.e., the microscopically inhomogeneous distribution of charge density on atoms or molecules) within the material and the electric component of the wave.

The most common and perhaps most intuitive model for this behavior is the mass-spring model where the electric field exerts a force on the electric charge which in turn moves a coupled mass (e.g., the molecule or parts of it). At its resonance frequency (i.e., when action and reaction take place in a 180° phase relation) the process stores (and for several reasons dissipates) a maximum of energy. Far below this frequency, the process reacts to the field but mostly in-phase and thus with low dissipation of energy. It does, on the other hand, influence the electric field and thus the propagation of the electromagnetic wave. Far above this frequency, it is too slow to react to the exciting field and therefore does not constitute an effect.

This is referred to as dielectric relaxation and described by various relaxation response expressions, most prominently the Debye relaxation but other responses are also used for dielectrics [70]. It is defined as

$$\varepsilon(\omega) = \varepsilon_\infty + \frac{\Delta\varepsilon}{1 + \mathrm{i}\omega\tau} \tag{8.14}$$

with ε_∞ being the limit permittivity above the resonance frequency and $\varepsilon_\infty + \Delta\varepsilon$ the limit permittivity below the resonance frequency. τ is called the relaxation time.

There are various ways of looking at this expression. In microwave engineering, dielectric properties are described by real part permittivity and the loss angle

$$\tan\delta = \frac{\mathrm{Im}(\varepsilon)}{\mathrm{Re}(\varepsilon)} \tag{8.15}$$

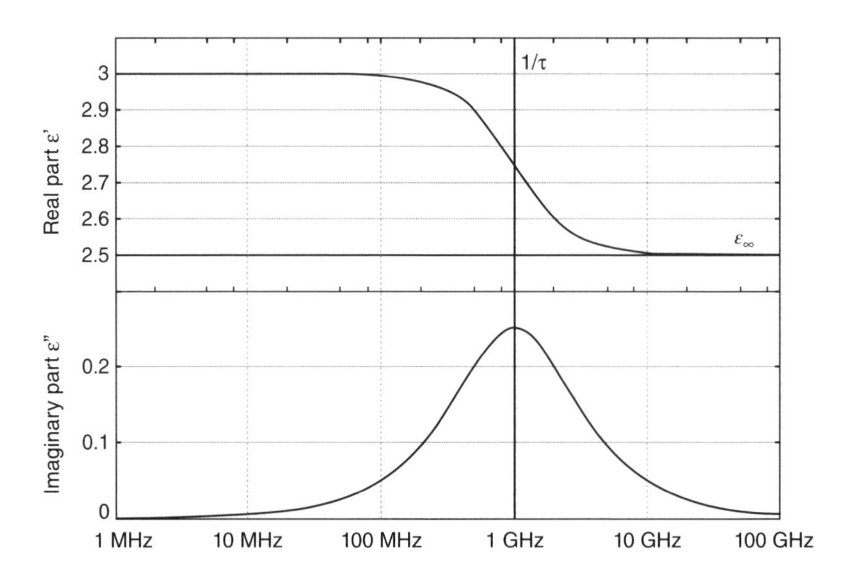

Figure 8.9 Example of a dielectric with a relaxation frequency at 1 GHz.

We can therefore look at an example dispersion curve (using $\varepsilon_{\infty} = 2.5$, $\Delta\varepsilon = 0.5$ and $1/\tau = 1$ GHz) which is shown in Figure 8.9. Obviously, the loss angle is at maximum at the resonance frequency. About one decade below and above it, it has almost vanished. Another observation is that the slope of the permittivity is near zero when the resonance frequency is sufficiently far from the region we are interested in.

Several relaxation processes can be present in one material. Typically, LCs show very high permittivity values near DC (i.e., in the range of 1 kHz). LCs for optical application will have low losses (or virtually none at all) in the range of 1.5 μm wavelength (which corresponds to 200 THz in frequency), but usually show rather high losses in the GHz and low THz range. This is because the imaginary part of the permittivity for relaxation processes extends far into the microwave range. However, it has been shown in the past by Merck KGaA, that LCs can be tailored to show low losses even in this frequency range.

A high DC permittivity, however, is of great use for tuning purposes, in this case losses do not play an important part.

It can be concluded that, if properly designed, LC mixtures can exhibit a near flat permittivity in a given frequency range and equally have a very low loss angle in the range of $\tan\delta = 10^{-3}$ which is comparable to the majority of low-loss polymers and even better than many glasses. Finally, a high DC permittivity enhances tunability.

8.2.1.4 Material Phases

In the previous section the properties of common dielectrics were discussed. At a fixed temperature and pressure (and orientation, but we will come to this later) this also holds true for LCs. Let us now look at a fixed frequency and change the temperature (but not orientation) for a given LC mixture.

Under atmospheric pressure, many materials exhibit three well known phases when transiting from low to high temperatures. So, while increasing temperature one would expect the material to take two transitions: Starting in the solid state, a melting point and further a boiling point are expected (Figure 8.10).

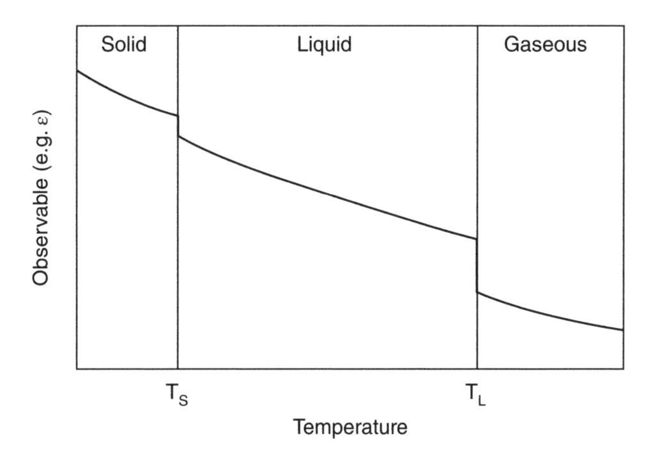

Figure 8.10 Schematic dependence of an observable, such as the real part of the permittivity ε, on temperature for a material exhibiting three phases – solid, liquid, and gaseous.

In the case of LCs this is different. Between solid and liquid one finds at least one additional transition. For the LC BL111, the first occurs at the melting point, where the formerly crystalline material becomes a milky liquid. It has oil-like texture and shows perturbations visible to the bare eye, referred to as *schlieren.*[1] At the second transition it remains liquid but loses its milky appearance and becomes clear. When lowering the temperature again, the process reverses with the transitions not necessarily at the same temperature (hysteretic effect). Many LCs exhibit such hysteretic behavior.

Other types of LC change from clear liquid through milky liquid to a wax-like state before they start to show crystalline features. They are commonly referred to as smectic LCs. Care needs to be taken since there are several smectic phases, the reader is advised to refer to deGennes' work [69]. The properties of such types of LCs are not the subject of this book.

The given measurements can be carried out applying a magnetic field to the sample. Depending on its orientation with respect to the electric component of the radio frequency (RF) field, the results look entirely different in a given temperature range. This is shown in Figure 8.11 where two orientations have been schematically superimposed. In the solid, that is, crystalline phase, due to a high number of defects the material is poly-crystalline and solid, which leads to a random grain orientation and thus quasi-isotropic behavior. The bias field has little impact here.

At the melting point T_{m}, the permittivity changes according to the orientation of the magnetic bias field with respect to the incident electric RF field. This is the nematic phase with a high permittivity and typically low losses in one direction (ordinary axis, the magnetic field is aligned with the RF E-field) and low permittivity and high losses in the other direction (magnetic bias field orthogonal to RF E-field).

The transition from nematic to isotropic takes place at T_{ni}. Here, the molecular order of the material collapses and both orientations show the same permittivity. This transition is reversible, although effects such as supercooling have been observed that cause a shift in T_{ni} when the temperature is varied from high to low values.

At very high temperatures (or low atmospheric pressures), LCs may go into a gaseous phase. This, however, is rather a question of handling problems when LC devices are designed for instance for space applications and will therefore not be considered here.

[1] From German "Schliere", meaning streak or flow mark.

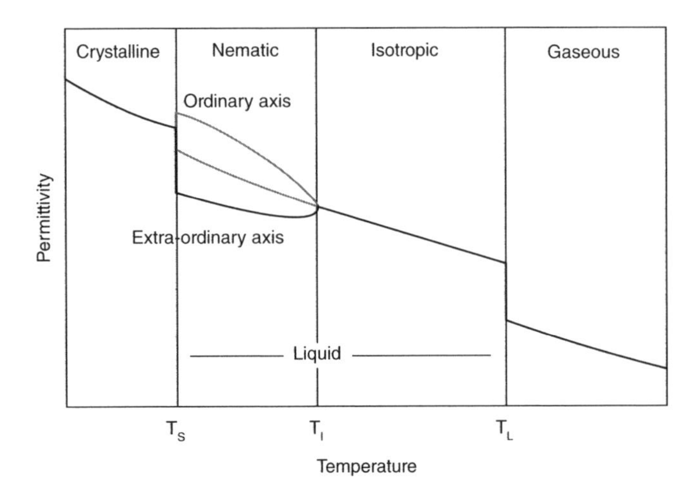

Figure 8.11 Schematic dependence of the permittivity of a liquid crystal material on temperature. In the nematic phase the effective permittivity depends on the orientation of the LC, that is, the material is anisotropic.

8.2.1.5 Description of Anisotropic Properties

As discussed before, a simple experiment illustrated the anisotropic behavior of LCs. In this section, a commonly accepted approach to model this behavior will be presented: the Frank–Zocher–Oseen continuum theory.

Single LC molecules exhibit polarizability which depends on the direction of the polarizing field. The LCs we consider here always have one optical axis along which polarizability is usually large; orthogonal to this axis, the polarizability is significantly lower.

In the isotropic phase, molecule movements and arrangement are sufficiently uncorrelated that while for the single molecule anisotropy still holds, the whole ensemble appears isotropic – hence the name of the phase.

Below a certain temperature, called the *clearing point* T_{ni} (subscript ni for transition from nematic to isotropic), a molecule is able to influence the alignment of other molecules in its vicinity. Neighboring molecules' orientations become correlated and a given volume can now show anisotropic behavior. Bear in mind that this does not mean that *all molecules are aligned in parallel* but rather *there is a common direction*, called the director, which can be seen as the average orientation of all molecules in the vicinity. Associated with this is an order parameter S. It is zero in the isotropic phase and becomes 1 for a fully ordered volume, that is, when all molecules have the same orientation. It is related to the distribution of orientations via the integral

$$S = \int f(\theta) \cdot \frac{1}{2}(3\cos^2\theta - 1)\, \mathrm{d}\Omega \tag{8.16}$$

where θ is the angle of a single molecule's axis with respect to the director, $f(\theta)$ the probability of a molecule having an orientation θ, and the integration is carried out over a full sphere $\mathrm{d}\Omega$.

There are several approaches to determining S that are discussed in detail in Ref. [69]. The order parameter S will only be used symbolically here in order to explain the two main characteristics of LC switching: phase transition and the change of anisotropy with temperature.

Assuming an order parameter of $S = 1$ was possible. This means all molecules are oriented along the director $\vec{n}$. Therefore, maximum polarizability and hence permittivity is obtained along the director. Orthogonally, the perpendicular permittivity value is obtained. The dielectric tensor of the material therefore is

$$\hat{\varepsilon} = \begin{pmatrix} \varepsilon_\perp & 0 & 0 \\ 0 & \varepsilon_\perp & 0 \\ 0 & 0 & \varepsilon_\parallel \end{pmatrix} \tag{8.17}$$

with $\varepsilon_\perp$ and $\varepsilon_\parallel$ related to the polarizability of the perfectly aligned molecules, hence the maximum respective permittivities.

If the order parameter decreases (with rising temperature or due to local field effects), the expression for the effective anisotropic dielectric tensor still holds, though now with values $\varepsilon'_\perp > \varepsilon_\perp$ and $\varepsilon'_\parallel < \varepsilon_\parallel$, that is, closer to a value in between. There are, in general, models that describe the behavior of S, but for the purpose of application it is a matter of characterization to obtain the effective values. The effect of the ordering parameter is clearly visible in Figure 8.12. The measured values for the effective permittivities in parallel and perpendicular to the director converge on each other with increasing temperature. The anisotropy then collapses at the phase transition temperature T_{ni} at the edge of the isotropic phase.

It should be kept in mind that S is foremost a function of temperature but not of electric or magnetic fields. Measurements show that high bias fields can speed up the orientation process but they cannot increase anisotropy. For weak bias fields only long-range disorder can cause lower effective anisotropy.

8.2.1.6 Mechanical and Dynamic Properties

In nematic LCs there is a short-range order, which was previously introduced and described by the LC director. The correlation length associated with this short-range order depends on several dimensions but is usually determined mainly by temperature.

It is in the range of micrometers. External (electric or magnetic) fields and wall effects can have an effect. Their main influence though lies in the long-range order. The director field itself can vary significantly over longer distances, that is, the LC orientation can nevertheless be a function of space.

A central concept for microwave and THz LC devices is the mechanics of this director field $\vec{n}$ as a function of biasing fields, intrinsic interactions, and boundary conditions, called the director dynamics.

In order to calculate the director field $\vec{n}$ of a macroscopic volume of LC, the concept of Landau free energy is used. The problem consists in minimizing the free energy f defined by its three

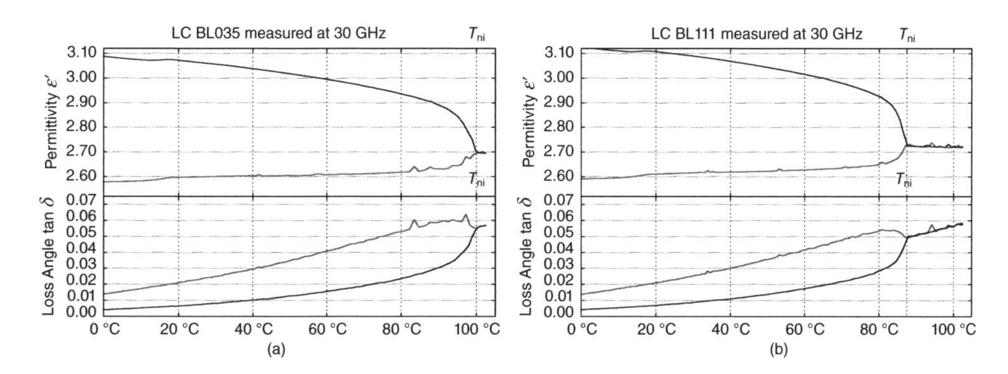

Figure 8.12 Permittivity and loss angle for two common LC blends ((a) BL035 and (b) BL111). Note the different clearing points T_{ni}.

components for the steady state – elastic free energy, surface (or interface) free energy, and electric field energy:

$$f = f_{\text{elast}} + f_{\text{surf}} + f_{\text{field}} \tag{8.18}$$

$$f_{\text{elast}} = \frac{K_{11}}{2}(\nabla\vec{n})^2 + \frac{K_{22}}{2}(\vec{n}\cdot\nabla\times\vec{n})^2 + \frac{K_{33}}{2}(\vec{n}\times\nabla\times\vec{n})^2 \tag{8.19}$$

$$f_{\text{surf}} = -\frac{K_{24}}{2}(\nabla\cdot((\nabla\cdot\vec{n})\vec{n} + \vec{n}\times\nabla\times\vec{n})) \tag{8.20}$$

$$f_{\text{field}} = -\frac{1}{2}\varepsilon_0\vec{E}^2\cdot(\Delta\varepsilon_{\text{r}}(\vec{E}\cdot\vec{n})^2 + \varepsilon_{\text{r},\perp}) - \frac{1}{2}\mu_0\vec{H}^2\cdot(\Delta\mu_{\text{r}}(\vec{H}\cdot\vec{n})^2 + \mu_{\text{r},\perp}) \tag{8.21}$$

In the Lagrangian formulation

$$\frac{d}{dt}\left(\frac{\partial f}{\partial q_i}\right) - \frac{\partial f}{\partial q_i} = Q_i \tag{8.22}$$

holds. The external forces (or the external moments) Q_i contain dissipative moments of the form

$$\vec{Q}_{\text{diss}} = \gamma_{\text{rot}}\frac{\text{d}\vec{n}}{\text{d}t} \tag{8.23}$$

Depending on the problem, certain terms in Eqs. (8.18)–(8.21) can be omitted. The resulting expression may for instance not contain the magnetic component of f_{field} if no magnetic fields are used. For large structures, surface effects may be neglected, resulting in the omission of Eq. (8.20). A derivation is carried out in Ref. [71].

Equation (8.19) shows the relation of the spatial director field to a set of four mechanical constants (γ, K_{11}, K_{22}, and K_{33}), external bias fields (H and E), wall forces and the free energy f. This is the commonly used continuum equation. The system tends to minimize its free energy.

The three constants K_{11}, K_{22}, and K_{33} are related to the three possible distortions of the director field (cf. Figure 8.13): bend, twist, and splay respectively. The constant γ_{rot} is the LC's viscosity in Pa s. Optical LC mixtures are optimized for very low viscosities in the range of 1–10 kPa s, that is, comparable to water. Common mixtures optimized for microwave application show viscosities of several hundred or thousand kPa s. Visible from Eq. (8.21) is the relation with the electric (or magnetic) field: while γ_{rot} inhibits or damps movement, its counterparts are the external fields. Increasing their strength can therefore decrease the switching time.

The presented equations are solved numerically. An approach which takes into account the relevant effects for microwave applications is discussed in Ref. [72].

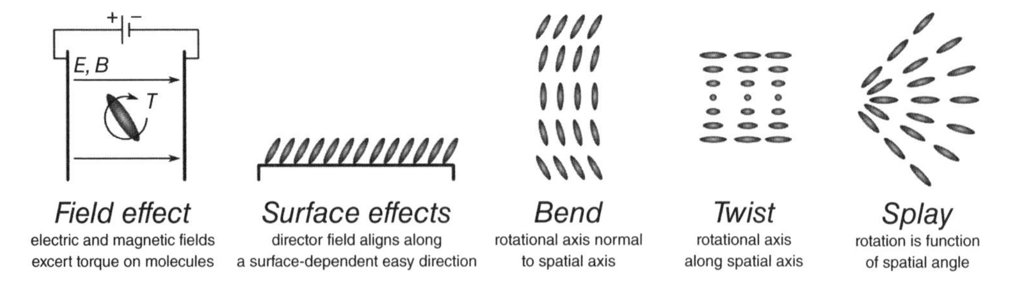

Figure 8.13 Illustration of free energy terms and their geometric representations.

Orientation Mechanisms

As discussed above, three mechanisms can be used to obtain orientation in LCs

1. Magnetic fields
2. Electric fields
3. Surface effects.

Their specific application depends strongly on the topology one desires to use: bulk material is best oriented with magnetic or electric fields. If electrodes cannot be applied to the device, only the magnetic field may be a solution. For planar structures very often a combination of electric field and surface effect is used: A main orientation is imposed by a rubbed wall (e.g., a rubbed polyimide film), which acts as an anchor for the LC.

This is a convenient way of establishing one orientation in a planar structure. In the case of an inverted microstrip line (IMSL) it is convenient to apply an electric field between the signal and the ground electrode. This establishes a vertical orientation. The orthogonal, horizontal orientation can then be provided by an orientation layer: If the electric field is strong enough, it will orient the LC vertically (from signal to ground); if it is switched off, the only remaining force on the LC is the surface force. The LC will therefore orient horizontally.

Which of the three effects is used depends strongly on the application. In the case of characterization, which will be discussed now, external magnetic and electric fields are often used for convenience. Compactness is rarely an issue and the orientation speed is usually much higher for these mechanisms.

8.2.2 Characterization

A number of standard characterization techniques specially suited for characterization of LCs are presented – starting from an approach well known in optics because it may be the most intuitive. Furthermore, it allows introduction of the line model which will help in understanding common microwave techniques that some readers may be unfamiliar with. As a last step, terahertz characterization which employs methods from optics *or* the microwave range, depending on the specific frequency or set-up that is available, will be discussed.

8.2.2.1 The Transmission Method

In an optical set-up, the material properties $\varepsilon_{\perp}$ and $\varepsilon_{\parallel}$ can be determined using a slab model. For solid materials, an etalon is fabricated with a thickness that allows on the one hand a compromise between feasibility (very thin layers will be deposited on a substrate which in turn has to be taken into account) and precision. On the other hand, a compromise has to be found between narrow-band measurements that allow for high sensitivity, especially with respect to losses, and wide-band measurements which resolve permittivity and losses reasonably well for lossy materials but over a wide frequency range. Using a Fabry–Pérot etalon we can achieve both, depending on the parameters we choose. The methods presented here are expressed in terms that should address people from microwave engineering as well as those working in optics and photonics. Saleh and Teich use a very similar transmission matrix method extensively in "Fundamentals of Photonics" [73]. It is similar in the sense that subsequent structures' matrices can be multiplied and provide the overall transmission matrix. The matrices used here, however, are determined differently and converted into S-parameters in a different way. The two methods should not be confused.

To start with, the set-up depicted in Figure 8.14 is considered. A slab of the material under test (MUT) is surrounded by two semi-infinite spaces of permittivity ε_{env}. The slab is of thickness d_{MUT} and unknown permittivity ε_{MUT}. Measurements are usually taken with a collimated laser beam and the samples are

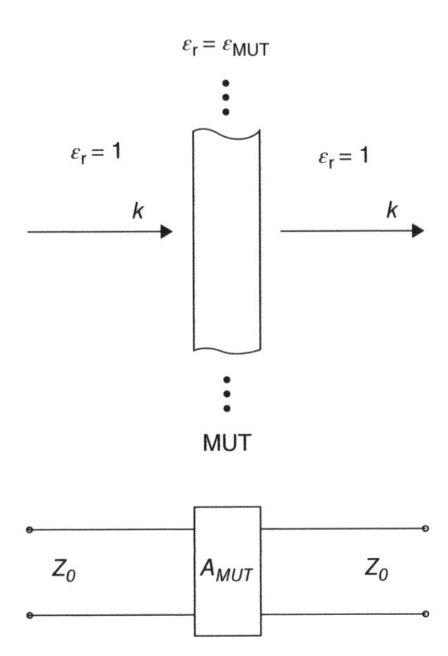

Figure 8.14 Transmission model for a single, solid material slab surrounded by air or vacuum.

much wider than the beam waist. This means a plane wave propagating from left to right with wave vector $\vec{k}$ pointing in the same direction[2] can be assumed.

In order to formulate the analytic model an ABCD-Parameter approach is used. The matrix for the slab of MUT is similar to that of a line [74]. It takes the form

$$A_{\mathrm{MUT}} = \begin{pmatrix} A & B \\ C & D \end{pmatrix} = \begin{pmatrix} \cos \beta_{\mathrm{MUT}} l_{\mathrm{MUT}} & jZ_{\mathrm{W,MUT}} \sin \beta_{\mathrm{MUT}} l_{\mathrm{MUT}} \\ \frac{j}{Z_{\mathrm{W,MUT}}} \sin \beta_{\mathrm{MUT}} l_{\mathrm{MUT}} & \cos \beta_{\mathrm{MUT}} l_{\mathrm{MUT}} \end{pmatrix} \tag{8.24}$$

$$\text{with } Z_{\mathrm{W,MUT}} = \frac{Z_0}{\sqrt{\varepsilon_{\mathrm{r,MUT}}}} \text{ and } \beta_{\mathrm{MUT}} = \sqrt{\varepsilon_{\mathrm{r,MUT}}} \cdot k_0 \tag{8.25}$$

Now, the overall transfer function is obtained by transforming A_{MUT} into an S-parameter matrix. For this, the transformation [74] is used

$$S = [A + B/Z_0 + CZ_0 + D]^{-1} \cdot \begin{pmatrix} A + B/Z_0 - CZ_0 - D & 2\,(AD - BC) \\ 2 & -A + B/Z_0 - CZ_0 + D \end{pmatrix} \tag{8.26}$$

The parameter $T = |S_{12}|^2$ carries the transmission information and can be fitted numerically onto the results. Z_0 refers to the impedance of the embedding medium (i.e., free space impedance in the case of air or vacuum).

In a real set-up for instance, a laser source would be used as source and a photodiode as a detector. Sweeping the laser wavelength and recording the diode response produces a transfer response that only needs to be normalized using an empty measurement as a reference.

The above expression can now be fitted to the measurement data yielding a complex value for β_{MUT} that can be transformed into a complex permittivity.

[2] It is actually a Gaussian beam, but near its focal plane it has a flat wavefront. In the case of a collimated beam, the curved wavefront of a Gaussian beam is sufficiently flat to be modeled by a plane wave.

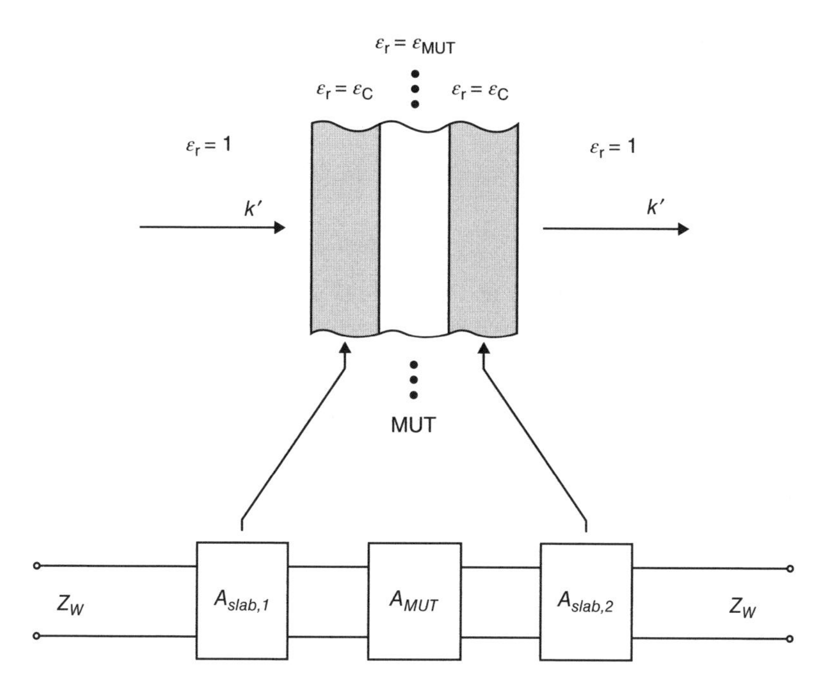

Figure 8.15 Transmission model of a more realistic characterization setup – the sample material (MUT, white) is embedded between two slabs of container material (gray).

The presented set-up is only reasonable for solid samples that can be fabricated at a given thickness. It does not take into account a sample container, which is necessary for liquid materials. A more realistic set-up would be to use three slabs surrounded by air or vacuum (Figure 8.15). Using line theory this can now be easily established. Instead of a single ABCD matrix A_{MUT} two more matrices need to be defined that each represent one slab of the container. Their expression is essentially the same except that β now depends on the container material and its permittivity. The multiplication is carried out from right to left.

$$A_{\mathrm{total}} = A_{\mathrm{slab,2}} \cdot A_{\mathrm{MUT}} \cdot A_{\mathrm{slab,1}} \tag{8.27}$$

The three matrices are multiplied and the resulting expression (only T in the case of a simple transmission measurement, or more S-parameters if available) is fit onto the measurement results. Both the algebra and the fitting can today easily, reliably, and quickly be done in software. In this way, we can even obtain an error estimation.

The obtained expressions hold for an isotropic dielectric. Under certain conditions they are applicable for anisotropic material as well. In a uniaxial anisotropic dielectric, which is the most complex material treated here, there is a single optical axis with one permittivity along the axis and another permittivity perpendicular to it.

Without going into too much detail, it can be seen that for two cases (optical axis parallel to the incident polarization and optical axis perpendicular to the polarization) the dielectric tensor may be simplified. Here, the material can be treated as an isotropic dielectric because the electric field does not "experience" the other tensor element. Assuming perfect alignment of the LC directors in the whole sample volume would allow the permittivities of the two orientations to be obtained in two measurements with the previously presented, very simple expression.

This is, in fact, the usual way of characterizing LC in the optical domain: A slab, sufficiently thick, is irradiated by a laser in the desired wavelength range and the transmission function is fit to an analytic expression. All this while orienting the LC using a strong, homogeneous magnetic field – parallel to the polarization of light and then perpendicular to it.

We have presented a way to treat this problem mathematically using the line theory with the free space expression for the propagation constant β.

8.2.2.2 Microwave Methods

The Line Method

A number of line topologies can be used to characterize LCs. The expression for a piece of line was presented above in Eq. (8.24). Its wave vector more generally depends on the shape of the field which is imposed by the line topology [42].

Propagation constants and line impedances are a function of the line type. As an example, the three types, best suited for LC characterization are given below: hollow waveguide (HL), microstrip line (MSL), and coaxial line (coax).

- **Microstrip Line (MSL)**

$$\beta_{\mathrm{MSL}} = k_0 \cdot \sqrt{\frac{\varepsilon_\mathrm{r}+1}{2} + \frac{\varepsilon_\mathrm{r}-1}{2\sqrt{1+12d/W}}}$$
$$Z_{\mathrm{MSL}} = \begin{cases} \dfrac{60}{\sqrt{\varepsilon_\mathrm{r}}} \ln\left(\dfrac{8d}{W} + \dfrac{W}{4d}\right) & \text{for } W/d \le 1 \\ 120\pi \cdot \left[\sqrt{\varepsilon_\mathrm{r}}\left[W/d + 1.393 + 0.667\ln\left(\dfrac{W}{d} + 1.444\right)\right]\right]^{-1} & \text{for } W/d > 1 \end{cases} \tag{8.28}$$

- **Coaxial Line (coax)**

$$\beta_{\mathrm{coax}} = k$$
$$Z_{\mathrm{coax}} = \frac{\sqrt{\mu/\varepsilon}}{2\pi} \ln\frac{D}{d} \tag{8.29}$$

- **Rectangular Hollow Waveguide (HL), transverse electric (TE) modes**

$$\beta_{HL} = k\sqrt{1-(f_{\mathrm{co}}/f)^2}$$
$$Z_{\mathrm{HL,TE}} = \frac{\eta}{\sqrt{1-(f_{\mathrm{co}}/f)^2}} \tag{8.30}$$

$$\text{with } f_{\mathrm{co}} = \frac{c_0}{2\pi\sqrt{\mu_\mathrm{r}\varepsilon_\mathrm{r}}}\sqrt{\left(\frac{m\pi}{a}\right)^2 + \left(\frac{n\pi}{b}\right)^2}, \eta \approx 377\ \Omega \text{ and } k = \frac{2\pi f}{c_0}\sqrt{\mu_\mathrm{r}\varepsilon_\mathrm{r}} \tag{8.31}$$

MSLs are to be used with great caution when looking at anisotropic media. It is easy to see in Figure 8.16 that in the case of horizontal orientation a considerable portion of the RF field is not parallel to the LC directors. Such results, although presented frequently, can only represent an estimate of LC permittivity.

An efficient characterization method in coaxial topology is presented in Ref. [75]. The outer conductor of a semi-rigid coaxial line and the dielectric are removed and replaced by a fabricated cavity of the same dimension. It is filled with air and has holes on both ends that allow filling of the empty cavity with LC (or any other liquid). The cavity is fabricated in brass and heated or cooled from both sides.

Using bias-tees it is possible to apply a DC voltage between the inner and outer conductors thus providing a cylindrical electric field that aligns the LC in parallel to the electric field of the fundamental coaxial mode (see Figure 8.17). As discussed in earlier sections, this allows determination of the permittivity along the optical axis of the LC. Using an external magnetic field and removing the electrical bias,

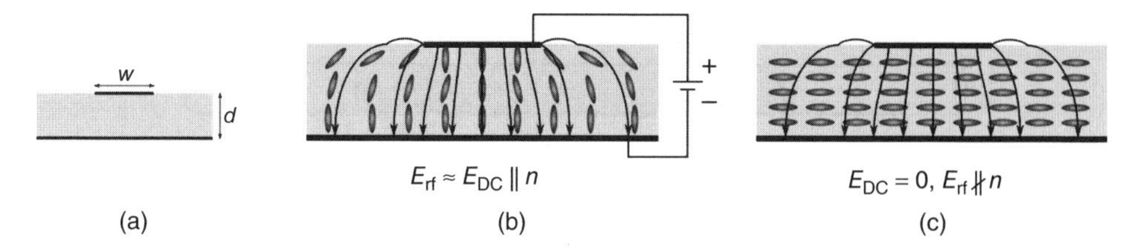

Figure 8.16 Microstrip line. (a) Dimensions of a MSL, (b) DC, RF and director field orientations in the electrically tuned state (on), and (c) RF and director field in the non-tuned state (off).

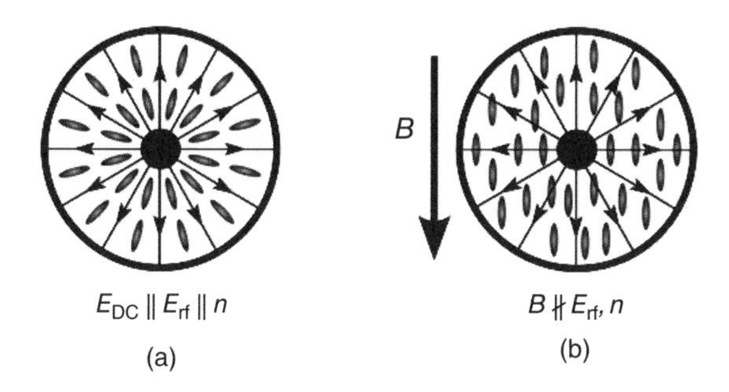

Figure 8.17 LC characterization in a coaxial line set-up. (a) Orientation for the parallel state biased using a voltage between the center conductor and shield. (b) Indirect determination of the perpendicular state using an external magnetic bias field.

it is possible to align the LC such that its uniform direction points across the cross-section of the coaxial line (see Figure 8.17b). The propagating wave now experiences different effective permittivities for each angle around the center conductor. Assuming the mode itself is only slightly disturbed, this allows calculation of an effective permittivity. As is clear from the expression, it contains both the parallel and the perpendicular permittivity.

$$\varepsilon_{\mathrm{r,eff}} = \frac{\varepsilon_{\mathrm{r},\perp} + \varepsilon_{\mathrm{r},\parallel}}{2} \tag{8.32}$$

Using the scattering matrix presented above, it is now possible to solve the analytical *S*-parameter expression for $\varepsilon_{\mathrm{r,eff}}$ in both states (electric bias and magnetic bias). The analytical solution for the *S*-parameters and the data obtained in a broadband vector network analyzer (VNA) measurement are fitted using a simple Newton–Raphson-based algorithm and permittivity and loss are obtained. Note that it is necessary in order to know the perpendicular permittivity $\varepsilon_{\mathrm{r},\perp}$ to carry out the electrical biasing experiment and obtain $\varepsilon_{\mathrm{r},\parallel}$ as well. Furthermore the margin of error of $\varepsilon_{\mathrm{r},\parallel}$ adds to the margin of error of $\varepsilon_{\mathrm{r},\perp}$. This may seem inconvenient but – as we will see later – for actual devices it is not an obstacle because in most cases perfect orientation of the LC is impossible. For very accurate characterization, especially for improving LC mixtures, it is crucial to have more precise results. This is true in particular for the losses which show an uncertainty of $\tan\delta < 1.7 \times 10^{-3}$, which is in the range of the $\tan\delta$ for current high-performance mixtures.

It should be pointed out that while microstrip and coplanar waveguide (CPW) lines can be used for characterization as well, the expressions for determining the permittivity become a lot more complex than in the aforementioned cases. Moreover, the higher line losses in general may also be considered, as they make characterization of low-loss LCs very difficult or even impossible. These topologies, however, are interesting for characterization of small amounts of a LC. Depending on the desired

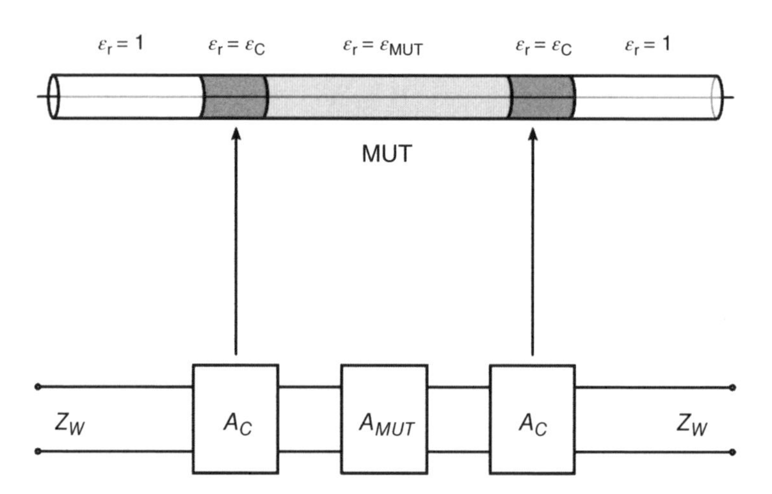

Figure 8.18 Transmission line model for a coaxial line characterization set-up. It is strictly identical with the three slab model except that the propagation constant is chosen to be of the form of a coaxial line propagation constant.

frequency range, only several microliters may suffice. It is also possible to obtain both orientations with magnetic biasing.

The Resonant Method

Looking at the last section one may notice that broadband measurements based on a line approach, while an interesting approach, show several major inconveniences. The first is certainly the low accuracy when determining dielectric loss. However, on closer inspection it becomes clear that the real part of the permittivity itself is subject to uncertainties of the order of percents. When, from the broadband investigation, the overall dielectric response of the material is found, one could settle for single frequency points and describe the dielectric function using these sampling points to get a broader view. Meanwhile, the accuracy at which both parts (real and imaginary) of the permittivity are being measured, may be greatly enhanced using a resonant approach.

The optical characterization method discussed before, can be made resonant by selecting the slab thickness using the Fabry–Pérot effect. Its FSR and finesse $\mathcal{F}$ are functions of permittivity and loss. Using microwave terminology, one would call this a multimode resonator with a high number of modes and a high Q-factor (which corresponds to a high finesse). Both finesse and Q-factor translate to the width of the resonance peaks and are very sensitive to material losses. The position of the resonance peak is a function of the geometry of the resonator, that is, its electrical (or optical) dimensions. The peaks therefore change position depending on the dielectric permittivity. Having said this, we will treat two limiting cases: an entirely filled (i.e., fully loaded) resonator and an almost empty (i.e., perturbed) resonator with only a small portion of sample.

Resonant Line Approach

Yet another method to determine material properties is to use a resonant line [76]. At first sight, and in particular concerning the mathematical approach, this is very similar to the free space approach. On further investigation, one notices that while this is a very convenient approach for isotropic materials, since there are expressions to obtain the material permittivity from the effective permittivity of the propagating mode, it is rather difficult to obtain an expression for anisotropic media. Electric biasing is, in this case, not an option as it makes the determination of the material properties unnecessarily complicated. A strong, homogeneous magnetic field creates a homogeneously oriented

LC layer, which, while anisotropic, can still be analyzed mathematically and makes an analytical separation of $\varepsilon_{r,\perp}$ and $\varepsilon_{r,\parallel}$ possible.

The error using this method is rather large for the above reasons. Nevertheless, it is acceptable for qualitative studies [77] and it allows investigation of structures like filters or phase shifters. It is therefore suggested to use a cavity perturbation method for characterization.

Cavity Perturbation Method

Unperturbed cavity modes can provide Q-factor values of several thousands, even around 10 000. While, until now, only methods have been discussed where a cavity is entirely filled with the MUT, for cavities this is not an option for two reasons: (i) In general, the amount of material would be too large (several milliliters are necessary for a 30 GHz cavity). (ii) The Q-factor of such a fully loaded cavity will be extremely low.

Several geometries for a partially loaded cavity have been proposed, depending on material type and sample geometry. For solid, machinable bulk material for instance, a variety of shapes can be sought and fabricated. In the case of LCs, the limiting problem is the container for the liquid. It should be of very low loss so as not to contribute to the loading of the cavity. It should also not be of a too large permittivity for it will disturb the mode without providing any insight. Last but not least, it should have a very well defined geometry or a geometry which can be easily described in terms of an analytical or semi-analytical model, so variations can be accounted for by running a reference measurement of an empty container.

In Refs. [78, 79] a rectangular cavity with one dominant high-Q mode is proposed. In this way, applying an auxiliary magnetic field in two directions, $\varepsilon_{r,\perp}$ and $\varepsilon_{r,\parallel}$ (and the respective $\tan\delta$) can be determined in two separate measurements. Both are determined independently of each other. The error in this procedure only depends on the unloaded quality factor of the cavity, variation of the container from the ideal cylindrical shape and perturbations induced by the coupling windows and the holes through which the sample container is inserted. However, for a mostly automatic measurement procedure the fact that two magnetic field orientations are necessary is inconvenient. It requires a usually rather bulky mechanical construction. Furthermore, moving the set-up may disturb calibration conditions and thus introduce further, unknown errors.

Therefore, in Ref. [80] a rectangular triple-mode cavity is proposed. It allows measurements of three independent material axes (which in the case of the LC would be $\varepsilon_{r,\parallel}$ along the optical axis and $\varepsilon_{r,\perp}$ twice perpendicular to it) without the need to change the external magnetic field orientation (cf. Figure 8.19). Obtaining two results for $\varepsilon_{r,\perp}$ may seem unnecessary. It should be pointed out though, that because of the relatively large freedom in choosing the modes to be used, the second result can be obtained for a

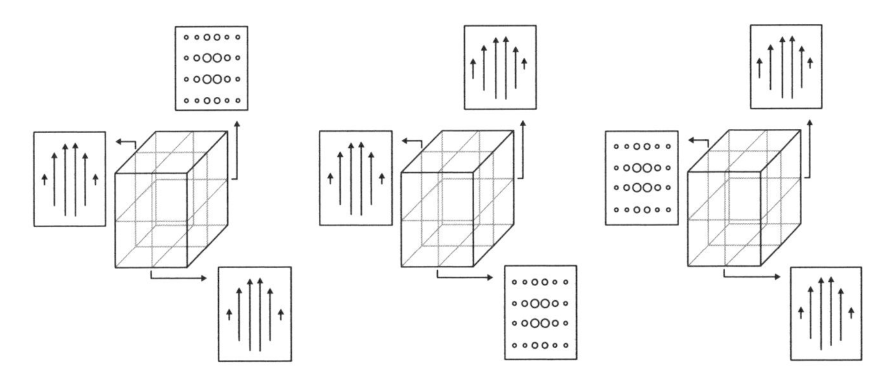

Figure 8.19 Fundamental modes of a rectangular cavity resonator. A sample placed in the center or along one of the central axis can be characterized in three directions.

significantly different frequency and therefore allow broader characterization. The authors also point out that instead of measuring $\epsilon_{r,\perp}$, the mode may be used to determine $\mu_{r,\perp}$ or $\mu_{r,\parallel}$.

The main issue with the rectangular cavity is fabrication. Inner radii are limited by mechanical constraints which become harder to meet with increasing frequency. Furthermore, it is impossible to place the filling hole in a place which is current free for all three modes. A way to circumvent this is by using a cylindrical rather than a rectangular cavity. At the expense of a third independent mode (which is unnecessary for biaxial media like LCs, Figure 8.20) both shortcomings of the previous approach can be solved. As a cylindrical structure, the cavity is very easy to fabricate. It is also rather easy to polish the cavity, which is not the case for a rectangular cavity. The cavity is closed by a flat piece which is advantageous as it avoids a seam that would otherwise perturb the resonant modes and break the symmetry. At last, it is now possible to place the sample very precisely in the very center of the cavity (Figure 8.21). Combined with the generally lower tolerances of this simple cylindrical structure, it allows for a very high repeatability.

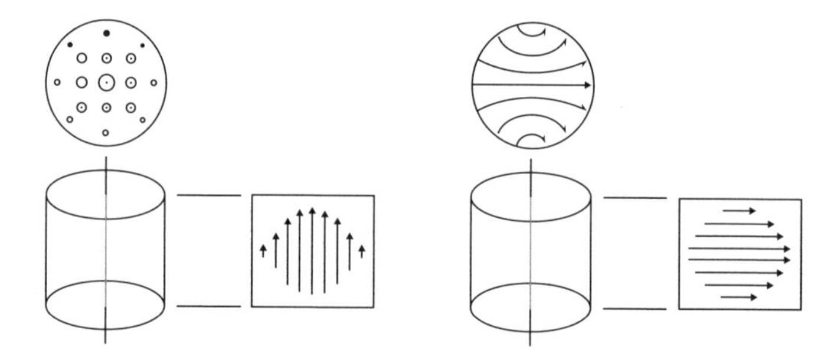

Figure 8.20 Fundamental modes of a cylindrical cavity resonator. A small sample can be placed along the center line.

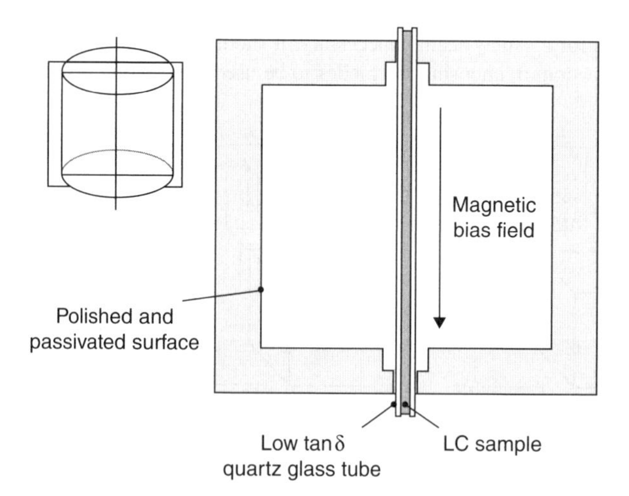

Figure 8.21 Cylindrical cavity resonator for characterization of liquid crystals. The glass tube in the center holds the LC sample, a magnetic bias field orients the LC along the axis of revolution of the cavity.

8.2.2.3 Terahertz Methods

The major challenge of terahertz technology is that wavelengths require dimensions that are generally extremely small for conventional fabrication techniques present in microwave engineering but extremely large with respect to optical (even far-infrared) wavelengths.

The methods previously presented are still generally applicable, even in the terahertz domain. It is possible to build a three slab sample holder – somewhat the standard procedure – as well as it is technically possible to build a cavity resonator. The latter is, however, incredibly delicate and it is impossible to get high quality factors (high in microwave technology refers to orders of $10^3 - 10^4$ or higher). This makes this method inferior to the simple transmission method, especially if Fabry–Pérot is considered. Given smooth surfaces, Bragg gratings provide quite high reflectivities and therefore high finesses (i.e., quality factors).

Most characterization methods for solids rely upon the simple slab method presented above. Duvillaret *et al.* presented a rather general and often cited numerical method to extract data from such set-ups [81]. Unfortunately, most publications are limited to the determination of permittivity or refractive index. Material loss is often neglected (for instance in [82] or [83]). Various works show why this is the case: In the non-resonant set-up, loss angles of $\tan\delta < 10^{-2}$ (corresponding to similar numerical values of κ in the case of organic materials) are impossible to resolve (cf. [63]).

8.2.3 Applications

Over the years, a considerable number of LC-based devices have been proposed. There are devices, based on waveguides, first conceived for the microwave range and steadily approaching the terahertz regime. Another approach is based on free space optics where phase shifting is an interesting opportunity to remove mechanical components from beam lines [66].

One of the great advantages of LCs over several other tuneable materials is their high linearity. Goelden *et al.* [84] have determined the IP3 of 5CB to be at least 52 dBm. Commercial mixtures reached values of the linearity range of at least 60 dBm.

8.2.3.1 Phase Shifters

Phase shifters are an integral part of many microwave applications. They serve as variable delay elements in phased array antennas, they can be used to compensate delays or, as in a Mach–Zender type interferometer, they can be used as switches. In terahertz set-ups they may potentially replace mechanical delay lines that are necessary for phase sensitive (heterodyne) measurements.

The first devices adopted well-known principles from LC display devices. The MSL phase shifter first published in Ref. [85] showed good performance. The figure of merit (FoM) for phase shifters, that is, the ratio of maximum phase shift over maximum insertion loss as defined by

$$\mathrm{FoM} = \frac{\Delta\varphi_{\mathrm{max}}}{IL_{\mathrm{max}}} \tag{8.33}$$

took values of 40–80°/dB up into the lower millimeter-wave region of 35 GHz. This was quickly extended to the W-band by Fritzsch *et al.* [86].

When it became clear that the true advantage of LC devices over other technologies were low losses (rather than switching speed, where ferrites, semiconductors or even micro-electromechanical structures (MEMS) are faster by at least 1 order of magnitude), the topology of choice became the hollow wave guide.

Gäbler *et al.* [87] demonstrated 150°/dB phase shift at 35 GHz in a hollow waveguide phase shifter with a three slab dielectric filling. Hoefle *et al.* [88] combined waveguide and planar topology to integrate a phase shifter and a Vivaldi antenna, namely using a finline structure in the W-band. FoM-values of 70°/dB in the well-matched range of 70–75 GHz were obtained, attaining 100°/dB for higher frequencies at the expense of matching. This concept was successfully extended to an array [89, 90].

The inhomogeneously filled hollow waveguide approach by Gäbler *et al.* was picked up several times. Low-weight Ka-band phase shifters have been demonstrated in Refs. [91, 92] and another W-band approach was presented in Ref. [93]. Combining classical micro-milling and micro-fabrication approaches, frequencies beyond 110 GHz come within reach.

Many rather simple free space phase shifter elements are of transmission type, that is, an LC slab is enclosed by two quartz glass plates and oriented either using a strong magnetic field or using the surface orientation effect and an electric bias (cf. [66, 83, 94, 95]).

8.2.3.2 Polarizers

Another way of driving the aforementioned structure is a polarizer. Unlike in a phase shifter, the polarizer is a structure where both fundamental modes TE01 and TE10 are propagating (degenerate mode). Such an LC-filled structure can be used to couple between TE01 and TE10 and vice versa. This principle is depicted in Figure 8.22.

Strunck *et al.* have presented such a structure in Ref. [97]. It works very much like a polarizing LC layer in an LC display. The optical axis of the LC is continuously turned. Under certain conditions, the light polarization follows. Optimal suppression of the orthogonal polarization is obtained when the Gooch–Terry criterion is met. While this was derived for plane wave propagation through a slab of LC, in modified form it still holds for propagation in waveguides. The main difference lies in the change in the propagation constant. In a waveguide it not only depends on the dielectric constant but also on the cut-off frequency of the waveguide.

8.2.3.3 Beam Steering

If an antenna array is used, the antenna beam pattern is a function of the phases of the individual radiating elements (i.e., patch or horn antennas). LC phase shifters can be used to efficiently set the phase, for example, in an array as shown in Figure 8.23. This principle can be applied using LC up to very high

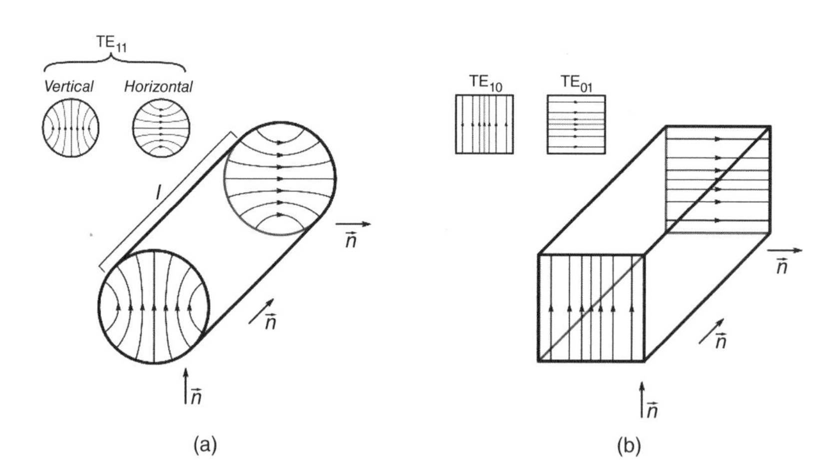

Figure 8.22 Principle of a LC hollow waveguide polarizer. The LC director field orientation n is continuously rotated to make the RF field follow. (a) Circular and (b) quadratic cross-section.

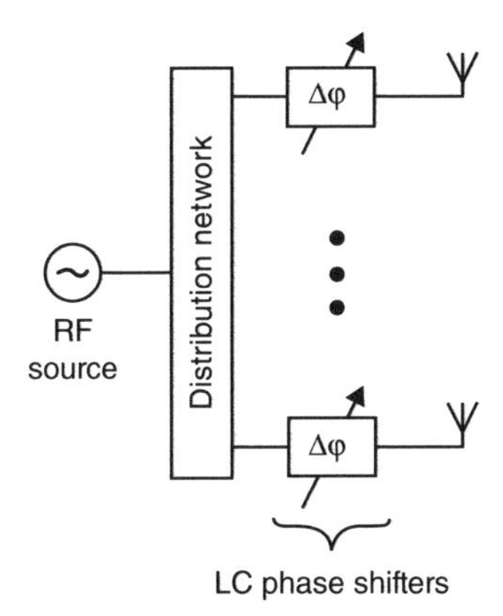

Figure 8.23 Phased transmit array with LC phase shifters.

frequencies. It was shown that an array of Vivaldi antennas with slotted guide phase shifters works at 94 GHz with a tuning range of ± 15° [90]. At the current state of research, the limit is rather the machining and fabrication of the antenna and/or phase shifter structure than the capabilities of the LC materials.

There are numerous techniques on the market to implement a phased array as depicted in Figure 8.23. Most prominently, in the past years MEMS and ferroelectrics (such as barium-strontium titanate, BST) have been used for phased arrays. In contrast to LC technologies however, there are distinct drawbacks: the performance of ferroelectrics-based systems degrades with frequency. While very good performance is observed in the Ka-band, most publications in the W-band show only half the performance in terms of FoM. MEMS perform better and only degrade as the line topology degrades but they are, in general, non-continuously tunable devices.

Another general advantage of LC technology is that it can be applied to hollow waveguide structures relatively easily. This is not the case for MEMS and BST structures. While for instance ferrite-based waveguide phase shifters are available, they usually require strong magnetic fields and thus high current (i.e., high power).

Several phased arrays based on LC have been demonstrated: planar devices (as presented, for example, in Ref. [98]) as well as 2.5/3D devices (such as [77, 90, 91, 99, 100].

One drawback of those arrays is the relatively slow tuning speed in comparison with the quasi-instantaneous tuning of semiconductor devices or the nanosecond to microsecond speed of ferrites or MEMS. LC phase shifters are in the range of milliseconds (planar structures) to tens of seconds (waveguide with current LCs).

This, however, does not necessarily turn out to be a problem: if the application is tracking a transmitter station at low relative angular speeds, then slow switching is acceptable and the advantages (low insertion losses in general, high relative phase shifts, that is, high FoMs and high-power acceptance) predominate.

8.3 Graphene for THz Frequencies

8.3.1 Theory and Material Properties

Single crystal carbon, diamond, is used in the THz-application field as an efficient heat conductor for heat sinking of high-power devices. Diamond has also been used as a field-effect transistor for high-voltage applications since this material has a very high breakdown field.

The first two-dimensional carbon lattices were the so-called buckyballs. They are used as a nanometric powder for numerous applications, even in drug-delivery concepts in the human body.

The next type of component was the carbon nanotube [101]. A large number of electronic components, from field-effect transistors to sensors, were experimentally realized, including applications in the THz frequency range.

Theoretically it was initially considered impossible to obtain graphene sheets, until the experimentalists Geim and Novoselov at the University of Manchester published their experimental realization of graphene by the special approach of mechanical exfoliation from pyrolytic graphite [102]. The Nobel Prize was awarded to them for this achievement.

Since then, a good number of fabrication techniques, often by epitaxial processes on such materials as SiC and by chemical vapor deposition, have been developed so that graphene is offered by a number of companies and universities for a reasonably low price.

It is now therefore possible to consider applications of graphene for the THz area. The most important specific properties of this material are summarized in the following.

8.3.1.1 Band Structure

The conduction band and valence band touch at the K-point of the Brillouin zone. Graphene is therefore a zero band gap semiconductor or a semi-metal. The point where valence and conduction band touch is called the Dirac point. Due to the hexagonal honeycomb structure in real space, there are 6 K points in momentum space. Two lattice vectors are required to describe the lattice. Therefore, the 6 K points are divided into two groups, namely the K and K′ points. The quantity to differentiate the two groups is called the pseudo-spin. Most 3D semiconductors show parabolic conduction and valence bands in the *E*-*k* space. For graphene, instead, the bands are linear, as illustrated in Figure 8.24 with

$$E_{C,V}(\vec{k}) = \pm\hbar v_F|\vec{k}| = \pm v_F p \tag{8.34}$$

where $p = \hbar k$ is the momentum measured relative to the Dirac points and $v_F = 10^6$ m/s is the Fermi velocity, 1/300 of the speed of light. The "+" sign accounts for the conduction band, E_C, the "−" sign in Eq. (8.34) is for the valence band, E_V. The conduction and valence band form cones with their tips touching at the Dirac point. These cones are often referred to as Dirac cones.

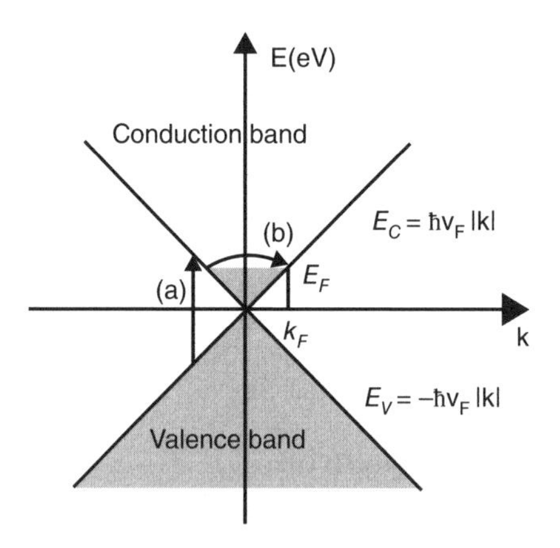

Figure 8.24 Energy band diagram of n-doped graphene at the K point. (a) Interband excitation of an electron from the valence band into the conduction band, for instance by absorption of a photon (high energy, little momentum transfer). (b) intraband transfer of an electron in the conduction band, for example, by emission or absorption of a phonon (little energy of a few millielectronvolts, large momentum transfer).

The linear dispersion relation for both electrons and holes results in a very unusual behavior of the carriers. In usual metals or semiconductors, electrons behave like ordinary particles with an effective mass, m^*. Their kinetic energy is given by $E = p^2/(2m^*) \sim k^2$. Clearly, this is not the case in graphene since carriers in graphene obey a linear dispersion. Therefore, the carriers cannot be described by the Schrödinger equation and do not behave like classical particles. The carriers obey the Dirac equation [103], actually describing relativistic Physics,

$$E = \sqrt{(m_0c^2)^2 + p^2c^2}. \tag{8.35}$$

Comparing Eq. (8.35) with Eq. (8.34), reveals that the rest mass, m_0, of carriers in graphene must be zero and that the velocity of light must be exchanged by the Fermi velocity, v_F, in the Dirac equation. This means that electrons and holes behave like relativistic particles with zero rest mass that move at about 1/300 of the speed of light. The 300 times smaller velocity of electrons compared to the speed of light makes graphene an excellent test bed for relativistic Physics. Although the rest mass of the electron is zero, the effective mass at finite energy is non-zero and is calculated via the De Broglie relation [104]

$$p = m^* v_F. \tag{8.36}$$

With Eq. (8.34) the effective mass is

$$m^* = E/v_F^2, \tag{8.37}$$

in analogy to the relativistic mass. That is, the effective mass of the electron is zero at the Dirac point and increases linearly above and below. For ordinary semiconductors, in contrast, the effective mass at the top of the valence band and at the bottom of the conduction band are not energy-dependent.

8.3.1.2 Charge Carrier Density in Graphene

The electron ($n^{(2D)}$) and hole ($h^{(2D)}$) concentrations in a graphene sheet are given by the following expressions:

$$n^{(2D)}(E_F) = \int_0^\infty \rho(E) f(E - E_F) dE = N_g \Im(E_F/k_B T) \tag{8.38}$$

$$h^{(2D)}(E_F) = \int_0^\infty \rho(E)[1 - f(E - E_F)] dE = N_g \Im(-E_F/k_B T) \tag{8.39}$$

here, $f(E)$ is the Fermi–Dirac distribution function which describes the occupational probability of a state depending on its displacement in energy from the Fermi energy, E_F (measured relative to the Dirac point). Further, N_g, the effective graphene density of states, is given by

$$N_g = \frac{2(k_B T)^2}{\pi(\hbar v_F)^2} \tag{8.40}$$

and $\Im$ is the complete Fermi–Dirac integral:

$$\Im(E_F/k_B T) = \int_0^\infty \frac{\eta}{1 + e^{\eta - E_F/k_B T}} d\eta \tag{8.41}$$

Figure 8.25 shows the calculated sheet carrier density in graphene for various Fermi energies and temperatures.

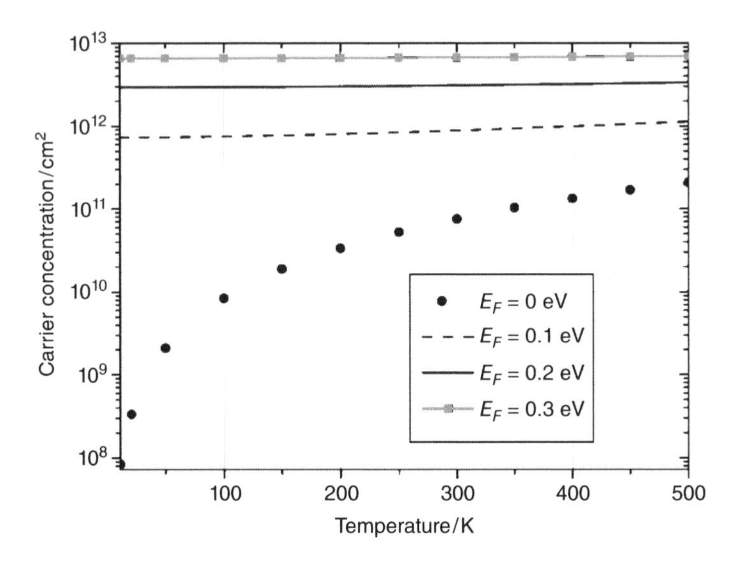

Figure 8.25 Carrier concentration versus temperature for several Fermi energies.

The Fermi energy in graphene depends on the doping level. It can be tuned by doping with adsorbates, substitution doping or by binding to a substrate. In as-grown epitaxial graphene on SiC, for instance, the dangling bonds of the carbon-rich buffer layer cause an n-type doping level in the range of $1.1 \times 10^{13}/\text{cm}^2$ [105]. The carrier concentration at 0 K can be calculated by integrating the Fermi circle in k-space. This yields

$$|n^{(2\text{D})}| = \pi k_F^2 / V_\text{k} = \frac{E_\text{F}^2}{\pi(\hbar v_\text{F})^2}, \tag{8.42}$$

where $V_\text{k} = \pi^2$ is the unit volume in k-space. Equation (8.34) was used to express the Fermi wave vector, k_F, in terms of the Fermi energy. Equation (8.42) holds for both electrons and holes; $n^{(2\text{D})}$ is either the electron or the hole density. The Fermi energy can now be expressed in terms of the carrier concentration as

$$E_\text{F} = \pm\hbar v_\text{F}\sqrt{\pi|n^{(2\text{D})}|}, \tag{8.43}$$

where the "+" sign has to be used for n-doping and the "−" sign for p-doping.

8.3.1.3 AC Conductivity and THz Transmission

The THz transmission and the (AC) conductance of graphene are dominated by an interband transition and an intraband transition of the charge carriers. The *interband* contribution corresponds to creation of an electron–hole pair by excitation of an electron from the lower Dirac cone into the upper Dirac cone at approximately the same k-vector as illustrated in Figure 8.24 by arrow a. The *intraband* contribution is due to transitions at the Fermi energy, transferring mainly momentum and only little energy (Figure 8.24, transition b).

Interband transitions are responsible for the commonly denoted transmission loss through graphene of 2.3% in the visible. In the THz range, however, intraband transitions dominate the transmission and the optical conductivity, even for only lightly doped graphene: The interband transition can only occur for excitation energies higher than $\sim 2E_\text{F}$ (see Figure 8.24, arrow a) since it requires an empty state in the upper Dirac cone and a filled state in the lower one (or vice versa). For terahertz radiation

($E = 4$ meV at 1 THz), the Fermi energy must therefore be smaller than $|E_F| < 2$ meV. Even for lightly doped graphene at low temperature with $n^{(2D)} > 3\times10^8/\text{cm}^2$, the Fermi energy is higher than 2 meV (see Eq. (8.43)) such that only intraband contributions to the conductivity have to be taken into account. Although there are some THz applications where interband transitions play a role, we focus in the following on the much more important intraband transitions. A derivation of the interband response can be found in Ref. [106].

In general, the conductivity of a two-dimensional electron gas (2DEG) is given by

$$\sigma^{(2D)}(\omega) = e|n^{(2D)}|\mu(\omega), \tag{8.44}$$

where $n^{(2D)}$ is the doping level of graphene which can be adjusted by doping with adsorbates or by coupling of graphene to its vicinity, as discussed in Section 8.3.1.2. The (collective) mobility can be derived from the equation of motion,

$$m^*\ddot{x}(t) + m^*\Gamma\dot{x}(t) = eE(t), \tag{8.45}$$

where m^* is the effective mass from Eq. (8.37), $\Gamma = 1/\tau$ is the damping factor resulting from scattering with an effective scattering time, τ, and $E(t)$ is the electric field that causes the carrier motion. The scattering time at room temperature is fairly short, with typical values in the tens of femtosecond range [106] and may reach much higher values at low temperatures. With harmonic excitation, $E(t) = E_0\exp(-\mathrm{i}\omega t)$, Eq. (8.45) becomes

$$(-\mathrm{i}\omega + \Gamma)\dot{x}(t) = \frac{e}{m^*}E(t). \tag{8.46}$$

Using $\dot{x}(t) = v(t) = \mu(\omega)E(t)$, the mobility is

$$\mu(\omega) = \frac{e}{m^*}\cdot\frac{1}{-\mathrm{i}\omega + \Gamma} = \mathrm{i}\frac{e}{m^*}\cdot\frac{1}{\omega + \mathrm{i}\Gamma}. \tag{8.47}$$

Finally, we obtain for the conductivity, by substituting the mobility of Eq. (8.47) into Eq. (8.44),

$$\sigma^{(2D)}(\omega) = \mathrm{i}\frac{e^2|n^{(2D)}|}{m^*}\cdot\frac{1}{\omega + \mathrm{i}\Gamma}. \tag{8.48}$$

This is equivalent to a classical particle motion with a Drude weight of

$$D = \pi\frac{e^2|n^{(2D)}|}{m^*}. \tag{8.49}$$

However, the effective mass is energy-dependent according to Eq. (8.37), since only the carriers at the Fermi energy have the option to move into free states. With Eq. (8.37), the effective mass can be expressed in terms of the carrier concentration as

$$m^* = \frac{\hbar}{v_F}\sqrt{\pi n^{(2D)}}, \tag{8.50}$$

and with Eq. (8.39), the Drude weight for graphene as [107]

$$D = \frac{e^2 v_F}{\hbar}\sqrt{\pi|n^{(2D)}|}. \tag{8.51}$$

The conductivity as a function of the carrier concentration becomes

$$\sigma^{(2D)}(\omega) = \mathrm{i}\frac{e^2 v_F}{\hbar\sqrt{\pi}}\sqrt{|n^{(2D)}|}\cdot\frac{1}{\omega + \mathrm{i}\Gamma}. \tag{8.52}$$

Also the damping factor, Γ, may depend on the carrier concentration, $n^{(2D)}$. Carrier concentration or doping-dependent origins of scattering are, for instance, impurity scattering by the dopants [108] and electron–electron scattering. Carrier-concentration independent scattering can arise from strong coupling of graphene to its vicinity, for example, to phonons of a substrate or a dielectric layer covering graphene, or by scattering with defects in the graphene film.

Note that the equation of motion in Eq. (8.45) is actually a classical description, despite the relativistic, Dirac-like nature of graphene's carriers. However, this classical description of the conductivity of graphene turns out to reproduce experimental results well [107, 109], since it describes the *average* motion of the charge ensemble rather than the motion of *individual* carriers. This average motion is taken into account by the effective mass and the scattering time. A more sophisticated derivation is given in Ref. [106] (and references therein), however, leading to the same result. For structured graphene, a resonant term $\omega_0^2 x(t)$ may have to be included on the left-hand side of Eq. (8.45), providing a Lorentz–Drude equation of motion.

According to Eq. (8.52), the (intraband) conductivity approaches zero at zero carrier concentration. However, we have neglected interband transitions so far. Even for zero carrier concentration, graphene remains conductive with a minimum conductivity of $\sigma_0 = e^2/(4\hbar) = 60.7\ \mu\text{S}$ [110] due to the interband term. Large area graphene can therefore never be turned into a high resistive state, which results in some difficulties in constructing graphene transistors or diodes. In Section 8.3.1.4 we will explain that this problem can be resolved by implementing nano-constrictions.

Equation (8.52) shows that the (intraband) conductivity of graphene varies as $\sim \sqrt{|n^{(2D)}|}$, whereas the conductivity of classical semiconductors and metals scales as $\sim |n^{(2D)}|$. This feature displays an important property of graphene that also impacts the THz transmission. The transmission through a 2DEG, situated at a surface between air and a material with refractive index, n, (not to be confused with the carrier density, $n^{(2D)}$) is given by [106]

$$T(\omega) = T_0 \cdot \frac{1}{\left|1 + \dfrac{\sigma^{(2D)}(\omega) Z_0}{1+n}\right|^2} = T_0 \cdot T_G(\omega), \tag{8.53}$$

where $T_0 = 1 - R_0 = 1 - [(n-1)/(n+1)]^2$ is the transmission through the interface without the 2DEG, and $Z_0 = 377\ \Omega$ is the free space wave impedance (see also Chapter 2, Section 2.2.4.2., Eq. (2.74) and discussion thereof). For suspended, undoped graphene at low temperature ($T_0 = 1, n_{\text{Air}} = 1$, only interband transitions with $\sigma = \sigma_0 = 60.7\ \mu\text{S}$), Eq. (8.53) yields $T = 97.7\%$. This is the well known transmission loss through graphene of 100–97.7% = 2.3%. In the THz range, however, the AC conductivity of graphene is usually (much) larger than σ_0 due to the intraband contribution. Particularly highly doped samples show strongly reduced THz transmission through a single graphene layer. For instance, graphene on SiC ($n_{\text{SiC}} = 3.2$) with typical carrier concentrations in the range of $n^{(2D)} = 1.1\times10^{13}/\text{cm}^2$ and DC mobilities of the order of $900\ \text{cm}^2/\text{Vs}$ [105] (corresponding to a damping constant of $\Gamma = 2\pi \cdot 4.6\ \text{THz}$), shows a fairly constant transmission of only $T_G = 76.7\%$ (corresponding to a transmission loss of 23.3%), from DC to frequencies $\upsilon_{\text{THz}} < \Gamma/2\pi = 4.6\ \text{THz}$. The transmission loss of 23.3% is more than one order of magnitude higher than the 2.3% caused by interband transitions only.

8.3.1.4 Semiconducting Graphene

The vanishing effective mass of graphene's charge carriers at the Dirac point implies that graphene is attractive for highest mobility applications. Several applications, however, require a finite band gap. An example is graphene transistors: even at the Dirac point, the conductivity is *not* zero (despite a vanishing carrier density). The transistor cannot be pinched off. A bandgap, however, would allow for pinch off: it switches off the low energy interband transitions. In bilayer graphene, a tunable band gap can be formed by applying a perpendicular electric field. Another case of semiconducting graphene is reported by sandwiching it between boron nitride [111]. Mono-layer graphene can be made into a semiconductor by reducing the width of narrow strips [112]. A bandgap can be formed by confining the graphene width

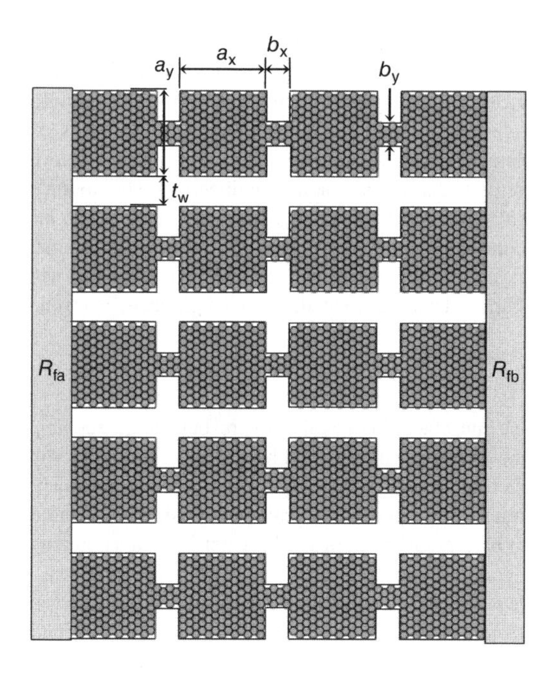

Figure 8.26 The graphene sheet on an insulating substrate like SiC is electron-lithographically structured such that a wide structure with a narrow band gap is followed by a narrow structure with a wide band gap.

in nanoribbon or nanoconstriction structures. For example, the induced bandgap by a 50 nm nanoribbon is about 6 meV, while for a nanoconstriction with a 30 nm constriction width, this can be about 14 meV [113]. This is mainly attributed to the extra confinement in the longitudinal direction. Such narrow constrictions require electron-beam lithography for fabrication. In order to obtain acceptable power levels of such components, one has to use multiple-layer graphene and matrices of components in series and in parallel. An example is given in Figure 8.26, where ballistic resonators are proposed for THz generation and other signal processing applications. Such matrix surfaces have to be incorporated into suitable THz resonating or transmission structures [114].

8.3.2 *Applications*

The special properties of graphene make it very attractive for a manifold of applications. These properties include the potential for extremely high carrier mobility, its thickness of one atomic layer, compatibility to standard lithography processes, high transmission in the visible (97.7%), and carrier-density controllable transmission coefficient in the THz. This section lists some recent applications and future perspectives for THz applications of graphene. Due to the vast amount of publications on graphene, this brief section is far from exhaustive, it just gives a glance at THz graphene optics and devices.

8.3.2.1 Graphene Field Effect Transistors

As described in Section 5.1.3 field effect transistors (FETs) can rectify THz radiation, generating a DC bias that is proportional to the incident THz power. The rectification remains efficient even if the transistor is operated far above its respective maximum frequency for amplification, f_T and f_{max}. Such transistors require ohmic source-drain contacts and a gate Schottky contact to a 2DEG. Graphene with its excellent material parameters can be used as the 2DEG material in FETs, offering mobilities much higher than those of silicon devices and being compatible with any kind of high resistivity substrate and complementary metal oxide semiconductor (CMOS) processes. In terms of device processing, it is fairly simple to achieve ohmic source and drain contacts to graphene. However, obtaining high quality, low leakage Schottky gate contacts requires some engineering. The absence of a band gap prevents formation of natural Schottky contacts to graphene. An oxide or dielectric insulation layer is required. For efficient rectification, the thickness of this layer, d_{Ox}, should be as small as possible in order to increase the control of the gate bias, U_{G}, on the carrier density in graphene, which follows a simple plate capacitor model. The 2D carrier density for gated, undoped graphene is given by

$$en^{(2D)} = \varepsilon_0 \varepsilon_{\mathrm{r}} \frac{1}{d_{\mathrm{Ox}}} U_{\mathrm{G}}. \tag{8.54}$$

Further, the gate dielectric must be of high quality to prevent charge trapping in the dielectric. Trapped charges lead to a drift in the effective gate bias and hysteresis. Thermally evaporated dielectrics show a large amount of chargeable traps and are a poor choice. Atomic layer-deposited (ALD) materials, such as Al_2O_3, have excellent material properties and can be deposited in atomically thin layers. However, growth of a closed film on bare graphene is difficult: water is used as precursor for Al_2O_3 growth, but the hydrophobic nature of graphene prevents adhesion of water. Two solutions to this problem have been found: Lin *et al.* [115] have processed a graphene FET with an $f_{\mathrm{T}} = 26$ GHz by using a NO_2-TMA precursor prior to ALD growth. Another solution is the implementation of a few nanometer thick organic seed layer [116]. Further, ALD materials other than Al_2O_3, without the requirement of water as precursor, represent a solution. Vicarelli *et al.* [117] have fabricated a THz graphene rectifier by contacting a graphene flake with a HfO_2-insulated Cr/Au gate. Some hysteretic behavior remained. The maximum THz responsivity is in the range of $R = 0.1$–0.15 V/W at 300 GHz and the room temperature noise equivalent power (NEP) is in the range of $30\,\mathrm{nW}/\sqrt{\mathrm{Hz}}$ for bilayer graphene. These values are about 3 orders of magnitude worse than the state-of-the-art in classical semiconductor FETs based on Si or III–V materials but already allow for some THz applications. The reasons for the weaker performance are (i) less optimization of the devices (such as impedance matching, etc.) and (ii) the absence of a band gap in graphene. The latter is responsible for a low on/off ratio in the range of 5–10 and a fairly small resistance of the device at the Dirac point without strong threshold behavior. However, for rectification, operation close to the threshold usually allows the highest responsivities. Formation of a band gap would, for instance, require nano-constrictions. To the knowledge of the authors, no graphene FET rectifiers with nano-constrictions have yet been demonstrated. A band gap may also be formed by breaking the coupling of multi-layer graphene [116].

First attempts to build THz amplifying graphene FETs have also been undertaken. Liao *et al.* [118] have developed a self-aligned process with very short ungated graphene regions where the gate is formed by a GaN nanowire. The authors did not measure the THz performance directly but from the transit-time, they estimated an $f_{\mathrm{T}} = 0.7$–1.4 THz.

The highest (extrapolated) f_{T} from RF measurements on graphene-based transistors are in the range of 300 GHz [119].

8.3.2.2 Switchable THz Filters

In the visible domain, the transmission loss through graphene is dominated by interband absorption. Due to the linear band diagram, it is fairly constant at a value of only 2.3%. Transmission in the THz domain, however, is dominated by intraband absorption of electrons (or holes) at the Fermi energy. The transmission through graphene can be controlled by the charge density in graphene: more carriers result in higher losses; the smallest losses are given for graphene at the Dirac point. The transmission through

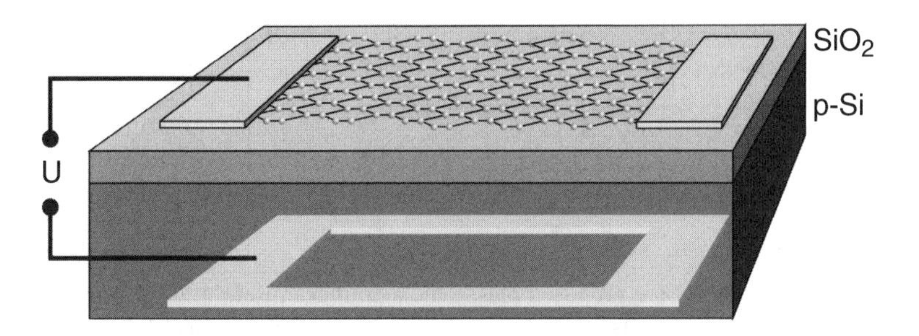

Figure 8.27 Schematic of the THz modulator in Ref. [120].

a continuous (doped) graphene film can be well described by a Drude model for the (AC) conductivity, as discussed in Section 8.3.1.3, Eqs. (8.52) and (8.53).

Several groups have designed structures where the carrier concentration (and hence the conductivity) in graphene is controlled by a gate in order to alter the transmission through the active graphene layer. Sensale-Rodriguez *et al.* [120] used a graphene layer on a thin, slightly doped semiconductor layer which acted as gate (see Figure 8.27). The semiconductor was biased by a gold ring back-gate. At specific frequencies, they achieved a transmission modulation depth (defined as $(T_{max} - T_{min})/T_{max}$) of 64%. The 3 dB switching speed was in the range of 4 kHz. The insertion loss is 2 dB.

8.3.2.3 THz Plasmonics

By adding structure to graphene, additional performance features can be generated. One of the simplest structures is a grating of graphene strips. The constriction gives rise to surface plasmon-polaritons (SPPs) which alter the THz transmission for the electric field along the grating. A Drude model is insufficient to describe the THz transmission, since it cannot reproduce the SPP resonance. For very thin strips, a Lorentz–Drude model can be used as a very coarse model. A more accurate description has been derived by Bludov *et al.* [121]. A first experimental demonstration is shown in Ref. [122].

A series of plasmonic structures for manipulating light has been proposed in Ref. [123] for the infrared range, which may be extended to the THz. This includes waveguides and splitters. Further, the chemical potential of a graphene sheet can be periodically altered in order to generate additional functionality by plasmons. This includes circular n–p junctions that act as a lens. Formation of n–p junctions within a (single) graphene layer can be achieved either by gating or by local intercalation with hydrogen [124]. The latter bears the advantage that no further metallization is required, allowing obstacle-free optical access to the graphene layer.

The first THz meta-materials with an active graphene layer have been presented [125].

8.3.2.4 Perspectives for THz Amplification and Lasing

A few groups have yet worked on theoretically investigating potential THz amplification using graphene. Two different operation modes are discussed:

Ryzhii *et al.* [126] have discussed an optically pumped graphene laser with a pump photon energy (much) above that of the THz wavelength to be amplified. The pump laser causes inter-band absorption, populating the upper Dirac cone with electrons and the lower one with holes, causing population inversion. They have shown that lasing is theoretically possible, however, with several obstacles to overcome. First, a single graphene sheet absorbs only 2.3% of the incident laser light via an interband transition. In order to overcome the low absorption, the authors suggested to use graphene multilayers. However,

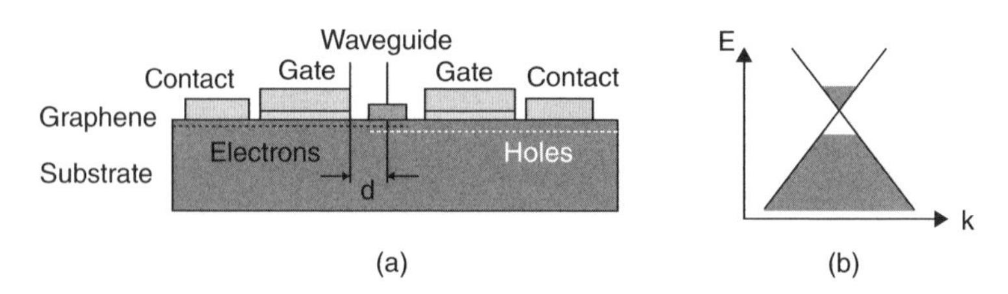

Figure 8.28 (a) Schematic of the proposed plasmonic amplifier in Ref. [127]. Electrons and holes in the graphene layer are vertically offset for clarity. (b) Ideal carrier energy distribution with population inversion at the center of the structure.

electronic coupling between the layers must be circumvented in order to preserve the properties of monolayer graphene, such as a linear band structure, low effective mass, and high mobility. Second, carrier relaxation and scattering can be extremely fast in graphene with values in the tens of femtoseconds to a few picoseconds for substrate-coupled graphene on SiC [126]. For instance, carriers can emit phonons, scatter with impurities, scatter on fluctuations of the Fermi-energy due to external fields or adsorbates, or scatter with other carriers. The latter becomes very important for high carrier concentrations, that is, high optical pumping, and may limit the maximum pump power. However, strong inversion (and hence, extreme pumping) will be required in order to achieve timescales for stimulated emission that are shorter than the relaxation or scattering time. Further, highest quality material is required in order to prevent impurity scattering. Graphene is very sensitive to adsorbates, which act as dopants or scatterers. This may require protecting layers. These protecting layers, however, can already alter the properties of graphene.

Rana [127] proposed an electrically pumped THz plasmon amplifier. A graphene sheet is connected with two ohmic contacts followed by (insulated) gates. The gate potentials determine the carrier type under the gate and are chosen such that holes are injected from one side and electrons from the other side into the active zone between the gates, as illustrated in Figure 8.28. At the center of the structure, electrons and holes can recombine, emitting photons or plasmons. A plasmonic waveguide is situated on top of the recombination zone where the amplified wave is guided. It is crucial that the carriers do not relax while transported from the gate to the recombination zone. This requires very long scattering times in order to cover the distance of $d = 1\ \mu m$ (or longer). Similar constraints on material properties and transport are apparent, as for the laser proposed in Ref. [126]. It is reasonable to expect that graphene-based amplifiers and lasers will have to be operated at low temperatures, similar to quantum cascade lasers. Rana [127] calculated the gain versus THz frequency. Reasonable gain is only achieved above 3 THz, even for 77 K operation. This also shows that lasers and amplifiers are unlikely to work in the low THz range but may be feasible in the higher THz range.

8.3.2.5 Further Potential Applications

There is another useful property of large graphene sheets, namely that they are conducting up to and beyond THz frequencies, but optically transparent [128, 129]. This is a property which makes graphene attractive for replacing the conductive, transparent semiconductors such as indium tin oxide [130]. For THz generation, such properties have been successfully employed for efficiency enhancement of THz signal generation by optical mixing in materials such as low-temperature fabricated GaAs [131], where graphene acted as a transparent electrode. A sometimes relevant effect is the plasmon dispersion in graphene [132].

Graphene stripes are open to a large number of component ideas. Tunneling structures can be produced by separating two graphene layers by an insulating sheet, such as SiO_2 or hexagonal boron nitride (h-BN)

[133]. Similarly, this can be realized by short gaps along a line. Also structures of varying strip width can be used to heat the charge carriers, which can be then made to drop energetically in a narrow-gap stretch, similarly to the quantum cascade laser, and thus emitting THz photons.

Graphene has been employed as a flexible component such as transparent heaters [134]. Graphene can also be doped with various types of impurities so that many more ideas can be attempted.

The field of THz components based on graphene is therefore very wide. Many of the basic engineering concepts, such as the characteristic impedance of graphene transmission lines, have not yet been evaluated systematically. Other important questions concern *1/f* noise contributions. Similarly, life-time limitations of components are unknown since such effects as electromigration of the material are not yet known. This section is therefore just an encouragement of further research work.

References

[1] Crocker, A., Gebbie, H.A., Kimmitt, M.F., and Mathias, L.E.S. (1964) Stimulated emission in the far infra-red. *Nature*, **201**, 250.

[2] Chang, T.Y. and Bridges, T.J. (1970) Laser action at 452, 496, and 541 μm in optically pumped CH3F. *Optics Communications*, **1**, 423.

[3] Andronov, A.A., Zverev, I.V., Kozlov, V.A. *et al.* (1984) Stimulated emission in the long-wavelength IR region from hot holes in Ge in crossed electric and magnetic fields. *JETP Letters*, **40**, 804.

[4] Köhler, R., Tredicucci, A., Beltram, F. *et al.* (2002) Terahertz semiconductor-heterostructure laser. *Nature*, **417**, 156.

[5] Mahler, L., Tredicucci, A., Beltram, F. *et al.* (2009) Vertically emitting microdisk lasers. *Nature Photonics*, **3**, 46.

[6] Richtmyer, R.D. (1939) Dielectric resonators. *Journal of Applied Physics*, **10** (6), 391.

[7] Mie, G. (1908) Beiträge zur Optik trüber Medien, speziell kolloidaler Metallösungen. *Annalen der Physik*, **330**, 377.

[8] Debye, P. (1909) Der Lichtdruck auf Kugeln von beliebigem Material. *Annalen der Physik*, **335**, 57.

[9] Garrett, C.G.B., Kaiser, W., and Bond, W.L. (1961) Stimulated emission into optical whispering modes of spheres. *Phys. Rev.*, **124**, 1807.

[10] Cohen, D.A., Hossein-Zadeh, M., and Levi, A.F.J. (2001) High-Q microphotonic electro-optic modulator. *Solid-State Electronics*, **45**, 1577.

[11] Ilchenko, V.S., Savchenkov, A.A., Matsko, A.B., and Maleki, L. (2003) Whispering-gallery-mode electro-optic modulator and photonic microwave receiver. *Journal of the Optical Society of America B*, **20**, 333.

[12] Savchenkov, A.A., Liang, W., Matsko, A.B. *et al.* (2009) Tunable optical single-sideband modulator with complete sideband suppression. *Optics Letters*, **34**, 1300.

[13] Strekalov, D.V., Schwefel, H.G.L., Savchenkov, A.A. *et al.* (2009) Microwave whispering-gallery resonator for efficient optical up-conversion. *Physical Review A*, **80**, 033810.

[14] Annino, G., Bertolini, D., Cassettari, M. *et al.* (2000) Dielectric properties of materials using whispering gallery dielectric resonators: experiments and perspectives of ultra-wideband characterization. *Journal of Chemical Physics*, **112**, 2308.

[15] Akay, M.F., Prokopenko, Y., and Kharkovsky, S. (2004) Resonance characteristics of whispering gallery modes in parallel-plates-type cylindrical dielectric resonators. *Microwave and Optical Technology Letters*, **40**, 96.

[16] Yariv, A. (1997) *Optical Electronics in Modern Communications*, 5th edn, Oxford University Press, New York.

[17] Preu, S., Schmid, S.I., Sedlmeir, F., Evers, J. *et al.* (2013) Directional emission of dielectric disks with a finite scatterer in the THz regime. *Optics Express*, **21**, 16370.

[18] Rivera-Lavado, A., Preu, S., García-Muñoz, L.E. *et al.* (2015) Dielectric rod waveguide antenna as THz emitter for photomixing devices. *IEEE Transactions on Antennas and Propagation*, **63**, 882.

[19] Oraevsky, A.N. (2002) Whispering-gallery waves. *Quantum Electronics*, **32**, 377.

[20] Breunig, I., Sturman, B., Sedlmeir, F. *et al.* (2013) Whispering gallery modes at the rim of an axisymmetric optical resonator: analytical versus numerical description and comparison with experiment. *Optics Express*, **21**, 30683.

[21] Comsol Group (2010) COMSOL Multiphysics.

[22] Humphries, S. (1997) *Finite-Element Methods for Electromagnetics*, Taylor & Francis.
[23] Oxborrow, M. (2006) Configuration of COMSOL Multiphysics for Simulating Axisymmetric Dielectric Resonators: Explicit Weak-Form Expressions Axisymmetric Electromagnetic Resonators.
[24] Yang, J., Lee, S.-B., Shim, J.-B. *et al.* (2008) Enhanced nonresonant optical pumping based on turnstile transport in a chaotic microcavity laser. *Applied Physics Letters*, **93**, 061101.
[25] Shao, L., Wang, L., Xiong, W. *et al.* (2013) Ultrahigh-Q, largely deformed microcavities coupled by a free-space laser beam. *Applied Physics Letters*, **103**, 121102.
[26] Vyatchanin, S.P., Gorodetskii, M.L., and Il'chenko, V.S. (1992) Tunable narrow-band optical filters with modes of the whispering gallery type. *Journal of Applied Spectroscopy*, **56**, 182.
[27] Gorodetsky, M.L. and Ilchenko, V.S. (1999) Optical microsphere resonators: optimal coupling to high-Q whispering-gallery modes. *Journal of the Optical Society of America B*, **16**, 147.
[28] Rowland, D.R. and Love, J.D. (1993) Evanescent wave coupling of whispering gallery modes of a dielectric cylinder. *IEE Proceedings Journal of Optoelectronics*, **140**, 177.
[29] Boriskina, S.V. (2006) Theoretical prediction of a dramatic Q-factor enhancement and degeneracy removal of whispering gallery modes in symmetrical photonic molecules. *Optics Letters*, **31**, 338.
[30] Schwefel, H.G.L. and Poulton, C.G. (2009) An improved method for calculating resonances of multiple dielectric disks arbitrarily positioned in the plane. *Optics Express*, **17**, 13178.
[31] Preu, S., Schwefel, H.G.L., Malzer, S. *et al.* (2008) Coupled whispering gallery mode resonators in the Terahertz frequency range. *Optics Express*, **16**, 7336.
[32] Benyoucef, M., Shim, J.-B., Wiersig, J., and Schmidt, O.G. (2011) Quality-factor enhancement of supermodes in coupled microdisks. *Optics Letters*, **36**, 1317.
[33] Shim, J.-B. and Wiersig, J. (2013) Semiclassical evaluation of frequency splittings in coupled optical microdisks. *Optics Express*, **21**, 24240.
[34] Pance, K., Viola, L., and Sridhar, S. (2000) Tunneling proximity resonances: interplay between symmetry and dissipation. *Physics Letters A*, **268**, 399.
[35] Witzany, M., Liu, T.-L., Shim, J.-B. *et al.* (2013) Strong mode coupling in InP quantum dot-based GaInP microdisk cavity dimers. *New Journal of Physics*, **15**, 013060.
[36] Gmachl, C., Capasso, F., Narimanov, E.E. *et al.* (1998) High-power directional emission from microlasers with chaotic resonators. *Science*, **280**, 1556.
[37] Kim, S.-K., Kim, S.-H., Kim, G.-H. *et al.* (2004) Highly directional emission from few-micron-size elliptical microdisks. *Applied Physics Letters*, **84**, 861.
[38] Schwefel, H.G.L., Rex, N.B., Tureci, H.E. *et al.* (2004) Dramatic shape sensitivity of directional emission patterns from similarly deformed cylindrical polymer lasers. *Journal of the Optical Society of America B*, **21**, 923.
[39] Song, Q., Ge, L., Redding, B., and Cao, H. (2012) Channeling chaotic rays into waveguides for efficient collection of microcavity emission. *Physical Review Letters*, **108**, 243902.
[40] Jiang, X.-F., Xiao, Y.-F., Zou, C.-L. *et al.* (2012) Highly unidirectional emission and ultralow-threshold lasing from on-chip ultrahigh-Q microcavities. *Advanced Materials*, **24**, OP260.
[41] Danylov, A.A., Waldman, J., Goyette, T.M. *et al.* (2007) Transformation of the multimode terahertz quantum cascade laser beam into a Gaussian, using a hollow dielectric waveguide. *Applied Optics*, **46**, 5051.
[42] Hentschel, M., Kwon, T.-Y., Belkin, M.A. *et al.* (2009) Angular emission characteristics of quantum cascade spiral microlasers. *Optics Express*, **17**, 10335.
[43] Wiersig, J. and Hentschel, M. (2006) Unidirectional light emission from high-Q modes in optical microcavities. *Physical Review A: Atomic, Molecular, and Optical Physics*, **73**, 031802.
[44] Dettmann, C.P., Morozov, G.V., Sieber, M., and Waalkens, H. (2009) Unidirectional emission from circular dielectric microresonators with a point scatterer. *Physical Review A*, **80**, 063813.
[45] Rubin, J.T. and Deych, L. (2010) Ab initio theory of defect scattering in spherical whispering-gallery-mode resonators. *Physical Review A*, **81**, 053827.
[46] Hales, R.F.M., Sieber, M., and Waalkens, H. (2011) Trace formula for a dielectric microdisk with a point scatterer. *Journal of Physics A: Mathematical and Theoretical*, **44**, 155305.
[47] Backes, S.A., Cleaver, J.R.A., Heberle, A.P. *et al.* (1999) Threshold reduction in pierced microdisk lasers. *Applied Physics Letters*, **74**, 176.
[48] Cai, X., Wang, J., Strain, M.J. *et al.* (2012) Integrated compact optical vortex beam emitters. *Science*, **338**, 363.
[49] Hakki, B.W. and Coleman, P.D. (1960) A dielectric resonator method of measuring inductive capacities in the millimeter range. *IRE Transactions on Microwave Theory and Techniques*, **8**, 402.

[50] Kobayashi, Y. and Katoh, M. (1985) Microwave measurement of dielectric properties of low-loss materials by the dielectric rod resonator method. *IEEE Transactions on Microwave Theory and Techniques*, **33**, 586.
[51] Fiedziuszko, S.J. and Heidmann, P.D. (1989) Dielectric resonator used as a probe for high T/sub c/ superconductor measurements. IEEE MTT-S International Microwave Symposium Digest, 1989, 555.
[52] Hanumegowda, N.M., Stica, C.J., Patel, B.C. *et al.* (2005) Refractometric sensors based on microsphere resonators. *Applied Physics Letters*, **87**, 201107.
[53] Chiou, W.C. and Pace, F.P. (1972) Parametric image upconversion of 10.6-μm illuminated objects. *Applied Physics Letters*, **20**, 44.
[54] Abbas, M.M., Kostiuk, T., and Ogilvie, K.W. (1976) Infrared upconversion for astronomical applications. *Applied Optics*, **15**, 961.
[55] Albota, M.A. and Wong, F.C. (2004) Efficient single-photon counting at 1.55 μm by means of frequency upconversion. *Optics Letters*, **29**, 1449.
[56] Karstad, K., Stefanov, A., Wegmuller, M. *et al.* (2005) Detection of mid-IR radiation by sum frequency generation for free space optical communication. *Optics and Lasers in Engineering*, **43**, 537.
[57] Temporão, G., Tanzilli, S., Zbinden, H. *et al.* (2006) Mid-infrared single-photon counting. *Optics Letters*, **31**, 1094.
[58] VanDevender, A.P. and Kwiat, P.G. (2007) Quantum transduction via frequency upconversion (Invited). *Journal of the Optical Society of America B*, **24**, 295.
[59] Ding, Y.J. and Shi, W. (2006) Efficient THz generation and frequency upconversion in GaP crystals. *Solid-State Electronics*, **50**, 1128.
[60] Khan, M.J., Chen, J.C., and Kaushik, S. (2007) Optical detection of terahertz radiation by using nonlinear parametric upconversion. *Optics Letters*, **32**, 3248.
[61] Ilchenko, V.S., Savchenkov, A.A., Byrd, J. *et al.* (2008) Crystal quartz optical whispering-gallery resonators. *Optics Letters*, **33**, 1569.
[62] Matsko, A.B., Strekalov, D.V., and Yu, N. (2008) Sensitivity of terahertz photonic receivers. *Physical Review A*, **77**, 043812.
[63] Weickhmann, C., Jakoby, R., Constable, E., and Lewis, R. (2013) Time-domain spectroscopy of novel nematic liquid crystals in the terahertz range. 38th International Conference on Infrared, Millimeter, and Terahertz Waves (IRMMW-THz).
[64] Nose, T., Sato, S., Mizuno, K. *et al.* (1997) Refractive index of nematic liquid crystals in the submillimeter wave region. *OSA Applied Optics*, **36**, 6383.
[65] Reuter, M., Vieweg, N., Fischer, B.M. *et al.* (2013) Highly birefringent, low-loss liquid crystals for terahertz applications. *APL Materials*, **1**, 012107.
[66] Koeberle, M., Hoefle, M., Gaebler, A. *et al.* (2011) Liquid crystal phase shifter for Terahertz frequencies with quasi-orthogonal electrical bias field. 2011 36th International Conference on Infrared, Millimeter and Terahertz Waves (IRMMW-THz).
[67] Demus, D., Goodby, J., Gray, G.W. *et al.* (eds) (1998) *Handbook of Liquid Crystals – Fundamentals*, Wiley-VCH Verlag GmbH, Weinheim.
[68] Tsai, T.-R., Chen, C.-Y., Pan, C.-L. *et al.* (2003) Terahertz time-domain spectroscopy studies of the optical constants of the nematic liquid crystal 5CB. *OSA Applied Optics*, **42**, 2372.
[69] de Gennes, P.-G. and Prost, J. (1995) *The Physics of Liquid Crystals*, Oxford University Press.
[70] Debye, P. (1913) *Zur Theorie der anomalen Dispersion im Gebiete der langwelligen elektrischen Strahlung*, Verhandlungen der Deutschen Physikalischen Gesellschaft, 777.
[71] Anderson, J., Watson, P.E., and Bos, P.J. (2001) *LC3D: Liquid Crystal Display 3-D Director Simulator Software and Technology Guide*, Artech House (Optoelectronics Library).
[72] Gäbler, A. (2013) Synthese steuerbarer Hochfrequenzschaltungen und Analyse Flüssigkristall-basierter Leitungsphasenschieber in Gruppenantennen für Satellitenanwendungen im Ka-Band, Fachbereich Elektrotechnik und Informationstechnik, Technische Universität Darmstadt.
[73] Saleh, B. and Teich, M. (2007) *Fundamentals of Photonics*, Wiley-Interscience, New York.
[74] Pozar, D.M. (1998) *Microwave-Engineering*, John Wiley & Sons, Inc., New York.
[75] Mueller S., Penirschke A., Damm C., *et al.* (2005) Broad-band microwave characterization of liquid crystals using a temperature-controlled coaxial transmission line (2005), *IEEE Transactions on Microwave Theory and Techniques*, **53**, 1937.
[76] Tonouchi, M. (2007) Cutting-edge terahertz technology. *Nature Photonics*, **1**, 97.
[77] Nagel, M., Bolivar, P.H., Brucherseifer, M. *et al.* (2002) Integrated planar terahertz resonators for femtomolar sensitivity label-free detection of DNA hybridization. *Applied Optics*, **41**, 2074.

[78] Penirschke, A., Mueller, S., Scheele, P. *et al.* (2004) Cavity perturbation method for characterization of liquid crystals up to 35 GHz. 34th European Microwave Conference, 545.
[79] Goelden, F., Lapanik, A., Gäbler, A. *et al.* (2007) Systematic investigation of nematic liquid crystal mixtures at 30 GHz. Digest of the IEEE/LEOS Summer Topical Meetings, 202.
[80] Gäbler, A., Goelden, F., Mueller, S., and Jakoby, R. (2008) Triple-mode cavity perturbation method for the characterization of anisotropic media. 38th European Microwave Conference, 909.
[81] Duvillaret, L., Garet, F., and Coutaz, J.-L. (1996) A reliable method for extraction of material parameters in terahertz time-domain spectroscopy. *IEEE Journal of Selected Topics in Quantum Electronics*, **2**, 739.
[82] Pogson, E.M., Lewis, R.A., Köberle, M., and Jakoby, R. (2010) Terahertz time-domain spectroscopy of nematic liquid crystals. *Proceedings of SPIE*, **7728**, 7728.
[83] Chandrasekhar, S. and Madhusudana, N.V. (1972) Spectroscopy of liquid crystals. *Applied Spectroscopy Reviews*, **6**, 189.
[84] Goelden, F., Mueller, S., Scheele, P. *et al.* (2006) IP3 measurements of liquid crystals at microwave frequencies. 36th European Microwave Conference, 971.
[85] Mueller, S., Scheele, P., Weil, C. *et al.* (2004) Tunable passive phase shifter for microwave applications using highly anisotropic liquid crystals. IEEE MTT-S International Microwave Symposium Digest, 1153.
[86] Fritzsch, C., Giacomozzi, F., Karabey, O.H. *et al.* (2011) Continuously tunable W-band phase shifter based on liquid crystals and MEMS technology. 41st European Microwave Conference (EuMC), 1083.
[87] Gäbler, A., Goelden, F., Manabe, A. *et al.* (2009) Investigation of high performance transmission line phase shifters based on liquid crystal. European Microwave Conference, 594.
[88] Hoefle, M., Koeberle, M., Penirschke, A., and Jakoby, R. (2011), Millimeterwave vivaldi antenna with liquid crystal phase shifter for electronic beam steering. 6th ESA Workshop on Millimetre-Wave Technology and Applications.
[89] Hoefle, M., Koeberle, M., Chen, M. *et al.* (2010) Reconfigurable Vivaldi antenna array with integrated antipodal finline phase shifter with liquid crystal for W-Band applications. 35th International Conference on Infrared Millimeter and Terahertz Waves (IRMMW-THz).
[90] Hoefle, M., Koeberle, M., Penirschke, A., and Jakoby, R. (2011) Improved millimeter wave Vivaldi antenna array element with high performance liquid crystals. 36th International Conference on Infrared, Millimeter and Terahertz Waves (IRMMW-THz).
[91] Weickhmann, C., Nathrath, N., Gehring, R. *et al.* (2013) Recent measurements of compact electronically tunable liquid crystal phase shifter in rectangular waveguide topology. *Electronics Letters*, **49**, 1345.
[92] Weickhmann, C., Nathrath, N., Gehring, R. *et al.* (2013) A light-weight tunable liquid crystal phase shifter for an efficient phased array antenna. 43rd European Microwave Conference.
[93] Jost, M., Weickhmann, C., Strunck, S. *et al.* (2013) Liquid crystal based low-loss phase shifter for W-band frequencies. *Electronics Letters*, **49**, 1460.
[94] Tsai, T.-R., Chen, C.-Y., Pan, R.-P. *et al.* (2004) Electrically controlled room temperature terahertz phase shifter with liquid crystal. *IEEE Microwave and Wireless Components Letters*, **14**, 77.
[95] Chen, C.-Y., Hsieh, C.-F., Lin, Y.-F. *et al.* (2004) Magnetically tunable room-temperature 2 pi liquid crystal terahertz phase shifter. *Optics Express*, **12**, 2625.
[96] Wu, H.-Y., Hsieh, C.-F., Tang, T.-T. *et al.* (2006) Electrically tunable room-temperature 2pi liquid crystal terahertz phase shifter. *IEEE Photonics Technology Letters*, **18**, 1488.
[97] Strunck, S., Karabey, O.H., Gäbler, A., and Jakoby, R. (2012) Reconfigurable waveguide polariser based on liquid crystal for continuous tuning of linear polarisation. *Electronics Letters*, **48**, 441.
[98] Karabey, O.H., Gäbler, A., Strunck, S., and Jakoby, R. (2012) A 2-D electronically steered phased-array antenna with 2×2 elements in LC display technology. *IEEE Transactions on Microwave Theory and Techniques*, **60**, 1297.
[99] Müller, S., Köberle, M., Mössinger, A. *et al.* (2008) Liquid crystal based electronically steerable 4×4 antenna array with single horn feed at Ka-Band. IEEE AP-S Antennas and Propagation Society International Symposium.
[100] Strunck, S., Karabey, O.H., Weickhmann, C. *et al.* (2013) Continuously tunable phase shifters for phased arrays based on liquid crystal technology. IEEE International Symposium on Phased Array Systems Technology, 82.
[101] Dragoman, M., Hartnagel, H.L., Tuovinen, J., and Plana, R. (2005) Microwave applications of carbon nanotubes. *Journal Frequenz*, **59**, 251.
[102] Novoselov, K.S., Geim, A.K., Morozov, S.V. *et al.* (2004) Electric field effect in atomically thin carbon films. *Science*, **306**, 666.

[103] Avouris, P. (2010) Graphene: electronic and photonic properties and devices. *Nano Letters*, **10**, 4285.
[104] Novoselov, K.S., Geim, A.K., Morozov, S.V. *et al.* (2005) Two-dimensional gas of massless Dirac fermions in graphene. *Nature*, **438**, 197.
[105] Emtsev, K.V., Bostwick, A., Horn, K. *et al.* (2009) Towards wafer-size graphene layers by atmospheric pressure graphitization of silicon carbide. *Nature Materials*, **9**, 203.
[106] Dawlaty, J.M., Shivaraman, S., Strait, J. *et al.* (2008) Measurement of the optical absorption spectra of epitaxial graphene from terahertz to visible. *Applied Physics Letters*, **93**, 131905.
[107] Docherty, C.J. and Johnston, M.B. (2012) Terahertz properties of graphene. *Journal of Infrared, Millimeter and Terahertz waves*, **33**, 797.
[108] Das Sharma, S., Adam, S., and Hwang, E.W. (2011) Electronic transport in two dimensional graphene. *Reviews of Modern Physics*, **83**, 407.
[109] Horng, J., Chen, C.-F., Geng, B. *et al.* (2011) Drude conductivity of Dirac fermions in graphene. *Physical Review B*, **83**, 165113.
[110] Nair, R.R., Blake, P., Grigorenko, A.N. *et al.* (2008) Fine structure constant defines visual transparency of graphene. *Science*, **320**, 1308.
[111] Quhe, R., Zheng, J., Luo, G. *et al.* (2012) Tunable and sizable band gap of single-layer graphene sandwiched between hexagonal boron nitride. *NGP Asia Materials*, **4**, e6.
[112] Li, X., Wang, X., Zhang, L., Lee, S., and Dai, H. (2008) Chemically derived, ultrasmooth graphene nanoribbon semiconductors. *Science*, **319**, 1229.
[113] Han, M.Y., Özyilmaz, B., Zhang, Y., and Kim, P. (2007) Energy band-gap engineering of graphene nanoribbons. *Physical Review Letters*, **98**, 206805.
[114] Hartnagel, H.L., Ong, D.S., and Al-Daffaie, S. (2013) Proposal of a THz signal generator by ballistic electron resonance device based on graphene. Proceedings of Conference WOCSDICE, Warnemünde, Germany 2013, ISBN: 978-3-00-041435-0.
[115] Lin, Y.-M., Jenkins, K.A., Valdes-Garcia, A. *et al.* (2009) Operation of graphene transistors at gigahertz frequencies. *Nano Letters*, **9**, 422.
[116] Xia, F., Farmer, D.B., Lin, Y.-M., and Avouris, P. (2010) Graphene field-effect transistors with high on/off current ratio and large transport band gap at room temperature. *Nano Letters*, **10**, 715.
[117] Vicarelli, L., Vitiello, M.S., Coquillat, D. *et al.* (2012) Graphene field-effect transistors as room-temperature terahertz detectors. *Nature Materials*, **11**, 865.
[118] Liao, L., Bai, J., Cheng, R. *et al.* (2010) Sub-100 nm channel length graphene transistors. *Nano Letters*, **10**, 3952.
[119] Wu, Y., Jenkins, K.A., Valdes-Garcia, A. *et al.* (2012) State-of-the-art graphene high-frequency electronics. *Nano Letters*, **12**, 3062.
[120] Sensale-Rodriguez, B., Yan, R., Rafique, S. *et al.* (2012) Extraordinary control of Terahertz beam reflectance in graphene electro-absorption modulators. *Nano Letters*, **12**, 4518.
[121] Bludov, Y.V., Ferreira, A., Peres, N.M.R., and Vasilevskiy, M.I. (2013) A primer on surface plasmon-polaritons in graphene. *International Journal of Modern Physics B*, **27**, 1341001.
[122] Ju, L., Geng, B., Horng, J., Girit, C. *et al.* (2011) Graphene plasmonics for tunable terahertz metamaterials. *Nature Nanotechnology*, **6**, 630.
[123] Vakil, A. and Engheta, N. (2011) Transformation optics using graphene. *Science*, **332**, 1291.
[124] Hertel, S., Waldmann, D., Jobst, J. *et al.* (2012) Tailoring the graphene/silicon carbide interface for monolithic wafer-scale electronics. *Nature Communications*, **3**, 957.
[125] Lee, S.H., Choi, M., Kim, T.-T. *et al.* (2012) Switching terahertz waves with gate-controlled active graphene metamaterials. *Nature Materials*, **11**, 936.
[126] Ryzhii, V., Ryzhii, M., Satou, A. *et al.* (2009) Feasibility of terahertz lasing in optically pumped epitaxial multiple graphene layer structures. *Journal of Applied Physics*, **106**, 084507.
[127] Rana, F. (2008) Graphene terahertz plasmon oscillators. *IEEE Transactions on Nanotechnology*, **7**, 91.
[128] Bonaccorso, F., Sun, Z., Hasan, T., and Ferrari, A.C. (2010) Graphene photonics and optoelectronics. *Nature Photonics*, **4**, 611.
[129] Lee, M.-S., Lee, K., Kim, S.-Y. *et al.* (2013) High-performance, transparent, and stretchable elektrodes using graphene-metal nanowire hybrid structures. *Nano Letters*, **13**, 2814.
[130] Hartnagel, H.L., Dawar, A.L., Jain, A.K., and Jagadish, C. (1995) *Semiconducting Transparent Thin Films*, Institute of Physics Publishing, Bristol and Philadelphia, PA.

[131] Al-Daffaie, S., Yilmazoglu, O., Küppers, F., and Hartnagel, H. (2013) Graphene LTG-GaAs photomixer for reliable continous wave Terahertz generation. 38th International Conference on Infrared; Terahertz and Millimeter Waves (IRMMW-THz), Mainz, Germany.
[132] Abergel, D.S.L., Apalkov, V., Berashevich, J. *et al.* (2010) Properties of graphene: a theoretical perspective. *Advances in Physics*, **59**, 261.
[133] Mikhailov, S.A. (2013) Graphene-based voltage-tunable coherent terahertz emitter. *Physical Review B*, **87**, 115405.
[134] Kang, J., Kim, H., Kim, K.S. *et.al.* (2011) High performance graphene-based transparent flexible heaters. *Nano Letters*, **11**, 5154.

Index

Semiconductor Terahertz Technology: Devices and Systems at Room Temperature Operation, First Edition.
Edited by Guillermo Carpintero, Luis Enrique García Muñoz, Hans L. Hartnagel, Sascha Preu and Antti V. Räisänen.